Frank
Bickford
TC3

*Quantitative Analysis*

# *Quantitative Analysis*

*Charles T. Kenner*
SOUTHERN METHODIST UNIVERSITY
DALLAS, TEXAS

*Kenneth W. Busch*
BAYLOR UNIVERSITY
WACO, TEXAS

*Macmillan Publishing Co., Inc.*
NEW YORK

*Collier Macmillan Publishers*
LONDON

A portion of this material has been adapted from *Analytical Separations and Determinations*, by C. T. Kenner, copyright © 1971 by Macmillan Publishing Co., Inc., and from *Instrumental and Separation Analysis*, by C. T. Kenner (Columbus: Charles E. Merrill, 1973), copyright © 1973 by Bell & Howell Company.

Macmillan Publishing Co., Inc.
866 Third Avenue, New York, New York 10022

Collier Macmillan Canada, Ltd.

**Library of Congress Cataloging in Publication Data**

Kenner, C. T.
Quantitative analysis.

Bibliography: p.
Includes index.
1. Chemistry, Analytic—Quantitative.
I. Busch, Kenneth W., joint author. II. Kenner, C. T. Analytical separations and determinations.
III. Title.
QD101.2.K46 545 78-4160
ISBN 0-02-362490-6
Printing: 1 2 3 4 5 6 7 8 Year: 9 0 1 2 3 4 5

*Dedicated to Bess Harrison Kenner*

# *Preface*

Modern analytical chemistry is the science of chemical measurements. Chemical measurements play an important role in many diverse disciplines, such as agriculture, geology, metallurgy, medicine, psychology, engineering, and environmental science, by providing information essential in solving practical problems. In its widest sense, analytical chemistry embraces the theory and practice of all the diverse means by which information of all forms may be obtained on the composition of matter. The content and scope of analytical chemistry is continually expanding as new techniques are developed to solve ever more challenging problems. To meet the need for increasingly more sophisticated knowledge regarding the composition of matter, analytical scientists are studying all facets of the chemical measurement process. Indeed, research on the development of new means of obtaining chemical information on the composition of matter may more appropriately be termed analytical science rather than analytical chemistry, because its progress depends on a combination of disciplines, including chemistry, physics, engineering, and applied mathematics.

The present work is an attempt on the part of the authors to meet the needs of students and teachers for an introductory textbook covering some of the more important aspects of modern analytical chemistry. Every effort has been made to produce a balanced, readable text flexible enough to be adaptable to the wide range of analytical programs currently being offered at colleges and universities throughout the country. In this respect, the text should be appropriate for programs where only a single course in analytical chemistry is offered and where the instructor wishes to discuss some basic instrumental methods as well as gravimetric and volumetric principles. Furthermore, we anticipate that this text will also prove useful at institutions that offer a separate course for the large number of students in related disciplines, such as biology, geology, medicine, clinical chemistry, and science education. Such students need some basic knowledge of instrumentation but not at the level or depth needed by chemistry majors who plan to become practicing chemists. Our philosophy in writing the instrumental portion of the book has been to emphasize the principles and capabilities of instruments to solve analytical problems rather than to stress their design and construction.

Because of the wide scope of modern analytical chemistry, it is difficult to select the subjects and instruments to discuss as well as to decide on the amount and level of material that should be included in an introductory text. Comprehensive coverage of basic quantitative analysis, separation science, and instrumental methods in one book is neither desirable nor practical; thus, this text is necessarily a compromise. Although there is no formal separation into parts, the first fifteen chapters in the text deal with material ordinarily covered in beginning analytical courses, stressing volumetric and gravimetric analysis. Chapters 16 through 27 cover instrumental methods and introductory separation science. Chapter 28 discusses and illustrates the most widely used manual laboratory techniques. The detailed table of contents gives a complete outline of the text, which will be helpful both to students and to instructors. Each chapter is written so that the material may be covered in any sequence desired by the instructor.

We believe that we have included sufficient material for the student to understand each technique and instrument without having a detailed knowledge of all of its operational aspects. We have selected subjects and instruments that are in widest use in commercial, educational, and research laboratories. As examples, we have given preference to volumetric analysis over gravimetric, and have included radiochemical techniques and electrophoresis instead of thermal and high-frequency methods. A short introduction to organic chelates and a short review of inorganic complexes are included for those students who have not completed a course in organic chemistry. This material is needed to understand extraction, which, in turn, is needed to understand liquid–liquid and gas–liquid chromatography. We have stressed all types of chromatography and all types of absorption photometry, together with flame atomic spectroscopy, potentiometric measurement of hydrogen and other ions, and automated analyses by giving more details of these techniques than is given for subjects such as x-ray, nuclear magnetic resonance, mass spectrometry, and emission spectroscopy.

No laboratory directions are given in the text, although the theoretical basis of typical determinations is discussed. The authors believe that most instructors prefer to design their own experiments, and that they normally modify the experiments in any text by issuing complete mimeographed instructions suited to their own needs. The students prefer to use the inexpensive or free instructions so that they will not have to bring their textbooks into the laboratory.

We wish to express our appreciation to the many students who have aided in the preparation of the book through their comments and class notes during the preparative stages. Thanks are expressed to Dr. George Howard Luttrell, Jr., who prepared Chapter 25 on automated analysis, and Rev. Walter Ross Purkey, who is responsible for Chapter 27 on microelectronics. We are deeply grateful to Mrs. Shirley McLean, who typed the manuscript with a minimum of errors. Our thanks also go to our wives, Bess and Marianna, for their forbearance during the preparatory stages and for sharing in the task of proofreading the manuscript.

C. T. K.

K. W. B.

# Contents

## 10 *Neutralization* *149*

# 13 *Potentiometric Measurements* 244



# 14 *Compleximetry* 267

# 22 Column and Plane Chromatography 436



# 23 Gas Chromatography 465

## 24 *Ion Exchange* 485



## 25 *Automated Analysis* 494



## 26 *Miscellaneous Methods* 512

## 28 Laboratory Techniques 562



## Appendixes



## Index 613

CHAPTER

# 1 Introduction

Analytical chemistry deals with the determination of the composition of materials and substances. The word **substance** in chemistry is usually used to refer to a pure element or compound, whereas a **material** is considered to be a mixture. Thus the compound sodium chloride is referred to as a substance, but sodium chloride as it occurs in nature and as it is mined from salt domes is referred to as a material, since it contains substances other than sodium chloride as impurities. As a result, a chemist practically never analyzes a substance but deals primarily with materials as they occur in nature or as they are manufactured or prepared from other materials.

Analytical chemistry has made great progress in the determination of smaller and smaller amounts of the substances found in materials. The increased interest in the quality of the environment and the concern over the effects of trace amounts of chemicals on life processes have led analytical chemists to develop practical methods that allow detection and determination of components in the part per million (ppm) and part per billion (ppb) ranges. This has called for the introduction of new procedures as well as the modification of older methods to take advantage of the new and improved techniques and instruments now available. The beginning student in quantitative chemistry should realize, however, that essentially all instrumental methods require some or all of the basic "wet" techniques (drying, weighing, dissolution, transfer, extraction, precipitation, filtration, volumetric measurements, aliquoting, etc.) in the preparation of the sample before the instrumental measurements can be made.

## TYPES AND METHODS OF ANALYSIS

The two main classes of analysis are qualitative and quantitative. The determination of the actual substances or constituents present in a material, without regard to the amount of each present, is known as **qualitative** analysis, whereas the determination of the actual amount of a constituent present is consider to be **quantitative** analysis. Thus qualitative analysis is concerned only with what materials or substances are present, whereas quantitative analysis deals with how much of each constituent determined is present. The result of a quantitative analysis is usually reported as a number, often the percentage by weight.

Analyses may be classified in several different ways. One method of classification is concerned with the number and type of constituents determined and reported. An **ultimate** or **complete** analysis accounts for each constituent present and should represent 100% of the sample. If all the constituents are not determined, the analysis is called a **partial** analysis, and if only one constituent is determined, it is often referred to as **single-constituent** analysis. Since the properties of materials are often affected by one or a few of the major, and sometimes the minor, constituents, partial analyses are run much more often than complete analyses.

Often similar materials may be determined and reported together if it is not necessary to know the value for each individually. Such analyses are called **proximate** analyses. An example is the determination of iron and aluminum in rocks. Both of these elements precipitate upon addition of ammonia to form hydrous oxides, which are filtered, washed, and ignited to a mixture of ferric oxide ($Fe_2O_3$) and aluminum oxide ($Al_2O_3$), which is then weighed. The report would show the combined percentages of $Fe_2O_3$ and $Al_2O_3$. Since this particular precipitate is known as the $R_2O_3$ precipitate (where R represents the metals), the report often simply shows percent $R_2O_3$. If it is necessary to know either or both of the metals individually, other methods would have to be used to determine one or both.

Analyses may also be classified on the basis of the type of material being analyzed. Thus, we refer to inorganic analysis and to organic analysis. Similarly, we speak of rock analysis, nonferrous alloy analysis, steel analysis, petroleum product analysis, fuel analysis, and so on, to indicate the type of material being analyzed.

Analyses may also be classified as to the amount of sample taken for analysis. In the beginning course in quantitative analysis, and in many industrial, research, and government laboratories, the majority of samples are 0.1 g or more and are classified as **macroanalyses**. If the sample weight is between 1 and 10 mg (0.001–0.010 g), the analysis is said to be a **microanalysis**. Sample weights between micro and macro are known as **semimicroanalyses**. Recently, methods that use a sample weight of less than 1 mg have been developed. Such methods are known as **ultramicroanalyses**.

The classification of analytical methods is based upon the final measurement step in the procedure and includes gravimetric, volumetric, and physicochemical methods. Methods in which the final determination is the accurate measurement of a weight of a substance of known purity are known as **gravimetric** methods, while those in which the final measurement deals with a volume are called **volumetric** methods. If the final volume measurement is made with a burette, the procedure is called a titration, and the method is classified as a **titrimetric** analysis. Neither gravimetric nor volumetric methods measure individual properties of the substance being determined or measured. Methods based upon the measurement of some physical or chemical property are known as **physicochemical** methods. Such methods often require special instruments to make the measurement and are known as **instrumental** methods. Physicochemical or instrumental methods include the measurement of optical properties (transmittance, absorbance, reflectance, emission, rotation), electrical properties (conductance, potential, resistance, etc.), chromatographic properties, magnetic properties, thermal properties, radioactivity, crystalline structure, and others.

## STEPS IN AN ANALYSIS

To analyze a material, the following steps are usually required: (1) the sample must be obtained, (2) the method of analysis must be selected, (3) the sample must be measured (usually weighed), (4) the sample must be dissolved, and (5) the constituent being determined must be separated and measured. Each of these will be discussed separately.

### OBTAINING THE SAMPLE

The sample used in the actual analytical determination (called the **determination** sample) must be a representative sample; that is, the determination sample must have the same average composition as the larger sample from which it is obtained and which it is supposed to represent. Although representative samples seldom create a problem in beginning courses in analytical chemistry, they are very important for laboratories that run analyses on large lots of material in order to check purchase specifications. The weight of the determination sample varies depending upon the method being used, but in the majority of cases it is 1 g or less. Yet this 1 g may have to represent as much as a trainload of coal or a boatload of ore. It is imperative that the determination sample that represents large amounts of material actually be a representative sample; otherwise, the analysis, no matter how carefully run, will be erroneous and may result in substantial loss or gain to the companies involved.

Correct methods of sampling large lots of materials are specified by various industrial and/or scientific associations concerned with the use of the materials. Such associations include the American Society for Testing Materials (ASTM), The Association of Official Analytical Chemists (AOAC), the American Oil Chemists Society (AOCS), the Association of the Petroleum Industry (API), and the National Bureau of Standards (NBS). Details of standard methods of sampling can be found in many reference books and in articles and books published by the associations involved.

In general, the sampling of large lots of material is performed in three steps. The first sample taken is called the **gross** sample and is usually obtained during the unloading process. A specified portion is retained, such as every twenty-fifth shovel, or some sort of riff is used on the unloading chute to separate a certain fraction of the material. The gross sample may range as high as several tons of material and must be obtained by a procedure that ensures it is representative of the large lot. The gross sample must be reduced to an amount suitable for transmission to the laboratory, usually 25 lb or less.

Reduction of the amount of the gross sample may be performed by the long pile and alternate shovel method or by the cone and quarter method. In the **long pile** method, the gross sample is arranged in a long pile during the unloading process. It is then separated into two equal piles using a shovel and throwing alternative shovels to opposite sides. One-half is discarded and the other half again separated in the same manner. This is continued until the remaining sample is of the proper weight. In the **cone and quarter** method, the gross sample is piled into a cone and the top flattened. The cone is then separated into quarters. Two opposite quarters are discarded and the other two opposite quarters retained and formed into a second cone. The process is repeated until the sample is of the correct amount to send to the laboratory.

During the reduction of the amount of the gross sample, it is often necessary to reduce the particle size to retain the proper proportion of large and small particles, to ensure uniformity, and to ensure that the sample sent to the laboratory is representative. Various types of crushers, grinders, ball mills, and so on, are used to reduce the particle size of the sample.

After receipt of the laboratory sample, the analyst must also be sure that the sample used in the actual analysis is uniform and is representative of the large lot. The analyst must often use the same methods (on a smaller scale, of course) for reduction of the amount and particle size of the sample. Sometimes the sample is such that it cannot be reduced to particles small enough to ensure uniform mixing. In such cases, the analyst will use a series of standard sieves to separate the sample into fractions of relatively equal particle size. The fraction of the total weight of the sample represented by each fraction is then calculated. The necessary portion of each fraction is then weighed individually and mixed to obtain the determination sample.

Standard sieves are made in the form of wire screens and are classified on the basis of the number of openings (meshes) per square inch. The larger the number of openings, the smaller is the mesh size and, as a consequence, the smaller a particle has to be to pass through a sieve. A particle retained by a number 40 screen would pass through any screen of a smaller number. Particle sizes of solids are usually specified in terms of the smallest number screen through which the particles will pass and the largest number screen upon which they will be retained. A sample classified as 40/80 will pass a number 40 sieve and be retained on a number 80 sieve. If only one sieve number is given in the particle-size specifications, it means that all the particles will pass that size screen. All the particles of a sample referred to as 100 mesh will pass through a number 100 screen. Standard screens vary in size from number 1 to number 300.

Some materials must be reported on an "as-received" basis and also on a "dry" basis. As a consequence, the analyst will take some rather large representative portion and weigh and dry it at 100–110°C to remove the adventitious moisture. **Adventitious moisture** is moisture that is loosely held by the solid and may be removed by simple heating to a temperature slightly above the boiling temperature of water. The determination sample itself must be dry, and, as a result, all that portion of the laboratory sample to be used for analytical determination is dried for at least 1 hr at a temperature between 100 and 110°C and is stored in a desiccator (see Figure 28-11) until the individual samples are weighed out. The results are then calculated on both the dry basis and the as-received basis.

## SELECTION OF THE METHOD OF DETERMINATION

Before deciding on the actual sample weight to be weighed, it is necessary to select the method of determination to be used. The first and primary consideration in the selection is the accuracy with which the constituent must be determined. The method selected must be capable of this requisite accuracy, since it is not economical to select a method that will not give adequate results or to select a method capable of accuracy beyond that which is required, especially if the more accurate method requires more time or more expensive equipment.

The method must also be selected on the basis of the percentage of the constituent in the sample and the size of sample available. Methods that are satisfactory for major components are usually not satisfactory to determine the same constituent if it is present in trace quantities. Also, a method that is satisfactory for a relatively large sample is usually not satisfactory for a sample size in the milligram range.

If several methods are available, the next consideration is the time involved in the analysis. This factor can be very important in control work in metallurgical and manufacturing plants when the product must be delayed until the analysis is complete. For instance, a steel mill prepares large quantities of special steels at one time and usually must keep the steel molten while components are being added. After the addition, the steel must be analyzed before the molten material can be poured into the forms desired. A few minutes of time saved in such an analysis may save the company a large sum of money over a comparatively short period of time. As a consequence, steel mills will often pay over $100,000 for an instrument that can save a few minutes in analysis time while the steel must be kept in the molten state.

For situations in which time is not critical, the actual time spent by the analyst on the determination must be considered. If one determination requires 1 hr of the analyst's time over a period of 2 days and another requires 2 hr during one 8-hr shift, the slower method will be selected if time is not important.

Other factors that are usually considered include costs other than analyst time, difficulty of procedure, availability and stability of reagents, special equipment needed and whether or not it can be justified by the number of analyses run, and other similar considerations.

## MEASUREMENT OF THE SAMPLE

### *Selection of Proper Sample Weight*

After selection of the method, the actual sample weight to be utilized must be chosen. In general, it is desirable to weigh the largest possible sample in order to decrease the relative error in weighing. However, the time and difficulties involved in later steps of the analysis must also be considered because these usually increase with sample weight. The optimum weight to be used will be the largest weight possible that will not cause undue difficulties in the later steps of the analysis. Only an experienced analyst can determine what the optimum weight will be for a given analysis, since the decision requires a knowledge of the procedure and the experimental difficulties that will be encountered in the course of the determination. Optimum weights are seldom expressed to more than two significant figures (0.50, 1.0, 10, etc.) because a sample weight within 5 or 10% of the optimum weight is usually satisfactory.

To select the actual weight of the sample to be weighed, it must be decided whether speed of weighing or ease of calculation is of primary importance. Weighing an exact weight of sample to the nearest part per thousand (ppt) can be a tedious and time-consuming operation if the sample weight is 1 g or less, since this means that a deviation of only 1 mg or less can be allowed. It is much faster to obtain an accurate weight of a sample if a range is allowed on either side of the optimum weight. As an example, weighing a sample of exactly 1.000 g takes much longer than weighing a sample of anywhere between 0.950 and 1.050 g, as long as the actual weight is measured to 1 ppt.

However, calculations are simplified by using weights that are exact fractions or multiples of 1 because most calculations involve dividing by the weight of the sample in at least one step. Thus, it is much easier to divide by 1.000 than by 0.979 or by 1.036. An even simpler method of calculation is involved if the sample weight chosen is the exact factor weight or an exact factor or multiple of the factor weight. The **factor weight** in a given determination is that weight of sample which will allow the percentage of the constituent being determined to be calculated by multiplying the weight of the precipitate produced (or the volume of a solution used) by 100 to obtain the percentage directly. If the factor weight deviates widely from the optimum weight, an exact fraction or multiple of the factor weight can be used so that the multiplier will be 10, 50, 200, and so on. Methods of calculating factor weights will be discussed later. Ease of calculation is not considered as necessary now as it was formerly because most modern laboratories are equipped with electric calculators.

### *Weighing by Difference*

If the sample changes weight upon exposure to the atmosphere, it is usually weighed by difference. When weighing a sample by difference, the material is weighed in a closed container, the sample to be analyzed is poured out of this container, and the container is weighed again. The weight of sample taken is then calculated from the difference in the weights before and after removal of the determination sample. Because it is essentially impossible to weigh a sample of an exact predetermined weight by this method, a range of weight must be used.

### *Aliquoting Procedure*

Sometimes it is preferable to weigh a relatively large sample so that it can be dissolved and diluted to a definite, accurately known volume. A known accurate fraction of this large volume is then used for the determination sample. This procedure is known as aliquoting, and the sample used is called an **aliquot**. An aliquot may be defined as an accurately known fraction of a larger volume from which it is taken. The aliquoting procedure may be used if the determination sample is too small to allow accurate weighing or if the sample changes weight upon exposure. In either case, the larger sample size can be weighed more accurately and more rapidly, especially if a range of weight is allowable. The aliquoting procedure to be used must be planned before the sample is weighed so that the aliquots used will contain the optimum weight of sample.

The analyst usually has available various sizes of volumetric (transfer) pipettes and various sizes of volumetric flasks. The volumetric flasks are calibrated to contain (TC) and thus can be used to prepare an accurately known volume of a solution. The volumetric pipettes are usually calibrated to deliver (TD) and are used to transfer an accurately known volume of a solution. The actual amount of sample to be weighed will depend upon the volumetric flask and the volumetric pipette used and upon the optimum weight of the determination sample. These values are related by the equation

$$\text{wt sample to be weighed} = \text{optimum weight determination sample} \times \frac{\text{volume flask}}{\text{volume pipette}} \qquad \textbf{(1-1)}$$

As an example, calculate the weight of sample to be weighed if the optimum weight is 0.25 g, the sample is to be diluted to 500.0 mL, and a 25.00-mL aliquot is to be used. Substitution in eq. (1-1) gives

$$\text{wt sample} = 0.25\ \text{g} \times \frac{500.0\ \text{mL}}{25.00\ \text{mL}} = 5.0\ \text{g}$$

In weighing the 5-g sample, a range of 5–10% may be used without seriously affecting the analysis; so the actual sample weight could be any value between 4.500 and 5.500 g, as long as it was known accurately.

It should be noted that the calculation of the actual weight of the determination sample is the reverse of the calculation in eq. (1-1). For example, assume that the actual weight of the sample in the example given above turned out to be 5.012 g; calculate the weight of the actual determination sample. The actual weight of that determination sample will be the weight of solute in the aliquot or will equal the weight weighed multiplied by the fraction of the total weight found in the aliquot. This relationship is expressed by the equation

$$\text{wt determination sample} = \text{wt weighed} \times \frac{\text{volume aliquot}}{\text{volume solution}} \qquad \textbf{(1-2)}$$

The calculation then becomes

$$\text{wt determination sample} = 5.012\ \text{g} \times \frac{25.00\ \text{mL}}{500.0\ \text{mL}} = 0.2506\ \text{g}$$

## SOLUTION OF THE SAMPLE

In practically all analyses, the first step after weighing the sample is to dissolve the weighed sample. The solvent used should dissolve the sample completely in the least amount of time possible and should cause the least possible interference with the later steps of the analysis. The general rule is that the solvent should be as gentle as possible and should dissolve the sample at a reasonable rate as close to room temperature as possible. For inorganic materials solvents are usually tried in the following order.

### *Water*

Samples that are composed of soluble salts may be dissolved in water at room temperature or at elevated temperatures. If the sample is composed of a mixture of soluble and insoluble substances, the residue from the water solution may be dissolved in a stronger solvent (such as a strong acid or a strong base) or may be fused with an acidic or basic flux. Some simple rules for judging solubility are given in Appendix 3.

### *Nonoxidizing Acids*

If the sample is not soluble in water, the next solvent to be tried is a nonoxidizing acid such as hydrochloric acid. Other acids which are usually considered nonoxidizing include dilute sulfuric acid and dilute perchloric acids. Concentrated sulfuric and perchloric acids can act as oxidizing acids at elevated temperatures. Hydrofluoric acid may be used as a nonoxidizing acid for many silicate samples.

### *Oxidizing Acids*

Oxidizing acids such as nitric, concentrated sulfuric, and perchloric acids at elevated temperatures, and aqua regia (which is a mixture of hydrochloric and nitric acids) are often used.

### *Fusion*

If the sample is not soluble in acids, it usually must be fused with an acidic or basic flux. The most popular **acidic flux** is potassium bisulfate ($KHSO_4$), which decomposes to potassium pyrosulfate ($K_2S_2O_7$) when melted. **Basic fluxes** include sodium carbonate or a mixture of sodium and potassium carbonates. If simple acidic or basic fusion will not render the sample soluble, **oxidizing fluxes** can be used. Sodium peroxide is often used as an oxidizing basic flux. Sodium carbonate plus potassium chlorate is also used as an oxidizing flux. Fusion is accomplished by mixing the sample intimately with the flux and heating until the mixture is molten. The melt is then cooled and dissolved in water or in acid.

In many cases, the sample is dissolved as completely as possible in an acid and the insoluble residue is then fused with a basic flux such as sodium carbonate.

### *Organic Solvents*

Some samples may be soluble in organic solvents such as chloroform, carbon tetrachloride, alcohols, ethers, and so on. These samples are usually organic materials that are insoluble in inorganic solvents such as water and acids.

The solvents usually used to dissolve various types of materials are summarized in Table 1-1.

**TABLE 1-1.** Solvents Used to Dissolve Inorganic Substances

| *Substances* | *Solvent* |
|---|---|
| Very active metals | Water |
| Moderately active metals | Nonoxidizing acids |
| Inactive metals | Oxidizing acids |
| Soluble salts | Water |
| Carbonates, hydroxides, oxides, phosphates, sulfides | Nonoxidizing acids |
| Silicates | Hydrofluoric acid or fusion |
| Various ores | Acidic, basic, or oxidizing fusion |

In some cases the type of determination being performed must be considered in the choice of solvent. For example, in the analysis of certain types of brasses and bronzes, tin may be determined either by a gravimetric or a volumetric method. For the gravimetric method, nitric acid is chosen as the solvent because metastannic acid, which is insoluble in nitric acid and can be filtered and ignited, is formed. A mixture of hydrochloric acid and nitric acid is used for the volumetric procedure because the hydrochloric acid will keep the tin in solution and the nitric acid will dissolve the copper.

## SEPARATION AND MEASUREMENT OF THE CONSTITUENT(S) BEING DETERMINED

After the sample has been dissolved, the constituent or constituents being determined must be separated from interfering substances before actual measurement. Separation methods include precipitation, extraction, chromatography, ion exchange, distillation, and electrodeposition. Most of these are discussed in later chapters. In many cases, an interfering substance can be masked to prevent the interference. **Masking** in analytical chemistry means conversion to a form that will not enter into the reaction being performed. This usually involves conversion to a soluble complex or to a different state of oxidation.

# PROBLEMS

**1** Arrange the following list of samples in order of decreasing particle size: 20/50, 4/8, 60/100, 300, 10/30, 100/200.

**2** Define and state the difference between (a) qualitative and quantitative analysis; (b) gross and determination sample; (c) micro- and ultramicroanalysis; (d) micro- and macroanalysis; (e) complete and partial analysis; (f) oxidizing and nonoxidizing acids; (g) acidic and basic flux; (h) aliquot and total volume; (i) organic and inorganic analysis; (j) representative sample and laboratory sample; (k) optimum weight sample and factor weight sample.

**3** Calculate the sample weight to be weighed and the range of weight that would be allowed for the following aliquoting procedures:

| *Optimum Weight* | *Volumetric Flask Used (mL)* | *Transfer Pipette Used (mL)* |
|---|---|---|
| (a) 0.015 | 1000 | 25 |
| (b) 0.50 | 250 | 50 |
| (c) 0.25 | 100 | 25 |
| (d) 0.33 | 500 | 100 |
| (e) 0.0035 | 500 | 10 |

*Ans.* (b) 2.500; 2.250–2.750 g

**4** Calculate the weight of the determination sample for the following aliquoting procedures:

| *Grams of Sample Weighed* | *Volumetric Flask Used (mL)* | *Volumetric Pipette Used (mL)* |
|---|---|---|
| (a) 1.036 | 100 | 25 |
| (b) 2.356 | 500 | 25 |
| (c) 8.579 | 1000 | 20 |
| (d) 2.624 | 250 | 10 |
| (e) 0.1039 | 500 | 50 |

*Ans.* (b) 0.1178

CHAPTER

# 2 The Analytical Balance

## HISTORY OF WEIGHING

Nomadic societies had little need for tools of measurement. However, with the advent of agrarian groups, length measurements became necessary to denote field boundaries, and weight measurements became necessary for the exchange of grains[1] and other products of land cultivation.

Although some agrarian societies had scholars of science, especially mathematicians, there existed little exchange between these men of science and the common man of the field. As a consequence, the system of weights and measures that evolved only provided the degree of precision needed to satisfy the comparisons for everyday commerce.

Allied to the development of modern science is the development of technology. The empirical consideration of physical reality must use reliable and suitable tools of investigation. A hoe would not be practical for the preparation of a 40-acre field, nor would a tractor be suitable for the preparation of a 4-$ft^2$ backyard garden. Tools of measurement are continually being developed and perfected to expand the body of scientific knowledge.

Our chief consideration as analytical chemists, at this stage, is the analytical balance. The principle of the analytical balance can be found in the ancient Greeks' laws of mechanics. Balances of rough construction features were used for hundreds of years for weight comparisons. However, the birth of this instrument as a reliable and suitable tool for precise investigative work occurred in the eighteenth century, when Antoine Lavoisier (1743–1794) recognized the necessity of precision in weighing for the clear understanding of the behavior of matter.

Lavoisier's experiments with the oxidation of metals required very precise weighings, a precision not possible on available balances. Lavoisier perfected the design of the analytical balance, and the accurate data he obtained led

[1] Grain continues to be a unit in the English system of weights; it is derived from the weight of a grain of wheat; 437.5 grains = 1 oz (avoirdupois).

him to a successful theory of combustion. Lavoisier left his scientific descendants with a very important legacy because chemistry has been equated with the quantification of data since his time. As a result of his far-reaching influence on the direction of chemistry, Lavoisier is often referred to as the father of modern chemistry.

## MASS AND WEIGHT

The analytical balance is so designed that objects of unknown mass are compared with objects of known mass until any variation in the two masses is less than the limits of detection of the balance. The concept of mass is sometimes obscure to the beginning student. **Mass** is the quantity of matter contained by an object and is constant. It is a measure of the inertia of a body, that is, the tendency of a body to remain at rest or in motion in a straight line.

By contrast, **weight** is a force measurement and is the result of the earth exerting a gravitational force on an object. It varies with geographical location, since the force of attraction depends on the distance of the object from the center of the earth.

These terms, *mass* and *weight*, might be illustrated in the following manner. When the astronauts of the Apollo 11 space project embarked from Cape Kennedy, they possessed several pencils, one of which was not used. This pencil had a certain mass and weight on the earth, but from the time of blast-off its weight changed continually until it finally reached zero (i.e., weightlessness). On the moon's surface its weight was approximately one-sixth the value on earth, because of the much smaller mass of the moon. During the same period, the mass, or quantity of matter of the pencil, was constant.

### *In Summary*

Mass is constant over the universe and depends only on the quantity of matter in the object; weight is not constant over the universe and depends on the quantity of matter and the relationship of the object to a much larger body that exerts a gravitational attraction on it.

Having drawn this careful distinction between mass and weight, we should add that they are frequently used interchangeably because the analytical balance is designed to compare an unknown weight with a known mass under conditions of equal gravitational attraction. The geographical location, and also the gravitational attraction, of the object and the known mass is essentially the same during the comparison, and, as a result, the weights are equal when the masses are equal. For convenience, the term *weight* is used as being identical with the term *mass*, and the distinction between the two is made only when necessary for clarity.

## THEORY OF WEIGHING

The two main types of analytical balances in use today in analytical laboratories are the substitution (single-pan) mechanical balance and the electronic balances, which are either single-pan or top-loading. The double-pan or equal-arm balance, similar to that used by Lavoisier, was the most popular until around 1950, when the single-pan mechanical balance was introduced and soon became the most widely used. The development and availability of electronic top-loading and single-pan balances around 1970 has led to increased use of this type of balance, especially in industrial and research installations. Examples of single-pan mechanical balances are shown in Figures 2-1 and 2-2.

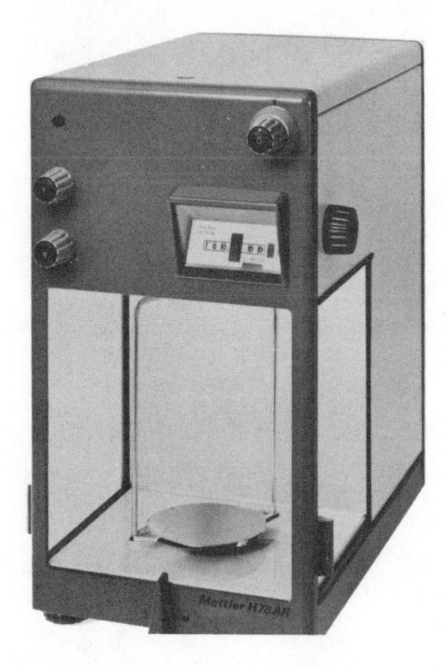


**Figure 2-1.** Substitution (single-pan) balance. (Courtesy of Mettler Instrument Corporation, Princeton, N.J.)

The theory of the mechanical balances is simply the theory of a lever of the first class, which has the fulcrum between the ends of the lever. We will explain the theory of the equal-arm mechanical balance first because the operation of the single-pan mechanical balance is easier to explain if the principle of the equal-arm lever is first understood. Electronic balances will be covered later.

### THE EQUAL-ARM MECHANICAL BALANCE

With the double-pan balance, the object to be weighed is placed on the left-hand pan, represented by $M_x$ in Figure 2-3, and known weights, $M_c$, are placed on the right. $M_x$ and $M_c$ would in reality represent not only the object

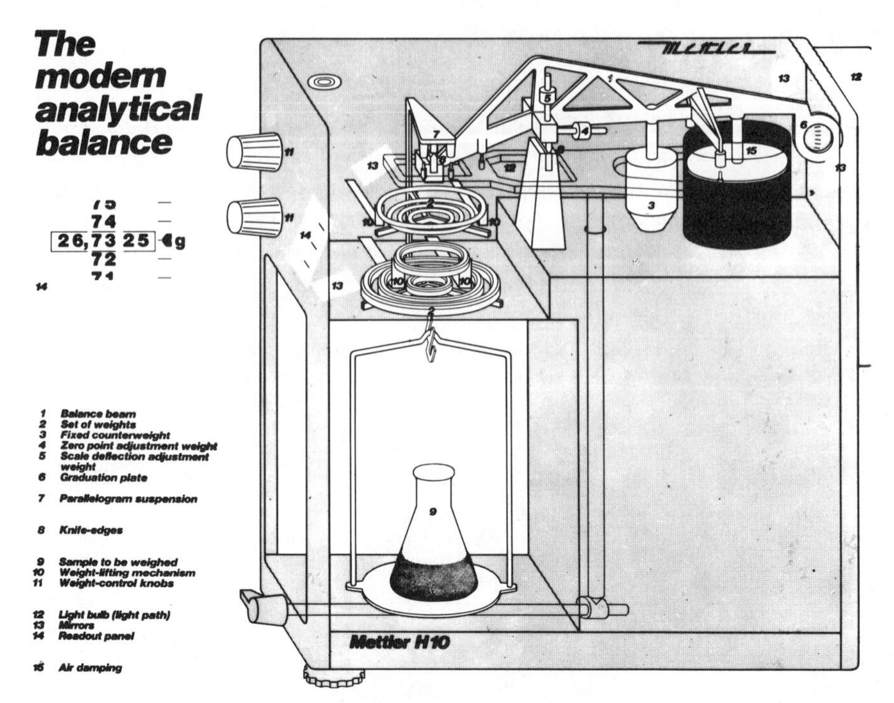


**Figure 2-2.** Side view of substitution balance. (Courtesy of Mettler Instrument Corporation, Princeton, N.J.)

and weights, respectively, but also the stirrups and pans suspended from the terminal knife edges. For simplicity, we shall assume that the two terminal suspension systems are matched, allowing us to consider only the additional objects on the pans.

The forces acting on the two lever arms under load conditions are

$$F_x = M_x g \qquad F_c = M_c g \tag{2-1}$$

where $g$ is the acceleration due to gravity.

In Figure 2-3, $AOB$ represents the equilibrium position of the balance. In this position the moments are equal, and, using the law of moments, we may write

$$F_x L_x = F_c L_c \tag{2-2}$$

Substituting eq. (2-1) into (2-2) yields

$$M_x g L_x = M_c g L_c \tag{2-3}$$

Since $L_x = L_c$ by design and $g$ is constant, this becomes

$$M_x = M_c$$

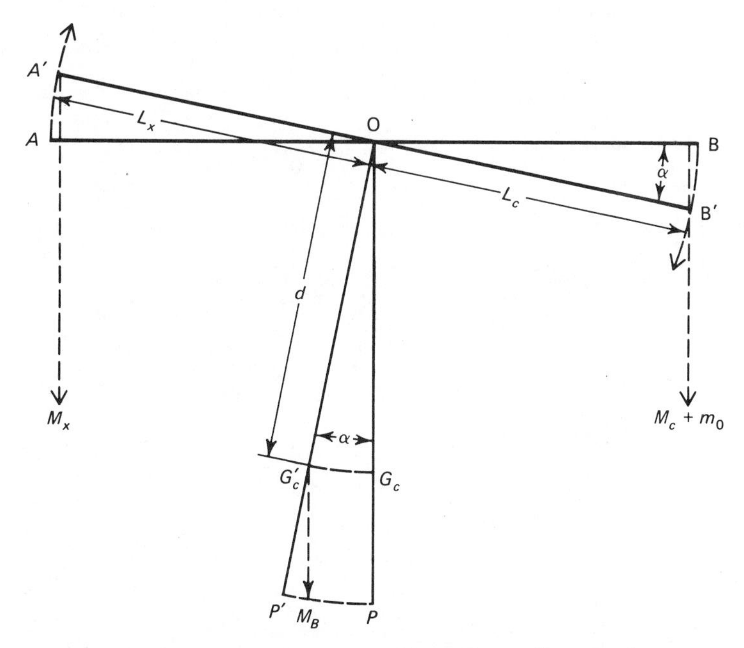


**Figure 2-3.** Displacement of beam by small overload.

When there is an excess mass on one of the pans, for example, $m_0$ in Figure 2-3, the additional force $m_0 L_c$ causes a deflection of the beam in a clockwise manner about the fulcrum $O$. There is a restoring torque due to the mass of the beam $M_B$ acting at its center of gravity $G_c'$ that causes the beam to stop and reverse its motion whenever the force of the torque exceeds the force causing the deflection. The beam would finally cease motion at the position $P'$ so that the addition of the mass $m_0$ actually causes a displacement of the center of gravity through the arc $G_c G_c'$. The sensitivity of the balance is a measure of the amount of weight necessary to cause a stated displacement of the center of gravity of the beam system. The sensitivity of the equal-arm balance varies with the amount of load on the pans, probably because of frictional effects that change as the weight on the pans increases.

## THE SUBSTITUTION MECHANICAL BALANCE

The substitution or single-pan balance is a modification of the equal-arm balance that uses a fixed weight on one arm of the balance and removable weights on the other, with the arm lengths and the total weight of the beam system adjusted so that the moments of the two arms are equal. A simplified diagram is given in Figure 2-4 and a more complete picture is shown in Figure 2-2.

For the design in Figure 2-4, eq. (2-2) becomes

$$F_f d_f = F_p d_p \tag{2-4}$$

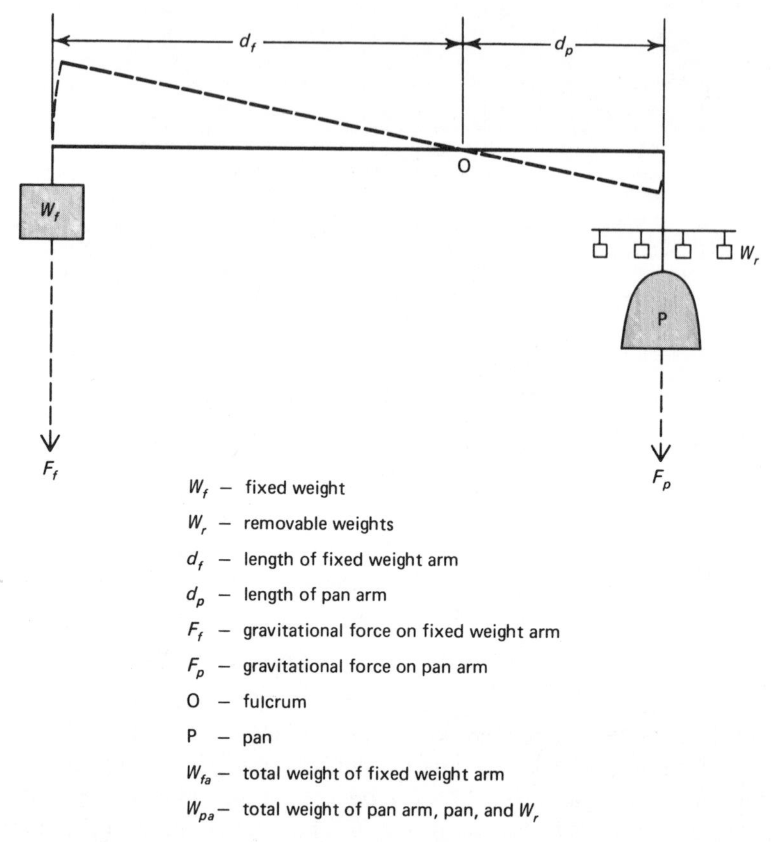


**Figure 2-4.** Substitution-balance-beam principle.

for which

$$F_p = W_{pa}g \tag{2-5}$$

and

$$W_{pa} = W_a + W_p + W_r \tag{2-6}$$

where $W_a$ is the weight of the beam arm, $W_p$ is the weight of the pan and suspension system, and $W_r$ is the sum of the removable weights. Since $F_f d_f$ is constant, $F_p d_p$ must always be the same when the balance is in equilibrium. The only variable in $F_p d_p$ is $W_r$, which can be changed by the removal of weights. As a consequence, weights are removed in the substitution balance rather than added as in the equal-arm balance.

If an object that has the exact weight of one of the removable weights is placed on the pan and the weight is removed, the beam will remain in equilibrium, since there is no effective change in the weight or moment of the beam arm.

However, it is seldom that the object and a removable weight (or weights) are exactly equal, since the smallest removable weight is 1 g on most laboratory balances. In these cases, the beam will be displaced from its equilibrium position as a result of the difference between the weight of the object and the sum of the weights removed and will assume the position of the dashed line in Figure 2-4. An optical scale projection is used to measure this displacement.

The essential characteristic of the substitution balance is that it is a constant-load balance. Like the equal-arm balance, the substitution balance is a lever of the first class, with the weights and pan on one side of the fulcrum and the counterweight on the other side, as shown in Figures 2-2 and 2-4. Note that the weighing pan and the set of weights share a common suspension from the beam. When an object is placed on the pan, weights that correspond to the weight of the object are removed. The beam is therefore under a constant load; thus, the balance operates at constant sensitivity. The beam is damped, and optical scale projection is used to determine that fraction of the mass less than the smallest weights represented by the dials. The degree of deflection of the beam from its rest-point position is translated on the optical scale to the mass required to restore the beam to its rest-point position. By reading the weights dialed and the optical scale projection, one has a direct measure of the mass of the object.

## ELECTRONIC BALANCES

Examples of electronic balances are shown in Figures 2-5 and 2-6. In this class of balance the position sensor is some type of electronic device and the restoring force some type of linear force generator or torque motor. Thus, the unknown weight is determined by comparison to a variable electromagnetic force that has been calibrated against a known standard weight. Most electronic balances use null-restoring devices in that the deflection caused by the unknown weight is sensed by the electronic position sensor and a proportionate force of opposite polarity is applied through an electronic force-generating device to restore the system to null. The current in this device can then be amplified and displayed as an analog of the weight.

**Figure 2-5.** Top-loading electronic balances. (Courtesy of Mettler Instrument Corporation, Princeton, N.J.)

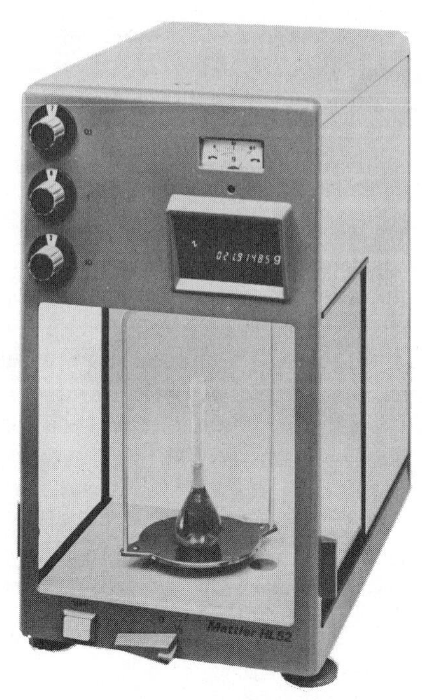


**Figure 2-6.** Upright electronic balance. (Courtesy of Mettler Instrument Corporation, Princeton, N.J.)

The first successful electronic balances could be considered as hybrids, since they were combinations of mechanical and electronic systems. Many of these used a conventional-type beam with removable weights on the sample-pan portion of the beam, balanced by a counterweight as in the single-pan mechanical balance. A portion of the restoring force is furnished, however, by a torque motor mounted at the fulcrum or by a variable-force coil. This allows both small and large samples to be weighed with equal ease. A block diagram of a top-loading hybrid electronic balance is shown in Figure 2-7.

Beamless balances are also available that use capacitive and photoelectric detectors and some type of force-coil compensation system, as shown in Figure 2-8.

Electronic balances suffer from a price disadvantage compared to the mechanical balances, but they are faster and much easier to operate because they have very few controls (often only one). Serial or repetitive weighings can be made much faster and with fewer errors because the weight of an object is indicated on the readout almost instantaneously after the object is placed on the pan. Also, most electronic balances have analog or digital outputs that can be connected to printers, calculators, recorders, and so on, to obtain permanent, statistically processed, printed records. One example is the use by pharmaceutical firms to count and record the number of pills placed in each bottle from the weight data. An excellent discussion of electronic balances can be found in the article by Ralph O. Leonard in the September 1976 issue of *Analytical Chemistry* [**48**, 879A (1976)].

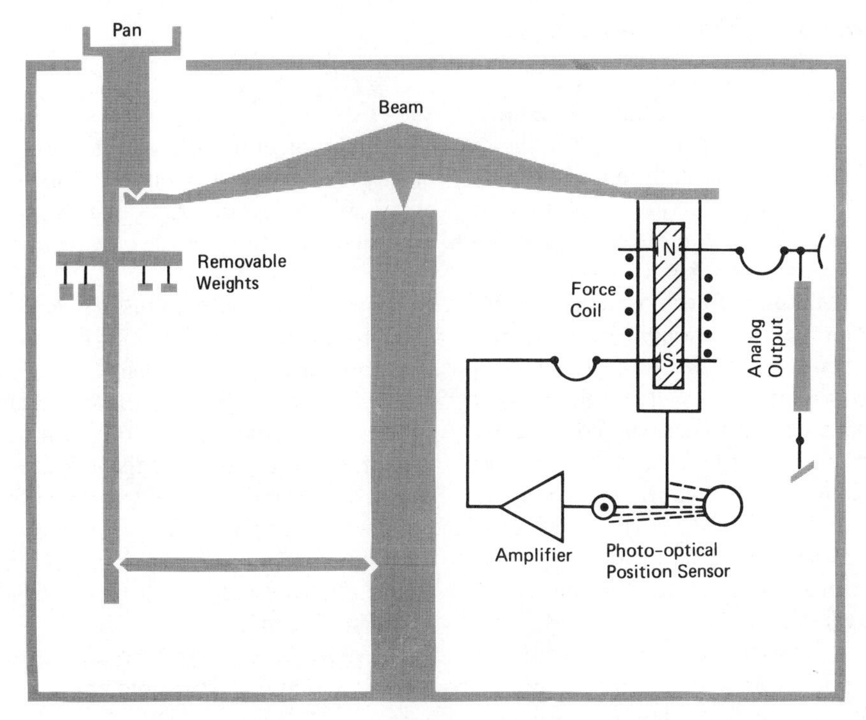


**Figure 2-7.** Schematic representation of top-loading hybrid balance.

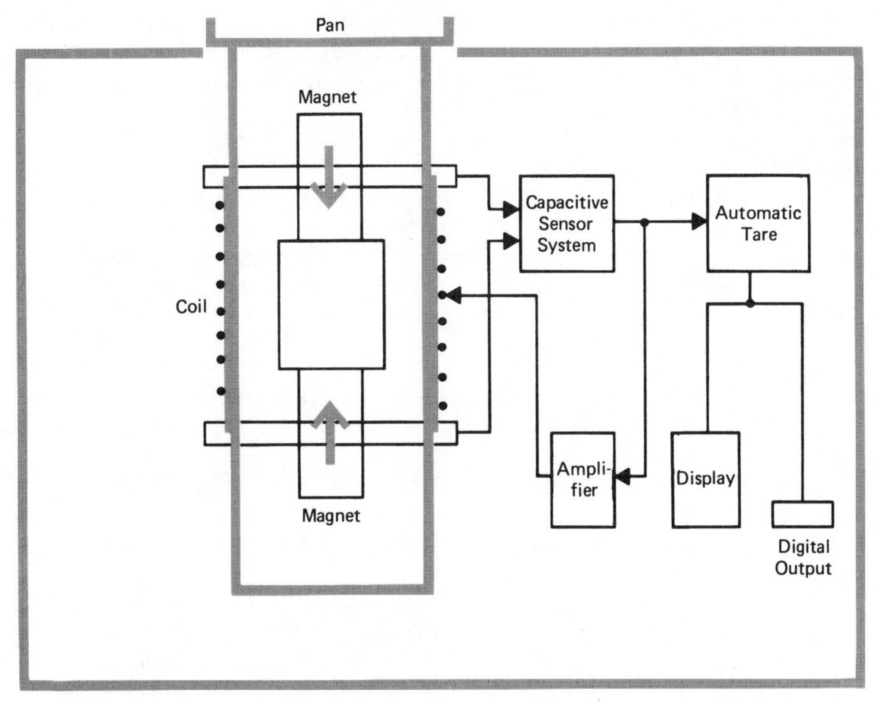


**Figure 2-8.** Schematic representation of a beamless force-coil balance system.

## ANALYTICAL WEIGHTS

The double-pan balance requires a set of comparison masses (weights) or analytical weights. A set for the chainomatic balance usually consists of the following denominations: 100, 50, 20, 10, 10, 5, 2, 2, and 1 g. A nonchainomatic balance would require additional weights of 500, 200, 200, 100, 50, 20, 20, and 10 mg. Other combinations that would allow a complete sequence of values are, of course, possible.

Manufacturers of weights have two very important considerations in selecting materials. The weight must be made of relatively dense material so that it is of convenient size, and the weight must be made of very noncorrosive material so that its mass will remain constant. Several metals satisfy these requirements to a degree that is directly related to their values; that is, gold or platinum would be the best metals to use except for their very high cost. Compromises are therefore necessary. The National Bureau of Standards (NBS) provides standards of construction and tolerances for weights. Specifications for various classes of weights in use in quantitative analysis are included in the National Bureau of Standards Circular 547, *Precision Laboratory Standards of Mass and Laboratory Weights.* The NBS tolerances are shown in Table 2-1.

Classes of analytical weights are identified by descriptive title and letter designation and comprise major classes M, S, S-1, and P. Principal characteristics of these classes are as follows.

**Class M**, high-precision scientific standards: designed for use as reference standards, for analytical work of the highest precision, and for specialized work requiring the most accurate determinations of mass. Weights of this class are made in one piece so that their effective density may be very exactly determined. The gram weights are normally made of brass and are plated with gold, platinum, or rhodium.

**TABLE 2-1.** Tolerances in Milligrams Applying to National Bureau of Standards Weight Classes

| *Denomination* | *Class M* | *Class S* | *Class S-1* | *Class P* |
|---|---|---|---|---|
| 1 kg | 5.0 | 2.5 | 10 | 20 |
| 500 g | 2.5 | 1.2 | 5.0 | 10 |
| 300 g | 1.5 | 0.75 | 3.0 | 6.0 |
| 200 g | 1.0 | 0.50 | 2.0 | 4.0 |
| 100 g | 0.50 | 0.25 | 1.0 | 2.0 |
| 50 g | 0.25 | 0.12 | 0.60 | 1.2 |
| 30 g | 0.15 | 0.074 | 0.45 | 0.90 |
| 20 g | 0.10 | 0.074 | 0.35 | 0.70 |
| 10 g | 0.050 | 0.074 | 0.25 | 0.50 |
| 5 g | 0.034 | 0.054 | 0.18 | 0.36 |
| 3 g | 0.034 | 0.054 | 0.15 | 0.30 |
| 2 g | 0.034 | 0.054 | 0.13 | 0.26 |
| 1 g | 0.034 | 0.054 | 0.10 | 0.20 |

**Class S**, scientific working standards and precise analytical weights: for use as working standards in the calibration of other weights, for exact analytical weighing without necessity for the application of correction factors, and for keyboard or dial-controlled weights in rapid weighing mechanized balances. Class S weights of 1 g and larger are normally made in two pieces, the knob being separable and serving to close an adjusting cavity beneath it. Lead may not be used as the adjusting material. The gram weights are made of brass, bronze, or stainless steel, brass or bronze being plated with platinum or rhodium or lacquered.

**Class S-1**, routine analytical weights: for general application as precise balance weights in quantitative analysis. With respect to design and materials, this class is identical in specifications with class S but is held to somewhat less exact adjustment tolerances to facilitate economy of manufacture.

**Class P** (formerly class S-2), rough analytical weights: for use as student weights with analytical balances and for routine analytical weighings in connection with methods not requiring or not capable of high precision. General requirements are similar to those of classes S and S-1 but less discriminatory in respect to materials of construction, finish, and packaging.

## WEIGHING ERRORS

There are quite a few errors in weighing that must be avoided or corrected for to obtain correct weights on an analytical balance. Such errors are grouped under four main classes: instrumental, static electrical, buoyant effect, and atmospheric effect.

1. **Instrumental errors** include any errors due to construction or operation of the balance or the weights used. Errors due to construction could be caused by unequal length of balance arms, chipped knife edges, corrosion on beam, distortion of beam due to excessive load, dust on beam surfaces, and contact between the vanes on the beam and the magnetic dampers. Most errors due to the weights are related to the difference between the actual weight and the labeled weight. For instance, there are usually two 10-g weights in each weight box, and it is unusual for them both to weigh within 0.05 mg of 10 g. This can be corrected for by calibration of the weights. Such calibrations should be checked periodically because handling or dropping the weights will often cause a change in the actual weight.
2. **Static electrical effects** are produced on glass when it is rubbed with a cloth or some grades of paper. When such a charged piece of glass is placed on a balance pan, part of the charge is slowly dissipated into the atmosphere but part of it conducted by the metal portions of the balance to create two or more zones of like charge on the balance. Since like charges repel each other, force will be exerted on the pans, causing an error in the weight observed. To offset such an error, a suitable length of time must elapse between cleaning the glass and the weighing operation in order for the charge to be completely dissipated. A small amount of a radioactive substance placed in the balance case

will dissipate the charge more rapidly, owing to the ionization of the air by the radioactive substance. This error is related to humidity and is more pronounced under conditions of low humidity.

3. **Buoyant effect errors** are due to the force that buoys up any object placed in a fluid. An object will be buoyed up by the difference in the density of the object and the density of the fluid it displaces—a cork floats on water because its density is less than that of water, whereas a stone sinks in water because its density is more than that of water. However, the weight of the stone in the water will be less than its weight in air, since air is less dense than water.

Such errors occur when the volume and density of the object being weighed differ from the volume and density of the weights used. These errors can be offset in some cases by using a counterpoise, or the correct weight can be determined by calculating the weight of the substance in vacuo. The weight in vacuo may be calculated by use of the equation

$$W_{vac} = W_{air} + W_{air}\left(\frac{0.0012}{d_o} - \frac{0.0012}{d_w}\right) \quad \textbf{(2-7)}$$

where $W_{vac}$ = weight in a vacuum
$W_{air}$ = weight in air
0.0012 = density of air
$d_o$ = density of the object
$d_w$ = density of the weights

An example will show how a weight in air can be corrected to the corresponding weight in vacuo.

• **Example.** What is the weight in vacuo of a sample of aluminum (density = 2.7 g/mL) that weighs 24.3567 g in air against brass weights (density = 8.4 g/mL)?

$$W_{vac} = 24.3567 + 24.3567\left(\frac{0.0012}{2.7} - \frac{0.0012}{8.4}\right)$$

$$= 24.3567 + 24.3567(0.000444 - 0.000143)$$

$$= 24.3567 + 24.3567(0.000301)$$

$$= 24.3567 + 0.0073 = 24.3640 \text{ g}$$

It should be noted that the greater the difference in the densities of weights and object, the greater will be the correction. If the density of the object is less than the density of the weights, as in the preceding example, the buoyancy correction will be positive. If the density of the object is more than the density of the weights, the buoyancy correction will be negative.

If the object being weighed is a closed vessel, the air inside must be at the same pressure as atmospheric pressure when weighed. At times it is necessary to weigh a closed container that contains gases other than air. An error will be involved unless correction is made for the difference in the weight of the gas and an equal volume of air. Another buoyant effect error is caused by weighing objects that are not at the temperature of the balance. A hot object will appear to weigh less than its true weight, whereas a cold object will appear to weigh more.

4. **Atmospheric effects.** Some materials will gain weight when exposed to the atmosphere, whereas others will lose weight. This may be due to pickup or loss of water, oxygen, carbon dioxide, and so on.

## PROBLEMS

**1** Calculate the sensitivity of the following in divisions per milligram, given the pointer deflections with the masses required to bring about those deflections.

(a) 5.6, 2.9 mg
(b) 2.5, 1.3 mg
(c) 3.7, 2.1 mg
(d) 4.0, 1.5 mg
(e) 1.8, 1.2 mg
(f) 4.5, 3.2 mg

*Ans.* (b) 1.9 div/mg

**2** If brass weights are used to weigh the following substances in air, calculate their weight in vacuo. (The density of brass is 8.4 g/mL.)

(a) 5.2092 g of aluminum (*d* = 2.7)
(b) 14.6270 g of gold (*d* = 19.3)
(c) 22.0000 g of iron (*d* = 7.9)
(d) 37.2945 g of brass
(e) 48.7234 g of mercury (*d* = 13.6)
(f) 27.5216 g of platinum (*d* = 21.4)

*Ans.* (b) 14.6258 g

**3** Calculate the vacuum weights of each of the substances given in Problem 2 if they were taken in air against platinum weights.

*Ans.* (b) 14.6271 g

**4** Calculate the densities of each of the following solids if the first number given is their weight in air against brass weights and the second number is their weight in vacuo.

(a) 20.0000 g, 20.0030 g
(b) 18.5000 g, 18.5010 g
(c) 9.0000 g, 8.9995 g
(d) 45.6250 g, 45.6225 g

*Ans.* (b) 6.1 g/mL

CHAPTER

# 3 Methods of Expressing Concentrations of Solutions

All the methods for expressing the concentration of a solution state in some way the amount of solute present in a given amount of the solution or the solvent. The amount may be stated as either weight or volume. In most analytical laboratories, the concentrations are expressed upon the basis of the weight of solute present in a stated volume of the solution, because accurate volume measurements are easily made. It is important for the analyst to understand the various ways in which concentration is expressed in order that he may convert from one method to another when necessary.

Concentrations may be expressed in qualitative form or in more exact terms. Thus, a solution may be referred to as a **concentrated** solution, meaning that it has relatively large amounts of solute per unit volume, or as a **dilute** solution, meaning that it contains relatively smaller amounts of solute per unit of volume. These terms are used mainly with acids and bases to designate the usual article of commerce (concentrated) or dilutions of it. However, the actual concentration of concentrated acids varies widely, as shown in Table 3-1. Other terms that have no quantitative meaning except with specifically stated systems are saturated, unsaturated, supersaturated, strong, weak, pure, contaminated, and so on. In each of these, the value must be defined or stated for the system being considered and will vary from system to system.

A solution for which the concentration is known with the degree of accuracy required for a quantitative determination is known as a **standard** solution. In most quantitative work this means that the analytical concentration must be within 0.2% (2 ppt) of the stated value.

It is often necessary to distinguish between the analytical and the equilibrium concentration of a solution. The **analytical** concentration is that concentration which can be calculated from the volume of the solution and the weight or volume of the solute dissolved in it. It can also be considered to be a measure of the amount of solute that can be recovered by evaporation of the solvent. The **equilibrium** (sometimes referred to as actual) concentration is the concentration of any particular form or forms of the solute present in the solution under the existing conditions. As an example, consider a solution of succinic acid

**TABLE 3-1.** Composition of Concentrated Reagents

| | HCl | $HNO_3$ | $H_2SO_4$ | $HC_2H_3O_2$ | $HClO_4$ | $H_3PO_4$ | $NH_4OH$ | NaOH |
|---|---|---|---|---|---|---|---|---|
| Molecular weight | 36.46 | 63.01 | 98.08 | 60.06 | 100.46 | 98.00 | 35.05 | 40.00 |
| Approx. specific gravity | 1.19 | 1.42 | 1.84 | 1.05 | 1.67 | 1.69 | 0.90 | 1.53 |
| Approx. % (w/w) | 37.0 | 69.5 | 96.0 | 99.6 | 70.0 | 85.0 | 57.6 | 50.0 |
| Approx. normality | 12.2 | 15.7 | 36.0 | 17.4 | 11.6 | 44.0 | 14.8 | 19.1 |
| Approx. milliliters per liter to prepare 1 *M* solution | 83.0 | 64.0 | 56.0 | 57.5 | 86.0 | 69.0 | 67.5 | 52.4 |

($H_2C_4H_4O_4$) prepared by dissolving 118 g (1 molecular weight) of the acid in sufficient water to make 1 liter of solution. The analytical concentration of such a solution is 1 *M* (1 molar). After being dissolved, however, the acid ionizes stepwise to form hydrogen ions, bisuccinate ions ($HC_4H_4O_4^-$), and succinate ions ($C_4H_4O_4^{2-}$), so that the solution contains these ions as well as the non-ionized molecules of the acid. As a consequence the actual concentration of molecular succinic acid is not 1 *M* but some smaller value. Further information (such as the ionization constants of the acid and the ionic strength) is needed to determine the exact concentration of succinic acid molecules. This exact concentration, if it is actually determined by calculation or other methods, is considered to be the equilibrium concentration of succinic acid molecules.

The methods described in the following section are used to express analytical concentration and include essentially all the methods in use for qualitative and quantitative expression of analytical concentration. The methods used to calculate equilibrium concentrations from analytical concentrations are described in later sections.

## METHODS OF EXPRESSING ANALYTICAL CONCENTRATIONS

### WEIGHT–VOLUME (w/v)

Concentration may be expressed as the weight of solute in a given volume of the solution such as grams per liter (g/liter) or milligrams per milliliter (mg/mL). The usual or clinical laboratory method of expressing the percentage concentration of a solution is in reality a weight–volume method, because it represents the weight in grams of the solute present in 100 mL of the solution. Also, the concentrations of the components of blood are often expressed as mg %(mg/100 mL of blood). Neither of these is a true percentage method but is based upon the fact that the density of many dilute solutions is approximately 1. The symbol (w/v) should be shown after the % sign to ensure no misunderstanding of the concentration. Another popular weight–volume method is the expression of the concentration as parts per million (ppm) or as parts per billion (ppb); 1 ppm equals 1 $\mu$g per $10^6$ $\mu$g, which is 1 $\mu$g/g. For solutions whose densities are approximately 1, 1 ppm is equivalent to 1 $\mu$g/mL or 1 mg/liter. Also, 1 ppb equals 1 nanogram (ng) per $10^9$ ng, which is 1 ng/g. For solutions whose densities are approximately one, 1 ppb is equivalent to 1 ng/mL or 1 $\mu$g/liter.

• **Example 1.** How many grams of sodium chloride are contained in (or are necessary to prepare) 500 mL of normal saline, which is 0.85% (w/v)?

$$\text{g} = \%\left(\frac{\text{g}}{100\text{ mL}}\right) \times \text{volume (mL)}$$

$$= \frac{0.85\text{ g}}{100\text{ mL}} \times 500\text{ mL} = 4.25\text{ g} \qquad \textbf{(3-1a)}$$

• **Example 2.** (a) One liter of a 5.0% (w/v) solution of glucose contains how many grams of glucose?

$$g = \frac{5.0\text{ g}}{100\text{ mL}} \times 1000\text{ mL} = 50\text{ g}$$

(b) One gallon (3800 mL) of a 10-ppm solution of ethyl alcohol contains how many milligrams of ethyl alcohol?

$$\text{mg} = \text{ppm}\left(\frac{\text{mg}}{1000\text{ mL}}\right) \times \text{mL}$$

$$= \frac{10\text{ mg}}{1000\text{ mL}} \times 3800\text{ mL} = 38\text{ mg} \qquad \textbf{(3-1b)}$$

## PERCENTAGE METHODS

The concentration of solutions may be expressed in terms of the percentage by weight or the percentage by volume in the case of a liquid solute. **Percent by weight** (w/w) is the weight in grams of solute present in 100 g of the solution, **percent by volume** is the number of milliliters of solute present in 100 mL of the solution. Since percentage is parts per hundred, the weight percent of a solution is expressed by

$$\%\text{ (w/w)} = \frac{\text{weight solute (g)}}{\text{weight solution (g)}} \times 100 \qquad \textbf{(3-2)}$$

Volume percent is

$$\%\text{ (v/v)} = \frac{\text{volume solute (mL)}}{\text{volume solution (mL)}} \times 100 \qquad \textbf{(3-3)}$$

• **Example 3.** Express the percent by weight of a solution that weighs 200 g and contains 25.0 g of sodium sulfate.

$$\%\,(\text{w/w}) = \frac{25.0\ \text{g}}{200\ \text{g}} \times 100\% = 12.5\%\,(\text{w/w})$$

• **Example 4.** Calculate the percent by volume of a solution prepared by adding 50.0 mL of methanol to 200 mL of water. Assume that the volumes are additive.

$$\text{volume solution} = 50.0 + 200 = 250\ \text{mL}$$

$$\%\,(\text{v/v}) = \frac{50.0\ \text{mL}}{250\ \text{mL}} \times 100\% = 20.0\%\,(\text{v/v})$$

• **Example 5.** Calculate the percent by weight of a solution prepared by dissolving 5.00 g of silver nitrate in 100 mL of water. Assume that the density of water is 1.00. (Note that this problem cannot be solved unless the density of the solvent is known.)

$$\begin{aligned} \text{wt solvent} &= 100\ \text{mL} \times 1.00\ \text{g/mL} = 100\ \text{g} \\ \text{wt solution} &= 100\ \text{g solvent} + 5.00\ \text{g}\ AgNO_3 \\ &= 105\ \text{g} \end{aligned}$$

$$\%\,(\text{w/w}) = \frac{5.00\ \text{g}}{105\ \text{g}} \times 100\% = 4.76\%\,(\text{w/w})$$

## DILUTION RATIO

The concentration of dilute acids and bases is often stated as the dilution ratio of the concentrated acid or base with water. The ratio is expressed as two numbers separated by a colon, with the first number representing the volume of concentrated acid or base and the second representing the number of volumes of water to which the acid or base is added. The ratio is usually given in parentheses before or after the name of the solute. Examples are HCl (1 : 1), $H_2SO_4$ (1 : 4), and (2 : 3) $H_3PO_4$. To prepare these particular solutions, 1 volume of concentrated sulfuric acid is added to 4 times its volume of water, and 2 volumes of concentrated phosphoric acid are added to 3 volumes of water. The compositions of the concentrated reagents used to prepare solutions expressed as dilution ratios are shown in Table 3-1. Concentrations expressed in this way may be converted easily to volume percent if the volumes are assumed to be additive. Organic solvent mixtures are usually designated using + instead of :, as in acetone–ether (2 + 3).

• **Example 6.** What volumes of concentrated acetic acid and of water would be used to prepare 300 mL of 1 : 5 solution? To solve such a problem, first obtain the total of the volumes shown in the ratio. Then divide the volume of solution to be prepared by this number to get the value of 1 volume.

$$1 + 5 = 6 \text{ total volumes in ratio}$$
$$300/6 = 50 \text{ mL} = 1 \text{ volume} = \text{volume acid used}$$
$$50 \times 5 = 250 \text{ mL} = \text{volume water used}$$

To prepare the solution, add 50 mL of concentrated acetic acid to 250 mL of water.

## MOLAR (*M*) AND FORMAL (*F*) CONCENTRATIONS

The **molar** concentration states the number of moles (gram molecular weight, MW) of the solute present in 1 liter of the solution; the **formal** concentration states the number of gram formula weights present in 1 liter of solution. From these definitions it might appear that the two methods are identical, and for many solutions they are. The use of formal concentration in place of molar concentration was suggested originally because there is no such thing as a "molecule" of an ionic substance; ionic substances exist as individual ions even in the pure or solid state. Also, in the case of organic compounds, which do exist as molecules, association or dissociation often occurs after the molecules are dissolved. These two methods could be used to distinguish between the analytical and the equilibrium concentration, with formal representing the analytical and molar being used for the equilibrium concentration. This particular distinction, however, has never come into general use, and as a consequence some authors emphasize molar, some emphasize formal, and some use both. For our purposes in this text, we shall assume that the two methods are identical and shall use molar rather than formal concentrations. One of the main reasons for the preference of molar over formal is the general use of the term *mole* in all fields of chemistry to represent one Avogadro's number of the entity in question, whether it be an atom, a molecule, an ion, an electron, or a compound formed by association of two or more molecules.

The relationship between the weight of solute and the number of moles of solute in a given volume of a solution of a given molar concentration may be developed as follows:

$$\text{moles solute dissolved} = \text{moles solute in solution}$$

$$\text{moles solute dissolved} = \frac{\text{g solute}}{\text{MW solute (g/mole)}}$$

$$\text{moles of solute in solution} = M \text{ (moles/liter)} \times \text{liters}$$

so that

$$M \text{ (moles/liter)} \times \text{liters} = \frac{\text{g solute}}{\text{MW solute (g/mole)}}$$
$$= \text{moles solute in solution} \qquad \textbf{(3-4)}$$

Equation (3-4) can be rearranged to give

$$\mathrm{MW}\left(\frac{\mathrm{g}}{\mathrm{mole}}\right) \times M\left(\frac{\mathrm{moles}}{\mathrm{liter}}\right) \times \mathrm{liters} = \mathrm{g\ solute} \qquad \textbf{(3-5)}$$

which indicates how the units cancel.

In the analytical laboratory, measurements are made in milliliters more often than in liters. Substitution of milliliters per 1000 for liters in eq. (3-5) gives

$$\mathrm{MW}\left(\frac{\mathrm{g}}{\mathrm{mole}}\right) \times M\left(\frac{\mathrm{moles}}{\mathrm{liter}}\right) \times \frac{\mathrm{mL}}{1000\ (\mathrm{mL/liter})} = \mathrm{g\ solute} \qquad \textbf{(3-6)}$$

Similarly, when needed, milligrams per 1000 can be substituted for grams, millimoles per 1000 for moles, and millimolecular weights per 1000 for molecular weight. One millimole (mmole) equals 0.001 mole and 1 mMW (millimolecular weight) equals 0.001 MW or molecular weights per 1000. Equations (3-5) and (3-6) can be rearranged to calculate any one of the terms as long as the other three are known. The two most common forms of the calculations used in the analytical laboratory are the calculation of the molarity of a solution prepared by dissolving a known weight in grams of solute in a given volume of solution and the calculation of the weight in milligrams of solute present in a given volume of a solution of known molar concentration. The equations for these are

$$M = \frac{\mathrm{g\ solute}}{\mathrm{MW} \times \mathrm{liters}} = \frac{\mathrm{g\ solute}}{\mathrm{MW} \times \mathrm{mL}/1000} = \frac{\mathrm{g\ solute} \times 1000}{\mathrm{MW} \times \mathrm{mL}}$$

$$= \frac{\mathrm{mg\ solute}}{\mathrm{MW} \times \mathrm{mL}} \qquad \textbf{(3-7)}$$

and

$$\mathrm{MW} \times M \times \mathrm{mL} = \mathrm{mg\ solute} \qquad \textbf{(3-8)}$$

For formal concentrations, substitute formula weight (FW) for molecular weight and formal ($F$) for molar ($M$) in the expressions (3-7) and (3-8).

• **Example 7.** A solution of sulfuric acid (MW, 98.0) contains 4.90 g of $H_2SO_4$ in 400 mL. Calculate the molar concentration. Using eq. (3-7)

$$M = \frac{\mathrm{g\ solute} \times 1000}{\mathrm{MW} \times \mathrm{mL}} = \frac{4.90\ \mathrm{g} \times 1000\ \mathrm{mL/liter}}{98.0\ \mathrm{g/mole} \times 400\ \mathrm{mL}} = 0.125\ M$$

• **Example 8.** How many grams of silver nitrate (MW, 169.9) are necessary to prepare 500 mL of 0.1250 $M$ solution?

$$\mathrm{g} = \mathrm{MW} \times M \times \mathrm{liters} = 169.9\ \mathrm{g/mole} \times 0.1250\ \mathrm{mole/liter} \times \frac{500}{1000}\ \mathrm{liter}$$

$$= 10.62\ \mathrm{g}$$

• **Example 9.** What is the molar concentration of a 0.85% (w/v) solution of sodium chloride (MW, 58.4)? Since %(w/v) represents grams per 100 mL, 1 liter contains 0.85 g × 1000 mL/100 mL, or 8.5 g.

$$M = \frac{\text{g/liter}}{\text{MW}} = \frac{8.5\ \text{g}}{58.4\ \text{g}} = 0.146\ M$$

• **Example 10.** How many millimoles are present in 150 mL of a 0.025 $M$ solution?

$$\text{mmoles} = \text{mL} \times M = 150 \times 0.025 = 3.75\ \text{mmoles}$$

## NORMAL (*N*) CONCENTRATION

Another commonly used method of expressing the concentration of a solution in analytical chemistry is the normal concentration. The normality of a solution expresses the concentration in terms of the number of gram equivalent weights of solute per liter of solution. Thus, a 1 $N$ solution contains 1 gram equivalent weight of solute per liter of solution. It is clear from the preceding discussion that an understanding of normality depends on an understanding of the term *equivalent weight* (eq wt). This will be discussed in detail in the following section, entitled Equivalent Weights.

Since the normality of a solution is the number of gram equivalent weights of solute per liter of solution, the relationship between the weight of solute and the number of equivalents in a given volume of a solution of stated normality is given by

$$N\ (\text{eq/liter}) \times \text{liter} = \text{number equivalents} = \frac{\text{g solute}}{\text{eq wt solute (g/eq)}} \qquad \textbf{(3-9)}$$

Because volume measurements in milliliters are more common than those in liters, this is easily converted to

$$N \times \text{mL} = \text{number meq} = \frac{\text{g} \times 1000}{\text{eq wt}} = \frac{\text{mg}}{\text{eq wt}} = \frac{\text{g}}{\text{meq wt}} \qquad \textbf{(3-10)}$$

One milliequivalent weight (meq wt) is usually referred to as a **milliequivalent (meq)**.

It will be shown in the discussion of equivalent weights that, for a balanced chemical reaction, the number of equivalents of each reactant and product must be identical. This statement is the basis of all quantitative work and explains the reason most analytical chemists prefer to use normal rather than other methods of expressing concentration.

• **Example 11.** How many grams of sodium sulfate (MW, 142.0; eq wt, 71.00) are necessary to prepare 200.0 mL of a 0.5000 $N$ solution for precipitation reactions?

$$\begin{aligned} g &= \text{eq wt} \times N \times \text{mL}/1000 \\ &= 71.00 \text{ g/eq} \times 0.5000 \text{ eq/liter} \times \frac{200}{1000} \text{ liter} \\ &= 7.100 \text{ g} \end{aligned}$$

• **Example 12.** Calculate the normality for neutralization reactions of a solution of sodium hydroxide (MW, 40.00; eq wt, 40.00) prepared by dissolving 100.0 g of sodium hydroxide in sufficient water to make 1000 mL of solution.

$$N = \frac{\text{g} \times 1000}{\text{eq wt} \times \text{mL}} = \frac{100.0 \times 1000}{40.00 \times 1000} = 2.500\ N$$

## TITER METHODS

The **titer** of a solution is defined as the weight of a pure substance that corresponds to 1 mL of the solution. "Corresponds to" in this case is taken to mean (a) will react exactly with, (b) is chemically equivalent to, (c) is contained in, (d) may be obtained from, or (e) may be substituted for. This method is used primarily to state the weight of some substance *other than the solute* with which 1 mL of the solution will react exactly. The important factor here is that the concentration is expressed in terms of some substance other than the solute and that the relationship is a weight–volume relation, so that volumes may be converted directly to weights by a simple multiplication.

As an example, the concentration of a silver nitrate solution that is to be used to determine the amount of sodium chloride by a titration could be expressed as the number of milligrams of sodium chloride with which 1 mL of the solution will react and to which it is chemically equivalent. Thus, 1 mL of a 1 $N$ solution of silver nitrate will react exactly with 58.44 mg of sodium chloride, so that the concentration could be expressed as being equivalent to 58.44 mg of sodium chloride per milliliter.

One symbolization for titer is to use a capital $T$ preceded by the formula of the substance specified (which we shall call the reactant) and followed by the formula of the solute written as a subscript. For the silver nitrate solution mentioned this would be

$$\text{NaCl}\ T_{\text{AgNO}_3} = 58.44 \text{ mg}$$

and would be read "the sodium chloride titer of the silver nitrate solution is 58.44 mg." Another method of expressing the titer is to give the name or formula of the solute followed by a statement indicating the number of milligrams of the reactant with which 1 mL of the solution will react. For the silver nitrate solution described, these could be shown as

silver nitrate, 1 mL = 58.44 mg NaCl

or

silver nitrate, 58.44 mg NaCl per mL

The calculation of the weight of solute present in a solution for which the concentration is expressed in the titer system involves the determination of the weight relationship of the reaction between the solute and the reactant. Since substances always react in the ratio of their equivalent weights, the normal concentration of the solution will be identical to the normal concentration of a solution of the reactant containing the titer weight in 1 mL. As a consequence, the first step is to calculate the normal concentration of a solution of the reactant containing the titer weight in 1 mL and then to calculate the weight of solute in a given volume of a solution of this normality. For example,

$$N_{\text{react}} = \frac{\text{titer (mg/mL)}}{\text{eq wt (react)}} \tag{3-11}$$

$$g_{\text{solute}} = \text{eq wt (solute)} \times N \text{ (reactant)} \times \text{liters (solution)} \tag{3-12}$$

These can be combined to give

$$g_{\text{solute}} = \frac{\text{eq wt (solute)} \times \text{titer (mg/mL)} \times \text{mL (solution)}}{\text{eq wt (reactant)} \times 1000} \tag{3-13}$$

which can be used for calculations involved in the preparation of solutions that utilize the titer system for concentration.

• **Example 13.** How many grams of silver nitrate (MW, 169.9) are necessary to prepare 500.0 mL of a solution that has a chloride (at wt, 35.45) titer of 0.5000 mg?

$$g_{AgNO_3} = \frac{169.9 \times 0.5000 \times 500.0}{35.45 \times 1000} = 1.198 \text{ g}$$

• **Example 14.** Calculate the sodium carbonate (eq. wt, 53.00) titer of 0.1037 *N* hydrochloric acid.

$$\begin{aligned} \text{titer} &= N \text{ (solution)} \times \text{eq wt (reactant)} \\ &= 0.1037 \times 53.00 = 5.496 \text{ mg/mL.} \end{aligned} \tag{3-14}$$

• **Example 15.** Calculate the $SO_3$ (MW, 80.06) titer of a $BaCl_2$ solution that was prepared by dissolving 24.43 g of $BaCl_2 \cdot 2H_2O$ (MW, 244.3) in sufficient water to prepare 1000 mL of solution

$$\text{titer} = \frac{\text{g (solute)} \times \text{eq wt (reactant)} \times 1000}{\text{eq wt (solute)} \times \text{mL (solute)}} \qquad \textbf{(3-15)}$$

$$= \frac{24.43 \times 40.03 \times 1000}{122.15 \times 1000}$$

$$= 8.006 \text{ mg/mL}$$

• **Example 16.** How many milligrams of $Fe_2O_3$ are present in an iron ore sample if the sample required 35.37 mL of a $K_2Cr_2O_7$ solution, which has a $Fe_2O_3$ titer of 5.000?

$$\text{mg (reactant)} = \text{titer} \times \text{mL (solution)}$$

$$\text{mg } Fe_2O_3 = \text{titer} \times \text{mL} = 5.000 \times 35.37 = 176.9 \text{ mg} \qquad \textbf{(3-16)}$$

The use of the titer concept is especially favored in quality control laboratories where the same reagent is used for the same procedure many times a day. Reading volumes and reporting weights directly in this manner short-circuits the calculations otherwise needed, saves time, and avoids errors.

## EQUIVALENT WEIGHTS

Students often have trouble grasping the concept of equivalent weight. Although there is nothing mysterious about the concept of equivalent weight, perhaps a clearer understanding of the term may be gained by considering its historical origin. The term *electrochemical equivalent weight* was first applied in connection with the famous electrolysis experiments performed in 1833 by Michael Faraday. Faraday assembled a simple electrolysis cell, shown in Figure 3-1, that allowed him to observe the effect of passing an electrical current

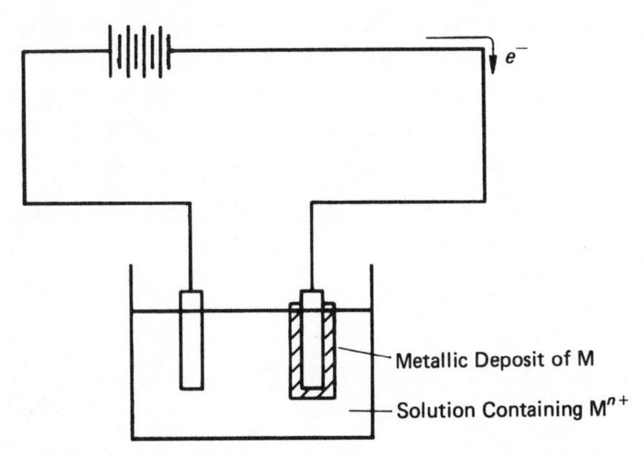


**Figure 3-1.** Schematic diagram of the simple electrolysis apparatus used by Faraday.

through various solutions. One of the observations with this electrolysis cell was that solutions containing dissolved salts could be decomposed by the passage of an electric current. At the same time, it was observed that a metallic deposit of the cation of the salt formed on one of the electrodes. The weight of the metallic deposit formed on the electrode for the passage of the same amount of current varied with the cation in solution. The weights of the metallic deposits observed for a series of cations for the passage of *equivalent amounts* of electricity through the cell became known as the electrochemical equivalent weights of the cations. Over the years, the term has been shortened to simply *equivalent weight*.

The quantity of electricity that suffices to deposit 1 electrochemical gram equivalent weight is 96,487 coulombs, or 1 faraday. Although unknown to Faraday at the time, this amount of electricity corresponds to exactly 1 mole of electrons, or Avogadro's number of electrons. From your previous study of general chemistry, it is apparent that the phenomenon Faraday was observing was the electrochemical reduction of various cations by 1 mole of electrons. That is,

$$\frac{1}{n}M^{n+} + 1e^- = \frac{1}{n}M^0 \tag{3-17}$$

where 1 mole of electrons will reduce $1/n$ moles of a cation whose charge is $n+$. Thus, the electrochemical equivalent weight of $M^{n+}$ is the atomic weight of $M$ divided by the charge on the ion. We can therefore define an equivalent weight as that weight of a substance which may in some way be related to 1 mole of electrons.

It is apparent from the preceding discussion that the concept of equivalent weight is basically an electrochemical concept, and consequently it is easy to see how this concept may be related to direct electron transfer reactions in solution. Consider the following half-reaction for the reduction of dichromate ion in acid solution; the equation describes the passage of 1 equivalent of electricity (i.e., 1 mole of electrons),

$$\tfrac{1}{6}Cr_2O_7^{2-} + \tfrac{7}{3}H^+ + 1e^- = \tfrac{1}{3}Cr^{3+} + \tfrac{7}{6}H_2O \tag{3-18}$$

Since the half-reaction has been written corresponding to 1 equivalent of electricity, the stoichiometric coefficients represent the fractional amount of a formula weight for each species that is equivalent to 1 mole of electrons. Thus, the weight of $K_2Cr_2O_7$ equivalent to 1 mole of electrons (i.e., the equivalent weight) would be one-sixth of the formula weight of $K_2Cr_2O_7$. If this half-reaction is combined with a second half-reaction capable of furnishing 1 equivalent of electrons, a balanced chemical reaction will result:

$$\tfrac{1}{2}Zn^0 = \tfrac{1}{2}Zn^{2+} + 1e^- \qquad \text{one equivalent of electricity furnished}$$

$$\tfrac{1}{6}Cr_2O_7^{2-} + \tfrac{7}{3}H^+ + 1e^- = \tfrac{1}{3}Cr^{3+} + \tfrac{7}{6}H_2O \qquad \text{one equivalent of electricity consumed}$$

$$\tfrac{1}{6}Cr_2O_7^{2-} + \tfrac{1}{2}Zn^0 + \tfrac{7}{3}H^+ = \tfrac{1}{3}Cr^{3+} + \tfrac{1}{2}Zn^{2+} + \tfrac{7}{6}H_2O \tag{3-19}$$

where the stoichiometric coefficients of each of the products and reactants represent the fractional amount of a formula weight corresponding to 1 equivalent of electricity. Thus, each of the fractional amounts of a formula weight for each product and reactant represents 1 equivalent weight.

Because the number of equivalents of electricity furnished or consumed by each half-reaction comprising the final balanced redox reaction must be equal, it is apparent that the number of equivalents of each species produced or consumed in the balanced reaction must also be equal. This fact may be used to advantage in simplifying many analytical stoichiometric calculations, even in the absence of the balanced chemical equation. For example, suppose that, in a particular experiment involving the reaction between dichromate ion and zinc metal, 3 equivalents of dichromate are consumed. At the same time, 3 equivalents of zinc metal must also have been consumed, and 3 equivalents of zinc ion produced. On a weight basis, 3 equivalents of zinc corresponds to 98.0 g, since from the half-cell for zinc, each equivalent represents $\frac{1}{2}$ a gram formula weight. Even in a sequence of stoichiometrically related reactions, 1 equivalent of one species will always consume or produce 1 equivalent of a second species.

The concept of equivalents may be extended to include other reaction types, such as acid–base, precipitation, and complexation reactions. Since

$$H^+_{aq} + 1e^- = \tfrac{1}{2}H_2 \tag{3-20}$$

it is apparent that 1 mole of hydrogen ions may be considered equivalent to 1 mole of electrons. Thus in an acid–base reaction, any substance furnishing or reacting with 1 mole of $H^+_{aq}$ represents 1 equivalent.

In ionic precipitation or complexation reactions, the charge on the ion involved in the particular precipitation or complex formation reaction is related to the number of equivalents. In the precipitation reaction involving barium ion and sulfate ion,

$$Ba^{2+} + SO_4^{2-} \longrightarrow BaSO_4 \tag{3-21}$$

the equivalent weight of $Na_2SO_4$ will be determined by the ionic charge. Thus, 1 mole of $Ba^{2+}$ represents 2 equivalents, since 2 faradays are required to reduce 1 mole of $Ba^{2+}$ ion. Since 1 mole of barium ion reacts with 1 mole of sulfate ion, 1 mole of sulfate must be 2 equivalents as well. Since 1 mole of $Na_2SO_4$ furnishes 1 mole of sulfate ion to the solution, 1 mole of sodium sulfate must represent 2 equivalents, and the equivalent weight of sodium sulfate in this reaction will be the formula weight divided by 2.

In the complex formation reaction involving cyanide ion and silver ion,

$$Ag^+ + 2CN^- \longrightarrow Ag(CN)_2^- \tag{3-22}$$

the equivalent weight of KCN will be determined by the amount of KCN that reacts with 1 mole of silver ion. In the particular reaction under discussion, 1 mole of silver ion, which represents 1 equivalent since 1 mole of electrons is required to reduce it, will react with 2 moles of cyanide ion in the form of KCN.

Therefore, the equivalent weight of KCN will be twice the formula weight for this reaction. (See Chapter 14 for a discussion of complexes.)

Even though there is no electron transfer involved in acid–base or precipitation reactions, the number of equivalents of each species involved in the reaction will always be equal. This is due to the conservation of charge in a balanced equation. Thus,

$$\begin{array}{ccccc} 2Sb^{3+} & + & 3S^{2-} & \longrightarrow & Sb_2S_3 \\ 6\text{ eq} & & 6\text{ eq} & & 6\text{ eq} \\ 2\text{ moles} \times 3\dfrac{\text{eq}}{\text{mole}} & & 3\text{ moles} \times 2\dfrac{\text{eq}}{\text{mole}} & & \end{array} \quad \textbf{(3-23)}$$

It is apparent from the preceding discussion, that the equivalent weights of substances involved in acid–base, precipitation/complexation reactions, and redox reactions are all based, either directly or indirectly, on the weight associated with 1 mole of electrons.

Now that we have some historical perspective and understand the origin of the concept of equivalent weights, we shall consider in detail how this concept can be applied to various classes of chemical reactions. The two main classes of chemical reactions are those in which there is an electron transfer (oxidation–reduction, redox) and those in which there is no electron transfer. The majority of the reactions in which there is no electron transfer may be classified as being neutralization (acid–base) or precipitation reactions.

In considering the various criteria in the following discussion of the determination of the equivalent weights for various classes of chemical reactions, the student should keep in mind that they are all ultimately based in some manner on the electrochemical equivalence with 1 mole of electrons.

## EQUIVALENT WEIGHTS FOR REACTIONS IN WHICH THERE IS NO ELECTRON TRANSFER

### *Neutralization*

**Acids.** The equivalent weight of an acid in a neutralization reaction is the weight of the acid that contains 1 gram-atom (g-atom) of replaceable hydrogen, since 1 g-atom of hydrogen is electrochemically equivalent to 1 mole of electrons. For a reaction in which all the replaceable hydrogens of the acid take part in the reaction, the equivalent weight equals the molecular weight divided by the number of replaceable hydrogens in one molecule. If only a portion of the replaceable hydrogens is involved because of stepwise ionization or neutralization of the acid, the equivalent weight equals the molecular weight divided by the number actually involved. Examples for complete reaction of available hydrogens are

$$\frac{HCl}{1} \quad \frac{HNO_3}{1} \quad \frac{HC_2H_3O_2}{1} \quad \frac{H_2SO_4}{2} \quad \frac{H_2CO_3}{2}$$

and for reaction of only a portion of available hydrogens are

$$H_3PO_4 + 2NaOH \longrightarrow Na_2HPO_4 + 2HOH \qquad \frac{H_3PO_4}{2} \qquad (3\text{-}24)$$

$$H_2CO_3 + NaOH \longrightarrow NaHCO_3 + HOH \qquad \frac{H_2CO_3}{1} \qquad (3\text{-}25)$$

**Bases.** The equivalent weight of a base is that weight that will react with 1 equivalent of hydrogen ions or 1 mole of hydrogen ions. Since 1 mole of hydrogen ions will react with 1 mole of hydroxyl ions, according to the reaction

$$\underset{1\ \text{eq}}{H^+_{aq}} + \underset{1\ \text{eq}}{OH^-} \longrightarrow HOH \qquad (3\text{-}26)$$

the equivalent weight of a base containing or producing hydroxyl ions will be that weight which contains or produces 1 gram-ion (g-ion) of hydroxyl. For example,

$$\frac{NaOH}{1} \quad \frac{Ba(OH)_2}{2} \quad \frac{Al(OH)_3}{3}$$

**Substances That React with Acids or Bases or Are Produced by Reaction of Acids and Bases.** The equivalent weight is the molecular weight divided by the number of hydrogens or hydroxyls with which 1 mole of the substance reacts or by the number of hydrogens or hydroxyls involved in the reaction that produced the substance. For example,

$$\begin{array}{lcl} Na_2CO_3 + HCl & \longrightarrow & NaHCO_3 + NaCl \\ \dfrac{Na_2CO_3}{1} & & \dfrac{NaHCO_3}{1} \quad \dfrac{NaCl}{1} \end{array} \qquad (3\text{-}27)$$

$$\begin{array}{lcl} Na_2CO_3 + 2HCl & \longrightarrow & H_2CO_3 + 2NaCl \\ \dfrac{Na_2CO_3}{2} & & \dfrac{2NaCl}{2} \quad \text{or} \quad \dfrac{NaCl}{1} \end{array} \qquad (3\text{-}28)$$

$$\begin{array}{lcl} 2NaOH + H_2SO_4 & \longrightarrow & Na_2SO_4 + 2HOH \\ & & \dfrac{Na_2SO_4}{2} \end{array} \qquad (3\text{-}29)$$

$$\begin{array}{lcl} (NH_4)_2SO_4 + 2NaOH & \longrightarrow & Na_2SO_4 + 2HOH + 2NH_3 \\ \dfrac{(NH_4)_2SO_4}{2} \quad \dfrac{NaOH}{1} & & \dfrac{Na_2SO_4}{2} \quad \dfrac{HOH}{1} \quad \dfrac{NH_3}{1} \end{array} \qquad (3\text{-}30)$$

### *Precipitation*

In precipitation reactions the equivalent weight is that weight of a substance that will be electrochemically equivalent to 1 mole of electrons. For substances furnishing cations involved in a precipitation reaction, the equivalent weight will be the formula weight divided by the charge on the cation. Thus, the equivalent weight of $Pb(NO_3)_2$ in the precipitation reaction

$$Pb^{2+} + 2Cl^- \longrightarrow PbCl_2 \qquad \textbf{(3-31)}$$

will be $Pb(NO_3)_2/2$, since 2 equivalents of electricity are required to reduce 1 mole of $Pb^{2+}$. For substances furnishing anions involved in a precipitation reaction, the equivalent weight will be based on the cation with which it reacts. If NaCl is used to provide the $Cl^-$ in the example described by eq. (3-31) the equivalent weight of NaCl would be the formula weight divided by 1, since 2 moles of chloride ion react with 2 equivalents of lead. Other examples are

$$\begin{array}{llcll} AgNO_3 + NaCl & & \longrightarrow & NaNO_3 + AgCl & \\ \dfrac{AgNO_3}{1} & \dfrac{NaCl}{1} & & \dfrac{NaNO_3}{1} & \dfrac{AgCl}{1} \end{array} \qquad \textbf{(3-32)}$$

$$\begin{array}{llcll} BaCl_2 + K_2SO_4 & & \longrightarrow & BaSO_4 + 2KCl & \\ \dfrac{BaCl_2}{2} & \dfrac{K_2SO_4}{2} & & \dfrac{BaSO_4}{2} & \dfrac{2KCl}{2} \text{ or } \dfrac{KCl}{1} \end{array} \qquad \textbf{(3-33)}$$

$$\begin{array}{llcll} 3Cu(NO_3)_2 + 2Na_3PO_4 & & \longrightarrow & Cu_3(PO_4)_2 + 6NaNO_3 & \\ \dfrac{Cu(NO_3)_2}{2} & \dfrac{Na_3PO_4}{3} & & \dfrac{Cu_3(PO_4)_2}{6} & \dfrac{NaNO_3}{1} \end{array} \qquad \textbf{(3-34)}$$

These considerations apply to analytical reactions involving the formation of coordination complexes. For example, consider the reaction between cupric ion and cyanide,

$$Cu^{2+} + 4CN^- \longrightarrow Cu(CN)_4^{2-} \qquad \textbf{(3-35)}$$

The number of equivalents involved in the reaction is determined by the charge on the cation, which in this case is 2+. Therefore, 1 mole of cupric ion represents 2 equivalents, and all the other species involved in the reaction must also be 2 equivalents. Thus, the equivalent weight of KCN in this reaction will be 2(KCN) because each mole of KCN furnishes 1 mole of cyanide ion to the solution, and 4 moles of cyanide ion must react to provide 2 equivalents.

### *General Method*

For any quantitative reaction, the total number of equivalents of one substance present in the reaction must equal the total number of equivalents of each other substance in the reaction. Consequently, if the total number of equivalents of one substance can be determined, this number may be used to calculate the equivalents of the other substances.

Thus, in reaction (3-33), the number of equivalents of barium sulfate present is 2, since the ionic charge of barium is 2+. Because there are 2 equivalents of barium sulfate, there must also be 2 equivalents of barium chloride, 2 of potassium sulfate, and 2 of potassium chloride. In reaction (3-34) there are 3 moles of copper nitrate in the balanced reaction. Since each mole represents 2 equivalents (ionic charge of copper is 2), the 3 moles must represent 6 equivalents, and there must also be 6 equivalents of sodium phosphate, of copper phosphate, and of sodium nitrate present. Since there are 2 moles of sodium phosphate, each mole must represent 3 equivalents.

## EQUIVALENT WEIGHTS FOR OXIDATION–REDUCTION REACTIONS

The equivalent weight for a substance in an electron transfer or oxidation–reduction reaction depends upon the electron change occurring in the reaction. Details of methods of balancing oxidation reactions are given in Chapter 11 and will not be discussed in this section. Balanced equations will be used to illustrate the method of determining the equivalent weights for reactions in which electron transfer occurs.

The equivalent weight is defined as the atomic, ionic, or molecular weight divided by the electron change per unit. It is important to know the exact reaction that takes place because many substances can show more than a single equivalent weight, depending upon the conditions under which the reaction takes place. Potassium permanganate is a good example of an oxidizing agent that can show several equivalent weights. In acid solution, the product is manganous [manganese(II)—$Mn^{2+}$] ion; in solutions that are nearly neutral, the product is manganese dioxide ($MnO_2$); and in strongly basic solutions, the product is manganate ion ($MnO_4^{2-}$).

To determine the total electron change and thus the equivalent weights of reactants and products, it is necessary that a balanced equation be written. The equation used can be the balanced oxidation–reduction reaction or the balanced half-cell reaction of the oxidizing agent or the reducing agent. As an example, consider the reaction of permanganate with oxalate in sulfuric acid solution. The molecular reaction is

$$2KMnO_4 + 5H_2C_2O_4 + 3H_2SO_4 \longrightarrow 2MnSO_4 + 10CO_2 + 8H_2O + K_2SO_4 \quad \textbf{(3-36)}$$

The ionic reaction is

$$2MnO_4^- + 5HC_2O_4^- + 11H^+ \longrightarrow 2Mn^{2+} + 10CO_2 + 8H_2O \qquad \textbf{(3-37)}$$

and the ionic half-cell reactions are

$$MnO_4^- + 8H^+ + 5e^- \longrightarrow Mn^{2+} + 4H_2O \qquad \textbf{(3-38)}$$

$$HC_2O_4^- \longrightarrow 2CO_2 + H^+ + 2e^- \qquad \textbf{(3-39)}$$

If the molecular reaction is utilized, the total electron change can be calculated from either the manganese or the oxalate. For manganese the change is from 7+ to 2+, which requires five electrons per atom, and since there are two atoms of manganese in the two molecules of potassium permanganate, the total number of electrons is ten. Similarly, the apparent charge on carbon is 3+ in oxalate and 4+ in carbon dioxide. Since ten carbons are involved, and each carbon requires one electron, a total of ten electrons are required in the reaction. The equivalent weight of the permanganate ion is one-fifth the ionic weight, since five electrons are utilized; similarly the equivalent weight of the potassium permanganate equals the molecular weight divided by 5. For the oxalate, the equivalent weight will be the molecular weight divided by 2. Identical results can be calculated from the half-cell reactions.

As a second example, consider the reaction of sodium thiosulfate with iodine (actually the soluble form of molecular iodine in aqueous solution is the triiodide ion, $I_3^-$) to form iodide ion and sodium tetrathionate. The molecular reaction is

$$2Na_2S_2O_3 + NaI_3 \longrightarrow Na_2S_4O_6 + 3NaI \qquad \textbf{(3-40)}$$

The ionic reaction is

$$2S_2O_3^{2-} + I_3^- \longrightarrow S_4O_6^{2-} + 3I^- \qquad \textbf{(3-41)}$$

and the half-cell reactions are

$$I_3^- + 2e^- \longrightarrow 3I^- \qquad \textbf{(3-42)}$$

$$2S_2O_3^{2-} \longrightarrow S_4O_6^{2-} + 2e^- \qquad \textbf{(3-43)}$$

From the molecular reaction it is seen that the equivalent weight of the sodium thiosulfate is the same as the molecular weight, since two molecules of the compound require two electrons. The equivalent weight of the sodium triiodide equals the molecular weight divided by 2 and that of the triiodide ion equals the ionic weight divided by 2. For iodide ion the equivalent weight equals $\frac{3}{2}$ the atomic weight.

Probably the best way to avoid becoming confused as to the equivalent weight of any substance in a particular reaction is to apply the general method described on page 39. To do this, first be sure to have a correctly balanced equation and determine the total number of electrons involved for either the oxidizing or the reducing agent. Then place this number under each reactant and product, and the equivalent weight will be the molecular weight multiplied by the ratio of the number of molecules and the number of electrons. For example, in the reaction of $K_2Cr_2O_7$ with $FeSO_4$ in the presence of sulfuric acid, the equation is

$$K_2Cr_2O_7 + 6FeSO_4 + 7H_2SO_4 \longrightarrow Cr_2(SO_4)_3 + 3Fe_2(SO_4)_3 + 7H_2O + K_2SO_4 \quad \textbf{(3-44)}$$

and the number of electrons involved is six (see eq. 3-18). The equivalent weights are

$$\frac{K_2Cr_2O_7}{6} \qquad \frac{6FeSO_4}{6} \text{ or } \frac{FeSO_4}{1}$$

$$\frac{Cr_2(SO_4)_3}{6} \qquad \frac{3Fe_2(SO_4)_3}{6} \text{ or } \frac{Fe_2(SO_4)_3}{2}$$

It should be remembered that the analyst is actually interested in the equivalent weights of only two substances in any reaction used as the basis of an analysis. These are the substance being used as the standard and the substance being determined. The equivalent weights of all the other reactants and of the products are not used in the calculation of the results and thus are not necessary in the analysis. In the example of iron and dichromate, which is used for the determination of iron, the equivalent weight of iron is utilized in the final calculation and the equivalent weight of dichromate must be known in order to determine the normality of the solution. These are the only two that need be known; the others are of academic interest only.

## THE CONCEPT OF THE ANALYTICAL REACTION

The equivalent weight of a substance is not necessarily fixed like the formula weight of the substance because it depends on how the substance reacts. To determine the equivalent weight of a substance unambiguously, the chemical behavior of the substance must be carefully specified. Thus, those substances that can react in more than one way will have more than one equivalent weight. The appropriate equivalent weight of the substance will be based on its behavior in the particular reaction under consideration. For an analytical procedure based on a sequence of reactions, the equivalent weights of the substances involved are all based on the *analytical reaction* that takes place. The analytical reaction is the reaction upon which the actual determination is based. For a volumetric analysis, this is the titration reaction where a standard solution is used; for a gravimetric determination, it is the reaction involving the precipitate that is actually weighed.

The importance of basing the equivalent weights of substances on the analytical reaction may be illustrated by the following examples. Suppose that the chromium content in chromium oxychloride ($CrO_2Cl_2$) is to be determined by dissolving the sample and oxidizing the chromium to dichromate, followed by titration with a standard solution of ferrous ion, according to the reaction

$$Cr_2O_7^{2-} + 14H^+ + 6Fe^{2+} \longrightarrow 2Cr^{3+} + 6Fe^{3+} + 7H_2O \quad \textbf{(3-45)}$$

The actual analytical reaction is the titration between dichromate ion and ferrous ion, and the equivalent weight of $CrO_2Cl_2$ will be based on this reaction. Since the titration is an oxidation–reduction reaction and since the balanced chemical reaction given by eq. (3-45) corresponds to 6 moles of electrons, 1 mole of dichromate ion must be 6 equivalents. Regardless of the details of the intermediate steps involved, it is clear that 2 moles of $CrO_2Cl_2$ is required to produce 1 mole of $Cr_2O_7^{2-}$,

$$2 \text{ moles } CrO_2Cl_2 \equiv 1 \text{ mole } Cr_2O_7^{2} \equiv 6 \text{ eq}$$

Therefore, the equivalent weight of $CrO_2Cl_2$ is the formula weight divided by 3. Notice that the oxidation state of chromium in $CrO_2Cl_2$ was not needed or employed in determining its equivalent weight because $CrO_2Cl_2$ was not directly involved in the analytical reaction.

Suppose now that the chloride content in $CrO_2Cl_2$ is to be determined by dissolving the sample and precipitating the chloride ion by silver, according to the analytical reaction

$$Ag^+ + Cl^- \longrightarrow AgCl \quad \textbf{(3-46)}$$

The equivalent weight of $CrO_2Cl_2$ will be the formula weight divided by 2. This may be determined as follows. Regardless of the specific details of the transformation involved, it is clear that 1 mole of $CrO_2Cl_2$ will ultimately produce 2 moles of chloride ion. Each mole of chloride ion represents 1 equivalent, since each mole of chloride ion is chemically equivalent to 1 mole of silver ion, and 1 mole of silver ion is 1 equivalent because 1 mole of electrons is required to reduce it. Thus,

$$1 \text{ mole } CrO_2Cl_2 \equiv 2 \text{ moles } Cl^- \equiv 2 \text{ moles } Ag^+ \equiv 2 \text{ eq}$$

Notice that the equivalent weight of $CrO_2Cl_2$ was not based on the charge on chromium in $CrO_2Cl_2$ because, in this case, the chromium was not involved in the actual analytical reaction.

## EQUIVALENT WEIGHTS IN ANALYTICAL SEQUENCES

It is not uncommon for an analytical procedure to consist of several chemical transformations prior to the actual analytical reaction. The question then arises as to how to find the equivalent weight of the substance being determined at the start of the procedure. For purposes of illustration, let us consider the **Winkler method** for the determination of dissolved oxygen in solutions. Oxygen may be determined from the following sequence of reactions. The sample is treated with excess Mn(II), NaOH, and NaI. $Mn(OH)_2$ forms immediately, and reacts further with dissolved oxygen to give $Mn(OH)_3$ according to eq. (3-47).

$$4Mn(OH)_2 + O_2 + 2H_2O \longrightarrow 4Mn(OH)_3 \tag{3-47}$$

Upon acidification of the solution, the $Mn(OH)_3$ reacts with the excess NaI that was added.

$$2Mn(OH)_3 + 3I^- + 6H^+ = I_3^- + 3H_2O + 2Mn^{2+} \tag{3-48}$$

The liberated triiodide is then determined by titration according to the analytical reaction (3-49).

$$I_3^- + 2S_2O_3^{2-} = 3I^- + S_4O_6^{2-} \tag{3-49}$$

We must know, of course, what the equivalent weight of $O_2$ is in this sequence. Inspection of the sequence reveals that oxygen is not involved in the actual analytical reaction. The equivalent weight ascribed to oxygen must therefore be determined indirectly from the stoichiometry of the sequence. In this case, the analytical reaction is a redox reaction involving triiodide ion. The half-cell for triiodide may be obtained from the table of reduction potentials in Appendix 7

$$I_3^- + 2e^- = 3I^- \tag{3-50}$$

Thus, 1 mole of $I_3^-$ is 2 equivalents. Going back through the sequence, we see that 1 mole of $I_3^-$ is produced by 2 moles of $Mn(OH)_3$, and 4 moles of $Mn(OH)_3$ are produced by 1 mole of $O_2$. Therefore,

$$1 \text{ mole } I_3^- \equiv 2 \text{ eq} \equiv 2 \text{ moles } Mn(OH)_3 \equiv \tfrac{1}{2} \text{ mole } O_2$$

In this sequence, the equivalent weight of $O_2$ will be $O_2/4$. Thus, for an analytical sequence, the number of equivalents involved in the analytical reaction is determined, and this number of equivalents is then stoichiometrically related back through the sequence to the initial step.

## CONVERSION FROM ONE METHOD TO ANOTHER

### WEIGHT–VOLUME TO MOLAR OR NORMAL

Calculate the concentration in grams per liter (g/liter) from the concentration as given and then divide by the molecular weight for molar or the equivalent weight for normal.

• **Example 17.** A solution of sodium iodide contains 5.00 mg/100 mL. Calculate (a) the molar concentration, (b) the normal concentration for the formation of iodate ion ($IO_3^-$), and (c) the normal concentration for precipitation of $PbI_2$.

$$\text{g/liter} = \frac{\text{mg/1000}}{\text{mL/1000}} = \text{mg/mL} \tag{3-51}$$

$$M = \frac{\text{g/liter}}{\text{MW}} \tag{3-52}$$

and

$$N = \frac{\text{g/liter}}{\text{eq wt}} \tag{3-53}$$

(a) The molarity is

$$\text{g/liter} = \frac{5.00 \text{ mg}}{100 \text{ mL}} = 0.0500 \text{ mg/mL}$$

$$M = \frac{0.0500}{149.9} = 0.000334\ M \quad \text{or} \quad 3.34 \times 10^{-4}\ M$$

(b) The half-cell equation is

$$I^- + 3H_2O \longrightarrow IO_3^- + 6H^+ + 6e^-$$

and the equivalent weight is 149.9/6, or 24.98.

$$N = \frac{0.0500}{24.98} = 0.00200\ N \quad \text{or} \quad 2.00 \times 10^{-3}\ N$$

(c) For precipitation of $PbI_2$ the equivalent weight is the same as the molecular weight, since 2 moles of iodide react with 2 equivalents of lead.

$$N = \frac{0.0500}{149.9} = 0.000334\ N \quad \text{or} \quad 3.34 \times 10^{-4}\ N$$

## MOLAR TO NORMAL AND VICE VERSA

One mole usually contains 1 or more equivalents, so the normal concentration is ordinarily equal to or greater than the molar concentration. Since the equivalent weight is equal to the molecular weight divided by the number of equivalents per mole,

$$N = M \times \text{number eq/mol} \tag{3-54}$$

or

$$M = \frac{N}{\text{number eq/mole}} \qquad \textbf{(3-55)}$$

• **Example 18.** What is the molar concentration of a solution of phosphoric acid that is 0.250 $N$ for the formation of phosphate ion? The formation of phosphate from phosphoric acid requires the neutralization of all three hydrogens so that the equivalent weight is one-third the molecular weight.

$$M = \frac{N}{3} = \frac{0.250}{3} = 0.0833\ M$$

• **Example 19.** What is the normal concentration of 0.100 $M$ sulfuric acid for the precipitation of barium sulfate? The equivalent weight for precipitation of sulfate equals one-half the molecular weight, or $H_2SO_4/2$, so

$$N = M \times 2 = 0.100 \times 2 = 0.200\ N$$

## MOLAR OR NORMAL TO TITER

To convert from molar to titer, first convert from molar to normal and then use eq. (3-14) to calculate the titer. To convert normal to titer, use eq. (3-14) directly.

• **Example 20.** Calculate the barium oxide titer of a 0.0500 $M$ solution of barium chloride. Since 1 mole of barium chloride produces 1 mole of barium oxide, the equivalent weights of both barium oxide and barium chloride equal one-half the molecular weight, so

$$N = M \times 2 = 0.0500 \times 2 = 0.100\ N$$

$$\text{titer} = N \times \text{eq wt} = 0.100 \times 153/2 = 7.65\ \text{mg/mL}$$

## WEIGHT PERCENT (w/w) TO MOLAR OR NORMAL

Because the percentage by weight does not specify a volume, the density $d$ (or specific gravity, sp gr) of the solution must also be known to convert the concentration to either molar or normal. The concentration as given is converted to grams per liter and then divided by the molecular weight or the equivalent weight. The grams of solute per liter is determined by multiplying the weight of 1 liter by the fraction of the weight due to the solute (percent per 100). Since the weight of 1 liter equals the density times 1000, the calculation becomes

$$\text{g/liter} = d \times 1000 \times \%/100 = 10 \times d \times \% \qquad \textbf{(3-56)}$$

and

$$M = \frac{10 \times d \times \%}{\text{MW}} \tag{3-57}$$

$$N = \frac{10 \times d \times \%}{\text{eq wt}} \tag{3-58}$$

• **Example 21.** Concentrated sulfuric acid has a density of 1.84 g/mL and contains 96.0% (w/w) of $H_2SO_4$. Calculate (a) the molar concentration and (b) the normal concentration.

(a) The molarity is $M = \frac{10 \times 1.84 \times 96.0}{98.0} = 18.0\ M.$

(b) The normality is $N = \frac{10 \times 1.84 \times 96.0}{49.0} = 36.0\ N.$

## PROBLEMS

1 A 23.7% (w/w) solution of $KHSO_4$ has a density of 1.15 g/mL. Express the concentration of this solution in the following ways: (a) molar; (b) normal for reactions in which (1) neutralization occurs, (2) $BaSO_4$ is precipitated, and (3) $SO_4^{2-}$ is reduced to $S^0$; (c) the following titers: (1) NaOH, (2) $SO_3$, and (3) KC1.

*Ans.* (a) 2.00 $M$; (b) (1) 2.00 $N$, (2) 4.00 $N$, (3) 12.00 $N$; (c) (1) 80.00, (2) 160, (3) 149

2 Repeat problem 1 for a 14.0% (w/w) solution of $KHSO_4$ with a density of 1.10 g/mL.

3 A 5.00% (w/w) solution of $NaHCO_3$ has a density of 1.04 g/mL. Calculate (a) the molar concentration; (b) the normal concentration for reactions in which (1) a base is added, (2) an acid is added, (3) $BaCO_3$ is precipitated, (4) $CO_2$ is produced, and (5) $Na_2UO_4$ is precipitated; (c) the following titers: (1) HCl, (2) NaOH, (3) $CO_2$, and (4) the $CO_2$ titer expressed as milliliters of gas at STP instead of milligrams.

4 Calculate the equivalent weights of each substance in the following balanced reactions:

(a) $2NaOH + H_4C_{10}H_{12}O_8N_2 \rightarrow Na_2H_2C_{10}H_{12}O_8N_2 + 2H_2O$

(b) $4NaOH + H_4C_{10}H_{12}O_8N_2 \rightarrow Na_4C_{10}H_{12}O_8N_2 + 4HOH$

(c) $Na_3PO_4 + 2HCl \rightarrow NaH_2PO_4 + 2NaCl$

(d) $Fe_2(NH_4)_2(SO_4)_4 + 4BaCl_2 \rightarrow 4BaSO_4 + 2FeCl_3 + 2NH_4Cl$

(e) $Fe(NH_4)_2(SO_4)_2 + 2BaCl_2 \rightarrow 2BaSO_4 + FeCl_2 + 2NH_4Cl$

(f) $Fe_2(NH_4)_2(SO_4)_4 + 6NaOH \rightarrow 2Fe(OH)_3 + (NH_4)_2SO_4 + 3Na_2SO_4$

(g) $6Fe(NH_4)_2(SO_4)_2 + K_2Cr_2O_7 + 7H_2SO_4 \rightarrow Cr_2(SO_4)_3 + 3Fe_2(SO_4)_3 + 6(NH_4)_2SO_4 + K_2SO_4 + 7H_2O$

(h) $KBrO_3 + 9KI + 3H_2SO_4 \rightarrow KBr + 3KI_3 + 3K_2SO_4 + 3H_2O$

(i) $2CuSO_4 + 5KI \rightarrow Cu_2I_2 + KI_3 + 2K_2SO_4$

(j) $2KMnO_4 + 5H_2O_2 + 3H_2SO_4 \rightarrow 2MnSO_4 + 5O_2 + K_2SO_4 + 8H_2O$

**5** Calculate the molar and normal concentrations for neutralization and/or precipitation reactions for each of the following solutions:
(a) 18.194 g of $NaH_2AsO_4 \cdot H_2O$ in 250 mL of solution
(b) 4.904 g of $H_2SO_4$ in 200 mL of solution
(c) 5.207 g of $BaCl_2$ in 500 mL of solution
(d) 61.08 g of $BaCl_2 \cdot 2H_2O$ in 2000 mL of solution
(e) 8.0423 g of $AgNO_3$ in 500 mL of solution
(f) 1.714 g of $Ba(OH)_2$ in 333 mL of solution
(g) 3.155 g of $Ba(OH)_2 \cdot 8H_2O$ in 333 mL of solution

*Ans.* (a) 0.4000 *M*, 0.8000 *N* for neutralization

**6** Calculate the molar and normal concentrations for the following solutions:
(a) 3.167% (w/v) NaCl
(b) 5.326% (w/v) $BaCl_2$
(c) 10.0% (v/v) HCl
(d) 25.0% (v/v) $H_2SO_4$
(e) 25.0% (w/v) $H_2SO_4$
(f) 5% (w/v) $AgNO_3$

*Ans.* (a) 0.5419 *M*, 0.5419 *N*

**7** Express the concentrations of the following solutions by the titer method:
(a) The $Cl^-$ titer of 0.500 *N* $AgNO_3$
(b) The $Cl^-$ titer of a solution of 8.4950 g of $AgNO_3$ in 500.0 mL of water
(c) The NaCl titer of 3.523% (w/v) $AgNO_3$
(d) The $Na_2O$ titer of 0.1067 *N* HCl
(e) The $CaCO_3$ titer of 0.02000 *M* EDTA
(f) The $CaCl_2$ titer of an EDTA solution that has a $CaCO_3$ titer of 1.000 mg/mL

*Ans.* (a) 17.73

**8** Calculate the normality, the molarity, the silver oxide titer, and the silver nitrate titer of a solution that contains 4.00 g of sodium chloride in 100 g of water and has a density of 1.02 g/mL.

CHAPTER

# 4 Reliability of Measurements

Much of the information we have about the nature and behavior of chemical systems is the direct result of careful measurements of various chemical and physical properties that characterize the particular chemical system. The final result of any measurement is a number that expresses the magnitude of the particular property of the system being measured in terms of some established unit for that quantity. Frequently, the desired chemical information must be calculated, using measured values for several quantities. The question naturally arises, therefore, of how reliable the numbers expressing the measured or calculated values are. This chapter will discuss the reliability of measurements.

## SIGNIFICANT FIGURES

Any measurement represents a comparison between the quantity being measured and some standard reference quantity that expresses the magnitude of the units being used in the comparison. For example, in measuring length, the length of the object is compared with a standard length that represents the desired unit of measurement. If the object is $2\frac{1}{2}$ times larger than a standard unit of length (which might be a centimeter, for example), we would say that the object is $2\frac{1}{2}$ cm long. It is important in scientific work to distinguish a casual comparison from a rigorous comparison. In other words, was the length merely about $2\frac{1}{2}$ cm long, or was the length determined with great care to the nearest 0.0001 cm?

The rigor with which a particular measurement (i.e., comparison) is made may be indicated by the number of digits used to express the result. Most careful measurements are reported just beyond the limit of the calibrations on the device used to make the comparison. Thus, the last digit of the measurement represents an estimate of the fractional amount of the particular property between the two closest calibrations. Since the last digit in a measurement is only an estimate of the distance between the two bracketing calibrations, it will be uncertain by a particular amount. It is normally assumed, if there is no information to the contrary, that the last digit in a measured result has been estimated to the nearest tenth of the distance between the two bracketing calibrations. Thus, if a length were reported as 2.516 cm, it would be assumed that measurements were made with a device graduated in hundredths of a centimeter and that the length of the object extended beyond 2.51 by six-tenths of the distance between 2.51 and 2.52 cm. Furthermore, it would be assumed that the absolute uncertainty in this measured value is $\pm 0.001$ cm. Thus, the value 2.516 cm implies a great deal more information than might be apparent at first glance.

At this point it would be helpful to distinguish between the quality of various measured quantities. The value 2.51600 cm implies absolute uncertainty of 0.00001 cm, which represents a more exacting measurement than 2.516 cm. The absolute uncertainty in a measurement is not, by itself, a completely reliable guide to the quality of the measured value, because it does not give any indication of how the absolute uncertainty compares with the magnitude of the quantity being measured. For example, an absolute uncertainty in length of 0.1 cm may imply a high-quality measurement in the length of a football field, but would not indicate an extremely exacting measurement of an object 1.0 cm long. Thus, the quality of a measurement is often expressed by the **relative uncertainty**, which is defined as the absolute uncertainty divided by the magnitude of the measured quantity. Since the absolute uncertainty and the magnitude of the measured quantity both have the same units, their ratio, the relative uncertainty, is dimensionless because the units cancel.

Naturally, it is important when dealing with measured quantities to express the results of the measurement so that the absolute and relative uncertainties implied by the number of digits used reflect the true rigor that went into the measurement. This is especially important in calculations involving measured quantities, not only because the measured values entering into the calculation may have different uncertainties but also because normal mathematical operations increase or decrease the number of digits in the answer. The digits required to express a measurement or a calculated result to the correct uncertainty are referred to as **significant figures**.

To determine the number of significant figures in any measured value, start counting from a digit other than zero to the left and count through the last digit, including zero, to the right. Note that the position of the decimal has nothing to do with the number of significant figures. Thus, the following numbers each have four significant figures: 136.7; 13.67; 1006; 1000; 70.70; and 0.001367.

At this point it is important to discuss the role of significant figures in mathematical calculations involving measured quantities. We have seen that significant figures represent digits that are experimentally meaningful. In many purely mathematical operations with numbers, the numbers of digits in the final answer is limited only by the extent to which the calculation is carried out. If these mathematical operations involve numbers that represent measured values, it is clear that the calculation cannot be carried out indefinitely before experimentally meaningless digits in the results are obtained. For example, if the density of water is calculated from a measured mass of 25.483 g and a measured volume of 25.39 mL, the following result can be obtained from the numbers:

$$\frac{25.483 \text{ g}}{25.39 \text{ mL}} = 1.0036628 \text{ g/mL}$$

Clearly, the density of water cannot be determined to a precision of 1 part in 100 million when the volume was determined only to about 1 part in 2500. Therefore, some of the digits in the number 1.0036628 are experimentally meaningless and do not represent significant figures. In calculations involving multiplication and division, the result should have a relative uncertainty of comparable magnitude with the factor with the worst relative uncertainty entering into the calculation. In the preceding example, the volume is known with a relative uncertainty of 1/2539, whereas the mass is known with a relative uncertainty of 1/25,483. Since the volume measurement is the limiting factor, the density should have a relative uncertainty comparable to 1/2539. The calculated density should therefore be expressed as 1.004 g/mL.

Notice that in rejecting the insignificant or experimentally meaningless digits, the digit in the thousandths place was increased from 3 to 4. When experimentally meaningless digits are rejected, treat the excess digits as a decimal fraction. If the decimal fraction is greater that 0.5, increase the last figure retained by 1. If the decimal fraction is less than 0.5, the last figure retained is left unchanged. If the decimal fraction is 0.5, increase the last retained figure by 1 only if it is an odd number.

Because the relative uncertainty in a measured value is determined by the number of significant figures used to express it, the relative-uncertainty requirements for multiplication and division may be adequately satisfied by retaining as many significant figures in the result as are present in the factor with the *least* number of significant figures. The simplest way to offset roundoff errors in calculations involving several numbers is to retain, in each of the more reliable factors, one more significant figure than is to be used to express the answer. Thus, in the problem

$$\%\,\text{Cu} = \frac{3.5 \text{ meq} \times 63.54 \text{ mg/meq}}{987.3 \text{ mg}} \times 100\%$$

the number of significant figures justifiable in the result is limited to two by the factor with the least number of significant figures (i.e., 3.5 meq). To avoid errors in the rounding off of each factor, the other factors can be rounded off to one more than the required number of significant figures. Thus,

$$\%\,\text{Cu} = \frac{3.5 \text{ meq} \times 63.5 \text{ mg/meq}}{987 \text{ mg}} \times 100\% = 23\%$$

and the final result has been rounded off to two significant figures. Notice that the factor 100% is not a measured quantity (this is evident because there are no units expressed), and therefore is a pure number with unlimited precision.

In considering the number of experimentally meaningful digits required in a problem involving addition and subtraction, it is the absolute uncertainty in the values that is important. Thus, the answer can be expressed with no more certainty than the value with the greatest absolute uncertainty involved in the calculation. In terms of significant figures, this means, in addition or subtraction, retain in each number only as many figures to the right of the decimal as there are in the number having the least number of figures to the right of the decimal. For addition of the numbers 516.3, 34.36, and 1.349, the correct sum would be expressed as 552.0.

Finally, when dealing with calculations involving logarithms, there should be the same number of decimal places in the mantissa of the logarithm as there are significant figures in the number being expressed. Thus, the log of 1.2300 is 0.08991, the log of 1.230 is 0.0899, the log of 1.23 is 0.090, and the log of 1.2 is 0.08.

## ACCURACY, PRECISION, AND ERRORS

Philosophers have long debated over the existence of absolute truth. Regardless of the existence or nonexistence of absolute truth, it can be said that truth may be approached asymptotically through the process of experience. In the preceding section, comparison with a standard was presented as the fundamental basis for all measurements. Since comparison requires standards, the truth of a series of measurements rests on the integrity of the standards. Because of the importance in the reliability of standards, every modern government has some agency responsible for the control of standards. Standards may be classified according to their reliability as primary, secondary, and so on. Thus, the meter is defined in terms of a particular wavelength emitted by krypton, which is a fundamental realizable primary standard of length. This wavelength may, by the process of interferometry, be used to produce a secondary standard in the form of a distance between two etchings on a metal bar. In turn the calibrated metal bar may be used to produce other tertiary standards, and so on.

Suppose that a series of bars exactly 1 meter (m) long is to be produced by comparison with a secondary standard metal bar described above. Even being as careful as possible, it would be found that, although the majority of the metal bars produced were very close to being 1 m long, some bars would be shorter than 1 m and others would be longer than 1 m. If in this hypothetical example an infinite number of metal bars were produced, a frequency distribution of lengths would be found similar to that shown in Figure 4-1. In this particular example, the distribution shown is symmetrical and is known as a Gaussian distribution, after the famous mathematician Carl Friedrich Gauss. The existence of a particular distribution of experimental results is of great importance in expressing the reliability of experimental data.

How can a distribution of lengths be explained when only one length was sought? The answer is that, even for a well-characterized measurement where errors have been reduced to a minimum, at some point the comparison process of measurement is only an estimate; therefore, at some point in the measurement

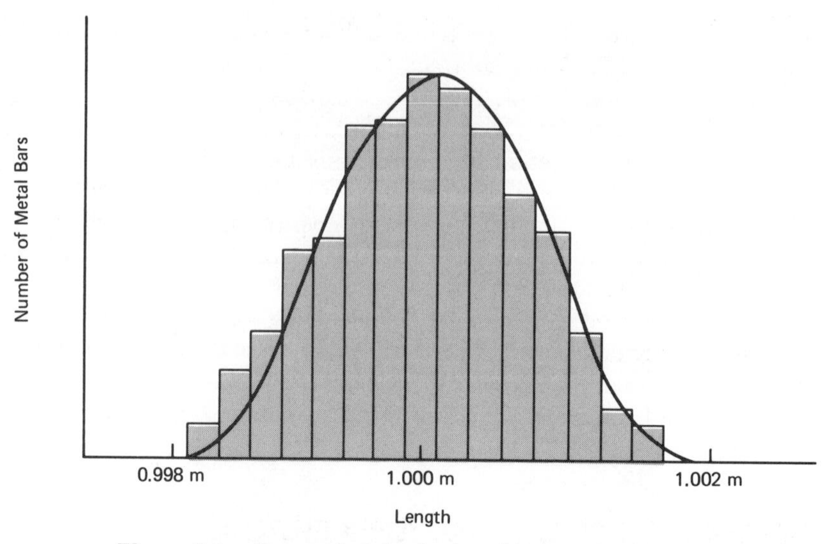


**Figure 4-1.** Symmetrical distribution of indeterminate errors.

a judgment on the part of the experimenter must be made. These experimental judgments on the part of the experimenter give rise to errors in estimation of the correct value. Such errors are known as **random errors** because they are equally likely to be positive or negative. The more the final result depends on estimation by the experimenter, the wider will be the distribution of results.

Random errors are not the only errors that can affect the validity of the results. For example, it is often found that factors initially considered unimportant in an experiment are later found by experience to influence the result. In the case of the metal bars, suppose that in their production the factor of temperature was not considered, and the lengths of the bars produced were compared at a different temperature than that used during the calibration of the secondary standard. If the metal bars being produced are of a different alloy than the calibration standard and if the temperature at which the bars are produced is different from that used during the calibration of the secondary standard, all of the lengths of the bars produced will be in error by the difference in the extent of expansion between the calibration standard and the bars. Such an error is known as a **systematic error** because, for a given set of conditions, the error influences the results only in one direction. Thus whereas random errors are equally likely to be positive or negative, a particular systematic error will always be either positive or negative for a given set of conditions.

The existence of these two types of errors enables one to categorize experimental results in two ways. First, one can consider how a series of measurements agree with one another without reference to their absolute validity. The agreement of successive measurements among themselves is referred to as the **precision** of the results and is a measure of the width of the distribution for a large number of measurements. If some representative value can be found that summarizes a series of measurements, this value may be compared with some more reliable standard. The agreement between the present measurements and the more reliable standard is referred to as the **accuracy** of the results. Accurate results can be achieved only to the extent that systematic errors can be discovered by experience and eliminated. Naturally, before one can talk about accuracy, a more reliable value for the quantity being measured must be available. Great pains must be taken to assure the absence of systematic error in a reference standard, and it is not uncommon for new sources of systematic error to be discovered in what was thought to be a well-characterized measurement. Absolute accuracy is therefore an ephemeral quantity that is approached but never completely realized.

The terms *accuracy* and *precision* refer to two distinctly different concepts, and should not be used interchangeably. Thus, the precision of a set of measurements implies nothing about their accuracy. Normally, however, an accurate set of results is precise as well.

When one speaks of differences among members of a set of measurements, **deviation** is used. Deviations may be expressed as absolute deviations or relative deviations. Relative deviations are obtained by dividing the absolute deviation by the magnitude of the particular quantity. When the difference between a measured value and some more reliable standard (which is taken as the "true" value) is considered, **error** is used. Errors may be expressed as absolute errors or as relative errors. The absolute error ($E$) is the difference between the observed ($O$) and the true value ($T$) and is always expressed in the units of the measurement. The relative error ($RE$) expresses the relationship of the error to the true value and is usually expressed as percentage error or as parts per thousand. Thus,

$$E = O - T \qquad \text{(in units of measurement)} \qquad \textbf{(4-1)}$$

$$RE = \frac{E}{T} \times 1000 \qquad \text{(in ppt)} \qquad \textbf{(4-2)}$$

For example, assume that two values are measured with the same absolute error, as follows:

| *Measurement (g)* | *True Value (g)* | *Absolute Error (g)* | *Relative Error (ppt)* |
|---|---|---|---|
| 10.545 | 10.544 | 0.001 | 0.094 |
| 0.105 | 0.104 | 0.001 | 9.6 |

Although the absolute errors are equal, the relative errors show that the first measurement is relatively more accurate than the second.

## CLASSIFICATION OF ERRORS

In the preceding section, errors were classified into two broad categories, random errors and systematic errors. Another name for random errors is **indeterminate** errors, indicating that it is not possible to determine their origin.

Indeterminate errors follow the symmetrical distribution shown in Figure 4-1, which shows that small errors are more probable than large errors and that positive and negative errors are equally probable. Another name for systematic errors is **determinate** errors, indicating that their origin can often be determined by a systematic variation of experimental conditions. Determinate errors can be either variable or constant, but, for a particular set of experimental conditions, a given determinate error will be always either positive or negative. Variable determinate errors will affect the precision of the results, but constant determinate errors do not.

A constant error that occurs in each and every measurement does not affect the precision of the set but does affect the magnitude of the average of the set. Constant errors of this type are called **bias**, and, if the extent of the bias can be determined, the results can be corrected for bias by addition or subtraction of the constant error from each and every result. The effect of a bias on a determination is shown by the dashed curve in Figure 4-2.

The shape of the distribution curve for a set of measurements is affected by the precision, as shown in Figure 4-3. A narrow distribution curve indicates good precision (small standard deviation), whereas a broad distribution is caused by poor precision (large standard deviation).

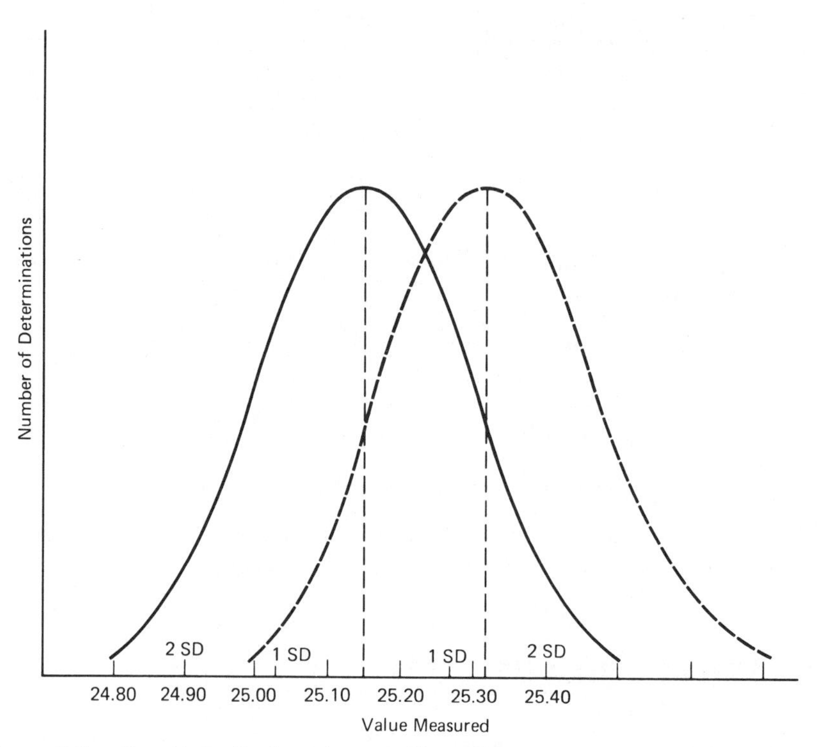


**Figure 4-2.** Normal distribution of errors. The solid curve represents measurements for which the average value is 25.15 with a standard deviation of 0.125. The dashed curve represents measurements with the same degree of precision but with a bias of 0.17 unit.

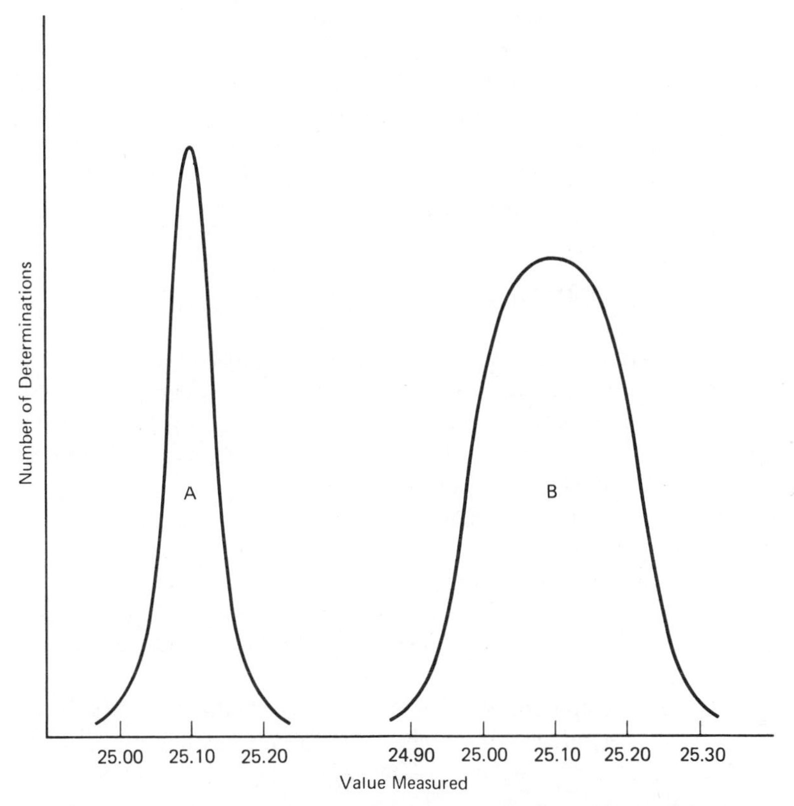


**Figure 4-3.** Effect of precision upon shape of distribution curve. (A) Good precision with a standard deviation of approximately 0.03. (B) Poorer precision with a standard deviation of approximately 0.10, even though the averages are identical.

## TYPES OF DETERMINATE ERRORS

Determinate errors are classified as personal, instrumental, or method errors.

**Personal errors** are those due to the carelessness of the observer and include such things as misreading a burette (29.38 instead of the correct value of 28.62), listing weights incorrectly (listing a 2-g weight when a 1-g weight was used), misinterpreting directions (adding concentrated acid instead of 1 : 1 or adding 10 mL when directions call for 1.0 mL), overshooting an end point in a titration, and use of dirty equipment. It should be noted that these errors probably would be eliminated in a careful rerun of the determination.

**Instrumental errors** are those due to the instruments themselves and include the use of uncalibrated equipment (weights, pipettes, burettes, etc.), construction errors in the instruments (balance arms not equal), and damaged equipment (chipped knife edge).

**Method errors** are those inherent in the method itself, which normally would not be eliminated in a rerun but often may be discovered and corrected for by calculation or procedural changes. These include impurities in reagents (correct by using a blank determination), solubility of precipitates in precipitation medium or wash solution (determine and correct mathematically or revise procedure to decrease solubility), contamination in or on precipitates (dissolve and reprecipitate or remove contaminating substances by prior treatment), indicator errors (run blank), and reaction not stoichiometric (can often be corrected by careful standardizations).

### TYPES OF INDETERMINATE OR RANDOM ERRORS

Indeterminate errors are classified as personal, as instrumental, or as being due to variations in external conditions. Note that for all the random errors listed below, essentially none can be discovered or corrected for and few, if any, would be eliminated in a repeat of the determination. This type of error is just as likely to occur during the second run as in the first.

Personal errors are due to the inability of the observer to distinguish between slightly different measurements or colors. Examples are misestimation of the position of the meniscus (estimating 25.56 instead of the correct value of 25.57), assuming two colors are identical when there is a slight difference in hue, misestimation of balance rest point, and misestimation of the position of a meter needle.

Indeterminate instrumental errors are due to random variations in the response of instruments, variations in line voltage, wandering rest points during a weighing, irreversible reactions caused by momentary nonhomogeneity of solutions, and so on.

Variations of external conditions over which the observer has no control also cause random errors. Included here would be errors due to variations of barometric pressure, variations of humidity, slight variations of temperature, and momentary air currents.

## STATISTICAL TREATMENT OF DATA

Whenever a large number of measurements on a particular quantity are made, it is especially convenient if the results can be summarized in some way without recourse to listing all the measured values. What is desired is a measure of the central tendency of the measurements, as well as some indication of the width of the distribution. The Gaussian distribution shown in Figure 4-2 is commonly encountered, and, although it is not the only possible distribution of experimental results, it is known as the **normal distribution of error**. The mathematical expression of the curve for the normal distribution of error is a function of only two quantities, $\mu$ and $\sigma$, where $\mu$ is known as the **population mean**, and $\sigma$ is the **standard deviation** from the mean. For a symmetrical distribution, the mean or average value of a large number of measurements is an appropriate measure of the central tendency of the results because the equally probable positive and negative random errors tend to cancel one another in the calculation. The quantity known as the standard deviation is a measure of the width of the distribution, as shown in Figure 4-3. Thus, simply by specifying these two quantities, the results of a large number of measurements may be summarized conveniently.

## MEAN AND STANDARD DEVIATION

The quantities $\mu$ and $\sigma$ represent the mean and the standard deviation of a hypothetically infinite set of measurements. Since an infinite set of measurements is not realizable in actual situations, these quantities are estimated for a limited set of data. Thus, $\mu$, the population mean, can be estimated by the mean of a limited set of data using eq. (4-3)

$$\bar{X} = \sum_{i=1}^{n} \frac{X_i}{n} \tag{4-3}$$

where $\bar{X}$ is the mean of the $n$ individual observations $X_i$. The standard deviation $\sigma$ may be estimated from the expression

$$s = \sqrt{\frac{\sum (X_i - \bar{X})^2}{n - 1}} \tag{4-4}$$

where $s$ represents the estimate of the standard deviation. The standard deviation is sometimes referred to as the **root-mean-square deviation**, by analogy with eq. (4-4), which is used for calculating it.

Students are often puzzled as to why the mean of the deviations squared is calculated by dividing by $n - 1$ instead of $n$. The reason concerns the number of independent pieces of information available. If $n$ observations are made, there are $n$ independent pieces of information. In calculating the standard deviation, each observation $X_i$ is required, as is the mean $\bar{X}$. However, the mean is calculated from the sum of the individual observations; thus, it is not an independent piece of information. The number of independent deviations $(X_i - \bar{X})$ available from the data is $n - 1$ because the $n$th or final deviation is fixed by the remaining $n - 1$ deviations, and the fact that $\sum (X_i - \bar{X}) = 0$. Statisticians refer to the number of independent pieces of information as the **degrees of freedom**. Although the value of the standard deviation obtained from eq. (4-4) is only an estimate, mathematicians have devised ways of estimating what the true $\sigma$ would have been had an infinite number of measurements been made. These procedures, as well as other methods for estimating $s$, will be discussed later in the chapter.

## CONFIDENCE LIMITS

Still more information characterizing the behavior of the results of a series of measurements may be determined if the results are known to follow a normal distribution. Notice that, for a normal distribution of errors, large deviations from the population mean are exceedingly infrequent. Since any measured value must lie somewhere between $-\infty$ and $+\infty$, we can say that the probability of obtaining a value between $-\infty$ and $+\infty$ must be 100%. In other words, there can be no chance that the value can be outside this range. This 100% probability may be associated with the area under the normal curve of error from $-\infty$ to $+\infty$.

If the total area under the normal curve of error is taken to represent the 100% probability level, fractional areas of the total area must represent fractional probability levels. Thus, a region symmetrically distributed about the mean and including 95% of the total area under the curve must represent a 95% probability level. The limits enclosing this region are known as the **confidence limits** for the given probability or confidence level. A 95% confidence level means that 95 measurements out of 100 will be within the symmetrical bounds enclosing 95% of the total area under the curve. Another way of stating this is that there will be only a 5% risk that a valid measurement will occur outside these limits. The interval enclosing 95% of the area can be referred to as the **confidence interval** for a 95% confidence level.

Because the interval enclosing 95% of the total area under the curve represents a 95% probability level, we can be 95% confident that, for a well characterized measurement from which all sources of systematic error have been eliminated, the true value must be somewhere within that interval. Mathematically, the confidence limit for a single measurement may be expressed as

$$\text{confidence limit for } \mu = X \pm z\sigma \qquad \textbf{(4-5)}$$

where $z$ is a factor that depends only on the confidence level desired. For example, $z = 1.00$ for a 68% probability level, $z = 1.96$ for a 95% probability level, and $z = 3.29$ for a 99.9% probability level.

Generally, we are more interested in the confidence limits of the mean of several measurements. Since the mean is a more reliable estimate of the true value than any single measurement (because the random errors associated with each value tend to cancel one another in the calculation of the mean), the confidence limits of the mean for a given confidence level are smaller than for a single measurement. Hence,

$$\text{confidence limit for } \mu = \overline{X} \pm \frac{z\sigma}{\sqrt{n}} \qquad \textbf{(4-6)}$$

where $n$ is the number of observations in the mean. This equation tells us that the confidence limits of the mean for any confidence level may be reduced to any desired value by increasing the number of observations. Although this is certainly true, a point of diminishing returns is rapidly reached. Thus, whereas the confidence interval associated with the mean of four observations is half the corresponding interval for a single observation, 12 more observations (for a total of 16) must be made to again reduce the confidence interval of the mean by an additional factor of 2.

## STUDENT'S $t$

The equations developed in the previous section for the confidence limits apply only if the value of $\sigma$ is known. The true standard deviation $\sigma$ is reliably estimated by eq. (4-4) for 20 or more observations. If the number of observations made is less than 20, the value calculated by eq. (4-4) tends to underestimate the true value of $\sigma$. To correct for the fact that the true value of $\sigma$ is unknown for a small set of measurements, the factor $z$ is replaced by the factor $t$, known as Student's $t$[1]:

$$\text{confidence limit for } \mu = X \pm ts \qquad \textbf{(4-7)}$$

$$\text{confidence limit for } \mu = \overline{X} \pm \frac{ts}{\sqrt{n}} \qquad \textbf{(4-8)}$$

where $t$ is a function not only of the confidence level desired but also of the number of degrees of freedom used in the calculation of $s$. The values of $t$ are shown in Table 4-1 for several confidence levels.

**TABLE 4-1.** The $t$ Distribution

| *D.F.** | $\alpha = 0.25$ | $\alpha = 0.025$ | $\alpha = 0.005$ |
|---|---|---|---|
| 1 | 1.0 | 12.7 | 63.6 |
| 2 | 0.81 | 4.3 | 9.9 |
| 3 | 0.76 | 3.2 | 5.8 |
| 4 | 0.74 | 2.8 | 4.6 |
| 5 | 0.73 | 2.6 | 4.0 |
| 9 | 0.70 | 2.3 | 3.2 |
| 19 | 0.67 | 2.1 | 2.9 |

*D.F. = $n - 1$; probability level = $1 - 2\alpha$.

## COMPARISON OF AVERAGES

We have seen that all experimental measurements are uncertain and vary over a range determined by the presence and extent of random error in the measurement. Therefore, in comparing experimental results, we must ask ourselves whether the observed differences could be due to the presence of random error or to some other cause. In other words, are the results the same or different statistically? Thus, even where two results should be clearly different for theoretical reasons, this difference may be obscured by the presence of random error and no confidence can be placed in the apparent difference observed. The concepts developed in the previous sections can be used to answer

[1] This function was first described by W. S. Gosset, who, because his employer (a well-known Irish brewery) would not allow research publications, adopted the pen name Student.

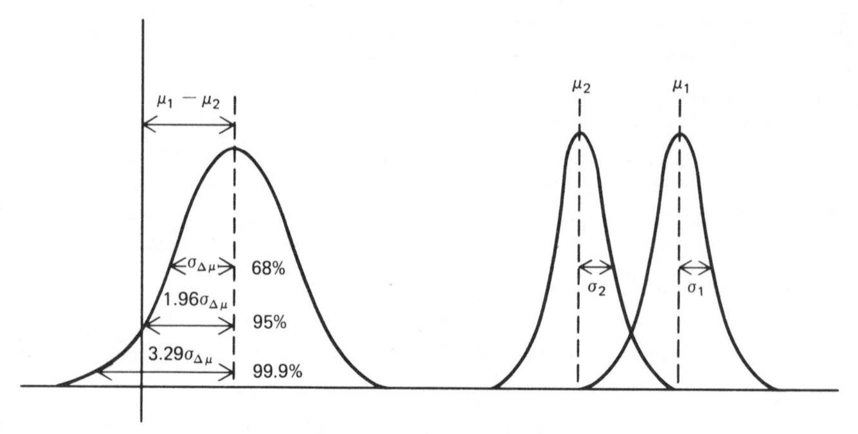


**Figure 4-4.** Comparison of averages.

the question as to whether a statistically significant difference exists between two means.

Consider the two population means $\mu_1$ and $\mu_2$ shown in Figure 4-4. The distribution about $\mu_1$ is characterized by a standard deviation $\sigma_1$, whereas the distribution about $\mu_2$ is characterized by a standard deviation $\sigma_2$. The difference, $\Delta\mu = \mu_1 - \mu_2$, will be characterized by a somewhat wider distribution with a standard deviation $\sigma_{\Delta\mu}$, shown in Figure 4-4, because $\Delta\mu$ will be simultaneously subject to the random errors about both $\mu_1$ and $\mu_2$. A significant difference will be said to exist between the two population means if $\mu_1 - \mu_2$ is greater than the confidence limits established for the distribution about $\mu_1 - \mu_2$. The confidence limits about $\mu_1 - \mu_2$ are shown for three different confidence levels. Notice that, at the 68% confidence level, the length of the line segment $\mu_1 - \mu_2$ from the origin is greater than $1\sigma_{\Delta\mu}$, and the means are significantly different at the 68% confidence level. However, the difference between the means $\mu_1$ and $\mu_2$ is not statistically significant at the 99.9% confidence level, since the length represented by $3.29\sigma_{\Delta\mu}$ is greater than the length of the line segment represented by $\mu_1 - \mu_2$. Thus, at the 99.9% confidence level, there is a chance that $\mu_1 - \mu_2 = 0$, or that $\mu_1 = \mu_2$.

For a limited set of data, where $\overline{X}_1$ has been computed from $n_1$ measurements and $\overline{X}_2$ has been computed from $n_2$ measurements, the confidence limits for $\Delta\overline{X}$ must be established from a knowledge of $t$ and $s_{\Delta\overline{X}}$ as described previously. The standard deviation for the difference of the two means $s_{\Delta\overline{X}}$ may be estimated by the sum of the individual variances for $\overline{X}_1$ and $\overline{X}_2$ [2]. Thus, $s_{\Delta\overline{X}}$ will be given by

$$s_{\Delta\overline{X}} = \sqrt{\frac{s_1^2}{n_1} + \frac{s_2^2}{n_2}} \tag{4-9}$$

[2] The **variance** is defined as the square of the standard deviation. The variance for a single observation is $\sigma^2$ (or $s^2$ if the true value for the standard deviation is estimated). The variance of the mean of several observations is $\sigma^2/n$ (or $s^2/n$), which is simply the square of the standard deviation of the mean of $n$ observations, $(\sigma/\sqrt{n})^2$. The variance $V_{\Delta\overline{X}}$ associated with the difference $\overline{X}_1 - \overline{X}_2$ will be $V_1 + V_2$, where $V_1$ is the variance of $\overline{X}_1$ and $V_2$ is the variance of $\overline{X}_2$.

and the confidence limits for the desired confidence level may be established with the aid of the Table 4-1. Thus,

$$\text{confidence limits for } \Delta\overline{X} = ts_{\Delta\overline{X}} = t\sqrt{\frac{s_1^2}{n_1} + \frac{s_2^2}{n_2}} \qquad \textbf{(4-10)}$$

where the value of $t$ is selected for the desired confidence level for $n_1 + n_2 - 2$ degrees of freedom (two degrees of freedom were lost in calculating $s_1$ and $s_2$). For the case where $s_1 = s_2$, which would occur in the analysis of two potentially different materials by the same analytical procedure, eq. (4-10) reduces to

$$\text{confidence limits for } \Delta\overline{X} = ts\sqrt{\frac{n_1 + n_2}{n_1 n_2}} \qquad \textbf{(4-11)}$$

Thus, if the quantity $ts\sqrt{(n_1 + n_2)/n_1 n_2}$ exceeds $\Delta\overline{X}$ for the given confidence level, the two means are statistically the same and the material may be said to contain experimentally equivalent amounts of $X$.

## REJECTION OF DATA

The question often arises in a series of measurements as to whether a certain measurement should be rejected from consideration because it appears to be different from the other measurements. If a series of well-characterized measurements are known to follow a normal distribution and if, from previous experience, reliable estimates of $\mu$ and $\sigma$ are available, we can determine the frequency with which a particular deviation from the mean is likely to occur. Thus, we know that 95% of the time, the deviation from the mean for a single measurement, $\mu - X$, will be less than $1.96\sigma$. Any deviation $\mu - X$ greater than $1.96\sigma$ will be due to random error only 5% of the time. For such situations, statistical predictions are reasonably reliable.

This is not the case for a small set of measurements for which $\mu$ and $\sigma$ are estimated by $\overline{X}$ and $s$, respectively. The random values that make up the frequency distribution shown in Figure 4-1 arise by chance and do not follow any particular order. Thus, it is perfectly possible to make three successive measurements that are higher than the mean, in much the same manner as it is possible to obtain heads on three successive tosses of a coin. If the fourth measurement turns out by chance to be lower than the mean, it appears to be divergent from the previous three. This conclusion would be immediately recognized as erroneous if the true mean and standard deviation were known. In the absence of this information, however, prejudice on the part of the investigator develops, with the concurrent desire to reject the offending value. In this example, to do so is clearly dangerous because the mean of the four results will be closer to the true mean than the mean of the three higher results. Clearly, simple inspection of the data is not a reliable guide to the rejection of spurious results, and statistical methods should be used. However, extreme care should be used even when employing these methods with a limited set of data. No mathematical procedure available can alter the worthlessness of shoddy measurements.

### The Q Test

One commonly used statistical method for discarding deviant result in sets of 10 or less is known as the $Q$ test. This method entails calculating the difference between the deviant figure and the closest value to it when the results are arranged in increasing or decreasing order. This difference is then divided by the range to obtain the $Q$ values. The calculated $Q$ value is compared to $Q$ values for 90% confidence as given by Table 4-2. The questionable value may be discarded if the calculated value of $Q$ exceeds the value given in the table. This discarding of a deviant result will be correct nine times out of ten. Consider the data of two students, R and S, obtained in the analysis of the same unknown (see p. 64).

| *Examples for data of* R *and* S | R | S |
|---|---|---|
| Number of measurements | 4 | 6 |
| Range | 0.24 | 0.18 |
| Difference | 0.14 | 0.10 |
| Calculated $Q$ value | 0.583 | 0.556 |
| $Q_{90}$ from table | 0.76 | 0.56 |

Since the calculated $Q$ value is definitely below the table value for the R data, the value 59.75 should be retained. Since the calculated $Q$ value for the data of S is very close to the table value, the value 59.27 can be either retained or discarded. In such cases, it is usually best to use some other criterion for discarding a given value.

**TABLE 4-2.** Values for Statistical Evaluation of Sets of Two to Ten Observations

| *Number of Observations,* $n$ | *Deviation,* $K_w$ | *Confidence Factors* $t_{w0.95}$ | $t_{w0.99}$ | *Rejection Quotient* $Q_{0.90}$ |
|---|---|---|---|---|
| 2 | 0.89 | 6.4 | 31.83 | |
| 3 | 0.59 | 1.3 | 3.01 | 0.94 |
| 4 | 0.49 | 0.72 | 1.32 | 0.76 |
| 5 | 0.43 | 0.51 | 0.84 | 0.64 |
| 6 | 0.40 | 0.40 | 0.63 | 0.56 |
| 7 | 0.37 | 0.33 | 0.51 | 0.51 |
| 8 | 0.35 | 0.29 | 0.43 | 0.47 |
| 9 | 0.34 | 0.26 | 0.37 | 0.44 |
| 10 | 0.33 | 0.23 | 0.33 | 0.41 |

*Source:* R. B. Dean and W. J. Dixon, *Anal. Chem.*, **23**, 636 (1951).

# PRACTICAL CONSIDERATIONS

The following section will apply the considerations developed in the previous sections to various examples and will discuss some practical considerations in the applications of statistical methods to analytical results.

## METHODS OF EXPRESSING PRECISION

Precision may be expressed as the **average deviation** (AD), the **relative average deviation** (RAD), or as the **standard deviation** (SD or *s*). The square of the SD is known as the **variance**. Even though the relative average deviation is used more often in the laboratory for a first course in quantitative analysis, the standard deviation is a better estimate of the precision of the results.

### *Average Deviation and Relative Average Deviation*

The **absolute deviation** (D) and the average deviation are expressed in the units of the measurement, whereas the **relative deviation** (RD) and the relative average deviation are expressed as parts per thousand or percent. The deviations are calculated as the difference between individual measurements, and the average or mean of the set and may be averaged to obtain the average deviation. The relative deviations are calculated by dividing the individual deviations by the average of the set and multiplying by 1000. These are averaged to obtain the relative average deviation. Following is an example of these calculations:

| *Measurements of % of* Cl *in Unknown* | *D* | RD (*ppt*) |
|---|---|---|
| 45.23 | 0.05 | $\dfrac{0.05 \times 1000}{45.28} = 1.1$ |
| 45.31 | 0.03 | $\dfrac{0.03 \times 1000}{45.28} = 0.7$ |
| 45.30 | 0.02 | $\dfrac{0.02 \times 1000}{45.28} = 0.4$ |
| Av = 45.28 | AD = 0.03 | RAD = 0.7 |

Note that AD is expressed to the same number of decimal places as the measurements and that RD and RAD are expressed to the closest 0.1 of 1 ppt.

Since accuracy and precision are both relative values, it is necessary to define the conditions under which a set of results will be considered accurate and/or precise. For the purpose of this text, a set of three results will be considered to be accurate if the RAD from the *true value* does not exceed 2.0 ppt, and the set will be considered to be precise if the RAD from the *average of the set* does not exceed 2.0 ppt and if any individual RD does not exceed 3.0 ppt.

### *Methods of Calculating Standard Deviations*

**Sum-of-Squares Method.** In this method, the following operations are performed:

1 Square each individual measurement and sum these squares. Do not round off.

2 Sum the individual measurements and square this sum. Do not round off.

3 Divide the square of the sum (step 2) by the number of measurements in the series, and then subtract from the sum of the squares of the individual measurements (step 1) to obtain the sum of the squares.

4 Divide the sum of squares (step 3) by the degrees of freedom (one less than the number of measurements) to obtain the variance.

5 Extract the square root of the variance to obtain the standard deviation.

*Examples* (columns R and S represent values obtained by different students running the same unknown for which the true value is 59.55):

| | R | | S | |
|---|---|---|---|---|
| | 59.51 | 3,541.4401 | 59.09 | 3,491.6281 |
| | 59.75 | 3,570.0625 | 59.17 | 3,501.0889 |
| | 59.61 | 3,553.3521 | 59.27 | 3,512.9329 |
| | 59.60 | 3,552.1600 | 59.13 | 3,496.3569 |
| | | | 59.10 | 3,492.8100 |
| | | | 59.14 | 3,497.5396 |
| Total | 238.47 | 14,217.0147 | 354.90 | 20,992.3564 |
| Average | 59.62 | | 59.15 | |
| $\frac{\text{Total}^2}{n}$ | $\frac{(238.47)^2}{4}$ | = 14,216.9852 | $\frac{(354.90)^2}{6}$ | = 20,992.3350 |
| Difference of sum of squares | | 0.0295 | | 0.0214 |
| Degrees of freedom | | 3 | | 5 |
| Variance | | 0.00983 | | 0.00428 |
| SD | | 0.0991 (0.10 rounded off) | | 0.0654 (0.07 rounded off) |

**Deviations-from-Average Method.** In this method, the following operations are performed:

1 Sum the individual measurements and divide by the number of measurements to obtain the average.

2 Subtract the average from each individual measurement to obtain the deviation of each.

3 Square each of these deviations and then sum the squares.

4 Divide the sum of the squares by the degrees of freedom to obtain the variance.

5 Extract the square root of the variance to obtain the SD.

| | R | | | S | | |
|---|---|---|---|---|---|---|
| | 59.51 | −0.11 | 0.0121 | 59.09 | −0.06 | 0.0036 |
| | 59.75 | +0.13 | 0.0169 | 59.17 | +0.02 | 0.0004 |
| | 59.61 | −0.01 | 0.0001 | 59.27 | +0.12 | 0.0144 |
| | 59.60 | −0.02 | 0.0004 | 59.13 | −0.02 | 0.0004 |
| | | | | 59.10 | −0.05 | 0.0025 |
| | | | | 59.14 | −0.01 | 0.0001 |
| Total | 238.47 | | 0.0295 | 354.90 | | 0.0214 |
| Average | 59.62 | | | 59.15 | | |
| Variance | $\frac{0.0295}{3} = 0.00983$ | | | $\frac{0.0214}{5} = 0.00428$ | | |
| SD | 0.0991 (0.10 rounded off) | | | 0.0654 (0.07 rounded off) | | |

**From-the-Range Method.**[3] In this method, the range of a set of measurements is determined and multiplied by a deviation factor to obtain the standard deviation. The range is the difference between the smallest and largest observation, and the deviation factor may be obtained from the Dean and Dixon table, which is reproduced in Table 4-2. The table includes values for sets that contain from two to ten measurements. The standard deviation calculated from the range is usually larger than the values obtained by the better known methods already given but is considered to be a better estimate of the actual SD of the set when small numbers of measurements are made. The range method also has the advantage of simplicity of calculation. Applied to the data of R and S,

| | R | S |
|---|---|---|
| Range | 59.75 − 59.51 = 0.24 | 59.27 − 59.09 = 0.18 |
| SD | 0.24 × 0.49 = 0.1176 (0.12 rounded off) | 0.18 × 0.40 = 0.0720 (0.07 rounded off) |

## CONFIDENCE LIMITS

There are two methods of calculating confidence intervals for a given set of determinations. One of these ultilizes the $t$ value of Student and the standard deviation of the average (the standard deviation of the individual measurements

[3] See R. B. Dean and W. J. Dixon, "Simplified Statistics for Small Numbers of Observations," *Anal. Chem.*, **23**, 636 (1951).

divided by the square root of the number of measurements). The other method utilizes the range and the $t_w$ value given by Dean and Dixon for sets containing ten or fewer measurements. Both of these methods will be illustrated using the data of R and S.

### *Method Using Student's t Value*

| | R | S |
|---|---|---|
| Degrees of freedom | 3 | 5 |
| $t$ for 95% limits | 3.2 | 2.6 |
| SD of individual determinations | 0.0991 | 0.0654 |
| Standard deviation of average ($SD/\sqrt{n}$) | 0.0496 | 0.0267 |
| 95% limits ($t \times$ SD of average) | 0.16 | 0.07 |

Thus, the true average value for R should be between 59.46 and 59.78 and that for S should be between 59.08 and 59.22.

### *Method Using Range and $t_w$*

| | R | S |
|---|---|---|
| Range | 0.24 | 0.18 |
| $t_w$ for 95% for number of measurements | 0.72 | 0.40 |
| 95% limits (range $\times t_w$) | 0.17 | 0.07 |

Thus, the true average value for R should be between 59.45 and 59.79 and that for S should be between 59.08 and 59.22. It should be noted that these values agree closely with those calculated by the first method.

## PROBLEMS

1 How many significant figures are there in each of the following measurements?

(a) 123.45 (d) $1.69 \times 10^{-5}$ (g) 987
(b) 0.01120 (e) $0.62 \times 10^{10}$ (h) 3,000,058
(c) 30.30 (f) $9.00 \times 10^{3}$ (i) 1.6295

*Ans.* (b) 4

**2** Add or subtract after rewriting the following additions and subtractions in vertical columns. Express each number (including the difference or the sum) in the proper number of significant figures.

(a) $12.67 + 357.8 + 1.349$
(b) $0.0025 + 2.5 \times 10^{-3} + 0.1025$
(c) $0.543 + 5.43 + 54.3 + 543$
(d) $1.2 \times 10^3 + 5605 + 630.0$
(e) $35.6982 - 13.4397$
(f) $1.634 - 0.29$
(g) $107.97 - 23$
(h) $6.53 \times 10^{-2} - 6.53 \times 10^{-3}$

*Ans.* (b) 0.1075; (f) 1.34

**3** Perform the indicated multiplications and divisions expressing each number, including the result, in the proper number of significant figures.

(a) $\dfrac{3.50 \times 0.1563}{35.07 \times 0.562}$

(b) $\dfrac{5.735 \times 0.565}{27.40 \times 6.8164}$

(c) $\dfrac{25.67 \times 0.1123}{1.023 \times 0.5000}$

(d) $\dfrac{3.5 \times 10^{-6} \times 7.66 \times 10^{-6}}{0.253 \times 35}$

(e) $\dfrac{5.43 \times 543}{0.543 \times 54.3}$

*Ans.* (b) 0.0174

**4** For each of the following sets of results, calculate (1) the average, (2) the deviations, (3) the average deviation, (4) the relative deviations in parts per thousand, (5) the relative average deviation in parts per thousand, (6) the standard deviation by the deviation from average method, (7) the standard deviation from the range, and (8) the 95% confidence limits.

(a) 35.47, 35.49, 35.42, 35.46
(b) 25.10, 25.20, 25.00
(c) 63.92, 63.75, 63.90, 63.86. 63.84
(d) 6.050, 6.048, 6.068, 6.054, 6.056
(e) 50.00, 49.96, 49.92, 50.15

*Ans.* (b) (1) 25.10, (3) 0.07, (5) 2.7, (6) 0.10, (7), 0.12, (8) $\pm 0.26$

**5** For each set in Problem 4, calculate the $Q$ value for the deviant result and state whether or not the deviant value should be discarded.

*Ans.* (b) 0.50; no.

**6** For each set in Problem 4 for which a deviant value can be discarded, recalculate the values requested in Problem 4 after the deviant value has been discarded.

**7** Assuming that the true values are (a) 35.53, (b) 25.06, (c) 63.80, (d) 6.064, and (e) 49.80, state whether the sets in Problem 4 are (1) both accurate and precise, (2) accurate but not precise, (3) precise but not accurate, or (4) neither accurate nor precise.

*Ans.* (b) (4)

**8** Classify the following errors as being (1) determinate-personal, (2) determinate-instrumental; (3) determinate-method, (4) indeterminate-personal, (5) indeterminate-instrumental, or (6) indeterminate-conditions:

(a) End point and equivalence point do not coincide
(b) Reading a burette as 15.63 instead of correct value of 15.65
(c) Reading a balance weight as 13.5692 instead of the correct weight of 13.6592
(d) Misestimation of equilibrium point of balance
(e) Reading a burette as 15.63 instead of the correct value of 14.37
(f) Use of uncalibrated pipette

(g) Variation of humidity during weighing of a hygroscopic solid
(h) Variation of temperature during a titration
(i) Solubility of a precipitate in the precipitation medium
(j) Using 1 : 10 HCl instead of the correct 1 : 1 HCl
(k) Use of nitric acid in place of sulfuric acid
(l) Failure to distinguish between two shades of orange
(m) Overshooting an end point by 0.03 mL
(n) Overshooting an end point by 10 drops
(o) Variation of the equilibrium point of a balance during a weighing
(p) Weighing a warm crucible
(q) The 5-g weight does not weigh exactly 5.000 g

CHAPTER 5

# Calculations of Gravimetric Analysis

In gravimetric analyses, the constituent to be determined is isolated in some relatively pure solid form that can be dried and weighed. This is one of the oldest, yet most accurate, methods of quantitative analysis and is still widely used by practicing analytical chemists. The constituent may be precipitated unchanged by change of solvent or by the addition of a compound that will form an insoluble addition complex. In most cases, however, an ion is converted to a compound, as in the precipitation of the sulfate ion ($SO_4^{2-}$) as barium sulfate ($BaSO_4$) and the nickel ion ($Ni^{2+}$) as nickel dimethylglyoximate

$$[Ni(C_4H_5N_2O_2)_2].$$

Gravimetric methods have the advantages of simplicity and accuracy but tend to be tedious and time-consuming in the laboratory. Also, the precipitates formed may be seriously contaminated (see Chapter 8). The typical analysis involves (1) the drying and weighing of the sample, (2) the dissolution of the sample, (3) preliminary adjustments, such as pH, (4) the addition of the precipitating agent and (5) the aging, filtration, washing, drying, and weighing of the precipitate. Some of these steps can be very tedious and exacting.

## GRAVIMETRIC (CHEMICAL) FACTORS

Gravimetric analysis is based upon the **law of constant composition** (the proportion by weight of the constituent elements in any pure compound is always the same) and upon the **law of definite proportions** (the masses of substances taking part in a chemical change always show a definite and invariable ratio to each other). In gravimetric analysis, the desired constituent is separated from interfering substances and either weighed directly or converted to a compound of known composition and purity that can be dried and weighed. Consequently, one of the most common calculations in gravimetric analysis is the conversion of a given weight of one substance (usually the precipitated compound) to the chemically equivalent weight of a second substance (usually the constituent being determined). Such a calculation is based upon the laws mentioned and upon the quantitative relationship expressed by a completely balanced equation.

The balanced equation for the reaction of sodium sulfate and barium chloride to produce barium sulfate and sodium chloride,

$$Na_2SO_4 + BaCl_2 \longrightarrow BaSO_4 + 2\,NaCl \qquad \textbf{(5-1)}$$

states that 142.0 g of sodium sulfate (1 molecular weight) reacts with 208.3 g of barium chloride (1 molecular weight) to produce 233.4 g of barium sulfate (1 molecular weight) and 116.9 g of sodium chloride (2 molecular weights). Because these substances always react and are produced in this ratio by weight, the weight of any one of them produced from (or reacted with) a given weight of any one of the others may be calculated by simple ratio. Thus, the weight of barium sulfate produced from 28.40 g of sodium sulfate is given by

$$\frac{X}{233.4} = \frac{28.40}{142.0} \qquad \textbf{(5-2)}$$

or, symbolically,

$$\frac{X}{BaSO_4} = \frac{28.40}{Na_2SO_4} \qquad \textbf{(5-3)}$$

Solving these gives

$$X = 28.40 \times \frac{BaSO_4}{Na_2SO_4} = 28.40 \times \frac{233.4}{142.0}$$
$$= 28.40 \times 1.643 = 44.66 \text{ g} \qquad \textbf{(5-4)}$$

Note that the ratio $BaSO_4/Na_2SO_4$ will occur in any calculation involving the weight of $BaSO_4$ produced from any given weight of $Na_2SO_4$. The value of this ratio may be calculated once and then used for any subsequent similar calculation. A ratio such as this, which states the relationship between chemically equivalent amounts of two substances, is called a **gravimetric factor**.

Many calculations in gravimetric analysis involve the determination of the weight of an element (constituent being determined) present in a given weight of a pure compound (substance precipitated). Chloride, for example, is often precipitated as silver chloride, which is weighed, and the weight of chloride is calculated from the weight of silver chloride. If a given sample produced 1.000 g of silver chloride, what weight of chloride was present?

$$\frac{\text{g Cl}}{\text{g AgCl}} = \frac{X}{1.000}$$

and the gravimetric factor for Cl in AgCl is

$$\frac{\text{Cl}}{\text{AgCl}} = \frac{35.45}{143.3} = 0.2474 \tag{5-5}$$

then

$$X = 1.000 \times 0.2474 = 0.2474 \text{ g}$$

Problems of this type can also be solved by analogy, as shown next. Since there is 1 atomic weight (at wt) of Cl in each formula weight (FW) of AgCl,

$$\text{no. at wts Cl} = \text{no. of FWs of AgCl}$$

$$\text{no. FWs AgCl} = \frac{\text{g AgCl}}{\text{FW AgCl}} = \frac{1.000}{143.3}$$

$$\text{no. at wts Cl} = \frac{\text{g Cl}}{\text{at wt Cl}} = \frac{\text{g Cl}}{35.45}$$

and

$$\frac{\text{g Cl}}{35.45} = \frac{1.000}{143.3}$$

$$\text{g Cl} = 35.45 \times \frac{1.000}{143.3} = 0.2474 \text{ g}$$

or, using symbols,

$$\text{g Cl} = \text{at wt Cl} \times \frac{\text{g AgCl}}{\text{FW AgCl}}$$

It is often necessary to determine the weight of a compound that is chemically equivalent to a given weight of another compound even though neither can be converted to the other by a single direct reaction. In this case, the two substances must be related to each other by a series of sequential reactions. For example, consider the determination of the weight of ferrous carbonate in a given sample of a carbonate iron ore that is to be analyzed by a gravimetric procedure. The ore is dissolved in a strong acid, after which the iron is oxidized to the ferric state and precipitated as the hydrous oxide, which is then ignited to $Fe_2O_3$. The completely balanced set of sequential reactions (a necessary condition) is

$$2FeCO_3 + 4H^+ \longrightarrow 2Fe^{2+} + 2H_2CO_3 \tag{5-6}$$

$$2Fe^{2+} + \text{oxidizing agent} \longrightarrow 2Fe^{3+} \tag{5-7}$$

$$2Fe^{3+} + 6OH^- \longrightarrow 2Fe(OH)_3 \tag{5-8}$$

$$2Fe(OH)_3 + \text{heat} \longrightarrow Fe_2O_3 + 3HOH \tag{5-9}$$

Consequently, 2 moles of $FeCO_3$ are necessary to produce 1 mole of $Fe_2O_3$, and the calculation is

$$\text{wt } FeCO_3 = \text{wt } Fe_2O_3 \frac{2\,FeCO_3}{Fe_2O_3} \tag{5-10}$$

In this case, both the original compound and the final compound contain the same element. However, it is also possible to calculate the weight of a compound chemically equivalent to another compound that does not contain the same element, provided the series of balanced reactions relating the two is known. In the analysis of arsenates, the arsenate may be quantitatively precipitated as silver arsenate, but this compound is not a weighable form. Consequently, it is dissolved in nitric acid, and the silver is precipitated as silver chloride, which is a weighable form. The weight of arsenate originally present can then be calculated from the weight of silver chloride produced. The equations are

$$Na_3AsO_4 + 3\,AgNO_3 \longrightarrow Ag_3AsO_4 + 3\,NaNO_3 \tag{5-11}$$

$$Ag_3AsO_4 + 3\,HNO_3 \longrightarrow 3\,AgNO_3 + H_3AsO_4 \tag{5-12}$$

$$3\,AgNO_3 + 3\,HCl \longrightarrow 3\,AgCl + 3\,HNO_3 \tag{5-13}$$

Thus, 1 mole of sodium arsenate produces 3 moles of silver chloride, and the calculation is

$$\text{wt } Na_3AsO_4 = \text{wt AgCl} \times \left(\frac{Na_3AsO_4}{3\,AgCl}\right) \tag{5-14}$$

The important points to remember in setting up gravimetric factors are

1 The number of molecules of each of the two substances in a gravimetric ratio must be such that they contain the same number of atoms of the particular element involved.

2 The molecular weight (or a proper multiple) of the substance weighed (precipitate) is always placed in the denominator, and the molecular weight (or a proper multiple) of the sought substance (constituent being determined) is always placed in the numerator.

3 In setting up the factor, only the substance weighed and the substance sought are used.

In following rule 1, it should be remembered that the number of oxygen atoms in the two formulas is only incidental to the analysis and need not be equal in the numerator and denominator.

## CALCULATION OF PERCENTAGE BY WEIGHT (GENERAL)

Since percent represents parts per hundred, the fraction of the total weight of a material due to a particular constituent is multiplied by 100 to obtain the percentage of the constituent. The fraction due to the constituent is determined by dividing the weight due to the part by the weight of the entire sample, or

$$\% = \frac{\text{wt constituent}}{\text{wt sample}} \times 100 \tag{5-15}$$

All methods of analysis are designed to measure in some way the weight of the desired constituent present in a given sample.

## CALCULATION OF PERCENT OF CONSTITUENT FOR GRAVIMETRIC ANALYSIS

Since the percent of the constituent is equal to 100 times the weight of the constituent divided by the weight of the sample, and since the weight of the constituent is equal to the weight of the precipitate times the factor, the calculation of percent of constituent is

$$\% \text{ constituent} = \frac{\text{wt ppt} \times \text{factor} \times 100}{\text{wt sample}} \tag{5-16}$$

The most common error made by beginning students in such a calculation is the use of an incorrect factor.

• **Example 1.** Calculate the percent $SO_3$ in a sample of gypsum if a 0.7560-g sample produced a precipitate of $BaSO_4$ weighing 0.9875 g.

$$\% \, SO_3 = \frac{\text{wt ppt} \times \text{factor} \times 100}{\text{wt sample}}$$

$$\text{factor} = \frac{SO_3}{BaSO_4} = \frac{80.06}{233.42} = 0.3430 \tag{5-17}$$

$$\% \, SO_3 = \frac{0.9875 \times 0.3430 \times 100}{0.7560} = 44.80\%$$

The student can check the expression to be sure that the correct factor has been used by showing the units and substances next to each number to see that the units and substances cancel satisfactorily. Equation (17) would be shown in the following manner to make such a check:

$$\%\,SO_3\left(\frac{g\ SO_3 \times 100}{g\ sample}\right) = \frac{0.9875(\cancel{g\ BaSO_4}) \times 0.3430\left(\frac{g\ SO_3}{\cancel{g\ BaSO_4}}\right) \times 100}{0.7560\ (g\ sample)}$$

$$= 44.80\,\%\left(\frac{g\ SO_3 \times 100}{g\ sample}\right) \qquad \textbf{(5-18)}$$

If the inverse factor had been used, the equation would show

$$\%\,SO_3\left(\frac{g\ SO_3 \times 100}{g\ sample}\right) = \frac{0.9875(g\ BaSO_4) \times 2.915\left(\frac{g\ BaSO_4}{g\ SO_3}\right) \times 100}{g\ sample}$$

$$= 380.8\,\%\left(\frac{(g\ BaSO_4)^2 \times 100}{g\ SO_3 \times g\ sample}\right) \qquad \textbf{(5-19)}$$

Not only is the setup incorrect since $BaSO_4$ does not cancel, but the answer of 380.8 % is obviously wrong.

In solving problems of this type, which require multiplication of several numbers and division by one number or the product of several numbers, a calculator is a useful aid. However if a calculator is not available, it would be best to use logarithms in place of longhand multiplication and division. Slide rules can be used for some calculations but are not capable of the part-per-thousand accuracy required in many analytical calculations.

## FACTOR WEIGHT SAMPLES

For any particular determination, a sample weight may be selected so that the percentage of the desired constituent may be calculated from the weight of the precipitate simply by moving the decimal. Such a weight is called a **factor weight** and is often used when a large number of samples are to be run, to simplify the calculation of the results. In the calculation of percentages, the numerical value of the factor appears in the numerator and the weight of the sample appears in the denominator. If the sample weight taken is equal to the factor, these two values cancel each other and the calculation becomes

$$\% = \frac{\text{wt ppt} \times \cancel{\text{factor}} \times 100}{\cancel{\text{wt sample}}} = \text{wt of ppt} \times 100 \qquad \textbf{(5-20)}$$

Thus, it is necessary only to weigh the precipitate and to move the decimal two places to the right to determine the percent of the constituent being determined. In this example, each 0.01 g of the precipitate represents 1 % of the constituent. If the factor weight so calculated does not approximate the optimum weight, a whole-number multiple or an exact fraction of the factor weight should be used so the calculation will consist of moving the decimal and multiplying or dividing by a small number.

• **Example 2.** Calculate the factor weight for the determination of the percentage of iron in an iron ore by precipitating the iron and igniting to ferric oxide.

$$\% \text{ Fe} = \frac{\text{wt } Fe_2O_3 \times (2Fe/Fe_2O_3) \times 100}{\text{wt sample}}$$

$$\text{factor} = 0.6994 \qquad \textbf{(5-21)}$$

$$\text{factor weight sample} = 0.6994 \text{ g}$$

If the optimum weight of sample is found to be 1.5 g, the sample to be weighed would be 2 × 0.6994, or 1.3988 g, and the calculation would be

$$\% \text{ Fe} = \frac{\text{wt } Fe_2O_3 \times 100}{2} \qquad \textbf{(5-22)}$$

In this case, each 0.02 g of the precipitate represents 1% of iron. It should be remembered that these samples have to be weighed to the exact weight as calculated and cannot be weighed within a given range.

Any problem that states the weight of a precipitate equivalent to a stated percentage of constituent is considered to be a factor weight problem. To calculate the weight of sample necessary, rearrange the usual percentage calculation to calculate the weight of sample instead of the percentage.

$$\text{wt sample} = \frac{\text{wt ppt} \times \text{factor} \times 100}{\%} \qquad \textbf{(5-23)}$$

• **Example 3.** Calculate the weight of a gypsum sample to be weighed if each milligram of $BaSO_4$ is to represent 0.05% $SO_3$. The factor $SO_3/BaSO_4$ equals 0.3430.

$$\text{g sample} = \frac{0.001 \times 0.3430 \times 100}{0.05} = 0.6860 \text{ g} \qquad \textbf{(5-24)}$$

## QUANTITY OF REAGENT REQUIRED FOR A GIVEN REACTION

It is important to add the proper amount of reagent when performing a precipitation, since too small an amount will cause incomplete precipitation and a large excess will often cause errors through secondary reactions. An example of a harmful secondary reaction occurs in the precipitation of chloride as silver chloride by the addition of silver nitrate to a chloride solution. In the presence of an excess of silver ion, the silver chloride forms a soluble complex by the reaction

$$AgCl + Ag^{+}(XS) \longrightarrow Ag_2Cl^{+} \quad \textbf{(5-25)}$$

The formation of the complex ion decreases the amount of the silver chloride precipitate and thus introduces a negative error in the determination. Similarly, silver chloride also reacts with an excess of chloride ion to form the soluble $AgCl_2^-$ ion. Slight to moderate excesses will not cause appreciable error, but large excesses of the precipitant ion can cause major errors in gravimetric and other analyses that include precipitation of a solid.

The amount of reagent to be added is normally the stoichiometric amount plus a small excess. The stoichiometric amount is that weight which will react exactly with the constituent being determined. In solving such a problem it should be remembered that stoichiometric quantities represent equal numbers of equivalents (or milliequivalents) of each of the reactants, and thus the number of milliequivalents of the precipitating agent must equal the number of milliequivalents of the substance with which it reacts. The number of milliequivalents present in a given volume of a solution of given normality equals the product of the normal concentration and the number of milliliters used. For solids the number of milliequivalents present is equal to the weight in grams of the substance divided by the milliequivalent weight in grams. Thus, the calculation becomes

$$\text{mL} \times N = \frac{\text{g solid}}{\text{meq wt solid}} \quad \textbf{(5-26)}$$

This can be solved for any one of the four quantities, provided that the other three are known.

• **Example 4.** How many milliliters of 0.5000 $N$ barium chloride are required to react exactly with 1.000 g of sodium sulfate? In solving such a problem, first calculate the equivalent weight (or milliequivalent weight) of the sodium sulfate (the substance for which a weight is given). The equivalent weight will be one-half the molecular weight because the valence of sulfate is 2, or symbolically, $Na_2SO_4/2$. The molecular weight of sodium sulfate is 142.05, so the equivalent weight is 71.03 and the milliequivalent weight is 0.07103.

$$\text{mL} = \frac{\text{wt solid}}{\text{meq wt (solid)} \times N\text{ (solution)}} = \frac{1.000}{0.07103 \times 0.5000} = 28.12 \text{ mL}$$

This type of problem can also be solved by calculating the weight of barium chloride necessary to react with 1.000 g of sodium sulfate and then calculating the volume of 0.5000 $N$ solution that contains this weight.

# INDIRECT ANALYSIS

In actual practice it is sometimes advisable to resort to indirect methods of analysis, that is, methods that utilize arithmetic or algebraic calculations in place of actual separations and determinations for one or more constituents. There are two principal methods of indirect analysis:

1 Two substances are weighed together and are then converted to some other substances, which are also weighed. From these data, the percent of each original substance can be calculated by the use of two simultaneous equations.

2 Two substances are weighed together and one of them is then determined by a separate method. From these data, the percent of each original substance can be calculated. It should be noted that this type of problem cannot be solved by the use of simultaneous equations.

## *Example of Method 1*

A sample containing only sodium chloride and potassium chloride weighs 0.2040 g. The sample is dissolved and precipitated as silver chloride, which weighs 0.4250 g. Calculate the percentage of sodium chloride in the sample.

*Step 1.* Let the weight of the substance for which the percentage is to be calculated equal $x$; let the weight of the other equal $y$; and set up an equation in $x$, $y$, and the weight of the original sample.

$$x + y = 0.2040 \tag{5-27}$$

*Step 2.* Using the values $x$ and $y$ for the weights of the two substances, calculate the weight of the precipitate produced from each by use of the correct factor.

$$\begin{aligned}\text{wt AgCl produced from } x \text{ g NaCl} &= x(\text{AgCl}/\text{NaCl}) \\ &= 2.452x\end{aligned} \tag{5-28}$$

$$\begin{aligned}\text{wt AgCl produced from } y \text{ g KCl} &= y(\text{AgCl}/\text{KCl}) \\ &= 1.923y\end{aligned} \tag{5-29}$$

Since these two values represent the total weight of AgCl produced, the second simultaneous equation can be set up using these and the weight of AgCl.

$$2.452x + 1.923y = 0.4250 \tag{5-30}$$

*Step 3.* Solve these two simultaneous equations for $x$ by multiplying eq. (5-27) by the amount of $y$ in eq. (5-30) and then subtracting the two equations to eliminate $y$.

$$\begin{array}{l} 2.452x + 1.923y = 0.4250 \\ 1.923x + 1.923y = 0.3923 \\ \hline 0.529x \qquad\qquad\quad = 0.0327 \end{array}$$

$$x = 0.0327/0.529 = 0.0618 \text{ g NaCl}$$

From eq. (5-27),

$$\text{wt KCl is } 0.2040 - 0.0618 = 0.1422.$$

*Step 4.* Calculate the percent of one or both constituents as directed.

$$\% \text{ NaCl} = \frac{0.0618 \times 100}{0.2040} = 30.29\%$$

$$\% \text{ KCl} = \frac{0.1422 \times 100}{0.2040} = 69.71\%$$

### *Example of Method 2*

A clay sample weighing 0.7500 g produced a mixture of NaCl and KCl weighing 0.3025 g. The KCl was converted to $K_2PtCl_6$, which weighed 0.3874 g. Calculate the percentage of $Na_2O$ and $K_2O$ in the sample.

*Step 1.* From the data given and the proper gravimetric factor, calculate the weight of the component determined separately.

$$\text{wt KCl} = \text{wt } K_2PtCl_6 \left( \frac{2KCl}{K_2PtCl_6} \right) = 0.3874 \times 0.3067 = 0.1188 \text{ g}$$

*Step 2.* Calculate the weight of the other component by subtracting the weight of KCl from the total weight of the two compounds.

$$\begin{array}{l} 0.3025 \text{ g (wt KCl and NaCl)} \\ 0.1188 \text{ g (wt KCl)} \\ \hline 0.1837 \text{ g (wt NaCl)} \end{array}$$

*Step 3.* Using the proper factors, convert each of these weights to the chemically equivalent weights of the substances for which percentages are to be calculated.

$$\text{wt } K_2O = \text{wt KCl}\left(\frac{K_2O}{2KCl}\right) = 0.1188 \times 0.6317 = 0.0750 \text{ g}$$

$$\text{wt } Na_2O = \text{wt NaCl}\left(\frac{Na_2O}{2NaCl}\right) = 0.1837 \times 0.5303 = 0.0975 \text{ g}$$

*Step 4.* Calculate the percentage of each component.

$$\% K_2O = \frac{0.0750 \times 100}{0.7500} = 10.00\%$$

$$\% Na_2O = \frac{0.0975 \times 100}{0.7500} = 13.00\%$$

## PROBLEMS

**1** Calculate the following gravimetric factors:

| *Substance Sought* | *Substance Weighed* |
|---|---|
| (a) NaCl | AgCl |
| (b) Cl | AgCl |
| (c) $AlCl_3$ | AgCl |
| (d) $SO_3$ | $BaSO_4$ |
| (e) $Fe_2(SO_4)_3$ | $BaSO_4$ |
| (f) $Fe_2(SO_4)_3$ | $Fe_2O_3$ |
| (g) $K_2O$ | $K_2PtCl_6$ |
| (h) $Fe_3O_4$ | $Fe_2O_3$ |

(*Ans.* (b) 0.2474)

**2** Calculate the percentage of the sought substance for the following gravimetric analyses:

| *Substance Sought* | *Sample Material* | *Sample Weight (g)* | *Substance Precipitated* | *Weight of Precipitate (g)* |
|---|---|---|---|---|
| (a) Cl | Table salt | 1.000 | AgCl | 2.344 |
| (b) $SO_3$ | Gypsum ore | 0.7560 | $BaSO_4$ | 0.9875 |
| (c) $FeCO_3$ | Iron ore | 0.5000 | $Fe_2O_3$ | 0.2124 |
| (d) CaO | Rock | 0.9958 | $CaCO_3$ | 0.2936 |
| (e) $K_2O$ | Rock | 1.023 | $K_2PtCl_6$ | 0.1016 |

(*Ans.* (b) 44.80%)

**3** Calculate the percentage of $CaCO_3$ in a sample of limestone if a 0.7560-g sample lost 0.3301 g when ignited. (The loss in weight represents loss of $CO_2$.)

**4** For each of the following, calculate the volume of the solution necessary to react exactly with the given weight of the stated solid for the reaction specified:

| Solution | Concentration (*N*) | Reactant | Weight of Reactant (*g*) | Product Formed |
|---|---|---|---|---|
| (a) $FeCl_3$ | 0.1250 | $Na_3PO_4$ | 0.3285 | $FePO_4$ |
| (b) $AgNO_3$ | 0.2078 | $Na_3AsO_4$ | 0.2678 | $Ag_3AsO_4$ |
| (c) $CaCl_2$ | 0.2500 | $Na_3PO_4$ | 0.1964 | $Ca_3(PO_4)_2$ |
| (d) $Na_2C_2O_4$ | 0.1056 | $CaCl_2$ | 0.2931 | $CaC_2O_4$ |
| (e) $H_2SO_4$ | 0.0505 | $BaCl_2$ | 0.1310 | $BaSO_4$ |

(*Ans.* (b) 18.60)

**5** Calculate the weight of sample to be weighed if each stated unit of precipitate is to represent the stated percentage of the constituent.

| Type of Sample | Weight of Precipitate | Percent of Constituent | Factor |
|---|---|---|---|
| (a) Iron ore | 0.0010 g $Fe_2O_3$ | 0.10% Fe | 0.7000 |
| (b) $Cu_2Cl(OH)_3$ | 0.0050 g AgCl | 0.20% Cl | 0.2474 |
| (c) $Cu_2Cl(OH)_3$ | 0.0005 g $Cu_2(SCN)_2$ | 0.10% Cu | 0.5226 |
| (d) Gypsum | 0.0020 g $BaSO_4$ | 0.25% $SO_3$ | 0.3430 |
| (e) Lead ore | 0.0010 g $PbSO_4$ | 0.50% PbO | 0.7360 |

(*Ans.* (b) 0.6185)

**6** A sample containing $BaCl_2$ and $SrCl_2$ weighed 0.3705 g. The salts were converted to the carbonates, which weighed 0.3485 g. Calculate the percentage of Ba and Sr in the sample.

*Ans.* Ba, 39.84%

**7** A rock sample weighing 0.9000 g produced a mixture of NaCl and KCl weighing 0.1025 g. This mixture was dissolved and produced a precipitate of $K_2PtCl_6$ weighing 0.2815 g. Calculate the percentage of $Na_2O$ and $K_2O$ in the rock.

**8** A sample containing $CuSO_4 \cdot 5H_2O$ and $NiSO_4 \cdot 6H_2O$ weighs 1.2500 g. The sample was dissolved, and 0.2800 g of copper was plated from the solution. Calculate the percent of each component.

**9** A 1.500-g sample of a rock produced a mixture of CaO and BaO, which weighed 0.5193 g. The oxides were converted to the sulfates, which weighed 0.9072 g. Calculate the percentage of Ca and Ba in the rock.

**10** A mixture of $Fe_2O_3$ and $Al_2O_3$ weighs 0.7100 g. On ignition in $H_2$ only the $Fe_2O_3$ is affected and is reduced to Fe. The mixture then weighs 0.6318 g. Calculate the percent of Al in the original mixture.

**11** A mixture of $BaCl_2 \cdot 2H_2O$ and LiCl weighs 0.6000 g and produces a precipitate of AgCl weighing 1.440 g. Calculate the percentage of Ba and of Li in the sample.

**12** A mixture of NaCl and $(NH_4)_2SO_4$ weighing 0.7970 g produced a precipitate of AgCl, which weighed 1.632 g. Calculate the percentage of $SO_3$, $NH_3$, and Na in the sample.

*Ans.* $SO_3$, 9.997%

**13** In rock analysis, a 10.00-g sample weighed 9.93 g after drying to constant weight at 110°C. A 1.250-g sample of the dried material produced a mixture of CaO and MgO, which weighed 0.5273 g. The oxides were converted to the carbonates, which weighed 1.010 g. Calculate the percentage of Ca and of Mg in the dried sample and in the rock as received.

**14** A 2.000-g sample of limestone lost 0.0420 g when dried to constant weight at 110°C. A 0.2016-g sample of the dried material produced a precipitate of magnesium oxinate, which weighed 0.1552 g, and a 0.9651-g sample of the dried material produced a mixture of CaO and MgO, which weighed 0.4838 g. Calculate the percentage of Mg and of Ca in the dried material and in the limestone as received.

*Ans.* Mg 5.99% in dried sample; 5.86% in sample as received.

**15** A 1000-mL sample of water was evaporated and treated to obtain a mixture of NaCl and KCl weighing 0.4498 g. The potassium was precipitated as $KClO_4$, which weighed 0.2011 g. Calculate the parts per million (ppm; mg/liter) of Na, NaCl, K, and KCl in the water.

**16** A 1.0320-g sample of a nickel ore produced 0.0254 g of $CoSO_4$ and 0.4325 g of nickel dimethylglyoximate $[Ni(C_4H_5N_2O_2)_2]$. Calculate percentage of NiO and $Co_3O_4$ in the sample.

CHAPTER

# 6 Calculations of Titrimetric Analysis

In a titrimetric analysis, a solution (the **titrant**) is added from a burette into a titration vessel that contains the substance being titrated. The purpose of the titration is to determine the volume of the titrant that will react quantitatively (exactly, stoichiometrically) with the substance being titrated. Consequently, titration may be defined as the determination of the volume of a solution that will react exactly with a given volume of another solution or with a given weight of another substance.

The **stoichiometric (equivalence) point** in a titration is attained when exactly equivalent quantities of the titrant and the substance being titrated have been brought into reaction or, in other words, when a sufficient volume of titrant has been added to react exactly with the amount of the substance being titrated that is present in the titration vessel. In many titrations the stoichiometric point is accompanied by some detectable change caused by one of the reactants or products. The detectable change may be visual (such as a change in color or the formation or disappearance of a precipitate) or it may be a change in a physical or chemical property (such as a change in potential, conductivity, capacitance, energy absorption or emission, or density), which usually requires the use of a special instrument for detection.

Often the detectable change does not appear at the exact equivalence point, since it may be caused by a slight excess or deficiency of the titrant. For this reason, the appearance of the detectable change is known as the **end point**, not as the equivalence point. The difference in the volume of titrant necessary to reach the equivalence point and that necessary to reach the end point is known as the **end-point error**.

In some titrations a detectable change is not furnished by the reactants or products. In this case, another substance must be added that will show a detectable change at or near the equivalence point. Such a substance is called an **indicator** and usually furnishes a visual change.

At the stoichiometric point the number of equivalents (or milliequivalents) of the titrant added to the titration vessel must equal the number of equivalents (or milliequivalents) of the titrated substance originally placed in the vessel.

This is the basis of all calculations involved in titrimetric analysis; that is, at the equivalence point

$$\text{meq titrant} = \text{meq titrated substance} \qquad \textbf{(6-1)}$$

The titration data must furnish a means of calculating the number of milliequivalents of either the titrant or the titrated substance in order that the number of milliequivalents of the other may be determined. For solutions, the number of milliequivalents of the solute equals the product of the volume (milliliters) and normality; for solids, the number of milliequivalents equals the weight of the solid divided by the milliequivalent weight. Consequently, the titration data must include an accurately known volume of a standard solution or an accurately known weight of a pure substance.

Many titrimetric procedures are utilized daily in industrial, governmental, academic, and research laboratories, especially since the introduction of automated titration instruments. Titrimetry is usually more rapid and less tedious (although less accurate) than gravimetry, even though it requires the preparation and standardization of the solutions used in the analyses. Titrimetric methods are often the procedure of choice for many routine determinations in industrial laboratories, including vinegar, antiacids, etching baths, plating solutions, bleaches, sulfur in gasoline, soda ash, and water hardness.

## STANDARDS

Standards are substances or solutions of known composition and are classified as primary and secondary. Usually the word *standard* by itself is applied only to solids, and the word *solution* is added when referring to a solution of known composition.

A **primary** standard is a solid of such purity that it can be used to prepare a standard solution by dissolving an accurately known weight in sufficient solvent to make an accurately known volume of solution. A **secondary** standard is a substance whose composition has been determined against a primary standard by some method of analysis. A suitable primary standard will meet the following requirements:

1 It should react with the substance being standardized or determined by a single, definite, quantitative chemical reaction. A reaction is considered to be quantitative if only a negligible amount of at least one reactant is left unreacted at the equivalence point.

2 The purity must be known definitely, and the types and amounts of impurities, if present, should be determinable by relatively simple procedures.

3 It should not change weight or composition upon exposure to the atmosphere. A solid that changes weight upon exposure is difficult to weigh accurately and consequently is not satisfactory as a primary standard.

4 It should be capable of being dried at normal temperatures, usually 105–110°C, without change of composition. This characteristic rules out most

hydrates, since hydrates tend to lose all or part of their water of hydration at or below these temperatures.

5 It should have a high equivalent weight so that relatively large samples can be used in standardization procedures.

Typical primary standards include potassium hydrogen phthalate ($KHC_8H_4O_4$, also known as potassium acid phthalate and potassium biphthalate and often abbreviated as KHP), sulfamic acid ($HSO_3NH_2$), and sodium carbonate ($Na_2CO_3$) for acids and bases; potassium dichromate ($K_2Cr_2O_7$), arsenious oxide ($As_2O_3$), potassium iodate ($KIO_3$), and sodium oxalate ($Na_2C_2O_4$) for oxidation–reduction reactions; and sodium chloride (NaCl) and silver nitrate ($AgNO_3$) for precipitation reactions. Primary standard grade chemicals may be purchased from the usual suppliers, and some may be obtained from the National Bureau of Standards. For drugs, the *United States Pharmacopeia* reference standards are usually used.

Secondary standards may be preferable in some cases, especially for procedures for which a suitable primary standard is not readily available. As an example, the standard samples of rocks, alloys, metals, and other materials furnished by the National Bureau of Standards are considered to be the equal of primary standards, even though the composition is determined by analysis.

## STANDARD SOLUTIONS

The concentration of a standard solution must be known within the limits of accuracy necessary in the actual analysis for which it is being used. In the beginning course in quantitative analysis, a solution will be considered a standard solution if its concentration is known accurately to within two parts per thousand of the stated value. Standard solutions are also classified as primary or secondary. The concentration of a secondary standard solution is determined by titration against a known weight of a standard solid or a known volume of a standard solution.

There are two methods of preparing a primary standard solution. The simplest is to dissolve a known weight of a primary standard solid in sufficient solvent to make a known volume of the solution. The second method involves determining the concentration by a gravimetric procedure of an accurately known volume of the solution. It should be noted in both of these methods that the two requirements are an accurately known weight of a solid of known purity and an accurately known volume of the solution.

An example of the first method is the preparation of standard silver nitrate in the laboratory determination of chloride. A known weight of silver nitrate is dissolved in water, transferred to a volumetric flask, and made to volume in the flask. The number of milliequivalents of silver nitrate (as determined by the weight dissolved) equals the number of milliequivalents of silver nitrate in the solution. The calculation is

$$N = \frac{\text{g solute}}{\text{mL solution} \times \text{meq wt solute}} \qquad \textbf{(6-2)}$$

If 8.5938 g of silver nitrate (eq wt, 169.9) is dissolved and diluted to 500.0 mL, the normality may be calculated as follows:

$$N = \frac{8.5938 \text{ g}}{500.0 \text{ mL} \times 0.1699 \text{ (g/meq)}} = 0.1012\ N \text{ (meq/mL)}$$

Often it is impossible to prepare a solution by weighing a primary standard solid because none is available. In this case the solution is prepared by dissolving an approximate weight of the solid (or volume of liquid) in enough solvent to make an approximate volume of the solution. An aliquot of this solution is then pipetted and the weight of solute determined by precipitation of a compound from which the weight of the solute may be calculated. Thus, a solution of potassium chloride may be standardized by precipitating the chloride in an aliquot as silver chloride and weighing the silver chloride. A solution so standardized would be considered a primary standard. The calculation of the normality is

$$N = \frac{\text{wt precipitate}}{\text{volume (mL) aliquot} \times \text{meq wt precipitate}} \tag{6-3}$$

Note that the milliequivalent weight of the precipitate, not the solute, is used in this calculation. If a 25.00-mL aliquot of a solution of potassium chloride produces a precipitate of silver chloride (eq wt, 143.32) weighing 0.3583 g, the normality is calculated by

$$N = \frac{0.3583 \text{ g}}{25.00 \text{ mL} \times 0.14332 \text{ (g/meq)}} = 0.1000\ N$$

The concentration of a secondary standard solution is usually determined by titration against a standard solid or a standard solution. An example of titration against a primary standard solid is the standardization of sodium hydroxide using potassium acid phthalate (KHP). In this procedure, samples of KHP (eq wt, 204.23) are weighed, dissolved, and then titrated with the sodium hydroxide solution using phenolphthalein as the indicator. If a 0.7546-g sample of biphthalate required 34.79 mL of the sodium hydroxide, the normality of the sodium hydroxide may be calculated by

$$N = \frac{\text{wt solid}}{\text{mL solution} \times \text{meq wt solid}} \tag{6-4}$$

or

$$N = \frac{0.7546 \text{ g}}{34.79 \text{ mL} \times 0.20423 \text{ (g/meq)}} = 0.1062\ N$$

An example of a titration against a standard solution is the determination of the concentration of a thiocyanate solution against standard silver nitrate solution using ferric alum indicator. Usually, an aliquot of the silver nitrate is pipetted and titrated with the thiocyanate from the burette. However, both solutions could be measured by burette. In this case both reactants are solutions and

$$\text{mL} \times N \text{ (solution 1)} = \text{mL} \times N \text{ (solution 2)} \qquad \textbf{(6-5)}$$

If 10.00 mL of 0.1015 $N$ silver nitrate required 20.30 mL of the thiocyanate, the $N$ of thiocyanate may be calculated by

$$N = \frac{\text{mL} \times N \text{ (of AgNO}_3\text{)}}{\text{mL thiocyanate}} \qquad \textbf{(6-6)}$$

or

$$N = \frac{10.00 \times 0.1015}{20.30} = 0.0500\ N$$

## CALCULATION OF PERCENT OF CONSTITUENT FROM TITRIMETRIC ANALYSIS

The two main types of titrimetric analysis are direct titration and indirect titration. In a **direct** titration the volume of a standard solution required to react with the constituent is determined by adding the titrant to the sample until the end point is reached. In the **indirect** or **back-titration** method, an excess of a standard solution is added to the sample, and the amount of excess is determined by titration with a second standard solution. The sample being determined may be a weighed sample of the solid material dissolved in water or may be an aliquot of an accurately prepared solution of the unknown. If the end point and stoichiometric point do not coincide, a blank should be run and the volume of titrant required by the blank should be subtracted from the volume of titrant used in the titration. A blank is a determination performed without an actual sample being present. The blank solution for a titration is a volume of water equal to that used to dissolve the sample. The blank solution is then carried through the analysis, and the amount of titrant necessary to produce the end point is determined.

### DIRECT TITRATION

The basis of the calculation of direct titration is the fact that at the equivalence point the milliequivalents of titrant added (milliliters × $N$) must equal the milliequivalents of the constituent present in the sample. Since the number of milliequivalents of the titrant is known, the weight of the constituent may be found by multiplying by the milliequivalent weight. Thus, the calculation of percentage becomes

$$\%\,\text{constituent} = \frac{\text{wt constituent} \times 100}{\text{wt sample}} \tag{6-7}$$

$$\text{wt constituent} = \text{number meqs} \times \text{meq wt constituent} \tag{6-8}$$

$$\text{number meq} = \text{mL} \times N\ \text{titrant} \tag{6-9}$$

or

$$\%\,\text{constituent} = \frac{\text{mL} \times N\,(\text{titrant}) \times \text{meq wt (constituent)} \times 100}{\text{wt sample}} \tag{6-10}$$

As an example, calculate the percent of chloride in an unknown if a 0.4179-g sample required 34.67 mL of 0.1012 $N$ silver nitrate by the direct titration method. The atomic weight of Cl is 35.453.

$$\%\,Cl^- = \frac{34.67 \times 0.1012 \times 0.035453 \times 100}{0.4179} = 29.77\,\%$$

If an aliquot of an accurately prepared solution is used in place of the weighed sample, the weight of the sample in the aliquot may be calculated and used in the calculation. For example, suppose that 8.3580 g of the unknown was dissolved and diluted to 500.0 mL and a 25.00-mL aliquot of this solution was used as the sample. The calculation of the weight of the sample used is

$$\text{wt sample} = 8.3580 \times \frac{25.00}{500.0} = 0.4179\ \text{g} \tag{6-11}$$

The Mohr method for chloride uses potassium chromate as an indicator and requires a blank because the end point and equivalence point are not identical. As an example of the calculation for such a determination, find the percentage of chloride in a sample if a 25.00-mL aliquot of a solution of 5.5920 g in 500.0 mL required 33.47 mL of 0.1098 $N$ $AgNO_3$ by the Mohr method. In a blank determination, 0.12 mL of the $AgNO_3$ was required.

$$\%\,Cl^- = \frac{(33.47 - 0.12) \times 0.1098 \times 0.035453 \times 100}{5.592(25.00/500.0)}$$

$$= \frac{33.35 \times 0.1098 \times 0.035453 \times 100}{5.592 \times 0.0500} = 46.43$$

Calculations for replicate determinations that use the same-size aliquots of a given solution so that the sample weights are identical may be simplified by calculating the value of all the terms except milliliters. This value, often represented by $Q$, may then be multiplied by the volume of titrant to obtain the percentage. For example, assume that three titrations of 25.00-mL aliquots of the unknown in the last example required 33.47, 33.52, and 33.39 mL, respectively. $Q$ may be calculated for all the samples as

$$Q = \frac{0.1098 \times 0.035453 \times 100}{5.592 \times 0.05} = 1.392 \qquad \textbf{(6-12)}$$

The calculation for the percentage for the three replicates then becomes

$$\% \text{ Cl}^- = 33.35 \times 1.392 = 46.43$$
$$\% \text{ Cl}^- = 33.40 \times 1.392 = 46.50$$
$$\% \text{ Cl}^- = 33.27 \times 1.392 = 46.32$$

## INDIRECT TITRATION

In indirect methods, an excess of one standard solution is added to a solution of the sample and the amount of excess is determined by titration with a second standard. An example is the **Volhard method** for chloride, in which an excess of standard silver nitrate is added to the sample, and the amount of excess silver nitrate is determined by titration with a standard solution of thiocyanate using ferric alum as the indicator. Since some of the silver nitrate reacts with all the chloride present and the remainder reacts with the thiocyanate, the total number of milliequivalents of the silver nitrate must be equal to the sum of the milliequivalents of chloride plus the milliequivalents of thiocyanate. Since the milliequivalents of silver nitrate are given by the volume and the normality of the standard solution used and the milliequivalents of thiocyanate are determined by the volume and normality of the solution used in the back titration, the milliequivalents of chloride may be found by subtraction of the milliequivalents of thiocyanate from the milliequivalents of silver nitrate.

$$\text{meq } AgNO_3 = \text{meq Cl}^- + \text{meq KSCN} \qquad \textbf{(6-13)}$$

or

$$\text{meq Cl}^- = \text{meq } AgNO_3 - \text{meq KSCN} \qquad \textbf{(6-14)}$$

or

$$\text{meq Cl}^- = (\text{mL} \times N \ AgNO_3) - (\text{mL} \times N \text{ KSCN}) \qquad \textbf{(6-15)}$$

and

$$\% \text{ Cl}^- = \frac{\text{meq Cl}^- \times \text{meq wt Cl}^- \times 100}{\text{wt sample}} \qquad \textbf{(6-16)}$$

or

$$\% \text{Cl}^- = \frac{[(\text{mL} \times N\ \text{AgNO}_3) - (\text{mL} \times N\ \text{KSCN})] \times \text{meq wt Cl}^- \times 100}{\text{wt sample}}$$

**(6-17)**

As an example, calculate the percent of $Cl^-$ in an unknown if a 0.3469-g sample required 15.56 mL of 0.0509 *N* KSCN to titrate the excess $AgNO_3$ remaining from 25.00 mL of 0.1018 *N* solution added to the sample.

$$\% \text{Cl}^- = \frac{[(25.00 \times 0.1018) - (15.56 \times 0.0509)] \times 0.035453 \times 100}{0.3469}$$

$$= \frac{(2.545 - 0.792) \times 0.035453 \times 100}{0.3469}$$

$$= \frac{1.753 \times 0.035453 \times 100}{0.3469} = 17.92\%$$

## OPTIMUM WEIGHT OF SAMPLE FOR STANDARDIZATIONS

The optimum weight of standard to be used in the standardization of solutions for which only the approximate concentration is known is the weight that will use the largest feasible volume of titrant but will not require a volume of titrant that exceeds the total readable volume (capacity) of the burette. Using the largest feasible volume will decrease the relative error due to misestimation of the positions of the meniscus before and after delivery of titrant, but using more than one burette-full increases the meniscus error because the position must be read four times instead of two. Since the maximum meniscus error in two estimations of the meniscus is normally 0.04 mL, the error in four estimations will be 0.08 mL, which will double the relative error unless a large portion of the burette volume is used both times. If a volume close to two burette volumes is used, the titration will require an excessively long time to perform. For a 50-mL burette, volumes between 20 and 50 mL are considered satisfactory.

The error made in preparation of solutions of approximate normalities can be either positive or negative. It is often large but is seldom greater than 30%. That is, the concentration of an approximately 0.1 *N* solution will seldom be less than 0.0700 *N* or greater than 0.1300 *N*.

If the solution is prepared so that the concentration is approximately 0.1 *N* and the weight of standard used is that weight which would require exactly 50.00 ml of exactly 0.1000 *N* solution, more than one burette-full will be needed if the concentration turns out to be less than 0.1000 *N*. Consequently, the weight of standard to be used is that which will require no more than one burette-full if the maximum negative error was made in the preparation. This is approximately 70% of the weight required to react with exactly one burette-full of a solution of the estimated normality. The calculation of the optimum weight then becomes

$$\text{optimum wt standard} = \text{volume burette} \times \text{estimated } N \times \text{meq wt standard} \times 0.7 \qquad \textbf{(6-18)}$$

As an example, solutions of sodium hydroxide are prepared to an approximate normality by preparing an approximately 50% (w/w) solution and diluting a calculated volume of this solution to the volume required with carbon dioxide free water. The solution is then standardized against potassium biphthalate (KHP). If a solution that is approximately 0.1 $N$ is prepared in this way, the optimum weight of KHP for the standardization using a 50-mL burette is

$$\text{wt KHP} = 50 \times 0.1 \times 0.20423 \times 0.7 = 0.7148 \text{ g}$$

If the maximum negative error had been made in the solution and 0.7148 g of KHP had been used, the volume of titrant required would be

$$\text{mL} = \frac{\text{wt KHP}}{N \times \text{meq wt}} = \frac{0.7148}{0.0700 \times 0.20423} = 49.99 \text{ mL}$$

If the maximum positive error had been made in the solution and 0.7148 g of KHP had been used, the volume of titrant required would be

$$\text{mL} = \frac{0.7148}{0.1300 \times 0.20423} = 26.92 \text{ mL}$$

Thus, the use of the 0.7 safety factor in the calculation of the weight of standard to be weighed will ensure that a satisfactory volume of titrant will be used in the titration, provided that the error in preparation of the solution does not exceed 30%.

## MAXIMUM WEIGHT OF UNKNOWN SAMPLE TO BE WEIGHED

No titration should require more titrant than the capacity of the burette; thus, the maximum weight of an unknown sample is that weight which requires no more than one burette-full of the titrant. However, it is also necessary that the sample weight be large enough to require a substantial portion of the capacity of the burette in order that the relative meniscus error be as small as possible. Since standard solutions are used in the titrations of unknowns, the problem becomes one of estimating the actual percentage of constituent in the sample. Although it is obvious that the maximum possible percentage of the constituent in the sample is 100%, remember that the actual percentage due to the constituent may be very small; thus, a sample weight calculated on the basis that the constituent is 100% of the sample could cause the use of volumes of titrant that are too small for satisfactory accuracy in the determination. That is, the relative error in estimating the volume would be greater than can be allowed to obtain the requisite accuracy needed in the analysis.

The weight to be used can be estimated readily only if the analyst has prior information concerning the composition of the unknown material. Often the maximum and minimum percentages of the constituent in such samples are known from previous analyses on similar materials. Sometimes the actual compound containing the constituent is known, even though the fraction of the weight of the unknown material due to this compound is not known.

If the maximum percentage of the constituent is known from previous analysis, the calculation of the maximum weight can be made in the following manner. The usual calculation of percentage from a direct titration is given by eq. (6-10):

$$\%\,\text{constituent} = \frac{\text{mL} \times N \times \text{meq wt (constituent)} \times 100}{\text{wt sample}} \qquad \textbf{(6-10)}$$

which, upon rearranging and using general terms, becomes

$$\text{max wt sample} = \frac{\text{volume burette} \times N\ \text{titrant} \times \text{meq wt constituent} \times 100}{\text{max}\ \%\ \text{constituent in sample}} \qquad \textbf{(6-19)}$$

As an example, an analyst is required to determine the percentage of iron in an iron ore, and his previous experience indicates that this type of sample will seldom contain more than 30% iron. What is the maximum weight of sample he should weigh for the determination if a 50-mL burette and 0.1000 *N* standard solution is to be used and the equivalent weight of iron in the titration selected is 55.85?

$$\text{max wt sample} = \frac{50.00 \times 0.1000 \times 0.05585 \times 100}{30}$$

$$= 0.9308\ \text{g}$$

If the compound containing the constituent is known, the maximum weight can be calculated assuming that this compound represents 100% of the unknown material by the following method:

$$\text{max wt sample} = \frac{\text{volume burette} \times N\ \text{titrant} \times \text{meq wt compound} \times 100}{100} \qquad \textbf{(6-20)}$$

As an example, a chloride unknown containing potassium chloride (eq wt, 74.56) and inert materials is to be analyzed by a direct method using a 25.00-mL burette and 0.2000 *N* silver nitrate. Calculate the maximum weight of sample to be weighed.

$$\text{max wt sample} = \frac{25.00 \times 0.2000 \times 0.07456 \times 100}{100}$$

$$= 0.3728 \text{ g}$$

If no satisfactory method of calculating the maximum weight of sample is available, an analyst will usually take an approximate weight of sample, dissolve and titrate rapidly, neglecting interferences, to an approximate end point. From the approximate data thus determined, the analyst can calculate the maximum weight of sample by simple ratio. That is,

$$\frac{\text{approx wt sample used}}{\text{approx vol titrant used}} = \frac{\text{wt sample to be weighed}}{\text{volume burette to be used}} \qquad \textbf{(6-21)}$$

## PROBLEMS

**1** Calculate the normality of a solution prepared by dissolving 8.5938 g of $AgNO_3$ and diluting to 500.0 mL.

*Ans.* 0.1012 *N*

**2** Repeat problem 1 for (a) 2.5439 g of NaCl in 250.0 mL; (b) 1.3216 g of $BaCl_2$ in 100 mL; (c) 3.9274 g of NaOH in 1000 mL; (d) 2.1696 g of $K_2Cr_2O_7$ in 500.0 mL.

**3** Calculate the normality of a NaCl solution if a 25.00-mL aliquot produced a precipitate of AgCl weighing 0.4015 g.

*Ans.* 0.1121 *N*

**4** Repeat problem 3 for the following solutions:

| *Solute* | *Aliquot* | *ppt* | *Wt of ppt (g)* |
|---|---|---|---|
| (a) $BaCl_2$ | 20.00 | $BaSO_4$ | 0.4783 |
| (b) $CaCl_2$ | 15.00 | $CaCO_3$ | 0.2222 |
| (c) KOH | 10.00 | $K_2PtCl_6$ | 0.4275 |
| (d) HCl | 25.00 | AgCl | 0.6539 |

**5** A 0.7546-g sample of KHP required 34.79 mL of a NaOH solution. Calculate the normality of the solution.

*Ans.* 0.1062 *N*

**6** Repeat problem 5 for the following weights of standards and volume in solutions:
(a) 0.4263 g of $AgNO_3$ required 25.37 mL of NaCl solution
(b) 0.1065 g of $Na_2CO_3$ required 38.83 mL of $H_2SO_4$ solution
(c) 0.1021 g of NaCl required 43.69 mL of $AgNO_3$ solution

**7** Calculate the percentage of chloride in an unknown if a 0.4179-g sample required 34.67 mL of 0.1012 *N* $AgNO_3$.

*Ans.* 29.77%

**8** Calculate the percent of chloride in an unknown if a 3.6380-g sample was dissolved and diluted to 500.0 mL. A 25.00-mL aliquot of this solution required 23.92 mL of 0.1069 *N* $AgNO_3$ by the Mohr method. A blank of 0.12 mL was necessary.

*Ans.* 49.50%

**9** For the solution prepared in problem 8, the $AgNO_3$ remaining after addition of 25.00 mL of 0.1258 *N* $AgNO_3$ to a 25.00-mL aliquot required 9.54 mL of 0.0629 *N* KSCN. Calculate the percent Cl.

**10** For the solution prepared in problem 8, a 20.00-mL aliquot required 21.62 mL of 0.0943 *N* $AgNO_3$ by a method that does not require a blank. Calculate the percent of Cl.

**11** Calculate the parts per million (ppm) of chloride and sodium chloride in a coastal bay water if a 50.00-mL sample required 27.38 mL of 0.0785 *N* $AgNO_3$ by the Mohr method. Assume a blank of 0.15 mL was necessary.

*Ans.* 1516 ppm Cl

**12** A 25.00-mL sample of the water in problem 11 required 20.76 mL of 0.0515 *N* $AgNO_3$ by a method that does not require a blank. Calculate the ppm of Cl and NaCl.

**13** The $AgNO_3$ remaining after the addition of 25.00 mL of 0.1030 *N* $AgNO_3$ to 25.00 mL of the water in problem 11 required 28.69 mL of 0.0525 *N* KSCN by the Volhard method. Calculate the ppm of Cl and NaCl.

**14** Calculate the optimum weight of standard for the following standardizations:

| *Standard* | *Solution* | *Normality* | *Volume of Burette* |
|---|---|---|---|
| (a) KHP | KOH | 0.2 | 25.00 |
| (b) KHP | NaOH | 0.5 | 50.00 |
| (c) $AgNO_3$ | NaCl | 0.05 | 10.00 |
| (d) $Na_2CO_3$ | HCl | 0.1 | 50.00 |
| (e) $C_4H_{10}O_2NOH$ | $H_2SO_4$ | 0.25 | 25.00 |

(*Ans.* (a) 0.7147)

**15** Calculate the maximum weight of sample to be weighed for an unknown atacamite ore [$Cu_2Cl(OH)_3$] in which the percent Cl is to be determined by using 0.1250 *N* $AgNO_3$ and a 25.00-mL burette.

*Ans.* 0.6677

**16** Calculate the maximum weight of sample for the unknown in problem 15 if the copper was to be determined iodometrically by titration with 0.0500 *N* $Na_2S_2O_3$ using a 50.00-mL burette. In this titration the equivalent weight of copper is the same as the atomic weight.

**17** Calculate the maximum weight of sample to be weighed in the analysis of a stassfurtite ore ($Mg_7Cl_2B_{16}O_{30}$) if the chloride is to be determined by the Mohr method using a 50.00-mL burette and 0.0575 *N* $AgNO_3$. Assume that a blank of 0.12 mL is necessary.

*Ans.* 1.282 g

**18** Calculate the maximum weight of sample to be weighed of the ore in problem 17 if the magnesium is to be determined with a 25.00-mL burette using 0.2075 *N* EDTA.

CHAPTER

# 7 Equilibrium

## THE ACTIVITY CONCEPT

When Arrhenius first proposed his theory of ionization it was believed that all substances existed as neutral molecules in the pure state and that these molecules dissociated or broke apart after dissolving to form charged particles called **ions**. **Strong electrolytes** were those compounds that dissociated to a great extent, and **weak electrolytes** were those that were ionized only to a slight extent after being dissolved. One of the main arguments for the dissociation into ions was the fact that solutions of electrolytes conducted an electric current and that solutions of strong electrolytes were better conductors than solutions of weak electrolytes. Another factor was that solutions of strong electrolytes lowered the freezing point of water more than solutions of equal concentrations of nonelectrolytes. Since the lowering was attributed to the number of particles in solution, strong electrolytes evidently existed as ions in solution. On this basis it was predicted that the freezing points of solutions of strong electrolytes should be lower by a factor of the number of ions produced; that is, the freezing point of a sodium chloride (NaCl) solution should be two times lower than that of a solution of sugar of the same concentration, and the freezing point of a solution of aluminum chloride ($AlCl_3$) should be four times lower than that of the sugar solution. Since the freezing-point depression of such solutions turned out to be a little less than predicted, Arrhenius postulated that strong electrolytes were not completely dissociated in water solution.

We know now that most strong electrolytes are completely ionic even in the pure or solid state; thus the Arrhenius theory must be modified to take this into account. Evidently, the ions in moderately concentrated solutions of strong electrolytes are not as effective or efficient in taking part in physical or chemical reactions as they are in very dilute solutions. The concept of active concentration (activity, active mass) has been introduced to explain the behavior of solutions of strong electrolytes. Strong electrolytes are assumed to be completely ionic in solution but are not completely effective (efficient), so that the active concentration is less than the actual concentration. The two concentrations are related by the expression

$$A = fM \tag{7-1}$$

where $A$ is the active concentration, $M$ the molar concentration, and $f$ a dimensionless number known as the activity coefficient. For simplicity in designating concentrations, it is customary to enclose the formula of the solute ion or molecule in parentheses to indicate activity or active concentration and in brackets to indicate molar concentration. Using the general formula for an acid as the solute, eq. (7-1) can be written as

$$(HA) = f_{HA}[HA] \tag{7-2}$$

A complete explanation of the activity concept is beyond the scope of this text, but a simplified explanation will be attempted to give the student some concept of effective concentration. The effective concentration of an ion is a function of the number of like ions present (the analytical molar concentration of the ion) and also of the number of other ions present in the solution. As an ion moves about in solution, it is attracted by ions of opposite charge and repelled by ions of like charge. These attractive and repulsive effects keep the ion from moving freely and thus decrease the efficiency of the ion. The effect can be compared to the plight of a student who is to meet a friend at the classroom door after class is dismissed. If there are only a small number of students in the room, his progress toward the door is not hindered to a great extent. If, however, the room is filled with students and he sits in the front of the room, his progress toward a door at the back of the room is seriously impeded by the other students; as a consequence, he cannot be completely effective, timewise, in reaching the door. The larger the number of students, the greater the interference with his motion and the less effective he can be. The same situation exists for an ion in solution—the greater the number of ions, the less effective any single ion can be.

Debye and Hückel related the activity coefficient to the number and charge of all the ions in the solution by using the concept of ionic strength. The ionic strength $\mu$ is defined by the equation

$$\mu = \tfrac{1}{2}\sum (M_i Z_i^2) \tag{7-3}$$

where $M_i$ is the analytical molar concentration of an ion, $Z_i$ is the charge on the ion and the $\sum$ is read "the sum of." In words, the ionic strength equals one-half the sum of the products of the molar concentration and the ionic charge (valence) squared, for all the ions in the solution. Since the ions are placed in solution in the form of compounds, the ionic strength due to any one solute compound may be calculated. Using the general formula $A_x^{m+}B_y^{n-}$ to represent a solute compound, the ionic strength will be

$$\mu = \frac{M(xm^2 + yn^2)}{2} \tag{7-4}$$

The total ionic strength of the solution is the sum of the ionic strengths due to the individual compounds. Equation (7-4) shows that the ionic strength due to a compound is either equal to the molar concentration or is a multiple of the molar concentration. The multiplier is dependent upon the ionic charge of the ions present in the compound: for 1 : 1 types (NaCl), the multiplier is 1; for 2 : 1 or 1 : 2 [$Na_2SO_4$ or $Ca(NO_3)_2$], the multiplier is 3; for 2 : 2 ($CaSO_4$), the multiplier is 4; for 1 : 3 or 3 : 1 ($K_3PO_4$ or $FeCl_3$), the multiplier is 6; for 3 : 2 ($Fe_2S_3$), the multiplier is 15.

As an example of the calculation of ionic strengths, determine the ionic strength of a solution that is 0.08 *M* in NaCl, 0.04 *M* in potassium sulfate ($K_2SO_4$), and 0.10 *M* in acetic acid ($HC_2H_3O_2$). In the case of acetic acid, remember that only ions affect the ionic strength and that since acetic acid is a weak electrolyte, which exists mainly as molecules in the solution, it will not contribute appreciably to the ionic strength of the solution.

| *Compound* | *Calculation* | $\mu$ |
|---|---|---|
| NaCl | $\frac{0.08(1 \times 1^2 + 1 \times 1^2)}{2}$ | 0.08 or $M \times 1$ |
| $K_2SO_4$ | $\frac{0.04(2 \times 1^2 + 1 \times 2^2)}{2}$ | 0.12 or $M \times 3$ |
| $HC_2H_3O_2$ | | Negligible |
| Total | | 0.20 |

Debye and Hückel also showed that the activity coefficient was related to the ionic strength by the equation

$$\log f_i = \frac{-0.5Z^2\sqrt{\mu}}{1 + \sqrt{\mu}} \tag{7-5}$$

Values calculated by this equation agree well with the experimental values up to an ionic strength of approximately 0.10 and can be used for approximations up to an ionic strength of about 0.25. Accurate activity coefficient values for ionic strengths above 0.1 must be determined experimentally.

• **Example 1.** As an example of the use of eq. (7-5), calculate the activity coefficients of the sodium, potassium, chloride, and sulfate ions in the solution given in the preceding table, in which the ionic strength is 0.20.

For sodium, potassium, and chloride.

$$\log f_i = \frac{-0.5 \times 1^2 \times \sqrt{0.20}}{1 + \sqrt{0.20}}; \qquad \sqrt{0.20} = 0.45$$

$$= \frac{-0.5 \times 1^2 \times 0.45}{1.45} = -0.16$$

$$f_i = 10^{-0.16} = 10^{0.84} \times 10^{-1} = 6.9 \times 10^{-1}$$

$$= 0.69$$

**TABLE 7-1.** Average Values of Activity Coefficients

| *Ionic Charge* | *Activity Coefficient at an Ionic Strength of:* | | | |
|---|---|---|---|---|
| | *0.01* | *0.025* | *0.05* | *0.10* |
| 1 ($\mathring{a}$* = 4) | 0.90 | 0.86 | 0.82 | 0.77 |
| 2 ($\mathring{a}$ = 5) | 0.67 | 0.56 | 0.47 | 0.37 |
| 3 ($\mathring{a}$ = 5) | 0.40 | 0.27 | 0.18 | 0.12 |
| 4 ($\mathring{a}$ = 6) | 0.21 | 0.105 | 0.055 | 0.027 |

*Source*: I. M. Kolthoff, E. B. Sandell, E. J. Meehan, and S. Bruckenstein, *Quantitative Chemical Analysis*, 4th ed. New York; Macmillan, 1969, by permission.

* $\mathring{a}$ average ionic radius for the charge specified in angstrom units.

For sulfate the calculation is

$$\log f_i = \frac{-0.5 \times 2^2 \times \sqrt{0.20}}{1 + \sqrt{0.20}}$$

$$= \frac{-0.5 \times 4 \times 0.45}{1.45} = -0.62$$

$$f_i = 10^{-0.62} = 10^{0.38} \times 10^{-1} = 2.4 \times 10^{-1}$$
$$= 0.24$$

Equation (7-5) may be used for the estimation of activity coefficients, but the calculated values are not precise because they do not include an ion-size parameter. The extended Debye–Hückel equation does include the ion-size parameter as well as the dielectric constant of the solution and the absolute temperature. The values given in Table 7-1 were calculated from the extended equation using 25°C and average values for the ionic radii based on the charge on the ion. These values are a better estimate of the true activity coefficients and can often be used in place of values calculated by eq. (7-5).

## EQUILIBRIUM CONSTANTS

If substances A and B react to form products C and D, and if C and D when mixed under proper conditions will react to form A and B, the reaction is said to be **reversible**; that is, it can take place in either direction, depending upon the conditions. Most reactions are reversible to some extent, even though for some the reverse reaction is so slow that it would take eons of time for detectable quantities to react.

All reversible reactions will attain equilibrium if left in a closed system without change of conditions for a long enough period of time, and the concentrations of the reactants and products at equilibrium are consequently a measure of the extent of the reaction. Some reactions attain equilibrium almost instantaneously, whereas others may require extremely long periods of time.

The state of equilibrium occurs in a chemical reaction when the rate of the forward reaction equals the rate of the reverse reaction. That is, in the general reaction

$$A + B \rightleftharpoons C + D \tag{7-6}$$

equilibrium is attained when the rate at which A and B react to form C and D is exactly equal to the rate at which C and D react to form A and B. Since the rates of formation and decomposition are equal, the concentrations will remain unchanged.

The equilibrium constant expression shows the relationship between the concentrations of reactants and products at equilibrium and may be written

$$K_{\text{eq}} = \frac{(\text{C}) \times (\text{D})}{(\text{A}) \times (\text{B})} \tag{7-7}$$

This expression is known as the **thermodynamic equilibrium constant** and states that the product of the activities of the products divided by the product of the activities of the reactants is a constant at a constant temperature. By convention, the products are shown in the numerator for the reaction as written. $K_{\text{eq}}$ for the reverse reaction is the inverse of $K_{\text{eq}}$, or $1/K_{\text{eq}}$. If the concentrations are expressed in terms of molar concentrations, this expression becomes

$$K_{\text{eq}} = \frac{f_c \times [\text{C}] \times f_d \times [\text{D}]}{f_a \times [\text{A}] \times f_b \times [\text{B}]} = \frac{f_c \times f_d}{f_a \times f_b} \times \frac{[\text{C}] \times [\text{D}]}{[\text{A}] \times [\text{B}]}$$

or

$$\frac{f_a \times f_b}{f_c \times f_d} \times K_{\text{eq}} = \frac{[\text{C}] \times [\text{D}]}{[\text{A}] \times [\text{B}]} = K_{\text{prac}} = K_M \tag{7-8}$$

The product of the activity coefficient term and the thermodynamic constant is known as the **practical** or **molar equilibrium constant**. Because the activity coefficients vary with ionic strength, which varies with concentration, the values at any given concentration can be calculated by use of eq. (7-5) or may be estimated from Table 7-1 and used to obtain the value of the practical constant. By convention, pure liquids, pure solids, and ideal gases at 1 atm partial pressure are assumed to have unit activity and need not be shown in equilibrium constant expressions.

In the dilute solutions normally used in quantitative analysis, the activity coefficients are often almost equal to 1 and usually are almost equal to each other. Assuming this to be true, the activity coefficient term becomes unity, and the equilibrium constant expression may be stated as

$$K_{\text{eq}} = \frac{[\text{C}] \times [\text{D}]}{[\text{A}] \times [\text{B}]} \tag{7-9}$$

This form is used in the majority of beginning courses in quantitative analysis. The error actually involved in disregarding activity coefficients is appreciable but is assumed to be negligible in order to simplify calculations.

From thermodynamic considerations, it can be shown that the concentration of each product or reactant in the equilibrium constant must be raised to the power of its multiplier in the balanced equation of the reaction. Thus, $K_{eq}$ for the reaction

$$2A + 3B \rightleftharpoons C + 4D$$

is

$$K_{eq} = \frac{[C] \times [D]^4}{[A]^2 \times [B]^3} \qquad \textbf{(7-10)}$$

The equilibrium constant expression may be applied to any equilibrium whether it be homogeneous (all reactants and products in one phase) or heterogeneous (reactants and products in more than one phase). Most homogeneous equilibria involved in analysis are those in which all reactants and products occur in a solution; most heterogeneous equilibria apply to a precipitate in contact with a solution.

The value of the equilibrium constant, and thus the concentrations of reactants remaining when equilibrium is established, is a measure of the completeness of the reaction. A reaction is said to be **complete** if only negligible quantities of at least one of the reactants remain at equilibrium. Since *negligible* is a relative term that expresses the relationship of one number to another larger number, it must be defined quantitatively for any particular system. In analytical work, a **negligible amount** is defined as an amount that is difficult to detect and to determine and is taken to be 0.1 mg or less. This amount in 100 mL represents a concentration of approximately $10^{-6}$ or $10^{-7}$ $M$. A reaction that involves equal numbers of moles of each reactant and product is assumed to be quantitatively complete if $K_{eq}$ has a value of $10^6$ or more. The $K_{eq}$ value for quantitative completeness must be higher than this for equilibria that involve differing numbers of moles of reactants and/or products.

## FACTORS THAT AFFECT AN ESTABLISHED EQUILIBRIUM

After equilibrium has been established in a reaction under a given set of conditions, the equilibrium may be upset or shifted by variations in these conditions. According to LeChâtelier's law, the system will react in such a way as to minimize the effect of the change; that is, the system will shift in the direction that tends to relieve the stress placed upon it because of the change.

Any variation of concentration, temperature, or pressure (if a gas is involved) will cause a change in the equilibrium. Addition of a catalyst has no measurable effect upon an established equilibrium, since it either has no effect upon the rates at equilibrium or else it affects both the forward and reverse reactions equally. A catalyst it used primarily to speed the attainment of equilibrium.

Of the factors noted, variation in temperature will affect the numerical value of the equilibrium constant as well as causing a shift in the equilibrium. Variation of the other factors will cause a shift in the equilibrium but will not affect the numerical value of the equilibrium constant.

The effect of variation of these factors can be evaluated if one considers the fact that concentrations of reactants and products remain constant until conditions are changed. If the change causes an increase in the concentration of one reactant, there will be an increase in the rate of reaction to use up the added substance. This will shift the equilibrium so that the momentarily increased concentration will be decreased. Consider the general reaction given ($A + B \rightleftharpoons C + D$) at equilibrium and assume that $K_{eq}$ has a value of 1 and that the concentration of each substance present is 0.5 *M*. Thus,

$$\frac{[C][D]}{[A][B]} = 1 = \frac{0.5 \times 0.5}{0.5 \times 0.5} \qquad \textbf{(7-11)}$$

Assume that enough of substance C is added to the system to increase its concentration momentarily to 1 *M*. Under these conditions the rate of the reaction of C with D is increased due to the law of mass action, which states that the rate of a chemical reaction is directly proportional to the product of the activities of the reactants. This tends to decrease the concentration of both C and D and to increase the concentration of both A and B. A new equilibrium will be established when the rates again become equal. The value of the equilibrium constant will remain the same, but the concentrations of the substances will differ in the new equilibrium. To show this mathematically, assume that the concentration of C is reduced to $1 - x$ at the new equilibrium. The concentration of D will be $0.5 - x$ and the concentration of A and B will both be $0.5 + x$. Substituting these values in $K_{eq}$, we have

$$\frac{(1 - x)(0.5 - x)}{(0.5 + x)(0.5 + x)} = 1$$

$$(1 - x)(0.5 - x) = (0.5 + x)(0.5 + x)$$

$$0.5 - 1.5x + x^2 = 0.25 + x + x^2$$

$$-2.5x = -0.25$$

$$x = 0.1$$

The concentrations at the new equilibrium are

$$A = 0.5 + 0.1 = 0.6$$
$$B = 0.5 + 0.1 = 0.6$$
$$C = 1.0 - 0.1 = 0.9$$
$$D = 0.5 - 0.1 = 0.4$$

Substitute these in $K_{eq}$ and

$$\frac{0.9 \times 0.4}{0.6 \times 0.6} = \frac{0.36}{0.36} = 1$$

Note that the equilibrium has been shifted in the direction of the reactants A and B; that is, the concentrations of these substances are larger in the new equilibrium than in the original. Note also that the concentration of D is decreased, whereas the concentration of C is larger than the original concentration due to the addition of more C but is less than the momentary concentration attained upon addition of more C.

## APPLICATIONS OF THE EQUILIBRIUM CONSTANT

### IONIZATION CONSTANTS

Electrolytes are substances that conduct an electric current in water solution and are classified as strong, moderate, and weak. Since the conduction of current through a solution depends upon the number and type of ions present, those substances that furnish relatively large numbers of ions are considered to be strong and those that furnish relatively small numbers are weak. Those in between are classified as moderate or moderately strong.

Most strong electrolytes are soluble compounds that are completely ionic even in the solid or pure state. The act of solution simply separates these ions and allows them to move freely in the solution. A few strong electrolytes, such as HCl, are molecular in the pure state but react with water upon dissolving to become essentially completely ionic. Since there will be relatively large numbers, the ions are not completely effective in reactions because of interference with each other; as a consequence, their activity coefficients will be less than 1. Soluble weak electrolytes exist as uncharged molecules in the pure or solid state, and after solution a relatively small fraction decomposes to form ions. This reaction to form ions from molecules is called **ionization** and is a reversible reaction.

Strong electrolytes include the strong acids (HCl, $HClO_3$, $HClO_4$, $H_2SO_4$, $HNO_3$, etc.), the strong bases (NaOH, KOH, etc.), and practically all soluble salts. Weak electrolytes include the slightly soluble ionic substances (barium sulfate, silver chloride, ferric hydroxide, etc.), the weak acids (organic, boric, carbonic, nitrous, hypochlorous, etc.), the weak bases ($NH_4OH$, organic amines, etc.), and a very few soluble salts (mercuric chloride, lead acetate, complex ions, etc.). The moderate electrolytes include the secondary ions of some diprotic acids (bisulfate), some acids (phosphoric), and the hydroxides of the alkaline earth metals, which are completely ionic but only moderately soluble.

Since the ionization of a weak electrolyte is a reversible reaction that will attain equilibrium, the equilibrium constant can be applied and, in this case, is called the **ionization constant** ($K_i$). The ionization constants of acids are often symbolized as $K_a$ and the ionization constants of bases as $K_b$. The ionization constants of weak electrolytes by convention relate to the reaction that produces ions; thus for the weak acid HA, the reaction is written

$$HA \rightleftharpoons H^+ + A^- \quad \textbf{(7-12)}$$

and the ionization constant is

$$K_a = \frac{[H^+][A^-]}{[HA]} \quad \textbf{(7-13)}$$

### *Calculation of Ionization Constants from Percent of Ionization and Molar Concentration*

If the very small concentration of hydrogen ion produced by the ionization of water is neglected, the concentration of the hydrogen ion in a solution of a weak acid HA will equal the molar concentration ($M_a$) of the compound multiplied by the fraction that has been ionized:

$$[H^+] = M_a \times \text{fraction ionized} \quad \textbf{(7-14)}$$

The fraction ionized equals the percent of ionization divided by 100:

$$[H^+] = \frac{M_a \times \%}{100} \quad \textbf{(7-15)}$$

Since the hydrogen ion and the anion are produced in equal numbers, the anion concentration will equal the hydrogen ion concentration if the hydrogen ion concentration from the ionization of water is negligible compared to that from the ionization of the acid. The concentration of the nonionized molecules of the acid will be equal to the molar concentration minus the fraction that has ionized:

$$[HA] = M_a - \frac{M_a \times \%}{100} = M_a - [H]^+ \quad \textbf{(7-16)}$$

Substituting these values in the ionization constant,

$$K_a = \frac{(M_a \times \%/100) \times (M_a \times \%/100)}{M_a - (M_a \times \%/100)} \quad \textbf{(7-17)}$$

• **Example 2.** Calculate the ionization constant of benzoic acid, which is 2.50% ionized in 0.100 $M$ solution. The student should review the use of logarithms and exponents in calculations before trying to follow the steps in this example.

$$[H^+] = [A^-] = \frac{M_a \times \%}{100} = \frac{0.100 \times 2.50}{100} = 0.0025$$

$$= 2.50 \times 10^{-3} = 10^{0.398} \times 10^{-3} = 10^{-2.602}$$

$$[HA] = M_a - \frac{M_a \times \%}{100} = 0.100 - 0.00250 = 0.0975$$

$$= 9.75 \times 10^{-2} = 10^{0.989} \times 10^{-2} = 10^{-1.011}$$

$$K_a = \frac{10^{-2.602} \times 10^{-2.602}}{10^{-1.011}} = \frac{10^{-5.204}}{10^{-1.011}} = 10^{-4.193}$$

$$= 10^{0.807} \times 10^{-5} = 6.41 \times 10^{-5}$$

or

$$K_a = \frac{(2.50 \times 10^{-3})^2}{9.75 \times 10^{-2}} = \frac{6.25 \times 10^{-6}}{9.75 \times 10^{-2}} = 6.41 \times 10^{-5}$$

In calculations involving equilibrium constants in analytical chemistry, it is seldom necessary to express exponents to more than three places, and in many instances two places are satisfactory. At present three places will be used in all such calculations.

## *Calculation of Ionic Concentrations from K and M*

As indicated in the previous section, the concentrations of the hydrogen ion and the anion in a solution of a weak acid are essentially equal, and the concentrations of the nonionized molecules equal the molar concentration of the acid ($M_a$) minus the concentration of the hydrogen ion, since each dissociation of the acid produces one such ion. Thus,

$$[H^+] = [A^-]$$
$$[HA] = M_a - [H^+]$$

and

$$K_a = \frac{[H^+]^2}{M_a - [H^+]} \qquad \textbf{(7-18)}$$

or

$$[H^+]^2 + K_a[H^+] - K_a M_a = 0 \qquad \textbf{(7-19)}$$

Equation (7-19) is in the general quadratic form of $x^2 + bx + c = 0$, for which

$$x = \frac{-b \pm \sqrt{b^2 - 4c}}{2} \tag{7-20}$$

in which $b = K_a$ and $c = -K_a M_a$. The usual quadratic form for the general equation $ax^2 + bx + c = 0$ is

$$x = \frac{-b \pm \sqrt{b^2 - 4ac}}{2a} \tag{7-21}$$

However, eq. (7-21) reduces to eq. (7-20) when $a = 1$, which is the general case for ionization calculations.

If the ionization constant is $10^{-4}$ or smaller, that is, if the percent of ionization in a 1 $M$ solution is less than 1 %, the number of molecules that have dissociated is negligible compared to the number that have not, so that $M_a - [H^+]$ is approximately equal to $M_a$. In such cases, the use of $M_a$ is permissible and the expression for [H] becomes

$$[H^+] = \sqrt{K_a M_a} \tag{7-22}$$

• **Example 3.** Calculate the molar hydrogen ion concentration in 0.100 $M$ benzoic acid for which $K_a$ is $6.61 \times 10^{-5}$:

$$6.61 = 10^{0.820}$$

Substitution in eq. (7-22) gives

$$[H^+] = \sqrt{10^{0.820} \times 10^{-5} \times 10^{-1}} = \sqrt{10^{0.820} \times 10^{-6}}$$

and, since dividing an exponent by 2 is equivalent to extracting the square root,

$$\begin{aligned} [H^+] &= 10^{0.820/2} \times 10^{-(6/2)} \\ &= 10^{0.410} \times 10^{-3} \\ &= 2.57 \times 10^{-3}\ M \qquad (\text{since } 10^{0.410} = 2.57) \end{aligned}$$

• **Example 4.** However, to calculate the hydrogen ion concentration in a 0.100 $M$ solution of chloroacetic acid, for which $K_a$ equals $1.41 \times 10^{-3}$, it is necessary to solve the quadratic equation, since in this case $M_a$ is not approximately equal to $M_a - [H^+]$. Thus,

$$\begin{aligned} b &= K_a = 1.41 \times 10^{-3} = 0.00141 = 10^{0.149} \times 10^{-3} = 10^{-2.851} \\ b^2 &= (1.41 \times 10^{-3})^2 = (10^{-2.851})^2 = 10^{-5.702} = 10^{0.298} \times 10^{-6} \\ &= 1.99 \times 10^{-6} = 0.00000199 \\ c &= -K_a M_a = -1.41 \times 10^{-3} \times 0.100 = -1.41 \times 10^{-4} = -0.000141 \\ -4c &= -(4 \times -1.41 \times 10^{-4}) = 5.64 \times 10^{-4} = 0.000564 \end{aligned}$$

Substituting these values in eq. (7-20)

$$[H^+] = \frac{-b \pm \sqrt{b^2 - 4c}}{2} = \frac{-0.00141 \pm \sqrt{0.00000199 + 0.000564}}{2}$$

$$= \frac{-0.00141 \pm \sqrt{0.000566}}{2} = \frac{-0.00141 \pm \sqrt{10^{-3.247}}}{2}$$

$$= \frac{-0.00141 \pm 10^{-1.624}}{2} = \frac{-0.00141 \pm 0.0238}{2}$$

$$= \frac{0.0224}{2} = 0.0112 = 1.12 \times 10^{-2}\ M$$

If the simplified formula (7-22) had been used in Example 4 the calculated value for $[H^+]$ would be $1.16 \times 10^{-2}$ $M$, which represents an error of approximately 3.6%, or 36 ppt. Such errors increase as the value of $K$ increases, so a larger error will be involved for $K_a$ values in the range of $10^{-2}$ and $10^{-1}$.

Equations (7-20) and (7-22) can also be used to calculate the $[H^+]$ in solutions of many polybasic (polyprotic) weak acids using $K_1$ for $K_a$, since the numbers of hydrogen ions produced by the second and subsequent ionization steps will be negligible in most cases compared to those produced by the primary ionization provided that $K_1$ and $K_2$ differ by at least $10^4$.

## SOLUBILITY PRODUCT

In any saturated solution of an ionic compound in contact with undissolved solute, an equilibrium exists between the undissolved solute and the dissolved ions because the rate of dissolution equals the rate of crystallization.

$$MA_{solid} \rightleftharpoons M^+ + A^- \qquad \textbf{(7-23)}$$

Pure solids are considered to be at unit activity and thus need not be shown in equilibrium constant expressions; therefore, the equilibrium constant for the heterogeneous equilibrium between the solid MA and its ions would be

$$K_{eq} = [M^+] \times [A^-] = K_{sp} \qquad \textbf{(7-24)}$$

Thus, the product of the concentrations of the ions in a saturated solution, each raised to the appropriate power, is a constant that is known as the solubility product and given the symbol $K_{sp}$. In analytical work, the solubility product is important because it is related to the maximum concentrations that may remain unprecipitated in solution. Whenever the momentary concentrations upon mixing exceed the solubility product, precipitation will occur and will continue until the product of the ion concentrations again equals the solubility product. (See Chapter 8 for a complete discussion of the theory of precipitate formation.)

### Calculation of $K_{sp}$ from Solubility

The steps involved in this calculation consist of converting the solubility to the molar solubility ($M_{sol}$), calculating the ionic concentrations from $M_{sol}$, and substituting these values in the solubility product expression.

• **Example 5.** Calculate the solubility product of nickel(II) hydroxide (MW, 92.7) if the solubility is $1.82 \times 10^{-2}$ mg/100 mL. First, convert the solubility to moles per liter ($M_{sol}$) by expressing the solubility as grams per liter and dividing by the molecular weight.

$$M_{sol} = \frac{1.82 \times 10^{-2}\ (\text{mg/100 mL}) \times 1000\ \text{mL/1000 mL/L} \times 1\ \text{g/1000 mg}}{92.7\ \text{g/mole}}$$

$$= \frac{1.82 \times 10^{-4}\ (\text{g/liter})}{92.7\ (\text{g/mole})} = 1.97 \times 10^{-6}\ M\ (\text{moles/liter}) \qquad \textbf{(7-25)}$$

The concentration of an ion is equal to $M_{sol}$ multiplied by the number of those ions in the formula of the compound, or

$$[\text{ion}] = M_{sol} \times \text{number of these ions in the formula} \qquad \textbf{(7-26)}$$

In our case,

$$[Ni^{2+}] = M_{sol} \times 1 = 1.97 \times 10^{-6} = 10^{0.294} \times 10^{-6} = 10^{-5.706}$$
$$[OH^-] = M_{sol} \times 2 = 2 \times 1.97 \times 10^{-6} = 3.94 \times 10^{-6}$$
$$= 10^{0.596} \times 10^{-6} = 10^{-5.404}$$

Substituting these values in $K_{sp}$, we obtain

$$K_{sp} = [Ni^{2+}] \times [OH^-]^2 = 10^{-5.706} \times (10^{-5.404})^2$$
$$= 10^{-5.706} \times 10^{-10.808} = 10^{-16.514} = 10^{0.486} \times 10^{-17}$$
$$= 3.06 \times 10^{-17}$$

### Calculation of Solubility from $K_{sp}$

This is the reverse of the preceding calculation and involves calculating $M_{sol}$ from $K_{sp}$ and converting this to the solubility desired. For compounds in which the ions have the same valence, $M_{sol}$ is simply the square root of $K_{sp}$ because the ion concentrations are equal to each other.

• **Example 6.** To calculate the solubility in milligrams per 100 mL of lead carbonate for which $K_{sp}$ is $2.50 \times 10^{-11}$, express the ion concentrations in terms of $M_{sol}$,

$$[Pb^{2+}] = [CO_3^{2-}] = M_{sol} \times 1$$

and substitute in the $K_{sp}$ expression to solve for $M_{sol}$,

$$K_{sp} = [Pb^{2+}] \times [CO_3^{2-}] = (M_{sol})^2 = 2.50 \times 10^{-11}$$
$$= 25.0 \times 10^{-12} = 10^{1.398} \times 10^{-12} = 10^{-10.602}$$
$$M_{sol} = \sqrt{25.0 \times 10^{-12}} = 5.00 \times 10^{-6}\ M$$

or

$$M_{sol} = \sqrt{10^{-10.602}} = 10^{-5.301} = 10^{0.699} \times 10^{-6} = 5.00 \times 10^{-6}$$

Then convert $M_{sol}$ to the solubility in milligrams per 100 mL,

$$\text{MW} \times M_{sol} \times \text{mL} = \text{mg}$$

or

$$267 \times 5.00 \times 10^{-6} \times 10^2 = 1335 \times 10^{-4}\ \text{mg} = 1.34 \times 10^{-1}\ \text{mg/100 mL}$$

• **Example 7.** For compounds in which the ions do not have the same ionic charge, the calculation of $M_{sol}$ is more complicated because the ionic concentrations have to be raised to the power of the number of ions involved. Thus, to calculate solubility in grams per 100 mL of cupric hydroxide for which $K_{sp}$ is $8.52 \times 10^{-20}$,

$$[Cu^{2+}] = M_{sol} \times 1 \qquad [OH^-] = M_{sol} \times 2$$

$$K_{sp} = [Cu^{2+}] \times [OH^-]^2 = M_{sol} \times (2M_{sol})^2 = 4(M_{sol})^3$$
$$= 8.52 \times 10^{-20}$$

$$(M_{sol})^3 = \frac{8.52 \times 10^{-20}}{4} = 2.13 \times 10^{-20} = 10^{0.328} \times 10^{-20}$$
$$= 10^{-19.672}$$

$$M_{sol} = \sqrt[3]{10^{-19.672}} = 10^{-19.672/3} = 10^{-6.557} = 10^{0.443} \times 10^{-7}$$
$$= 2.77 \times 10^{-7}\ M$$

$$\text{MW} \times M_{sol} \times \text{liter} = 97.6 \times 2.77 \times 10^{-7} \times 10^{-1}$$
$$= 2.69 \times 10^{-6}\ \text{g/100 mL}$$

For $Ca_3(PO_4)_2$,

$$K_{sp} = [3M_{sol}]^3[2M_{sol}]^2 = 108M_{sol}^5$$

### *Calculation of Amount of Ion or Compound Left Unprecipitated*

One way of estimating the effectiveness of a precipitation is to calculate the weight of the original ion left unprecipitated under the conditions of precipitation. Most precipitations are performed by adding a definite volume of the precipitant to a known volume of solution containing the sample being analyzed. The majority of precipitations in any method of analysis are planned in conjunction with the sample weight and with the usual range of percentage of the substance being determined to give a slight to moderate excess of the precipitating agent. If the final volume of the solution and the final concentration of the precipitating ion is known, the concentration and thus the weight of the sought ion or substance that is left unprecipitated can be calculated.

• **Example 8.** Calculate the milligrams of chromate ion left unprecipitated if sufficient lead nitrate is added to a solution of sodium chromate to make the final lead ion concentration equal to 0.01 $M$ and the final volume equal to 100 mL.

$$K_{sp} = [Pb^{2+}] \times [CrO_4^{2-}] = 1.82 \times 10^{-14}$$

or, rearranging and solving for $CrO_4^{2-}$,

$$[CrO_4^{2-}] = \frac{K_{sp}}{[Pb^{2+}]} = \frac{1.82 \times 10^{-14}}{0.01} = 1.82 \times 10^{-12}\ M$$

and the milligrams of chromate ion left unprecipitated are

$$116 \times 1.82 \times 10^{-12} \times 10^2 = 2.11 \times 10^{-8} \text{ mg of } CrO_4^{2-}$$

In working these problems, it should be noted that the $K_{sp}$ of the precipitate and the concentration of the precipitant is used to calculate the concentration of the sought ion left unprecipitated and that the ionic or molecular weight of the sought substance or ion is used to calculate the actual weight left unprecipitated. Thus, in Example 8, 116 is the ionic weight of $CrO_4^{2-}$, not the molecular weight of sodium chromate.

In a saturated solution of lead chromate made by dissolving lead chromate in water, the concentration of the lead ion and also of the chromate ion would be $(K_{sp})^{1/2}$ or $1.35 \times 10^{-7}\ M$, and 100 mL of the solution would contain $1.57 \times 10^{-3}$ mg of chromate ion. Thus, the addition of the excess of lead ion decreased the amount of chromate ion left in the solution to a value that is approximately $1.3 \times 10^{-5}$ times as large as the amount left in a saturated solution. This is an example of the **common ion effect**, which can be stated as: the solubility of a slightly soluble compound can be decreased by the addition of a common ion in excess. However, too large an excess may cause complex formation, with resultant increase in solubility. It is important to remember that the common ion effect actually works properly only with slight to moderate excesses of the precipitant ion.

## CONCENTRATIONS OF SPECIES PRESENT AT A GIVEN HYDROGEN ION CONCENTRATION

It is often necessary to calculate the equilibrium concentration of a molecular acid or one of the ions formed by stepwise ionization of a polyprotic acid. Calculations of this type are also necessary in finding the relative concentrations of the two forms of an indicator at a definite hydrogen ion concentration. A general formula for calculating the concentration of any species is derived, in which the total concentration of all the related species in the solution is symbolized as $M_t$ and the fraction due to any one species is symbolized by $\alpha$, with subscripts denoting the particular species. Thus, the molecular form is indicated by a subscript of zero and the ion produced by the primary ionization by a subscript 1. The actual or equilibrium concentration of any species is then obtained as the product of $M_t$ and $\alpha$. Using the general formula for a triprotic acid, $H_3A$, the ionization steps with corresponding constants are

$$H_3A \rightleftharpoons H^+ + H_2A^- \qquad K_1 = \frac{[H^+] \times [H_2A^-]}{[H_3A]} \tag{7-27}$$

$$H_2A^- \rightleftharpoons H^+ + HA^{2-} \qquad K_2 = \frac{[H^+] \times [HA^{2-}]}{[H_2A^-]} \tag{7-28}$$

$$HA^{2-} \rightleftharpoons H^+ + A^{3-} \qquad K_3 = \frac{[H^+] \times [A^{3-}]}{[HA^{2-}]} \tag{7-29}$$

and the total concentration is given by

$$M_t = [H_3A] + [H_2A^-] + [HA^{2-}] + [A^{3-}] \tag{7-30}$$

Equations (7-27), (7-28), and (7-29) may be rearranged to solve for each equilibrium concentration in terms of the equilibrium concentration of $H_3A$ as follows:

$$[H_2A^-] = \frac{K_1[H_3A]}{[H^+]} \tag{7-31}$$

$$[HA^{2-}] = \frac{K_2 \times [H_2A^-]}{[H^+]} = \frac{K_2 \times (K_1 \times [H_3A]/[H^+])}{[H^+]}$$

$$= \frac{K_1K_2[H_3A]}{[H^+]^2} \tag{7-32}$$

$$[A^{3-}] = \frac{K_3[HA^{2-}]}{[H^+]} = \frac{K_3 \times (K_1K_2[H_3A]/[H^+]^2)}{[H^+]}$$

$$= \frac{K_1K_2K_3[H_3A]}{[H^+]^3} \tag{7-33}$$

Substitution of these values in eq. (7-30) gives

$$M_t = [H_3A] + \frac{K_1[H_3A]}{[H^+]} + \frac{K_1K_2[H_3A]}{[H^+]^2} + \frac{K_1K_2K_3[H_3A]}{[H^+]^3} \tag{7-34}$$

The values for $\alpha$ can now be determined as

$$\alpha_0 = \frac{[H_3A]}{M_t} = \frac{[H^+]^3}{[H^+]^3 + K_1[H^+]^2 + K_1K_2[H^+] + K_1K_2K_3} \tag{7-35}$$

$$\alpha_1 = \frac{[H_2A^-]}{M_t} = \frac{K_1[H^+]^2}{[H^+]^3 + K_1[H^+]^2 + K_1K_2[H^+] + K_1K_2K_3} \tag{7-36}$$

$$\alpha_2 = \frac{[HA^{2-}]}{M_t} = \frac{K_1K_2[H^+]}{[H^+]^3 + K_1[H^+]^2 + K_1K_2[H^+] + K_1K_2K_3} \tag{7-37}$$

$$\alpha_3 = \frac{[A^{3-}]}{M_t} = \frac{K_1K_2K_3}{[H^+]^3 + K_1[H^+]^2 + K_1K_2[H^+] + K_1K_2K_3} \tag{7-38}$$

The equilibrium concentration of each species may now be calculated by multiplying $M_t$ by the proper $\alpha$, or

$$[\text{species}] = M_t\alpha_{\text{species}} \tag{7-39}$$

It should be noted in eqs. (7-35) through (7-38) that the denominator is the same in all calculations for $\alpha$ and that the numerator is identical to the term in the denominator that has a number one higher than the subscript on $\alpha$. Thus, in eq. (7-37) for $\alpha_2$ the numerator is the third term in the denominator. Utilizing this fact and the more general formula for a polyprotic acid $H_nA$, the denominator for calculation of $\alpha$ will be

$$[H^+]^n + [H^+]^{n-1}K_1 + [H^+]^{n-2}K_1K_2 + \cdots + [H^+]K_1K_2\cdots K_{n-1} + K_1\cdots K_n \tag{7-40}$$

and the value for a particular $\alpha$ will use this denominator and the proper term of the denominator as the numerator.

• **Example 9.** Calculate the molar concentration of $HPO_4^{2-}$ in a 0.100 $M$ solution of phosphate in which the hydrogen ion concentration is $10^{-7}$ $M$

$$K_1 = 10^{-2.12} \qquad K_2 = 10^{-7.21} \qquad K_3 = 10^{-12.32}$$

At $[H^+] = 10^{-7}$, the denominator will be

$$\text{denominator} = (10^{-7})^3 + (10^{-7})^2 \times 10^{-2.12} + 10^{-7} \times 10^{-2.12} \times 10^{-7.21} + 10^{-2.12} \times 10^{-7.21} \times 10^{-12.32}$$

or

$$\text{denominator} = 10^{-21} + 10^{-16.12} + 10^{-16.33} + 10^{-21.65}$$

Disregarding the first and last terms because the quantity represented by $10^{-21}$ is insignificant in comparison with that represented by $10^{-16}$

$$\text{denominator} = 7.6 \times 10^{-17} + 4.7 \times 10^{-17} = 12.3 \times 10^{-17}$$

so that

$$\alpha_2 = \frac{4.7 \times 10^{-17}}{12.3 \times 10^{-17}} = 0.382$$

and

$$[HPO_4^{2-}] = M_t\alpha_2 = 0.100 \times 0.382 = 0.0382\ M$$

## FACTORS THAT AFFECT SOLUBILITY AT A GIVEN TEMPERATURE

### *Effect of Diverse Ions on Solubility*

In general, an increase in the concentration of ions not common to the precipitate increases the solubility of a precipitate, since an increase in the ionic strength increases the value of the practical ionization constant and thus the molar concentrations of the ions of the precipitate in the saturated solution. As an example, the solubility of barium chromate is $4.8 \times 10^{-5}\ M$ in a solution with ionic strength of 0.25 and is $1.6 \times 10^{-5}\ M$ in pure water.

### *Effect of Hydrogen Ion Concentration on Solubility*

Although the solubility of salts of strong acids such as silver chloride is not greatly affected by moderate concentrations of strong acids such as nitric acid, the salts of weak acids may be completely dissolved by small to moderate concentrations of strong acids.

• **Example 10.** Calculate the solubility of calcium oxalate in a 0.05 $M$ sodium oxalate solution to which enough hydrochloric acid has been added to make the hydrogen ion concentration equal to 0.1 $M$. We must first consider the equilibrium existing among the various forms of oxalate ($H_2C_2O_4$, $HC_2O_4^-$, and $C_2O_4^{2-}$), since only $C_2O_4^{2-}$ appears in the solubility product of calcium oxalate and affects the solubility. The ionization constants for oxalic acid are

$$H_2C_2O_4 \rightleftharpoons H^+ + HC_2O_4^-$$

for which

$$K_1 = \frac{[H^+][HC_2O_4^-]}{[H_2C_2O_4]} = 5.6 \times 10^{-2} \quad \textbf{(7-41)}$$

and

$$HC_2O_4^- \rightleftharpoons H^+ + C_2O_4^{2-}$$

for which

$$K_2 = \frac{[H^+][C_2O_4^{2-}]}{[HC_2O_4^-]} = 5.2 \times 10^{-5} \quad \textbf{(7-42)}$$

To calculate the ratio of $[H_2C_2O_4] : [HC_2O_4^-] : [C_2O_4^{2-}]$ in a solution with a total oxalate concentration of 0.05 $M$, let $[H_2C_2O_4]$ be 1, rearrange eq. (7-41) and (7-42) to calculate $[HC_2O_4^-]$ and $[C_2O_4^{2-}]$, and calculate $[HC_2O_4^-]$ and $[C_2O_4^{2-}]$ at the hydrogen ion concentration specified. Thus, for the 0.1 $M$ hydrogen ion,

$$[HC_2O_4^-] = \frac{[H_2C_2O_4] \times K_1}{[H^+]} = \frac{1 \times 5.6 \times 10^{-2}}{10^{-1}} = 0.56 \quad \textbf{(7-43)}$$

$$[C_2O_4^{2-}] = \frac{[HC_2O_4^-] \times K_2}{[H^+]} = \frac{0.56 \times 5.2 \times 10^{-5}}{10^{-1}} = 0.00029 \quad \textbf{(7-44)}$$

The ratio of $[H_2C_2O_4] : [HC_2O_4^-] : [C_2O_4^{2-}]$ is thus 1 : 0.56 : 0.0003, and the fraction due to oxalate is 0.0003/1.5603, or 0.0002. Since the total concentration is 0.05 $M$, the actual $[C_2O_4^{2-}]$ is $1.0 \times 10^{-5}$ $M$. Using the $K_{sp}$ of calcium oxalate as $2.0 \times 10^{-9}$, the concentration of calcium ions and thus the solubility of $CaC_2O_4$ is

$$[Ca^{2+}] = \frac{2.0 \times 10^{-9}}{1.0 \times 10^{-5}} = 2.0 \times 10^{-4}\ M$$

By a similar calculation the solubility at a hydrogen ion concentration of 1.0 $M$ is $1.4 \times 10^{-2}$ $M$, at a hydrogen ion concentration of 0.001 $M$ it is $8.3 \times 10^{-7}$ $M$, and at a hydrogen ion concentration of $10^{-5}$ $M$ it is $4.8 \times 10^{-8}$ $M$. It should be noted that the 100-fold increase in hydrogen ion from 0.001 to 0.1 $M$ caused more than a 200-fold increase in the solubility, and the 10-fold increase from 0.1 $M$ to 1.0 $M$ caused a 70-fold increase. Another way of showing this effect is to calculate the milligrams of calcium left unprecipitated when the pH is raised slowly by addition of ammonia to 100 mL of a strong acid solution that contains 2.0 mmoles of calcium and 5 mmoles of oxalate. The calcium does not start to precipitate until the pH reaches a value between zero and 1, and 0.80 mg of calcium ($40 \times 2.0 \times 10^{-4} \times 10^2$) is left unprecipitated at a pH of 1. At a pH of 3, only 0.0033 mg ($40 \times 8.3 \times 10^{-7} \times 10^2$) of calcium is left in solution, and the precipitation is quantitatively complete. These values have been obtained neglecting the effect of ionic strength on the ionization constants and solubility product. By utilizing the correct values of the practical constants at an ionic strength of 0.16, the solubility of calcium oxalate at pH 1 turns out to be $5.0 \times 10^{-4}$ *M*, or approximately three times as large as the value obtained neglecting the effect of diverse ions.

## *Effect of Complex Formation on Solubility*

As mentioned previously, most precipitation procedures are designed to add only a slight to moderate excess of the precipitant in order to offset secondary reactions. The majority of the harmful secondary reactions are caused by the reaction of the precipitate with the large excess of precipitant to form a soluble complex. This causes some of the solid precipitate to dissolve or prevents its formation. A good example of such a compound is AgCl, which can react with excess chloride to form $AgCl_2^-$ and with excess silver ion to form $Ag_2Cl^+$. Such reactions can cause appreciable errors in gravimetric procedures and separations.

• **Example 11.** Calculate the silver ion concentration (which is a measure of the solubility) in a 0.1 *M* solution of NaCl in contact with solid AgCl: (1) assuming no complex formation, and (2) assuming complex formation. The $K_{sp}$ of AgCl is $1.56 \times 10^{-10}$ and the stability (formation) constant of $AgCl_2^-$ is $1.91 \times 10^5$.

For (1)

$$[Ag^+] = \frac{K_{sp}}{[Cl^-]} = \frac{1.56 \times 10^{-10}}{10^{-1}} = 1.56 \times 10^{-9}\ M$$

For (2), the $[Ag^+]$ must satisfy both $K_{sp}$ and $K_{stab}$. The equilibrium constants for complexes are designated as stability constants ($K_{stab}$) for the formation of the complex and as instability constants ($K_{instab}$) for the ionization of the complex. For the dichloroargentate(I) ion in this example, the formation equation and constant would be

$$Ag^+ + 2Cl^- \rightleftharpoons AgCl_2^- \qquad K_{stab} = \frac{[AgCl_2^-]}{[Ag^+] \times [Cl^-]^2} \tag{7-45}$$

and the ionization or instability equation and constant would be

$$AgCl_2^- \rightleftharpoons Ag^+ + 2Cl^- \qquad K_{instab} = \frac{[Ag^+] \times [Cl^-]^2}{[AgCl_2^-]} \tag{7-46}$$

The numerical values of $K_{stab}$ and $K_{instab}$ are the inverse of each other. Since many tables list stability constants and many list instability constants, the student must be careful to use the correct value. One simple method of distinguishing the two is that the stability constant will ordinarily be a very large number, whereas the instability constant will be a small number. In terms of exponents, the exponents for stability constants are positive, whereas the exponents for instability constants are negative. Solving eq. (7-45) for $[Ag^+]$ gives

$$[Ag^+] = \frac{[AgCl_2^-]}{K_{stab}[Cl^-]^2} \tag{7-47}$$

and since $[Ag^+]$ must satisfy both $K_{sp}$ and $K_{stab}$,

$$[Ag^+] = \frac{K_{sp}}{[Cl^-]} = \frac{[AgCl_2^-]}{K_{stab}[Cl^-]^2} \quad \textbf{(7-48)}$$

We now solve eq. (7-48) for $[AgCl_2^-]$ and obtain

$$\begin{aligned}[AgCl_2^-] &= K_{sp}K_{stab}[Cl^-] \\ &= 1.56 \times 10^{-10} \times 1.91 \times 10^5 \times 10^{-1} = 3.00 \times 10^{-6}\ M\end{aligned}$$

This value represents more than a 1000-fold increase in the solubility of silver chloride over that calculated from $K_{sp}$ only and points out the danger of adding too large an excess of the precipitant.

Another factor affecting complex formation is the presence in the solution of another complexing agent that will prevent quantitative precipitation of the desired solid. Silver ion forms a soluble complex with thiosulfate ions, and this fact is used in photography to remove unactivated silver halides from the exposed film in the development and fixing process. To illustrate the effect of the presence of such agents in the precipitation medium, we can use Example 12.

• **Example 12.** Calculate the solubility of AgCl in 0.01 *M* sodium thiosulfate. The instability or dissociation constant for $Ag(S_2O_3)_2^{3-}$ is $6.0 \times 10^{-14}$. The equation for the reaction involved is

$$AgCl_{(s)} + 2S_2O_3^{2-} \rightleftharpoons Ag(S_2O_3)_2^{3-} + Cl^- \quad \textbf{(7-49)}$$

and the final chloride ion concentration is a measure of the solubility. Since the silver ion concentration in the solution must satisfy both $K_{sp}$ and $K_{instab}$

$$[Ag^+] = \frac{K_{sp}}{[Cl^-]} = \frac{K_{instab}[Ag(S_2O_3)_2^{3-}]}{[S_2O_3^{2-}]^2} \quad \textbf{(7-50)}$$

Neglecting the small free silver ion concentration in the solution, the concentration of the complex equals the concentration of chloride, and

$$\begin{aligned}[Cl^-]^2 = [Ag(S_2O_3)_2^{3-}]^2 &= \frac{K_{sp}[S_2O_3^{2-}]^2}{K_{instab}} \\ &= \frac{1.56 \times 10^{-10} \times 10^{-4}}{6.0 \times 10^{-14}} = 0.26\end{aligned}$$

$$[Cl^-] = [Ag(S_2O_3)_2^{3-}] = \sqrt{0.26} = 0.51\ M$$

Thus, the solubility of AgCl in 0.01 *M* thiosulfate is approximately 0.51 *M*, and AgCl could not be precipitated quantitatively from this solution. This aspect of complex formation is often used to offset the interference of an ion in an analytical procedure by addition of a complex-forming ion to tie up the interfering ion in a slightly ionized complex. This removal of interference is called masking, discussed in Chapter 1.

## SEPARATIONS BY FRACTIONAL PRECIPITATION

If two ions in a solution form slightly soluble compounds with the same ion and the two compounds have sufficiently different solubilities, one compound may be precipitated quantitatively before the other starts to precipitate. The effectiveness of such a separation depends on the solubility products and the ionic concentrations in saturated solutions of the two compounds. If a solution contains $A^-$ and $B^-$ ions that form slightly soluble compounds with $M^+$ ions, slow addition of $M^+$ will cause the solubility product of the less soluble compound to be exceeded and that compound will precipitate before the solubility product of the more soluble compound is attained. To determine which compound is precipitated first, it is necessary to calculate the minimum concentration of $M^+$ that will cause precipitation of each. Since the lowest of these two values will be exceeded first as the concentration of $M^+$ is increased from zero, the compound that requires the lowest minimum concentration of $M^+$ will precipitate first. A quantitative separation is theoretically possible if the concentration of this ion remaining when the second starts to precipitate has been reduced to a negligible value.

The **Mohr method** for chloride is a good example of the application of this principle to an analytical method. In this method, chloride ion is titrated with silver nitrate using potassium chromate as the indicator. Since silver chloride is white and silver chromate is a reddish brown, the formation of the first silver chromate produces a change of color that is utilized as the end point.

• **Example 13.** In the usual procedure the chromate ion concentration is about 0.002 *M* and the chloride ion concentration is usually assumed to be approximately 0.02 *M*. The solubility product of silver chloride is $1.56 \times 10^{-10}$; that of silver chromate is $9.0 \times 10^{-12}$. Thus, the minimum concentrations of $Ag^+$ for precipitation of AgCl and $Ag_2CrO_4$ are

$$[Ag^+]_{min} \text{ for AgCl} = \frac{K_{sp}}{[Cl^-]} = \frac{1.56 \times 10^{-10}}{2 \times 10^{-2}} = 7.8 \times 10^{-9}\ M$$

$$[Ag^+]_{min} \text{ for } Ag_2CrO_4 = \sqrt{\frac{K_{sp}}{[CrO_4^{2-}]}} = \sqrt{\frac{9 \times 10^{-12}}{2 \times 10^{-3}}}$$

$$= \sqrt{10^{0.653} \times 10^{-9}} = 10^{-(8.347/2)}$$

$$= 10^{-4.174} = 10^{0.826} \times 10^{-5}$$

$$= 6.70 \times 10^{-5}\ M$$

and silver chloride will precipitate first. The concentration of chloride left unprecipitated when silver chromate starts to precipitate will be that which is present in a solution having the silver ion concentration necessary to cause precipitation of silver chromate and may be calculated as

$$[Cl^-] = \frac{1.56 \times 10^{-10}}{6.70 \times 10^{-5}} = 2.33 \times 10^{-6}\ M$$

This represents a chloride ion concentration of approximately 0.0083 mg/100 mL left unprecipitated, and so the separation is theoretically quantitative. Remember that such calculations indicate only that the separation is theoretically possible and do not prove that the actual separation is possible or even probable under laboratory conditions.

• **Example 14.** As a second example, calculate whether or not the separation of $Pb^{2+}$ and $Ba^{2+}$ by slow addition of $K_2CrO_4$ is theoretically possible if the concentration of each ion is approximately 0.10 $M$. The solubility product for $PbCrO_4$ is $1.8 \times 10^{-14}$ and for $BaCrO_4$ it is $2.4 \times 10^{-10}$.

$$[CrO_4^{2-}]_{min} \text{ for } PbCrO_4 = \frac{K_{sp}}{[Pb^{2+}]} = \frac{1.8 \times 10^{-14}}{10^{-1}} = 1.8 \times 10^{-13}\ M$$

$$[CrO_4^{2-}]_{min} \text{ for } BaCrO_4 = \frac{K_{sp}}{[Ba^{2+}]} = \frac{2.4 \times 10^{-10}}{10^{-1}} = 2.4 \times 10^{-9}\ M$$

Lead chromate will be precipitated first, and the concentration left unprecipitated when barium chromate starts to precipitate will be

$$[Pb^{2+}] = \frac{K_{sp}}{[CrO_4^{2-}]} = \frac{1.8 \times 10^{-14}}{2.4 \times 10^{-9}} = 7.5 \times 10^{-6}\ M$$

and the milligrams of $Pb^{2+}$ left in 100 mL would be

$$207 \times 7.5 \times 10^{-6} \times 100 = 0.16 \text{ mg}$$

and so the separation is theoretically possible.

## PROBLEMS

**1** Calculate the ionic strength of the following solutions:
(a) 0.100 $M$ NaCl
(b) 0.050 $M$ $K_2SO_4$
(c) 0.150 $M$ $Al(C_2H_3O_2)_3$
(d) 0.250 $M$ $Ca_3(C_6H_5O_7)_2$ (citrate)
(e) A solution that is 0.010 $M$ in $K_3PO_4$, 0.020 $M$ in $K_2HPO_4$, 0.030 $M$ in $KHSO_4$, and 0.40 $M$ in $Al_2(SO_4)_3$ (assume that no precipitate forms)

*Ans.* (b) 0.150

**2** Calculate the activity coefficient for the stated ion in a solution of the stated ionic strength.

| | *Ion* | *Ionic Strength* |
|---|---|---|
| (a) | $Na^+$ | 0.025 |
| (b) | $Ca^{2+}$ | 0.250 |
| (c) | $H^+$ | 0.100 |
| (d) | $Al^{3+}$ | 0.010 |
| (e) | $Cl^-$ | 0.225 |

(*Ans.* (b) 0.21)

**3** Calculate the ionization constants for the following monobasic acids:
(a) An acid that is 0.900% ionized in 0.100 *M* solution
(b) An acid that is 1.05% ionized in 0.0100 *M* solution
(c) An acid that is 2.50% ionized in 1.00 *M* solution
(d) An acid that is 0.0250% ionized in 0.0100 *M* solution
(e) An acid that is 1.23% ionized in 10.00 *M* solution

*Ans.* (b) $1.11 \times 10^{-6}$

**4** Calculate the molar hydrogen ion concentration in each of the following solutions:
(a) 0.125 *M* acetic acid ($pK_a = 4.74$)
(b) 0.100 *M* arsenic acid ($H_3AsO_4$) ($pK_1 = 2.22$)
(c) 0.250 *M* hypochlorous acid ($pK_a = 7.53$)
(d) 0.500 *M* phosphoric acid ($pK_1 = 2.12$)
(e) 1.00 *M* benzoic acid ($pK_a = 4.18$)

*Ans.* (b) 0.0217

**5** Calculate the $K_{sp}$ of each of the following compounds from the solubility data given:

| | *Compound* | *Solubility* |
|---|---|---|
| (a) | $Ag_2C_2O_4$ | $3.30 \times 10^{-2}$ g/liter |
| (b) | $Ag_2S$ | $5.70 \times 10^{-16}$ mg/mL |
| (c) | $Cd(OH)_2$ | $2.60 \times 10^{-2}$ mg/100 mL |
| (d) | CdS | $1.00 \times 10^{-9}$ mg/liter |
| (e) | $Bi_2S_3$ | $3.10 \times 10^{-18}$ g/100 mL |

(*Ans.* (b) $4.9 \times 10^{-53}$)

**6** Calculate the solubility of the following compounds:

| | *Compound* | *Express Solubility as:* | $pK_{sp}$ |
|---|---|---|---|
| (a) | $Cu(OH)_2$ | Milligrams per liter | 19.07 |
| (b) | $Co(OH)_2$ | Milligrams per 100 mL | 15.60 |
| (c) | $Ca_3(PO_4)_2$ | Grams per liter | 28.70 |
| (d) | AgCl | Milligrams per mL | 9.81 |
| (e) | $Cr(OH)_3$ | Molar | 30.22 |

(*Ans.* (b) 0.037)

**7** Calculate the number of milligrams of $Mn^{2+}$ left unprecipitated in 100 mL of a 0.100 *M* solution of $MnSO_4$ to which enough sodium sulfide has been added to make the final $S^{2-}$ ion concentration equal 0.001 *M*. Assume no change in volume due to addition of the sodium sulfide.

*Ans.* $6.04 \times 10^{-9}$

**8** Repeat problem 7 for the following:

| *Ion Left* | *Compound Present* | *Compound Added* | *Final Ion Concentration (M)* |
|---|---|---|---|
| (a) $SO_4^{2-}$ | $Na_2SO_4$ | $BaCl_2$ | $Ba^{2+} = 0.005$ |
| (b) $CrO_4^{2-}$ | $K_2CrO_4$ | $PbCl_2$ | $Pb^{2+} = 0.010$ |
| (c) $Fe^{3+}$ | $FeCl_3$ | NaOH | $OH^- = 1.0 \times 10^{-4}$ |
| (d) $Ag^+$ | $AgNO_3$ | KI | $I^- = 0.033$ |

**9** Calculate the molar solubility of silver oxalate in a 0.020 *M* solution of sodium oxalate to which enough strong acid has been added to make the final hydrogen ion concentration equal to 0.10 *M*.

*Ans.* $2.0 \times 10^{-3}$ *M*

**10** Repeat Problem 9 for the following solubilities:

| $M_{sol}$ *of:* | *In:* | *Final Molar Hydrogen Ion Concentration* |
|---|---|---|
| (a) $PbF_2$ | 0.050 *M* NaF | 0.010 |
| (b) $CaC_2O_4$ | 0.250 *M* $Na_2C_2O_4$ | $2.00 \times 10^{-4}$ |
| (c) $BaCrO_4$ | 0.100 *M* $K_2CrO_4$ | $5.00 \times 10^{-5}$ |
| (d) $CrPO_4$ | 0.500 *M* $Na_3PO_4$ | $1.00 \times 10^{-9}$ |

**11** Calculate the silver ion concentration that is a measure of the molar solubility of AgCN in a 0.050 *M* solution of NaCN in contact with solid AgCN (a) assuming no complex formation and (b) assuming complex formation.

*Ans.* (a) $3.2 \times 10^{-13}$; (b) $1.0 \times 10^{6}$

**12** Repeat Problem 11 for the molar solubility of the compound listed in the solution specified:

| $M_{sol}$ *of:* | *In:* |
|---|---|
| (a) AgCl | 0.010 *M* $Na_2S_2O_3$ |
| (b) AgCl | 0.500 *M* $NH_3$ |
| (c) AgI | 1.00 *M* $NH_3$ |
| (d) $BaSO_4$ | 0.010 *M* $Na_2$EDTA |

**13** Calculate the molar concentration of $H_2PO_4^-$ in a 0.050 *M* solution of $Na_3PO_4$ to which enough strong acid has been added to make the final hydrogen ion concentration equal $1.00 \times 10^{-8}$.

*Ans.* 0.0070 *M*

**14** Repeat problem 13 for the molar concentrations of the following ions:

| | *Ion* | *Compound* (0.050 *M*) | *Final Molar Hydrogen Ion Concentration* (*M*) |
|---|---|---|---|
| (a) | $HC_2O_4^-$ | $Na_2C_2O_4$ | $1.00 \times 10^{-3}$ |
| (b) | $H_2AsO_4^-$ | $Na_3AsO_4$ | $2.50 \times 10^{-6}$ |
| (c) | $HC_8H_4O_4^-$ (phthalate) | $K_2C_8H_4O_4$ | $1.00 \times 10^{-4}$ |
| (d) | $HC_6H_5O_7^{2-}$ (citrate) | $Na_3C_6H_5O_7$ | $5.00 \times 10^{-5}$ |

**15** If hydrogen sulfide is bubbled into 100 mL of an acidified solution that is 0.100 *M* in lead nitrate and 0.100 *M* in manganous nitrate, which sulfide will precipitate first? Calculate the number of milligrams of this ion remaining unprecipitated in the solution when the second ion starts to precipitate.

*Ans.* $1.34 \times 10^{-10}$ mg of $Pb^{2+}$

**16** Repeat Problem 15 for the following solutions:
(a) 0.100 *M* lead nitrate and 0.100 *M* nickel nitrate
(b) 0.050 *M* cupric sulfate and 0.050 *M* manganous sulfate

**17** If 15 *M* NaOH is added dropwise to 100 mL of a solution which is 0.050 *M* in $FeSO_4$ and 0.050 *M* in $Fe_2(SO_4)_3$, which ion will precipitate first? Calculate the molar concentration of this ion remaining when the second starts to precipitate, assuming no change in volume due to addition of the NaOH.

**18** Repeat problem 17 for the following solutions:
(a) 0.005 *M* $CuSO_4$ and 0.100 *M* $Ce_2(SO_4)_3$
(b) 0.250 *M* $CoCl_2$ and 0.005 *M* $CrCl_3$

CHAPTER

# 8 Theory of Precipitation

In Chapter 7, the factors that affect the solubility of a solid in a liquid were studied without consideration of the factors that affect the actual formation and purity of a precipitate in a solution. In this chapter, the steps involved in the formation of a precipitate, the factors that affect particle size, and the types of impurities that contaminate the solid will be discussed.

It will be beneficial to review the terms *saturated* and *supersaturated* as applied to solutions of solids before beginning the discussion. If an ionic solid is placed in a solution, the solid begins to dissolve from the surface, and thus the concentrations of the ions in the solution continually increase. However, as the ions move about in the solution, they may approach the surface close enough to be reincorporated into the crystal. The rate of dissolving is a function of the temperature and the surface area but can be assumed to be constant for a given solid and solution. The rate of reincorporation into the solid, however, is a function of the concentration of the ions in the solution and consequently increases as more solid dissolves. Under these circumstances, an equilibrium must be attained at which the rate of dissolving is just offset by the rate of reincorporation into the solid. This solution is said to be a **saturated** solution.

Thus, a saturated solution contains all the dissolved solid it could dissolve at that temperature in contact with undissolved solid and may be defined as a solution in which the dissolved solute will remain in equilibrium with solid solute. It should be pointed out that the solution would still be saturated if it were decanted or filtered from the solid, as long as the temperature remained constant. The product of the concentrations of the ions (raised to the appropriate powers) in a saturated solution equals the solubility product.

The solubility of most solids increases with increase in temperature, so that if a given amount of a solvent is saturated with a solid at an elevated temperature, any excess solid removed, and the solution cooled, it is possible that no solid would separate from the solution. Such a solution would be called **supersaturated** since it contains more dissolved solid than it could dissolve at the cooler temperature in contact with undissolved solute. Supersaturation is a metastable condition, and the supersaturation often may be broken by agitation or by addition of crystals of the solute or other small particles that can act as nucleation centers.

Let us also review the types of dispersions that are encountered in analysis, especially in relation to particle size. Dispersions are classified as being homogeneous or heterogeneous. Solutions are examples of **homogeneous dispersions** in which the ions or molecules of the solute are the dispersed phase and the molecules of the solvent are the dispersion medium. The dispersed phase in a solution is distributed homogeneously throughout the mixture, does not settle out, and cannot be removed by filtration. Since the maximum dimension of many ions or molecules is less than 10 Å (angstroms), the dispersed phase in solutions is usually considered to be 1 nm or less in size.

Suspensions of solids in liquids, such as sand in water, are examples of **heterogeneous dispersions**. In suspensions, the particles are large enough to settle out by gravity alone if left undisturbed a sufficient length of time. The minimum particle size that will settle out by gravity alone is usually considered to be about 100 nm. The dispersed phase in suspensions is particles composed of many atoms or molecules that are dispersed heterogeneously among the molecules of the dispersion medium. They will settle out and can be removed by ordinary filtration.

Dispersions in which the size of the dispersed phase is between those of solutions and suspensions are known as **colloids** or as **colloidal dispersions**. They are heterogeneous, since the dispersed phase consists of particles composed of a large number of atoms or molecules, however, colloids do not settle out and the dispersed phase cannot be separated by ordinary filtration since the particles are small enough to pass through the finest filters. Colloids are stable and exhibit the Tyndall effect and Brownian movement. The **Tyndall effect** refers to the fact that the path of a beam of light through a colloid can be seen because particles are large enough to reflect light even though they are too small to be seen by the human eye. The unordered or random motion of the colloidal particles is known as **Brownian movement**. The particles appear to move in straight lines with sudden changes of direction and velocity. The stability of colloids results, at least in part, from the fact that colloidal particles are charged and thus repel each other due to the electrical repulsion between like charges. Colloidal particles also exhibit a very large surface area or have a large surface/mass ratio.

## FORMATION OF PRECIPITATES

If two soluble compounds containing ions that form an insoluble solid are mixed in solution, the insoluble compound may or may not separate from the solution depending upon the concentrations of the ions. A precipitate will form only if the instantaneous concentrations of the ions upon mixing exceed the solubility product of the insoluble substance. The actual steps necessary for the formation of the precipitate can be very involved, but for a compound formed from two ions, the ions first come together to form ion pairs, which are held together loosely by electrostatic attraction. Such ion pairs can grow by addition of ions or ion pairs to form ion clusters, which can be separated easily until some minimum size cluster is attained. The minimum size is a critical size above which the cluster will continue to grow and form a particle with a definite crystalline pattern. Below this critical size, the clusters will separate more rapidly than they grow. The critical size varies for individual substances but is often between six and ten ions. The cluster is called a **nucleus** after attaining the critical size for growth.

Thus, the steps involved in the formation of the solid particles are first, the formation of ion pairs, then the formation of ion clusters, followed by the formation of nuclei capable of growing in the presence of concentrations that exceed the solubility product. The ion pairs and clusters can form in solutions in which the concentrations are not great enough to form a precipitate, but nuclei will not form until the solubility product is exceeded. The period of time involved in the formation of nuclei of a size sufficient to continue growth is called the **induction period**. It varies in length according to the precipitate and the conditions; for example, the induction period is short for silver chloride, moderate for barium sulfate, and long for magnesium ammonium phosphate.

## PARTICLE SIZE OF PRECIPITATES

The final particle size is an important property of the precipitate; it governs, at least in part, both the purity and the filterability of the precipitate. One of the main factors that affects particle size is the speed of precipitation. Von Weimarn studied the effect of rate of formation of the precipitate upon the particle size and found that the faster the precipitation, the smaller the particle size. He also found that the rate was dependent upon the extent of supersaturation and defined relative supersaturation as $Q - S$, where $Q$ is the molar concentration of the mixed reagents before precipitation occurs (the instantaneous concentrations upon mixing) and $S$ is the molar solubility of the precipitate when the system has come to equilibrium. According to von Weimarn, the rate of formation is governed by the ratio of the relative supersaturation to the molar solubility, or

$$\text{rate} = \frac{Q - S}{S} \tag{8-1}$$

Actually, the rate of formation of nuclei differs from the rate of crystal growth, and it has been found that final particle size is a function of both rates. These rates depend upon the extent of supersaturation, with the rate of formation of nuclei being governed by $(Q - S)^n$ (where $n$ may be as large as 4) and the rate of crystal growth being affected by the surface area of the particles also, or $A(Q - S)$ (where $A$ is the surface area). The values of these two rates are such that, after the first group of nuclei are formed, crystal growth predominates over nuclei formation provided that $Q - S$ is small. If $Q - S$ is large, nuclei formation predominates over crystal growth and many small particles will be formed.

The variation in particle size can also be explained by use of supersolubility curves (temperature–concentration curves that plot the concentrations above which supersaturation may not exist) compared to ordinary solubility–temperature curves. Theoretically, a supersaturated solution at a constant temperature may remain indefinitely without precipitation provided that no crystals or nucleation sites are added. However, at any given temperature, there is an instantaneous concentration upon mixing for any solid above which nucleation will occur whether or not nucleation sites are present. A typical graph is shown in Figure 8-1. The upper or dashed curve represents the supersolubility curve and the lower the regular solubility curve. The area above the dashed curve is called the **labile region**. Whenever the instantaneous concentrations are in this region, nucleation occurs rapidly. The area between the two curves is known as the **metastable region**. In this region, no new nuclei will be formed, and the particles will grow until the concentrations equal those in a saturated solution. The area below the lower curve is the **unsaturated region**.

If the concentration of a precipitant at temperature $T$ is raised slowly and uniformly throughout the solution, no precipitation will occur until the concentration represented by point $P$ on the upper curve is attained. At this point particles will form, and the particles will grow until the concentration has been

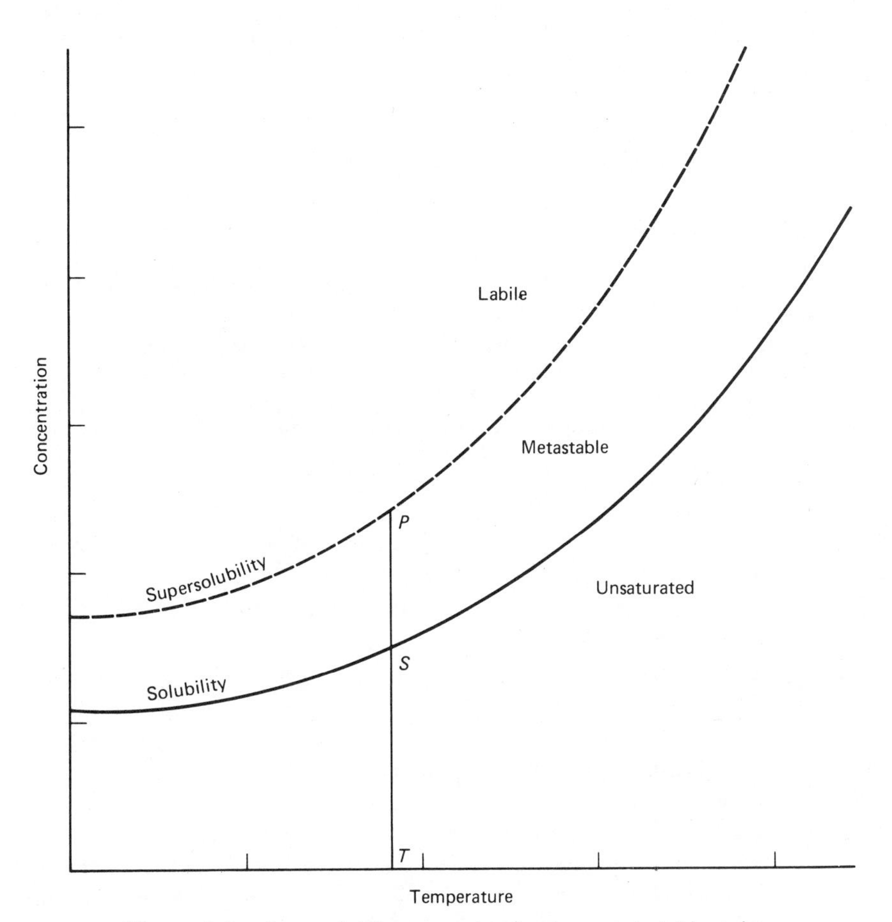


**Figure 8-1.** Supersolubility curve showing large metastable region.

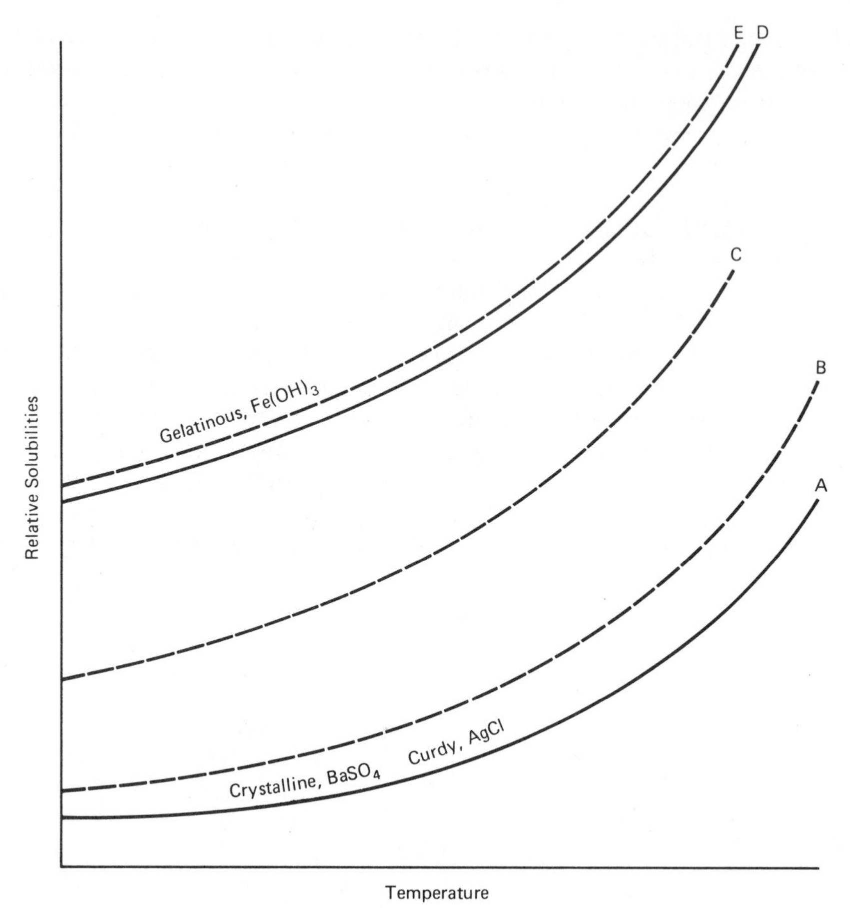


**Figure 8-2.** Relative sizes of metastable regions for various types of precipitates. (A) Solubility curve for AgCl and $BaSO_4$. (B) Supersolubility curve for AgCl. (C) Supersolubility curve for $BaSO_4$. (D) Solubility curve for $Fe(OH)_3$. (E) Supersolubility curve for $Fe(OH)_3$. Particle size for gelatinous precipitates is usually less than 0.02 $\mu$m, for curdy precipitates from 0.02 to 0.1 $\mu$m, and for crystalline precipitates 0.1 $\mu$m or more.

lowered to the value at point *S* on the lower curve. If, however, during mixing, local instantaneous concentrations again exceed the value represented by point *P*, more small particles will be formed.

The size of the metastable region, or the difference in solubility and supersolubility curves, affects the length of the induction period and thus has a great effect upon the size of the precipitate particles. As shown in Figure 8-2, the metastable area for barium sulfate is much larger than the one for silver chloride, and, as a result, barium sulfate particles are larger than silver chloride particles. The metastable area is even smaller for ferric hydroxide and other compounds that tend to form gelatinous or flocculant-type precipitates.

It is apparent that any factor that affects the solubility of the precipitate will affect the particle size and that slow and uniform mixing is important if a small number of large crystals (rather than a large number of small crystals) is desired.

# GROWTH OF AN INDIVIDUAL PARTICLE OF A PRECIPITATE

To understand the incorporation of impurities with a precipitate, it is necessary to understand the growth of individual particles in the solution. Most precipitates are formed by addition of a solution of the precipitant ion to a solution of the ion to be precipitated so that particles actually form and grow in the presence of an excess of one of the ions of the precipitate. Let us now consider this using as an example calcium oxalate when it is formed by the slow addition of dilute sodium oxalate to a solution of calcium chloride. As we have seen, the ions form ion pairs, then ion clusters, and finally nuclei that can continue growth. Calcium oxalate forms in cubic arrangement so that after a short while a particular crystal contains many ions arranged in the general pattern shown in Figure 8-3. The crystal will grow by addition of both calcium and oxalate ions at particular points on the surface. Thus, an oxalate ion is attracted to a calcium ion on the surface, and a calcium ion is attracted to an oxalate ion on the surface. This is caused by the fact that each ion on the surface has a slight charge, since it is not completely surrounded by other ions of opposite charge. A calcium ion in the interior of the crystal is surrounded by six oxalate ions (four in the plane of the ion and one each in the two adjacent planes). As a consequence, its charge is equally distributed among six ions, or one-sixth of its charge to each ion. A calcium ion on the surface, however, is surrounded by only five oxalates, and as a consequence has a slight residual positive charge equal approximately to one-sixth of its charge. An oxalate ion is attracted to this positive area on the surface and takes its position in the crystal lattice. Similarly, calcium ions are attracted to the area of oxalate ions and cause continued growth.

Because the oxalate is being added to the calcium, the calcium will be in excess until equivalent quantities are added. Consider a calcium oxalate particle in the solution if the addition is stopped before an equivalent quantity has been added. It will still try to grow by addition of both calcium and oxalate, but the oxalate will soon give out, whereas calcium ions will still be present. Calcium ions continue to be attracted to and become attached to the surface so that the

| | | | | | | | | | |
|---|---|---|---|---|---|---|---|---|---|
| | | $Ca^{2+}$ | | | | $Ca^{2+}$ | | | |
| | $Ca^{2+}$ | $C_2O_4^{2-}$ | $Ca^{2+}$ | $C_2O_4^{2-}$ | $Ca^{2+}$ | $C_2O_4^{2-}$ | $Ca^{2+}$ | $C_2O_4^{2-}$ | |
| $Ca^{2+}$ | $C_2O_4^{2-}$ | $Ca^{2+}$ | $C_2O_4^{2-}$ | $Ca^{2+}$ | $C_2O_4^{2-}$ | $Ca^{2+}$ | $C_2O_4^{2-}$ | $Ca^{2+}$ | |
| | $Ca^{2+}$ | $C_2O_4^{2-}$ | $Ca^{2+}$ | $C_2O_4^{2-}$ | $Ca^{2+}$ | $C_2O_4^{2-}$ | $Ca^{2+}$ | $C_2O_4^{2-}$ | $Ca^{2+}$ |
| | $C_2O_4^{2-}$ | $Ca^{2+}$ | $C_2O_4^{2-}$ | $Ca^{2+}$ | $C_2O_4^{2-}$ | $Ca^{2+}$ | $C_2O_4^{2-}$ | $Ca^{2+}$ | |
| | | | $Ca^{2+}$ | | | | $Ca^{2+}$ | | |

**Figure 8-3.** One plane of a calcium oxalate crystal in the presence of excess $Ca^{2+}$ ions, showing the adsorption of $Ca^{2+}$ ions at the negative sites on the surface.

particle assumes a positive charge. Since the system as a whole must be electrically neutral, this charge will be neutralized by an excess of chloride ions in the immediate vicinity of the particle. These, however, will not be as firmly attached to the crystal as the calcium ions. The calcium ions that are attached to the crystal in excess of the stoichiometric amount are said to be **adsorbed** on the crystal. The depth of the solution around the crystal in which the charge is neutralized is the **counterion region**.

## IMPURITIES IN OR ON PRECIPITATES

Precipitate impurities may occur on the surface of the precipitate particle or inside the precipitate particle. Those on the surface may be adsorbed or postprecipitated, whereas those inside the particle may be occluded or coprecipitated. Each of these will be discussed in detail.

### ADSORPTION

The calcium ions discussed in the previous section are said to be adsorbed on the crystal, so adsorption is a surface or boundary phenomenon. The concentration of the material adsorbed on the surface will be different from its concentration in the body of either of the two phases that constitute the boundary.

Adsorption on ionic solids always occurs in two layers, called the **primary layer** and the **secondary layer**, as shown in Figure 8-4. In the case of the calcium oxalate particle discussed previously, the calcium ion is considered to be in the primary layer since it is held to the particle by some force of attraction, and the chloride ions in the counterion region would be considered to be in the secondary layer. The ions in the secondary layer are free to move about in the solution; those in the primary layer are not.

One of the main questions concerning adsorption that is of interest to the analyst is the type of ions that tend to be adsorbed. Considering first the primary layer, ions that are common to the precipitate and are in excess in the solution

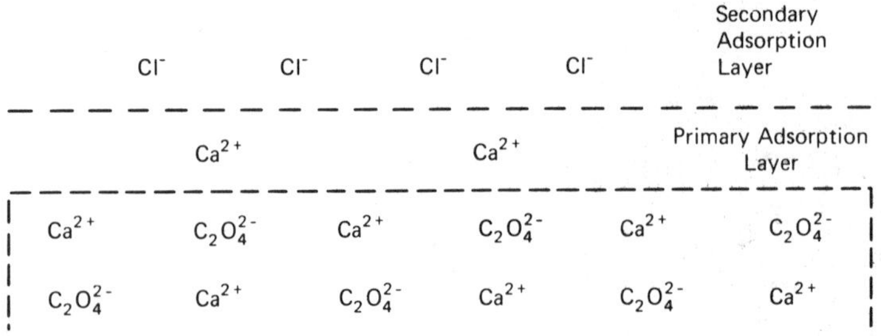

$Cl^-$ $Cl^-$ $Cl^-$ $Cl^-$ Secondary Adsorption Layer

$Ca^{2+}$ $Ca^{2+}$ Primary Adsorption Layer

$Ca^{2+}$ $C_2O_4^{2-}$ $Ca^{2+}$ $C_2O_4^{2-}$ $Ca^{2+}$ $C_2O_4^{2-}$

$C_2O_4^{2-}$ $Ca^{2+}$ $C_2O_4^{2-}$ $Ca^{2+}$ $C_2O_4^{2-}$ $Ca^{2+}$

**Figure 8-4.** Surface of a calcium oxalate particle in the presence of excess $Ca^{2+}$ ions, showing primary adsorption of $Ca^{2+}$ ions and secondary adsorption of $Cl^-$ ions.

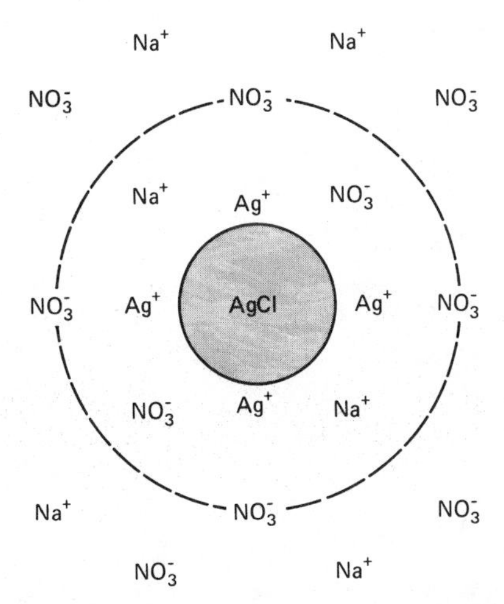


**Figure 8-5.** Colloidal particle of AgCl in the presence of excess $Ag^+$ ion, showing the primary adsorption of $Ag^+$ ions and the counterion region in which the charge is neutralized.

will be adsorbed more strongly than any other ions. If there are no common ions present in excess, any ion that forms an insoluble compound with either of the ions of the precipitate may be adsorbed. In case there is more than one such ion present, the ion that forms the most insoluble compound will tend to be adsorbed the most strongly. Considering calcium oxalate, a carbonate ion would be adsorbed in preference to a nitrate ion, and a barium ion would be adsorbed rather than a potassium ion.

One result of adsorption is that the particles become charged and thus repel each other because they all have the same charge. The force of repulsion is a function of the size of the charge and the distance of separation. For any given size of charge the repelling force increases as the particles come closer together. This effect does not extend beyond the counterion region since the charge is neutralized by the excess of oppositely charged ions in this region, as shown in Figure 8-5.

There are also **cohesive forces** (often called **van der Waals' forces**) between the particles that tend to attract the particles to each other so that they may form larger aggregates. This cohesive force is similar to gravity, in that it increases as the mass increases and as the distance of separation decreases. For two charged particles to be able to combine to form a larger particle or aggregate, the cohesive force must be greater than the repelling force owing to like charges as the particles come close together.

The combination of two or more charged small particles to form a larger particle is called **agglomeration** (also known as **flocculation** and as **coagulation**). Factors that favor agglomeration are those that tend to decrease the depth of the counterion region, and include both the concentration and charge of the counterions. An increase in concentration of the counterion will decrease the depth of the counterion region, and the larger the charge of the counterion, the greater the effect of any given concentration.

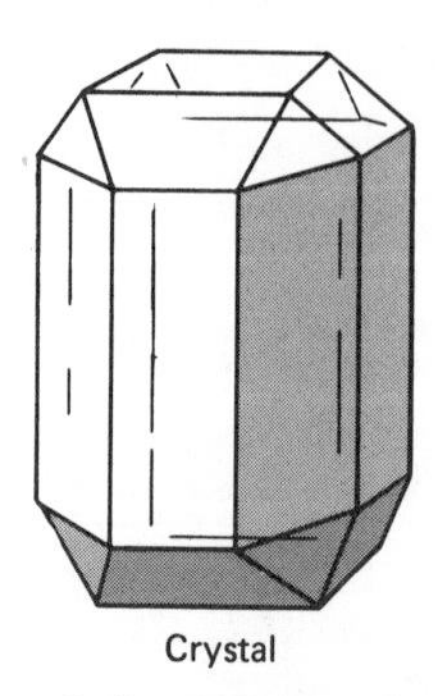

Crystal Curd

**Figure 8-6.** Contrast between crystalline and curdy precipitate particles.

The agglomeration of small particles differs from crystal growth because the agglomerated particles retain their original form and do not actually increase in size, as is the case when particles grow by addition of ions to the crystal lattice. The difference between crystalline and agglomerated particles is shown in Figure 8-6. The agglomerated particles carry both the primary layer and the secondary layer of adsorbed ions into the agglomerated particle, since they must remain neutral to be held together by the cohesive forces. Removal of the counterions by washing will cause the particles to redisperse or **peptize**. As a consequence, the wash solution for an agglomerated precipitate must contain an electrolyte to prevent peptization of the agglomerates.

## POSTPRECIPITATION

The second type of impurity often found on the surface of a precipitate comprises the **postprecipitated** impurities that form on the surface of a previously formed precipitate particle. The postprecipitation of zinc sulfide on copper(II) sulfide is an excellent example of this type of impurity. In slightly acid solution, cupric sulfide will be precipitated quantitatively by the addition of excess hydrogen sulfide, but zinc sulfide will not form under these conditions. If both copper and zinc ions are present, the copper sulfide will precipitate and, if filtered immediately, will be almost free of zinc ion. If allowed to remain in contact with the solution for an appreciable time, the copper sulfide particles will become coated with precipitated zinc sulfide and thus becomes contaminated by zinc. The extent of the contamination is governed, at least in part, by time because the amount of postprecipitated impurity will increase with time until a maximum is reached. The effect is the same whether the zinc ion is present before precipitation or added after the precipitate is formed.

Postprecipitation is probably an adsorption phenomenon, since the copper sulfide is precipitated in the presence of an excess of sulfide ion; thus, sulfide ion is adsorbed in the primary layer of the particles. This attracts zinc ion as the secondary ion so that the solubility product of the zinc sulfide is probably exceeded in the counterion region. As a result, zinc sulfide precipitates on the surface of the copper sulfide particles.

Another example of postprecipitation occurs in the oxalate separation of calcium and magnesium in limestone analysis. In this procedure, an excess of oxalate is used in the precipitation step. Calcium oxalate precipitates, but magnesium oxalate forms a complex and remains in solution. If the calcium is filtered immediately, little or no magnesium will be carried down with the calcium oxalate. If the solution is allowed to stand for some time (more than 1 or 2 hr) magnesium oxalate will postprecipitate and contaminate the precipitate.

Even though the primary adsorption of the common ion in excess is probably the main factor in postprecipitation, all cases cannot be explained by adsorption alone. One possible explanation is that a supersaturated solution of the second material is formed on the surface of the particle and that the surface offers nucleation sites to break the supersaturation and cause the secondary precipitate to form. Other factors that might be involved include active sites on the precipitate surface, which cause deformation of ions and exert a catalytic effect on the secondary precipitation.

## OCCLUSION

Occlusion is the physical entrapment of impurities in the crystal as the particle grows rapidly. The impurities may be trapped as mother liquor or as individual ions or molecules. Essentially, the particle is growing so fast that ions in the immediate vicinity are simply surrounded by the growing particle and cannot escape back into the mother liquor before being completely enclosed by the crystal. Such impurities may not be removed by washing, but they are decreased in some cases during aging of the precipitate.

## COPRECIPITATION

The process by which soluble impurities become incorporated into a precipitate particle during its formation is known as coprecipitation. Coprecipitation may occur either by **isomorphous replacement** (sometimes referred to as **mixed crystals** or as **solid solution formation**) or by adsorption followed by entrapment or occlusion.

Coprecipitation due to solid solution formation in which the impurity actually enters the crystal lattice of the precipitate may be illustrated by the coprecipitation of barium chromate with barium sulfate. The addition of barium chloride to a solution that is 0.02 *M* in chromate ion and 0.1 *M* in hydrogen ion causes no precipitation of barium chromate. Addition of barium chloride to a solution that is 0.02 *M* in sulfate ion and 0.1 *M* in hydrogen ion causes the quantitative precipitation of barium sulfate. Addition of barium chloride to a solution that is 0.02 *M* in chromate, 0.02 *M* in sulfate, and 0.1 *M* in hydrogen ion causes the precipitation of barium sulfate contaminated with approximately 5% by weight of barium chromate. Even though barium chromate alone would not precipitate under these conditions, the presence of the barium sulfate lattice allows the chromate ion to be incorporated into the precipitate.

There are relatively few cases of solid solution formation as the cause of impurities, since the two compounds must not only form in the same crystal pattern, but the radii of the ions of like charge must be within 10% of each other.

Coprecipitation caused by adsorption and entrapment is much more common in analytical procedures. For example, barium sulfate may contain either anions or cations coprecipitated in this way. The cations tend to predominate if barium is added to sulfate; the anions predominate if the sulfate is added to the barium solution. This can be explained by consideration of the adsorption of the common ion that is in excess and the counterions that would be attracted to the particle. Assume that less than the stoichiometric amount of barium is added to a sulfate solution. The particles will adsorb sulfate ions in the primary layer and cations in the secondary layer. If more barium is now added, the rapid growth of the particles in the immediate vicinity of mixing can entrap the secondary layer of cations in the precipitate particle. Similarly, addition of the sulfate to the barium solution will cause entrapment of the anions in the secondary layer. The extent of occlusion by adsorption and entrapment is affected by the solubility of the entrapped substance. Thus, calcium is entrapped more than sodium with barium sulfate, and nitrate is entrapped more than chloride. The extent of coprecipitation is also a function of the ratio of concentrations of the impurity ion and the ion being precipitated. The coprecipitation increases as the concentration of the ion being precipitated decreases, that is, as the ratio of the impurity to precipitated ion becomes large.

In some procedures, coprecipitation is desirable and actually the basis of the separation. As an example, manganese coprecipitates with the hydrous oxide of iron (ferric hydroxide) when precipitated at high pH with sodium hydroxide. Even though manganese is present in too small a concentration to precipitate itself, the high pH causes the hydrous oxide particles to become negatively charged and attract the manganese in the secondary layer. This particular process is known as **gathering** and is used in several analytical procedures.

## PURIFICATION OF PRECIPITATES

Precipitate impurities must often be removed or reduced to obtain a satisfactory result in a determination. Some impurities may be removed by simple washing, but most require more sophisticated treatment.

Adsorbed impurities may be removed by simple aqueous washing of crystalline precipitates, but the wash solution must contain an electrolyte to offset peptization for curdy or flocculent precipitates. In this case, the electrolyte ion actually replaces one of the adsorbed ions, either in the primary or the secondary layer. If the precipitate is to be dried, the compound formed and left on the precipitate must be volatile. In the washing of silver chloride precipitates that contain adsorbed silver ions in the primary layer and nitrate ions in the secondary layer, the wash solution is usually dilute nitric acid because the hydrogen ion will displace the silver ions from the primary layer to form nitric acid, which will volatilize when heated. The hydrous oxide precipitates cannot be washed with an acid solution, so ammonium compounds are usually used in the wash solutions because most ammonium compounds volatilize completely when heated.

Postprecipitated impurities cannot be removed by simple washing either with or without an electrolyte. Postprecipitated impurities may be reduced by dissolving the precipitate and reprecipitating, being careful to filter as promptly as possible after the precipitate is formed.

Occluded and coprecipitated impurities are not affected by washing, since they occur in the body of the precipitate particles. In some cases, they can be removed only by solution and reprecipitation. The amount of impurities occluded or coprecipitated will be less in the second precipitation because their concentrations will be lower.

Some precipitates may be partially purified by aging or digestion. **Aging** is allowing the precipitate to remain in contact with the mother liquor for extended periods of time; **digestion** is the same process, except at elevated temperatures for shorter lengths of time.

Very small particles of a precipitate are more soluble than larger particles. Since the mother liquor constitutes a saturated solution, small particles will tend to dissolve during aging or digestion and deposit on the larger crystals. Thus, the small particles will disappear and the larger particles will grow larger. The impurities trapped during the growth of the smaller particles will be returned to the solution and probably will not be trapped during the slow growth of the larger particles. Even though the particles are not small enough to have increased solubility, the smaller particles will dissolve during aging or digestion and reprecipitate on the larger particles, since the smaller surface area represents a more stable energy relationship.

## TYPES OF PRECIPITATES

Precipitates are classified as crystalline, curdy, and gelatinous or flocculent. **Crystalline** precipitates are characterized by easily filterable particles that are relatively pure. As a consequence, all analysts prefer to work with crystalline precipitates. **Curdy** precipitates are those that are formed as colloidal particles but agglomerate to form larger aggregates that are of filterable size. Curdy precipitates are usually readily filterable, somewhat more impure than crystalline precipitates, and must be washed with an electrolyte solution. **Gelatinous** precipitates are also flocculated colloids, but the particle size is smaller than the particles of a curdy precipitate. Gelatinous precipitates are difficult to filter, contain more impurities than curdy precipitates, and also must be washed with an electrolyte solution to offset peptization.

## RULES FOR PRECIPITATION

Some of the general rules for the preparation of a relatively pure and easy-to-filter precipitate include the following.

1 The precipitate should be formed under conditions that tend to increase the solubility. If necessary the conditions can be changed to decrease the solubility after the precipitate is formed. To do this, the precipitate should be formed at elevated temperatures and in the presence of compounds that will increase the solubility. Thus, barium sulfate is precipitated from solutions that contain small amounts of hydrochloric acid, since barium sulfate is more soluble in acid than in water, owing to the formation of some bisulfate ions.

2 The solutions should be as dilute as possible and should be mixed slowly with vigorous agitation.

3 A slight to moderate excess of the precipitant ion should be added to take advantage of the common ion effect. Large excesses should be avoided because they may cause complex formation.

4 If possible, the precipitate should be aged or digested under conditions of increased solubility. However, precipitates that are subject to postprecipitation should be filtered as soon as possible after formation.

5 If necessary, the precipitate should be dissolved and reprecipitated in order to decrease impurities.

## PRECIPITATION FROM HOMOGENEOUS SOLUTION

In ordinary precipitation, two solutions are mixed together and, as precipitation occurs, there are great changes in the concentrations of the precipitant ion and the ion being precipitated in the immediate vicinity of mixing. As a result, temporary supersaturation in excess of the supersolubility curve will often occur in the immediate vicinity of mixing before agitation causes the solution to become uniform. These are the conditions that lead to rapid formation of a large number of small particles and the consequent adsorption and entrapment of impurities in the precipitate. To understand what happens, let us consider a drop of barium chloride that has been placed without mixing in a solution of sodium sulfate. The barium ions will migrate outward into the sulfate solution, and the sulfate ions will migrate into the barium solution. At the juncture of the two solutions, barium sulfate will be precipitated and deplete the concentration of both ions. The barium sulfate particles closest to the barium solution will be in a region of excess barium ion and will adsorb barium ions in the primary layer to become positively charged. Similarly, the barium sulfate particles closest to the sulfate solution will adsorb sulfate ions and become negatively charged. Thus, some of the particles will tend to entrap cations and some will entrap anions, so both types of impurities will be present in the crystals being formed. These conditions continue as the ions migrate toward each other and form more particles. As a result, precipitates formed by the mixing of two solutions tend to be impure and to consist largely of small particles.

If the precipitant ion could be generated slowly and uniformly throughout the solution, the localized concentration effects that occur when solutions are mixed would be eliminated. In such cases there should be very few areas in which the instantaneous concentrations exceed the supersolubility curve, so a relatively small number of particles should be formed, and these should be able to grow slowly without adsorption and entrapment of counterions. In other words, excessive supersaturation with consequent small particles and occluded impurities should not occur because the concentration of the precipitant ion is raised slowly and uniformly throughout the solution by a chemical reaction or a physical change.

This process of generating the precipitating ion or condition uniformly throughout the solution is known as **precipitation from homogeneous solution**. The essential factor in this process is the gradual increase in the available concentration of the precipitant ion so that the solubility product is exceeded uniformly throughout the solution at a slow or moderate rate, which allows crystal growth to predominate over nucleation. Precipitation from homogeneous solution is used to improve separations, to study and reduce coprecipitation, to form larger crystalline particles, and to produce precipitates that are purer and easier to filter.

This type of precipitation can be applied to any system in which the necessary reagent can be generated slowly and uniformly throughout the solution by a chemical reaction or a physical change. Physical changes that may be used include the volatilization of an organic solvent or of ammonia. These changes either decrease the solubility or lower the pH. Chemical reactions can be used that will generate the actual ion or compound needed or will produce hydrogen or hydroxyl ions to raise or lower the pH.

For any of these, the rate must be satisfactory. A reaction that is too slow will cause too much time to be spent in the analysis whereas a reaction that is essentially instantaneous will cause undesired local concentration excesses. There are many examples of reactions that can be used for control of pH or generation of individual ions or molecules. A few typical ones will be discussed.

## TYPES OF REACTIONS

### *Control of* pH

In most cases, slow rise of pH is desired rather than slow increase in acidity. The procedure used most often is the hydrolysis of urea at elevated temperatures. At room temperature urea hydrolyzes to form ammonia and carbon dioxide so slowly that the necessary rise in pH would require an excessively long time.

The rate at 100°C is satisfactory, being neither too fast nor too slow. The hydrolysis reactions are often written

$$CO(NH_2)_2 + H_2O \longrightarrow 2NH_3 + CO_2 \quad \textbf{(8-2)}$$

$$2NH_3 + 2H^+ \rightleftharpoons 2NH_4^+ \quad \textbf{(8-3)}$$

but it has been shown that the hydrolysis is actually a rearrangement to form ammonium cyanate followed by hydrolysis and reaction with hydrogen ion as follows:

$$CO(NH_2)_2 \longrightarrow NH_4^+ + OCN^- \quad \textbf{(8-4)}$$

$$OCN^- + 2H^+ + H_2O \longrightarrow NH_4^+ + CO_2 \quad \textbf{(8-5)}$$

The ammonium cyanate does not accumulate below pH 5, and the rate of hydrolysis is almost independent of pH between pH 1 and 7. If the elevated temperature is not desired, cyanates can be used at room temperature to raise the pH, essentially by reaction (8-5). Other methods include the hydrolysis at elevated temperature of acetamide. hexamethylenetetramine, mixtures of halide and halate, and thiosulfate. The hydrolysis of 2-hydroxyethyl acetate can be used to lower pH slowly at elevated temperature, since the acetic acid formed by hydrolysis reacts with hydroxyl ions and since the ethylene glycol (which is the other product) is neutral. For ammoniacal solutions the pH can be lowered by simple volatilization of the ammonia.

### *Generation of Other Reagents*

Various substances can be used at room or at elevated temperatures for slow generation of particular ions. Sulfamic acid and methyl or ethyl sulfate are utilized to generate sulfate, ethylene or propylene chlorohydrin for chloride, methyl or ethyl oxalate for oxalate, methyl phosphate or metaphosphoric acid for phosphate, trichloroacetic acid for carbonate, oxidation of chromic ion by bromate to form chromate, and oxidation of iodine by chlorate to form iodate.

In the case of precipitation using organic reagents, the reagent may be formed by some suitable reaction in the solution. Examples are the slow formation of dimethylglyoxime by the reaction of biacetyl and hydroxylamine and the hydrolysis of 8-acetoxyquinoline to form 8-hydroxyquinoline. Since the chelate or metal salt of the organic compound is usually much more soluble in organic solvents than in water, this type of compound may be precipitated by slow volatilization of an organic solvent such as acetone.

## USES OF PRECIPITATION FROM HOMOGENEOUS SOLUTION

### *Precipitation of Hydrous Oxides*

Gelatinous precipitates such as the hydrous oxides are very difficult to filter and are usually rather impure. Precipitation from homogeneous solution produces precipitates that are denser, purer, and much easier to filter. The presence of an organic anion often improves the characteristics of the precipitate. The precipitation is usually effected by the adjustment of pH to a value just below incipient precipitation, urea is added, and the solution is maintained at or near the boiling point for 1 or 2 hr or until precipitation is complete. In the presence of some organic acids (such as formic acid), the iron precipitate contains basic salts [such as basic ferric formate—$FeOCHO_2$ or $Fe(OH)_2CHO_2$] and a ferric oxide monohydrate (FeOOH). These precipitates are much purer than those obtained by heterogeneous precipitation, and the purity can be improved even further by use of a two-step procedure in which the solution is cooled rapidly and filtered when the precipitation is approximately 90–95% complete. After filtration, the solution is again heated to complete the precipitation and the two precipitates are combined. These precipitates are much easier to filter and to wash than those produced by addition of ammonia or sodium hydroxide.

### *Production of Relatively Large Crystals*

Substances such as silver chloride, which form as colloidal particles and then agglomerate to form curdy precipitates, may be precipitated in the form of comparatively large crystalline particles by precipitation from homogeneous solution. Crystals with dimensions of as much as 0.2 mm have been produced by destruction of the silver amine complex [$Ag(NH_3)_2^+$] by the hydrolysis of 2-hydroxyethyl acetate and by the volatilization of ammonia from solutions that contain an excess of chloride ion. Nickel dimethylglyoximate, aluminum oxinate, thorium iodate, cadmium and other sulfides, and many other compounds that tend to form in small particles have been precipitated in crystalline form in much larger particles by precipitation from homogeneous solution.

### *Study of Coprecipitation*

Precipitation from homogeneous solution has been used to study coprecipitation, and it has been shown that coprecipitation often occurs during the later stages of the precipitation when the concentration of the ion being precipitated becomes small compared to the concentration of the coprecipitated impurity. In the coprecipitation of yttrium with ferric periodate, formed by liberation of periodate from paraperiodic acid by hydrolysis of acetamide, it was found that coprecipitation was essentially negligible until approximately 99.5% of the ion had been precipitated. Coprecipitation increased rapidly in the final stages of precipitation after 99.9% of the iron had been precipitated. This explains the increased purity of the precipitate in the two-step method.

## PROBLEMS

**1** Discuss the effect of the rate of precipitation upon the particle size of the precipitate, including:
(a) The von Weimarn principle
(b) The effect of the relative depth of the metastable region
(c) The effect of temperature
(d) The effect of precipitation from homogeneous solution

**2** Discuss the growth of an individual particle of a precipitate.

**3** Which types of impurities are found on the surface of a particle and which are found inside the particle? Explain.

**4** Assuming equimolar concentrations of $NO_3^-$, $Ba^{2+}$, $Li^+$, $OH^-$, and $C_2H_3O_2^-$ in a solution in which neutral colloidal particles of $SrSO_4$ are suspended, which ion will be adsorbed most strongly?

*Ans.* $Ba^{2+}$

**5** Repeat problem 4 for the following situation:

| | *Suspended Solid* | *Ions* |
|---|---|---|
| (a) | $CaCO_3$ | $PO_4^{3-}$, $Cl^-$, $Na^+$, $NH_4^+$, $OH^-$ |
| (b) | CuS | $Na^+$, $SO_4^{2-}$, $Pb^{2+}$, $C_2H_3O_2^-$, $Al^{3+}$ |
| (c) | AgCl | $Mg^{2+}$, $NO_3^-$, $Hg_2^{2+}$, $SO_4^{2-}$, $Ba^{2+}$ |
| (d) | $AlPO_4$ | $SO_4^{2-}$, $K^+$, $Cl^-$, $Li^+$, $OH^-$ |

**6** Which one of the following compounds would require the lowest molar concentration to be equally effective in decreasing the depth of the counterion region of colloidal lead sulfate suspended in a solution of sodium sulfate: sodium citrate, iron(II) acetate, tin(IV) chloride, aluminum nitrate, and magnesium malonate?

*Ans.* Tin(IV) chloride

**7** Repeat Problem 7 for the following situations

| | *Suspended Solid* | *In a Solution of* | *Compounds Added* |
|---|---|---|---|
| (a) | $PbSO_4$ | $PbCl_2$ | $Na_3C_6H_5O_7$, $Fe_4(EDTA)_3$, $SnCl_4$, $Al(NO_3)_3$, $MgC_4H_2O_4$ |
| (b) | CuS | $CuCl_2$ | NaCl, $MgSO_4$, $Al(NO_3)_3$, $Na_3C_6H_5O_7$, $Fe_2(SO_4)_3$ |
| (c) | $AlPO_4$ | $Al_2(SO_4)_3$ | $NH_4Cl$, $NaNO_3$, $KC_2H_3O_2$, $Li_2SO_4$, $NaClO_3$ |
| (d) | $MgCO_3$ | $Na_2CO_3$ | KCl, $(NH_4)_2SO_4$, $Al(NO_3)_3$, $SnCl_4$, $Na_4EDTA$ |

**8** Discuss the advantage and disadvantages of precipitation from homogeneous solution.

CHAPTER

# 9 Gravimetry and Precipitation Titrimetry

## GRAVIMETRY

A large number of gravimetric methods have been used in the analysis of materials since the beginning of careful analytical determinations. As satisfactory titrimetric methods and, more recently, instrumental methods are developed for many constituents, fewer and fewer gravimetric procedures are used. However, many gravimetric methods are still the method of choice for the analysis of a relatively large number of constituents determined in analytical laboratories. Often, if the results of the analysis are to be used as legal evidence in a court, a gravimetric procedure will be used because of its greater chance of being accepted and understood in a court of law.

Four representative gravimetric procedures that illustrate the three types of precipitates encountered in analytical methods together with one example of precipitation from homogeneous solution will be presented in this chapter. The basic theory and calculations of gravimetry dealing with the formation and solubility of precipitates, the completeness of precipitation, and the impurities found on or in precipitates have been discussed in previous chapters. The three types of precipitates encountered in gravimetric procedures are the crystalline precipitates exemplified by $BaSO_4$, the curdy precipitates, of which AgCl is the most common example, and the gelatinous precipitates, such as the hydrous oxides of iron and aluminum, usually called ferric hydroxide and aluminum hydroxide.

### DETERMINATION OF SULFATE AS BARIUM SULFATE

The precipitation of barium sulfate in the gravimetric determination of sulfate is performed by the addition of a barium chloride solution to a hot solution of sulfate that has been acidified with hydrochloric acid.

$$Ba^{2+} + SO_4^{2-} \longrightarrow BaSO_4 \qquad \textbf{(9-1)}$$

The hydrochloric acid prevents the precipitation of slightly soluble barium salts such as barium carbonate and barium phosphate. Since barium sulfate is more soluble in acid solutions than in water, because of the formation of bisulfate ($HSO_4^-$) ions, the hydrochloric acid also favors the formation of larger particles according to the von Weimarn rule.

The barium sulfate precipitate tends to be contaminated with occluded and coprecipitated impurities, and quantitatively exact determinations can be accomplished only on sulfuric acid solutions and sodium sulfate solutions under well-defined conditions. The extent of the contamination and the type of contamination depends in part upon which reagent is added to the other. If the barium chloride solution is added to the sulfate solution, coprecipitation of cations predominates; if the sulfate is added to the barium, coprecipitation of anions predominates. The extent of coprecipitation also depends upon the relative concentrations of the impurities in the solution and upon the solubility of the barium or sulfate compounds formed with the impurities.

Nitrate and chlorate are coprecipitated to a greater extent than other common anions such as the halides. Chloride is coprecipitated less than most other anions and is the anion of choice whenever possible. Almost all the heavy metals are coprecipitated with barium sulfate, with the ion of higher valance for any particular metal being coprecipitated more than ions of lower valence. Thus, iron(III) (ferric) ions are coprecipitated much more than iron(II) (ferrous) ions. Ferric iron is present more often than other heavy metal ions and consequently causes major interference. The interference of ferric ion can be offset to some extent by reduction to ferrous ion but cannot be completely eliminated except by prior removal of the ferric ion by precipitation or complexation.

Coprecipitation can cause either positive or negative errors, depending upon the relationship of the equivalent weights of the metals or anions in question and upon which component of the precipitate is present in excess. Thus, ferric ion causes a negative error for sulfate because iron has a lower equivalent weight than barium. In the analysis of barium, coprecipitation of ferric ion would cause a positive error, since it would increase the total weight of the precipitate in the presence of excess sulfate. Similarly, nitrate causes a positive error in the determination of sulfate but would cause a negative error in the determination of barium, since the nitrate is converted to the oxide upon ignition.

Cations that form insoluble sulfates in acid solutions, such as lead and strontium, and anions that form insoluble barium salts in acid solution, such as chromate, must be absent, since coprecipitation is extensive for such ions.

The precipitate is usually filtered through fine-porosity quantitative paper, washed with hot water until free of chloride, and ignited after the paper has been charred carefully and the carbon burned off. Oxidizing conditions must exist during the charring, burning, and ignition procedures, to prevent reduction of the precipitate to the sulfide.

## DETERMINATION OF CHLORIDE AS SILVER CHLORIDE

Chloride is often determined by precipitation as silver chloride by addition of a solution of silver nitrate to a chloride solution that has been acidified with nitric acid. After a suitable aging or digestion period (preferably in the dark), the precipitate is filtered, washed with very dilute nitric acid, dried, and weighed. The precipitate must be protected from the light because silver chloride is reduced by light. When the proper procedure is followed, practically all sources of error can be eliminated, and the method is one of the most accurate procedures among analytical determinations. In fact, much of the early work on the determination of accurate atomic weights involved the conversion of the element to the chloride followed by the analysis of the chloride by precipitation as silver chloride.

The solution of chloride is acidified with nitric acid to prevent the precipitation of insoluble silver salts of weak acids such as carbonates and phosphates. The nitric acid also speeds the coagulation of the precipitate, which forms as colloidal particles upon addition of the silver nitrate. The nitric acid further aids in the purification of the precipitate by displacing adsorbed silver ions to form adsorbed nitric acid, which is volatilized in the drying process.

A large excess of silver nitrate must be avoided because silver chloride forms a soluble complex ion, $Ag_2Cl^+$, in the presence of excess silver ion. Formation of such ions causes incomplete precipitation with a consequent negative error. A slight to moderate excess is desirable, however, to decrease the solubility of the precipitate by the common ion effect and thus decrease the amount of chloride ion left unprecipitated.

The precipitate forms immediately upon addition of the silver nitrate solution in colloidal particles that adsorb silver ions in the primary layer and nitrate ions in the secondary layer. The resulting particles are positively charged as a result of the adsorbed silver ions and tend to remain dispersed as colloidal particles too small to be of filterable size. Upon aging overnight in the presence of nitric acid, the particles coagulate to form filterable-sized particles called **curds**. The actual particles are still the same size but they have clumped together to form a relatively large conglomerate that can be filtered. The adsorbed silver ions are removed in part by displacement by hydrogen ions during the aging process. If sufficient time is not available to allow overnight aging, the precipitate can be coagulated by digestion for approximately 1 hr near the boiling point of water.

The precipitate is usually filtered through a Gooch crucible or through a sintered glass crucible of medium porosity. It must be washed with very dilute nitric acid to prevent peptization of the precipitate. If water is used as the wash liquid, the ions in the secondary layer are more easily removed than those in the primary layer. As a result, the particles again become charged, repel each other, and redisperse as colloidal particles. As previously mentioned, the process of redispersion is known as peptization. The precipitate must be washed until there is no test for $Ag^+$ ions in the filtrate. The precipitate is then dried at 110–130°C, at which temperature essentially all the water is removed.

The precipitate must be protected from the light, especially while wet, because it is photosensitive and is easily reduced by light. This is the same reaction that makes photography possible, since photographic film is essentially a dispersion of colloidal silver halides in a polymeric base. When light strikes the film, the silver halides are sensitized or partially reduced to the metal. The development process is simply the further reduction of all the sensitized halides to the metal by a weak reducing agent, that is not strong enough to reduce unsensitized silver halides. Unreacted halides are then removed by dissolving in a sodium thiosulfate solution.

The reactions involved in the reduction of the silver chloride precipitate in the solution before filtration may be shown as

$$2AgCl + light \longrightarrow 2Ag^0 + Cl_2^0 \qquad \textbf{(9-2)}$$

$$3Cl_2^0 + 3H_2O \longrightarrow 6H^+ + ClO_3^- + 5Cl^- \qquad \textbf{(9-3)}$$

$$5Cl^- + 5Ag^+ \longrightarrow 5AgCl \qquad \textbf{(9-4)}$$

The formation of chloride ions occurs in a solution that contains excess silver ion, and more silver chloride is precipitated. Since the metallic silver formed in the decomposition remains with the precipitate, a positive error occurs. If, however, the decomposition takes place after filtration, the error will be negative as a result of the loss of elementary chloride.

## DETERMINATION OF IRON AS FERRIC OXIDE

Iron is often determined by the addition of ammonia to a hot solution containing iron(III) (ferric) ions to form the hydrous oxide. The precipitate is a gelatinous mass of coagulated colloidal particles, which are often described as microcrystalline because they do not show a definite crystal pattern until after being aged for some time. The precipitate is a hydrous oxide of indefinite composition that may be represented as $Fe_2O_3 \cdot xH_2O$. For simplicity it is usually designated as $Fe(OH)_3$ and called ferric hydroxide, even though the ratio of $Fe_2O_3$ to $H_2O$ is seldom 1 : 3. Ferric hydroxide is very insoluble, with $K_{sp}$ near $10^{-37}$, so it may be quantitatively precipitated even at a pH of 4. The gelatinous precipitate rapidly clogs the pores of any filter paper so that a paper of coarse porosity must be used and the precipitate washed as much as possible by decantation before transfer to the filter. After filtration and washing with an electrolyte solution to prevent peptization, the paper is charred, the carbon removed by heating over a Bunsen flame, and the precipitate ignited at temperatures near 1000°C before weighing. Oxidizing conditions must be maintained during the charring, carbon removal, and ignition steps to prevent reduction of the precipitate to $Fe_3O_4$.

Any cation (such as aluminium and chromic) that forms insoluble hydroxides with ammonia will contaminate the precipitate. Many cations that will not form hydroxide precipitates with ammonia are coprecipitated with the iron, especially at high pH. At low pH (below approximately 8) the particles are positively charged and repel cations, but at higher pH the particles become negatively charged, as a result of adsorption of hydroxide ions, and thus attract cations into the secondary layer. Most procedures therefore call for slow addition of ammonia to the hot solution until a faint odor of ammonia persists after ammonia fumes have been dispelled by blowing across the surface. If the amount of iron is small, it is difficult to determine when precipitation starts upon addition of the ammonia. The best indication is a slight deepening of the light yellow or tan color of the solution owing to the formation of colloidal particles of the hydrous oxide. If the solution is then digested for a short period of time, the precipitate will coagulate into the usual gelatinous mass. Many procedures call for coagulation by gentle boiling for a few minutes after the formation of the precipitate.

The fact that cations are attracted to the precipitate at high pH values is used in the **gathering process** by which trace amounts of certain metals are coprecipitated with the hydrous oxide of iron. The procedure calls for the addition of a concentrated solution of NaOH, which raises the pH quickly to values in the range 12–14, followed by coagulation at elevated temperatures. Trace amounts of manganese can be precipitated in this way for determination by satisfactory methods.

Fluoride and pyrophosphate ions must be absent, since these two anions form very stable complexes with ferric ion, which prevent the precipitation of the hydrous oxide.

The precipitate is usually washed with warm solutions of an ammonium salt, either ammonium nitrate or ammonium chloride, to prevent peptization of the precipitate and to form a volatile compound in the primary and secondary layers, which will be removed during the ignition step.

## PRECIPITATION OF NICKEL DIMETHYLGLYOXIMATE FROM HOMOGENEOUS SOLUTION

Nickel(II) ion is often precipitated as dimethylglyoximate by the addition of ammonium hydroxide to an acidic solution containing nickel(II) ion and dimethylglyoxime (DMG). However, the precipitate thus formed is flocculent, fluffy, and difficult to filter. On the other hand, precipitation from homogeneous solution by the hydrolysis of urea in the presence of nickel and dimethylglyoxime produces a precipitate formed of relatively large particles that is easy to filter. The basic theory and advantages of precipitation from homogeneous solution are covered in Chapter 8.

Dimethylglyoxime is essentially specific for nickel, since it reacts only with nickel in solutions that are nearly neutral (pH 5 and above). It will also precipitate palladium from dilute hydrochloric acid solutions and bismuth from solutions of high pH (11–11.5). No other single element interferes, but cobalt and ferric iron together will hinder the nickel from precipitating in the usual bright red form. Dimethylglyoxime forms a chelate with nickel, which can be represented by the equation in Figure 9-1.

Chelates are derivatives of organic compounds that have two functional groups so arranged that a ring structure can be formed with a metal cation. The stable ring structure may be contained in a neutral compound or may exist as an ion. Chelating precipitation agents are usually much more specific than inorganic precipitants. Insoluble chelate compounds are also less contaminated by coprecipitation or adsorption of ionic impurities than inorganic precipitates. See Chapter 14 for a more complete discussion of chelate compounds.

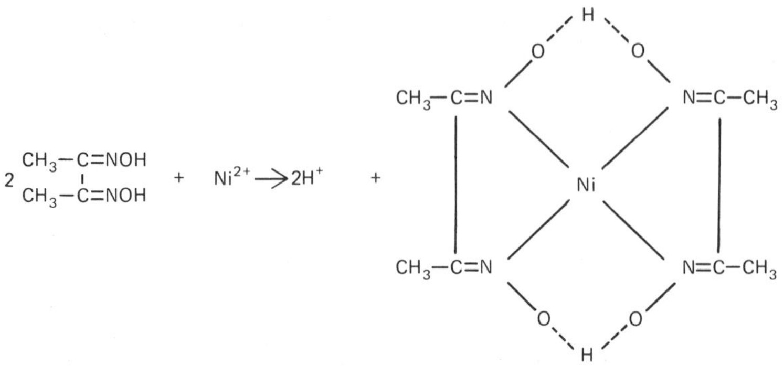

$$2\begin{array}{l}CH_3-C{=}NOH\\ \quad\;\,|\\ CH_3-C{=}NOH\end{array} + Ni^{2+} \longrightarrow 2H^+ + \begin{array}{ccccc} & O & \cdots H \cdots & O & \\ CH_3-C{=}N & & & & N{=}C-CH_3 \\ | & & Ni & & | \\ CH_3-C{=}N & & & & N{=}C-CH_3 \\ & O & \cdots H \cdots & O & \end{array}$$

**Figure 9-1.** Structural equation showing the formation of nickel dimethylglyoximate. The single line represents a single covalent bond, the double line represents a double covalent bond, and the dashed line represents a hydrogen bond.

In the determination by precipitation from homogeneous solution, the pH of the solution must be adjusted to a value just below incipient precipitation. A pH of approximately 2.6–2.8 is satisfactory. A pH greater than 2.8 will allow the precipitate to form immediately upon addition of dimethylglyoxime, whereas a pH much below 2.6 will require too long a time before precipitation begins. The dimethylglyoxime is dissolved in 1-propanol rather than ethanol to decrease loss by evaporation during the heating.

The pH is raised slowly and uniformly throughout the solution by the hydrolysis of urea (see Chapter 8). The hydrolysis is slow at room temperature but is moderate at temperatures near the boiling point of water, so the necessary pH rise occurs over a reasonable length of time.

Dimethylglyoxime is insoluble in water, so solutions in organic solvents must be used. Since the precipitate is also more soluble in organic reagents, the amount of reagent, and thus the amount of dimethylglyoxime added, are controlled so that the dimethylglyoxime itself will not precipitate; at the same time, loss of the nickel dimethylglyoxime precipitate as a result of solubility of the precipitate in the organic solvent is avoided. Use of a volatile solvent such as methanol or ethanol allows too much loss by evaporation, with consequent precipitation of the excess precipitating agent.

The gravimetric factor of 0.2032 is very favorable because it requires almost a 5-mg error in the precipitate to cause a 1-mg error in the amount of nickel.

## PRECIPITATION TITRIMETRY

Precipitation titrations are titrations in which at least one product is an insoluble solid. Although there are many reactions in which a substance is quantitatively precipitated only a few are known for which satisfactory indicators are available. Precipitation methods are used primarily in the titration of chloride with silver, although others are known and used. This section will deal only with the chloride titration methods.

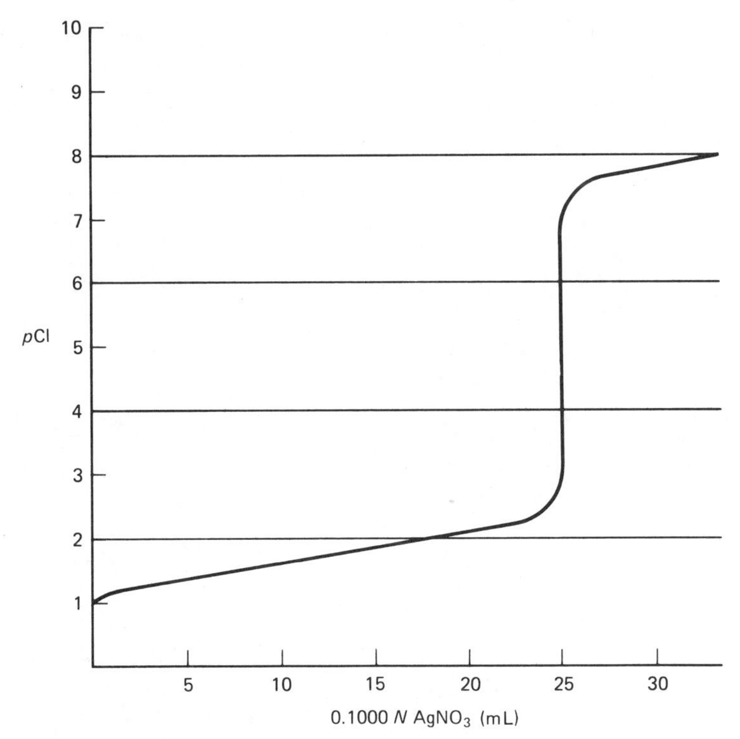


**Figure 9-2.** Titration curve for the titration of 25.00 mL of 0.1000 *N* NaCl with 0.1000 *N* $AgNO_3$; $M_{Cl^-} = 10^{-pCl}$.

There are three distinct methods for the titration of chloride with silver based upon the different types of indicators used in the methods. The three types of indicators are (1) the formation of a colored solid, as used in the Mohr method; (2) the formation of a colored soluble complex, as in the Volhard method; and (3) the change of color associated with the adsorption of the indicator on the surface of a solid particle, which is the basis of the Fajans method.

In any titration there must be a very great change in the concentration of at least one ion at the equivalence point, and to show such changes graphically, the p system of expressing concentrations is used. In this system, the molar concentration of an ion is expressed in exponential form, and the p value is the exponent with the sign changed. Thus, a molar concentration of chloride of $5.00 \times 10^{-3}$ is equal to $10^{-2.30}$, for which the pCl would be 2.30. (See Appendix 9 concerning the use of logarithms and the expression of numbers in exponential form.) A typical graph for the variation of pCl with volume of titrant is shown in Figure 9-2. It should be noted that there is a break of 4 pCl units from 3 to 7 with 2–4 drops of titrant. The molar concentration change represented by this break at the end point is from $10^{-3}$ to $10^{-7}$, which is a 10,000-fold change.

## THE MOHR METHOD

As discussed in Chapter 7, in the Mohr method silver nitrate is used to titrate the chloride solution to which potassium chromate has been added to act as the indicator. The end point is indicated by the change in color due to the formation of reddish-brown silver chromate ($Ag_2CrO_4$). As the titrant is added, the silver ion is removed by reaction with the chloride ion of the solution until the chloride has been quantitatively precipitated as AgCl. Then $Ag_2CrO_4$ begins to precipitate as the silver ion concentration increases due to the complete precipitation of chloride. The end point depends upon the relative solubility of AgCl and $Ag_2CrO_4$ or upon the fractional precipitation of chloride in the presence of chromate.

For the method to be sufficiently accurate for use in analysis, the silver chromate must not begin precipitation until the chloride has been reduced to a negligible value, as shown symbolically in eqs (9-5) and (9-6).

$$Ag^+ + \begin{matrix} Cl^- \\ CrO_4^{2-} \end{matrix} \longrightarrow AgCl \qquad \text{(until pCl is approximately 6)} \tag{9-5}$$

$$2Ag^+ + CrO_4^{2-} \longrightarrow Ag_2CrO_4 \qquad \text{(only after pCl is approximately 6)} \tag{9-6}$$

The calculations that show chloride is reduced to a neglible value are given in Example 13 of Chapter 7, and should be reviewed by the student.

A blank is necessary in this method because the $Ag_2CrO_4$ actually begins to precipitate slightly beyond the equivalence point, which occurs at pCl 5, and because some silver ion must be used to produce enough silver chromate to give a visible change. The latter error can be offset in part by using a potassium chromate indicator solution that is saturated with silver chromate.

The solution must be neutral or slightly basic (pH 7–9) because chromate reacts with hydrogen ion in acid solutions to form hydrogen chromate ions, which reduces the concentration of chromate

$$CrO_4^{2-} + H^+ \longrightarrow HCrO_4^- \tag{9-7}$$

At higher pH there are enough hydroxide ions present to cause the formation of silver hydroxide or oxide.

$$2Ag^+ + 2OH^- \longrightarrow 2AgOH \longrightarrow Ag_2O + H_2O \tag{9-8}$$

Even though chromic acid is a strong acid, the second ionization is definitely weak; $K_2 = 3.2 \times 10^{-7}$. Consequently, at pH values below 7, an appreciable fraction of the indicator is converted to the hydrogen chromate ion. Errors due to pH are offset by adjustment to the neutral point of phenolphthalein before titration. This is accomplished by adding phenolphthalein and making the solution basic with 0.5 *N* NaOH followed by addition of 1–2 drops of 0.5 *N* $H_2SO_4$ to remove the color.

Most directions also call for the volume at the end point to be approximately 100 mL in order to keep the chromate ion concentration essentially equal in all titrations.

## THE FAJANS METHOD (ADSORPTION INDICATOR METHOD)

The Fajans method is based upon the fact that some organic compounds adsorb strongly on certain precipitates and undergo a change in color upon adsorption so that they can be used as indicators for selected titrations. Fajans and his coworkers found that fluorescein and several of its derivatives are strongly adsorbed on silver chloride precipitates and, upon adsorption, change in color from yellow with a greenish fluorescence to pinkish red. Adsorption indicators can be used only for those titrations in which conditions are proper for the adsorption to occur or disappear at or near the equivalence point.

In the titration of chloride by silver, the precipitated silver chloride is formed in a solution that contains excess chloride and consequently adsorbs chloride ion in the primary layer to become negatively charged. These charged particles then attract cations in the secondary layer (see Figures 8-4, 5), which can be symbolized as $AgCl : Cl^- : Na^+$. The particles will remain negatively charged until the exact equivalence point at which they become neutral and tend to coagulate. With the first excess of silver, however, the particles adsorb silver ion strongly and becomes positively charged. The positive particles attract anions into the secondary layer, which can be symbolized as $AgCl : Ag^+ : NO_3^-$. As a consequence, there is a sharp change in charge on the particles just beyond the equivalence point. If the chloride is added to a silver solution, the particles are positively charged before the equivalence point and become negatively charged just beyond the equivalence point.

Fluorescein is a weak acid ($HC_{20}H_{11}O_5$, or HFl as a short abbreviated form), which is dissolved as the sodium salt, and exists in solution as the fluoresceinate ion $Fl^-$. The fluoresceinate ion is attracted to the positively charged silver chloride particles and repelled by the negatively charged particles. In the titration of chloride with silver, the indicator anion is repelled by the particles until just beyond the equivalence point, when the charge changes to positive. The attraction of the fluoresceinate ion into the secondary layer, which can be symbolized as $AgCl : Ag^+ : Fl^-$, causes a change in structure that produces a reddish pink color on the surface of the particles.

The color change can be explained as being due to the precipitation of silver fluoresceinate upon the surface because of the increased concentration of fluoresceinate ions in the counterion region. Alternatively, the pink color has been explained as a structural change in the fluoresceinate ion due to the energy relationships in the adsorption process on the particle surface. With either explanation the color change must be preceded by the adsorption of the indicator into the secondary layer. Since the molecular or nonionized form of fluorescein is not attracted to the particles and will not be adsorbed, the indicator must be in the ionic form to be effective.

The ratio of ionized to nonionized form for weak acids is a function of hydrogen ion concentration, and as a consequence the pH of the solution is important for the proper functioning of the indicator. For fluorescein, the solution must be essentially neutral or slightly basic (pH in the range 6–10) for sufficient ions to be present. Dichlorofluorescein, which is a stronger acid, can be used satisfactorily down to a pH of approximately 4.

Since the indicator works by adsorption, it is important that the surface area be kept as large as possible, which means that the particle size should be as small as possible. To offset coagulation of the particles before the equivalence point, a protective colloid such as dextrin is usually added to the solution being titrated. The dextrin forms a permeable coating around the particle that extends the depth of the counter ion region and retards coagulation.

Even though the end point occurs with an excess of titrant, there is no satisfactory way to determine the end-point error since there is no convenient way to run a blank determination. The end-point error is usually small and can be neglected in most cases; it can be estimated when necessary by titrating primary standard solutions of silver nitrate and sodium chloride of identical concentrations.

## ADSORPTION INDICATOR METHOD FOR SULFATE

Sulfate may also be determined by the adsorption indicator method, which usually requires less time than the gravimetric procedure. In this determination, a 1 : 1 mixture of 0.2% alizarin red S and 0.2% thorin is used as the indicator and the titration is run in a methanol–water solution using standard barium perchlorate as the titrant. The procedure gives better results with higher percentages of methanol but can be run satisfactorily with as low as 40% methanol in water. For best results the standard perchlorate should be made up with the same percentage methanol that will be used in the actual titration.

In the usual procedure, the sample is dissolved or an aliquot is pipetted and sufficient methanol is added to give the desired concentration. The solution is adjusted to a pH of 3.0–3.5 with 1 *M* perchloric acid or 1 *N* potassium hydroxide. Several drops of the mixed indicator are added, and the solution is titrated with the standard perchlorate to the end point, which is a sharp clear change from golden orange to orange-pink. Better results are obtained if approximately 90% of the titrant can be added rapidly before the addition of the indicator, and the titration finished slowly. Barium chloride may be substituted for the perchlorate, but the end points are not as sharp as with the perchlorate.

In the presence of methanol, the barium sulfate separates as a fluffy, highly adsorptive precipitate that is ideal for an adsorption indicator method because of its large surface area. The precipitation, however, suffers from the same types of cation and anion interference and contaminations as found in the gravimetric method. Good results may be obtained on highly contaminated solutions if the interferences are first separated by passage through ion-exchange columns before titration. For high ionic strength solutions, such as oil-well brines, the anions and cations may be removed by passing the acidified solution through a column of alumina, which retains the sulfate and allows other ions to pass on through. The sulfate is then eluted with ammonium hydroxide and the ammonium ions removed by passage through a cation-exchange column in the hydrogen form. Methanol is added, and the pH of the resulting sulfuric acid solution is adjusted to 3.0–3.5 with KOH before titration.

## THE VOLHARD METHOD

The Volhard method is an indirect or back-titration method in which an excess of a standard solution of silver nitrate is added to a chloride solution. The excess silver nitrate is then determined by titration with a standard solution of potassium or ammonium thiocyanate with ferric ion as indicator. The actual titration involves the reaction of silver ion with thiocyanate ion,

$$Ag^+ + SCN^- \longrightarrow AgSCN \qquad \textbf{(9-9)}$$

The first excess of thiocyanate reacts with the ferric ion to form the intensely colored $FeSCN^{2+}$ complex,

$$SCN^- + Fe^{3+} \longrightarrow FeSCN^{2+} \qquad \textbf{(9-10)}$$

The indicator is usually a concentrated [40%(w/w) or almost saturated] solution of ferric alum [ferric ammonium sulfate hydrate, $FeNH_4(SO_4)_2 \cdot 12H_2O$] in 20%(v/v) nitric acid. The nitric acid helps to bleach the color of the ferric ion and also prevents the hydrolysis of the ferric ion to form the hydrous oxide.

The standard silver solution is added to the acidified chloride solution to precipitate AgCl. The acid (usually 5 mL of $HNO_3$ for each 25 mL of chloride solution) speeds the coagulation of the particles and tends to displace adsorbed silver ions. Since AgSCN ($K_{sp}$, $1 \times 10^{-12}$) is less soluble (more insoluble) than AgCl ($K_{sp}$, $1 \times 10^{-10}$), the $SCN^-$ can react with and dissolve the precipitated AgCl, or

$$AgCl_{solid} + SCN^- \longrightarrow AgSCN_{solid} + Cl^- \qquad \textbf{(9-11)}$$

For this reason, the precipitated chloride must be removed from contact with the solution before titration with thiocyanate. The removal can be accomplished by filtration but the resultant dilution causes indistinct end points because of the large volume of solution. Also, some of the excess silver ion may not be completely washed out of the paper and would cause a negative error.

Silver chloride may also be prevented from dissolving by the addition of nitrobenzene, an insoluble organic liquid that will form a coating around the silver chloride particles and thus remove them from contact with the solution, since the coating is impermeable to water and the solution. Practically all procedures now call for the addition of nitrobenzene rather than filtration. Note that the compound used to prevent dissolution must be insoluble in water and must coat the particles with an impermeable layer to keep them from contact with the solution, even though they are still suspended in the titration vessel that contains the solution.

The Volhard method can also be used for determination of bromide and iodide without removal or protection of the precipitate, since silver bromide and silver iodide are less soluble than silver thiocyanate. In the titration of iodide, the indicator should not be added until after the precipitation of the silver iodide, to prevent oxidation of iodide ion by ferric ion. The Volhard method can also be used as a direct method for the determination of silver by titration with thiocyanate or for the determination of thiocyanate by titration with silver.

## PROBLEMS[1]

**1** A sample of rock salt weighing 3.133 g was dissolved and diluted to 500.0 mL. Calculate the percentage of sodium chloride and of chloride present by the following methods:

(a) A 50.00-mL aliquot produced a precipitate of silver chloride weighing 0.7562 g.

(b) A 25.00-mL aliquot required 27.99 mL of 0.09467 *N* silver nitrate by the Mohr method. The blank was 0.12 mL.

(c) A 40.00 mL aliquot required 20.29 mL of 0.1975 *N* silver nitrate using an adsorption indicator.

(d) The excess silver ion remaining from the addition of 20.00 mL of 0.0626 *N* silver nitrate to a 10.00-mL aliquot required 6.49 mL of 0.0306 *N* thiocyanate using the Volhard method.

*Ans.* (a) NaCl, 98.43%

**2** A sample of gypsum ore weighing 1.635 g was dissolved and divided into two equal portions. One portion produced a precipitate of $BaSO_4$ that weighed 0.9614 g, and the other portion required 23.54 mL of 0.1750 *N* barium perchlorate using an adsorption indicator. Calculate the percentage of $SO_3$, of $CaSO_4$, and of $CaSO_4 \cdot 2H_2O$.

**3** A sample of butter weighing 1.000 lb was treated with hydroxylamine and nickel chloride and produced a precipitate of nickel dimethylglyoximate weighing 0.5053 g. Calculate the percentage of biacetyl in the butter.

*Ans.* 0.033%

**4** An enriched magnetite ore sample weighing 1.056 g produced an ignited iron(III) oxide precipitate that weighed 0.6886 g. Calculate the percentage of magnetite and of iron in the ore.

[1] See also the Problems for Chapters 5 and 6.

CHAPTER

# 10 Neutralization

## ACIDS AND BASES

For many years acids and bases were defined in terms of the typical properties that are associated with them. Thus, acids are substances that turn litmus red, whereas bases turn this dye blue; acids react with bicarbonates (or carbonates) to form carbon dioxide, whereas bases react with carbon dioxide to form bicarbonates (or carbonates); acids have a sour taste, whereas bases feel slippery.

With the advent of the Arrhenius theory of ionization, the fact that acids react with metals to produce hydrogen and a salt led to the definition of acids as substances that dissociate into hydrogen ions and anions when dissolved in water, or as substances that contain "replaceable" hydrogen. Since acids react with bases to form a salt and water, bases must be substances that dissociate into hydroxyl ions and cations when dissolved in water. These definitions are not completely satisfactory, because they neglect the role played by the solvent and do not explain why solutions of salts often are acidic or basic rather than neutral.

Brønsted and Lowry independently proposed a more general definition in 1923 in which an acid is defined as any substance that can give up hydrogen ions (protons) to the solvent or to any base present and a base is any substance that can combine with hydrogen ions. Thus, an **acid** is a proton donor, whereas a **base** is a proton acceptor. However, hydrogen ions or bare protons do not exist in water solution because the hydrogen ion is hydrated or combined with a molecule of water to form the hydronium ion, $H_3O^+$. For simplicity, since it makes no major difference in most calculations or conceptions, the hydronium ion will be used only when necessary for clarity.

According to the Brønsted definition, acids and bases are related by the equation

$$\text{acid} \longrightarrow H^+ + \text{base} \quad \textbf{(10-1)}$$

Examples are

$$H_2O \longrightarrow H^+ + OH^- \quad \textbf{(10-2)}$$

$$HCl \longrightarrow H^+ + Cl^- \quad \textbf{(10-3)}$$

$$HC_2H_3O_2 \longrightarrow H^+ + C_2H_3O_2^- \quad \textbf{(10-4)}$$

$$H_3PO_4 \longrightarrow H^+ + H_2PO_4^- \quad \textbf{(10-5)}$$

$$H_2PO_4^- \longrightarrow H^+ + HPO_4^{2-} \quad \textbf{(10-6)}$$

$$NH_4^+ \longrightarrow H^+ + NH_3 \quad \textbf{(10-7)}$$

$$Al(H_2O)_6^{3+} \longrightarrow H^+ + AlOH(H_2O)_5^{2+} \quad \textbf{(10-8)}$$

These acids and bases are known as **conjugate pairs**, chloride being the conjugate base of hydrochloric acid and the ammonium ion being the conjugate acid of the base ammonia.

Acids may be classified according to charge as (1) neutral molecular acids such as acetic, (2) anion acids such as primary phosphate ($H_2PO_4^-$), and (3) cation acids such as the ammonium ion. Acids may also be classified, according to the number of replaceable hydrogens, as **monoprotic** (acetic, hydrochloric), **diprotic** (sulfuric), **triprotic** (phosphoric), and **polyprotic** (any acid with more than one replaceable hydrogen). Those substances that can act as either acid or base (primary phosphate, bicarbonate) are said to be **amphiprotic**.

To emphasize the role of the solvent, the general acid–base reaction can be written

$$\text{acid}_1 + \text{base}_2 \longrightarrow \text{acid}_2 + \text{base}_1 \quad \textbf{(10-9)}$$

and the reaction will take place quantitatively as written only if it is much easier to transfer a proton from acid$_1$ to base$_2$ than it is to transfer a proton from acid$_2$ to base$_1$. For example, the ionization of water takes place only to a limited extent, whereas the reverse of this reaction (the neutralization of a strong acid with a strong base) attains essential completion. These reactions may be shown as

$$H_2O + H_2O \longrightarrow H_3O^+ + OH^- \quad \text{(slight extent)} \quad \textbf{(10-10)}$$

$$H_3O^+ + OH^- \longrightarrow H_2O + H_2O \quad \text{(essential completion)} \quad \textbf{(10-11)}$$

Similarly, the ionization of hydrochloric acid or the dissolving of hydrogen chloride gas in water,

$$HCl + H_2O \longrightarrow H_3O^+ + Cl^- \quad \textbf{(10-12)}$$

takes place to a great extent because it is much easier to remove a proton from a chloride ion (or hydrogen chloride molecule) in water than it is to remove a proton from a hydronium ion. Another way of stating this is that the chloride ion has a much smaller affinity for a hydrogen ion than does a water molecule.

Reactions such as these are not confined only to water solutions but may occur in any solvent that has acid or basic properties. The dissolving of hydrogen chloride gas in ethyl alcohol and in glacial acetic acid is shown by

$$HCl + C_2H_5OH \longrightarrow C_2H_5OH_2^+ + Cl^- \qquad \textbf{(10-13)}$$

$$HCl + HC_2H_3O_2 \longrightarrow H_2C_2H_3O_2^+ + Cl^- \qquad \textbf{(10-14)}$$

The extent to which such reactions take place depends upon several factors, including the intrinsic strength of the acid, the intrinsic basicity of the solvent, and the dielectric constant.

The intrinsic strength of the acid is a function of its ability to react with amphiprotic solvents (ionize) to form a solvated proton. Hydrogen chloride acts as a strong acid in water whereas hydrogen acetate (acetic acid) is a weak acid. Hydrogen chloride acts as a weak acid, however, in glacial acetic acid, since it is not much more difficult to remove a proton from an acetic acid molecule than from a hydrogen chloride molecule, or, stated in another way, the acetate ion in glacial acetic acid has only a slightly greater attraction for a proton than does a chloride ion. The intrinsic basicity of the solvent has an effect because the more basic solvents promote greater ionization due to the acid–base type of reaction. Since acetic acid is a weaker base (stronger acid) than water, the ionization of hydrogen chloride is less in acetic acid than in water.

The dielectric constant also plays a part, since it is a measure of the insulating ability of the solvent. Because of the freedom of movement of the ions that is allowed by the very slight attraction and repulsion of the other ions, water has a high dielectric constant and promotes ionization. Organic solvents usually have lower dielectric constants, and as a consequence most ionic substances dissolved in organic solvents exist as ion pairs rather than as free individual ions. In glacial acid, for example, even strong acids such as perchloric acid exist primarily as the ion pair ($H_2C_2H_3O_2^+ \cdot ClO_4^-$) rather than as free $H_2C_2H_3O_2^+$ (acetonium) and $ClO_4^-$ (perchlorate) ions.

Because we are primarily concerned with aqueous solutions in analytical work, we shall confine our discussion to the strength of acids and bases in water. As previously stated, the value of the ionization constant is a measure of the strength of the acid or base, with small values indicating slight dissociation or weak acids and bases. For conjugate acid and base pairs, the strength of the base varies inversely as the strength of the acid. Thus, Brønsted classifies the chloride ion as an infinitely weak base and the acetate ion as a strong base. Similarly, the sodium ion is classified as an infinitely weak acid and the aluminum ion as a strong acid.

## ACIDITY AND BASICITY

Pure water is dissociated into hydrogen and hydroxyl ions so slightly that 1 liter of water contains only 0.0000001 g of hydrogen ion and 0.0000017 g of hydroxyl ion. As a result, the concentration of nonionized molecules of water is a constant in pure water and in the dilute solutions ordinarily used in analysis. For this reason the ionization constant of water is usually stated as the ion product constant $K_w$. At 25°C (room temperature),

$$K_w = (H^+) \times (OH^-) = 1.0 \times 10^{-14} \quad \textbf{(10-15)}$$

Because we are disregarding the effect of ionic strength and assuming that active and molar concentrations are equal, this becomes

$$[H^+] \times [OH^-] = 1.0 \times 10^{-14} \quad \textbf{(10-16)}$$

In pure water, the hydrogen ion concentration equals the hydroxyl ion concentration since they are produced in equal numbers by ionization and

$$[H^+] = [OH^-] = \sqrt{K_w} = 1.0 \times 10^{-7}\ M \quad \textbf{(10-17)}$$

Pure water is said to be neutral because there is no excess of either hydrogen or hydroxyl ions. Dissolving an acid in water produces hydrogen ions, and as a result the hydrogen ion concentration in an acid solution must be greater than it is in pure water, or greater than $10^{-7}$ $M$. Similarly, a solution of a base contains more hydroxyl ions that pure water, and the hydroxyl ion concentration is greater than $10^{-7}$ $M$. Because of the equilibrium that exists between the water molecules and the ions, an increase in the concentration of either ion causes a corresponding decrease in the other. Consequently, the hydroxyl ion concentration in an acid solution and the hydrogen ion concentration in a basic solution are both less than $10^{-7}$ $M$.

The molar hydrogen ion concentration in neutral or basic solutions is a very small number and consequently difficult and tedious to use. To simplify the expression of hydrogen ion concentrations. Sørensen suggested the use of the term pH, which he defined as the negative logarithm of the molar hydrogen ion concentration:

$$\text{pH} = -\log[H^+] \quad \textbf{(10-18)}$$

Similarly, pOH is the negative logarithm of the molar hydroxyl ion concentration and p$K$ the negative logarithm of the ionization constant. The measurement of hydrogen ion concentrations by potentiometric methods gives the hydrogen activity rather than the molar concentration; as a result, different symbolizations have been suggested to distinguish the type of concentration referred to. Some of the symbolizations found in the literature are $p_aH$ and pH for active concentrations, $p_cH$ and $p_MH$ for molar concentrations, and $p_sH$ for potentiometric measurements. The practical pH scale in use today is defined in terms of measurements against pH standards set by the National Bureau of Standards in Washington, D.C. For our purposes, we shall use the symbols $p_cH$ and $p_cOH$ to emphasize the fact that we are neglecting the effect of ionic strength on the effective concentration of an ion.

If the term *acidity* is used to denote the hydrogen ion concentration in a solution, it may be misleading because the hydrogen ion concentration and the concentration of the acid in a solution are seldom identical except in the case of strong monoprotic acids. For this reason, the acidity of a solution is usually thought to be of three types: momentary, titratable (or total), and reserve (or residual) acidity. The **momentary acidity** is the actual hydrogen ion concentration or the $p_cH$ of the solution. The **titratable acidity** is that which can be neutralized by addition of base. For strong acids the momentary and titratable acidity are essentially the same, since the acid is completely ionized; for weak acids, which exist mainly as molecules in the solution, the momentary and total acidities differ greatly. The weak acids ionize to furnish more hydrogen ions as the base neutralizes those present and consequently constitute a reservoir of hydrogen ions that are available only after ionization. This acidity, which is available only after ionization, is known as the **reserve** or **residual acidity**. The sum of the momentary and reserve acidities equals the total or titratable acidity.

## RELATIONSHIPS AMONG $p_cH$, $[H^+]$, $p_cOH$, $[OH^-]$, $K_w$ AND $pK_w$

The molar concentration and the $p_cH$ are related by the expression

$$[H^+] = 10^{-p_cH} \qquad \textbf{(10-19)}$$

or the $p_cH$ is the exponent of the hydrogen ion concentration with its sign changed. Thus, the $p_cH$ of pure water is 7, since the hydrogen ion concentration is $10^{-7}$ *M*. For acid solutions in which the hydrogen ion concentration is greater than $10^{-7}$ *M*, the $p_cH$ is a number less than 7; whereas for basic solutions in which the hydrogen ion concentration is less than $10^{-7}$ *M*, the $p_cH$ is a number greater than 7.

To convert molar hydrogen ion concentration to $p_cH$, express the concentration as 10 raised to the correct power. The $p_cH$ is this exponent with the sign changed. The calculation of the $p_cH$ of a solution in which the hydrogen ion concentration is $3.45 \times 10^{-9}$ *M* would be

$$[H^+] = 3.45 \times 10^{-9} = 10^{0.53} \times 10^{-9} = 10^{-8.47}$$
$$p_cH = 8.47$$

The reverse, or conversion of $p_cH$ to hydrogen ion concentration, is performed as follows:

$$p_cH = 5.18$$
$$[H^+] = 10^{-p_cH} = 10^{-5.18} = 10^{0.82} \times 10^{-6} = 6.61 \times 10^{-6}\ M$$

The relationship between the $p_cH$, the $p_cOH$, and $pK_w$ may be shown by converting the $K_w$ expression to negative logarithms,

$$[H^+] \times [OH^-] = K_w = 10^{-14} \qquad (10\text{-}20)$$

$$-\log[H^+] + (-\log[OH^-]) = -\log K_w = -\log 10^{-14} \qquad (10\text{-}21)$$

so that, by definition,

$$p_cH + p_cOH = pK_w = 14 \qquad (10\text{-}22)$$

and the sum of $p_cH$ and $p_cOH$ equals 14 in water and aqueous solutions.

## ACID–BASE TITRATIONS

### CONDITIONS EXISTING IN TITRATED SOLUTION AT THE EQUIVALENCE POINT

At the equivalence or stoichiometric point, equivalent quantities of the reactants have been brought into reaction so that the equivalents of base must be equal to the equivalents of acid. Another way of stating this is to say that the milliequivalents of titrant must be equal to the milliequivalents of the titrated substance at the equivalence point. Since the general reaction may be shown as

$$\text{acid} + \text{base} \longrightarrow \text{salt} + \text{water} \qquad (10\text{-}23)$$

the solution at the equivalence point is a solution of the salt formed and is identical to a solution prepared by dissolving the salt in water. As a consequence, the calculation of the $p_cH$ at the equivalence point in a titration becomes the calculation of the $p_cH$ of a solution of the salt formed.

#### *Concentration of Salt at the Equivalence Point*

The $p_cH$ at the equivalence point in many cases is affected by the concentration of the salt. The concentration of the salt can be calculated from the concentration and volume (milliequivalents) of the acid or base and the final volume of the titrated solution at the equivalence point. The normal concentration must then be converted to molar concentration before the calculation for $p_cH$ is made. As an example, calculate the molar concentration of the salt at the equivalence point in the titration of 40.00 mL of 0.1025 *N* hydrochloric acid with 0.2075 *N* barium hydroxide (1) assuming no dilution with water, (2) assuming dilution with 20.25 mL of water, and (3) assuming dilution to 100 mL with water. To solve such a problem, first determine the number of milliequivalents of the acid or of the base used, since this will equal the number of milliequivalents of salt formed:

$$\text{meq salt} = \text{mL} \times N \text{ (of either acid or base)}$$
$$= 40.00 \times 0.1025 = 4.100 \text{ meq}$$

Next, calculate the total volume of the solution if it is not given. For (1) the volume will be 40.00 mL for the acid plus 19.75 mL for the base. The milliliters of base added can be found by dividing the milliequivalents of the acid by the normality of the base (in this case, 4.100/0.2075 = 19.75). For (2) the volume will be 59.75 for the acid and base plus the 20.25 mL of water added, or 80.00 mL. The third step is to calculate the normal concentration of the salt as shown below:

$$N_{\text{salt}} = N_s = \text{meq/mL} \tag{10-24}$$

for

$$\text{(1)} \quad N_s = \frac{4.100}{59.75} = 0.0686\ N$$

$$\text{(2)} \quad N_s = \frac{4.100}{80.00} = 0.0513\ N$$

$$\text{(3)} \quad N_s = \frac{4.100}{100.0} = 0.0410\ N$$

Then convert the normal concentration to molar concentration by

$$M_{\text{salt}} = M_s = \frac{N_s}{\text{eq/mole}} \tag{10-25}$$

for

$$\text{(1)} \quad M_s = \frac{0.0686}{2} = 0.0343\ M$$

$$\text{(2)} \quad M_s = \frac{0.0513}{2} = 0.0257\ M$$

$$\text{(3)} \quad M_s = \frac{0.0410}{2} = 0.0205\ M$$

## SOLUTIONS OF SALTS

### *Hydrolysis*

When some salts are dissolved in water, the cation or the anion or both will react with water to some extent. This reaction is known as hydrolysis and is the opposite of neutralization. Since neutralization reactions go to essential completion, the reverse or hydrolysis reaction usually takes place only to a slight extent. For hydrolysis to occur, a weak acid or a weak base or both must be formed; thus, only the salts of weak acids and/or weak bases undergo hydrolysis. If acids and bases are classified as strong and weak, the four possible classifications of salts are

1 Salt of strong acid and strong base, such as NaCl.
2 Salt of strong acid and weak base, such as $NH_4Cl$.
3 Salt of weak acid and strong base, such as $NaC_2H_3O_2$.
4 Salt of weak acid and weak base, such as $NH_4C_2H_3O_2$.

Of these, the first does not undergo hydrolysis because no weak acid or weak base can be formed, and as a consequence the solution is neutral with $p_cH$ equal to 7. In the second case, the cation of the weak base reacts with water,

$$NH_4^+ + HOH \rightleftharpoons NH_4OH + H^+ \quad \textbf{(10-26)}$$

to form the weak base and hydrogen ions, so the solution is acidic with $p_cH$ less than 7. For the third class, the anion of the weak acid reacts with water,

$$C_2H_3O_2^- + HOH \rightleftharpoons HC_2H_3O_2 + OH^- \quad \textbf{(10-27)}$$

to form the weak acid and hydroxyl ions, so the solution is basic with $p_cH$ greater than 7. The salts of weak acids and weak bases react with water,

$$NH_4^+ + C_2H_3O_2^- + HOH \rightleftharpoons NH_4OH + HC_2H_3O_2 \quad \textbf{(10-28)}$$

to form both the weak base and the weak acid, and the $p_cH$ will depend upon the relative strengths of the acid and base. In most cases, the $p_cH$ will be close to 7 but not necessarily exactly 7.

The same conclusions as to acidity or basicity of salt solutions can be deduced by use of the Brønsted theory, since conjugate acids and bases have inverse strengths and since anions are bases and cations are acids. In the case of NaCl, the $Na^+$ ion is an infinitely weak acid because $OH^-$ is the strongest possible base in aqueous solution; also, the $Cl^-$ ion is an infinitely weak base because hydrochloric acid is completely dissociated in solution and $H^+$ (or $H_3O^+$) is the strongest possible acid in aqueous solutions. As a consequence, the solution is neutral, since both the acid and base are weaker than water.

In the case of $NH_4Cl$, the ammonium ion is stronger as an acid than chloride is as a base, consequently the solution is acidic; whereas for sodium acetate, the solution is basic because acetate ion is a strong base. For salts of weak acids and weak bases such as $NH_4C_2H_3O_2$, the solution will contain a strong Brønsted acid ($NH_4^+$) and a strong Brønsted base ($C_2H_3O_2^-$); thus, the solution will be close to neutral, and the hydrogen ion concentration will depend upon the relative strengths of the strong acid and base present.

## $p_cH$ *of Solutions of Salts*

**Salts of Strong Acids and Strong Bases.** As previously explained, the $p_cH$ of such solutions is 7.00 because it is the same as water. If the effect of ionic strength on $K_w$ is taken into account, the $p_cH$ will deviate slightly from 7 in solutions of high ionic strength.

**Salts of Strong Bases and Weak Acids.** Consider a solution of NaA, the sodium salt of the weak acid HA. The anion will react with water as follows:

$$A^- + HOH \rightleftharpoons OH^- + HA \tag{10-29}$$

for which

$$K_h = \frac{[OH^-][HA]}{[A^-]} \tag{10-30}$$

If this equation is multiplied by $[H^+]/[H^+]$, it becomes

$$K_h = \frac{[H^+][OH^-][HA]}{[H^+][A^-]} = \frac{K_w}{K_a} \tag{10-31}$$

Neglecting the small hydroxyl ion concentration produced from ionization of water, the hydroxyl ion concentration and the concentration of nonionized HA are equal, since they are produced in the same amounts. Also, since the amount of anion that reacts is relatively small compared to the total amount present, the concentration of $A^-$ can be considered to be equal to the molar concentration of the salt ($M_s$). Making these substitutions, eq. (10-30) may be written

$$K_h = \frac{K_w}{K_a} = \frac{[OH^-]^2}{M_s} \tag{10-32}$$

or

$$[OH^-]^2 = \frac{K_w M_s}{K_a} \tag{10-33}$$

$$[OH^-] = \sqrt{\frac{K_w M_s}{K_a}} \tag{10-34}$$

Taking negative logs, this becomes

$$-\log [OH^-] = \tfrac{1}{2}(-\log K_w - \log M_s + \log K_a)$$

or

$$p_cOH = \tfrac{1}{2}(pK_w - pK_a - \log M_s) \tag{10-35}$$

If $K_w/[H^+]$ is substituted for $[OH^-]$ in eq. (10-33), it becomes

$$\frac{K_w^2}{[H^+]^2} = \frac{K_w M_s}{K_a} \tag{10-36}$$

or

$$[H^+]^2 = \frac{K_w^2 K_a}{K_w M_s} = \frac{K_w K_a}{M_s} \tag{10-37}$$

and

$$[H^+] = \sqrt{\frac{K_w K_a}{M_s}} \tag{10-38}$$

which becomes, upon taking negative logs,

$$p_cH = \tfrac{1}{2}(pK_w + pK_a + \log M_s) \tag{10-39}$$

Any of these equations (10-34, 10-35, 10-38, or 10-39) may be used to calculate the $p_cH$ of a solution of a salt of a monoprotic weak acid, the secondary salt of a diprotic weak acid, or the tertiary salt of a triprotic weak acid.

• **Example 1.** Calculate the $p_cH$ of a 0.020 $M$ solution of sodium benzoate. Benzoic acid is monoprotic with $pK_a$ equal to 4.20.

(a) Using eq. (10-34),

$$[OH^-] = \sqrt{\frac{10^{-14} \times 2 \times 10^{-2}}{10^{-4.20}}} = \sqrt{\frac{10^{0.30} \times 10^{-16}}{10^{-4.20}}}$$

$$= \sqrt{\frac{10^{-15.70}}{10^{-4.20}}} = \sqrt{10^{-11.50}} = 10^{-5.75}\ M$$

$$p_cOH = 5.75$$

$$p_cH = 14.00 - 5.75 = 8.25$$

(b) Using eq. (10-35),

$$p_cOH = \tfrac{1}{2}[14.00 - 4.20 - (-1.70)] = \frac{11.50}{2} = 5.75$$

$$p_cH = 14.00 - 5.75 = 8.25$$

(c) Using eq. (10-38),

$$[H^+] = \sqrt{\frac{10^{-14} \times 10^{-4.20}}{10^{-1.70}}} = \sqrt{10^{-16.50}} = 10^{-8.25}$$

$$p_cH = 8.25$$

(d) Using eq. (10-39),

$$p_cH = \tfrac{1}{2}(14.00 + 4.20 - 1.70) = 8.25$$

It can be seen from Example 1 that the use of eq. (10-39) involves the fewest calculations; consequently, it will be used in the remaining problems.

• **Example 2.** Calculate the $p_cH$ at the equivalence point in the titration of 25.00 mL of 0.1000 $N$ acetic acid with 0.1000 $N$ sodium hydroxide. The $pK_a$ of acetic acid is 4.76. Assume no dilution with water. First, calculate $M_s$, using eqs. (10-24) and (10-25)

$$N_s = \frac{25.000 \times 0.1000}{50.00} = 0.0500\ N$$

$$M_s = \frac{0.0500\ N}{1} = 0.0500\ M = 10^{-1.30}$$

Then, using eq. (10-39),

$$p_cH = \tfrac{1}{2}(14.00 + 4.76 - 1.30) = \frac{17.46}{2} = 8.73$$

Salts of Weak Bases and Strong acids. The cation of the salt MA of a weak base and strong acid will react with water,

$$M^+ + HOH \rightleftharpoons H^+ + MOH \qquad \textbf{(10-40)}$$

for which $K_h$ equals

$$K_h = \frac{[H^+][MOH]}{[M^+]} \qquad \textbf{(10-41)}$$

If this equation is multiplied by $[OH^-]/[OH^-]$, it becomes

$$K_h = \frac{[H^+][OH^-][MOH]}{[M^+][OH^-]} = \frac{K_w}{K_b} \qquad \textbf{(10-42)}$$

Neglecting the small hydrogen ion concentration from the ionization of water, the hydrogen ion concentration and the concentration of nonionized molecules of the weak base MOH must be equal, since they are produced in the same amounts by the reaction of the cation with water. Also, the concentration of the cation may be assumed to be equal to the molar concentration of the salt, since the salt is completely ionic and the hydrogen ion concentration is small compared to the salt concentration. Making these substitutions, eq. (10-42) may be written as

$$K_h = \frac{K_w}{K_b} = \frac{[H^+]^2}{M_s} \qquad \textbf{(10-43)}$$

or

$$[H^+] = \sqrt{\frac{K_w M_s}{K_b}} \qquad \textbf{(10-44)}$$

and, taking negative logs,

$$-\log [H^+] = \tfrac{1}{2}[-\log K_w - (-\log K_b) + (-\log M_s)]$$
$$p_cH = \tfrac{1}{2}(pK_w - pK_b - \log M_s) \qquad \textbf{(10-45)}$$

Either eq. (10-44) or (10-45) may be used to calculate the $p_cH$ of solutions of salts of weak bases and strong acids.

• **Example 3.** Calculate the $p_cH$ of a 0.100 *M* solution of silver nitrate. The $pK_b$ of silver hydroxide is 3.96.

(a) Using eq. (10-44),

$$[H^+] = \sqrt{\frac{10^{-14} \times 10^{-1}}{10^{-3.96}}} = \sqrt{10^{-11.04}} = 10^{-5.52}$$

$$p_cH = 5.52$$

(b) Using eq. (10-45),

$$p_cH = \tfrac{1}{2}(14.00 - 3.96 + 1.00) = \frac{11.04}{2} = 5.52$$

• **Example 4.** Calculate the $p_cH$ at the equivalence point in the titration of 40.00 mL of 0.0750 *N* $NH_4OH$ with 0.1063 *N* hydrochloric acid. Assume that the solution has been diluted to 100 mL. The $pK_b$ of $NH_4OH$ is 4.74.

$$N_s = \frac{40.00 \times 0.0750}{100} = 0.0300\ N$$

$$M_s = \frac{N_s}{1} = 0.0300\ M = 10^{-1.52}\ M$$

$$p_cH = \tfrac{1}{2}(14.00 - 4.74 + 1.52) = \frac{10.78}{2} = 5.39$$

**Salts of Weak Bases and Weak Acids.** The titration of weak bases with weak acids is avoided in actual analysis because there is no sharp change in $p_cH$ at the equivalence point. The $p_cH$ of solutions of salts of weak acids and weak bases is close to 7 and may be calculated by

$$p_cH = \tfrac{1}{2}(pK_w + pK_a - pK_b) \qquad \textbf{(10-46)}$$

This equation will not be derived, but note that it indicates that the $p_cH$ is independent of the concentration of the salt.

• **Example 5.** Calculate the $p_cH$ of a 0.250 *M* solution of ammonium benzoate if $pK_a$ is 4.20 and $pK_b$ is 4.74.

$$p_cH = \tfrac{1}{2}(14.00 + 4.20 - 4.74) = \frac{13.46}{2} = 6.73$$

**Amphiprotic Salts.** Amphiprotic salts are formed in the titration of polyprotic acids with strong bases when dibasic acids are one-half neutralized and tribasic acids are one-third or two-thirds neutralized. The $p_cH$ of the equivalence point solution is controlled by the amphiprotic ion because the salt is completely ionized into the amphiprotic ion and the cation, which is an infinitely weak Brønsted acid. Considering a solution of NaHX, the amphiprotic salt of the diprotic acid $H_2X$, the $HX^-$ anion can undergo two separate reactions:

$$HX^- \rightleftharpoons H^+ + X^{2-} \qquad \textbf{(10-47)}$$

$$HX^- + H^+ \rightleftharpoons H_2X \qquad \textbf{(10-48)}$$

which may be summed to give

$$2HX^- \rightleftharpoons H_2X + X^{2-} \qquad \textbf{(10-49)}$$

The equilibrium constants for these equations are

$$K_{(10\text{-}47)} = \frac{[H^+][X^{2-}]}{[HX^-]} = K_2$$

$$K_{(10\text{-}48)} = \frac{[H_2X]}{[H^+][HX^-]} = \frac{1}{K_1}$$

$$K_{(10\text{-}49)} = \frac{[H_2X][X^{2-}]}{[HX^-]^2} = \frac{K_2}{K_1}$$

To get $[H^+]$ into the expression for $K_{(10\text{-}49)}$, we can substitute the value of $[HX^-]$ from the expression for $K_{(10\text{-}48)}$:

$$[HX^-] = \frac{K_1[H_2X]}{[H^+]} \qquad \text{or} \qquad \frac{1}{[HX^-]} = \frac{[H^+]}{[H_2X]K_1}$$

and

$$\frac{1}{[HX^-]^2} = \frac{[H^+]^2}{[H_2X]^2K_1^2}$$

to obtain

$$[H_2X][X^{2-}]\left(\frac{[H^+]^2}{[H_2X]^2K_1^2}\right) = \frac{[H^+]^2[X^{2-}][H_2X]}{[H_2X]^2K_1^2} = \frac{K_2}{K_1}$$

or

$$\frac{[H^+]^2[X^{2-}]}{[H_2X]} = \frac{K_2K_1^2}{K_1}$$

which may be simplified to

$$[H^+] = \sqrt{\frac{[H_2X]}{[X^{2-}]}}\sqrt{K_1K_2} \qquad \textbf{(10-50)}$$

Since reactions (10-47) and (10-48) usually take place to approximately the same extent, $[H_2X] \simeq [X^{2-}]$ and eq. (10.50) becomes

$$[H^+] = \sqrt{K_1K_2} \qquad \textbf{(10-51)}$$

and, upon taking negative logs,

$$p_cH = \tfrac{1}{2}(pK_1 + pK_2) \qquad \textbf{(10-52)}$$

Both eqs. (10-51) and (10-52) are applicable to solutions of primary salts of weak diprotic acids and the primary and secondary salts of weak triprotic acids. The hydrogen ion concentration is independent of the molar concentration of the salt provided the molar concentration is large enough to be greater than $3K_1$ and also greater than $2K_w/K_2$. These conditions are necessary for the concentrations of the nonionized acid and the secondary anion to be approximately equal.

• **Example 6.** Calculate the $p_cH$ of 0.068 $M$ potassium hydrogen malate. Malic acid is diprotic with $pK_1$ equal to 3.40 and $pK_2$ equal to 5.05.

To solve such a problem, first check to see if the simplified formula (10-51) can be used, that is, that $M_s$ is greater than $3K_1$ and greater than $2K_w/K_2$.

$$M_s = 0.068 = 10^{-1.17}$$

$$3K_1 = 3 \times 10^{-3.40} = 10^{-2.92}$$

$$\frac{2K_w}{K_2} = 2 \times \frac{10^{-14}}{10^{-5.05}} = \frac{10^{-13.70}}{10^{-5.05}}$$

$$= 10^{-8.65}$$

$M_s$ satisfies both necessary conditions so that eq. (10-51) or (10-52) may be used

$$p_cH = \tfrac{1}{2}(pK_1 + pK_2) = \tfrac{1}{2}(3.40 + 5.05)$$
$$= \frac{8.45}{2} = 4.23$$

## $p_c$H OF SOLUTIONS OF SALTS CONTAINING EXCESS ACID OR BASE

The presence of a salt of a strong acid or base has no effect on the $p_c$H of the solution of a strong acid or base other than the effect of ionic strength on the practical $pK_w$. The presence of a salt of a weak acid or a weak base in a solution of the acid or base will affect the $p_c$H. Since slight to moderate additions of strong acids or strong bases do not appreciably affect the $p_c$H of such solutions, they are known as buffered solutions and will be discussed in more detail later.

### *Solutions of a Weak Acid and One of Its Salts*

The hydrogen ion concentration in a solution containing a weak acid is governed by the ionization constant of the weak acid and the ratio of concentrations of the nonionized molecules and of the anion of the acid:

$$[H^+] = K_a \frac{[HA]}{[A^-]} \qquad \textbf{(10-53)}$$

Because the salt is completely ionic and the anions of the salt react with hydrogen ions to form nonionized molecules, the hydrogen ion concentration is less in such a solution than in a solution of the acid itself. In such a solution, when $pK_a$ is greater than 4, the molar concentrations of the nonionized acid and of the anion are both very large numbers compared to the hydrogen ion concentration and also very large compared to the amount of the salt that reacts with hydrogen ions to form nonionized molecules. As a result, the concentration of nonionized molecules is essentially equal to the molar concentration of the acid ($M_a$), and the anion concentration is essentially equal to the molar concentration of the salt ($M_s$). Using these terms, eq. (10-53) becomes

$$[H^+] = K_a \frac{M_a}{M_s} \qquad \textbf{(10-54)}$$

Substituting millimoles per milliliter for molar concentration and canceling the milliliters because the volume is the same for both the acid and the salt,

$$[H^+] = K_a \frac{\text{mmoles acid/mL}}{\text{mmoles salt/mL}} = K_a \frac{\text{mmoles acid}}{\text{mmoles salt}} \qquad \textbf{(10-55)}$$

Converting to negative logs, eqs. (10-54) and (10-55) become

$$\begin{aligned}-\log [H^+] &= -\log K_a + (-\log M_a) - (-\log M_s)\\ &= -\log K_a + (-\log \text{mmoles acid}) - (-\log \text{mmoles salt})\end{aligned}$$

or

$$\begin{aligned}p_cH &= pK_a - \log M_a + \log M_s\\ &= pK_a - \log \text{mmoles acid} + \log \text{mmoles salt}\end{aligned} \quad \textbf{(10-56)}$$

Either molar concentrations or millimoles of acid and salt can be used in the calculation. It is often preferable to use millimoles rather than molar concentrations because the logarithms of millimoles will be positive numbers, whereas the logarithms of the molar concentrations are usually negative numbers.

• **Example 7.** Calculate the $p_cH$ of a solution that is 0.50 $M$ in propionic acid and 0.25 $M$ in sodium propionate. The $pK_a$ for propionic acid is 4.87. First find the logs of acid and salt concentration.

For acid

$$\begin{aligned}0.50 &= 5.0 \times 10^{-1} = 10^{0.70} \times 10^{-1} = 10^{-0.30}\\ \log 0.50 &= -0.30\end{aligned}$$

For salt

$$\begin{aligned}0.25 &= 2.5 \times 10^{-1} = 10^{0.40} \times 10^{-1} = 10^{-0.60}\\ \log 0.25 &= -0.60\end{aligned}$$

Then

$$\begin{aligned}p_cH &= 4.87 - \log 0.50 + \log 0.25 = 4.87 - (-0.30) + (-0.60)\\ &= 4.87 + 0.30 - 0.60 = 4.57\end{aligned}$$

• **Example 8.** Calculate the $p_cH$ of a solution prepared by addition of 20.00 mL of 0.1000 $M$ sodium hydroxide to 30.00 mL of 0.1500 $M$ acetic acid. The $pK_a$ for acetic acid is 4.76.

(a) Calculate the millimoles of salt formed. The millimoles of salt formed equals the millimoles of base added since acetic acid and sodium hydroxide are both univalent compounds, so that

$$\text{mmoles salt} = 20.00 \times 0.1000 = 2.000$$

(b) Calculate the millimoles of acid left unreacted.

$$\begin{aligned}&\text{mmoles acid at start } (30.00 \times 0.1500) &&= 4.500\\ &\text{mmoles acid that reacted (mmoles base added)} &&= 2.000\\ &\text{mmoles acid left unreacted} &&= 2.500\end{aligned}$$

(c) Calculate $p_cH$.

$$\begin{aligned} p_cH &= pK_a - \log \text{mmoles of acid} + \log \text{mmoles of salt} \\ &= 4.76 - \log 2.50 + \log 2.00 \\ &= 4.76 - 0.40 + 0.30 = 4.66 \end{aligned}$$

• **Example 9.** Calculate the $p_cH$ of a solution that contains 2.72 g of $KH_2PO_4$ and 3.48 g of $K_2HPO_4$ in 100 mL of solution. The $pK_2$ of phosphoric acid is 7.21. MWs: $KH_2PO_4$, 136, $K_2HPO_4$, 174.

(a) Calculate mmoles of acid and salt.

$$\text{mmoles acid } (KH_2PO_4) = \frac{2.72}{0.136} = 20.0$$

$$\text{mmoles salt } (K_2HPO_4) = \frac{3.48}{0.174} = 20.0$$

(b) Calculate $p_cH$.

$$\begin{aligned} p_cH &= pK_a - \log \text{mmoles acid} + \log \text{mmoles salt} \\ &= 7.21 - \log 20.0 + \log 20.0 = 7.21 \end{aligned}$$

### *Solutions of a Weak Base and One of Its Salts*

The hydroxyl ion concentration (and thus the hydrogen ion concentration) in a solution of a weak base and one of its salts will be governed by the ionization constant ($K_b$) of the base and the ratio of concentrations of the nonionized molecules and of the cation of the base:

$$[OH^-] = \frac{K_b[MOH]}{[M^+]} \qquad \textbf{(10-57)}$$

As in solutions of weak acids and salts, the hydroxyl ion concentration in solutions of weak bases and salts is less than in the solution of the base alone. If the $pK_b$ of the base is 4 or more, the molar concentration of the nonionized molecules may be assumed to be equal to the molar concentration of the base, and the molar concentration of the cation may be assumed to be equal to the molar concentration of the salt. Making these substitutions, eq. (10-57) becomes

$$[OH^-] = \frac{K_b M_b}{M_s} \qquad \textbf{(10-58)}$$

which, upon taking negative logarithms, becomes

$$p_cOH = pK_b - \log M_b + \log M_s \qquad \textbf{(10-59)}$$

and millimoles can be substituted for molar concentration since the two compounds are contained in the same solution, so eq. (10-59) may be written

$$p_cOH = pK_b - \log \text{mmoles base} + \log \text{mmoles salt} \tag{10-60}$$

• **Example 10.** Calculate the $p_cH$ of a solution that is 0.50 $M$ in $NH_4OH$ and 0.30 $M$ in $NH_4Cl$. The $pK_b$ of $NH_4OH$ is 4.74.

$$\begin{aligned} p_cOH &= 4.74 - \log 0.50 + \log 0.30 \\ &= 4.74 - (-0.30) + (-0.52) \\ &= 4.74 + 0.30 - 0.52 = 4.52 \\ p_cH &= 14.00 - 4.52 = 9.48 \end{aligned}$$

• **Example 11.** Calculate the $p_cH$ in a solution that contains 2.00 mmoles of aconitine and 4.00 mmoles of aconitine chloride in 50.0 mL. The $pK_b$ of aconitine is 5.89.

$$\begin{aligned} p_cOH &= 5.89 - \log 2.00 + \log 4.00 \\ &= 5.89 - 0.30 + 0.60 = 6.19 \\ p_cH &= 14.00 - 6.19 = 7.81 \end{aligned}$$

• **Example 12.** How many grams of solid ammonium chloride (MW, 53.5) should be added to 100 mL of 1.00 $M$ ammonium hydroxide to produce a solution with $p_cH$ equal to 9.30? Assume no change in volume due to addition of the solid $NH_4Cl$. The $pK_b$ of $NH_4OH$ is 4.74. Since it is easier to calculate $p_cOH$ than $p_cH$ in solutions of a weak base and salt, first convert to $p_cOH$ by subtraction from 14.00:

$$p_cOH = 14.00 - p_cH = 14.00 - 9.30 = 4.70$$

Using eq. (10-58), calculate the molar concentration of the salt $M_s$.

$$[OH^-] = K_b \frac{M_b}{M_s}$$

or

$$\begin{aligned} M_s &= M_b \frac{K_b}{[OH^-]} = 1.00 \frac{10^{-4.74}}{10^{-4.70}} = 10^{-0.04} \\ &= 10^{0.96} \times 10^{-1} = 9.12 \times 10^{-1} = 0.912 \end{aligned}$$

Next, calculate the weight of $NH_4Cl$ in 100 mL of a solution of this concentration:

$$g = MW \times M \times \text{liters} = 53.5 \times 0.912 \times 0.100 = 4.88 \text{ g}$$

## TITRATION CURVES

The $p_cH$ of the titrated solution may be calculated after the addition of various amounts of titrant and the results graphed as $p_cH$ versus milliliters of titrant added. Such graphs are called **titration curves**. The actual shape of a titration curve depends upon the concentration and p$K$ values of the acid and base used and upon which is the titrant.

### *Titration Curve for Strong Acid by Strong Base*

As an example, we shall use the titration of 25.00 mL of 0.1000 $N$ HCl with 0.1000 $N$ NaOH and calculate the $p_cH$ at various points on the titration curve. In each case, we shall assume no dilution with water.

**Before Addition of Any NaOH.** The $p_cH$ will be that of 0.1 $N$ HCl, which is completely ionic and monoprotic, so that $N$ equals $M$ and the $p_cH$ at the beginning of the titration will be 1.00.

$$[H^+] = 10^{-1.00}\ M$$

$$p_cH = 1.00$$

**After Addition of 5.00 mL of NaOH.** The added base will neutralize an equal number of milliequivalents of acid to produce a solution containing unneutralized acid and NaCl. Since NaCl is the salt of a strong acid and strong base, it will have no effect on the $p_cH$, so that the calculation involves only the determination of the concentration of the unneutralized acid and the conversion to $p_cH$. In our example, milliequivalents equal millimoles since both the acid and base are uni-univalent compounds and the millimoles of base added will be equal to the millimoles of acid reacted.

mmoles acid at start $= 25.00 \times 0.1000 = 2.500$
mmoles acid reacted (mmoles base added) $= 5.00 \times 0.1000 = 0.500$
mmoles acid left unreacted $= 2.500 - 0.500 = 2.000$ mmoles

$$\text{volume solution} = 25.00 + 5.00 = 30.00 \text{ mL}$$

$$[H^+] = \frac{2.00}{30.0} = 0.067 = 10^{-1.17}\ M$$

$$p_cH = 1.17$$

### After Addition of 20.00 mL of NaOH

$$\begin{aligned}
\text{mmoles acid at start} &= 2.500\\
\text{mmoles acid reacted (base added)} &= 2.000\\
\text{mmoles acid left unreacted} &= 0.500
\end{aligned}$$

$$\text{volume solution } (25.00 + 20.00) = 45.00 \text{ mL}$$

$$[H^+] = \frac{0.500}{45.00} = 0.0111 = 10^{-1.95}$$

$$p_cH = 1.95$$

Note that the change in $p_cH$ in going from one-fifth to four-fifths neutralized is less than 1 unit.

### After Addition of 24.00 mL of NaOH

$$\begin{aligned}
\text{mmoles acid at start} &= 2.500\\
\text{mmoles acid reacted} &= 2.400\\
\text{mmoles acid left unreacted} &= 0.100
\end{aligned}$$

$$\text{volume } (25.00 + 24.00) = 49.00 \text{ mL}$$

$$[H^+] = \frac{0.100}{49.00} = 0.00204 = 10^{-2.69}$$

$$p_cH = 2.69$$

### After Addition of 24.50 mL of NaOH

$$\begin{aligned}
\text{mmoles acid at start} &= 2.500\\
\text{mmoles acid reacted} &= 2.450\\
\text{mmoles acid left unreacted} &= 0.050
\end{aligned}$$

$$\text{volume } (25.00 + 24.50) = 49.50 \text{ mL}$$

$$[H^+] = \frac{0.050}{49.50} = 0.00101 = 10^{-3.00}$$

$$p_cH = 3.00$$

After Addition of 24.90 mL (2 Drops Before Equivalence Point)

mmoles acid at start = 2.500
mmoles acid reacted = 2.490
mmoles acid left unreacted = 0.010

volume (25.00 + 24.90) = 49.90 mL

$$[H^+] = \frac{0.010}{49.90} = 0.000200 = 10^{-3.70}$$

$p_cH = 3.70$

After Addition of 24.95 mL of NaOH (1 Drop from Equivalence Point)

mmoles acid at start = 2.500
mmoles acid reacted = 2.495
mmoles acid left unreacted = 0.005

volume (25.00 + 24.95) = 49.95 mL

$$[H^+] = \frac{0.005}{49.95} = 0.000100 = 10^{-4.00}$$

$p_cH = 4.00$

After Addition of 25.00 mL of NaOH (the Equivalence Point). Since there is no acid or base left unreacted, this solution is the same as a 0.05 *M* solution of NaCl, in which the $p_cH$ is 7.00. Note that there is a jump of 3 $p_cH$ units due to addition of this 1 drop of base.

After Addition of 25.05 mL of NaOH (1 Drop Beyond Equivalence Point). Beyond the equivalence point the solution will contain sodium chloride and the excess base. As a consequence, calculate the molar hydroxyl ion concentration and $p_cOH$ instead of the molar hydrogen ion concentration.

mmoles base added (25.05 × 0.1000) = 2.505
mmoles base reacted (mmoles acid at start) = 2.500
mmoles base unreacted = 0.005

volume (25.00 + 25.05) = 50.05 mL

$$[OH^-] = \frac{0.005}{50.05} = 0.0000999 = 10^{-4.00}\ M$$

$p_cOH = 4.00 \qquad p_cH = 14.00 - 4.00 = 10.00$

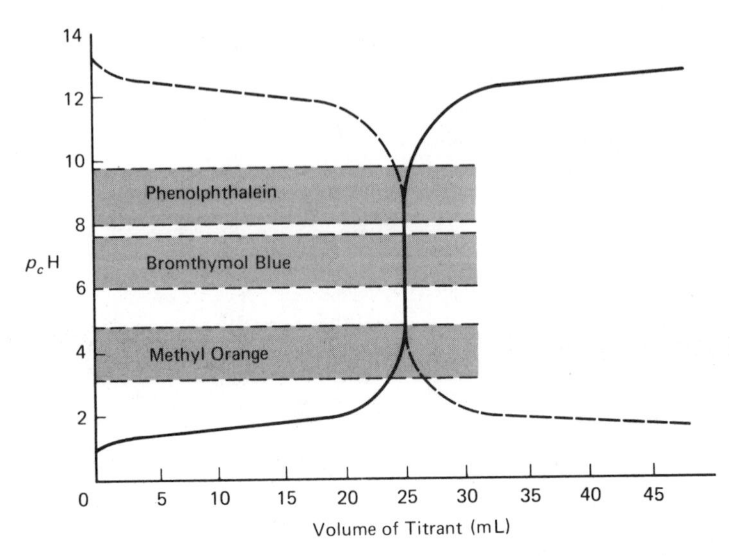


**Figure 10-1.** Titration of 25.00 mL of 0.1000 *N* HCl with 0.1000 *N* NaOH. The dashed line represents the reverse titration, in which the acid is added to the base.

This represents a change of 3 $p_cH$ units due to addition of the 1 drop of base in excess beyond the equivalence point, or a change of 6 $p_cH$ units within 1 drop on either side of the equivalence point.

**Continuing with Similar Calculations.** The $p_cH$ at 25.10 mL is 10.30, at 25.50 mL is 11.00, at 26.00 mL is 11.29, at 30.00 mL is 11.95, and at 45.00 mL is 12.46. A graph of these results is shown in Figure 10-1.

## *Titration Curve for Weak Acid by Strong Base*

As an example, we shall use the titration of 25.00 mL of 0.1000 *M* acetic acid with 0.1000 *M* NaOH and shall assume that the volumes are additive and that there has been no dilution with water. The $pK_a$ of acetic acid is 4.76.

**Before Addition of any Base.** This will be a solution of a weak acid, since $pK_a$ is greater than 4:

$$[H^+] = \sqrt{K_a M_a} \qquad \text{or} \qquad p_cH = \tfrac{1}{2}(pK_a - \log M_a)$$

$$\log M_a = \log 10^{-1.00} = -1.00$$

$$p_cH = \tfrac{1}{2}(4.76 + 1.00) = \frac{5.76}{2} = 2.88$$

**After Addition of 5.00 mL of Base.** The solution now contains unreacted acid and the salt formed by the reaction of the base, so it is necessary to calculate the millimoles of salt and acid present.

$$\begin{aligned}
&\text{mmoles acid at start } (25.00 \times 0.1000) &&= 2.500\\
&\text{mmoles salt formed (or base added) } (5.00 \times 0.1000) &&= 0.500\\
&\text{mmoles acid left unreacted} &&= 2.000
\end{aligned}$$

$$\begin{aligned}
\mathrm{p_cH} &= \mathrm{p}K_a - \log \text{mmoles acid} + \log \text{mmoles salt}\\
&= 4.76 - \log 2.0 + \log 0.5 = 4.76 - 0.30 + (-0.30) = 4.16
\end{aligned}$$

**After Addition of 12.50 mL of NaOH.** Since the acid is half neutralized, the concentration of the acid and salt are equal, so that $\mathrm{p_cH} = \mathrm{p}K_a$:

$$\mathrm{p_cH} = 4.76$$

This is an important fact to remember, that is, that

$$\mathrm{p_cH} = \mathrm{p}K_a$$

at the half-neutralization point.

**After Addition of 20.00 mL of NaOH.** This would be calculated as for the first addition of base.

$$\begin{aligned}
&\text{mmoles acid at start} &&= 2.500\\
&\text{mmoles salt formed} &&= 2.000\\
&\text{mmoles acid left unreacted} &&= 0.500
\end{aligned}$$

$$\begin{aligned}
\mathrm{p_cH} &= \mathrm{p}K_a - \log 0.5 + \log 2.0\\
&= 4.76 + 0.30 + 0.30 = 5.36
\end{aligned}$$

**After Addition of 24.00 mL of NaOH**

$$\begin{aligned}
&\text{mmoles salt formed} &&= 2.400\\
&\text{mmoles acid left unreacted} &&= 0.100
\end{aligned}$$

$$\mathrm{p_cH} = 4.76 - (-1.00) + 0.38 = 6.14$$

**After Addition of 24.90 mL of NaOH (2 Drops Before Equivalence Point)**

$$\begin{aligned}
&\text{mmoles salt formed} &&= 2.490\\
&\text{mmoles acid unreacted} &&= 0.010
\end{aligned}$$

$$\mathrm{p_cH} = 4.76 - (-2.00) + 0.40 = 7.16$$

**After Addition of 25.00 mL of NaOH (the Equivalence Point).** The solution now contains only sodium acetate, so the $p_cH$ can be calculated from eq. (10-39).

$$M_s = \frac{25.00 \times 0.1000}{25.00 + 25.00} = \frac{2.500}{50.00} = 0.05\ M = 10^{-1.30}$$

$$p_cH = \tfrac{1}{2}(pK_w + pK_a + \log M_s) = \tfrac{1}{2}(14.00 + 4.76 - 1.30)$$

$$= \frac{17.46}{2} = 8.73$$

**After Addition of 25.10 mL of NaOH (2 Drops Beyond Equivalence Point).** The solution now contains sodium acetate and excess strong base. Since the salt has no effect on the $p_cH$ in a solution of excess base, the calculation is the same as that for the titration of a strong acid by a strong base:

$$\text{mmoles base in excess} = 2.510 - 2.500 = 0.010$$

$$\text{volume} = 25.00 + 25.10 = 50.10\ \text{mL}$$

$$[OH^-] = \frac{0.010}{50.10} = 0.0001996 = 10^{-3.70}$$

$$p_cOH = 3.70$$

$$p_cH = 10.30$$

The $p_cH$ values in the titrated solution for further additions of base are identical with those calculated for the titration of a strong acid with a strong base. The titration curve for this type of titration is shown in Figure 10-2.

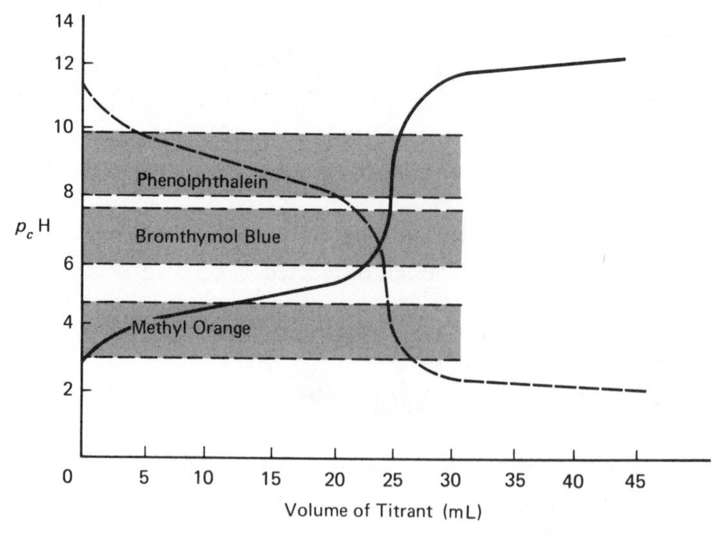


**Figure 10-2.** Titration of 25.00 mL of 0.1000 *N* $HC_2H_3O_2$ with 0.1000 *N* NaOH. The dashed line represents the titration of 25.00 mL of 0.1000 *N* $NH_4OH$ with 0.1000 *N* HCl.

### *Titration Curve for Weak Base by Strong Acid*

As our example we shall use the titration of 25.00 mL of 0.1000 $M$ $NH_4OH$ with 0.1000 $M$ HCl and assume that the volumes are additive and that there is no dilution with water. The $pK_b$ for $NH_4OH$ is 4.47.

**Before Addition of any HCl.** This will be a solution of a weak base so that

$$[OH^-] = \sqrt{K_b M_b} \qquad \text{or} \qquad p_cOH = \tfrac{1}{2}(pK_b - \log M_b)$$

$$M_b = 10^{-1.00}$$

$$p_cOH = \tfrac{1}{2}(4.74 + 1.00) = \frac{5.74}{2} = 2.87$$

$$p_cH = 11.13$$

**After Addition of 5.00 mL of HCl.** This solution will now contain unreacted $NH_4OH$ together with the $NH_4Cl$ formed by the reaction of the acid and base. The $p_cH$ may be calculated from eq. (10-60) as follows:

mmoles base at start (25.00 × 0.1000) = 2.500
mmoles salt formed (acid added) = 5.00 × 0.1000 = 0.500
mmoles base left unreacted = 2.000

$$\begin{aligned} p_cOH &= pK_b - \log \text{mmoles base} + \log \text{mmoles salt} \\ &= 4.74 - \log 2.00 + \log 0.5 \\ &= 4.74 - 0.30 + (-0.30) = 4.74 - 0.60 \\ &= 4.14 \end{aligned}$$

$$p_cH = 14.00 - p_cOH = 14.00 - 4.14 = 9.86$$

All other $p_cH$ values for additions of acid before the equivalence point is reached can be calculated in the same manner.

**After Addition of 25.00 mL of Acid (the Equivalence Point).** At the equivalence point, the solution will contain the salt with no excess of either acid or base, and the $p_cH$ can be calculated by use of eq. (10-45) as follows:

$$M_s = \frac{25.00 \times 0.1000}{25.00 + 25.00} = 0.05\ M = 10^{-1.30}$$

$$\begin{aligned} p_cH &= \tfrac{1}{2}(pK_w - pK_b - \log M_s) \\ &= \tfrac{1}{2}[14.00 - 4.74 - (-1.30)] = \frac{10.56}{2} = 5.28 \end{aligned}$$

**After Addition of 30.00 mL of HCl.** Beyond the equivalence point, the solution contains the salt and excess strong acid. Since the salt of a strong acid does not affect the $p_cH$ of solutions of that acid, the $p_cH$ is calculated from the millimoles of strong acid and volume of the solution as follows:

| | |
|---|---|
| mmoles strong acid added (30.00 × 0.1000) | = 3.000 |
| mmoles acid reacted (mmoles base at start) | = 2.500 |
| mmoles strong acid unreacted (in excess) | = 0.500 |

$$\text{volume } (25.00 + 30.00) = 55.00 \text{ mL}$$

$$[H^+] = \frac{0.500}{55.00} = 0.00909 = 10^{-2.04}$$

$$p_cH = 2.04$$

## FEASIBILITY OF TITRATIONS

For a titration to be feasible in the laboratory, there must be a sufficiently sharp break in the pH at the equivalence point and a suitable indicator must be available. The pH break depends upon the ionization constant of the weak acid or base and upon the concentration of the titrant. For most acid–base titrations, there must be a sharp break of between 1 and 2 pH units with 0.2–0.4% (usually 2–4 drops) of the volume of titrant needed to attain the equivalence point. For titrations using 0.1 *N* strong acid or base as the titrant, such a break will occur if the ionization constant is greater than $10^{-8}$. For other concentrations of titrant, the product of the concentration of titrant and ionization constant ($MK_a$) must exceed $10^{-9}$, as shown in Figure 10-3. This is only a

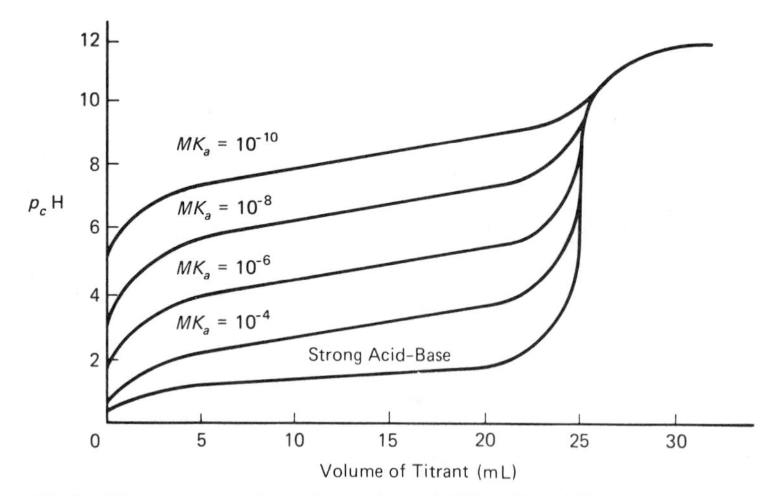


**Figure 10-3.** Titration curves for various values of $MK_a$, where $M$ is the molar concentration of the NaOH titrant and $K_a$ is the ionization constant of the weak acid.

general rule and should be applied with caution. Feasibility also depends upon the solubilities and upon the availability of a satisfactory indicator in the pH region of the equivalence point. Also, titrant concentrations less than 0.001 *N* are seldom used, even though more dilute solutions are theoretically satisfactory.

## TITRATION OF POLYPROTIC ACIDS AND MIXTURES OF ACIDS

Since polyprotic acids are neutralized by stepwise reaction with the hydrogen ions, two or more equivalence points should be possible in the titration of polyprotic acids. Thus, in the titration of maleic acid ($H_2C_4H_2O_4$), two equivalence points are theoretically possible according to the equations

$$H_2C_4H_2O_4 + NaOH \longrightarrow NaHC_4H_2O_4 + HOH \qquad \textbf{(10-61)}$$

$$NaHC_4H_2O_4 + NaOH \longrightarrow Na_2C_4H_2O_4 + HOH \qquad \textbf{(10-62)}$$

To be able to utilize such equivalence points in an actual titration there must be a sufficiently sharp break in $p_cH$ at both equivalence points. For titrations using 0.1 *N* strong base as titrant, satisfactory breaks will occur if the difference in the p$K$ values is approximately 4 or more and if both p$K$ values are 9 or less. For the case of maleic acid, for which p$K_1$ is 2.00 and p$K_2$ is 6.26, both equivalence points may be used; however, for carbonic acid, with p$K_1$ equal to 6.37 and p$K_2$ equal to 10.25, only one of the equivalence points would give a large enough break in $p_cH$, and this one is very seldom used, because the low solubility of carbonic acid in water. As shown in Figure 10-4, curve A, only two equivalence points can be utilized with phosphoric acid (p$K_1$, 2.12; p$K_2$, 7.21; p$K_3$, 12.32).

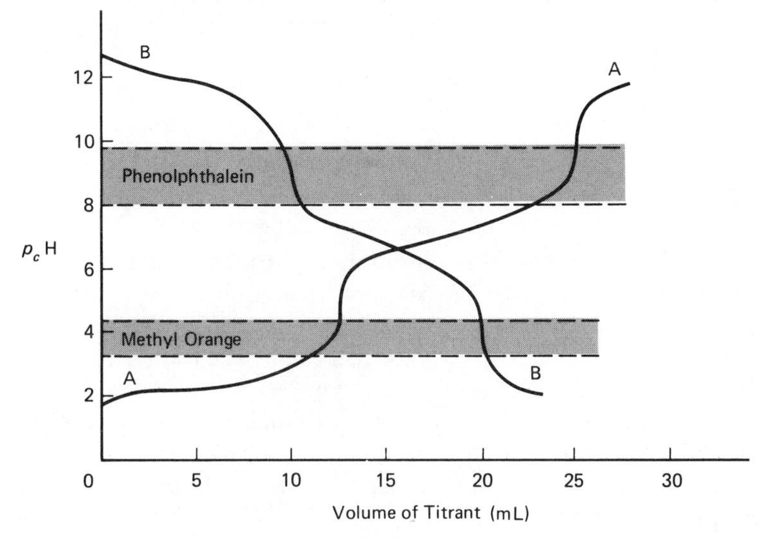


**Figure 10-4.** (A) Titration of 12.50 mL of 0.1000 *M* $H_3PO_4$ with 0.1000 *M* NaOH. (B) Titration of 10.00 mL of 0.1000 *M* $Na_2CO_3$ with 0.1000 *M* HCl.

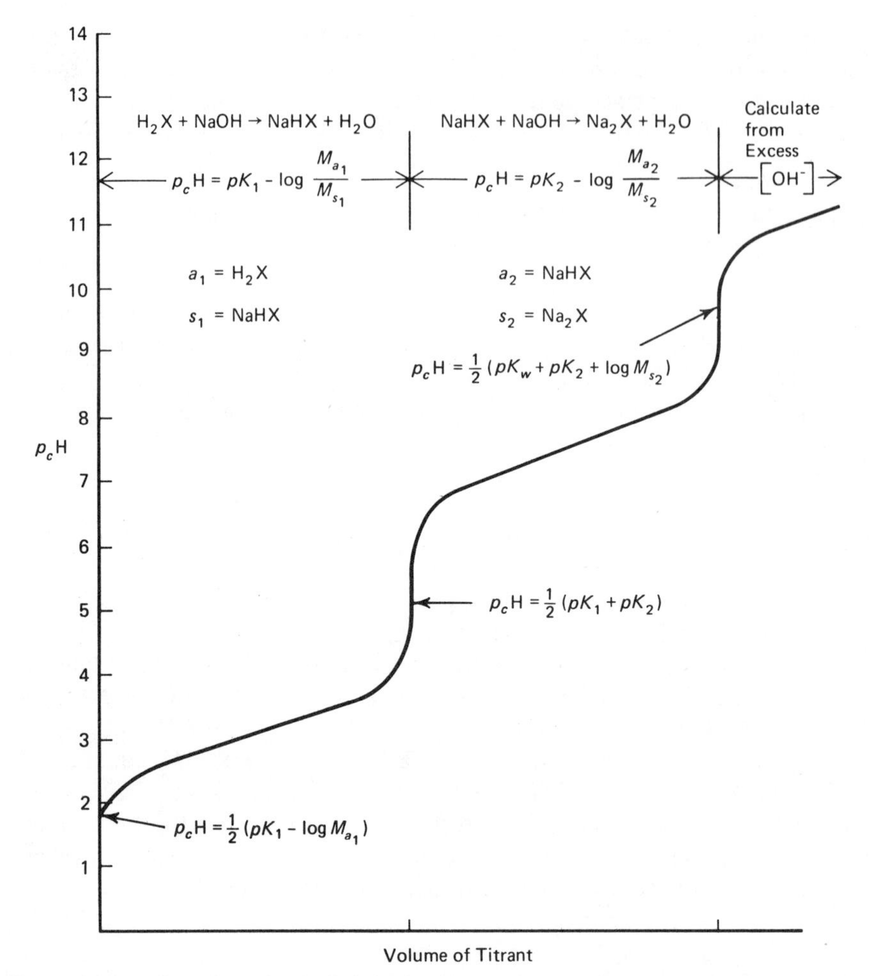


**Figure 10-5.** Summary of calculation of $p_cH$ at various points on the titration curve for the titration of a dibasic acid $H_2X$ with NaOH. $M_a$ is the analytical molar concentration of the acid, and $M_s$ the analytical molar concentration of the salt.

Similarly, only one end point is available with tartaric acid (p$K_1$, 3.02; p$K_2$, 4.54).

The same principles hold for the titration of mixtures of acids, and in general mixtures of strong and weak acids show two equivalence points, whereas mixtures of two strong acids or mixtures of two weak acids will show only one. Thus, both acetic acid and hydrochloric acid can be determined in the presence of each other but acetic and propionic cannot.

In many cases, the salts of diprotic weak acids that show only one equivalence point can be titrated with strong acid to give two equivalence points that are readily utilizable in actual titrations. Thus, $Na_2CO_3$ can be titrated with HCl according to the equations

$$Na_2CO_3 + HCl \longrightarrow NaHCO_3 + NaCl \qquad \textbf{(10-63)}$$

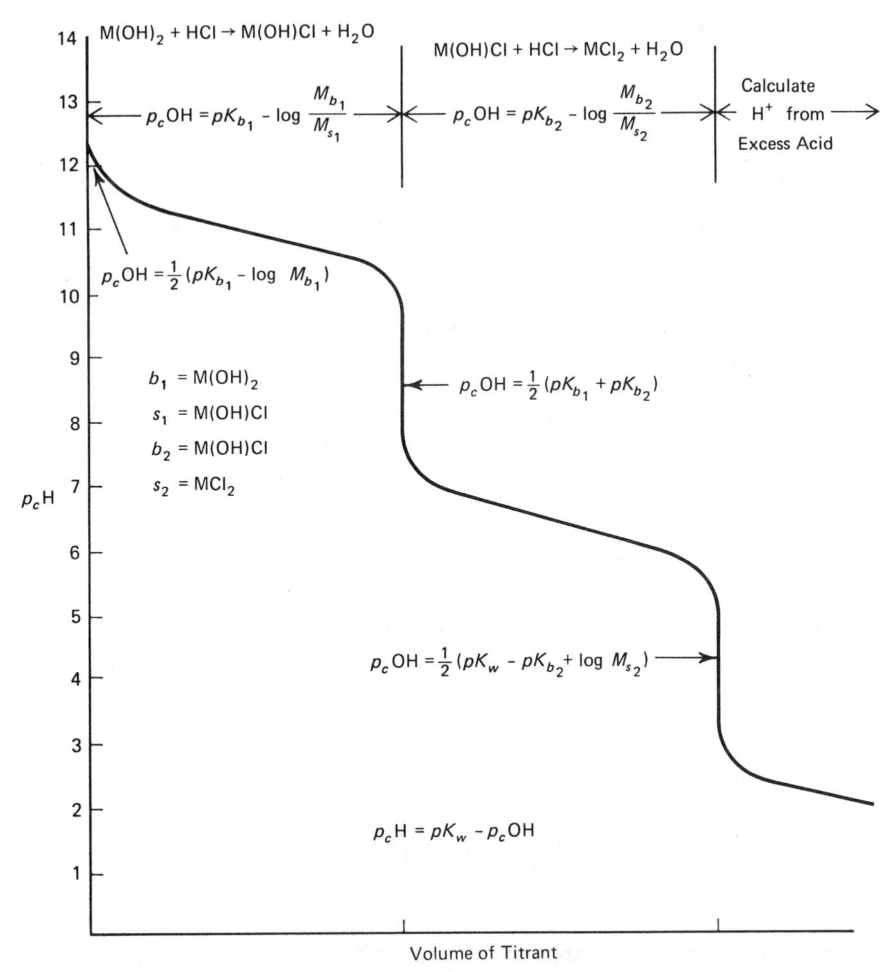


**Figure 10-6.** Summary of calculation of $p_cOH$ and $p_cH$ at various points on the titration curve for the titration of a diacid base, $M(OH)_2$, with HCl. $M_b$ is the analytical molar concentration of base, and $M_s$ the analytical molar concentration of salt.

$$NaHCO_3 + HCl \longrightarrow H_2CO_3 + NaCl \qquad \textbf{(10-64)}$$

As shown in Figure 10-4, curve B, the $p_cH$ at the equivalence point for reaction (10-63) is approximately 8, whereas that for reaction (10-64) is approximately 4. The $p_cH$ break at both equivalence points is sufficiently large that either or both may be utilized in actual titrations.

In those cases where multiple equivalence points exist and can be used, potentiometric titration methods may be utilized if satisfactory indicators are not available.

Typical titration curves for weak diabasic acids and for weak diacid bases are shown in Figures 10-5 and 10-6, respectively. These figures also show the formulas for calculation of $p_cH$ and $p_cOH$ in various regions of the titration.

## INDICATORS

Indicators for acid–base titrations are substances that change color over a definite $p_cH$ range. Most neutralization indicators are weak organic acids or weak organic bases for which the molecules and ions have different colors. One form may be colorless and the other colored, or both may be colored. The color that exists at the lower $p_cH$ is called the **acid color** and that at the upper $p_cH$ the **basic color**, irrespective of the actual hydrogen ion concentrations at which they occur. The color that is seen when both colored forms are present is called the **intermediate** or **transition color** or **virage**. Thus, for methyl orange the acid color is red and the basic color is yellow, so that, when both forms are present, the mixture of yellow and red appears to be orange.

### EXPLANATION OF INDICATOR ACTION

The reason indicators show different colors at different hydrogen ion concentrations is usually explained on the basis of the effect that variations in the hydrogen ion concentration have on the ratio of concentrations of nonionized molecules and indicator ion.

For weak acids such as phenolphthalein,

$$\mathrm{HIn} \rightleftharpoons \mathrm{H^+ + In^-} \qquad \textbf{(10-65)}$$

for which the ionization constant is

$$K_{\mathrm{In}} = \frac{[\mathrm{H^+}][\mathrm{In^-}]}{[\mathrm{HIn}]} \qquad \textbf{(10-66)}$$

the relationship may be expressed as

$$[\mathrm{H^+}] = K_{\mathrm{In}}\frac{[\mathrm{HIn}]}{[\mathrm{In^-}]} = K_{\mathrm{In}}\frac{[\text{acid form}]}{[\text{basic form}]} \qquad \textbf{(10-67)}$$

For weak bases such as methyl orange,

$$\mathrm{InOH} \rightleftharpoons \mathrm{In^+ + OH^-}$$

the ionization expression may be written

$$[\mathrm{OH^-}] = K_{\mathrm{In}}\frac{[\text{basic form}]}{[\text{acid form}]} \qquad \textbf{(10-68)}$$

Using the weak acids as an example, $K_{\mathrm{In}}$ is constant, and any variation in the hydrogen ion concentration will cause a corresponding change in the ratio $[\mathrm{HIn}]/[\mathrm{In^-}]$. At relatively large values of the hydrogen ion concentration, the indicator will exist mainly as molecules; whereas at relatively low values, the ions will predominate. Thus, for weak acids, the color of the nonionized molecules will predominate in acid solutions and the color of the ions will appear in basic solutions. It should be noted, however, that both forms must be present in all solutions that contain the indicator. For weak bases the basic form is the nonionized molecule, which predominates in basic solution.

The actual color changes of indicators are due to complex structural rearrangements that occur when the indicator reacts with an acid or base. Since most acid–base indicators are organic compounds that contain two or more benzene rings joined by one or more atoms of carbon or nitrogen, the most common change involves the interconversion of benzenoid and quinoid types of structure:

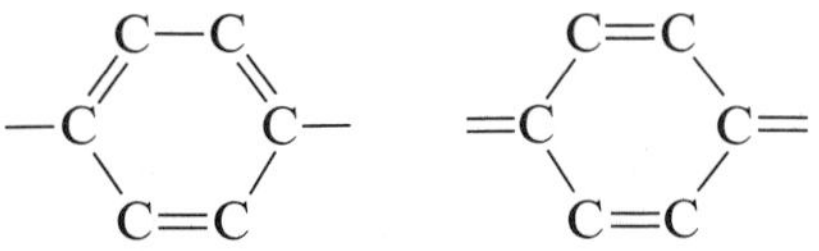

```
     C—C              C=C
    //   \\          /   \
  —C       C—     =C       C=
    \     /          \   /
     C=C              C=C
```

Other chromophoric groups are also associated with color and are involved in the color changes of some indicators.

## RANGE OF INDICATORS

If the hydrogen ion concentration equals the ionization constant of the indicator, the ratio of concentrations must be unity and the two concentrations will be equal, as shown by eq. (10-67). Under these conditions, the eye sees the mixture of colors or the intermediate color of the indicator. At any other hydrogen ion concentration the eye will see some variation of the intermediate color or will see only the acid color or only the basic color. If the two colored forms are equally easy to see, the eye can ordinarily distinguish the presence of one form with the other only if the concentration is at least one-tenth the total concentration of the indicator. Thus, the eye will see some type of intermediate virage when the concentration ratio is between 1 : 10 and 10 : 1 of the acid-to-basic form. Outside this range the eye sees only the predominant form. Writing eq. (10-67) in negative logarithmic form,

$$-\log\,[\mathrm{H}^+] = -\log K_{\mathrm{In}} + \left(-\log \frac{\text{acid form concentration}}{\text{basic form concentration}}\right) \tag{10-69}$$

$$\mathrm{p_cH} = \mathrm{p}K_{\mathrm{In}} - \log \frac{\text{acid form concentration}}{\text{basic form concentration}} \tag{10-70}$$

Substituting the ratios 1 : 10 and 10 : 1 in this equation, we obtain

$$\mathrm{p_cH} = \mathrm{p}K_{\mathrm{In}} - \log 10 = \mathrm{p}K_{\mathrm{In}} - 1$$

and

$$\mathrm{p_cH} = \mathrm{p}K_{\mathrm{In}} - \log 0.1 = \mathrm{p}K_{\mathrm{In}} - (-1) = \mathrm{p}K_{\mathrm{In}} + 1$$

**TABLE 10-1.** Acid–Base Indicators

| *Common Name* | *Chemical Name* | pH *Range* | $pK_{In}$ | *Wavelength of Maximum Absorbance* | *Color Change (Acid–Alkaline)* |
|---|---|---|---|---|---|
| Cresol red (acid range) | *o*-Cresolsulfonphthalein | 0.2–1.8 | | | Red to yellow |
| Methyl violet | | 0.5–1.5 | | | Yellow to blue |
| Thymol blue (acid range) | Thymolsulfonphthalein | 1.2–2.8 | 1.7 | 544 | Red to yellow |
| Methyl yellow | Dimethylaminoazobenzene | 2.8–4.0 | 3.3 | 508 | Red to yellow |
| Bromphenol blue | Tetrabromophenolsulfonphthalein | 3.0–4.6 | 3.8 | 592 | Yellow to blue |
| Methyl orange | Dimethylaminoazobenzene sulfonic acid | 3.1–4.4 | 3.5 | 506 | Red to orange-yellow |
| Bromocresol green | Tetrabromophenol-*m*-cresol sulfonphthalein | 4.0–5.6 | 4.7 | 614 | Yellow to blue |
| Methyl red | Dimethylaminoazobenzene-*o*-carboxylic acid | 4.2–6.2 | 5.0 | 533 | Red to yellow |
| Chlorphenol red | Dichlorosulfonphthalein | 4.8–6.4 | 6.0 | 573 | Yellow to red |
| Bromthymol blue | Dibromothymolsulfonphthalein | 6.0–7.6 | 7.1 | 617 | Yellow to blue |
| Phenol red | Phenolsulfonphthalein | 6.4–8.0 | 7.8 | 558 | Yellow to red |
| Cresol red (base range) | *o*-Cresolsulfonphthalein | 7.2–8.8 | 8.1 | 572 | Yellow to red |
| Thymol blue (base range) | Thymolsulfonphthalein | 8.0–9.6 | 8.9 | 596 | Yellow to blue |
| Phenolphthalein | | 8.0–9.8 | 9.3 | 553 | Colorless to red |
| Thymolphthalein | | 9.3–10.5 | 9.9 | 598 | Colorless to blue |
| Alizarin yellow | *p*-Nitroaniline azo sodium salicylate | 10.0–12.0 | 11.1 | | Yellow to violet |
| 2,4,6-Trinitrotoluene | | 12.0–14.0 | | | Colorless to orange |

Thus, the range of the color change of an indicator is approximately $pK_{In}$ plus or minus 1 pH unit or a total of 2 pH units. In actual practice, the range is anywhere from 1 to 3 pH units and is seldom exactly 2, as shown in Table 10-1. This table lists some selected commonly used indicators, giving the range, the acid and base colors, the $pK_{In}$ values, and the wavelength of maximum absorbance. Of these the sulfonphthaleins are the most popular, together with methyl orange, methyl red, and phenolphthalein. Phenolphthalein is used in more titrations than any other indicator, since it works well for most titrations in which sodium hydroxide is the titrant. Some dibasic indicators, such as thymol blue, have two color changes.

## DIFFERENTIAL TITRATION OF ALKALIES

Soda ash, which is often used as an inexpensive base in industry, is usually considered to be anhydrous sodium carbonate. It is seldom pure, since it is contaminated by water and often is a mixture of sodium carbonate with either sodium hydroxide or sodium bicarbonate rather than sodium carbonate alone. As a consequence, the total alkalinity is usually determined by titration with acid to the methyl orange end point, using an end-point comparison standard, or to a modified methyl orange end point. Sometimes it is necessary to know the composition as well as the total alkalinity. The composition can be determined if the sample is titrated to the phenolphthalein end point before the addition of methyl orange for the total alkalinity determination. It should be noted that the phenolphthalein end point for the titration of bases is the change from colored to colorless rather than the colorless to colored usually used. Colorless end points should be avoided when possible because the color change is usually gradual and there is no way to estimate how much the end point is overshot.

The basic reaction involved is the stepwise neutralization of sodium carbonate with hydrochloric acid shown in eqs. (10-63) and (10-64). The end point for the titration to bicarbonate occurs at approximately pH 8, at which point phenolphthalein becomes colorless, and the end point for the carbonic acid titration is near pH 4, which is in the methyl orange range. Note that for samples that contain sodium carbonate only, the volume of acid required to reach the phenolphthalein end point should exactly equal the volume required to go from the phenolphthalein end point to the methyl orange end point. If the sample contains either bicarbonate or hydroxide, the volumes required for the two end points will not be equal.

Figure 10-7 shows the titration curves for samples containing only sodium carbonate and for samples that contain sodium carbonate together with either hydroxide or bicarbonate. Upon addition of acid, any strong base present is neutralized first. After this, any sodium carbonate present is converted to sodium bicarbonate. Both these reactions are completed when the first end point is reached. Further addition of acid converts the sodium bicarbonate to carbonic acid. As a result, for mixtures of hydroxide and carbonate, the volume of acid required to reach the phenolphthalein end point will be greater than the volume required from the phenolphthalein to the methyl orange end point. For

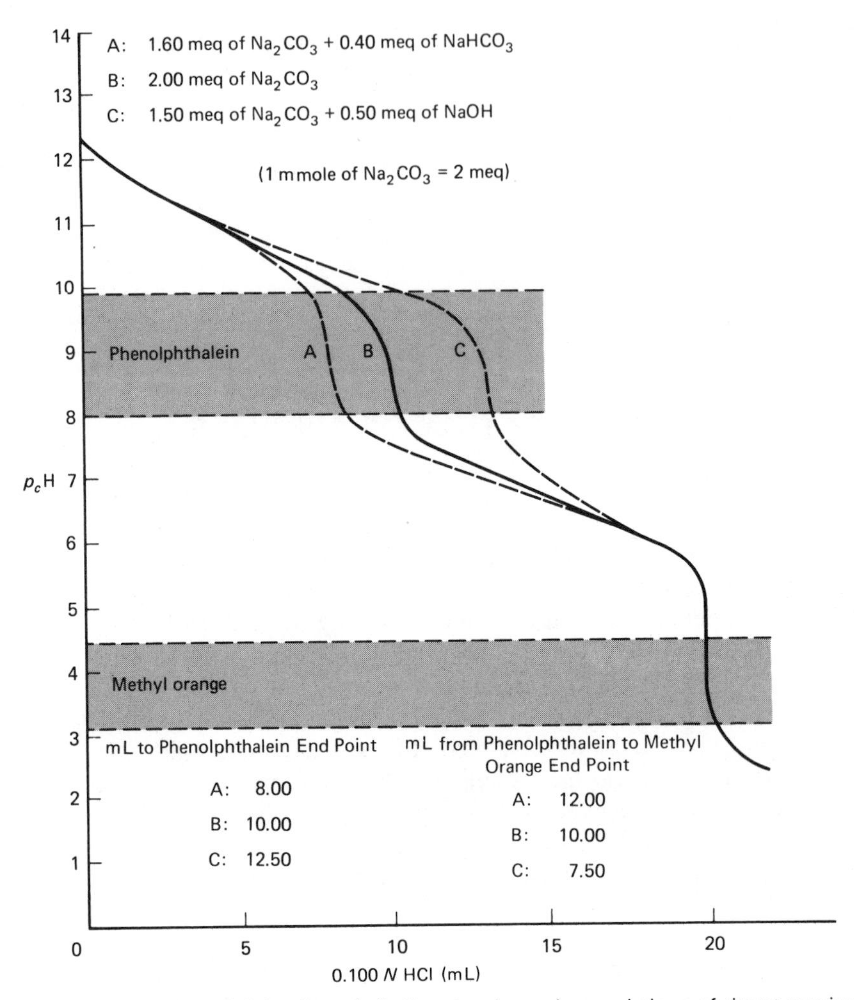


**Figure 10-7.** Differential titration of alkalies showing volume relations of titrant required to the phenolphthalein end point and from the phenolphthalein end point to the methyl orange end point for mixtures of various compositions.

mixtures of sodium carbonate and sodium bicarbonate, the volume required to go from the phenolphthalein end point to the methyl orange end point will exceed that necessary to attain the phenolphthalein end point. For samples containing only sodium carbonate, the two volumes will be equal. Another way of stating these differences is that, if more than half the total volume of acid required is utilized to reach the first end point, the sample contains sodium hydroxide and sodium carbonate; if exactly half the total volume is used in the first titration, the sample contains only sodium carbonate; if less than half the total volume is consumed in reaching the first end point, the sample contains sodium bicarbonate and sodium carbonate. Mixtures of sodium hydroxide and sodium bicarbonate are not possible because these two substances will react to form sodium carbonate and water. Examples of calculations for such titrations follow.

• **Example 13.** A 1.000-g sample of an industrial sodium carbonate alkali is dissolved and diluted to 100.0 mL. A 25.00-mL aliquot is titrated to the phenolphthalein end point and requires 24.76 mL of 0.1015 *N* hydrochloric acid. A second 25.00-mL aliquot is titrated to the methyl orange end point and requires 43.34 mL of the same acid. (a) State the components present and (b) calculate the percentage of each.

(a) Since the total volume required is 43.34 mL and the amount to the first end point is 24.76, or more than half the total, the sample contains sodium hydroxide and sodium carbonate.

(b) First calculate the volume of acid required to convert the sodium carbonate present to carbonic acid. This will be twice the volume of acid required to go from the phenolphthalein end point to the methyl orange end point:

$$\begin{aligned}\text{volume required} &= 2(43.34 - 24.76) = 2 \times 18.58\\ &= 37.16 \text{ mL}\end{aligned}$$

Then calculate the volume of acid required to neutralize the hydroxide. This will be the difference between the total volume required and the volume necessary for the sodium carbonate:

$$\text{volume required} = 43.34 - 37.16 = 6.18$$

This value may also be calculated by subtracting one-half the volume required for the sodium carbonate from the volume required in the titration to the phenolphthalein end point.

Next, calculate the percentage of sodium hydroxide (MW, 40.00) present.

$$\begin{aligned}\%\ \text{NaOH} &= \frac{\text{mL} \times N \times \text{meq wt} \times 100}{\text{wt sample}}\\ &= \frac{6.18 \times 0.1015 \times 0.04000 \times 100}{1.000 \times 25.00/100.0}\\ &= 10.04\%\end{aligned}$$

Finally calculate the percentage of $Na_2CO_3$ (MW, 106.0).

$$\begin{aligned}\%\ \text{Na}_2\text{CO}_3 &= \frac{\text{mL} \times N \times \text{meq wt} \times 100}{\text{wt sample}}\\ &= \frac{37.16 \times 0.1015 \times 0.05300 \times 100}{1.000 \times 25.00/100.0}\\ &= 79.96\%\end{aligned}$$

The percentage of sodium carbonate can also be calculated from the volume required to go from the first to the second end point as follows:

$$\% \, Na_2CO_3 = \frac{18.58 \times 0.1015 \times 0.1060 \times 100}{1.000 \times 25.00/100.0}$$

$$= 79.96\%$$

Note that in the second method of calculation of sodium carbonate that the volume of acid is one-half the volume for the first calculation and that the equivalent weight is twice the value used in the first calculation.

• **Example 14.** An alkaline unknown contains inert material plus any one or any combination of sodium carbonate, sodium bicarbonate, and sodium hydroxide. A sample weighing 1.023 g was dissolved, phenolphthalein was added, and the sample titrated to loss of color with 0.1008 *N* HCl. Methyl orange was then added, and the sample was titrated with the same acid to the methyl orange point. The first titration required 30.69 mL and the second required an additional 40.92 mL. (a) State the components present and (b) calculate the percentage of each in the sample.

(a) Since a larger volume of acid is required to go from the phenolphthalein end point to the methyl orange end point than was required to attain the phenolphthalein end point, the sample contains sodium carbonate and sodium bicarbonate.

(b) First calculate the percentage of sodium carbonate from the data for the phenolphthalein end point. Since this represents the conversion of sodium carbonate (MW, 106.0) to sodium bicarbonate (MW, 84.01), the equivalent weight of the sodium carbonate will be the same as the molecular weight.

$$\% \, Na_2CO_3 = \frac{\text{mL} \times N \times \text{meq wt} \times 100}{\text{wt sample}}$$

$$= \frac{30.69 \times 0.1008 \times 0.1060 \times 100}{1.023}$$

$$= 32.05\%$$

Next, determine the volume of acid required to react with the sodium bicarbonate present by subtracting the volume to the phenolphthalein end point from that required for the second titration:

$$\text{mL for } NaHCO_3 = 40.92 - 30.69 = 10.23$$

This value can also be calculated by substituting twice the volume to the phenolphthalein end point from the total volume required for both titrations:

$$\text{mL for } NaHCO_3 = 71.61 - 61.38 = 10.23$$

Finally, calculate the percentage of $NaHCO_3$ present.

$$\% \ NaHCO_3 = \frac{10.23 \times 0.1008 \times 0.08401 \times 100}{1.023}$$

$$= 8.47\%$$

• **Example 15.** A 2.000-g sample is known to contain inert material plus any one or any combination of sodium carbonate, sodium bicarbonate, and sodium hydroxide. The sample was dissolved and titrated in the cold with 0.5000 *N* HCl. When phenolphthalein was the indicator, the end point was reached with 24.00 mL of the acid. Methyl orange was then added and a total of 48.00 mL of the acid was required to attain the end point. (a) State the components present and (b) calculate the percentage of each.

(a) The volume of acid required to reach the phenolphthalein end point (24.00 mL) is exactly half the total volume (48.00 mL) required for the complete titration. As a consequence, sodium carbonate is the only basic component present.

(b) The percentage of sodium carbonate (MW, 106.0) may be calculated from either titration. Using the phenolphthalein end point,

$$\% \ Na_2CO_3 = \frac{24.00 \times 0.5000 \times 0.1060 \times 100}{2.000}$$

$$= 63.60\%$$

Using the methyl orange end point or the total titration,

$$\% \ Na_2CO_3 = \frac{48.00 \times 0.5000 \times 0.05300 \times 100}{2.000}$$

$$= 63.60\%$$

## BUFFERS

A solution that resists a change in the $p_cH$ of the solution upon addition of slight to moderate amounts of acid or base is known as a **buffered solution**. Or, stated in another way, a solution is considered to be buffered if the addition of slight to moderate amounts of strong acid or strong base will not cause an appreciable change in the $p_cH$.

There are several terms that are often used interchangeably but actually have slightly different meanings. Some of these terms are

**Buffer:** A buffer is a compound or a mixture of compounds that has the ability to maintain the $p_cH$ of a solution between narrow limits.

**Buffer system:** A buffer composed of two or more compounds is usually referred to as a buffer system.

**Buffer action:** The actual reaction by which a buffer system maintains the $p_cH$ upon addition of acid or base is known as the buffer action.

**Buffering solution:** A buffering solution is a concentrated solution of a buffer system that may be added in small amounts to another solution to cause it to be buffered.

## TYPES OF BUFFERS

There are two types of buffers that can cause a solution to be buffered. One of these is a strong acid or strong base at a concentration of 0.1 *N* or more. In this case, the solution is buffered simply because it contains relatively large amounts of strong acid or strong base. The second type of buffer is a mixture of a weak acid or a weak base together with a salt of the acid or base.

## EXPLANATION OF BUFFER ACTION

When 1 mmole of a strong acid is added to 100 mL of water, the hydrogen ion concentration is $10^{-2}$ *M* and the $p_cH$ is 2.00. For water, the addition of the 1 mmole to 100 mL causes a change of 5.00 $p_cH$ units, and the same type of change (from 7.00 to 12.00) would be expected if 1 mmole of strong base were added to water.

A 0.1 *N* solution of a strong acid contains 10.0 mmoles of hydrogen ion per 100 mL of solution and has a $p_cH$ of 1.00. If 1 mmole of hydrogen ion is added without appreciable change of volume, the concentration of the hydrogen ion is increased to 11.0 mmoles/100 mL and the $p_cH$ is 0.96. Similarly, if 1 mmole of strong base is added without appreciable change in volume, the solution would contain 9.00 mmoles of hydrogen ion in 100 mL and the $p_cH$ would be 1.05. Thus, the addition of slight to moderate amounts of strong acid or strong base does not cause an appreciable change in the $p_cH$ ,and the solution is considered to be buffered. The same situation exists in a 0.1 *N* solution of a strong base.

If 1 mmole of hydrogen ion is added without appreciable change of volume to 100 mL of 0.00001 *N* strong acid, the $p_cH$ changes from 5.00 to 2.00, and if 1 mmole of a strong base had been added instead, the $p_cH$ would change from 5.00 to 12.00. Thus, dilute solutions of strong acids or bases do not furnish enough hydrogen or hydroxyl ions to buffer a solution. If a buffered solution with $p_cH$ between 1.00 and 13.00 is needed, simple solution of strong acids or strong bases cannot be used. In this region, buffer mixtures of a weak acid and one of its salts or a weak base and one of its salts must be used to obtain a buffered solution.

If 1 mmole of strong acid is added to 100 mL of a solution that is 0.100 *M* in acetic acid and 0.100 *M* in sodium acetate, the $p_cH$ changes from 4.74 to 4.66; the addition of 1 mmole of strong base causes the $p_cH$ to change from 4.74 to 4.83. Evidently in solutions that contain a weak acid and its salt, some mechanism exists that tends to neutralize the effect of the addition of strong acid or strong base. This action can be explained if we consider that the weak acid is present primarily as molecules, whereas the salt is completely ionic. Thus, the solution contains a relatively large number of nonionized molecules and a relatively large number of a common ion.

The ionization of a weak acid may be shown as

$$HA \rightleftharpoons H^+ + A^- \quad \textbf{(10-71)}$$

for which the ionization constant is

$$K_a = \frac{[H^+][A^-]}{[HA]} \quad \textbf{(10-72)}$$

which can be written

$$[H^+] = \frac{K_a[HA]}{[A^-]} = K_a \frac{[HA]}{[A^-]} \quad \textbf{(10-73)}$$

For weak acids it can be assumed that the acid is primarily present in the molecular form and that the salt is completely ionic so that the [HA] may be considered to be equal approximately to the molar concentration (analytical concentration, $M_a$) of the acid and $[A^-]$ to the molar concentration (analytical concentration, $M_s$) of the salt. Substituting, eq. (10-73) becomes

$$[H^+] = K_a \frac{M_a}{M_s} = K_a \frac{\text{mmoles acid}}{\text{mmoles salt}} \quad \textbf{(10-74)}$$

Since the ionization of a weak acid is an equilibrium, addition of strong acid or strong base simply shifts the equilibrium. A strong acid will react with the salt to form more weak acid molecules, whereas the strong base will react with the molecular acid to form more anions. Thus,

$$\text{strong acid + anions} \longrightarrow \text{molecules weak acid} \quad \textbf{(10-75)}$$

and

$$\text{strong base + molecules weak acid} \longrightarrow \text{anions} \quad \textbf{(10-76)}$$

Applying this reasoning to the example of acetic acid and sodium acetate given, addition of strong acid converts acetate ion to acetic acid molecules, whereas addition of strong base changes acetic acid molecules to acetate ions. The calculations necessary to illustrate the $p_cH$ changes may be shown by use of eq. (10-74) using $1.8 \times 10^{-5}$ as the $K_a$ of acetic acid.

$$[H^+] = 1.8 \times 10^{-5} \times \frac{0.10}{0.10} = 1.8 \times 10^{-5}$$

$$= 10^{0.26} \times 10^{-5} = 10^{-4.74}$$

$$p_cH = 4.74$$

The addition of 1.0 mmole of strong acid such as hydrochloric acid would convert 1.0 mmole of acetate ion to 1.0 mmole of acetic acid and form 1 mmole of sodium chloride. Since there are 10.0 mmoles of each originally present ($100 \times 0.10$), after the addition there will be 11.0 mmoles of acetic acid and 9.0 mmoles of acetate ion, or the solution will be 0.11 *M* in acetic acid, 0.09 *M* in sodium acetate, and 0.01 *M* in sodium chloride. Applying these values in eq. (10-74),

$$[H^+] = 1.8 \times 10^{-5} \times \frac{0.11}{0.09} = 1.8 \times 10^{-5} \times 1.22$$

$$= 2.20 \times 10^{-5} = 10^{0.34} \times 10^{-5} = 10^{-4.66}$$

$$p_cH = 4.66$$

The addition of 1.0 mmole of a strong base such as sodium hydroxide would convert 1.0 mmole of acetic acid to 1.0 mmole of sodium acetate and 1.0 mmole of water, so that the concentration of acetic acid would be 0.09 *M* and that of sodium acetate would be 0.11 *M*. Substituting these values,

$$[H^+] = 1.8 \times 10^{-5} \times \frac{0.09}{0.11} = 1.8 \times 10^{-5} \times 0.82$$

$$= 1.48 \times 10^{-5} = 10^{0.17} \times 10^{-5} = 10^{-4.83}$$

$$p_cH = 4.83$$

The necessary conditions for a buffer, then, are either the presence of a large concentration of hydrogen or hydroxyl ions or the presence of large concentrations of a slightly ionized material and a common ion.

## LEVEL AND RANGE OF BUFFERS

The level of a buffer is the actual $p_cH$ of the buffered solution containing the buffer, and the range is the $p_cH$ region in which it is effective as a buffer.

The level of a buffer depends upon the ionization constant ($K_a$) of the weak acid or base of the buffer system and upon the ratio of concentrations of the slightly ionized material and salt, since it may be calculated using eq. (10-74) or eqs. (10-54) through (10-59).

Inspection of Figure 10-2 for the titration of acetic acid with sodium hydroxide shows that the $p_cH$ rises fairly rapidly at the beginning of the titration but then increases rather slowly until just preceding the equivalence point. This region of slow rise of $p_cH$ is known as the **buffer region**. The rise covers approximately two pH units from the $p_cH$ at 10% neutralization to the $p_cH$ at 90% neutralization. This can be shown using eq. (10-56) and calculating the $p_cH$ at these two partial neutralization values.

$$p_cH = pK_a - \log \frac{M_a}{M_s}$$

For 10% neutralization, $M_a = 0.90$ and $M_s = 0.10$, so

$$\mathrm{p_cH} = \mathrm{p}K_a - \log \frac{0.90}{0.10} = \mathrm{p}K_a - 0.95$$

For 90% neutralization, $M_a = 0.10$ and $M_s = 0.90$, so

$$\begin{aligned} \mathrm{p_cH} &= \mathrm{p}K_a - \log \frac{0.10}{0.90} = \mathrm{p}K_a - (-0.95) \\ &= \mathrm{p}K_a + 0.95 \end{aligned}$$

Consequently, the range of a buffer system is approximately $\mathrm{p}K_a \pm 1.0$ pH unit, or a total of two pH units, centered at the value of $\mathrm{p}K_a$.

## CAPACITY OF BUFFERS

The capacity of a buffer is a measure of the ability of the buffer system to neutralize added acid or base without a major change in $\mathrm{p_cH}$. To make the definition more quantitative, the capacity is often defined as the number of millimoles of strong acid required per milliliter to cause a unit change in $\mathrm{p_cH}$. The capacity is also defined in terms of capacity toward added base and called the **beta value** ($\beta$). The beta value is the number of millimoles of strong base required per milliliter to cause a unit rise in the $\mathrm{p_cH}$.

The capacity of a buffer depends upon the ratio of concentrations and upon the total concentration. For any given ratio, the larger the total concentration, the greater the capacity.

• **Example 16.** Calculate (a) the $\mathrm{p_cH}$ and (b) the capacity toward added acid for a solution that contains 12.0 mmoles of acetic acid ($\mathrm{p}K_a$, 4.74) and 8.0 mmoles of sodium acetate in 100 mL.

(a) Using eq. (10-56),

$$\begin{aligned} \mathrm{p_cH} &= \mathrm{p}K_a - \log \text{mmoles acid} + \log \text{mmoles salt} \\ &= 4.74 - \log 12.0 + \log 8.0 \\ &= 4.74 - 1.08 + 0.90 = 4.56 \end{aligned}$$

(b) To determine the capacity, first determine the ratio of millimoles of acid to salt at a $\mathrm{p_cH}$ 1 unit less than the $\mathrm{p_cH}$ of the solution

$$\text{new } \mathrm{p_cH} = 4.56 - 1.00 = 3.56$$

Using eq. (10-55),

$$[\mathrm{H^+}] = K_a \frac{\text{mmoles acid}}{\text{mmoles salt}}$$

rearrange to calculate the ratio of mmoles after addition of sufficient acid

$$\frac{\text{mmoles acid}}{\text{mmoles salt}} = \frac{[H^+]}{K_a} = \frac{10^{-3.56}}{10^{-4.74}} = 10^{1.18} = 15.1$$

or

$$\text{mmoles acid} = 15.1 \times \text{mmoles salt}$$

This value can then be substituted in the expression for the total number of millimoles present to calculate the millimoles of salt after the addition of sufficient acid to lower the $p_cH$ 1 unit:

$$\text{total mmoles} = \text{mmoles acid} + \text{mmoles salt} = 20.0$$

$$15.1 \times \text{mmoles salt} + \text{mmoles salt} = 20.0$$

$$16.1 \times \text{mmoles salt} = 20.0 \qquad \text{or} \qquad \text{mmoles salt} = \frac{20.0}{16.1}$$

$$\text{mmoles salt left after addition of sufficient acid} = 1.24$$

Since there were 8.0 mmoles of salt in the buffered solution before the addition of any acid, and the added acid reacted with the salt to form molecular acetic acid, the difference between the millimoles in the original solution and those left after the addition of the acid will be a measure of the millimoles of acid added.

$$\begin{aligned}\text{mmoles of acid added} &= \text{mmoles salt at start} - \text{mmoles salt left}\\ &= 8.0 - 1.2 = 6.8\end{aligned}$$

The capacity is defined as the number of millimoles of hydrogen ion per milliliter required to cause a unit change in $p_cH$, and the solution involved here contains 100 mL and required 6.8 mmoles of hydrogen ion, so that the capacity is

$$\text{capacity} = \frac{6.8}{100} = 0.068 \text{ mmole/mL}$$

Note that for a solution that contained ten times the concentration shown in this problem, the capacity would be greater. Since the total millimoles would equal 200, the difference in millimoles before and after would be 68 rather than 6.8, and the capacity thus would be 0.68 mmole/mL, or ten times higher.

• **Example 17.** Calculate (a) the $p_cH$ and (b) the beta value of a solution that contains 2.5 mmoles of $NaH_2PO_4$ and 3.5 mmoles of $Na_2HPO_4$ in 100 mL. The $pK_2$ of phosphoric acid is 7.21.

(a) Calculate the $p_cH$ using eq. (10-56).

$$p_cH = pK_2 - \log \text{mmoles acid} + \log \text{mmoles salt}$$

In this case, the $NaH_2PO_4$ is the acid and the $Na_2HPO_4$ is the salt, so that

$$\begin{aligned} p_cH &= 7.21 - \log 2.5 + \log 3.5 \\ &= 7.21 - 0.40 + 0.54 = 7.35 \end{aligned}$$

(b) To determine the beta value, first determine the ratio of millimoles of acid to millimoles of salt at a $p_cH$ 1 unit higher than the $p_cH$ of the solution.

$$\text{new } p_cH = 7.35 + 1.00 = 8.35$$

Using eq. (10-55) rearranged as in Example 16,

$$\begin{aligned} \frac{\text{mmoles acid}}{\text{mmoles salt}} &= \frac{[H^+]}{K_a} = \frac{10^{-8.35}}{10^{-7.21}} = 10^{-1.14} \\ &= 10^{0.86} \times 10^{-2} = 7.3 \times 10^{-2} \\ &= 0.073 \end{aligned}$$

so that

$$\text{mmoles acid} = 0.073 \times \text{mmoles salt}$$

The total number of millimoles of acid and salt is 6.0:

$$\text{mmoles acid} + \text{mmoles salt} = 6.0$$

Substituting 0.073 mmole of salt for millimoles of acid,

$$\begin{aligned} 0.073 \text{ mmole salt} + \text{mmoles salt} &= 6.0 \\ 1.073 \text{ mmoles salt} &= 6.0 \\ \text{mmoles salt} &= \frac{6.0}{1.073} = 5.6 \end{aligned}$$

The change in millimoles of salt will equal the amount of base added since 1 mmole of NaOH reacts with 1 mmole of acid to form 1 mmole of salt. Accordingly,

$$\text{increase in mmoles salt} = 5.6 - 3.5 = 2.1$$

or 2.1 mmoles of NaOH would be required per 100 mL to cause a 1-unit increase in $p_cH$. The beta value is, then,

$$\text{beta value} = \frac{2.1 \text{ mmoles}}{100 \text{ mL}} = 0.021 \text{ mmole/mL}$$

Each of the solutions in Examples 16 and 17 will show different capacities to added acid and added base since the number of millimoles of acid is not equal to the millimoles of salt. The maximum capacity to either added acid or base for any given total concentration occurs when the concentrations of the salt and acid are the same. Solutions that contain more acid than salt will have a higher capacity to added base than to added acid, whereas solutions containing more salt than acid will have a higher capacity to added acid than to added base.

## PREPARATION OF BUFFER SOLUTIONS OF DEFINITE $p_cH$ AND DEFINITE TOTAL MOLAR CONCENTRATION

It is often necessary to prepare a buffer of known $p_cH$ for use in a given determination. Since the capacity as well as the $p_cH$ is usually important in any procedure, the total molar concentration must also be specified so that the buffer system will have sufficient capacity to keep the $p_cH$ within narrow limits during the determination. For optimum capacity to either acid or base, the p$K$ of the weak acid or base must be close to the $p_cH$ desired so that the molar ratio of acid or base to salt will be close to unity. The acid or base and salt selected must also be sufficiently soluble to prepare a solution with the desired capacity. The calculations necessary for the preparation of a buffer of definite $p_cH$ and definite total molar concentration include the calculation of the molar ratio of the acid or base and the salt followed by use of this ratio and the total molar concentration to determine the actual concentration of each.

• **Example 18.** Calculate (a) the molar concentration of acid and salt, and (b) the volumes of 1.00 $M$ solutions of each that must be mixed and diluted to prepare 500 mL of a solution of sodium acetate and acetic acid that has a $p_cH$ of 4.65 and a total molar concentration of 0.500 $M$. The $pK_a$ of acetic acid is 4.74.

First, calculate the molar ratio of acid and salt using eq. (10-74) rearranged as follows:

$$\frac{[H^+]}{K_a} = \frac{\text{mmoles acid}}{\text{mmoles salt}} = \frac{10^{-4.65}}{10^{-4.74}} = 10^{+0.09}$$

$$\text{mmoles acid} = 1.23 \text{ mmoles salt}$$

The total number of millimoles equals 500 × 0.500 or 250, so

$$\text{mmoles acid} + \text{mmoles salt} = 250$$

Substituting the value for millimoles of acid in this expression gives

$$1.23 \text{ mmoles salt} + \text{mmoles salt} = 250$$

$$2.23 \text{ mmoles salt} = 250$$

$$\text{mmoles salt} = \frac{250}{2.23} = 112.1 = 112$$

$$\text{mmoles acid} = 250 - 112 = 138$$

Next, calculate the volumes of 1.00 $M$ solutions of acid and base that contain these numbers of millimoles.

$$\text{for the acid:} \quad \text{mL} = \frac{\text{mmoles}}{M} = \frac{138}{1.00} = 138 \text{ mL}$$

$$\text{for the salt:} \quad \text{mL} = \frac{\text{mmoles}}{M} = \frac{112}{1.00} = 112 \text{ mL}$$

To prepare the solution, mix 138 mL of 1.00 $M$ acetic acid with 112 mL of 1.00 $M$ sodium acetate and dilute to 500 mL.

• **Example 19.** How many grams of ammonium chloride (MW, 53.5) and how many milliliters of concentrated ammonium hydroxide (16.0 $M$) should be dissolved, mixed, and diluted to 100 mL to prepare a solution with a $p_cH$ of 9.00 and a total molar concentration of 5.00? The $pK_b$ of $NH_4OH$ is 4.76. Since the slightly ionized material is a base, it is preferable to work with $p_cOH$ rather than $p_cH$, so first convert $p_cH$ to $p_cOH$.

$$p_cOH = 14.00 - p_cH = 14.00 - 9.00 = 5.00$$

Next, determine the ratio of millimoles of base and salt using eq. (10-58).

$$\frac{[OH^-]}{K_b} = \frac{\text{mmoles base}}{\text{mmoles salt}} = \frac{10^{-5.00}}{10^{-4.76}} = 10^{-0.24}$$

$$= 10^{0.76} \times 10^{-1} = 5.76 \times 10^{-1} = 0.576$$

and

$$\text{mmoles base} = 0.576 \text{ mmole salt}$$

The total number of millimoles equals $100 \times 5.00$ or 500, so

$$\text{mmoles base} + \text{mmoles salt} = 500$$

Substituting in this expression,

$$0.576 \text{ mmole salt} + \text{mmoles salt} = 500$$

$$1.576 \text{ mmoles salt} = 500$$

$$\text{mmoles salt} = \frac{500}{1.576} = 317 \text{ mmoles}$$

$$\text{mmoles base} = 500 - 317 = 183 \text{ mmoles}$$

The grams of $NH_4Cl$ is given by

$$g = \text{mmoles} \times \text{mMW} = 317 \times 0.0535 = 17.0 \text{ g}$$

The milliliters of concentrated $NH_4OH$ is given by

$$\text{mL} = \frac{\text{mmoles}}{M} = \frac{183}{16.0} = 11.3 \text{ mL}$$

To prepare the solution, dissolve 17.0 g of $NH_4Cl$ in a small amount of water, add 11.3 mL of concentrated $NH_4OH$, and dilute to 100 mL.

## NONAQUEOUS TITRATIONS

### SOLVENTS

Nonaqueous solvents may be classified as amphiprotic, aprotic, and basic but not acidic. Most of the analytically important solvents are amphiprotic and can act either as an acid or as a base, depending upon the solute. The aprotic or inert solvents do not ionize and consequently are neither acidic nor basic. Examples of aprotic solvents are benzene, chloroform, and carbon tetrachloride. Some solvents, such as pyridine, acetone, and diethyl ether, can show weakly basic properties but no acidic properties.

Amphiprotic solvents undergo self-ionization or **autoprotolysis**, as shown in eqs. (10-77) and (10-78). The solvated proton is considered to be the strongest possible acid and the solvent anion the strongest possible base in that solvent. In water, the mineral acids, such as perchloric, hydrochloric, and nitric, are actually stronger proton donors than the hydronium ion but are leveled or reduced to the strength of the hydronium ion. Thus, water does not differentiate between the relative proton-donating ability or intrinsic strength of these acids. Water can differentiate between the intrinsic strengths of acetic acid and hydrochloric acid because acetic acid is a weaker proton donor than the hydronium ion. Since acetic acid is a stronger acid (or weaker base) than water, it can be used as a differentiating solvent for some of the strong acids, such as hydrochloric and perchloric. In acetic acid, perchloric acid acts as a strong acid whereas hydrochloric acts as a weak acid.

Similar considerations may be applied to the more basic solvents such as liquid ammonia, in which the amine ($NH_2^-$) ion is the strongest base and the ammonium ion ($NH_4^+$) is the strongest acid. In this solvent the stronger bases or proton acceptors are leveled to the strength of the amine ion. Water, with hydroxide as the strongest possible base, levels the strong bases such as sodium and potassium hydroxide and consequently cannot differentiate between their intrinsic strengths. Ammonia, owing to its stronger basicity, can act as a differentiating solvent for some of the stronger bases.

## TITRATIONS

Water is not a satisfactory solvent for the titration of many organic acids and bases because a large number of organic compounds are not soluble in water and those that are soluble are often very weak acids or bases with ionization constants less than $10^{-8}$. Since compounds that have very low ionization constants do not give a satisfactory break when titrated in water solution, only a very few organic acids and bases can be titrated in water. It has long been known that the strength of basic compounds, such as amines, is greater in glacial acetic acid than in water because acetic acid is less basic than water. Similarly, some strong aqueous acids such as hydrochloric are actually weak acids in acetic acid, whereas some acids such as perchloric are essentially completely ionized in both acetic acid and water.

In water, the strongest acid is the solvated proton known as the hydronium ion, which is formed by the ionization of water.

$$2H_2O \rightleftharpoons H_3O^+ + OH^- \qquad \textbf{(10-77)}$$

Similarly, in acetic acid the solvated proton formed by the ionization of acetic acid is also the strongest possible acid.

$$2HC_2H_3O_2 \rightleftharpoons H_2C_2H_3O_2^+ + C_2H_3O_2^- \qquad \textbf{(10-78)}$$

Since perchloric acid is essentially completely ionized in acetic acid, a solution of perchloric acid in acetic acid is usually chosen as the titrant solution for bases. Water interferes, and, since the usual concentrated perchloric acid contains approximately 30% water, acetic anhydride is added to such solutions to remove the water by converting it to acetic acid.

$$(CH_3CO)_2O + H_2O \rightleftharpoons 2HC_2H_3O_2 \qquad \textbf{(10-79)}$$

The general equation for the titration of a base in acetic acid is

$$H_2C_2H_3O_2^+ + B \longrightarrow BH^+ + HC_2H_3O_2 \qquad \textbf{(10-80)}$$

Nonaqueous titration in acetic acid is often used for aliphatic and aromatic amines, such as aniline, acridine, brucine, butylamine, tribenzylamine, and nicotinamide. Also, salts, such as choline chloride, codeine phosphate, thiamine nitrate, potassium biphthalate, sodium acetate, sodium bromide, sodium nitrate, sodium sulfate, and sodium tungstate, may be titrated in acetic acid using perchloric acid as the titrant. Potassium biphthalate is often used as a primary standard base in nonaqueous titrations, since it can be obtained readily in stated purity. In aqueous work also, this compound is probably the most popular primary standard acid used.

Other organic solvents that are used for titration of bases include dioxane, acetonitrile, nitrobenzene, chloroform, carbon tetrachloride, petroleum ether, and nitromethane. The perchloric acid titrant is usually dissolved in acetic acid for use with the inert solvents but may be dissolved in the solvent being used in some cases.

Weak organic acids may also be titrated in nonqueous media, but these titrations are not as popular as the weak base titrations. Common titrants include sodium methoxide or sodium ethoxide in methanol or ethanol, respectively, although in some cases benzene–methanol (1:6) may be used as a solvent. Other possible titrants are alcoholic potassium and sodium hydroxides. Phenols and other very weak acids may be titrated in anhydrous ethylenediamine with sodium aminoethoxide in ethylenediamine–ethanolamine as the titrant.

Most nonaqueous titrations are performed potentiometrically using silver–silver chloride and glass electrodes, with the glass electrode being the indicator electrode. The saturated calomel electrode may be utilized as the reference electrode but requires the use of a salt bridge. An ordinary pH meter is the usual measuring device. The pH reading should not be considered as the hydrogen ion concentration but only as a measure of the voltage or potential change during the titration.

Visual indicators may also be used. Crystal violet and methyl violet are the most popular for titrations in acetic acid, but 1-naphtholbenzein and benzoylauramine are also used. The indicator is usually dissolved in an inert solvent such as chlorobenzene, chloroform, nitrobenzene, or acetonitrile. The color change with crystal violet and methyl violet is from violet to blue to green, with blue being the end point. Most directions state that the end point is the disappearance of the last tinge of violet in the solution. The end point for most titrations is actually very sharp and can be obtained with less than 1 drop of titrant when using 0.1 *N* solutions. However, water interferes and must be absent because, in the presence of very small amounts of water, the end point is very difficult to determine. The water prolongs the disappearance of the violet tinge and makes the end point indistinct.

## PROBLEMS

**1** List the conjugate acids of the following bases:
(a) $H_2AsO_4^-$
(b) $CO_3^{2-}$
(c) $Fe(OH)(H_2O)_5^{2+}$
(d) $C_6H_5NH_2$ (aniline)
(e) $C_6H_5O^-$ (phenolate)
(f) $HC_6H_5O_7^{2-}$ (hydrogen citrate)

**2** List the conjugate bases for the following acids:
(a) $HNO_3$
(b) $H_2MoO_4$
(c) $H_2C_6H_5O_7^-$ (dihydrogen citrate)
(d) $NH_4^+$
(e) $Ni(H_2O)_6^{2+}$
(f) $HCrO_4^-$

**3** Calculate the $p_cH$ of the following solutions:
(a) 0.0025 $M$ $HNO_3$
(b) 0.0093 $N$ $H_2SO_4$
(c) 0.0153 $M$ HCl
(d) 0.00050 $M$ $H_2SO_4$
(e) $10^{-4.15}$ $N$ $HClO_4$

*Ans.* (b) 2.03

**4** Calculate the $p_cH$ of the following solutions:
(a) 0.036 *N* $NaOH$
(b) 0.250 *N* $Ba(OH)_2$
(c) 1.00 *N* $KOH$
(d) 0.0048 *M* $Ba(OH)_2$
(e) 0.0000077 *M* $Ca(OH)_2$

*Ans.* (b) 13.40

**5** How many millimoles of HCl would have to be added to 250 mL of water to produce solutions with the following $p_cH$ values: (a) 2.43; (b) 0.64; (c) 4.33; (d) 5.09; (e) 1.12.
*Ans.* (b) 57.3

**6** How many millimoles of NaOH would have to be added to 333 mL of water to produce solutions with the following $p_cH$ values: (a) 9.45; (b) 12.67; (c) 13.75; (d) 8.90; (e) 10.40.
*Ans.* (b) 15.6

**7** How many millimoles of NaOH would have to be added to 500 mL of water to produce solutions with the following $p_cOH$ values: (a) 1.49; (b) 4.73; (c) 6.10; (d) 0.50; (e) 2.50.
*Ans.* (b) 0.0093

**8** Calculate the molar concentration of hydrochloric acid solutions that have the following $p_cH$ values: (a) 1.23; (b) 2.34; (c) 3.45; (d) 4.56; (e) 5.67.
*Ans.* (b) 0.0046 *M*

**9** Calculate the molar concentration of sodium hydroxide solutions that have the following $p_cH$ values: (a) 7.89; (b) 13.00; (c) 8.97; (d) 12.23; (e) 10.12.
*Ans.* (b) 0.10 *M*

**10** Calculate the molar concentration of the salt at the equivalence point in the following titrations assuming: (1) no dilution with water, (2) dilution with water to 100.0 mL, and (3) dilution with 20.00 mL of water:
(a) 25.00 mL of 0.1017 *N* $HC_2H_3O_2$ with 0.0983 *N* $NaOH$
(b) 20.00 mL of 0.2408 *N* $H_2SO_4$ with 0.1013 *N* $KOH$
(c) 10.00 mL of 0.1525 *M* $H_3PO_4$ to $Na_2HPO_4$ with 0.1123 *M* $NaOH$
(d) 50.00 mL of 0.0505 *N* $HCl$ with 0.0894 *N* $NaOH$
(e) 25.00 mL of 0.2500 *N* $NH_4OH$ with 0.2075 *N* $HCl$
*Ans.* (b) (1) 0.0357 *M*, (2) 0.0241 *M*, (3) 0.0275 *M*

**11** State whether the $p_cH$ of the following solutions will be 7, less than 7, more than 7, or approximately 7:
(a) $NH_4NO_3$
(b) $Ca(NO_2)_2$
(c) $Na_2SO_4$
(d) $AlCl_3$
(e) $NH_4ClO_2$
(f) $Fe(C_2H_3O_2)_3$

**12** Calculate the $p_cH$ of the following solutions of salts:
(a) 0.0250 *M* $Ca(C_2H_3O_2)_2$
(b) 0.1050 *M* $NaClO$
(c) 0.2134 *M* $KC_7H_5O_2$ (potassium benzoate)
(d) 0.1000 *M* $LiC_3H_5O_2$ (lithium propionate)
(e) 0.1250 *M* $Cs_2C_4H_4O_5$ (cesium malate)

*Ans.* (b) 10.28

**13** Calculate the $p_cH$ of the following solutions of salts:
(a) 0.1000 *M* $NH_4NO_3$
(b) 0.0100 *M* $NH_4Cl$
(c) 0.2150 *M* $AgNO_3$
(d) 0.1250 *M* $(NH_4)_2SO_4$
(e) 0.0500 *M* $CuSO_4$

*Ans.* (b) 5.63

**14** Calculate the $p_cH$ at the equivalence points in the titrations for problem 10(a), (c), and (e).

*Ans.* (a) (1) 8.73, (2) 8.59, (3) 8.66

**15** Calculate the $p_cH$ of the following solutions of salts:
(a) 0.0125 *M* $NH_4C_7H_5O_2$ (ammonium benzoate)
(b) 0.0019 *M* $AgC_3H_5O_2$ (silver propionate)
(c) 0.5000 *M* $NH_4C_2H_3O_2$
(d) 0.2000 *M* $(NH_4)_2CO_3$
(e) 0.1033 *M* $AgNO_2$

*Ans.* (b) 7.46

**16** Calculate the $p_cH$ of the following solutions of salts:
(a) 0.2500 *M* $NaHCO_3$
(b) 0.0525 *M* $NaHC_4H_4O_5$ (sodium hydrogen malate)
(c) 0.074 *M* $NaH_2PO_4$
(d) 0.1000 *M* $Na_2HPO_4$
(e) 0.1250 *M* $Na_2H_2C_{10}H_{12}O_8N_2$ (disodium dihydrogen EDTA)

*Ans.* (b) 4.23

**17** Calculate the $p_cH$ for the following solutions:
(a) 0.50 *M* propionic acid and 0.50 *M* sodium propionate ($NaC_3H_5O_2$)
(b) 0.010 *M* sodium benzoate and 0.015 *M* benzoic acid ($HC_7H_5O_2$)
(c) 0.15 *M* disodium hydrogen phosphate ($Na_2HPO_4$) and 0.25 *M* sodium dihydrogen phosphate ($NaH_2PO_4$)
(d) 0.22 *M* disodium hydrogen phosphate ($Na_2HPO_4$) and 0.18 *M* trisodium phosphate
(e) 0.25 *M* $NH_4OH$ and 0.050 *M* $NH_4Cl$

*Ans.* (b) 4.02

**18** Calculate the $p_cH$ in the titrated solution assuming no dilution with water for the titration of 25.00 mL of 0.1000 *N* acid with 0.1000 *N* base after the addition of (1) 5.00 mL, (2) 12.50 mL, (3) 17.50 mL, and (4) 30.00 mL of the base for the following titrations: (a) hydrochloric acid with NaOH, (b) benzoic acid with NaOH, (c) ammonium hydroxide with hydrochloric acid, (d) sodium dihydrogen phosphate with NaOH to form disodium hydrogen phosphate, and (e) disodium hydrogen phosphate with HCl to form sodium dihydrogen phosphate.

*Ans.* (b) (1) 3.60, (2) 4.20, (3) 4.56, (4) 11.96

**19** Calculate the $p_cH$ of the following solutions:
(a) A solution that contains 7.5 g of $Na_2CO_3$ and 6.0 g of $NaHCO_3$ in 400 mL
(b) A solution that contains 6.85 millimoles of sodium acetate and 3.43 millimoles of acetic acid in 250 mL
(c) A solution prepared by adding 50.0 mL of 0.250 *M* NaOH to 50.0 mL of 0.150 *M* $H_3PO_4$
(d) A solution prepared by adding 10.0 mL of 3.00 *N* HCl to 90.0 mL of 1.00 *N* $NH_4OH$
(e) A solution that contains 48 millimoles of tartaric acid and 56 millimoles of potassium hydrogen tartrate in 330 mL

*Ans.* (b) 5.06

**20** Calculate (1) the capacity to added acid and (2) the beta value for the solutions listed in problem 17. Assume 100 mL of each solution.

*Ans.* (b) (1) 0.0084, (2) 0.01172

**21** Calculate (1) the capacity to added acid and (2) the beta value for the solutions listed in problem 19.

*Ans.* (b) (1) 0.0206, (2) 0.0118

**22** A 2.000-g sample of an industrial sodium carbonate alkali is dissolved and diluted to 100.0 mL. One 25.00-mL aliquot is titrated to the phenolphthalein end point with HCl and a second 25.00-mL aliquot is titrated with the same acid to the methyl orange end point. Using the volumes and normalities shown below, (1) state the compounds present, and (2) calculate the percentage of each present:

| *Milliliters Phenolphthalein End Point* | *Milliliters to Methyl Orange End Point* | *N of Acid* |
|---|---|---|
| (a) 35.48 | 49.93 | 0.1936 |
| (b) 24.63 | 49.26 | 0.1450 |
| (c) 15.34 | 35.42 | 0.2250 |

*Ans.* (b) (1) $Na_2CO_3$ only, (2) 75.71% $Na_2CO_3$

**23** An alkaline unknown contains inert material plus any one or any combination of sodium hydroxide, sodium carbonate, and sodium bicarbonate. A given weight of sample is dissolved and titrated to the phenolphthalein end point with acid. Methyl orange is then added and the titration continued to the methyl orange end point with the same acid. For the data given below, (1) state the components present, and (2) calculate the percentage of each present:

| *Weight of Sample (g)* | *Milliliters to Phenolphthalein End Point* | *Milliliters from Phenolphthalein End Point to Methyl Orange End Point* | *N of Acid* |
|---|---|---|---|
| (a) 0.998 | 25.27 | 20.63 | 0.1500 |
| (b) 1.002 | 27.89 | 35.69 | 0.2250 |
| (c) 0.500 | 23.33 | 23.33 | 0.2000 |

*Ans.* (b) (1) $Na_2CO_3$, $NaHCO_3$; (2) $Na_2CO_3$, 66.39%; $NaHCO_3$, 14.71%

**24** How many grams of each constituent must be dissolved and diluted to prepare the buffer solutions of given $p_cH$ and total molar concentration listed below:

| *Constituents* | *Volume of Solution (mL)* | $p_cH$ | *Total Molar Concentration (M)* |
|---|---|---|---|
| (a) $HC_2H_3O_2 + NaC_2H_3O_2$ | 400 | 4.35 | 0.250 |
| (b) $NH_4OH + NH_4Cl$ | 250 | 9.50 | 1.33 |
| (c) $NaHCO_3 + Na_2CO_3$ | 1000 | 10.50 | 0.500 |
| (d) $NaH_2PO_4 + K_2HPO_4$ | 500 | 7.00 | 1.25 |
| (e) $KHC_8H_4O_4 + K_2C_8H_4O_4$ (phthalate) | 100 | 5.10 | 1.00 |

*Ans.* (b) $NH_4OH$, 7.40 g; $NH_4Cl$, 6.49 g

CHAPTER

# 11 Oxidation–Reduction Theory

There are a very large number of oxidation–reduction reactions that are used as the basis for analytical methods. Oxidation was defined originally as the reaction of a substance with oxygen, and the substance that reacted was said to be oxidized. Thus, the rusting of iron, which was considered to be the reaction of iron with the oxygen in the air to form iron(III) oxide, was referred to as the oxidation of iron. Similarly, the term *reduction* was used to designate the removal of oxygen from a material, since many metal ores were oxides, and, in the metallurgical process, the ore was said to be reduced to the metal. As a consequence, the removal of oxygen became known as reduction. As the knowledge of the electronic structure of atoms increased, it was realized that, in the reaction with iron, oxygen atoms actually removed electrons from iron atoms to form charged particles called ions, so that oxidation was really the removal of electrons from an atom. Other reactions, such as the reaction of iron and chlorine, were also recognized to involve loss and gain of electrons, even though oxygen was not a reactant. Since these reactions caused an increase in positive charge or oxidation state, as did the reaction with oxygen, they were also termed oxidation. Similarly, the removal of oxygen from an oxide represented a gain of electrons by the metal ion and a decrease in oxidation state, and all such reactions were termed reductions, whether or not oxygen was involved.

**Oxidation** is now defined as the loss of electrons by a substance, and **reduction** is defined as the gain of electrons by a substance. Since electrons cannot be created or destroyed in ordinary chemical reactions, any oxidation must be accompanied by a corresponding reduction. The loss of electrons causes an increase in oxidation state, so oxidation is also defined as an increase in the oxidation state of an atom or ion. Similarly, the gain of electrons causes a decrease in oxidation state, and reduction is defined as the decrease in the oxidation state of an atom or ion.

A substance that causes oxidation to occur in another is called an **oxidizing agent** and, since it must accept electrons from the substance it oxidizes, is reduced during the reaction. Also, a substance that causes reduction to occur is called a **reducing agent** and is itself oxidized during the reaction. All oxidation–reduction reactions, then, can be considered to be the reaction of an oxidizing agent with a reducing agent. In all such reactions the oxidizing agent is reduced and the reducing agent is oxidized.

## BALANCING OXIDATION-REDUCTION REACTIONS

There are several satisfactory methods for balancing oxidation–reduction reactions, each of which has certain advantages over the other systems. Since analytical chemists are interested primarily in the actual species that react in a solution and in the net reaction that occurs, most analysts prefer the partial ion equation method, which is often called the **half-cell reaction**, **half-reaction**, or **electrode reaction method**. In this method only the actual ions or molecules that are involved in the reaction between the oxidizing agent and the reducing agent are shown. If a molecular equation is necessary, it can be obtained from the net ionic reaction, which is the result of the partial ion equation method of balancing reactions.

The steps taken to balance an equation include: (1) writing and balancing the ionic reaction of the oxidizing agent; (2) writing and balancing the ionic reaction of the reducing agent; (3) multiplication of one or both of these equations, if necessary, to balance the number of electrons; and (4) summation of the two equations to obtain the net ionic reaction. The following rules are utilized in balancing the half-reactions:

1 Show each substance as it actually occurs in the solution. Ionic substances (strong electrolytes) are shown as ions, and weak and nonelectrolytes are shown as molecules. The student should remember that the symbol of an ion includes its charge and that the symbol of an element respresents an atom, not an ion, of the element. Similarly, $Cr_2O_7^{2-}$ is the symbol of the dichromate ion, whereas $Cr_2O_7$ (without the charge) is the formula of chromium heptoxide.

2 The equations must be balanced both as to atoms and as to charges.

3 The number of electrons involved in each half-reaction must be equal, or the half-reactions must be multiplied by numbers that make the number of electrons involved equal.

4 The molecular equation may be obtained from the ionic equation by adding the ions that occur with the reactant ions and distributing them among the product ions. Any ions left over must represent whole numbers of molecules for the substances represented.

5 Water or its ions may be required to obtain material balance. Remember, however, that hydrogen ions cannot be shown as reactant or product in a basic solution, and hydroxide ions may not be shown as reactant or product in an acidic solution.

Several examples will be given as illustrations of this method of balancing equations.

• **Example 1.** Balance the reaction between $K_2Cr_2O_7$ and $FeCl_2$ in HCl solution to form $FeCl_3$, $CrCl_3$, and other products. Since salts and strong acids are completely ionic, the unbalanced ionic equation may be represented as

$$Cr_2O_7^{2-} + Fe^{2+} + H^+ \longrightarrow Cr^{3+} + Fe^{3+} + H_2O \qquad \textbf{(11-1)}$$

The unbalanced ionic half-reaction of the oxidizing agent is

$$Cr_2O_7^{2-} + H^+ \longrightarrow Cr^{3+} + H_2O \qquad \textbf{(11-2)}$$

To obtain material (atom) balance, there must be 2 $Cr^{3+}$ ions, and to balance the oxygen from the $Cr_2O_7^{2-}$, there must be 7 molecules of water, which, in turn, would require 14 $H^+$ ions. The half-reaction then becomes

$$Cr_2O_7^{2-} + 14H^+ \longrightarrow 2Cr^{3+} + 7H_2O \qquad \textbf{(11-3)}$$

The number of electrons involved may now be obtained by balancing the charge on both sides of the equation. That is, the total charge on the products must equal the total charge on reactants. The total charge on the reactants is 12 positive and on the products is 6 positive. To reduce 12 positive to 6 positive requires the addition of 6 electrons as reactant:

$$Cr_2O_7^{2-} + 14H^+ + 6e^- \longrightarrow 2Cr^{3+} + 7H_2O \qquad \textbf{(11-4)}$$

and the equation is balanced both as to atoms and as to charge.

The unbalanced ionic equation for iron is

$$Fe^{2+} \longrightarrow Fe^{3+} \qquad \textbf{(11-5)}$$

This equation is balanced as to atoms and requires only balancing as to charge. Since an increase in oxidation states represents a loss or removal of electrons, it will require the loss of 1 electron by the $Fe^{2+}$ to become $Fe^{3+}$:

$$Fe^{2+} - 1e^- \longrightarrow Fe^{3+} \qquad \textbf{(11-6)}$$

The electrons can actually be shown on either side of the equation and eq. (11-6) could be written

$$Fe^{2+} \longrightarrow Fe^{3+} + 1e^- \qquad \textbf{(11-7)}$$

Since the number of electrons lost by iron must equal the number of electrons gained by the dichromate, eq. (11-6) must be multiplied by 6 before the two equations can be added to obtain the net ionic reaction:

$$\begin{array}{llll} Cr_2O_7^{2-} + 14H^+ + 6e^- & \longrightarrow & 2Cr^{3+} + 7H_2O & \textbf{(11-4)} \\ 6Fe^{2+} \qquad\qquad - 6e^- & \longrightarrow & 6Fe^{3+} & \textbf{(11-8)} \\ \hline Cr_2O_7^{2-} + 6Fe^{2+} + 14H^+ & \longrightarrow & 2Cr^{3+} + 6Fe^{3+} + 7H_2O & \textbf{(11-9)} \end{array}$$

If the molecular equation is desired, the ions occurring with the reactants can be added and distributed among the products:

$$
\begin{array}{llllllll}
Cr_2O_7^{2-} & +6Fe^{2+} & +14H^+ & \longrightarrow & 2Cr^{3+} & +6Fe^{3+} & +7H_2O & \\
2K^+ & +12Cl^- & +14Cl^- & \longrightarrow & 6Cl^- & +18Cl^- & +2K^+ & +2Cl^- \\
\hline
\multicolumn{3}{l}{K_2Cr_2O_7 + 6FeCl_2 + 14HCl} & \longrightarrow & \multicolumn{4}{l}{2CrCl_3 + 6FeCl_3 + 7H_2O + 2KCl}
\end{array}
$$

**(11-9)**

**(11-10)**

As shown in this example, the ions left over after distribution among the products must represent a whole number of molecules of substances.

• **Example 2.** Balance the reaction for the titration of calcium oxalate by permanganate in the analysis of limestone by the oxalate–permanganate method. In this method, the precipitated calcium oxalate is dissolved in sulfuric acid and heated before being titrated with the permanganate. The oxalate will be present mainly as bioxalate due to the acidity of the solution. The unbalanced ionic equation then becomes

$$MnO_4^- + HC_2O_4^- + H^+ \longrightarrow Mn^{2+} + CO_2 + H_2O \quad \textbf{(11-11)}$$

The unbalanced ionic half-reaction for $MnO_4^-$ is

$$MnO_4^- + H^+ \longrightarrow Mn^{2+} + H_2O \quad \textbf{(11-12)}$$

To obtain material balance, 4 molecules of water will be formed, and this will require eight hydrogen ions, or

$$MnO_4^- + 8H^+ \longrightarrow Mn^{2+} + 4H_2O \quad \textbf{(11-13)}$$

To obtain charge balance, the total charge on the reactants is 7 positive whereas the charge on the products is 2 positive. To convert a charge of 7 positive to 2 positive requires the addition of 5 electrons, or

$$MnO_4^- + 8H^+ + 5e^- \longrightarrow Mn^{2+} + 4H_2O \quad \textbf{(11-14)}$$

The unbalanced ionic half-reaction for bioxalate is

$$HC_2O_4^- \longrightarrow CO_2 + H^+ \quad \textbf{(11-15)}$$

Material balance requires two molecules of $CO_2$:

$$HC_2O_4^- \longrightarrow 2CO_2 + H^+ \quad \textbf{(11-16)}$$

Charge balance requires the removal of 2 electrons:

$$HC_2O_4^- - 2e^- \longrightarrow 2CO_2 + H^+ \quad \textbf{(11-17)}$$

Before the addition of the two half-cell reactions, the minimum number of electrons that will satisfy both equations is 2 × 5, or 10, so eq. (11-14) must be multiplied by 2 and eq. (11-17) by 5:

$$2MnO_4^- + 16H^+ + 10e^- \longrightarrow 2Mn^{2+} + 8H_2O \quad \textbf{(11-18)}$$

$$5HC_2O_4^- - 10e^- \longrightarrow 10CO_2 + 5H^+ \quad \textbf{(11-19)}$$

$$2MnO_4^- + 5HC_2O_4^- + 16H^+ \longrightarrow 2Mn^{2+} + 10CO_2 + 8H_2O + 5H^+ \quad \textbf{(11-20)}$$

By convention, the same ion or molecule is not shown as both reactant and product, so the $5H^+$ are subtracted from both sides of the equation, to obtain

$$2MnO_4^- + 5HC_2O_4^- + 11H^+ \longrightarrow 2Mn^{2+} + 10CO_2 + 8H_2O \quad \textbf{(11-21)}$$

• **Example 3.** Balance the reaction between cupric ion and iodide ion in the determination of copper by the iodometric method. In this reaction, the products are triiodide ion and cuprous iodide. The unbalanced ionic reaction is

$$Cu^{2+} + I^- \longrightarrow Cu_2I_2 + I_3^- \quad \textbf{(11-22)}$$

The unbalanced ionic half-reaction for cupric ion is

$$Cu^{2+} \longrightarrow Cu_2I_2 \quad \textbf{(11-23)}$$

Material balance requires 2 cupric ions and 2 iodide ions as reactants:

$$2Cu^{2+} + 2I^- \longrightarrow Cu_2I_2 \quad \textbf{(11-24)}$$

Charge balance requires converting a 2 positive charge to 0 by the addition of 2 electrons:

$$2Cu^{2+} + 2I^- + 2e^- \longrightarrow Cu_2I_2 \quad \textbf{(11-25)}$$

The unbalanced ionic reaction for iodide is

$$I^- \longrightarrow I_3^- \quad \textbf{(11-26)}$$

Material balance requires 3 iodide ions:

$$3I^- \longrightarrow I_3^- \quad \textbf{(11-27)}$$

and charge balance requires conversion of a 3 negative charge to a 1 negative charge by the removal of 2 electrons:

$$3I^- - 2e^- \longrightarrow I_3^- \quad \textbf{(11-28)}$$

Addition of eqs. (11-25) and (11-28) gives the balanced net ionic reaction:

$$2Cu^{2+} + 2I^- + 2e^- \longrightarrow Cu_2I_2 \qquad \textbf{(11-25)}$$

$$3I^- - 2e^- \longrightarrow I_3^- \qquad \textbf{(11-28)}$$

$$2Cu^{2+} + 5I^- \longrightarrow Cu_2I_2 + I_3^- \qquad \textbf{(11-29)}$$

Note in Examples 1, 2, and 3 that there is both charge balance and material balance in the net ionic reaction. As a consequence, the student should check the charge balance of the net ionic reaction each time an oxidation–reduction reaction is balanced to be sure that the balancing has been done correctly.

## EQUIVALENT WEIGHTS IN OXIDATION-REDUCTION REACTIONS

The method of calculating the equivalent weights in any oxidation–reduction reaction was discussed in Chapter 3 and should be reviewed. It should be emphasized that the equivalent weight is a function of the electron change and not the oxidation state of the ion, atom, radical, or molecule. As a consequence, the student must know the actual products of the reaction in order to determine the electron change. Since most oxidizing agents and reducing agents can produce different products under different conditions, information as to the products to be expected is usually given in the beginning course in quantitative analysis. For the more common oxidants, $MnO_4^-$ is always reduced to $Mn^{2+}$ in acid solution, but $MnO_2$ is the product in solutions that are near neutrality, and manganate ($MnO_4^{2-}$) is the product in strongly basic solutions. Dichromate in acid solution is reduced to $Cr^{3+}$. Usually, the student is expected to know the product for dichromate and permanganate titrations under various conditions.

## ELECTROCHEMICAL THEORY OF OXIDATION-REDUCTION

The electric current that flows through wires and furnishes the energy for lighting, heating, and cooling homes and for industrial uses is simply a flow of electrons along the wire. The flow is characterized by three units—the volt, the ampere, and the ohm. The **volt** (V) is the unit of driving force known as the **potential** or the **electromotive force** (emf) that causes the electrons to flow. The **ampere** (A) is the unit of current and is a measure of the number of electrons that pass a given point in a given unit of time. The **ohm** (Ω) is the unit of resistance to the flow of the electrons through the conductor. These three factors are related by **Ohm's law**, which can be stated as

$$E = IR \qquad \textbf{(11-30)}$$

in which $E$ is the potential in volts, $I$ the current in amperes, and $R$ the resistance in ohms. Each of these quantities has an absolute definition, but they are usually defined in terms of each other: 1 V is the potential necessary to cause a current of 1 A to flow through a resistance of 1 Ω; 1 A is the amount of current that passes a given point when a potential of 1 V operates through a resistance of 1 Ω; 1 Ω is the amount of resistance offered to a current of 1 A under a potential of 1 V.

When an oxidation–reduction reaction takes place in a solution, there is a transfer of electrons through space from one atom or ion to another. In the reaction between metallic zinc and copper sulfate solution, which produces metallic copper and zinc sulfate solution, a copper ion must come in very close contact with a zinc atom for the transfer to take place, since electrons do not flow freely through a solution. The ionic reaction may be shown as

$$Cu^{2+} + Zn \longrightarrow Cu + Zn^{2+} \qquad \textbf{(11-31)}$$

and the individual half-reactions are

$$Cu^{2+} + 2e^- \longrightarrow Cu \qquad \textbf{(11-32)}$$

$$Zn - 2e^- \longrightarrow Zn^{2+} \qquad \textbf{(11-33)}$$

If some device could be designed that would separate the reactants and cause the electron transfer to take place along a wire, an electric current would be produced as the electrons flowed through the wire. A **galvanic** or **voltaic cell** is such a device. A galvanic cell is composed of two half-cells, each of which must contain the two forms of an oxidation–reduction pair and a metallic conductor called the **electrode**. The electrode can be one of the two forms of the oxidation–reduction pair, or it may be an inert conductor such as platinum, which does not enter into the half-cell reaction. A galvanic cell that takes advantage of the copper–zinc reaction of eq. (11-31) could be designed as shown in Figure 11-1. A container that has a bar of zinc metal immersed in a solution of zinc sulfate is connected by a U-tube filled with a concentrated solution of an electrolyte to a second container, which has a bar of copper immersed in a solution of copper sulfate. The two metal bars (electrodes) are connected by a wire. The U-tube, filled with electrolyte, is known as a **salt bridge** and allows ions to move from one container to the other to maintain electrical neutrality. It is usually filled with a concentrated solution of KCl embedded in a gelatin matrix to prevent free syphoning from one solution to the other. A voltmeter and/or an ammeter may be incorporated into the circuit to measure the potential created and/or the amount of current that flows.

Upon completion of the circuit, either by connecting the wire or by inserting the salt bridge, oxidation–reduction reactions occur at the electrodes and electrons flow through the wire. The direction of flow of electrons depends upon the composition of the two half-cells. Each half-cell exerts a potential or driving force that is related to the ease of reduction of the metal ions and causes electrons to flow from it to another half-cell. However, each half-cell can also accept

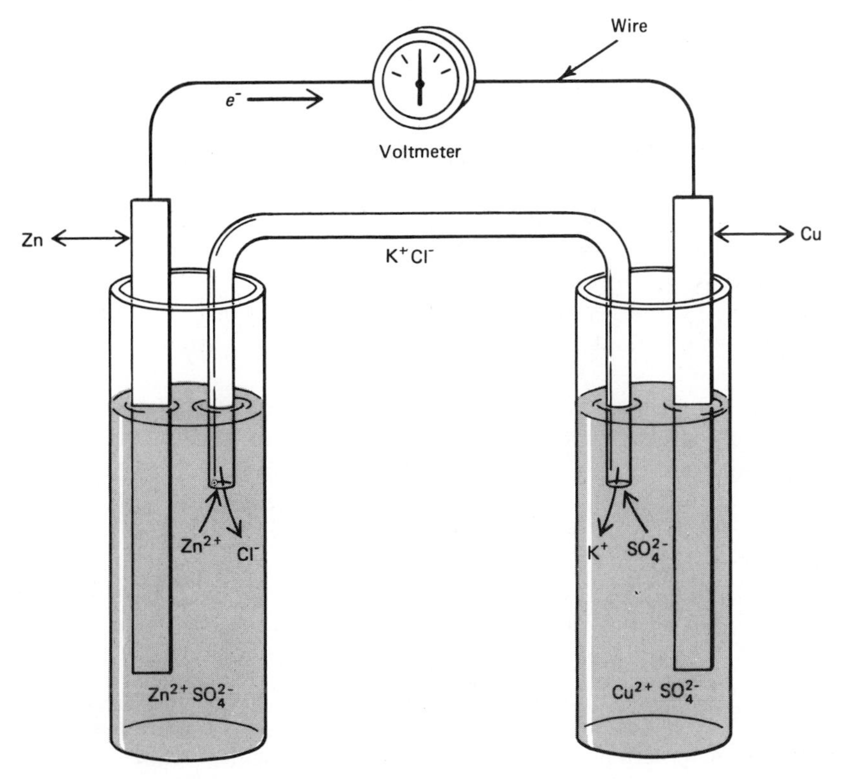


**Figure 11-1.** Galvanic cell showing a zinc–zinc sulfate half-cell and a copper–copper sulfate half-cell connected by a potassium chloride salt bridge.

electrons. As a consequence, the half-cell with the greater driving force or potential will force the other half-cell to accept electrons. In the zinc–copper cell, zinc has a greater driving force than copper, because copper ion is more easily reduced than zinc ion, and electrons flow through the wire from the zinc half-cell to the copper half-cell. In the zinc half-cell, zinc atoms go into solution to form zinc ions, leaving two electrons on the metal bar. These two electrons then flow through the wire to the copper bar where they are given to a copper ion in solution at the electrode interface, which then deposits on the electrode as elementary copper. As a consequence, the zinc bar will lose weight while the copper bar gains weight. The reactions that occur at the electrodes may be shown as

$$\mathrm{Zn} - 2e^- \longrightarrow \mathrm{Zn}^{2+}$$

$$\mathrm{Cu}^{2+} + 2e^- \longrightarrow \mathrm{Cu}$$

which are the same as eqs. (11-32) and (11-33). Summation of these two reactions gives

$$\mathrm{Zn} + \mathrm{Cu}^{2+} \longrightarrow \mathrm{Cu} + \mathrm{Zn}^{2+} \qquad \textbf{(11-34)}$$

which is the same as eq. (11-31) for the reaction that takes place when zinc is placed in a copper sulfate solution.

The zinc electrode in the preceding example is called the **negative pole** because it has a negative charge due to the electrons remaining on it as zinc atoms are converted to zinc ions. By comparison, the copper electrode is the **positive electrode**. The zinc electrode, having the higher driving force, is considered to be a region of high electron density, whereas the copper electrode, with a weaker driving force, has a lower electron density, and, because electrons flow freely along a metallic conductor from a region of high electron density to a region of lower electron density, the electrons flow through the wire from the negative half-cell to the positive half-cell, or from the zinc to the copper.

The salt bridge maintains electrical neutrality in the two solutions by transport of ions. The deposition of a copper ion leaves an excess of two negative charges in the copper solution, while the formation of a zinc ion causes an excess of two positive charges in the zinc solution. These charges are neutralized by the movement of ions into, out of, or through the salt bridge. It really makes no difference whether two $Cl^-$ ions move from the bridge into the zinc solution and two $K^+$ ions move into the copper solution or whether the zinc and sulfate ions migrate into the salt bridge, as long as electrical neutrality is preserved. In reality, probably both types of migration occur.

The driving force, or potential, of a half-cell cannot be measured alone, only by comparison to other half-cells. Another way of stating this is that differences in potential can be measured, but individual half-cell potentials cannot be measured. As a consequence, it is necessary to adopt a particular electrode as the standard electrode and measure all other potentials against it. The hydrogen half-cell has been designated as the **standard half-cell** and assigned a potential of zero when the hydrogen ion is at unit activity.

Half-cells that cause the hydrogen half-cell to accept electrons (reduce hydrogen ions to hydrogen gas) are assigned a negative potential, whereas those half-cells that accept electrons from the hydrogen half-cell (convert hydrogen gas to hydrogen ions) are assigned a positive potential. In this system, the strong oxidizing agents such as permanganate are assigned positive potentials and the strong reducing agents such as zinc are assigned negative potentials. This system has been adopted by the International Union of Pure and Applied Chemistry (IUPAC) to offset confusion in the assignment of positive and negative potentials to half-cells. There is another convention that assigns electrode potentials exactly opposite to the IUPAC convention and consequently causes confusion. In the IUPAC convention, the half-cell reaction is written to show the gain of electrons or to show reduction; tables that use the alternative convention write the half-cell reaction as the loss of electrons by the reactant or as oxidation of the reactant.

## THE NERNST EQUATION

The actual potential of any given half-cell depends not only upon the components of the half-cell reaction system but also upon the concentrations of all the ions involved in the half-cell reaction. For example, a cell composed of two zinc half-cells will produce a current if the concentrations of the zinc ions differ in the two half-cells.

The equation that shows the relationship between the actual potential of a half-cell and the active concentration of the reactants and products of the half-cell reaction is known as the **Nernst equation**. For the general half-cell reaction,

$$aA + bB + ne^- \rightleftharpoons cC + dD \quad \textbf{(11-35)}$$

the Nernst equation[1] is

$$E = E^0 + \frac{2.3RT}{nF} \log \frac{(A)^a(B)^b}{(C)^c(D)^d} \quad \textbf{(11-36)}$$

in which $E$ is the actual potential in volts produced by the half-cell; $E^0$ is the standard reduction potential of the half-cell; $R$ is the universal gas constant [8.31 V-coulombs $(°K)^{-1}(\text{mole})^{-1}$]; $T$ is the absolute temperature; $F$ is the faraday (96,500 coulombs); and $n$ is the number of electrons in the balanced half-cell reaction. The value of $2.3RT/F$ is 0.060 at 30°C and 0.059 at 25°C. It should be noted that the activity quotient expression in the logarithm term is identical to the reciprocal of the equilibrium constant for the reaction as written, so that eq. (11-36) may be shown as

$$E = E^0 + \frac{0.059}{n} \log \frac{1}{K_{eq}} \quad \textbf{(11-37)}$$

For a half-cell composed of a metal and metal ion such as the zinc–zinc ion half-cell, the equation becomes

$$E_{Zn} = E^0_{Zn} + \frac{0.059}{2} \log \frac{(Zn^{2+})}{(Zn_{solid})}$$

$$= E^0_{Zn} + \frac{0.059}{2} \log (Zn^{2+}) \quad \textbf{(11-38)}$$

since solids are considered to be at unit activity. Note that the concentrations specified in the Nernst equation are the active (effective) concentrations or

[1] The Nernst equation for reduction reactions of this type is often written in the form

$$E = E° - \frac{2.3RT}{nF} \log \frac{(C)^c(D)^d}{(A)^a(B)^b}$$

in which the symbols have the same meaning as given for eq. (11-36). The two expressions are equivalent since

$$\log \frac{(A)^a(b)^b}{(C)^c(D)^d} = -\log \frac{(C)^c(D)^d}{(A)^a(B)^b}$$

or

$$\log \frac{1}{K_{eq}} = -\log K_{eq}$$

activities. In many calculations, molar concentrations can be utilized in place of the activities without major error.

For a half-cell reaction such as permanganate in acid solution, for which the equation is

$$MnO_4^- + 8H^+ + 5e^- \rightleftharpoons Mn^{2+} + 4H_2O \qquad \textbf{(11-14)}$$

the Nernst equation, using molar concentrations, is

$$E_{MnO_4^-/Mn^{2+}} = E^0_{MnO_4^-/Mn^{2+}} + \frac{0.059}{5} \log \frac{[MnO_4^-][H^+]^8}{[Mn^{2+}]} \qquad \textbf{(11-39)}$$

Note that eq. (11-14) exemplifies reduction and that the logarithm term utilizes the reciprocal of the equilibrium constant for the reduction reaction as written, or the equilibrium constant for the reverse reaction.

Complete tables of standard reduction potentials, $E^0$, are listed in various handbooks and textbooks. An abbreviated table of $E^0$ values for certain half-cells is given in Appendix 7. These $E^0$ values are those obtained when the concentrations of all ions involved in the half-cell reaction are at unit activity, so that the logarithmic term becomes zero as the logarithm of 1 and has no effect on the half-cell potential. The conducting portion of the half-cell is assumed to be platinum or some other inert conductor for those half-cells that do not involve a metal as one member of the oxidation–reduction pair, such as the permanganate–magnanous ion half-cell and the ferric–ferrous ion half-cell.

• **Example 4.** Calculate the actual potential of the ferric–ferrous ion half-cell in the titration with dichromate before the addition of any dichromate. In most such titrations, the concentration of the ferrous ion is adjusted to approximately 0.1 *M* before the addition of dichromate and the ferric ion concentration is, of course, negligible. Analysts usually consider $10^{-6}$ *M* (or less) to be a negligible concentration, and $10^{-6}$ *M* will be used to represent a negligible concentration in this text whenever the actual value is not known.

The $E^0$ value for the ferric–ferrous ion half-cell is 0.77 V.

$$E_{Fe^{3+}/Fe^{2+}} = E^0_{Fe^{3+}/Fe^{2+}} + \frac{0.059}{n} \log \frac{[Fe^{3+}]}{[Fe^{2+}]} \qquad \textbf{(11-40)}$$

$$= 0.77 + \frac{0.059}{1} \log \frac{10^{-6}}{10^{-1}}$$

$$= 0.77 + 0.059 \log 10^{-5}$$

$$= 0.77 + 0.059\,(-5)$$

$$= 0.77 - 0.30 = 0.47 \text{ V}$$

• **Example 5.** Calculate the actual potential of the ferric–ferrous ion half-cell at the end of the titration with dichromate. In this example assume that the concentrations are the reverse of the concentrations in Example 4, since the ferrous ion has been quantitatively converted to ferric ion. The calculation is

$$E_{Fe^{3+}/Fe^{2+}} = 0.77 + 0.059 \log \frac{[Fe^{3+}]}{[Fe^{2+}]}$$

$$= 0.77 + 0.059 \log \frac{10^{-1}}{10^{-6}}$$

$$= 0.77 + 0.059 \log 10^{+5}$$

$$= 0.77 + 0.059\,(5)$$

$$= 0.77 + 0.30 = 1.07 \text{ V}$$

• **Example 6.** Calculate the actual potential in a solution in which the permanganate concentration is $10^{-1}$ *M*, the manganous concentration is $10^{-4}$ *M*, and the $P_cH$ is 1. Also, calculate the half-cell potential if the $p_cH$ is 3. The $E^0$ value for the permanganate–manganous ion half-cell is 1.51. Substitution in eq. (11-39) gives

$$E_{MnO_4^-/Mn^{2+}} = 1.51 + \frac{0.059}{5} \log \frac{10^{-1} \times (10^{-1})^8}{10^{-4}}$$

$$= 1.51 + \frac{0.059}{5} \log 10^{-5}$$

$$= 1.51 + \frac{0.059}{5} (-5)$$

$$= 1.51 - 0.060 = 1.45 \text{ V}$$

For $p_cH$ 3, the calculation becomes

$$E_{MnO_4^-/Mn^{2+}} = 1.51 + \frac{0.059}{5} \log \frac{10^{-1} \times (10^{-3})^8}{10^{-4}}$$

$$= 1.51 + \frac{0.059}{5} \log 10^{-21}$$

$$= 1.51 + \frac{0.059}{5} (-21) = 1.51 - 0.24$$

$$= 1.27 \text{ V}$$

## EFFECT OF $p_cH$ UPON THE POTENTIAL OF HALF-CELLS

Example 6 shows that the potential of certain half-cells will be affected by a variation in the hydrogen ion concentration, or $p_cH$. The extent of this effect can be estimated for various electrodes to see which is the most greatly affected by the variations in $p_cH$. Only those half-cells in which hydrogen or hydroxide is involved in the half-cell reaction are affected greatly by variations in $p_cH$. Other half-cells may be affected by the hydrogen ion concentration, but the extent cannot be predicted by the Nernst equation. Using the general oxidation–reduction half-cell reaction,

$$XO_{m/2}^{b-} + mH^+ + ne^- \longrightarrow X^{c+} + \frac{m}{2} H_2O \qquad \textbf{(11-41)}$$

the Nernst equation is

$$E = E^0 + \frac{0.059}{n} \log \frac{[XO_{m/2}^{b-}][H^+]^m}{[X^{c+}]} \qquad \textbf{(11-42)}$$

which may be written as

$$E = E^0 + \frac{0.059}{n} \log \frac{[XO_{m/2}^{b-}]}{[X^{c+}]} + \frac{0.059m}{n} \log [H^+] \qquad \textbf{(11-43)}$$

Consequently, at constant concentrations of $XO_{m/2}^{b-}$ and $X^{c+}$, the value of $E$ is governed by the variations in the value of the second logarithmic term, which contains $m/n$ as a multiplier. The ratio of $m/n$ can be calculated and used to predict the extent of the variation caused by a given change in $p_cH$.

As an example, determine which of the following half-cells will be most greatly affected and which will be least affected by a given change in $p_cH$: $Cr_2O_7^{2-}/2Cr^{3+}$; $MnO_4^-/Mn^{2+}$; $MnO_4^-/MnO_2$; $IO_3^-/I^-$; $AsO_4^{3-}/AsH_3$, for which the half-cell reactions are

$$Cr_2O_7^{2-} + 14H^+ + 6e^- \longrightarrow 2Cr^{3+} + 7H_2O \qquad \textbf{(11-4)}$$

$$MnO_4^- + 8H^+ + 5e^- \longrightarrow Mn^{2+} + 4H_2O \qquad \textbf{(11-14)}$$

$$MnO_4^- + 4H^+ + 3e^- \longrightarrow MnO_2 + 2H_2O \qquad \textbf{(11-44)}$$

$$IO_3^- + 6H^+ + 6e^- \longrightarrow I^- + 3H_2O \qquad \textbf{(11-45)}$$

$$AsO_4^{3-} + 11H^+ + 8e^- \longrightarrow AsH_3 + 4H_2O \qquad \textbf{(11-46)}$$

The $m/n$ values are

| | | |
|---|---|---|
| $Cr_2O_7^{2-}/2Cr^{3+}$ | : | $14/6 = 2.33$ |
| $MnO_4^-/Mn^{2+}$ | : | $8/5 = 1.60$ |
| $MnO_4^-/MnO_2$ | : | $4/3 = 1.33$ |
| $IO_3^-/I^-$ | : | $6/6 = 1.00$ |
| $AsO_4^{3-}/AsH_3$ | : | $11/8 = 1.38$ |

Of these five half-cells, $Cr_2O_7^{2-}/2Cr^{3+}$ will be the most greatly affected by a given change in $p_cH$, and $IO_3^-/I^-$ will be the least affected.

## RELATIONSHIP BETWEEN STANDARD POTENTIALS AND THE EQUILIBRIUM CONSTANT

The value of the equilibrium constant for an oxidation–reduction reaction depends upon the values of the standard potentials of the oxidizing and reducing agents and upon the number of electrons involved in the balanced equation for the reaction. To develop the expression that relates these factors, consider the general oxidation–reduction reaction

$$Ox_{(1)} + Red_{(2)} \longrightarrow Ox_{(2)} + Red_{(1)} \tag{11-47}$$

for which the equilibrium constant expression is

$$K_{eq} = \frac{[Ox_{(2)}][Red_{(1)}]}{[Ox_{(1)}][Red_{(2)}]} \tag{11-48}$$

Equation (11-47) should be read "the oxidized form of substance 1 reacts with the reduced form of substance 2 to form the reduced form of substance 1 and the oxidized form of substance 2."

In any oxidation–reduction reaction, there must be electrical equilibrium whenever chemical equilibrium exists. Stated another way, the potential of the oxidation half-cell must equal the potential of the reduction half-cell whenever chemical equilibrium exists in an oxidation–reduction system, or

$$E_{(1)} = E_{(2)} \quad \text{at equilibrium} \tag{11-49}$$

Using the subscript (1) to represent the oxidizing agent and (2) to represent the reducing agent, the Nernst equations for $E_{(1)}$ and $E_{(2)}$ are

$$E_{(1)} = E^0_{(1)} + \frac{0.059}{n} \log \frac{[Ox_{(1)}]}{[Red_{(1)}]} \tag{11-50}$$

$$E_{(2)} = E^0_{(2)} + \frac{0.059}{n} \log \frac{[Ox_{(2)}]}{[Red_{(2)}]} \tag{11-51}$$

Equations (11-50) and (11-51) can be equated, since $n$ is the same in both and equals the number of electrons in the balanced oxidation–reduction reaction.

$$E^0_{(1)} + \frac{0.059}{n} \log \frac{[Ox_{(1)}]}{[Red_{(1)}]} = E^0_{(2)} + \frac{0.059}{n} \log \frac{[Ox_{(2)}]}{[Red_{(2)}]} \quad \textbf{(11-52)}$$

Equation (11-52) may be rearranged to give

$$E^0_{(1)} - E^0_{(2)} = \frac{0.059}{n} \left( \log \frac{[Ox_{(2)}]}{[Red_{(2)}]} - \log \frac{[Ox_{(1)}]}{[Red_{(1)}]} \right) \quad \textbf{(11-53)}$$

or

$$\frac{n}{0.059} (E^0_{(1)} - E^0_{(2)}) = \log \frac{[Ox_{(2)}][Red_{(1)}]}{[Red_{(2)}][Ox_{(1)}]} \quad \textbf{(11-54)}$$

The logarithmic term is the same as eq. (11-48) for the equilibrium constant for the reaction, and $K_{eq}$ may be substituted to give

$$\log K_{eq} = \frac{n}{0.059} (E^0_{(1)} - E^0_{(2)}) \quad \textbf{(11-55)}$$

In eq. (11-55), the subscript (1) refers to the oxidizing agent and the subscript (2) refers to the reducing agent. As in other equilibria, the value of $K_{eq}$ is a measure of the completeness of the reaction or a measure of the concentrations of the reactants left at equilibrium. The larger the value of $K_{eq}$, the more complete the reaction.

The equilibrium constant for any oxidation–reduction reaction can be calculated using eq. (11-55). However, a high value of the constant does not indicate that the reaction is satisfactory for analytical determinations because the kinetics of the reaction are not indicated by the equilibrium constant. Even though very negligible amounts of reactants might be left at equilibrium, a reaction is not suitable for an analytical determination if the attainment of equilibrium requires an unduly long period of time.

• **Example 7.** Calculate $K_{eq}$ using eq. (11-55) for the reaction of potassium dichromate and ferrous iron represented by eq. (11-9). The $E^0$ values for the two half-cells are 1.33 for the dichromate–chromic ion half-cell and 0.77 for the ferric–ferrous ion half-cell. The number of electrons in the balanced reaction is 6, so

$$\log K_{eq} = \frac{6}{0.059} (1.33 - 0.77)$$

$$= \frac{6}{0.059} (0.56) = 56.95 \quad \textbf{(11-56)}$$

and

$$K = 10^{+56.95}$$

Since the value of the equilibrium constant is large, the reaction goes to completion and the reaction is quantitative.

• **Example 8.** Calculate the equilibrium constant for reaction (11-29) for copper(II) ion plus iodide ion. The $E^0$ values are 0.86 and 0.54, and $n$ equals 2.

$$\log K_{eq} = \frac{2}{0.059}(0.86 - 0.54)$$

$$= \frac{2}{0.059}(0.32) = 10.85 \qquad \textbf{(11-57)}$$

and

$$K_{eq} = 10^{10.85} = 7.1 \times 10^{10}$$

Even though the value for $K_{eq}$ for the copper–iodide reaction is much less than that for the dichromate–iron reaction, the copper–iodide reaction is quantitative and can be used satisfactorily in analytical determinations.

A reaction is assumed to be quantitative if the $K_{eq}$ is $10^6$ or greater, so the minimum difference in $E^0$ values that allows quantitatively complete reaction is $0.35/n$. This value is obtained from eq. (11-55) as follows:

$$0.059 \log 10^6 = n(E^0_{(1)} - E^0_{(2)})_{min}$$

$$(E^0_{(1)} - E^0_{(2)})_{min} = \frac{0.059 \times 6}{n} = \frac{0.35}{n} \qquad \textbf{(11-58)}$$

For any reaction to take place as written, the $K_{eq}$ value must be 1 or more, which means that the $E^0$ value of the oxidizing agent must be a larger positive value than the $E^0$ value of the reducing agent. A $K_{eq}$ value less than 1 indicates that the reverse reaction takes place and, if the value is $10^{-6}$ or less, that the reverse reaction is quantitative. Also, for any reaction to take place, both an oxidizing agent and a reducing agent must be present.

The value of $n(E^0_{(1)} - E^0_{(2)})$ can be used for several similar reactions to predict which will have the largest and which will have the smallest equilibrium constants. The reaction that has the largest value of $n$ times the difference in $E^0$ values will have the largest equilibrium constant.

## TYPICAL CURVE FOR OXIDATION-REDUCTION TITRATIONS

The change in potential during the course of an oxidation–reduction titration can be followed by immersing two electrodes in the solution during the titration. One of the two electrodes must be a reference electrode and the other must be an indicating electrode. A reference electrode produces a constant potential

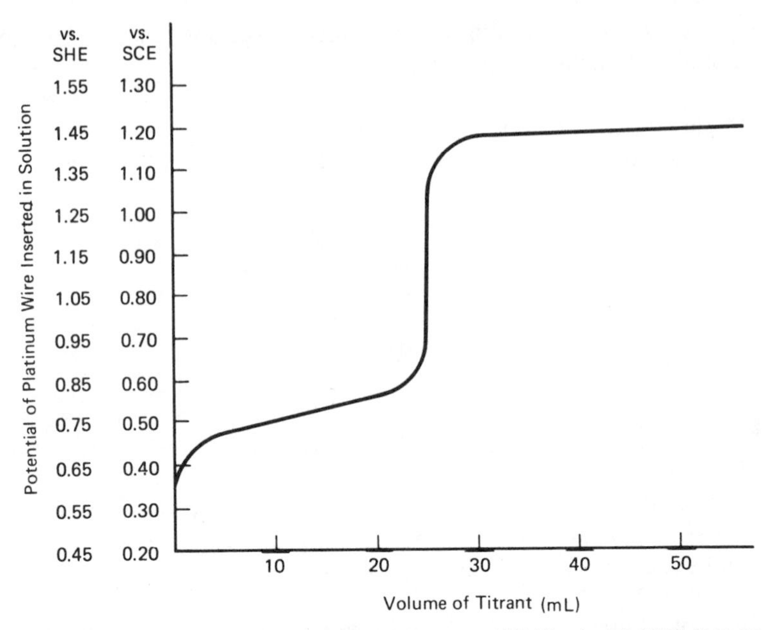


**Figure 11-2.** Variation of potential during the titration of 25.00 mL of 0.1000 *N* $FeSO_4$ with 0.1000 *N* $Ce(SO_4)_2$ in $H_2SO_4$ solution. SHE is the standard hydrogen electrode and SCE is the saturated calomel electrode.

irrespective of changes in the composition of the solution in which it is immersed, but the potential exerted by an indicator electrode must vary quantitatively as the concentration of the measured ion varies in the solution. Since the concentration of the titrated ion varies as it reacts with the titrant, the potential difference between the two electrodes will vary with the volume of titrant that is added.

A typical graph for the change in potential during the course of a titration is shown in Figure 11-2, which represents the titration of 25.00 mL of 0.1000 *N* ferrous sulfate with 0.1000 *N* ceric sulfate in sulfuric acid solution. The **saturated calomel electrode** (SCE) used in this titration is probably the most popular of all reference electrodes. The potential measurement against a **standard hydrogen half-cell** (SHE) is shown for comparison.

The potential at the beginning of the titration up to the end point is controlled by the potential of the titrated ion, and after the end point it is controlled by the potential of the titrant ion. As a consequence, the break at the equivalence point is a function of the difference in $E^0$ values. In the iron–ceric ion reaction, the break is approximately 0.4 to 0.5 V. As with neutralization titrations, there must be a sharp break in potential at the equivalence point for the tiration to be satisfactory.

# CALCULATION OF POTENTIALS AT VARIOUS POINTS ON THE TITRATION CURVE

## BEFORE THE EQUIVALENCE POINT

The ceric sulfate–ferrous sulfate titration will be used as the example in the calculation of the potential along the titration curve. Since it is difficult to estimate the actual ferric ion concentration in the solution at the beginning of the titration before the addition of any ceric sulfate, we shall calculate the potentials after the addition of 5.00, 12.50, 20.00, and 24.90 mL of the ceric sulfate.

### *After Addition of 5.00 mL of Titrant*

The equation for the reaction of ceric ion with ferrous ion is

$$Ce^{4+} + Fe^{2+} \longrightarrow Ce^{3+} + Fe^{3+} \qquad \textbf{(11-59)}$$

and there is an electron change of 1 on both ions. The $E^0$ value for ceric–cerous is 1.44 V, and for ferric–ferrous it is 0.77 V.

At the beginning of the titration there are 2.500 meq of ferrous iron in the solution. Since there is an electron change of 1, the equivalent weight is equal to the atomic weight for both cerium and iron and milliequivalents equal millimoles for both.

After addition of titrant, the solution will contain ferrous ions, ferric ions, cerous ions, and ceric ions. The ferrous and ferric concentrations can be calculated from the millimoles at the start and the millimoles that have reacted, but there is no simple way to estimate the ceric ion concentration left unreacted from the data of the titration. As a consequence, the ferric–ferrous system should be used to calculate the potential in the solution before the equivalence point is reached.

$$\begin{aligned}
&\text{mmoles } Fe^{2+} \text{ at start} = 25.00 \times 0.1000 && = 2.500\\
&\text{mmoles } Fe^{3+} \text{ and } Ce^{3+} \text{ formed} (= \text{mmoles } Ce^{4+} \text{ added})\\
&\qquad\qquad\qquad\qquad = 5.00 \times 0.1000 && = \underline{0.500}\\
&\text{mmoles } Fe^{2+} \text{ left unreacted} && = 2.000
\end{aligned}$$

Since the millimoles of ferrous and ferric ion are in the same solution, millimoles may be substituted for molar concentration in the Nernst equation, so that the calculation of the potential becomes

$$\begin{aligned}
E_{Fe^{3+}/Fe^{2+}} &= 0.77 + \frac{0.059}{1} \log \frac{0.500}{2.000}\\
&= 0.77 + 0.059 \log 0.25\\
&= 0.77 + 0.059 \log 10^{-0.60}\\
&= 0.77 + 0.059(-0.60) = 0.77 - 0.035 = 0.77 - 0.04\\
&= 0.73 \text{ V} \qquad \textbf{(11-60)}
\end{aligned}$$

### After the Addition of 12.50 mL of Titrant

mmoles $Fe^{2+}$ at start = 2.500
mmoles $Fe^{3+}$ and $Ce^{3+}$ formed (=mmoles $Ce^{4+}$ added) = 1.250
mmoles of $Fe^{2+}$ left unreacted = 1.250

$$E_{Fe^{3+}/Fe^{2+}} = 0.77 + 0.059 \log \frac{1.250}{1.250}$$

$$= 0.77 \text{ V}$$

### After Addition of 20.00 mL of Titrant

mmoles $Fe^{2+}$ at start = 2.500
mmoles $Fe^{3+}$ and $Ce^{3+}$ formed (=mmoles $Ce^{4+}$ added) = 2.000
mmoles $Fe^{2+}$ left unreacted = 0.500

$$E_{Fe^{3+}/Fe^{2+}} = 0.77 + 0.059 \log \frac{2.000}{0.500}$$

$$= 0.77 + 0.059 \log 4 = 0.77 + 0.059 \log 10^{0.60}$$

$$= 0.77 + 0.059(0.60) = 0.77 + 0.035 = 0.77 + 0.04$$

$$= 0.81 \text{ V}$$

### After Addition of 24.90 mL of Titrant

mmoles $Fe^{2+}$ at start = 2.500
mmoles $Fe^{3+}$ and $Ce^{3+}$ formed (=mmoles $Ce^{4+}$ added) = 2.490
mmoles $Fe^{2+}$ left unreacted = 0.010

$$E_{Fe^{3+}/Fe^{2+}} = 0.77 + 0.059 \log \frac{2.490}{0.010}$$

$$= 0.77 + 0.059 \log 249$$

$$= 0.77 + 0.059 \log 10^{2.40}$$

$$= 0.77 + 0.059(2.40) = 0.77 + 0.14$$

$$= 0.91 \text{ V}$$

## AT THE EQUIVALENCE POINT

The potential at the equivalence point for reactions of the general type

$$a\text{A} + b\text{B} \rightleftharpoons a\text{C} + b\text{D}$$

is given by

$$E_{eq} = \frac{n_A E_A^0 + n_B E_B^0}{n_A + n_B} \qquad \textbf{(11-61)}$$

where the subscript A refers to the half-reaction of the oxidizing agent and the subscript B refers to the half-reaction of the reducing agent. Applied to the ferrous–ceric titration, this becomes

$$E_{eq} = \frac{1.44 + 0.77}{1 + 1} = \frac{2.21}{2} = 1.105 \text{ V}$$

Equation (11-61) does not hold for more complicated titration reactions such as the permanganate–ferrous iron titration.

## BEYOND THE EQUIVALENCE POINT

Beyond the equivalence point, it is difficult to estimate the amount of ferrous ion left unreacted, but it is easy to calculate the concentrations of cerous and ceric ions if it is assumed that the millimoles of cerous formed equals the millimoles of ferrous at the start of the titration, that is, that the amount of ferrous ion left unreacted is negligible compared to the amount of cerous ion that is formed.

### *After the Addition of 25.10 mL of Titrant*

mmoles $Fe^{2+}$ at start = mmoles $Ce^{3+}$ formed = $25.00 \times 0.1000 = 2.500$
mmoles $Ce^{4+}$ added = $25.10 \times 0.1000 = 2.510$
mmoles $Ce^{4+}$ left unreacted $= 0.010$

$$E_{Ce^{4+}/Ce^{3+}} = E^0_{Ce^{4+}/Ce^{3+}} + \frac{0.059}{n} \log \frac{\text{mmoles } Ce^{4+}}{\text{mmoles } Ce^{3+}} \qquad \textbf{(11-62)}$$

$$= 1.44 + \frac{0.059}{1} \log \frac{0.010}{2.500}$$

$$= 1.44 + 0.059 \log 0.004$$

$$= 1.44 + 0.059 \log 10^{-2.40}$$

$$= 1.44 + 0.059(-2.40) = 1.44 - 0.14$$

$$= 1.30 \text{ V}$$

### *After Addition of 30.00 mL of Titrant*

$$\begin{aligned}
\text{mmoles } Ce^{3+} \text{ formed} &= 25.00 \times 0.1000 = 2.500 \\
\text{mmoles } Ce^{4+} \text{ added} &= 30.00 \times 0.1000 = 3.000 \\
\text{mmoles } Ce^{4+} \text{ left unreacted} &= 0.500
\end{aligned}$$

$$\begin{aligned}
E_{Ce^{4+}/Ce^{3+}} &= 1.44 + 0.059 \log \frac{0.500}{2.500} \\
&= 1.44 + 0.059 \log 0.200 \\
&= 1.44 + 0.059 \log 10^{-0.70} \\
&= 1.44 + 0.059(-0.70) \\
&= 1.44 - 0.04 = 1.40 \text{ V}
\end{aligned}$$

### *After Further Additions of Titrant*

By similar calculations the potential after 42.50 mL of titrant is 1.43, and after 50.00 mL it is 1.44.

## CALCULATION OF CONCENTRATIONS AT VARIOUS POINTS ON THE TITRATION CURVE

### BEFORE THE EQUIVALENCE POINT

Using the ceric–ferrous reaction as the example, the ferrous, ferric, and cerous ion concentrations can be calculated as shown in the preceding section. The ceric ion concentration left unreacted cannot be calculated directly from the titration data, and, as a consequence, the ferric–ferrous half-cell is used to calculate the potential before the equivalence point. Since the potential of the two half-cells must be equal, the calculated potential can be used with the Nernst equation for the ceric–cerous half-cell to calculate the concentration left unreacted. As an example, calculate the ceric ion concentration left unreacted after the addition of 12.50 mL of 0.1000 $N$ $Ce(SO_4)_2$ to 25.00 mL of 0.1000 $N$ $FeSO_4$ in sulfuric acid solution. Assume that the solution has been diluted to 100 mL. The potential in the solution is 0.77 V and the $E^0$ potential for the ceric–cerous half-cell is 1.44 V.

The milliequivalents of cerous ion produced, assuming that the concentration of ceric ion left unreacted is negligible compared to the cerous ion concentration, will be equal to the milliequivalents of ceric added:

$$\begin{aligned}
\text{meq } Ce^{3+} \text{ produced} &= \text{meq } Ce^{4+} \text{ added} \\
&= 12.50 \times 0.1000 = 1.250 \text{ meq}
\end{aligned}$$

and the normal concentration of $Ce^{3+}$ will be 1.250 meq/100.0 mL, or 0.0125 $N$.

Since there is an electron change of 1, the normal and molar concentrations will be the same, so the molar $Ce^{3+}$ concentration is 0.0125 *M*.

Substitution of these values in the Nernst equation,

$$E_{Ce^{4+}/Ce^{3+}} = E^0_{Ce^{4+}/Ce^{3+}} + \frac{0.059}{n} \log \frac{[Ce^{4+}]}{[Ce^{3+}]} \tag{11-62}$$

gives

$$0.77 = 1.44 + 0.059 \log \frac{[Ce^{4+}]}{0.0125}$$

$$= 1.44 + 0.059 \log \frac{[Ce^{4+}]}{10^{-1.90}}$$

which can be rearranged as

$$\log 10^{1.90}[Ce^{4+}] = \frac{0.77 - 1.44}{0.059} = -11.36$$

or

$$10^{1.90}[Ce^{4+}] = 10^{-11.36}$$

and

$$[Ce^{4+}] = \frac{10^{-11.36}}{10^{1.90}} = 10^{-13.26} = 5.5 \times 10^{-14}\ M$$

## AT THE EQUIVALENCE POINT

The concentrations of reactants left unreacted at the equivalence point is a measure of the completeness of the reaction and can be calculated from the titration data and the equilibrium constant or the equivalence point potential. The equilibrium constant method should be used for all reactions that involve different numbers of electrons in the half-reactions.

Examples of both methods of calculation will be shown for the ceric–ferrous reaction, and an example of calculation by the equilibrium constant method will be shown for the arsenious acid–potassium permanganate reaction.

In the ceric–ferrous reaction, the cerous and ferric ion concentrations can be calculated for both methods from the titration data if it is assumed that the reaction is quantitatively complete, that is, that the concentrations of reactants left unreacted at the equivalence point are negligible compared to the concentrations of the products formed. The milliequivalents of cerous ion formed will equal the milliequivalents of ceric ion added, and the milliequivalents of ferric ion formed will equal the milliequivalents of ferrous ion at the start of the titration. For the titration of 25.00 mL of 0.1000 $N$ $FeSO_4$ with 0.1000 $N$ $Ce(SO_4)_2$, assuming an equivalence point volume of 100.0 mL,

$$\begin{aligned}\text{meq } Ce^{3+} \text{ formed} &= \text{meq } Ce^{4+} \text{ added}\\ &= 25.00 \times 0.1000 = 2.500\end{aligned}$$

and

$$\begin{aligned}\text{meq } Fe^{3+} \text{ formed} &= \text{meq } Fe^{2+} \text{ at start}\\ &= 25.00 \times 0.1000 = 2.500\end{aligned}$$

The normal concentration of $Ce^{3+}$ and $Fe^{3+}$ will be

$$N = \frac{2.500}{100.0} = 0.025\ N$$

and, for both ions,

$$M = \frac{N}{\text{eq/mole}} = \frac{0.025}{1} = 0.025\ M$$

Since there is an electron change of 1 in both half-reactions, milliequivalents equal millimoles and normal concentration equals molar concentration for all reactants and products.

### *By the Equivalence Potential*

The equivalence potential for the ceric–ferrous reaction is 1.105 V, as calculated by eq. (11-61). Since the potential of each half-cell must be equal and must equal the equivalence potential at the equivalence point,

$$E_{eq} = 1.105 = E^0_{Fe^{3+}/Fe^{2+}} + \frac{0.059}{n} \log \frac{[Fe^{3+}]}{[Fe^{2+}]} \qquad \textbf{(11-63)}$$

and

$$E_{eq} = 1.105 = E^0_{Ce^{4+}/Ce^{3+}} + \frac{0.059}{n} \log \frac{[Ce^{4+}]}{[Ce^{3+}]}. \qquad \textbf{(11-64)}$$

Using the ferric–ferrous half-cell first,

$$1.105 = 0.77 + 0.059 \log \frac{0.025}{[Fe^{2+}]}$$

or

$$\log \frac{0.025}{[Fe^{2+}]} = \frac{1.105 - 0.77}{0.059} = \frac{0.335}{0.059} = 5.68$$

and

$$\frac{0.025}{[Fe^{2+}]} = 10^{5.68}$$

so

$$[Fe^{2+}] = \frac{0.025}{10^{5.58}} = \frac{10^{-1.60}}{10^{5.68}} = 10^{-7.28}$$

$$= 5.2 \times 10^{-8}\ M$$

Using the ceric–cerous half-cell, the concentration of ceric ion left unreacted is calculated by eq. (11-64) as

$$1.105 = 1.44 + 0.059 \log \frac{[Ce^{4+}]}{[Ce^{3+}]}$$

$$= 1.44 + 0.059 \log \frac{[Ce^{4+}]}{0.025}$$

or

$$0.059 \log \frac{[Ce^{4+}]}{0.025} = 1.105 - 1.44 = -0.335$$

and

$$\log 40[Ce^{4+}] = -\frac{0.335}{0.059} = -5.68$$

$$10^{1.60}[Ce^{4+}] = 10^{-5.68}$$

$$[Ce^{4+}] = \frac{10^{-5.68}}{10^{1.60}} = 10^{-7.28} = 5.2 \times 10^{-8}$$

### *By Use of the Equilibrium Constant*

The equilibrium constant for the ceric–ferrous reaction may be calculated by eq. (11-55) as

$$\log K_{eq} = \frac{1}{0.059}(1.44 - 0.77) = \frac{0.67}{0.059} = 11.36$$

$$K_{eq} = 10^{11.36}$$

The equilibrium constant expression for the ceric–ferrous reaction is

$$\frac{[Ce^{3+}][Fe^{3+}]}{[Ce^{4+}][Fe^{2+}]} = 10^{11.36} \tag{11-65}$$

$[Ce^{3+}] = [Fe^{3+}] = 0.025 = 10^{-1.60}$ *M* for the titration of 25.00 mL of 0.1000 *N* $FeSO_4$ with 0.1000 *N* $Ce(SO_4)_2$, assuming an equivalence point volume of 100 mL. Since the milliequivalents of $Ce^{4+}$ left unreacted must equal the milliequivalents of $Fe^{2+}$ left unreacted, $[Ce^{4+}] = [Fe^{2+}]$. Substitution of $[Ce^{4+}]$ for $[Fe^{2+}]$ in eq. (11-65) gives

$$\frac{10^{-1.60} \times 10^{-1.60}}{[Ce^{4+}]^2} = 10^{11.36}$$

or

$$[Ce^{4+}]^2 = \frac{10^{-3.20}}{10^{11.36}} = 10^{-14.56}$$

and, extracting the square root,

$$[Ce^{4+}] = 10^{-7.28} = 5.2 \times 10^{-8}\ M$$

and, since $[Ce^{4+}] = [Fe^{2+}]$,

$$[Fe^{2+}] = 5.2 \times 10^{-8}\ M$$

The small difference in the values calculated from the equivalence potential and from the equilibrium constant arises from the rounding off of the exponents to two decimal places.

### *Calculation of the Equivalence Point Concentrations for the Permanganate–Arsenious Acid Reaction*

Since both half-cells for the ceric–ferrous reaction have an electron change of 1, the calculation of the equivalence point concentrations is simpler than for reactions in which the electron changes of the half-cells differ, such as the permanganate–arsenious acid reaction. The balanced reaction for permanganate and arsenious acid in hydrochloric acid solution is

$$2MnO_4^- + 5H_3AsO_3 + 6H^+ \longrightarrow 2Mn^{2+} + 5H_3AsO_4 + 3H_2O \tag{11-66}$$

Since the number of electrons involved in the two half-reactions differs, the equilibrium constant should be used for the calculation of concentrations. The equilibrium constant, calculated by eq. (11-55), is

$$\log K_{eq} = \frac{10}{0.059}(1.53 - 0.56) = \frac{9.7}{0.059} = 164.41$$

For the titration of 25.00 mL of 0.1000 $N$ $H_3AsO_3$ with 0.1000 $N$ $KMnO_4$, assuming a $p_cH$ of 1 in an equivalence point volume of 100.0 mL, the concentrations of the $Mn^{2+}$ ion and $H_3AsO_4$ can be calculated from the titration data since the milliequivalents of these substances (assuming quantitative reaction) will be equal to the milliequivalents of the reactants in the titration:

$$\begin{aligned} \text{meq } Mn^{2+} \text{ formed} &= \text{meq } MnO_4^- \text{ added} \\ &= 25.00 \times 0.1000 = 2.500 \end{aligned}$$

and

$$\begin{aligned} \text{meq } H_3AsO_4 \text{ formed} &= \text{meq } H_3AsO_3 \text{ at start} \\ &= 25.00 \times 0.1000 = 2.500 \end{aligned}$$

The normal concentration of each of these in the equivalence point solution is

$$\frac{2.500 \text{ meq}}{100.0 \text{ mL}} = 0.0250\ N$$

Since molar concentrations are required in the equilibrium constant, the normal concentrations can be converted to molar by dividing by the number of equivalents per mole:

$$[H_3AsO_4] = \frac{N}{2} = \frac{0.0250}{2} = 0.0125 = 10^{-1.90}\ M$$

and

$$[Mn^{2+}] = \frac{N}{5} = \frac{0.0250}{5} = 0.0050 = 10^{-2.30}\ M$$

The equilibrium constant expression for the permanganate–arsenious acid reaction is

$$K_{eq} = \frac{[Mn^{2+}]^2[H_3AsO_4]^5}{[MnO_4^-]^2[H_3AsO_3]^5[H^+]^6} = 10^{164.41} \qquad \textbf{(11-67)}$$

The molar concentrations of arsenic acid and manganous ion are $10^{-1.90}$ and $10^{-2.30}$ $M$, respectively, as shown previously. There is no simple way to calculate the concentrations of permanganate and arsenious acid left unreacted from the titration data, and, as a consequence, the relationship between the molar concentrations of the two reactants must be determined so that one can be expressed in terms of the other for substitution in the equilibrium constant expression.

Since the milliequivalents of $MnO_4^-$ and $H_3AsO_3$ left unreacted are equal, the normal concentrations are equal:

$$N_{MnO_4^-} = N_{H_3AsO_3}$$

For permanganate, the molar concentration is

$$[MnO_4^-] = \frac{N_{MnO_4^-}}{5}$$

or

$$N_{MnO_4^-} = 5[MnO_4^-]$$

For arsenious acid,

$$[H_3AsO_3] = \frac{N_{H_3AsO_3}}{2}$$

or

$$N_{H_3AsO_3} = 2[H_3AsO_3]$$

and

$$5[MnO_4^-] = 2[H_3AsO_3]$$

or

$$[H_3AsO_3] = \tfrac{5}{2}[MnO_4^-] = 2.5[MnO_4^-] = 10^{0.40}[MnO_4^-]$$

Substitution in eq. (11-67) gives

$$\frac{[10^{-2.30}]^2 \times [10^{-1.90}]^5}{[MnO_4^-]^2 \times (10^{0.40}[MnO_4^-])^5 \times (10^{-1})^6} = 10^{164.41}$$

or

$$[MnO_4^-]^2 \times 10^{2.00}[MnO_4^-]^5 = \frac{10^{-4.60} \times 10^{-9.50}}{10^{-6} \times 10^{164.41}}$$

and

$$10^{2.00}[MnO_4^-]^7 = 10^{-172.51}$$

$$[MnO_4^-]^7 = \frac{10^{-172.51}}{10^{2.00}} = 10^{-174.51}$$

$$[MnO_4^-] = 10^{-174.51/7} = 10^{-24.93} = 1.2 \times 10^{-25}\ M$$

The arsenious acid concentration is

$$[H_3AsO_3] = 2.5[MnO_4^-] = 10^{0.40} \times 10^{-24.93}$$
$$= 10^{-24.53} = 3.0 \times 10^{-25}\ M$$

## OXIDATION-REDUCTION INDICATORS

In some oxidation–reduction titrations, a colored reactant or product can act as the indicator. For example, permanganate ion has a deep purple color, whereas the color of maganous ion is a very pale pink, so pale, in fact, that it is essentially colorless in the concentrations found at the equivalence point in permanganate titrations. As a result, the first excess of permangante colors the solution providing the other products are colorless. The appearance or disappearance of the color of the triiodide ion in procedures involving iodine can also serve as an end-point indicator.

For reactions in which the reactants or products do not indicate the end point, indicators must be added to the solution. Such indicators may form a colored complex with a reactant or product or else must be reversible oxidation–reduction systems in which the two members of the oxidation–reduction pair have different colors. An example of a colored complex indicator is the use of thiocyanate ion for the titration of iron(III) (ferric ion) with titanium(III). The deep red color of the $FeSCN^{2+}$ complex colors the solution until all the iron is reduced. Another example is the use of starch in titrations involving iodine. Starch forms a blue complex with iodine and provides very sharp end points.

There are several oxidation–reduction indicators that are oxidation–reduction systems. It can be shown that the color change of such indicators occurs over a range of approximately 0.12 V if both colored forms are equally easy to see. The range is centered at the $E^0$ value and is usually expressed as $E^0 \pm 0.06$ V. As a consequence, the $E^0$ value of the indicator must lie between the $E^0$ values of the oxidant and reductant and should be close to the equivalence point potential.

Diphenylamine and diphenylamine sulfonate are used as indicators for the titration of iron(II) (ferrous) ion with dichromate in the presence of phosphoric acid. The color change is from colorless to lavender or purple. The indicator does not change color at the true equivalence point unless phosphoric acid is present.

Ferroin (ferrous 1,10-phenanthroline) and modified ferroins such as nitroferroin (ferrous 5-nitro-1,10-phenanthroline) are often used as indicators for ceric titrations. The dark red ferrous complex is oxidized to the pale blue ferric complex. $E^0$ for ferroin is 1.06 V, and for the modified ferroins it ranges from 0.81 to 1.25 V.

## PROBLEMS

**1** Balance the following equations to obtain the ionic net reaction and the molecular equation.

(a) $K_2Cr_2O_7 + FeSO_4 + H_2SO_4 \rightarrow Cr_2(SO_4)_3 + Fe_2(SO_4)_3 + H_2O + K_2SO_4$
(b) $KMnO_4 + Fe(NH_4)_2(SO_4)_2 + H_2SO_4 \rightarrow MnSO_4 + Fe_2(SO_4)_3 + (NH_4)_2SO_4 + K_2SO_4 + H_2O$
(c) $Na_2S_2O_3 + KI_3 \rightarrow Na_2S_4O_6 + KI + NaI$
(d) $KIO_3 + KI + HC_2H_3O_2 \rightarrow KI_3 + KC_2H_3O_2 + H_2O$
(e) $KBrO_3 + SbCl_3 + HCl \rightarrow KBr + SbCl_5 + H_2O$
(f) $KMnO_4 + AsH_3 + H_2SO_4 \rightarrow H_3AsO_4 + MnSO_4 + H_2O + K_2SO_4$
(g) $H_3AsO_3 + KMnO_4 + H_2SO_4 \rightarrow H_3AsO_4 + MnSO_4 + K_2SO_4 + H_2O$
(h) $Ce(SO_4)_2 + SnSO_4 \rightarrow Ce_2(SO_4)_3 + Sn(SO_4)_2$
(i) $Ce(SO_4)_2 + H_2C_2O_4 \rightarrow Ce_2(SO_4)_3 + CO_2 + H_2SO_4$
(j) $KMnO_4 + K_4Fe(CN)_6 + H_2SO_4 \rightarrow MnSO_4 + K_3Fe(CN)_6 + K_2SO_4 + H_2O$

*Ans.* (b) $2KMnO_4 + 10Fe(NH_4)_2(SO_4)_2 + 8H_2SO_4 \rightarrow 2MnSO_4 + 5Fe_2(SO_4)_3 + 10(NH_4)_2SO_4 + K_2SO_4 + 8H_2O$

**2** Calculate the numerical values of the equivalent weights of the reactants and products involved in the electron change for the reactions in problem 1.

*Ans.* (b) $KMnO_4$, 31.61; $MnSO_4$, 30.20; $Fe(NH_4)_2(SO_4)_2$, 284.1; $Fe_2(SO_4)_3$, 199.9

**3** Balance the reactions and express the equivalent weights by showing the number of equivalents per mole for each of the substances listed on the right for the reactions listed on the left:

| | *Reaction* | *Substance* |
|---|---|---|
| (a) | $MnO_4^- + H^+ \rightarrow Mn^{2+} + H_2O$ | $KMnO_4$, $MnSO_4$, Mn |
| (b) | $MnO_4^- + H^+ \rightarrow MnO_2 + H_2O$ | $KMnO_4$, $MnO_4^-$, $MnO_2$, Mn |
| (c) | $H_2O_2 + H^+ \rightarrow H_2O$ | $H_2O_2$, O, $H_2O$ |
| (d) | $PbO_2 + H^+ \rightarrow Pb^{2+} + H_2O$ | Pb, $PbO_2$, $PbSO_4$, PbS, PbO |
| (e) | $H_3AsO_4 + H^+ \rightarrow H_3AsO_3 + H_2O$ | $Na_3AsO_4$, $As_2O_5$, $As_2O_3$, $H_3AsO_4$, As |
| (f) | $SO_4^{2-} + H^+ \rightarrow H_2SO_3 + H_2O$ | $K_2SO_4$, $Na_2SO_3$, S, $SO_3$, $SO_2$ |

$$\left( \textit{Ans.}\ (b)\ \frac{KMnO_4}{3}\ \frac{MnO_4^-}{3}\ \frac{MnO_2}{3}\ \frac{Mn}{3} \right)$$

**4** Calculate the actual potential of a $Cr_2O_7^{2-}/2Cr^{3+}$ half-cell at the following $p_cH$ and concentration values:

| | $p_cH$ | $[Cr_2O_7^{2-}]$ | $[Cr^{3+}]$ |
|---|---|---|---|
| (a) | −1 | $10^{-7}$ | 0.05 |
| (b) | 1 | 0.05 | 0.01 |
| (c) | 3 | $10^{-5}$ | $10^{-5}$ |
| (d) | 6 | $10^{-1}$ | $10^{-1}$ |
| (e) | 0 | $10^{-8}$ | 0.025 |

(*Ans.* (b) 1.22)

**5** Repeat problem 4 for the $MnO_4^-/Mn^{2+}$ half-cell, substituting $[MnO_4^-]$ for $[Cr_2O_7^{2-}]$ and $[Mn^{2+}]$ for $[Cr^{3+}]$.

*Ans.* (b) 1.42

**6** Repeat problem 4 for the $H_3AsO_4/H_3AsO_3$ half-cell, substituting $[H_3AsO_4]$ for $[Cr_2O_7^{2-}]$ and $[H_3AsO_3]$ for $[Cr^{3+}]$.

*Ans.* (b) 0.52

**7** Repeat problem 4 for the $IO_3^-/I^-$ half-cell, substituting $[IO_3^-]$ for $[Cr_2O_7^{2-}]$ and $[I^-]$ for $[Cr^{3+}]$.

*Ans.* (b) 1.04

**8** Repeat problem 4 for the $2BrO_3^-/Br_2$ half-cell, substituting $[BrO_3^-]$ for $[Cr_2O_7^{2-}]$ and $[Br_2]$ for $[Cr^{3+}]$.

*Ans.* (*b*) 1.47

**9** Which one of the four half-cells in problem 3(a), (b), (c), and (d) will be affected (1) most greatly and (2) the least by a given change in $p_cH$?

*Ans.* (1) $PbO_2/Pb^{2+}$, (2) $H_2O_2/H_2O$

**10** Which one of the half-cells in problems 4, 5, 7, and 8 will be most greatly affected and which will be least affected by a given change in $p_cH$? Prove your deduction using the answers to parts (c) and (e) of problems 4–8.

**11** Calculate the equilibrium constants for the reactions in problem 1, except (f).

*Ans.* (b) $4.2 \times 10^{62}$

**12** Balance the equations and calculate the equilibrium constants for the following reactions:

(a) Cerium(IV) (ceric) sulfate plus chromium(II) (chromous) sulfate to form cerium(III) (cerous) sulfate and chromium(III) (chromic) sulfate
(b) Potassium permanganate plus oxalic acid in the presence of sulfuric acid to form manganese(II) (manganous) sulfate and carbon dioxide
(c) Potassium iodate plus tin(II) (stannous) chloride in the presence of hydrochloric acid to form potassium iodide and tin(IV) (stannic) chloride
(d) Iron(III) (ferric) chloride plus potassium iodide to form potassium triodide and iron(II) (ferrous) chloride
(e) Arsenic acid plus sulfurous acid to form arsenious acid and sulfuric acid.

*Ans.* (b) $9.6 \times 10^{338}$

**13** Without calculating the actual equilibrium constant values, predict which one of the reactions in problem 1 will have the largest equilibrium constant and which one will have the smallest equilibrium constant.

**14** Without calculating the actual equilibrium constant values, predict which one of the reactions in problem 12 will have (a) the largest equilibrium constant and (b) the smallest equilibrium constant.

*Ans.* (a) b; (b) d

**15** Predict which of the following reactions will take place spontaneously as written, and without calculating the actual equilibrium constant values, predict which one will have the largest equilibrium constant. The equations are not balanced.

(a) $MnO_4^- + Fe^{2+} + H^+$
(b) $Ce^{4+} + Sn^{2+} + H_2SO_4$
(e) $Ce^{3+} + Fe^{3+} + H_2SO_4$
(d) $Cr_2O_7^{2-} + Ce^{4+} + H_2SO_4$
(e) $MnO_4^- + Sn^{2+} + H^+$
(f) $Mn^{2+} + Fe^{2+} + H^+$

**16** Calculate the potential in the titrated solution after the addition of the indicated volume of 0.1000 *N* $KMnO_4$ to 25.00 mL of 0.1000 *N* $Sn(SO_4)$ in $H_2SO_4$. Assume in each case that the volume is 100.0 mL and that the hydrogen ion concentration is 1 *M*

(a) 5.00 mL
(b) 12.50 mL
(c) 20.00 mL
(d) 24.95 mL
(e) 25.00 mL
(f) 25.10 mL
(g) 30.00 mL
(h) 40.00 mL
(i) 50.00 mL

*Ans.* (b) 0.15, (g) 1.50

**17** Repeat problem 16 for the titration of 0.2000 *N* $FeSO_4$ with 0.2000 *N* $K_2Cr_2O_7$ in sulfuric acid solution assuming that $[H^+]$ is 2 *M*.

**18** Repeat problem 16 for the titration of 1.8219 g of $Sb_2O_3$ which has been dissolved in 25.00 mL of HCl with 1.000 *N* $KIO_3$ assuming that the $p_cH$ is zero.

**19** Repeat problem 16 for the titration of 0.2793 g of iron wire which has been dissolved in hydrochloric acid and reduced to the iron(II) (ferrous) state before titrating with 0.2000 *N* cerium(IV) (ceric) chloride. Note that the $E°$ potential of the ceric–cerous half-cell is 1.28 in chloride solution.

**20** (a) Calculate the molar concentration of the titrant ion left unreacted in the titrated solution before the equivalence point in problem 16(a), (b), (c), and (d).
(b) Calculate the molar concentrations of the titrant ion and of the titrated ion left unreacted for problem 16(e).
(c) Calculate the molar concentration of the titrated ion left unreacted for problem 16(f), (g), (h), and (i).
*Ans.* (a) (b) $5.6 \times 10^{-118}$; (b) titrant ion, $5.9 \times 10^{-36}$ *M*; (c) (g) $1.9 \times 10^{-48}$ *M*.

**21** Repeat problem 20 using the titration in problem 17.

**22** Repeat problem 20 using the titration in problem 18.

**23** Repeat problem 20 using the titration in problem 19.

CHAPTER

# 12 Oxidation–Reduction Titrations

Oxidation–reduction titrations constitute a major portion of all titrimetric methods and have the advantage that the end points can often be determined potentiometrically for those titrations for which there is no satisfactory visual indicator. A representative example of each of the more important reagents will be discussed in this chapter. One of the most effective oxidizing agents used is potassium permanganate, which is a strong oxidizing agent that does not require the use of a separate indicator. The main disadvantages of permanganate are that it is difficult to prepare in stable form and that it cannot be used in acid solutions containing chloride unless special precautions are taken. Potassium dichromate is also used extensively because it can be obtained in primary standard grade and can be used in acid solutions containing chloride. Ceric sulfate is a strong oxidizing agent that can also be used in acid solutions containing chloride, but it has the disadvantage (or in some cases the advantage) that the strength of cerium(IV) solutions as oxidizing agents depends upon the anions present; the standard half-cell potential ranges from 1.28 V in chloride solutions to 1.70 V in perchlorate solutions. Ceric ion is usually used in sulfate solution, for which the standard potential is 1.44 V. Iodine (as triiodide) is used as a weak oxidizing agent with strong reducing agents in direct (iodimetric) titrations but has the disadvantage of losing strength slowly by volatilization of iodine from the solution. Many widely used methods are indirect (iodometric) in that iodide ion is first oxidized to iodine (or triiodide), which is then titrated with thiosulfate.

## PERMANGANATE METHODS

Permanganate ion is used for the determination of calcium as the oxalate, for the titration of iron, for the determination of arsenic as arsenite, and for other substances such as hydrogen peroxide, antimony, ferrocyanide, and sulfur dioxide. Permanganate is usually used in strong acid solutions (approximately

1–2 *N*) in which the product is manganese(II) (manganous) ion. In neutral and slightly acidic solutions the product is manganese dioxide, whereas in strongly basic solutions manganate ion ($MnO_4^{2-}$) is produced. The interference of chloride ion can be offset partially by adding a solution containing phosphoric acid, sulfuric acid, and manganous sulfate.

## STANDARDIZATION

Potassium permanganate is not a primary standard, and permanganate solutions must be standardized against primary standard substances. The most popular primary standards are sodium oxalate, arsenious oxide, and high-purity (99.99%) iron wire. Mohr's salt (ferrous ammonium sulfate hexahydrate, $Fe(NH_4)_2(SO_4)_2 \cdot 6H_2O$, has been used as a primary standard in the past. It has the advantage of a large equivalent weight, but the fact that it cannot be dried without decomposition is a distinct disadvantage. Selection of the primary standard depends upon the titration for which the solution will be used. In general, it is preferable to standardize a solution by the same reaction that will be utilized in the analysis. As an example, sodium oxalate is preferable for solutions that will be used in the calcium determination, whereas iron wire would be selected for solutions utilized in the determination of iron.

## DETERMINATION OF CALCIUM IN LIMESTONE BY THE OXALATE–PERMANGANATE METHOD

Magnesium usually occurs with calcium in natural materials such as limestone and other rock samples, so the method of analysis for calcium normally involves the separation of the magnesium before the actual determination of the calcium. One advantage of the oxalate method for calcium is the fact that oxalate can be used to separate the calcium from any magnesium that may be present, since magnesium forms a soluble complex with excess oxalate, and calcium does not. Other heavy metals interfere and must be removed prior to the oxalate separation.

The limestone sample is usually dissolved carefully and slowly in hydrochloric acid in a covered beaker to ensure that none of the sample is lost through effervescence. Most procedures also call for a digestion period in the acid solution to ensure complete solution of the carbonate fraction. The acid-insoluble residue left is composed mainly of silica and insoluble silicates. For samples in which the acid-insoluble residue is composed of insoluble silicates, especially clays, the residue is filtered, dried, and fused with sodium carbonate, and the fused material is dissolved in a small amount of acid. This solution is then added to the original solution of the sample before continuation of the analysis. The insoluble residue remaining from fusion is weighed and reported as silica.

Limestone samples often contain iron and aluminum in amounts large enough to cause interference in the oxalate separation. The iron and aluminum are usually removed by making the solution basic with ammonia and filtering the precipitated hydrous oxides. Care must be exercised in this step to offset negative errors caused by adsorption of calcium and magnesium ions on the hydrous oxides if the pH is too high. The hydrous oxide particles are colloidal in size and consequently are charged due to adsorption of ions in the primary region. The particles are positively charged at pH values below 8 for aluminum and below 8.5 for iron. At higher pH values the particles become negatively charged as a result of adsorption of hydroxide ions and attract positive ions such as calcium and magnesium into the secondary layer. Thus, most procedures call for the addition of a very slight excess of ammonia in the precipitation step. The oxides can be ignited and weighed or can be determined separately by photometric methods.

The filtrate from the hydrous oxides is made acidic with hydrochloric acid, and methyl red is added as an indicator. An excess of oxalate is then added (usually as ammonium oxalate), and the pH is raised to a value between 5 and 6 by the slow addition of ammonia until the intermediate orange color of the indicator is attained. This pH ensures the quantitative precipitation of calcium oxalate and offsets the coprecipitation of calcium hydroxide that will occur if the pH exceeds 7. The magnesium remains in solution as a soluble complex if there is sufficient excess oxalate. The particles of calcium oxalate are very small and must be digested for approximately 1 hr before attaining sufficient size to allow satisfactory filtration. The digestion period must not be too long, however, since magnesium oxalate will coprecipitate if the precipitate is left in contact with the solution for extended periods of time.

The transfer and washing of the precipitated calcium oxalate also requires careful attention to detail because the oxalate is to be titrated rather than the calcium. As a consequence, all the excess oxalate must be washed out of the precipitate. Since calcium oxalate is comparatively soluble, extended washing will cause a negative error as a result of dissolving the precipitate. As a consequence, most procedures call for washing by decantation a specified number of times with specified volumes of cold water, followed by transfer and washing in the filtration device (paper or suction crucible) with a specified number of washes of a specified volume.

In spite of the difficulties, results are precise if the washing directions are followed carefully. The accuracy is also satisfactory if directions are followed because the factors that can cause positive errors are carefully balanced against the factors that cause negative errors. Any deviation from the directions, however, can be the cause of major errors in the determination, even though the precision will still be excellent as long as all samples are treated exactly the same.

The precipitated calcium oxalate is dissolved in dilute sulfuric acid and titrated with potassium permanganate at elevated temperatures. The reaction involved in the titration may be represented by the ionic equation

$$2MnO_4^- + 5HC_2O_4^- + 11H^+ \longrightarrow 2Mn^{2+} + 10CO_2 + 8H_2O \quad \textbf{(12-1)}$$

and the half-reaction for bioxalate, which is equivalent to the calcium, is

$$HC_2O_4^- - 2e^- \longrightarrow 2CO_2 + H^+ \qquad \textbf{(12-2)}$$

Each bioxalate ion requires two electrons, and since each bioxalate represents one calcium ion, as shown by eq. (12-3), the equivalent weight for calcium in the titration equals one-half the atomic weight:

$$CaC_2O_4 + H^+ \longrightarrow Ca^{2+} + HC_2O_4^- \qquad \textbf{(12-3)}$$

The bioxalate solution must be heated almost to boiling and must be maintained at a temperature above 60°C during the titration to ensure that the reaction takes place at a reasonable rate. At room temperature, the reaction is so slow that the titration is not feasible. Even at elevated temperatures the reaction proceeds slowly until some manganous ion is formed. The manganous ion acts as a catalyst for the reaction, which will proceed rapidly at elevated temperatures in the presence of manganous ion. For this reason, the solution should be allowed to decolorize after the addition of the first few drops of titrant before the titration is continued.

Equation (12-1) indicates that hydrogen ions are used up in the reaction. If the titrant is allowed to run into the solution without agitation, the pH may be raised in the region of entry to a value at which manganese dioxide will be precipitated. Even though the manganese dioxide redissolves upon agitation, the reaction is not completely reversible and an error can result. Most procedures specify adequate agitation during the titration to offset this error.

The end point of the titration occurs when the first excess of permanganate colors the solution pink to lavender. Since the end point occurs beyond the equivalence point, a positive error is involved. This error can be offset by determination of the amount of permanganate required to produce the same color in a sulfuric acid solution of equal volume and by subtracting this blank from the titration values.

## DICHROMATE METHODS

Potassium dichromate is used primarily in the determination of iron in iron ores but is also used in the determination of uranium, sodium [after precipitation as $NaZn(UO_2)_3(C_2H_3O_2)_9 \cdot 6H_2O$], and organic compounds. Organic compounds are boiled with an excess of standard dichromate solution. The excess dichromate is then determined by addition of an excess of standard ferrous iron solution followed by titration of the excess ferrous iron by standard dichromate.

Potassium dichromate is obtainable in primary standard grade and standard solutions can be prepared by the direct method. The standard solutions so prepared are so stable that they can be boiled without decomposition and do not oxidize chloride ion in dilute (1–2 *N*) acid solution. The product in 1–2 *N* acid solution is always the $Cr^{3+}$ ion, since there are no other stable oxidation states between $Cr_2O_7^{2-}$ and $Cr^{3+}$. The half-cell reaction for dichromate in acid solution is

$$Cr_2O_7^{2-} + 14H^+ + 6e^- \longrightarrow 2Cr^{3+} + 7H_2O \qquad \textbf{(12-4)}$$

so the equivalent weight of potassium dichromate is one-sixth the molecular weight. A 0.1 *N* solution of potassium dichromate is transparent, and the meniscus is read easily. Diphenylamine sulfonate is a satisfactory indicator for most dichromate titrations.

## DETERMINATION OF IRON IN AN IRON ORE BY TITRATION WITH POTASSIUM DICHROMATE

Many iron ores are directly soluble in hydrochloric acid with little or no acid-insoluble residue. Some of the ores that contain materials insoluble in hydrochloric acid can be determined without removal of the acid-insoluble residue, but fusion must be used in those cases in which the acid-insoluble material contains iron.

After solution in hydrochloric acid, the iron must be reduced to the iron(II) (ferrous) state before titration with potassium dichromate. The prior reduction of iron is an important and critical step in the procedure and must be performed with care. Before reduction, the solution is evaporated to low volume (usually about 5 mL) on a steam or electric hot plate. The iron is present as a chloro complex, which has a dark brown color that is used as the indicator for the reduction. If the sample goes to dryness on the hot plate, it must be redissolved in hydrochloric acid before being reduced. In some samples, a precipitate will form in the concentrated solution, especially if the volume has been reduced too low. Such precipitates can usually be dissolved by addition of a few milliliters of warm water to the warm solution.

The iron is reduced to the ferrous state by reaction with tin(II) (stannous) chloride and, since ferrous iron is oxidized readily in acid solution by the oxygen of the air, rapid work is necessary after the reduction step to prevent a negative error. For this reason, most procedures require that all chemicals be measured and ready for addition and that the burette be filled with the titrant before the reduction step is started.

The reaction of stannous ion with ferric ion is usually written

$$Sn^{2+} + 2Fe^{3+} \longrightarrow Sn^{4+} + 2Fe^{2+} \qquad \textbf{(12-5)}$$

Since the iron and tin are present as chloro complexes, the reaction actually is

$$SnCl_4^{2-} + 2FeCl_4^- \longrightarrow SnCl_6^{2-} + 2Fe^{2+} + 6Cl^- \qquad \textbf{(12-6)}$$

The $FeCl_4^-$ ion has a yellow color (dark brown at this concentration), whereas the $Fe^{2+}$ ion is a pale green color and essentially colorless in the concentrations found in the reduced solution. As a consequence, the fading color of $FeCl_4^-$ ion can be used to follow the course of the reduction. The reaction is comparatively slow near the end point, and the stannous chloride must be added dropwise with a 10- to 15-sec interval between drops. It is important that all the iron be reduced to the ferrous state, but an excess of stannous chloride of more than 1 or 2 drops will cause errors and difficulty in the later steps in the analysis. The solution must be diluted before titration, and hydrochloric acid should be added before dilution to prevent precipitation of $Sn(OH)_4$.

Since potassium dichromate reacts with stannous tin, the excess stannous ion added in the reduction step must be removed before titration. The removal is accomplished by the addition of mercury(II) (mercuric) chloride. The stannous ion reacts with the mercuric ion to form mercury(I) (mercurous) chloride, a very insoluble substance that will separate from the solution as a silky, finely divided white precipitate. Since a finely divided precipitate is desired, the mercuric chloride is added rapidly with constant agitation. The desired reaction is

$$SnCl_4^{2-} + 2HgCl_2 \longrightarrow SnCl_6^{2-} + Hg_2Cl_2 \qquad \textbf{(12-7)}$$

The mercurous chloride is sufficiently insoluble that it will not be oxidized by the potassium dichromate and thus does not cause an error. If too large an excess of stannous ion is added in the reduction step, the mercuric chloride will react to form elementary mercury by

$$SnCl_4^{2-} + HgCl_2 \longrightarrow SnCl_6^{2-} + Hg^0 \qquad \textbf{(12-8)}$$

The elementary mercury formed by this reaction does react with potassium dichromate and will cause an error in the determination. As a consequence, most procedures call for a finely divided, shiny white precipitate to form within 3 min after the addition of the mercuric chloride. If no precipitate forms, insufficient stannous chloride was added and all the ferric ion was not reduced. If a heavy white to gray precipitate forms, too much stannous chloride was added, and the sample should be discarded because the mercury formed will react with the titrant to give a positive error.

Immediately after the reduction step and the formation of the mercurous chloride precipitate, sulfuric acid, phosphoric acid, and the diphenylamine indicator should be added and the solution titrated as rapidly as possible to prevent air oxidation of the ferrous iron.

The sulfuric acid is added to ensure the correct acidity of the solution and because the indicator works better in the presence of sulfate ions. The phosphoric acid forms a colorless complex [probably $Fe(H_2PO_4)_6^{3-}$] with the ferric ions as they are produced in the reaction. The formation of the colorless complex prevents the tan color of the $FeCl_4^-$ from obscuring the end point. The complex also serves to depress the potential of the ferric–ferrous half-cell and keeps the potential at a lower value before the end point is reached. This causes a sharper potential break at the end point and allows the use of diphenylamine sulfonate as the indicator. The effect of the phosphoric acid upon the potential change

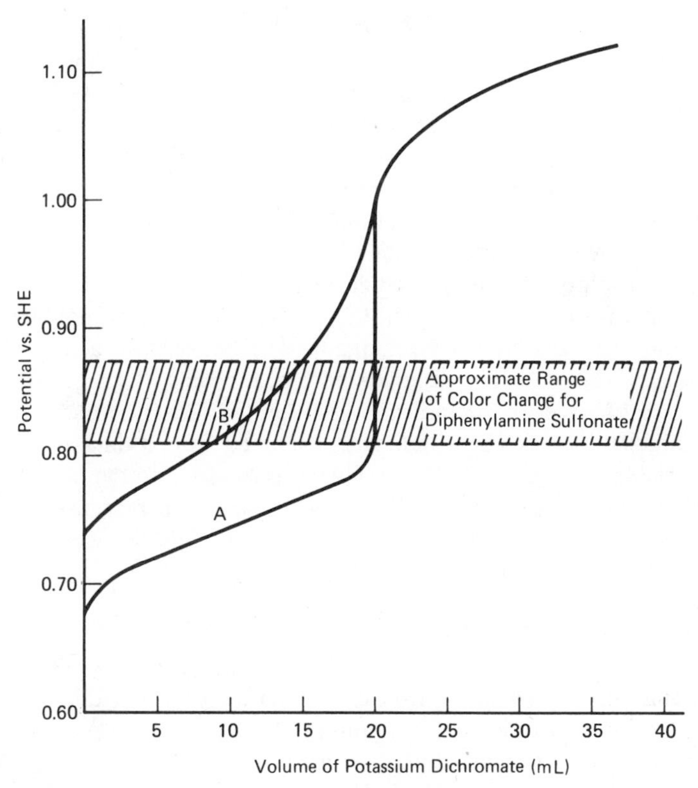


**Figure 12-1.** Effect of phosphoric acid on the potential during the titration of 20.00 mL of 0.1000 *N* $FeSO_4$ with 0.1000 *N* $K_2Cr_2O_7$ (A) in 3 *M* $H_3PO_4$ and 1 *M* $H_2SO_4$; (B) in 1 *M* $H_2SO_4$.

during the titration is shown in Figure 12-1. In the absence of phosphoric acid, the color changes appears before the equivalence point.

Diphenylamine sulfonate serves as the indicator and is oxidized from colorless to a lavender shade. In the actual titration the presence of the complexed ferric ions and the green color of the chromic ions combine with the color of the indicator to form a gray to black color just before the end point, and the end point is a dark purple color that will persist for at least 30 sec. Some dichromate is necessary to oxidize the indicator, and it is difficult or impossible to run a blank determination because the color change is not sharp in the absence of iron. A blank of 0.07 mL of 0.1000 *N* $K_2Cr_2O_7$ is usually subtracted from the titrant volume.

## CERIUM (IV) (CERIC) METHODS

Cerium(IV) (ceric) compounds act as strong oxidizing agents in acid solution and are reduced to colorless cerium(III) (cerous) ions. The strength of the cerium(IV) solutions is dependent upon the anions present: the standard half-cell potential is 1.28 V in hydrochloric acid, 1.44 V in sulfuric acid, 1.61 V in nitric acid, and 1.70 V in perchloric acid. The variation in the standard half-cell potential is sometimes explained as being due to complex formation because cerium(IV) forms complexes with sulfate, chloride, and nitrate ions but does not form complexes with perchlorate ions. Also, the stability constants of the complexes correlate with the standard potentials. Cerium(IV) is usually used in sulfuric or hydrochloric acid solutions for the titration of those substances that can be titrated with permanganate.

Cerium(IV) solutions must be standardized because the usual salts such as ceric ammonium sulfate [$(NH_4)_2Ce(SO_4)_3 \cdot 2H_2O$] are not primary standards. The solution is usually standardized against sodium oxalate, arsenious oxide, or high-purity iron wire. The disappearance of the yellow-orange color of the ceric ion can be used as the indicator, but ferroin [tris(1,10-phenanthroline) iron(II) sulfate] is an excellent indicator, changing from a dark red to a light blue (colorless in the actual titration) when it is oxidized to the iron (III) state.

Ceric solutions are stable indefinitely, can be used in the presence of chloride, and have the advantage that there is only one possible change in oxidation number, as shown by the half-cell equation,

$$Ce^{4+} + 1e^- \longrightarrow Ce^{3+} \qquad \textbf{(12-9)}$$

Ceric solutions may be used for the determination of iron, oxalate, hydrogen peroxide, arsenite, antimony(III), uranium (IV), vanadium(IV), tin(II), and plutonium(III). Ferroin or similar phenanthroline indicators are usually satisfactory for most procedures.

## METHODS INVOLVING IODINE

Elementary iodine is comparatively insoluble in water but reacts with iodide ion to form a soluble complex known as triiodide ion, which has a dark brown color.

$$I_2^0 + I^- \longrightarrow I_3^- \qquad \textbf{(12-10)}$$

As a consequence, the triiodide ion is the reactive species in all methods that involve iodine in aqueous solution rather than elementary iodine. In this discussion, the two terms (iodine and triiodide) will be used interchangeably and will be taken to mean the triiodide ion when an actual reaction is being discussed.

Iodine is a weak oxidizing agent that is used for direct titration of strong reducing agents such as arsenite. Indirect titration methods, in which an oxidizing agent is allowed to react with iodide ion to produce triiodide ion, which is then titrated with sodium thiosulfate, are probably more popular than the direct methods. Both direct and indirect methods must be used in neutral or acidic solutions; the reactions are not stoichiometric in basic solution because of side reactions. The appearance or disappearance of the color of the triiodide ion can be used to indicate the end point in both direct and indirect methods, but starch is usually used as the indicator except in strongly acid solution. Iodine forms a blue complex with starch, which constitutes a very sensitive method for the detection of the presence of iodine or triiodide ion. The actual color of the starch–iodine complex depends upon the relative concentration of triiodide ion. The color deepens from blue to dark blue to purple to black as the concentration of triiodide increases.

Iodine is limited in its use as a titrant in direct methods owing to the low standard half-cell potential of 0.54 V and to the fact that the concentration of iodine solutions changes slowly because of volatilization of the solute. Although iodine can be obtained in a very pure state, actually primary standard grade, by successive sublimations, the concentrations of standard solutions must be checked periodically, even though they were prepared originally by the direct method. The preparation of any standard iodine solution involves the solution of weighed amounts of iodine and potassium iodide in as small a volume of water as possible, followed by dilution with water to the desired volume.

Arsenic(III), antimony(III), tin(II), hydrogen sulfide, sulfite, thiosulfate, and ferrocyanide are often determined by the direct method of titration with standard iodine using starch as the indicator. The substances ordinarily determined by the indirect method include the halogens and oxyanions of the halogens, strong oxidizing agents such as permanganate and dichromate, organic compounds that can be brominated quantitatively, and copper. The more active halogens displace iodide from potassium iodide to form triiodide ion, which is then titrated with thiosulfate as indicated in

$$Cl_2^0 + 3KI \longrightarrow KI_3 + 2KCl \qquad \textbf{(12-11)}$$

$$KI_3 + 2Na_2S_2O_3 \longrightarrow Na_2S_4O_6 + KI + 2NaI \qquad \textbf{(12-12)}$$

The general ionic reactions for all indirect methods are, assuming a two-electron change for the oxidizing agent,

$$Ox + 3I^- \longrightarrow Red + I_3^- \qquad \textbf{(12-13)}$$

$$I_3^- + 2S_2O_3^{2-} \longrightarrow 3I^- + S_4O_6^{2-} \qquad \textbf{(12-14)}$$

The determination of copper will be discussed in detail as an example of indirect iodine methods.

## INDIRECT IODINE METHODS INVOLVING TITRATION WITH THIOSULFATE

### *Preparation and Standardization of Thiosulfate Solutions*

Thiosulfate solutions are usually prepared from sodium thiosulfate pentahydrate ($Na_2S_2O_3 \cdot 5H_2O$), which is not a primary standard, and, as a result, thiosulfate solutions must be standardized against suitable standard solids or solutions. The half-cell reaction of thiosulfate is represented by

$$2S_2O_3^{2-} - 2e^- \longrightarrow S_4O_6^{2-} \quad \textbf{(12-15)}$$

The $S_4O_6^{2-}$ ion is known as tetrathionate. Equation (12-15) indicates that each thiosulfate radical requires one electron to be converted to a tetrathionate radical so that the equivalent weight of sodium thiosulfate pentahydrate equals the molecular weight.

Thiosulfate solutions are unstable because they are readily decomposed by bacteria and by acid. The bacterial decomposition may be offset by using recently boiled and cooled water to prepare the solution or by the addition of a germicidal agent such as chloroform or mercuric chloride. The acid decomposition is avoided by adding a small amount of sodium carbonate to the solution. The acid decomposition involves the disproportionation of the hydrogen thiosulfate ion ($HS_2O_3^-$) to the hydrogen sulfite ion ($HSO_3^-$) and sulfur. Freshly prepared solutions of thiosulfate change concentration slowly for a period of time so that the solution should be prepared and allowed to stand several days before standardization.

Thiosulfate solutions may be standardized against many oxidizing agents, but potassium permanganate or potassium iodate solutions are preferable. Standardization against permanganate involves the addition of an aliquot of a standard solution of permanganate to an acidified solution of potassium iodide followed by titration with the thiosulfate solution using starch as the indicator. The reaction of permanganate with iodide is

$$2MnO_4^- + 15I^- + 16H^+ \longrightarrow 2Mn^{2+} + 5I_3^- + 8H_2O \quad \textbf{(12-16)}$$

Permanganate is often used simply because previously standardized solutions are available. The reaction is rapid, and the pale pink color of the manganous ion does not interfere with the color change at the end point.

Potassium iodate is also used for the standardization of thiosulfate. Potassium iodate may be obtained in primary standard grade and is readily soluble in water. In the standardization, an aliquot of a standard solution of iodate is added to a solution containing excess iodide. The solution is then acidified with HCl (1 : 10), and the iodate and iodide react to form triiodide ion by

$$IO_3^- + 8I^- + 6H^+ \longrightarrow 3I_3^- + 3H_2O \quad \textbf{(12-17)}$$

This reaction is unusual in that it can be used to determine any one of the three reactants. In the presence of excess iodide and hydrogen ions, it is quantitative for iodate; in the presence of excess iodate and hydrogen ions, it is quantitative for iodide; and in the presence of excess iodate and iodide ions, it is quantitative for hydrogen ions and can be used to standardize acid solutions. The triiodide formed is titrated with thiosulfate. The balanced sequence of reactions involved in the standardization of thiosulfate is

$$IO_3^- + 8I^- + 6H^+ \longrightarrow 3I_3^- + 3H_2O \qquad (12\text{-}17)$$

$$3I_3^- + 6S_2O_3^{2-} \longrightarrow 9I^- + 3S_4O_6^{2-} \qquad (12\text{-}18)$$

which may be summed to give

$$IO_3^- + 6S_2O_3^{2-} + 6H^+ \longrightarrow I^- + 3S_4O_6^{2-} + 3H_2O \qquad (12\text{-}19)$$

Equation (12-19) shows that one $IO_3^-$ reacts with six $S_2O_3^{2-}$ by the transfer of six electrons to form one $I^-$, three $S_4O_6^{2-}$, and three $H_2O$. As a consequence, the equivalent weight for potassium iodate in the standardization of thiosulfate equals the molecular weight divided by 6, or one-sixth the molecular weight. Most students feel that the equivalent weight should be one-fifth the molecular weight, since each iodate requires five electrons in the first reaction to form triiodide; however, in standardizations the complete balanced sequence of reactions must be utilized to determine the equivalent weight of a reactant or product.

Other substances used to standardize thiosulfate solutions include potassium bromate, potassium dichromate, oxalic acid, potassium hexacyanoferrate(III), iodine, and metallic copper.

### *Determination of Copper by the Indirect Iodine Method*

Copper ores and other materials containing copper are usually determined by the indirect iodine method or by electrodeposition of the copper from a nitrate and sulfate solution. The indirect iodine method is usually faster, but the electrodeposition method is capable of greater accuracy.

Many copper ores contain metallic copper as well as a salt or oxide of copper, and as a consequence the sample is usually dissolved in nitric acid. This procedure creates several problems, however, because nitric acid and the oxides of nitrogen both react with iodide to form triiodide ion and thus cause positive errors. In the solution step, metallic copper reacts with the nitric acid to form copper(II) (cupric) nitrate and oxides of nitrogen, as indicated by

$$Cu + 4HNO_3 \longrightarrow Cu(NO_3)_2 + 2NO_2 + 2H_2O \qquad (12\text{-}20)$$

and the copper salts or oxides also react to form soluble cupric nitrate, for example,

$$CuO + 2HNO_3 \longrightarrow Cu(NO_3)_2 + H_2O \qquad (12\text{-}21)$$

$$CuCO_3 + 2HNO_3 \longrightarrow Cu(NO_3)_2 + CO_2 + H_2O \qquad (12\text{-}22)$$

The excess nitric acid and the oxides of nitrogen must be removed prior to the determination of copper. Most procedures call for evaporation of the solution to low volume followed by dilution and further evaporation to remove the oxides of nitrogen and the excess nitric acid.

The oxides or salts of iron(III) often contaminate copper ores and will cause positive errors if allowed to remain in the solution because iron(III) (ferric) ions react with iodide in neutral or slightly acid solution to form triiodide ions. The interference due to iron is offset by the addition of ammonium bifluoride ($NH_4HF_2$), which reacts with ferric ions to form a soluble colorless complex,

$$Fe^{3+} + 6F^- \longrightarrow FeF_6^{3-} \qquad \textbf{(12-23)}$$

The iron often precipitates as the hydrous oxide during the pH adjustment step, but the formation constant of the hexafluoroferrate(III) ion is so large that any anhydrous oxide precipitated will be dissolved and converted to the complex.

The pH is usually adjusted before the addition of the fluoride to prevent or slow down the air oxidation of iodide ion, which is rapid in acid solution. The pH adjustment is usually accomplished by the addition of ammonium hydroxide or potassium biphthalate. If ammonia is used, it is added until a slight blue precipitate of copper(II) hydroxide is formed or until the color of the solution begins to deepen as a result of the formation of the tetrammine copper(II) complex, $Cu(NH_3)_4^{2+}$. Addition of the ammonium bifluoride to such a solution lowers the pH to below 4 and prevents the interference of arsenic and antimony, which are often present in copper ores.

After the addition of the bifluoride it is necessary to work rapidly to prevent etching the glassware by the hydrofluoric acid formed by the addition of the bifluoride. Potassium iodide is added and reacts with the copper(II) ion to form triiodide ion and copper(I) (cuprous) iodide, which is insoluble and separates as a brownish precipitate.

$$2Cu^{2+} + 5I^- \longrightarrow Cu_2I_2 + I_3^- \qquad \textbf{(12-24)}$$

The triiodide is then titrated with thiosulfate using starch as the indicator. The starch should be added just before the loss of the color due to triiodide because the formation of a black starch complex, when too large a concentration of triiodide is present, is not completely reversible. The end point is the change from the starch–iodine color to colorless.

The cuprous iodide formed upon addition of iodide ion adsorbs or occludes some of the triiodide ion. The extent of adsorption can be estimated from the color of the precipitate, since pure cuprous iodide is white. The occluded triiodide ion is released by the addition of potassium or ammonium thiocyanate. Cuprous thiocyanate is less soluble (more insoluble) then cuprous iodide, so the cuprous iodide dissolves and is converted to cuprous thiocyanate.

$$Cu_2I_2 + 2KSCN \longrightarrow Cu_2(SCN)_2 + 2KI \qquad \textbf{(12-25)}$$

As the precipitate dissolves, the adsorbed and occluded triiodide is released to the solution and produces a blue color with the starch. This color may be removed, usually by the addition of 1 or 2 drops of titrant.

The half-reaction for copper in the determination is

$$2Cu^{2+} + 2I^- + 2e^- \longrightarrow Cu_2I_2 \qquad \textbf{(12-26)}$$

Since there is an electron change of 1 for each copper ion, the equivalent weight of copper equals the atomic weight in this reaction.

CHAPTER

# 13 Potentiometric Measurements

## ORIGIN OF POTENTIAL DIFFERENCE

Matter may ultimately be considered to be composed of atoms, which are in turn composed of three fundamental particles: the proton, the electron, and the neutron. In a neutral atom, the positive charge due to the protons in the nucleus exactly balances the negative charge due to the surrounding electrons. In many forms of matter, most especially in metals, the atoms of the metallic element form a crystalline array in which the valence electrons are arranged in bonding orbitals. The higher energy filled orbitals extend over the entire crystal array, forming a conduction band in which mobile valence electrons, no longer confined to a single atom, are free to migrate. Whenever there is an excess of electrons in a particular region, that region becomes negatively charged; conversely, the region from which the excess electrons originated becomes positively charged.

Matter in its most stable state is electrically neutral, and work must be done to produce a separation of charge. The work in joules (J) required to separate 1 coulomb of charge is known as a volt (i.e., 1 V equals 1 J/coulomb). The separation of charge produces a force described by Coulomb's law, and under this force a charged particle, such as an electron, will move from a region of excess to a region of deficiency, thereby producing a flow of charged particles, or a **current**. In metals, the mobile charge carriers are electrons, and a current flow amounting to 1 coulomb of charge carriers per second is known as an ampere (i.e., 1 A equals 1 coulomb/sec). The separation of charge, the **potential difference**, provides the driving force for the flow of current, the current flow being directly proportional to the potential difference (or voltage or electromotive force). The proportionality constant between current flow and voltage is known as the resistance of the material, as indicated by Ohm's law,

$$E = IR \quad \textbf{(11-30)}$$

where $E$ is the potential difference in volts, $I$ the current in amperes, and $R$ the resistance in ohms.

## GALVANIC (VOLTAIC) CELLS

The electrochemical theory of oxidation-reduction, including galvanic cells and the Nernst equation, is covered in Chapter 11 and can be reviewed if necessary.

Complete tables of standard reduction potentials, $E^0$, are listed in various handbooks and textbooks. An abbreviated table of $E^0$ values for certain half-cells is given in Appendix 7. These $E^\circ$ values are those obtained when the concentrations of all ions which are involved in the half-cell reaction are at unit activity so that the logarithmic term becomes zero as the logarithm of one and has no effect on the half-cell potential. The conducting portion of the half-cell is assumed to be platinum or some other inert conductor for those half-cells which do not involve a metal as one member of the oxidation-reduction pair, such as the permanganate-manganous ion half-cell and the ferric-ferrous ion half-cell.

### CELL POTENTIALS

The potential exerted by any given voltaic or galvanic cell is the difference between the actual potentials of the two half-cells of which it is composed, or

$$E_{\text{cell}} = E_1 - E_2 \qquad \textbf{(13-1)}$$

in which $E_1$ and $E_2$ are the actual potentials of the two half-cells. Cells may be shown diagramatically as in Figure 11-1 or may be designated in type by using a single vertical line to represent the juncture between a solid and a liquid and a double vertical line to show the juncture between two liquids or solutions. Thus the cell composed of a zinc–zinc sulfate half-cell and a copper–copper sulfate half-cell connected by a saturated potassium chloride salt bridge and shown diagrammatically in Figure 11-1, may be represented as

$$\text{Zn}|\text{ZnSO}_{4(\text{solution})}\|\text{saturated KCl}\|\text{CuSO}_{4(\text{solution})}|\text{Cu} \qquad \textbf{(13-2)}$$

By convention the half-cell written on the left is designated as half-cell 2 and the one written on the right as half-cell 1. If the two cells are reversed, the numerical value of the potential will be the same but the sign will be changed. As written above the cell potential will be positive, but if written to show the copper half-cell on the left, the potential would be negative.

## CONCENTRATION CELLS

Since the half-cell potentials depend in part upon the active concentrations of the half-cell reactants, a potential will be developed between two half-cells which are identical except for the concentrations of the solutions. Such cells are known as concentration cells and may be used to determine the activity in a solution by measurement against a solution of known activity. For example, a galvanic cell composed of two silver–silver nitrate half-cells will produce a potential if the two solutions have different concentrations. The cell potential for the cell

$$Ag|(AgNO_3)_u\|\text{saturated } KNO_3\|(AgNO_3)_s|Ag \qquad \textbf{(13-3)}$$

where the subscripts $s$ and $u$ refer to the standard and unknown solutions, respectively, is given by

$$E_{\text{cell}} = E_{Ag_s} - E_{Ag_u} \qquad \textbf{(13-4)}$$

or

$$E_{cell} = \left[E^0_{Ag} + \frac{0.059}{1}\log(Ag^+)_s\right] - \left[E^0_{Ag} + \frac{0.059}{1}\log(Ag^+)_u\right] \qquad \textbf{(13-5)}$$

Equation (13-5) may be simplified by cancellation of $E^0_{Ag}$ and rearranged to give

$$E_{\text{cell}} = 0.059[\log(Ag^+)_s - \log(Ag^+)_u]$$

and

$$\log(Ag^+)_u = \log(Ag^+)_s - \frac{E_{cell}}{0.059} \qquad \textbf{(13-6)}$$

As an example, calculate the active silver ion concentration in the unknown solution for the following cell:

$$Ag|AgNO_3(\text{unknown})\|\text{saturated } KNO_3\|AgNO_3\ (0.100A)|Ag$$

for which the cell potential is 0.080 V. Note that it is not necessary to know the standard potential of the silver half-cell since the standard potential does not appear in eq. (13-6).

$$\log(Ag^+) = \log 0.100 - \frac{0.080}{0.059}$$

$$= -1.000 - 1.358 = -2.358$$

$$(Ag^+) = 10^{-2.358} = 10^{0.642} \times 10^{-3}$$

$$= 4.39 \times 10^{-3}\ \text{A}$$

If the molar instead of the active concentration of the standard is known, the active concentration may be determined by calculation of the activity coefficient ($f$) at the ionic strength of the solution. The activity is the product of the molar concentration and the activity coefficient.

Debye and Hückel related the activity coefficient to the number and charge of all the ions in the solution by using the concept of ionic strength, which was discussed in Chapter 7.

## POTENTIOMETRIC DETERMINATION OF pH

Even though pH or hydrogen ion activity can be determined using a concentration cell composed of two hydrogen electrodes, it is usually measured potentiometrically with a pH meter using a reference electrode and a hydrogen electrode or a substitute hydrogen electrode. A reference electrode shows the same potential irrespective of the composition (or changes in composition) of the solution in which it is immersed. Substitute hydrogen electrodes are electrode systems that respond quantitatively to variations in hydrogen ion activity, as does the hydrogen electrode.

### REFERENCE ELECTRODES

The most popular reference electrodes are the **calomel electrodes**, of which the *saturated* calomel electrode is used more than any other. The saturated calomel electrode is constructed as indicated in Figure 13-1. An inner chamber contains mercury and mercurous chloride (calomel) in contact with a solution that is saturated with mercurous chloride and potassium chloride. A conductor, usually platinum wire, connects the mercury to the connecting wire for the electrode. A fiber connector is used to make contact between the inner electrode and the outer chamber, which is filled with saturated potassium chloride. In some electrodes, a minute hole or a porous frit is used in place of the fiber. An asbestos fiber is used to make the connection between the outer chamber and the solution in which the electrode is immersed. Other calomel electrodes utilize molar or 0.1 $M$ KCl solutions in place of the saturated KCl solution and are known as the **molar** (or **normal**) and **tenth-molar calomel electrodes**. The descriptive adjective representing concentration refers to the concentration of the potassium chloride rather than the mercurous chloride. The potential of the electrode depends upon the KCl concentration, varying from 0.244 V for the saturated electrode to 0.335 V for the tenth-normal electrode at 25°C.

The reaction that takes place in the calomel electrode is

$$Hg_2Cl_2 + 2e^- \rightleftharpoons 2Hg + 2Cl^- \qquad \textbf{(13-7)}$$

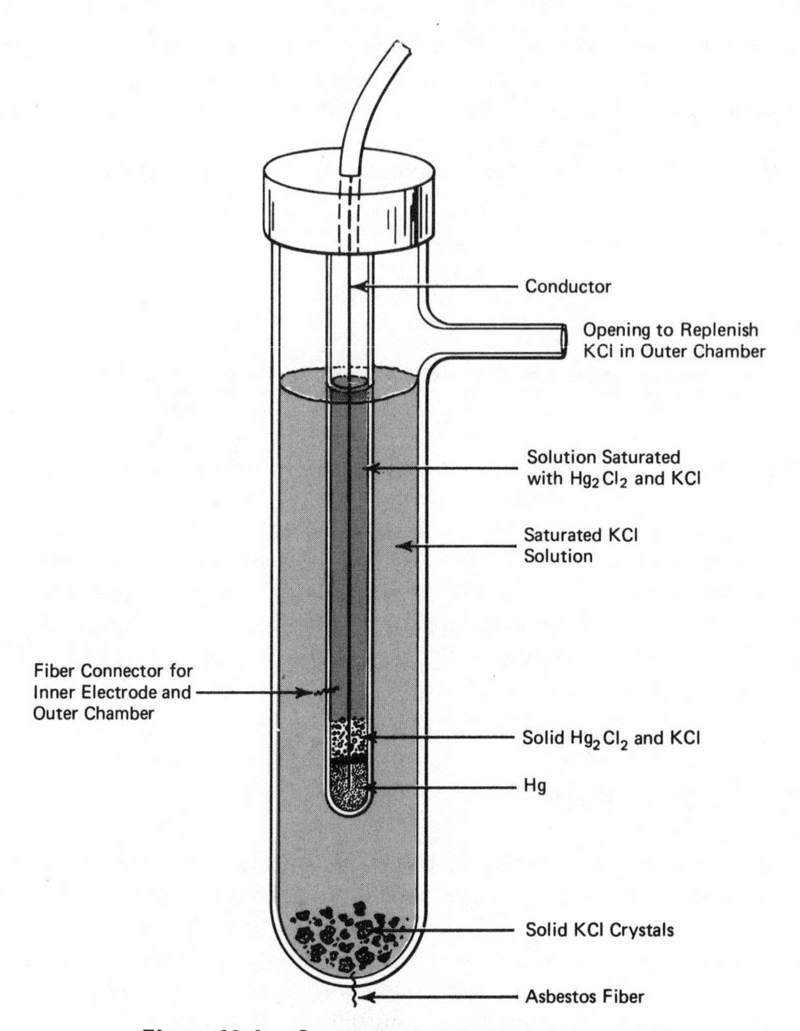


**Figure 13-1.** Saturated calomel reference electrode.

and the potential remains constant when only small amounts of current flow because the activity of the mercurous ion remains constant irrespective of the direction of flow. If the electrode accepts electrons, and thus converts some mercurous ion to mercury, some of the solid mercurous chloride dissolves to resaturate the solution; if the electrode produces electrons, some mercury is converted to mercurous ion, but, because the solution is saturated with mercurous chloride, the mercurous ions produced precipitate as mercurous chloride. Since there is a very high concentration of chloride ion, the small changes in chloride ion concentration do not affect the potential appreciably; the main effects of the chloride ion are to depress the solubility by the common ion effect and to maintain a constant ionic strength, and thus a constant value for the activity coefficient.

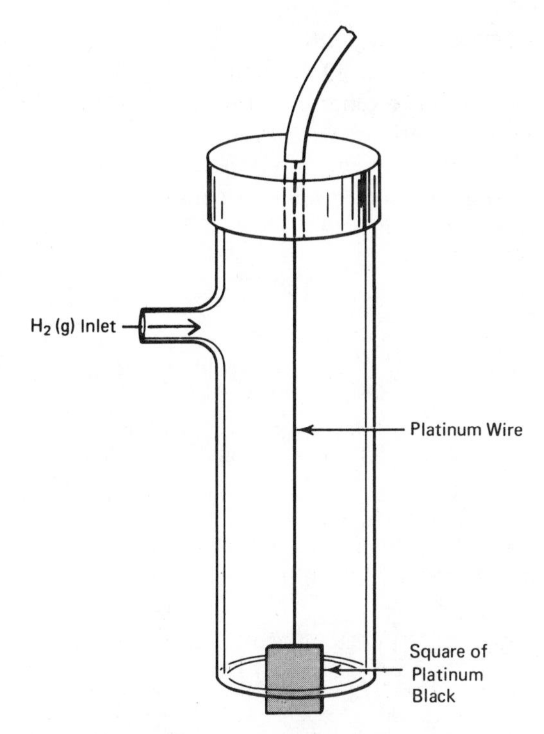


**Figure 13-2.** The hydrogen electrode.

## THE HYDROGEN ELECTRODE

Even though the hydrogen electrode is the standard electrode upon which the values of all standard potentials are based, it is actually used only occasionally, owing to the difficulty of operation. One form of the hydrogen electrode is shown in Figure 13-2. A square of platinum black foil is arranged so that half extends below the bottom of the glass tube. Hydrogen gas, under pressure, enters through the side arm and forces the liquid out of the tube so that the square is immersed half in the solution and half in the gaseous hydrogen. Under these conditions, the electrode acts like a metal–metal ion electrode and produces a potential that is dependent upon the hydrogen ion activity in the solution. The main objections to the hydrogen electrode are the difficulties encountered due to impure hydrogen, the fact that some ions may be reduced by the hydrogen, and that hydrogen sweeps out any other acidic or basic gases dissolved in the solution. The actual pH measured is the pH after all such reactions take place and not the pH of the solution before immersion of the electrode.

## SUBSTITUTE HYDROGEN ELECTRODES

### *The Quinhydrone Electrode*

Quinhydrone ($C_6H_4O_2 \cdot H_2C_6H_4O_2$) is a 1 : 1 addition compound of quinone ($C_6H_4O_2$) and hydroquinone ($H_2C_6H_4O_2$), which is slightly soluble in water and dissociates slightly into quinone and hydroquinone in solution. For convenience, we shall use Q to represent quinone, $H_2Q$ to represent hydroquinone, and $QH_2Q$ to represent quinhydrone. If a solution is saturated with quinhydrone and a platinum wire is immersed in the solution, the potential of the platinum wire will be dependent upon the hydrogen ion concentration in the solution. Quinone and hydroquinone constitute an oxidation–reduction pair, as shown by

$$Q + 2H^+ + 2e^- \rightleftharpoons H_2Q \tag{13-8}$$

The potential, as given by the Nernst equation, is

$$E_Q = E_Q^0 + \frac{0.059}{2} \log \frac{(Q)(H^+)^2}{(H_2Q)} \tag{13-9}$$

Substitution of the activity coefficient times the molar concentration for active concentration and rearrangement gives

$$E_Q = E_Q^0 + \frac{0.059}{2} \log \frac{[Q]}{[H_2Q]} + \frac{0.059}{2} \log \frac{f_Q}{f_{H_2Q}} + \frac{0.059}{2} \log (H^+)^2 \tag{13-10}$$

The second term on the right equals zero because the molar concentration of Q and $H_2Q$ are equal; they are produced in equal numbers by dissociation of quinhydrone. The activity coefficients are almost equal to 1, since both quinone and hydroquinone are molecular, so that the third term on the right is also essentially zero. As a consequence, eq. (13-10) reduces to

$$E_Q = E_Q^0 + 0.059 \log (H^+) \tag{13-11}$$

The quinhydrone electrode is easy to use because the only requirement is to saturate the solution with quinhydrone and to insert a platinum wire and a reference electrode. It has the disadvantage that it cannot be used above pH 9 because hydroquinone is a weak acid that ionizes at high pH; thus Q is no longer equal to $H_2Q$. Also, some oxidizing and reducing agents will react with quinone or hydroquinone and upset the equilibrium. Measurements must be made without unnecessary delay, since hydroquinone is oxidized slowly by air.

### *The Glass Electrode*

The most popular substitute electrode is the glass electrode, which has many advantages and only a few disadvantages when compared to the hydrogen electrode for the potentiometric determination of pH. Formerly, it was believed that the glass membrane electrode potential arose because the thin glass membrane was semipermeable, allowing hydrogen ions to diffuse through it.

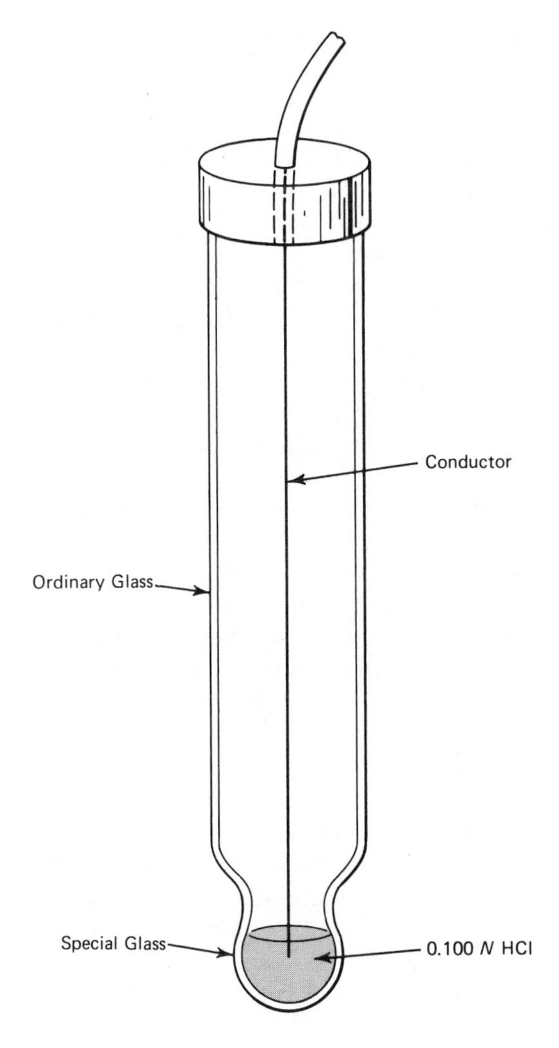


**Figure 13-3.** The glass electrode. The conductor may be silver coated with silver chloride to form a silver–silver chloride–chloride ion inner electrode or may be platinum immersed in an acid solution saturated with quinhydrone to form a quinhydrone inner electrode.

This hypothesis for the glass membrane potential has now been disproved experimentally; the membrane potential produced when two solutions of different hydrogen ion activity are separated by a thin membrane is now explained on the basis of an ion exchange process occurring at the membrane–solution interface. To respond to pH, the glass membrane must be soaked in an aqueous solution, which produces a hydrated layer on either side of the membrane. This hydrated layer is necessary for the ion exchange process to occur.

A diagram of a glass electrode is shown in Figure 13-3. The barrel of the electrode is usually prepared from ordinary glass with a thin bulb of special glass sealed on the end. The solution inside the electrode is usually 0.1 *N* HCl. A constant potential internal reference electrode must be enclosed in the glass electrode. This can be obtained by using a conductor of silver coated with silver chloride, which will produce a silver–silver chloride–chloride ion half-cell of constant potential. In other electrodes, the hydrochloric acid is saturated with quinhydrone and the conductor is a platinum wire.

## ORIGIN OF THE MEMBRANE POTENTIAL

We have said that the glass electrode potential arises because of an ion exchange process between ions in solution and ions in the hydrated glass layer formed at the solution–membrane boundary. Consider two solutions having different hydrogen ion activities separated by a thin glass ion exchange membrane as shown in Figure 13-4, where the hydrogen ions in each solution are in equilibrium with their respective membrane boundaries. It is certainly true that, for a system to be in equilibrium, all the constituents of the system must have the same potential energy. If this were not true, whichever constituent had the greater potential energy would provide a driving force for altering the state of the system until all the constituents of the system achieved the same potential energy. At equilibrium, the activity of the hydrogen ions in solution will be different from the activity of the hydrogen ions in the hydrated glass layer. In order for the hydrogen ions in solution to have the same potential energy as the hydrogen ions in the hydrated glass layer, a potential develops between the solution and the hydrated glass layer. This potential is known as a **junction potential**. Junction potentials arise whenever two dissimilar phases are brought into contact in order that the potential energies of constituents in each phase will be the same. At equilibrium, the total potential difference between solution 1 and solution 2 will be

$$V_1 - V_2 = J_1 + J_2 + (V_A - V_B) \tag{13-12}$$

where $J_1$ is the junction potential that develops between solution 1 and the left-hand edge of the membrane, $J_2$ is the junction potential that develops between the right-hand edge of the membrane and solution 2, and $V_A - V_B$ is the voltage difference across the membrane itself. For an ideal membrane, $V_A - V_B$ is zero.

It can be shown theoretically that for an ideal membrane,

$$V_1 - V_2 = \text{constant} + \frac{RT}{F} \ln \frac{a(H^+)_2}{a(H^+)_1} \tag{13-13}$$

which shows that the potential difference developed between solutions 1 and 2 is a function solely of the hydrogen ion activities in these solutions.

If a glass membrane is combined in an appropriate electrochemical cell, such as,

$$\underbrace{\text{Ag/AgCl, 0.1 } M \text{ Cl}^-, a(\text{H}^+)_1 \overset{V_1}{\Big|} \begin{matrix}\text{glass}\\ \text{membrane}\end{matrix}}_{\text{glass electrode}} \overset{V_2}{\Big|} \underbrace{a(\text{H}^+)_2 = \text{unknown}}_{\text{unknown solution}} \Big| \underbrace{\text{SCE}}_{\text{reference electrode}} \tag{13-14}$$

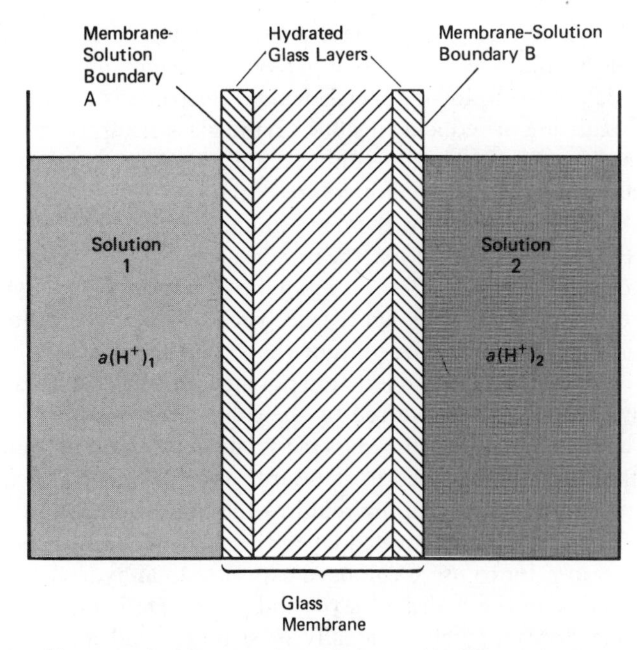


**Figure 13-4.** Glass membrane separating solutions with different hydrogen ion activities.

and if the hydrogen ion activity is maintained constant in solution 1,

$$E_{\text{cell}} = E_{\text{SCE}} + E_j - E_{\text{Ag/AgCl}} - (V_1 - V_2) \quad \textbf{(13-15)}$$

where $E_{\text{Ag/AgCl}}$ is the potential of the internal Ag/AgCl reference electrode; $E_{\text{SCE}}$ is the potential of the external reference electrode; and $E_j$ is the liquid junction potential between the SCE and the unknown solution.

Substituting eq. (13-13) into eq. (13-15) gives

$$E_{\text{cell}} = \text{constant} + \frac{RT}{F}\,\text{pH} = \text{constant} + 0.059\ \text{pH} \qquad (\text{at } 25°\text{C}) \quad \textbf{(13-16)}$$

which shows that the cell potential is a function of the pH of the unknown solution. Unfortunately, eq. (13-16) cannot be used to calculate the pH of a solution from the measured cell voltage because the magnitude of the liquid junction potential, $E_j$, is unknown. Consequently, the cell must be standardized using buffer solutions of known pH.

## FACTORS AFFECTING THE USE OF THE GLASS ELECTRODE

In the previous section, it was shown how a thin glass membrane incorporated into an appropriate electrochemical cell could produce a cell potential that varies directly with the pH of the unknown solution. The advantages and simplicity of such an arrangement have resulted in almost universal use of the glass electrode for the determination of pH. Among the many advantages of the glass electrode are the following: (1) The glass electrode is not affected by the presence of oxidizing or reducing agents; (2) the glass electrode may be used to measure the pH of small solution volumes; and (3) the glass electrode may be used to measure the pH of turbid or colored solutions.

Despite its wide applicability, there are limitations inherent in pH measurements with the glass electrode. First, the presence of the liquid junction potential, $E_j$, in the "constant" of eq. (13-16) is a potential source of error. Accurate pH measurement requires not only that the electrochemical cell be calibrated using buffer solutions but that the buffer solutions be similar enough to the composition of the unknown so that the liquid junction potential produced with the unknown will be identical with that produced by the calibration buffers. Naturally, the pH measurements made with the glass electrode cell can be no more accurate than the calibration buffers used to standardize the pH meter. Second, proper functioning of the glass membrane requires the maintenance of the hydrated glass layer. To provide this hydrated layer, the electrode must be soaked in an aqueous solution prior to use. Prolonged exposure to anhydrous solutions can cause dehydration of the hydrated layer and erratic performance. Finally, the glass electrode does not behave ideally in strongly acid and strongly basic solutions, giving rise to an acid error at a pH below zero and an alkaline error at a pH above 9. For pH measurements at pH values greater than 9, special glass electrodes are available that minimize the alkaline error.

## THE ALKALINE ERROR AND OTHER CATION-SENSITIVE GLASS MEMBRANES

In acid solutions, the hydrogen ion concentration is considerably higher than in basic solutions, and the glass electrode responds solely to changes in hydrogen ion activity. In very basic solutions (above pH 9) the hydrogen ion activity becomes very small, and the glass electrode begins to respond to other monovalent cations (such as sodium) that are present in high concentrations in basic solutions. The response of the glass membrane to the presence of high concentrations of other monovalent cations is due to the ability of these monovalent ions to enter into the ion exchange process at the glass membrane boundary in the same fashion as hydrogen ions. As a result of this response, an alkaline solution above pH 9 appears more acid than it actually is, giving rise to the alkaline error. Thus, for very alkaline solutions, the glass electrode becomes a cation-sensitive electrode because it responds to other monovalent cations, such as sodium. The magnitude of the effect is due to the cation-exchange properties of the glass membrane, which, in turn, depend on the glass composition.

The response of the glass membrane under alkaline conditions has been shown to follow an equation of the form

$$V_1 - V_2 = \text{constant} + \frac{RT}{F} \ln[a(\mathrm{H}^+)_2 + Ka(\mathrm{M}^+)_2] \qquad \textbf{(13-17)}$$

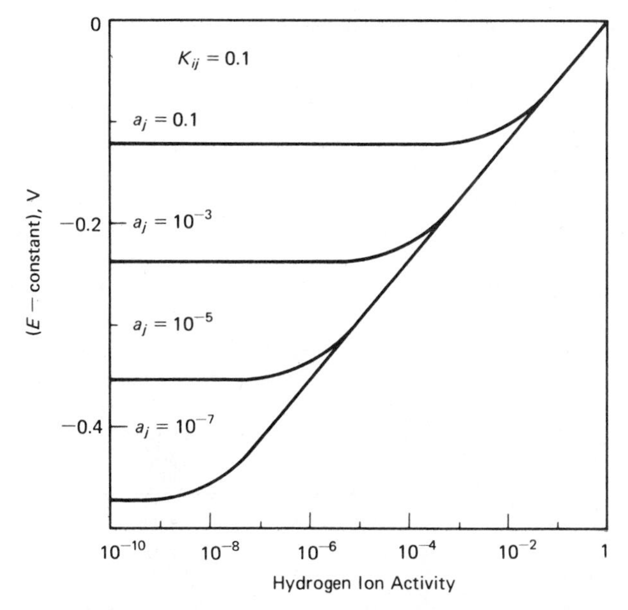


**Figure 13-5.** Plot of $[(V_1 - V_2) - \text{constant}]$ versus hydrogen ion activity. Graph shows response of electrode with a selectivity coefficient of $K = 0.1$ for various activities, $a_j$, of $M^+$. (Reprinted from H. A. Laitinen and W. E. Harris, *Chemical Analysis*. New York: McGraw-Hill, 1975, with permission.)

where $V_1 - V_2$ is the potential difference on opposite sides of the membrane, $a(H^+)_2$ is the hydrogen ion activity in solution 2, $a(M^+)_2$ is the activity of the cation $M^+$ in solution 2, and $K$ is the selectivity coefficient of the membrane for hydrogen ions in the presence of $M^+$ ions. Equation (13-17) shows that the electrode response will be primarily for hydrogen ions as long as the first term in the bracket is much larger than the second.

Figure 13-5 shows a plot of $V_1 - V_2$ versus pH in a solution containing cation $M^+$. Under acid conditions, the electrode responds in a linear manner to pH. As the pH of the solution increases and the hydrogen ion activity decreases, the response ceases to be a function of pH and the electrode begins to respond to the activity of $M^+$. As the concentration of $M^+$ increases, the onset of $M^+$ response shifts to more acid solutions, as predicted by eq. (13-17).

The observation and understanding of the alkaline error phenomenon with the glass membrane electrode has led to the development of special glass compositions purposely formulated to enhance this phenomenon, and thereby to produce cation-selective electrodes. Table 13-1 shows the composition and properties of various cation-enhanced sensitivity glasses. With the glass membrane electrode, the ion exchange sites are negatively charged $SiO^-$ sites, which, as a result of their high electrostatic field strength, have a strong affinity for hydrogen ions. By changing the composition of ordinary soda-lime glass by adding aluminum oxide, $AlOSi^-$ sites are formed; because of their lower electrostatic field strength, these have a greater affinity for cations over hydrogen ions.

**TABLE 13-1.** The Selectivity of Electrodes Varies with the Composition of the Glass

| | *Logarithm of Selectivity Ratio K* | | | | | | |
|---|---|---|---|---|---|---|---|
| *Glass Composition* | $Na^+$ *to* $H^+$ | $K^+$ *to* $H^+$ | $Li^+$ *to* $H^+$ | *Selectivity Order* | | | |
| $Na_2O$–$0.4Al_2O_3$–$6.3SiO_2$ | 0.04 | −1.04 | −2.01 | $Na^+$ | $H^+$ | $K^+$ | $Li^+$ |
| $Na_2O$–$0.75Al_2O_3$–$6.5SiO_2$ | 0.04 | −1.62 | −2.43 | $Na^+$ | $H^+$ | $K^+$ | $Li^+$ |
| $Na_2O$–$Al_2O_3$–$6SiO_2$ | −0.61 | −2.94 | −2.76 | $H^+$ | $Na^+$ | $Li^+$ | $K^+$ |
| $Na_2O$–$1.25Al_2O_3$–$5.5SiO_2$ | −1.07 | −3.78 | −3.23 | $H^+$ | $Na^+$ | $Li^+$ | $K^+$ |
| $Li_2O$–$0.5Al_2O_3$–$7SiO_2$ | 0.07 | −2.40 | −1.96 | $Na^+$ | $H^+$ | $Li^+$ | $K^+$ |
| $Li_2O$–$0.75Al_2O_3$–$6.5SiO_2$ | −0.20 | −3.89 | −2.43 | $H^+$ | $Na^+$ | $Li^+$ | $K^+$ |
| $Na_2O$–$0.5B_2O_3$–$7SiO_2$ | −1.82 | −3.04 | −3.35 | $H^+$ | $Na^+$ | $K^+$ | $Li^+$ |
| $Na_2O$–$0.75B_2O_3$–$6.65SiO_2$ | −2.82 | −3.99 | −3.58 | $H^+$ | $Na^+$ | $Li^+$ | $K^+$ |
| $1.25Na_2O$–$TiO_2$–$6.75SiO_2$ | −4.95 | −5.54 | −5.74 | $H^+$ | $Na^+$ | $K^+$ | $Li^+$ |
| $Na_2O$–$Al_2O_3$–$P_2O_5$–$4.0SiO_2$ | −0.76 | −3.30 | −2.82 | $H^+$ | $Na^+$ | $Li^+$ | $K^+$ |

*Source:* G. Rechnitz, *Chem. Eng. News,* **45**(25), 146 (1967); with permission.

## LIQUID ION EXCHANGE SYSTEMS

In addition to fixed-site ion exchange membranes such as glass membranes, where the negatively charged exchange sites are immobilized, mobile-site ion exchange materials, such as liquid ion exchangers, may also be used to produce ion selective electrodes. Indeed, the use of liquid ion exchangers permits the fabrication of anion-selective electrodes by using positively charged exchange groups. Figure 13-6 shows a schematic diagram of a typical liquid ion exchange electrode sensitive to calcium ions. In analogy with the glass electrode, the interior is filled with a solution of calcium chloride of constant composition into which dips a silver–silver chloride internal reference electrode. A porous

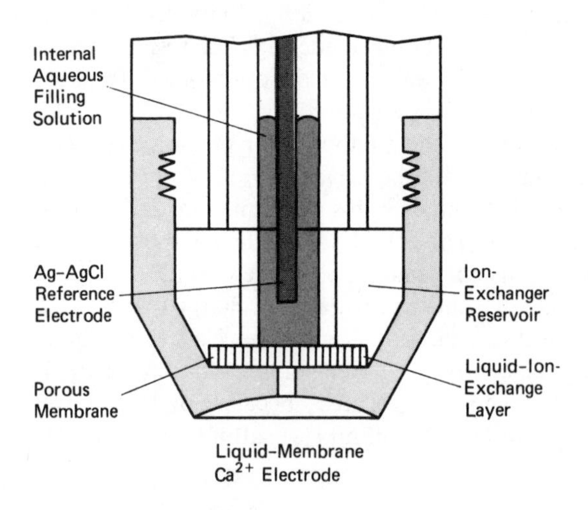


**Figure 13-6.** Liquid ion exchange electrode sensitive to calcium. (Reprinted from R. A. Durst, *Ion Selective Electrodes,* NBS Special Publication 314, Washington, 1969, with permission.)

membrane separates the interior electrode compartment from the external solution. This porous membrane is saturated with a water-immiscible liquid containing a calcium-selective ion exchanger. The liquid flows slowly under the influence of gravity from an adjacent reservoir so that the porous membrane is kept saturated with water-immiscible organic material.

In the case of the calcium-selective electrode, the calcium-selective ion exchange material is the calcium salt of didecylphosphoric acid, dissolved in the water-immiscible substance, di-*n*-octylphenylphosphonate. The ion exchange molecule used in this electrode is composed of two dissimilar portions—an organic portion, which is hydrophobic and therefore has an aversion for water, and a hydrophyllic, or water-seeking, ionic group containing a phosphorous atom. Because of the different affinities of the two portions of the molecule for water, the hydrophobic organic portion remains dissolved in the water-immiscible solvent (di-*n*-octylphenylphosphonate), whereas the hydrophyllic ionic group protrudes from the organic solvent into the aqueous solution. The insolubility of the organic tail of the molecule prevents the material from dissolving completely in the aqueous phase and thereby diffusing out into the bulk of the solution.

In the absence of any interfering species, the electrochemical cell formed between the calcium-selective electrode and an SCE obeys the expression

$$E_{\text{cell}} = \text{constant} + \frac{0.059}{2}\,\text{pCa} \qquad \text{(at 25°C)} \qquad \textbf{(13-18)}$$

which shows that the cell potential exhibits Nernstian behavior with respect to the calcium ion.

## SOLID STATE MEMBRANE ELECTRODES

One of the most successful ion-selective electrodes developed to date is the solid state fluoride electrode. The fluoride-sensitive component of this electrode consists of a membrane composed of a single crystal of $LaF_3$, to which has been added a small quantity of europium(II). The addition of the small quantity of europium(II) to the otherwise pure single crystal of $LaF_3$ creates disorders in the crystalline lattice known as **lattice defects**. These consist of vacancies in the crystal lattice, which, being of the appropriate size, may be filled by an adjacent lattice fluoride ion. The presence of these lattice defects throughout the crystal allows conduction of current to occur as mobile lattice fluoride ions move to fill lattice vacancies. The crystal, therefore, behaves as a semipermeable membrane, where the only permeable species is the fluoride ion. Figure 13-7 shows a schematic diagram of a solid state membrane electrode with a silver–silver chloride internal reference electrode dipping into a solution of constant composition containing sodium fluoride and sodium chloride.

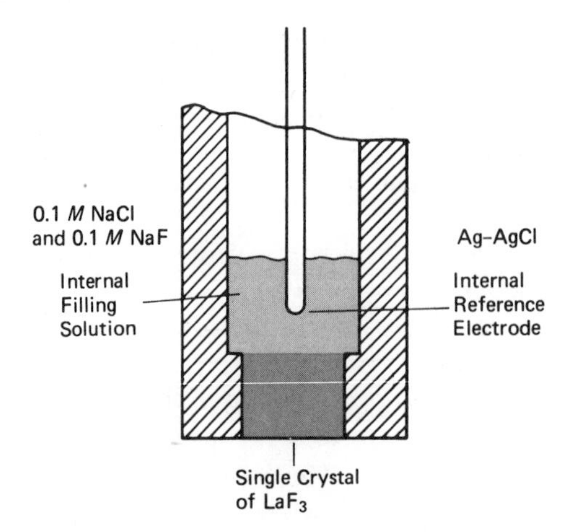


**Figure 13-7.** Solid state fluoride electrode. (Reprinted from R. A. Durst, *Ion Selective Electrodes*, NBS Special Publication 314, Washington, 1969, with permission.)

In addition to the use of single crystal membranes, various electrodes may be made from polycrystalline materials. In these polycrystalline materials, the membrane is formed either by pressing the material into a pellet to provide electrical continuity or by mixing the polycrystalline material with a binder such as silicone rubber. Electrical conduction usually involves the lattice ion with the smallest ionic radius and charge. Table 13-2 shows the various cations and anions that may be determined potentiometrically using solid state membrane electrodes.

**TABLE 13-2.** Applications of Solid State Membrane Electrodes

| *Test Ion* | *Solid State Membrane* | *Interfering Ions* |
|---|---|---|
| $F^-$ | $LaF_3$ | $OH^-$ |
| $Cl^-$ | $AgCl/Ag_2S$ | $Br^-$, $I^-$, $CN^-$, $S^{2-}$ |
| $Br^-$ | $AgBr/Ag_2S$ | $I^-$, $CN^-$, $S^{2-}$ |
| $I^-$ | $AgI/Ag_2S$ | $CN^-$, $S^{2-}$ |
| $CN^-$ | $AgI/Ag_2S$ | $I^-$, $S^{2-}$ |
| $SCN^-$ | $AgSCN/Ag_2S$ | $Br^-$, $I^-$, $CN^-$, $S^{2-}$ |
| $La^{3+}$ | $LaF_3$ | $OH^-$ |
| $Cd^{2+}$ | $CdS/Ag_2S$ | $Ag^+$, $Hg^{2+}$ |
| $Pb^{2+}$ | $PbS/Ag_2S$ | $Ag^+$, $Hg^{2+}$ |

## ANALYTICAL UTILIZATION

Analytical methods based on potentiometric measurements may be classified into two broad categories: direct potentiometry based on the Nernst equation and potentiometric titrations. Let us consider the direct potentiometric methods first.

## DIRECT POTENTIOMETRIC TECHNIQUES

Direct potentiometric techniques may be classified into the following categories: (1) Direct application of the Nernst equation, (2) null-point potentiometry, (3) use of calibration curves, and (4) standard addition methods.

Analytical methods based on the **direct application of the Nernst equation** consist of calculating the ionic activity of the desired species from the measured cell potential produced when a suitable indicator electrode and reference electrode are placed in electrical contact with an unknown solution. Although such methods may appear straightforward, they suffer several limitations. First, these methods assume ideal Nernstian response by the indicator electrode, which may not be true. Second, as indicated by eq. (13-15), the measured cell potential contains an unknown term, the liquid junction potential, $E_j$, which cannot be evaluated. Finally, a knowledge of activity coefficients is required to convert the calculated ionic activities into their corresponding concentrations.

Methods based on **null-point potentiometry** employ two identical indicator electrodes arranged to form a concentration cell composed of an unknown solution and a known solution. Figure 13-8 shows a concentration cell used to determine fluoride ion. The procedure consists of varying the concentration of the desired ion in the known solution until the cell potential becomes zero. When the cell potential is zero, the concentration of the given ion in the unknown solution must equal the concentration in the known solution. To avoid errors caused by the presence of liquid junction potentials at the salt bridge, a large excess of an inert electrolyte is generally added to both solutions. The presence of the large excess of inert electrolyte causes the liquid junction potentials at either end of the salt bridge to be equal; therefore, they cancel one another in the overall measured cell potential. Furthermore, ionic activity considerations are unnecessary, since the high ionic strength of the solution produced by the presence of the inert electrolyte causes the activity coefficients of the ion to be equal for both solutions.

The cell potential produced in this concentration cell will be given by

$$E_{\text{cell}} = \frac{0.059}{n} \log \frac{C_1}{C_2} \qquad \textbf{(13-19)}$$

where $C_2$ is the concentration of the given ion in the unknown solution and $C_1$ the concentration in the known solution. Thus, if $E_{\text{cell}}$ is plotted versus log $C_1$ for various values of $C_1$, a straight line will be obtained and the concentration $C_1$ for which $E_{\text{cell}} = 0$ may be determined from the graph.

In the **calibration-curve method** a graph of cell potential versus ionic activity of the given ion is prepared using a series of standard solutions. The cell potential produced with the unknown sample is then determined, and the concentration corresponding to this potential is obtained from the calibration curve. The presence of appreciable quantities of other ions in the unknown, which are not present in the standards, will alter the ionic strength of the unknown in comparison with the standards and will, therefore, affect the accuracy of the determination by indirectly altering the activity of the ion being determined. To avoid

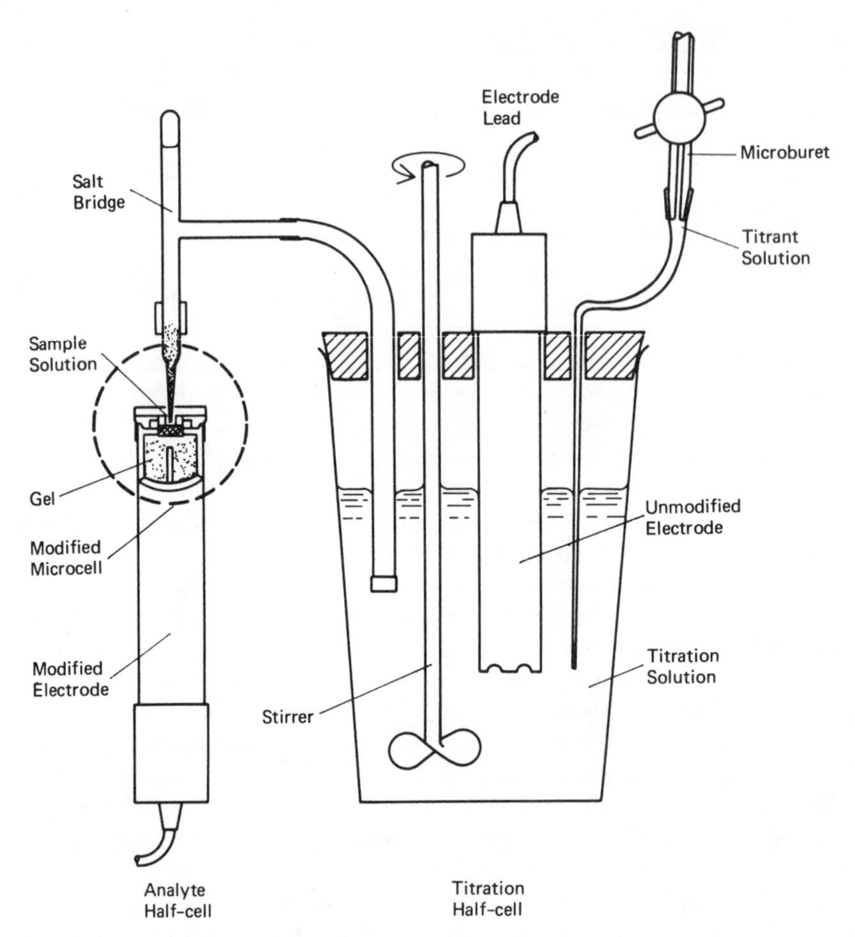


**Figure 13-8.** Concentration cell used to determine fluoride ion. (Reprinted from R. A. Durst, *Ion Selective Electrodes*, NBS Special Publication 314, Washington, 1969, with permission.)

errors of this type, the standards should approximate the composition of the unknown, or a large excess of inert electrolyte should be added to both the standards and unknown to swamp out ionic strength differences between solutions.

To avoid errors with potentiometric measurements on complex samples, the sample itself may be used in the standardization procedure, as in the **method of standard addition**. In this procedure, the cell potential obtained with the unknown sample is measured. The sample is then spiked with the ion of interest in such a way that the ionic strength and sample volume are essentially unaltered. If the initial potential, $E_{in}$, is given by

$$E_{in} = Q + \frac{RT}{nF} \ln f_X C_X + E_j \qquad \textbf{(13-20)}$$

where $Q$ is a constant; $E_j$ is the liquid junction potential; $f_X$ is the activity coefficient for ion X; and $C_X$ is the concentration of X in the sample, the potential after spiking will be

$$E_{\text{final}} = Q + \frac{RT}{nF} \ln (C_X + C_s)f_X + E_j \tag{13-21}$$

where $C_s$ is the concentration increase produced by adding the known incremental amount of X. Since $Q$, $f_X$, and $E_j$ will all be constant for both measurements, the difference, $\Delta E$, will be given by

$$\Delta E = E_{\text{final}} - E_{\text{in}} = \frac{2.3\ RT}{nF} \log \frac{C_X + C_s}{C_X} \tag{13-22}$$

Since the electrode may not behave in an ideally Nernstian fashion, the Nernst slope, $2.3RT/nF$, should be determined experimentally with a series of simple standard solutions. $C_X$ can then be determined by solving eq. (13-22) for $C_X$,

$$C_X = C_s(10^{\Delta E/S} - 1)^{-1} \tag{13-23}$$

where $S$ is the experimentally obtained slope of $E_{\text{cell}}$ versus log $C$ obtained with the standard solutions.

## POTENTIOMETRIC TITRATION

In potentiometric titrations, the potential existing between two electrodes immersed in the solution is measured as the titrant is added. One of the electrodes must be a reference electrode, and the other must be an indicating electrode, the potential of which will vary quantitatively with the activity of the measured ion in the solution. The most popular reference electrode is the saturated calomel electrode. The most popular indicating electrode for neutralization titrations is the glass electrode, whereas for oxidation–reduction titrations it is the platinum electrode. In any potentiometric titration there must be a sharp break in the measured potential at the equivalence point.

Potentiometric titrations are useful in a wide variety of situations, and can be used in those cases where a satisfactory visual indicator is unavailable or where substances are present in the titrated solution that would interfere with indicator action. Often two or more substances may be determined in the same solution by a single potentiometric titration. The main disadvantage of manual potentiometric titrations is the fact that they usually require more time and that the end point must be determined by graphical methods. Both of these disadvantages are eliminated by the use of an automatic titrator, of which several commercial models are available.

### *Graphical Methods of Selection of End Point*

A graph of the pH versus the milliliters of titrant added will produce a curve similar to that in Figure 13-9, in which there is a sharp break in pH (or potential) at the equivalence point. It can be shown mathematically that the equivalence point in any potentiometric titration coincides with the steepest point

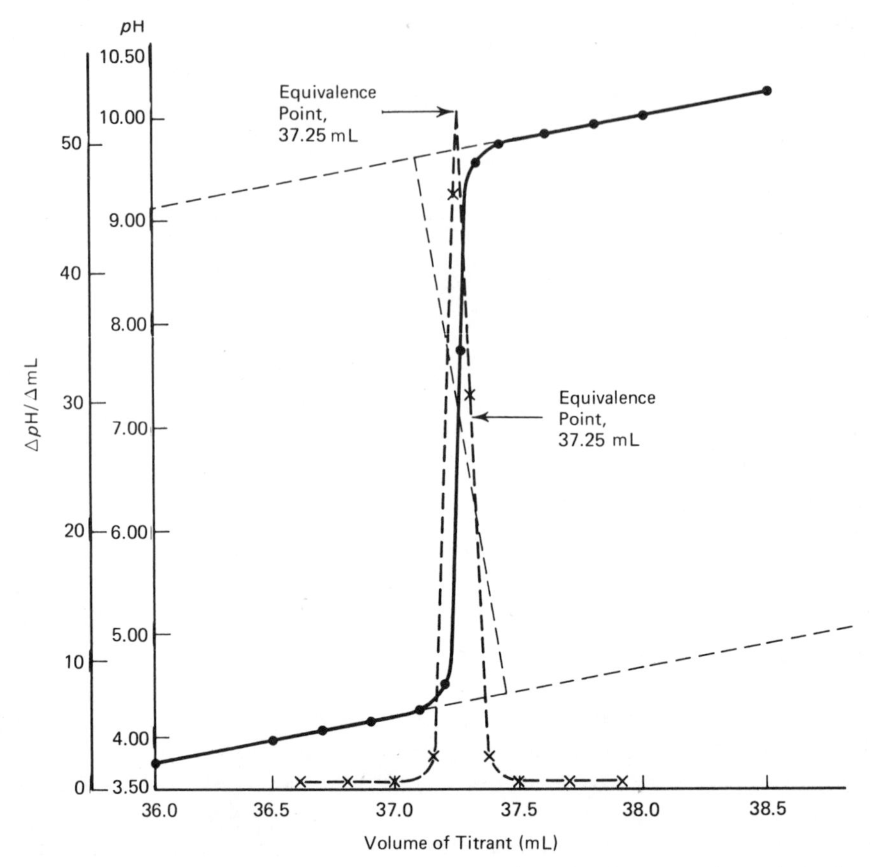


**Figure 13-9.** Expanded graph of the equivalence-point region for a potentiometric titration, showing graphical methods of selecting the end point.

on the graph so that the problem becomes one of determining the location of the steepest point. Since the desired accuracy in estimating the volume of titrant necessary is approximately 0.02 mL, the plot must be made so that the position of the steepest point can be estimated with this accuracy. This means that each small division of the paper must represent no more than 0.05–0.10 mL. Most titrations that involve a 50-mL burette are designed so that the volume necessary to reach the equivalence point is between 30 and 40 mL. A graph that shows the entire titration using 0.05- or 0.1-mL values for the smallest division of the paper will require a very large piece of paper. Selection of the equivalence point, however, requires only that portion of the graph in which the sharp break occurs. As a consequence, it is only necessary to plot the data for 1 or 2 mL on either side of the equivalence point. The steepest point of the curve usually is not obvious by inspection and must be located by some graphic method. Two such procedures are shown in Figure 13-9. The straight-line portions of the graph of pH versus milliliters (shown as a solid line) before and beyond the equivalence point break are extrapolated. Theoretically, these two lines should be parallel, but in actual practice they seldom are. The perpendicular between these two extrapolations should be bisected by the curve at the steepest point on the curve. As a consequence, the distance between the two extrapolations is measured, and the perpendicular is drawn at a point at which it will be bisected by the curve. The intersection of the curve and the perpendicular is the equivalence point.

The equivalence point can also be determined by the differential method. In this method the rate of change of pH with volume of titrant is plotted against the midpoint of each incremental addition of titrant. The difference between the pH values ($\Delta$pH) for a given increment of titrant is divided by the titrant increment ($\Delta$mL) to obtain the average rate of change of pH with volume over that particular increment. These values are then plotted against the milliliters of titrant. The differential value should be the greatest at the steepest point on the curve, and the curve should show very low values of $\Delta$pH/$\Delta$mL until close to the equivalence point. In the equivalence point region, the curve should rise sharply to a maximum and then decrease sharply. Such a graph is shown as the dashed curve in Figure 13-9 for the same titration. Note that the values are plotted at the midpoint of the increments. The equivalence point is the intersection of the lines obtained by extrapolating the best straight lines through the points on either side of the pH break. The values obtained by the two methods should agree.

If the same process of finding the slope is applied to the $\Delta$pH/$\Delta$mL versus volume curve, the slope of this curve versus volume may be found. The slope of the $\Delta$pH/$\Delta$mL versus volume curve represents the second derivative of the pH versus volume curve, or $\Delta^2\text{pH}/(\Delta\text{mL})^2$. From Figure 13-9, it can be seen that the slope of the $\Delta$pH/$\Delta$mL versus volume curve is highly positive before the equivalence point, highly negative after the equivalence point, and zero at the equivalence point. Thus, the point where the second derivative of the pH versus volume curve equals zero corresponds to the volume at the equivalence point.

## *Automatic Titrators*

As any quantitative analysis student can verify, potentiometric titrations are both tedious and time consuming. Therefore, in industrial situations where large numbers of determinations are performed, an apparatus that can automatically perform potentiometric titrations is an asset. These devices fall into three basic categories. The simplest consists of an electrochemical cell containing the unknown into which dip appropriate indicator and reference electrodes. The cell emf is measured with an electrometer, and the amplified signal is sent to a strip-chart recorder. With the recorder running, the titrant is delivered automatically at a constant rate, thereby causing the recorder pen to follow the cell emf, giving the familiar sigmoid titration curve. The end point is then determined manually by the operator from the inflection point of the titration curve. This arrangement is most advantageous with unknowns for which the equivalence point potential is not known or where the titration curve has several inflection points. This procedure assumes that the instantaneous potential of the electrochemical cell indicated by the recorder, and corresponding to a particular volume of titrant, is the equilibrium potential. This is rarely the case, since appreciable time is often required to attain chemical equilibrium because mixing is not instantaneous. The problem is particularly acute for slow reactions.

A more satisfactory approach involves stopping the automatic delivery of titrant at a predetermined calculated equivalence point potential. This approach mimics the manner in which a titration is carried out manually. Thus, in a manual titration, titrant is added rapidly at the start of the titration, but, as the equivalence point potential is approached, the delivery is slowed so that the equivalence point potential is approached but not exceeded. To accomplish this instrumentally requires proper placement of the electrodes with respect to the direction of stirring and the burette tip. The equivalence point potential may be anticipated if the electrodes are placed in close proximity to the burette tip, and so oriented with respect to the direction of stirring as to be immediately downstream from the burette tip. In this manner the cell potential will indicate a potential beyond the equivalence point potential because the titrant will not have had time to mix completely with the bulk solution. This excess potential will cause the automatic delivery of titrant to stop. As the titrant mixes with the bulk solution, the cell potential will drop below the equivalence point potential, causing the delivery of titrant to resume. In this manner, the equivalence point is approached by addition of small increments of titrant, and the instrument stops finally at the predetermined equivalence point potential.

The final category of automatic titrators are known as differential titrators. These instruments add titrant at a constant rate and electronically take the derivative of the potential versus time curve that is produced. As discussed previously, the slope (i.e., the first derivative) of a titration curve is a maximum at the equivalence point, whereas the second derivative of a titration curve is zero at the equivalence point. Thus, instruments utilizing the first derivative of the titration curve stop delivery of titrant when the derivative circuit produces a sharp maximum pulse. Instruments utilizing the second derivative terminate titrant delivery when the second derivative circuit produces a sharp change in sign. The advantage of differential titrators is that the equivalence point potential need not be known to employ these devices.

Finally, it should be stressed that automatic titrators are no more accurate than manually performed titrations. They are, therefore, most useful in those situations where speed is the predominant factor.

## PROBLEMS

**1** Calculate the pH for each of the following cell potentials obtained using a hydrogen electrode and a saturated calomel electrode at 25°C. $E_{Hg}$ at 25°C is 244 mV: (a) −0.666 V; (b) −0.446 V; (c) −0.303 V; (d) −0.999 V; (e) −0.990 V.

*Ans.* (b) 3.42

**2** Calculate the pH for each of the following cell potentials obtained using a quinhydrone electrode and a tenth-molar (normal) calomel electrode at 25°C. $E^0_Q = 699$ mV and

$E_{Hg}$ = 335 mV at 25°C: (a) −61 mV; (b) 100 mV; (c) 301 mV; (d) −205 mV; (c) 205 mV.

*Ans.* (b) 4.47

**3** Calculate the pH for each of the following cell potentials obtained using a glass electrode and a molar (normal) calomel electrode at 25°C. $E_g^0$ = 692 mV and $E_{Hg}$ = 278 mV at 25°C: (a) 0.325 V; (b) −0.385 V; (c) −0.015 V; (d) 115 mV; (e) 256 mV.

*Ans.* (b) 13.54

**4** Calculate the silver ion activity in the following cell for the potentials given:

$$Ag|\text{unknown } AgNO_3\|\text{saturated } KNO_3\|0.0440 \text{ A } AgNO_3|Ag$$

(a) −0.085 V; (b) 0.085 V; (c) −0.010 V; (d) 0.025 V; (e) −0.060 V.

*Ans.* (b) = $1.58 \times 10^{-3}$ A

**5** The following data were taken during the potentiometric titration of 25.00 mL of HCl with 0.1067 *N* NaOH. (a) Plot the data in the equivalence point region as pH versus milliliters of titrant; (b) select the end point and calculate the normality of the hydrochloric acid; (c) calculate the ΔpH/ΔmL values and plot against milliliters of titrant in the equivalence point region; (d) select the end point from this differential graph and calculate the normality of the acid:

| *Burette Readings* | pH | *Burette Readings* | pH | *Burette Readings* | pH | *Burette Readings* | pH |
|---|---|---|---|---|---|---|---|
| 0.55 | 1.70 | 25.70 | 3.45 | 26.00 | 7.50 | 26.40 | 10.52 |
| 24.50 | 3.00 | 25.80 | 3.50 | 26.05 | 9.95 | 26.50 | 10.56 |
| 25.00 | 3.17 | 25.85 | 3.60 | 26.10 | 10.20 | 27.00 | 10.74 |
| 25.50 | 3.37 | 25.90 | 3.75 | 26.20 | 10.35 | 27.50 | 10.92 |
| 25.60 | 3.41 | 25.95 | 5.00 | 26.30 | 10.47 | | |

**6** The following data were taken during the potentiometric titration of 25.00 mL of 0.1167 *N* HCl with NaOH. (a) Plot the data in the equivalence point region as pH versus milliliters of titrant; (b) select the end point from the graph and calculate the normality of the base; (c) calculate the ΔpH/ΔmL values and plot against milliliters of titrant in the equivalence point region; (d) select the end point from this differential graph and calculate the normality of the base:

| *Burette Readings* | pH | *Burette Readings* | pH | *Burette Readings* | pH | *Burette Readings* | pH |
|---|---|---|---|---|---|---|---|
| 0.25 | 1.60 | 25.40 | 3.35 | 25.70 | 7.40 | 26.10 | 10.42 |
| 24.20 | 2.90 | 25.50 | 3.40 | 25.75 | 9.85 | 26.20 | 10.46 |
| 24.70 | 3.07 | 25.55 | 3.50 | 25.80 | 10.10 | 26.70 | 10.64 |
| 25.20 | 3.27 | 25.60 | 3.65 | 25.90 | 10.25 | 27.20 | 10.82 |
| 25.30 | 3.31 | 25.65 | 4.90 | 26.00 | 10.37 | | |

**7** Calculate the silver ion activity in the following cell if the cell potential is −82 mV.

$$Ag|AgNO_3\ (0.044\ M)\|\text{sat.}KNO_3\|AgNO_{3(u)}|Ag$$

*Ans.* $1.79 \times 10^{-3}$

**8** Repeat problem 7 for the following silver concentration cells:

| *Molar Concentration of* *Standard* $AgNO_3$ | *Cell* *Potential* (V) |
|---|---|
| (a) 0.0036 *M* | 0.089 |
| (b) 0.1440 *M* | −0.089 |
| (c) 0.090 *M* | 0.118 |
| (d) 0.0250 *M* | −0.0118 |

**9** A chemist mixed 2.40 mmoles of NaHX and 2.00 mmoles of $Na_2X$ and diluted the resulting solution to 100 mL. When hydrogen and saturated calomel electrodes ($E = 244$ mV) were inserted in the solution, the cell potential was −658 mV. Calculate (a) the ionic strength of the solution; (b) the pH; (c) the activity coefficient; (d) the hydrogen ion activity; (e) the molar hydrogen ion concentration; (f) $K_2$ and $pK_2$ at this ionic strength.

*Ans.* (f) $K_2 = 1.03 \times 10^{-7}$; $pK_2 = 6.99$

**10** Repeat problem 9 for 4.00 mmoles NaHX plus 2.00 mmoles of $Na_2X$ in 100 mL and a cell potential of −599 mV.

**11** Repeat problem 9 for 8.1 mmoles of NaHX plus 4.4 mmoles of $Na_2X$ in 100 mL and a cell potential of −599 mV.

**12** An indicator turned yellow when added to a solution of hydrochloric acid. A combination glass and saturated calomel electrode was inserted and dilute base was added dropwise. As soon as a color change was noted, the cell potential was determined to be 194 mV. Addition of more base produced a pure blue color, at which point the potential was 88 mV. The $E°$ of the glass electrode was 692 mV. Calculate the pH range of the indicator.

*Ans.* 4.3 to 6.1

**13** Repeat problem 12 for the following cell potentials: (a) 271 mV and 177 mV; (b) 94 mV and −6 mV; (c) −24 mV and −118 mV; (d) 418 mV and 359 mV.

CHAPTER

# 14 Compleximetry

Coordination complexes comprise a large group of neutral and ionic substances that can play an important role in many analytical processes. For example, many coordination compounds are insoluble in water and form the basis of rather specific precipitation reactions in gravimetry. Complexing agents that form coordination complexes with certain specific metal ions may often be used as masking agents to prevent a metal ion from undergoing an undesired reaction that would result in an interference. Because coordination compounds are often highly colored, they are frequently used in many colorimetric and photometric procedures. Furthermore, the solubility of many neutral coordination compounds in organic solvents allows the extraction of metal ions from aqueous solution into the organic phase and thus forms the basis for some analytical separations that are not feasible otherwise. Because the formation of coordination complexes is so important to a large number of analytical applications, a knowledge of the factors that affect their formation and the properties of the resultant complexes is helpful in the design and understanding of analytical processes. Although a comprehensive discussion of coordination complexes is beyond the scope of this book, a short summary will be given. For more complete discussions, the student should consult recent general chemistry texts and more advanced books in analytical and inorganic chemistry.

## CHARACTERISTICS OF COORDINATION COMPOUNDS

Coordination complexes are neutral or ionic compounds that involve the formation of at least one coordinate covalent bond between the metal ion and a complexing agent. A **coordinate covalent bond** is formed between two atoms whenever one atom furnishes both electrons of the electron pair. After formation, a coordinate covalent bond is indistinguishable from other covalent bonds in which the electron pair is formed by donation of a single electron by each atom, since the source of the electrons has no effect on bond characteristics. Thus, to form a coordinate covalent bond with a metal ion, the complexing agent, or **ligand**, must have a lone pair of electrons available for sharing.

The formation of coordination complexes may be considered as an acid–base reaction in the Lewis sense. According to the Lewis convention, acid–base reactions involve the formation of a coordinate covalent bond between an acid, defined as an **electron pair acceptor**, and a base, defined as an **electron pair donor**. Since the metal ion is deficient in electrons and has empty orbitals available, the ligand can form a coordinate bond by furnishing the two electrons necessary for bond formation. Thus, the ligand is a base in the Lewis sense and the metal an acid. The formation of the complex is attributed to the attempt by the metal to attain the stable electronic structure of an inert or noble gas.

**Chelates** are coordination complexes that are formed between a metal ion and a ligand that contains at least two electron-donating groups arranged so that a ring structure can be formed upon coordination. Chelates containing five- or six-membered ring structures are especially stable. A ligand containing one electron-donating group is called a monodentate ligand, whereas ligands containing two, three, four, or six electron-donating groups in the same molecule are referred to as bi-, tri-, tetra-, or hexa-dentate ligands, respectively. The words *dentate* and *chelate* are of Greek origin, the former being derived from the word for *tooth* whereas the latter comes from the word for *claw*.

The number of bonds formed to a metal ion by its ligands is known as the **coordination number** or **ligancy** of the metal. Two typical four-coordinate complexes formed by copper (II) are

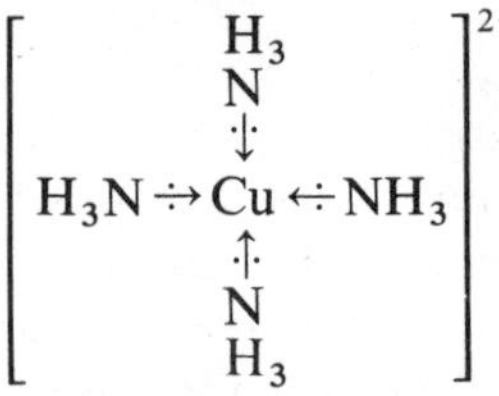

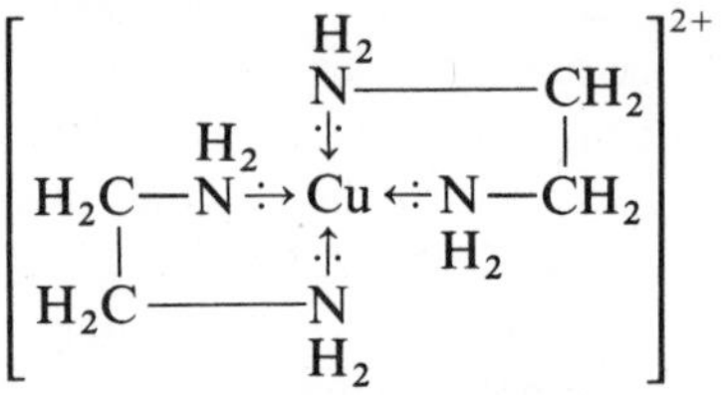

$$\left[\begin{array}{ccc} & H_3 & \\ & N & \\ & \downarrow & \\ H_3N \rightarrow & Cu & \leftarrow NH_3 \\ & \uparrow & \\ & N & \\ & H_3 & \end{array}\right]^{2+} \qquad \left[\begin{array}{ccccc} & & H_2 & & \\ & & N & \text{———} & CH_2 \\ & H_2 & \downarrow & & | \\ H_2C - & N \rightarrow & Cu & \leftarrow N - & CH_2 \\ | & & \uparrow & H_2 & \\ H_2C & \text{———} & N & & \\ & & H_2 & & \end{array}\right]^{2+}$$

The arrows used in these formulas represent coordinate covalent bonds formed by donation of a pair of electrons from the nitrogen. After formation there is no distinction between coordinate covalent bonds and other covalent bonds. Note that the ammonia is a monodentate ligand, whereas ethylenediamine [$H_2NCH_2CH_2NH_2$] is a bidentate ligand that forms five-membered chelate rings. In the actual formation of these complexes, the ligands displace the four molecules of water that are always coordinated to the copper (II) ion in aqueous solution. In both complexes the charge on the complex is the same as the charge on the metal ion, because the ligands themselves are neutral. In the case of negative ligands, however, the complex can have a negative charge as in $Ag(CN)_2^-$ [dicyanoargentate (I) ion].

The charge on the complex can be used analytically. For example, in extraction procedures, neutral complexes are often used as extraction species, whereas charged complexes are often used as masking agents to form nonextractable ionic species with some metal ions to increase the specificity of extraction procedures for other metals.

An example of an important chelating agent is dimethylglyoxime, a bidentate ligand.

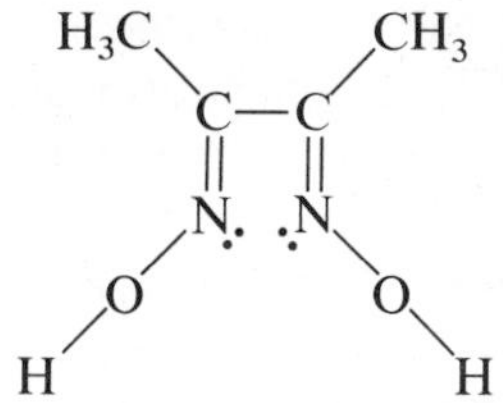

```
 H3C       CH3
    \     /
     C---C
     ||  ||
     N:  :N
    /      \
   O        O
  /          \
 H            H
```

Coordination by dimethylglyoxime occurs through the lone pairs of electrons on the nitrogens.

Nickel (II) ion reacts readily with dimethylglyoxime to form a red square-planar complex that is used in gravimetric, photometric, and extraction procedures for nickel because the reaction is essentially specific for nickel in solutions that are near neutrality. The structure of the bis(dimethylglyoximato) nickel (II) complex is

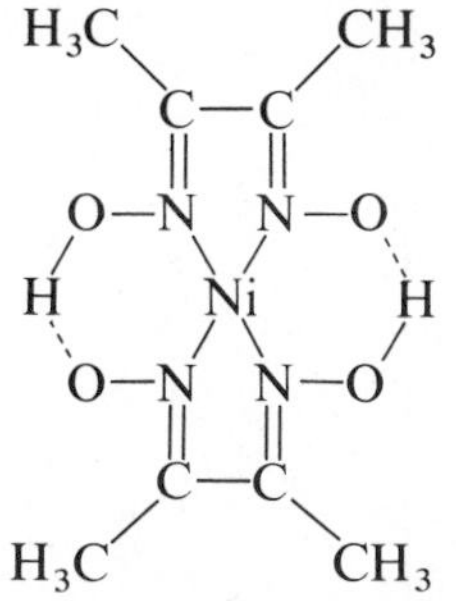

```
      H3C       CH3
         \     /
          C---C
          ||  ||
     O---N    N---O
    /     \  /     .
   H       Ni       H
    .     /  \     /
     O---N    N---O
          ||  ||
          C---C
         /     \
      H3C       CH3
```

The dashed lines in the structure indicate hydrogen bonds. Note that the complex is neutral owing to the fact that each dimethylglyoxime molecule loses a hydrogen ion in the formation of the complex; the dipositive charge on nickel is therefore balanced by the single negative charges on each dimethylglyoxime.

**TABLE 14-1.** Typical Functional Groups Involved in Chelation

| *Acidic or Anionic Groups* | | *Uncharged Groups* | |
|---|---|---|---|
| Carboxylic acids | $-C(=O)-OH$ or $-C(=O)-O^-$ | Amines | $-NH_2$, $-N(H)-$, $-N(|)-$ |
| Oximes | $=NOH$ or $=NO^-$ | Carbonyl | $-C(=O)-$ |
| Hydroxyls | $R-OH$ or $R-O^-$ | Thiocarbonyl | $-C(=S)-$ |
| Phenols | $\phi-OH$ or $\phi-O^-$ | Nitroso | $-NO$ |
| Sulfonic acids | $-S(=O)_2-OH$ or $-S(=O)_2-O^-$ | Cyclic amine | $\geqslant N$ (ring N with one double and one single bond) |
| Mercaptyl | $R-SH$ or $R-S^-$ | | |

A ligand involved in complexation can be either charged [as in the case of bis(dimethylglyoximato)Ni(II)] or uncharged [as in the copper(II) complexes illustrated]. With certain acidic functional groups, coordination requires displacement of hydrogen, and these ligands are therefore anionic. Uncharged ligands can coordinate without dissociation. Typical examples of both types of ligating functional groups are shown in Table 14-1.

Examples of other chelates encountered in analytical chemistry are

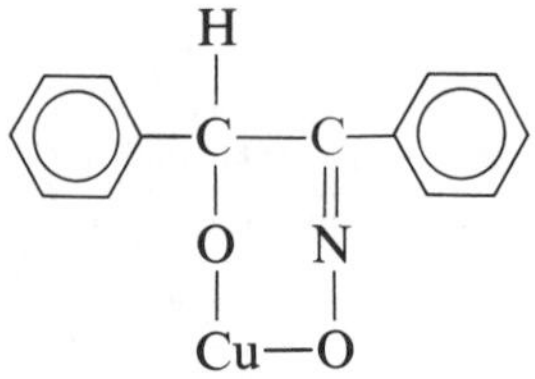

Cu(II) α-benzoin oxime complex

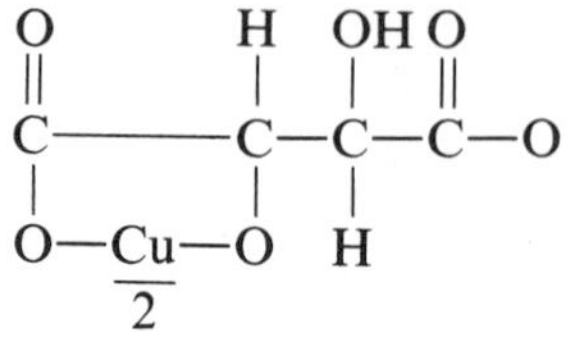

Cu(II) tartaric acid complex

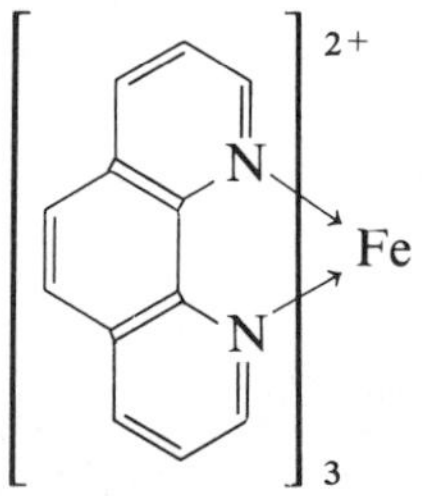

Fe(II) 1,10-phenanthroline complex

Chelating agents can be used as precipitants, extractants, or masking agents for metal ions. As precipitants they have the advantages of increased specificity, freedom from ionic coprecipitation, and low gravimetric factors. The fact that most chelates are colored allows their use in photometric procedures, and their solubility in organic reagents increases their adaptability to extraction procedures.

## STABILITY AND INSTABILITY CONSTANTS

Most complex ions contain more than one ligand coordinated to the metal, and the formation of the complex takes place in steps. Actually, practically all metal ions are hydrated in water, and the formation of the complex can be considered as the displacement of the coordinated water by a stronger ligand. Even though we often write the formula of the cupric ion in solution as $Cu^{2+}$, it is actually present as $Cu(H_2O)_4^{2+}$, and the blue color associated with cupric ions in solution is the color of the hydrated ion. The anhydrous ion is colorless, as is indicated by the fact that anhydrous cupric sulfate is white.

The formation of complexes in solution is illustrated in eqs. (14-1) and (14-2), in which M is assumed to be a metal ion with coordination number of *n* and L represents a ligand other than water, which can occupy one coordination position on the metal.

$$M(H_2O)_n \rightleftharpoons H_2O + M(H_2O)_{n-1}L \quad \textbf{(14-1)}$$

$$M(H_2O)_{n-1}L + L \rightleftharpoons H_2O + M(H_2O)_{n-2}L_2 \quad \textbf{(14-2)}$$

The reaction continues in steps until the highest complex, $ML_n$, is formed. For simplicity in writing equations, the water of the aquo complexes is usually omitted. Furthermore, since the metal ion can be mono- or polyvalent and since the ligands can be neutral or ionic, the ionic charges are omitted in the following general discussion.

Each stage in the formation of a complex is governed by the law of mass action and forms an equilibrium, so each step has its own formation or stability constant. Thus, for the formation of $ML_n$ from M and L, the stepwise sequence is

$$M + L \rightleftharpoons ML \qquad k_1 = \frac{[ML]}{[M][L]} \quad \textbf{(14-3)}$$

$$ML + L \rightleftharpoons ML_2 \qquad k_2 = \frac{[ML_2]}{[ML][L]} \quad \textbf{(14-4)}$$

$$ML_2 + L \rightleftharpoons ML_3 \qquad k_3 = \frac{[ML_3]}{[ML_2][L]} \quad \textbf{(14-5)}$$

$$ML_{n-1} + L \rightleftharpoons ML_n \qquad k_n = \frac{[ML_n]}{[ML_{n-1}][L]} \quad \textbf{(14-6)}$$

The overall formation constant for any given stage of complexation is given by the product of the successive $k$ values; thus,

$$M + 2L \rightleftharpoons ML_2 \qquad K_2 = k_1k_2 = \frac{[ML_2]}{[M][L]^2} \quad \textbf{(14-7)}$$

$$M + 4L \rightleftharpoons ML_4 \qquad K_4 = k_1k_2k_3k_4 = \frac{[ML_4]}{[M][L]^4} \quad \textbf{(14-8)}$$

The total molar concentration of the metal in a solution containing a metal ion and a complexing ligand is the sum of the equilibrium concentrations of each species present:

$$[M]_t = [M] + [ML] + [ML_2] + [ML_3] + \cdots + [ML_n] \quad \textbf{(14-9)}$$

Another way of stating this is that all the possible species will be present even though some are present in such small concentrations that they are negligible compared to the predominant species.

In many cases successive constants are almost equal so that there can be rather wide ranges of ligand concentration in which adjacent species will be present in almost equal amounts. Furthermore, in the presence of high ligand concentration, the highest complex is usually favored, and the presence of the intermediate species may be neglected in many calculations.

The reverse reaction or the ionization of the complex also occurs in steps from the complex with the greatest number of ligands to the free metal ion. The ionization or instability constants for these reactions are the inverse of the constants for the formation reactions. Tables of constants for both complex formation and ionization are available in the literature and are also known as stability constants ($K_{stab}$) and instability constants ($K_{instab}$). Note carefully which type the table is. In general, the easiest method of distinguishing is to examine the values of the constants. In most cases the values for the stability constants will be large numbers (exponents will be positive), whereas the values for the instability constants will be very small numbers (exponents will be negative).

## COMPLEXIMETRIC TITRATIONS

### TITRATION OF CHLORIDE WITH MERCURIC NITRATE

In reality, the titration of chloride with mercuric nitrate is an example of the formation of a slightly ionized compound rather than complex formation, since mercuric chloride is only slightly ionized in water solution (less than 1% in 0.1 *N* solution). Because it is very unusual for a metallic chloride to be nonionic, this titration is usually discussed with complex formation titrations.

Several indicators are available for the titration, the most popular being sodium nitroprusside [$Na_2Fe(CN)_5NO$] and diphenyl carbazone

$$[C_6H_5(NH)_2CON_2C_6H_5].$$

The nitroprusside forms a precipitate (turbidity) with excess mercuric ion and should not be added until very close to the end point. The diphenylcarbazone forms a soluble violet-colored complex with mercuric ions and is the indicator of choice. Correct pH is very important for the proper functioning of the diphenylcarbazone indicator and the pH must be maintained between 3.0 and 3.5 with nitric acid. For this reason, the indicator solution is often a mixture of diphenylcarbazone and bromphenol blue to aid in the adjustment of pH. The color of the bromphenol blue in this pH range also increases the sharpness of the end point.

The mercurimetric diphenylcarbazone method for chloride is suitable for titration in very dilute solutions in the range 1–100 ppm and is used for determination of chloride in natural waters and in blood. The method is also used in the determination of bromide, thiocyanate, and cyanide and is remarkably free of interference.

Most directions call for the mercuric nitrate to be made up in concentrations of 0.025 *N* or less in dilute nitric acid and standardized against primary standard sodium chloride solutions. The sample should not contain more than 20 mg of chloride. Before titration, the pH is adjusted with 0.05 *N* $HNO_3$ to the yellow color of bromphenol blue followed by the addition of 1 mL excess of the acid. The end point is the appearance of a blue-violet color that deepens as more titrant is added.

## TITRATION OF CYANIDE WITH SILVER NITRATE

The addition of a silver nitrate solution to a solution containing cyanide ion causes the formation of the soluble dicyanoargentate(I) ion, $Ag(CN)_2^-$. After sufficient silver has been added to react with all the cyanide by this reaction, the first excess will react to form insoluble silver dicyanorgentate(I), $Ag[Ag(CN)_2]$, which is often referred to and written as silver cyanide, AgCN. The ionic equations for the reactions are

$$Ag^+ + 2CN^- \rightleftharpoons Ag(CN)_2^- \qquad \textbf{(14-10)}$$

$$Ag^+ + Ag(CN)_2^- \rightleftharpoons Ag[Ag(CN)_2] \qquad \textbf{(14-11)}$$

As a consequence, the appearance of a permanent precipitate indicates the quantitative completion of the first reaction and can be taken as the equivalence point. It should be noted that the equivalent weight of cyanide in the reaction is 52.06, or twice the radical weight, since it requires two cyanides to react with each silver ion.

The end point is not sharp, owing to formation of solid silver dicyanoargentate(I) caused by local excess concentrations in the zone of mixing of the titrant and the solution. This precipitate is slow to redissolve in the equivalence point region. The end point can be sharpened by the addition of ammonia and iodide ion because the ammonia will prevent the precipitation of silver dicyanoargentate(I) but does not prevent the precipitation of silver iodide. As a consequence, the first excess of silver precipitates as silver iodide, which forms as a yellow opalescence in the solution and constitutes a very sharp end point. The silver iodide does not form before the equivalence point, because it is kept in solution by the excess cyanide in which it is soluble. The concentration of ammonia used is very important; too little will cause the end point to appear early, and too much will cause the end point to appear late. The concentration of iodide is also important for similar reasons. Most directions call for the solution to be approximately 0.3 *M* in ammonia and to contain approximately 0.2 g of potassium iodide per 100 mL.

Under the proper conditions, the reaction is accurate, and the end point is sufficiently sharp. The experiment is seldom carried out in beginning courses in analysis, however, owing to the danger associated with the use of extremely poisonous cyanide solutions.

## TITRATION WITH ETHYLENEDIAMINETETRAACETIC ACID (EDTA)

Ethylenediaminetetraacetic acid (also known as ethylenedinitrilotetraacetic acid and commonly referred to as EDTA) is the most important example of a class of reagents, introduced in 1945 by Schwartzenbach, that form soluble stable chelate complexes with many metals. This group of compounds is often called **complexones, chelons** or **versenes** as a general group name and the substances are usually aminopolycarboxylic acids. They can also be described as tertiary amines containing several carboxylic acid groups.

EDTA is a tetraprotic acid with the formula

$$(HOOCCH_2)_2N{-}CH_2{-}CH_2{-}N(CH_2COOH)_2$$

that forms soluble 1 : 1 metal complexes of widely different stabilities with metals. The acid, usually written as $H_4Y$ for convenience, is insoluble in water, and solutions are usually prepared with the disodium salt $Na_2H_2Y$, which is available in high purity as the dihydrate. The solution, however, is referred to as EDTA rather than as disodium EDTA. Since EDTA has both donor nitrogen and donor oxygen atoms, it is hexadentate and can form as many as five five-membered chelate rings; consequently, it forms complexes with practically all metals except the alkali metals of group I. However, the 1 : 1 mole reaction ratio makes equilibrium calculations simpler than for most complexes that are formed by stepwise equilibria.

EDTA is a weak acid that ionizes stepwise, as indicated in eqs. (14-12) through (14-15):

$$H_4Y \rightleftharpoons H^+ + H_3Y^- \qquad k_1 = \frac{[H^+][H_3Y^-]}{[H_4Y]} \tag{14-12}$$

$$H_3Y^- \rightleftharpoons H_2Y^{2-} + H^+ \qquad k_2 = \frac{[H^+][H_2Y^{2-}]}{[H_3Y^-]} \tag{14-13}$$

$$H_2Y^{2-} \rightleftharpoons H^+ + HY^{3-} \qquad k_3 = \frac{[H^+][HY^{3-}]}{[H_2Y^{2-}]} \tag{14-14}$$

$$HY^{3-} \rightleftharpoons H^+ + Y^{4-} \qquad k_4 = \frac{[H^+][Y^{4-}]}{[HY^{3-}]} \tag{14-15}$$

The p$K$ values for the stepwise constants are $pK_1 = 2.00$, $pK_2 = 2.67$, $pK_3 = 6.16$, and $pK_4 = 10.26$, and the cumulative constant is $pK_{instab} = 21.09$. The secondary ion, $H_2Y^{2-}$, reacts with metals by displacement of hydrogen ions, as exemplified by

$$Mg^{2+} + H_2Y^{2-} \rightleftharpoons MgY^{2-} + 2H^+ \tag{14-16}$$

$$Al^{3+} + H_2Y^{2-} \rightleftharpoons AlY^- + 2H^+ \tag{14-17}$$

Equations (14-16) and (14-17) indicate that the smaller the stability constant, the higher the pH of the solution must be to form the complex. In general, complexes of divalent ions are stable in alkaline or slightly acidic solution, complexes of the trivalent ions are stable at pH 1 or 2, and complexes of the tetravalent cations are stable in strong acid solutions. A number of exceptions to these general statements can be found.

EDTA is used primarily as a titrant in volumetric analysis but is also utilized as a masking agent in various reactions and extractions and as a method of dissolving insoluble salts. Our discussion will deal only with its use as a titrant.

## *Titration Methods*

Several titration methods are available because of the wide range of stabilities, the effect of pH, and the ability to use masking agents in the presence of the titrant. The most important of these are the direct and indirect titration procedures.

**Direct Titration.** Several metal ion indicators are available for use in direct titrations. In general, the solution of the metal ion is buffered to the proper pH, masking agents and the indicator are added, and the solution is titrated with EDTA to the color change at the end point.

**Indirect or Back Titration.** Metals that cannot be determined easily by direct titration may often be determined by the addition of an excess of a standard EDTA solution followed by the titration of the excess with a standard magnesium or zinc solution.

**Displacement of a Metal Ion.** Addition of a solution of zinc or magnesium EDTA complex to a solution of a metal that forms a more stable complex with EDTA than zinc or magnesium causes the displacement of zinc or magnesium from the complex. As an example, when a solution of magnesium EDTA is added to a solution of ferric ion, the ferric ion replaces the magnesium in the complex.

$$Fe^{3+} + MgY^{2-} \rightleftharpoons FeY^{-} + Mg^{2+} \qquad \textbf{(14-18)}$$

The liberated $Mg^{2+}$ is then titrated with standard EDTA and is a measure of the amount of $Fe^{3+}$ in the solution.

**Titration of Hydrogen Liberated.** According to eqs. (14-16) and (14-17), the reaction of a metal ion with EDTA releases two hydrogen ions. The hydrogen ion can be titrated with standard NaOH to determine the amount of cation indirectly.

**Titration of Anions.** Anions that form insoluble compounds with metal ions titratable by EDTA can be determined by adding an excess of a standard solution of the cation. The excess cation is then titrated with standard EDTA.

An example is the determination of sulfate by the addition of an excess of standard barium solution, which is then titrated by EDTA.

$$SO_4^{2-} + Ba^{2+} \rightleftharpoons BaSO_4 + Ba^{2+}(XS) \qquad \textbf{(14-19)}$$

$$Ba^{2+}(XS) + H_2Y^{2-} \rightleftharpoons BaY^{2-} + 2H^+ \qquad \textbf{(14-20)}$$

The end point has a tendency to fade because the $BaSO_4$ will dissolve in any excess EDTA added beyond the end point.

## *Indicators*

A number of visual indicators are available that form stable highly colored complexes with metal ions. The metal-indicator complex ion must be less stable than the corresponding EDTA complex, so that EDTA can remove the metal from the indicator complex. Using magnesium and Erio T (erio-chrome black T, also known as Indicator F241) as the example,

$$\text{Mg-Erio} + \text{EDTA} \rightleftharpoons \text{Mg-EDTA} + \text{Erio} \qquad \textbf{(14-21)}$$

The end point is the appearance of the color of the uncomplexed indicator.

Erio T, which was one of the first indicators to be used in EDTA titrations, works satisfactorily only in solutions in which the pH is approximately 8 to 10. At lower pH values, the end point is not sharp, owing to the fact that the dye has a purple red color similar to the color of the metal dye complexes, and at higher pH values it is converted to an orange form the does not function as an indicator. In the pH range 8–10, Erio T can be used to determine the sum of all the alkaline earth metals present but cannot be used for the determination of a single alkaline earth in the presence of any of the others. It is also used for titrations of Zn, Cd, and Pb.

CalRed (also known as the indicator of Patton and Reeder) can be used in highly basic solutions (pH 12–14) for the determination of calcium. At this high pH, magnesium precipitates as the hydroxide and does not interfere. The color change for CalRed is from red to pure blue, which is the same as the Erio T change.

Other indicators used, together with the useful pH range and chemical type, are listed in Table 14-2.

## *Standardization*

Disodium EDTA dihydrate can be purchased in high purity and can be used in some cases as a primary standard for the preparation of standard EDTA solutions. For most samples, however, it is preferable to standardize the solution against a standard solution of the metal to be determined. Many directions call for the standardization against a standard magnesium solution

**TABLE 14-2.** Indicators for EDTA Titrations

| *Indicator* | *Chemical Type* | *Useful* pH *Range* | *Typical Metals Determined* |
|---|---|---|---|
| Erio T | Azo | 8–10 | Mg, Zn, Cd, Pb, (Mg + Ca) |
| Erio R | Azo | 8–12 | Mg, Zn (Mg + Ca) |
| CalRed | Azo | 12–14 | Ca |
| Calmagite | Azo | 9–11 | Mg, Zn, Cd (Mg + Ca) |
| PAN | Azo | 2–11 | Cu, Pb |
| Murexide | Purine | 6–13 | Ca, Cu, Co, Ni |
| Calcein Blue | Iminodiacetate | 5–9, 13–14 | Cu, Ca |
| Salicylic acid | — | 2–3 | Fe |
| Zincon | Hydrazo | 5–10 | Zn, Cu, Ca |

using Erio T in a solution buffered to pH 9 with an ammoniacal buffer. The end point is the change from a wine red or purple to a pure sky blue. Since solutions of Erio T are not stable for extended periods of time, indicator solutions should be prepared fresh each week. To offset the trouble of preparing fresh indicator periodically, the indicator can be coated on crystals of a soluble salt, such as potassium chloride, potassium sulfate, or sodium sulfate, that will not interfere with a titration. The solid indicator is stable, so dispersions on solid crystals are also stable. Sufficient crystals are added to the solution being titrated to produce a satisfactory color.

Most standard solutions are 0.01 *M* (0.02 *N*) and are often also specified as having a calcium carbonate titer of 1.00 mg/mL, especially for solutions used to determine hardness of water.

## *Determination of Total Hardness in Water*

One of the main uses of EDTA is the determination of total hardness in water. Total hardness is due to the presence of both calcium and magnesium ions so that total hardness (or the sum of calcium and magnesium) is determined readily using Erio T as the indicator. The stability constant for the calcium–EDTA complex ($\log K = 11.00$) is larger than the stability constant ($\log K = 8.69$) for the magnesium complex. As a result, the addition of EDTA first converts the calcium to the EDTA complex; the magnesium is converted to the EDTA complex only after all the calcium is complexed. The stability constant of the magnesium–EDTA complex is larger than the stability constant for the magnesium–Erio T complex ($\log K = 7.00$), so the end point occurs when all the magnesium is converted from the Erio T complex to the EDTA complex. The color of Erio T is a pure sky blue at pH values between 8 and 10, whereas the color of the magnesium complex is wine red. Since total hardness is usually expressed as parts per million (milligrams per liter) of calcium carbonate, irrespective of the ion causing the hardness, the calculation becomes simply the multiplication of the volume of titrant by the calcium carbonate titer, followed by conversion to the number of milligrams in 1 liter of the water.

### Determination of Calcium and Magnesium in Limestone

Prior to the advent of EDTA as a titrant, calcium and magnesium were determined in limestone and other rocks by the precipitation of calcium as the oxalate followed by precipitation of magnesium as magnesium ammonium phosphate. The calcium oxalate was either converted to the oxide or carbonate and weighed, or dissolved and titrated with standard permanganate. The magnesium ammonium phosphate was ignited to magnesium pyrophosphate and weighed. Early work with EDTA using Erio T as the indicator showed that accurate results of the sum of the magnesium and calcium present could be obtained in rock samples, but there was no really satisfactory indicator for calcium or magnesium individually in the presence of each other. As a consequence, procedures to determine both calcium and magnesium depended upon the oxalate separation followed by titration of the magnesium in the filtrate with EDTA using Erio T as the indicator.

CalRed became available in 1956 as in indicator for the titration of calcium in the high pH range in which magnesium is quantitatively precipitated as the hydroxide and thus allowed the determination of calcium in a sample. Combination of the data for the calcium-only titration with the data from the titration for the sum of the calcium and magnesium makes possible the calculation of both calcium and magnesium individually. Other indicators are now available for titration of individual alkaline earths.

Many directions for calcium and magnesium in limestone call for the dilution of the filtrate from the $R_2O_3$ precipitate to a definite volume and the titration of aliquots of this solution with EDTA using Erio T and CalRed as indicators for separate aliquots.

**Titration for Sum of Calcium and Magnesium.** The aliquot to be titrated is buffered with ammoniacal buffer to a pH about 9, and masking agents are added. Ordinarily, small amounts of both hydroxylamine and potassium cyanide solutions are added. The hydroxylamine reduces ferric iron to the ferrous form, which does not interfere as much as ferric ion. The cyanide is added to form complexes with iron and other heavy metals such as copper to prevent their interference with the indicator action. The indicator is added as a solution or dispersed on crystals, and the solution is titrated to a pure sky blue end point. The theory of the titration is the same as the determination of total hardness discussed previously.

**Titration for Calcium only.** A concentrated (8 *N*) solution of potassium or sodium hydroxide is added to the aliquot to be titrated for calcium to cause the quantitative precipitation of magnesium as the hydroxide. In some cases, the high pH will also cause the precipitation of some calcium hydroxide, and the end point will be indistinct, since the calcium hydroxide dissolves slowly as the titrant is added. The calcium precipitation can be prevented by the addition of some of the standard EDTA before the addition of the base. The EDTA complexes part of the calcium and prevents its precipitation as the hydroxide but does not prevent precipitation of magnesium hydroxide. Care must be used to ensure that the amount added is not greater than the amount needed for the titration. Small amounts of hydroxylamine and potassium cyanide solutions are added to reduce and mask interfering metals. The CalRed indicator is added as a solution or dispersed on crystals, and the solution is titrated with standard EDTA to the end point, which is a change from wine red to pure sky blue.

Calculations. The amount of calcium present is obtained by direct calculations from the data of the CalRed titration for calcium. The amount of magnesium is determined by the difference between the volumes of EDTA used for identical aliquots in the Erio T and the CalRed titrations. The excess volume of EDTA required in the Erio T titration over that required in the CalRed titration is a measure of the amount of magnesium in the aliquot.

## PROBLEMS

**1** A 50.00-mL sample of river water required 25.33 mL of a $Hg(NO_3)_2$ solution with a chloride titer of 0.150 mg/mL to reach the diphenylcarbazone end point. Calculate the ppm of Cl in the water.

*Ans.* 76.00 ppm

**2** A 25.0-mL sample of water required 21.2 mL of a 0.00352 *N* solution of $Hg(NO_3)_2$ to reach the diphenylcarbazone end point. Calculate the ppm of Cl in the water.

**3** A 7.25-g sample of a silver ore was dissolved and after various separation steps the silver (as $AgNO_3$) was diluted to 50.0 mL. This solution was used as the titrant for a solution that contained 0.0980 g of NaCN, 0.050 g of KI, and 7.5 mmoles of $NH_3$. It required 37.5 mL to reach the AgI end point. Calculate the percentage of silver in the ore.

*Ans.* 1.98%

**4** Repeat problem 3 for a silver ore sample weighing 6.32 g that required 78.2% of the final silver solution to react with 25.0 mL of 0.0401 *N* NaCN containing the proper amounts of KI and $NH_3$.

**5** A 50.0-mL sample of city water required 20.4 mL of an EDTA solution with a $CaCO_3$ titer of 1.05 mg/mL to reach the Erio T end point. Calculate the ppm of total hardness based on $CaCO_3$.

*Ans.* 428 *ppm.*

**6** Repeat problem 5 for the following water samples:

| *Sample Size (mL)* | *EDTA Titer* | *EDTA Required (mL)* |
|---|---|---|
| 100 | 0.995 | 32.4 |
| 25.0 | 1.10 | 12.3 |
| 50.0 | 0.753 | 18.3 |

7 A 1.025-g sample of limestone was dissolved and, after removal of the acid-insoluble residue and $R_2O_3$ precipitate, was diluted to 500.0 mL. Calculate the percentages of CaO, MgO, $CaCO_3$, and $MgCO_3$ from the following data:
   (a) A 100.0-mL aliquot required 28.36 mL of 0.1037 *N* $KMnO_4$. A blank of 0.06 mL was required to reproduce the end-point color.
   (b) A 20.00-mL aliquot required 28.94 mL of an EDTA solution with a $CaCO_3$ titer of 1.015 mg/mL to reach the CalRed end point.
   (c) A 10.00-mL aliquot required 19.85 mL of the same EDTA solution to reach the Erio T end point.

*Ans.* (a) $CaCO_3$, 71.64%; (c) MgO, 10.73%

8 A 0.9924-g sample of a dolomitic limestone was dissolved and diluted to 500.0 mL after removal of the acid-insoluble residue, which weighed 0.0101 g. Calculate the percentages of calcite ($CaCO_3$), dolomite ($CaCO_3 \cdot MgCO_3$), CaO, MgO, and acid-insoluble residue from the following data:
   (a) A 20.00-mL aliquot required 20.96 mL of an EDTA solution with a $CaCO_3$ titer of 1.103 mg/mL to reach the CalRed end point.
   (b) A 10.00-mL aliquot required 19.15 mL of the same EDTA solution to reach the Erio T end point.

*Ans.* calcite, 10.06%; MgO, 19.41%

9 Give one example each from *your own laboratory experience* for the following uses of coordination compounds or ions in analytical chemistry.
   (a) Masking agent to remove interference due to color in an oxidation–reduction titration of iron
   (b) End-point indicator for a precipitation titration
   (c) Masking agent to remove interference in determination of copper by iodometric method
   (d) As a titrant
   (e) A *chelate* used as the basis of a colorimetric method

CHAPTER

# 15 Analytical Uses of Reaction Rates

Chemical reactions may be characterized in two ways: (1) in terms of the final equilibrium composition and (2) in terms of the speed with which a given reaction approaches equilibrium. The final equilibrium composition for a given reaction depends on the temperature of the reaction mixture and on the thermodynamic properties of the reactants and products. Chemical thermodynamics can tell us whether a particular reaction will occur but gives no information about the time required for the reaction to reach equilibrium. The study of the rates of chemical reactions is known as **chemical kinetics**. Chemical kinetics is an important branch of chemistry because it can provide information on the mechanisms of chemical reactions. This chapter will be devoted to the analytical applications of reaction rates.

The analytical use of reaction rates is based on the premise that some relationship exists between the rate of a reaction and the concentration of the substance being analyzed (i.e., the analyte) present in the sample. There are several advantages to the analytical use of reaction rates: (1) reactions that would ordinarily require excessive amounts of time to reach equilibrium may be amenable to reaction rate methods; (2) measurements may be made rapidly, before interfering side reactions have time to occur; (3) reaction-rate measurements are relative measurements and are therefore less subject to various systematic errors such as liquid junction potentials, turbidity, and dirty spectrophotometer cells; and (4) differential rate measurements often permit the determination of similar chemical compounds in mixtures without prior chemical separation.

Several practical considerations must be taken into account, however, in the analytical application of reaction rates. First, convenience and accuracy require that the rate of the analytical reaction neither be too fast (so that measurement of the rate is difficult) nor too slow (so that measurements require excessive time). Also, the rates of many chemical reactions are very sensitive to the prevailing experimental conditions. For this reason, it is important that parameters such as the temperature, pH, and ionic strength be rigorously controlled for all measurements.

## SOME ASPECTS OF CHEMICAL KINETICS

Before we can discuss the analytical application of reaction rates, we must turn our attention to some elementary aspects of chemical kinetics. The rate of a chemical reaction can be experimentally determined from the time dependence of the concentration of any species involved in the reaction. Figure 15-1 shows the time dependence of the concentration of R and Q for the hypothetical reaction,

$$3R + 2S = Q + P \qquad \textbf{(15-1)}$$

At the start of the reaction, R is present in large amounts whereas the concentration of Q is zero. As the reaction proceeds, the concentration of R decreases while the concentration of Q increases. As the reaction approaches equilibrium, the concentrations of R and Q become independent of time. Thus, the net rate of a reaction is given by

$$\text{net rate} = \text{forward rate} - \text{reverse rate} \qquad \textbf{(15-2)}$$

At equilibrium, the forward rate of the reaction equals the reverse rate, and the net rate is zero.

The rate of the production of Q at any time during the reaction may be determined from Figure 15-1 by determining the slope of the tangent to the curve at the given time. Since the slope of the tangent to a curve at any point is the derivative of the curve at that point, the rate of production of Q is given as

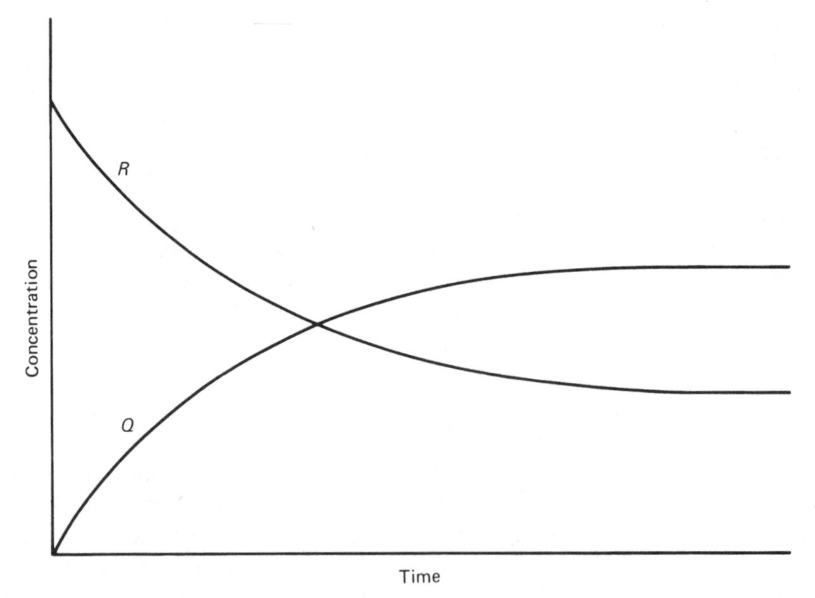


**Figure 15-1.** Time dependence of the concentration of R and Q for the hypothetical reaction 3R + 2S = Q + P.

$d[Q]/dt$. Since the progress of any reaction may be followed by measuring the growth of products or the loss of reactants, the experimenter is free to choose which substance he wishes to monitor in determining a reaction rate. Usually, the substance that can most easily be determined as a function of time is selected. It can be seen from the stoichiometry of reaction (15-1) that the rate of disappearance of R will be three times the rate of formation of Q. To avoid ambiguity, the rate of the reaction is defined as

$$\text{rate of reaction} = \pm\frac{1}{a}\frac{d[X]}{dt} \qquad \textbf{(15-3)}$$

where $a$ is the stoichiometric coefficient of substance X in the balanced chemical reaction, and the minus sign is used only when the substance X is a reactant. Thus for reaction (15-1),

$$\text{rate of reaction} = \frac{d[Q]}{dt} = \frac{d[P]}{dt} = -\frac{1}{3}\frac{d[R]}{dt} = -\frac{1}{2}\frac{d[S]}{dt} \qquad \textbf{(15-4)}$$

The rate of a reaction is influenced by several factors, including the temperature of the reaction mixture and the concentrations of the species involved in the reaction. Since the reaction rate varies with time, it is often convenient to eliminate the time dependence by determining the initial rate of the reaction from the tangent to the curve at zero time. There are several advantages to using the initial rate. First, the initial rate represents the maximum rate. Since the reaction has not proceeded to an appreciable extent, and the concentration of products is low, the net rate of the reaction is essentially the forward rate. Finally, the influence of side reactions may often be reduced by using the initial rate.

The mathematical expression that relates the rate of the reaction to the concentrations of reactants is known as the **differential rate law**. Consider reaction (15-5)

$$aA + bB = cC + dD \qquad \textbf{(15-5)}$$

The differential rate law for this reaction may frequently be expressed as

$$-\frac{1}{a}\frac{d[A]}{dt} = k[A]^p[B]^q \qquad \textbf{(15-6)}$$

The exponents $p$ and $q$ are frequently integral or half-integral. The exponent $p$ is referred to as the order of the reaction with respect to A, whereas $q$ refers to the order of the reaction with respect to B. The sum $p + q$ is the overall order of the reaction. Thus, if $p$ is 1 and $q$ is 2, the reaction is first order in A, second order in B, and third order overall. It should be stressed that the differential rate law must be *experimentally determined*, and that $p$ and $q$ are not necessarily equivalent to the stoichiometric coefficients $a$ and $b$ in the balanced chemical equation.

The constant $k$ in eq. (15-6) is the specific rate constant for the reaction. The specific rate constant is the rate of the reaction when A and B are present in unit concentration.

One way in which the differential rate law may be experimentally determined is by observing how the initial rate varies with changes in the reactant concentrations. Suppose that a series of solutions containing a constant amount of B but varying amounts of A is prepared. If the initial rate is found to vary directly with the concentration of A, the reaction is first order in A. If the initial rate is found to increase as the square of the concentration of A, the reaction is second order in A. Once the order of the reaction in A has been determined, the process may be repeated with a series of solutions containing a constant amount of A and varying amounts of B.

For some reactions, measurement of the initial rate may be difficult or impossible. For such reactions, the change in concentration of reactant may be determined as a function of time, and the time dependence observed may be compared with the expected behavior for various reaction orders. The expected time dependence of concentration for first- and second-order reactions may be determined from the integrated rate laws. Consider a reaction that is known to be first order in A. The disappearance of A can be written

$$-\frac{d[\mathrm{A}]}{dt} = k[\mathrm{A}] \qquad \textbf{(15-7)}$$

Integration of this equation gives

$$2.3 \log [\mathrm{A}] = 2.3 \log [\mathrm{A}]_0 - kt \qquad \textbf{(15-8)}$$

where [A] represents the concentration of A at time $t$, and $[\mathrm{A}]_0$ represents the initial concentration of A at $t = 0$. Thus, if a plot of log [A] versus $t$ gives a straight line with a negative slope, the reaction is first order in A. By a similar procedure, it can be shown that a reaction that is second order in A would follow an equation of the form

$$\frac{1}{[\mathrm{A}]} = \frac{1}{[\mathrm{A}]_0} + kt \qquad \textbf{(15-9)}$$

Thus, if the reaction is second order in A, a plot of 1/[A] versus time should produce a straight line.

From an analytical point of view, reactions with orders greater than first are of limited utility. Even reactions of high overall order, however, may sometimes be made to behave as if they were first order reactions by selecting the appropriate experimental conditions. Consider, for example, the reaction between Sn(II) and Fe(III) in hydrochloric acid,

$$2\,\mathrm{Fe(III)} + \mathrm{Sn(II)} = 2\,\mathrm{Fe(II)} + \mathrm{Sn(IV)} \qquad \textbf{(15-10)}$$

The differential rate law for this reaction has been experimentally determined to be

$$\frac{1}{2}\frac{d[\mathrm{Fe(II)}]}{dt} = k[\mathrm{Fe(III)}][\mathrm{Sn(II)}][\mathrm{Cl^-}]^4 \qquad \textbf{(15-11)}$$

which is sixth order overall. If, however, the concentrations of Sn(II) and $Cl^-$ are made sufficiently high, their concentrations will not change appreciably during the reaction, and eq. (15-11) may be written

$$\frac{1}{2}\frac{d[\mathrm{Fe(II)}]}{dt} = k'[\mathrm{Fe(III)}] \qquad \textbf{(15-12)}$$

where the concentrations of Sn(II) and $Cl^-$ have been incorporated with $k$ to give the pseudo-first-order rate constant $k'$. The analytical utility of a first or pseudo-first-order reaction is demonstrated by Figure 15-2, which shows a plot of reaction rate versus concentration of a substance Q known to follow first-order kinetics. It can be seen from the figure that the reaction rate is linearly dependent on the concentration of Q. If a plot of the initial reaction rate versus the concentration of Q is experimentally determined under carefully controlled experimental conditions, the plot can be used to determine the concentration of Q in a sample, provided that the reaction rate is determined under similar experimental conditions.

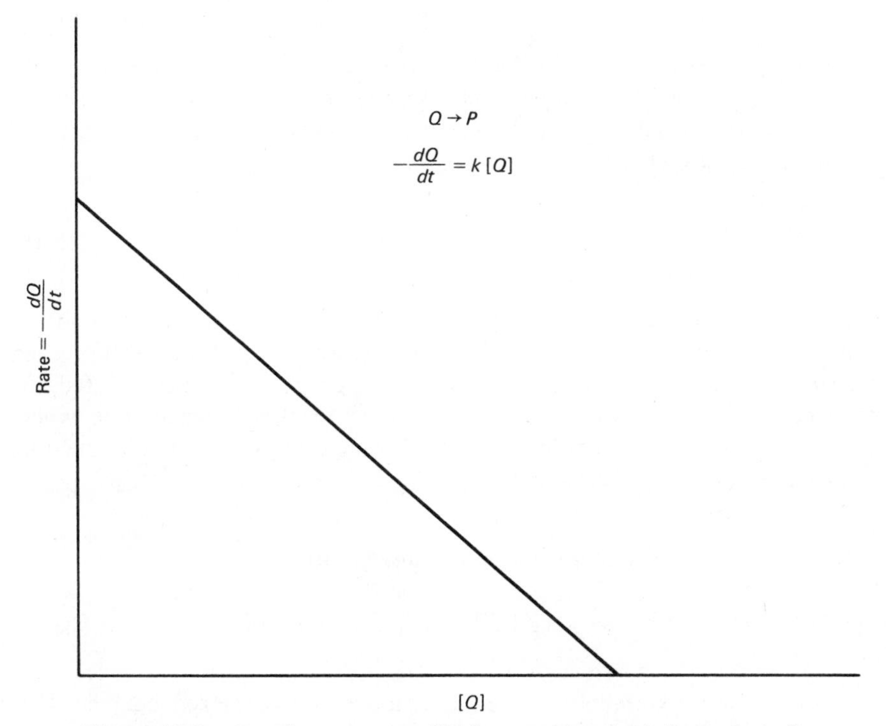


**Figure 15-2.** Reaction rate versus [Q] for a reaction that is first order in Q.

## ENZYME KINETICS

Molecules of biological and clinical significance can frequently be determined by reaction-rate procedures. Many of these reactions are catalyzed by special proteins known as **enzymes**. These enzymatic reactions typically follow **Michaelis–Menten kinetics**, which are based on the following simple three-step mechanism. A reaction mechanism is a detailed step-by-step description of how reactant molecules are converted to products. Each of the individual steps is called an **elementary process**. The sum of all the elementary processes involved in the mechanism should give the balanced chemical reaction. The first step in these enzyme-catalyzed reactions is the formation of an enzyme–substrate complex according to reaction (15-13):

$$\mathrm{E} + \mathrm{S} \underset{k_{-1}}{\overset{k_1}{\rightleftharpoons}} \mathrm{ES} \qquad \textbf{(15-13)}$$

where E is the enzyme, S represents the substrate, ES represents the enzyme–substrate complex, $k_1$ is the rate of the forward reaction, and $k_{-1}$ is the rate at which the complex dissociates. The second step involves the breakdown of the enzyme–substrate complex into the products, P, and the original enzyme. This may be represented as

$$\mathrm{ES} \xrightarrow{k_2} \mathrm{E} + \mathrm{P} \qquad \textbf{(15-14)}$$

where $k_2$ is the rate constant for the breakdown of the enzyme–substrate complex into the enzyme and products. According to this mechanism, the rate of appearance of products, $V$, should be related to the concentration of enzyme–substrate complex, [ES], by relation (15-15):

$$V = k_2[\mathrm{ES}] \qquad \textbf{(15-15)}$$

To relate the rate of formation of products to the concentration of enzyme and substrate, we must determine the relationship between these concentrations and the steady state concentration of ES. Shortly after the reaction is initiated, the concentration of the enzyme–substrate complex reaches a steady state, where the rate of formation of complex exactly balances the rate of its destruction. This may be expressed mathematically as

$$\text{rate of formation} = \text{rate of destruction}$$

$$k_1[\mathrm{E}][\mathrm{S}] = k_{-1}[\mathrm{ES}] + k_2[\mathrm{ES}] \qquad \textbf{(15-16)}$$

Equation (15-16) contains the concentration of uncomplexed enzyme, [E], which is not known. The total enzyme concentration must equal the sum of the free uncomplexed enzyme plus the enzyme present in the form of the enzyme–substrate complex. Thus,

$$[E]_0 = [E] + [ES]$$

or **(15-17)**

$$[E] = [E]_0 - [ES]$$

where $[E]_0$ is the total amount of enzyme present. Substitution of eq. (15-17) into eq. (15-16) and solving for [ES] gives

$$[ES] = \frac{k_1[E]_0[S]}{k_{-1} + k_2 + k_1[S]} \quad \textbf{(15-18)}$$

for the steady state concentration. The rate of appearance of products should be given by

$$V = \frac{k_1 k_2[E]_0[S]}{k_{-1} + k_2 + k_1[S]} \quad \textbf{(15-19)}$$

The Michaelis constant may now be defined as

$$K_m = \frac{k_{-1} + k_2}{k_1} \quad \textbf{(15-20)}$$

and substituted into eq. (15-19) to give

$$V = \frac{k_2[E]_0[S]}{K_m + [S]} \quad \textbf{(15-21)}$$

The limiting velocity $V_{max}$ is the reaction velocity that the reaction assumes when the substrate concentration becomes very large. As [S] becomes very large, eq. (15-21) reduces to

$$V = V_{max} = k_2[E]_0 \quad \textbf{(15-22)}$$

Substitution of eq. (15-22) into eq. (15-21) gives the normal form of the equation for the dependence of the reaction rate on substrate concentration,

$$V = \frac{V_{max}[S]}{K_m + [S]} \quad \textbf{(15-23)}$$

Figure 15-3 shows a plot of reaction rate versus substrate concentration for an enzymatic reaction obeying Michaelis–Menten kinetics. If the experimental

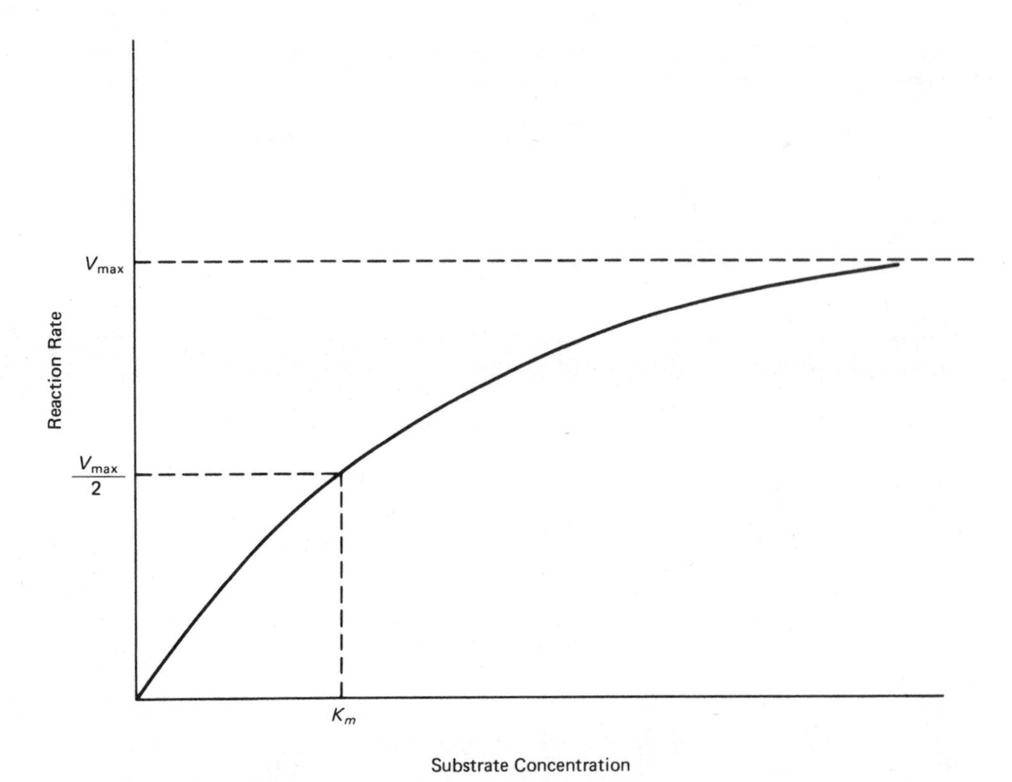


**Figure 15-3.** Reaction rate versus substrate concentration for an enzymatic reaction following Michaelis–Menten kinetics.

conditions can be adjusted so that the substrate concentration [S] can be made much smaller than $K_m$, eq. (15-23) may be simplified to

$$V = \frac{V_{max}[S]}{K_m} \qquad \text{when } K_m > 100[S]_0 \tag{15-24}$$

where $[S]_0$ is the initial substrate concentration. Equation (15-24) shows that the rate of formation of product is directly proportional to the substrate concentration whenever the initial substrate concentration is much less than $K_m$.

## ANALYTICAL APPLICATIONS

Analytical methods based on reaction kinetics may be broadly classified into two categories depending on the number of components determined: single-component determinations and simultaneous multicomponent determinations. Single-component determinations may be further classified as differential methods and integral methods.

## SINGLE-COMPONENT DETERMINATIONS

### *Differential Methods Utilizing the Initial Rate*

Single-component determinations utilizing the initial rate of the reaction are all predicated on establishing first- or pseudo-first-order behavior in the analyte concentration by proper selection of the reaction conditions. Once the appropriate reaction conditions have been established so that the reaction rate is directly proportional to the analyte concentration, an appropriate indicator substance involved in the reaction, whose concentration may be conveniently monitored as a function of time, is selected. For example, if the reaction involves the production or consumption of hydrogen ions, the reaction rate may be determined by monitoring the pH of the solution potentiometrically as a function of time. Alternatively, if the reaction involves the formation or disappearance of an absorbing species, the reaction rate may be determined by spectrophotometrically monitoring the concentration. In this respect, instrumental methods where the response is directly proportional to concentration and can monitor the concentration continuously as a function of time are especially useful in kinetic determinations. Figure 15-4 shows a typical recorder tracing

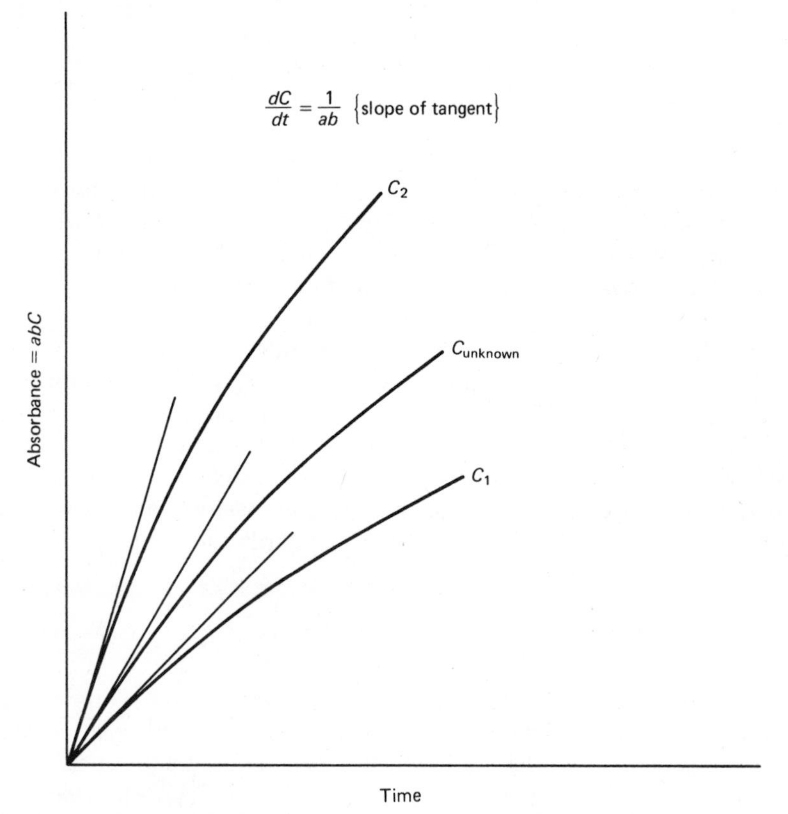


**Figure 15-4.** Absorbance versus time for various initial concentrations: $C_2 > C_1$.

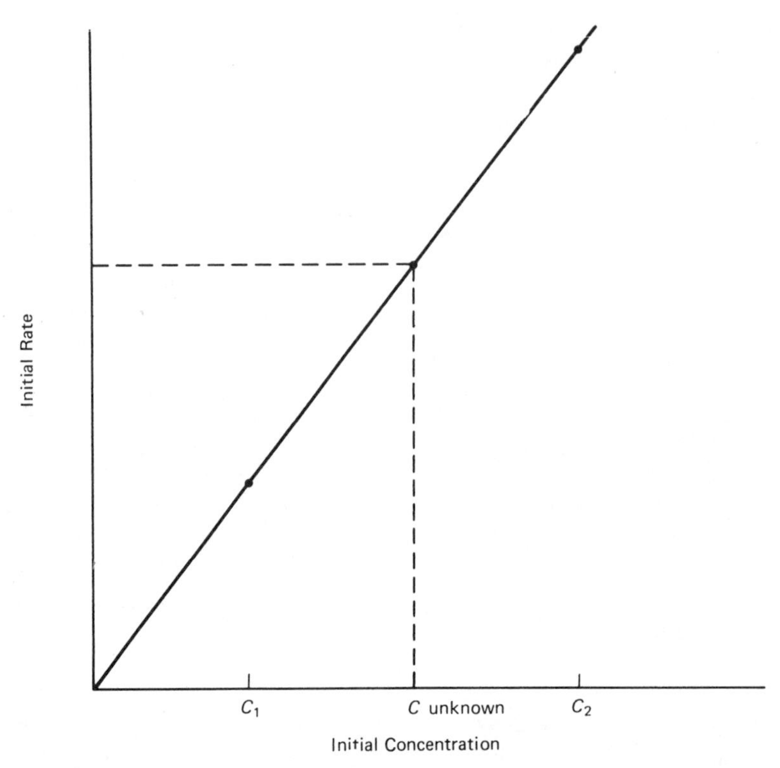


**Figure 15-5.** Initial reaction rate versus initial analyte concentration.

of the absorbance of a colored product versus time. Since the absorbance is directly proportional to the concentration of the absorbing species, the recorder tracing can be used to determine the rate of formation of the absorbing species. To determine the initial rate of formation, the slope of the tangent to the curve at $t = 0$ must be determined, as shown in Figure 15-4. If the initial rate is determined in this way for two different initial analyte concentrations, $C_1$ and $C_2$, where $C_2 > C_1$, a plot of initial rate versus analyte concentration may be established, as shown in Figure 15-5. The unknown concentration may then be determined by comparing the observed rate with the calibration plot of Figure 15-5. Naturally, for this procedure to be valid, the reaction conditions, such as the pH, temperature, ionic strength, and any other factor that might influence the observed rate must be the same for the standards and samples.

## *Integral Methods*

In contrast with differential methods, where the actual initial rate is measured, integral methods approximate the rate. To illustrate this difference, consider a reaction producing a substance I, which can be conveniently monitored by instrumental means. The rate of production of I is given by $d\mathrm{I}/dt$, which can be approximated by $\Delta\mathrm{I}/\Delta t$. In other words, rather than determine the slope of the concentration–time plot, the rate may be approximated by the ratio of the two finite differences, ΔI and $\Delta t$. The advantage to this approach lies in the fact that it opens the possibility of automatic instruments for determining the rate. Thus, it is relatively easy to visualize instruments that could be constructed that would automatically determine the rate in two potential ways: (1) the time required to produce a given change in the concentration of I could be determined; and (2) the change in I produced in a given time interval could be determined. For a first- or pseudo-first-order reaction,

$$\frac{\Delta \text{I}}{\Delta t} \simeq \frac{d\text{I}}{dt} = k[\text{A}] \tag{15-25}$$

where $k$ is the first-order rate constant and A the analyte concentration. If the change in I is kept constant,

$$\frac{1}{\Delta t} = k'[\text{A}] \tag{15-26}$$

and the reciprocal of the time required to produce the given change, ΔI, is directly proportional to the analyte concentration, where $k' = k/\Delta\text{I}$. Alternatively, if $\Delta t$ is kept constant,

$$\Delta \text{I} = k''[\text{A}] \tag{15-27}$$

and the corresponding change in I is proportional to the analyte concentration, where $k'' = k\,\Delta t$. These procedures are illustrated in Figure 15-6.

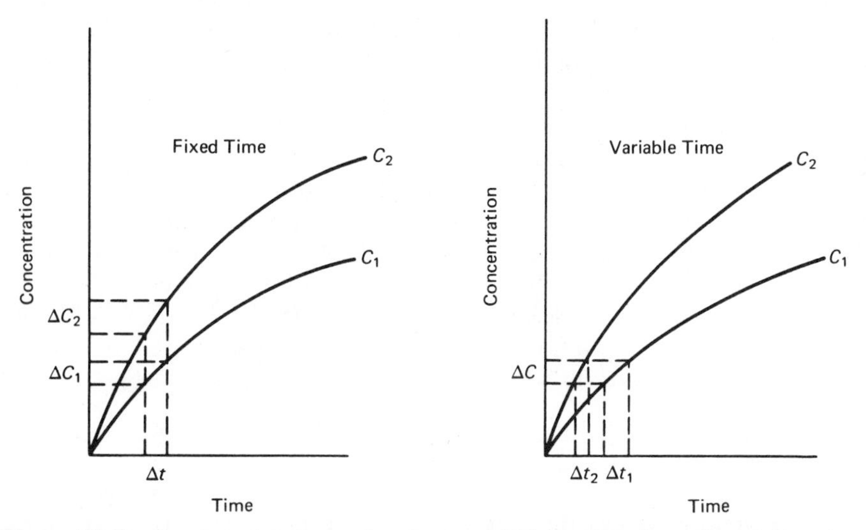


**Figure 15-6.** Concentration versus time for two initial concentrations, $C_1$ and $C_2$, where $C_2 > C_1$. It can be seen for the fixed-time integral method that $\Delta C$ is greater for the higher initial concentration $C_2$. Conversely, for the variable-time integral method, the higher initial concentration requires less time to produce the given $\Delta C$.

### *Integral Methods Maintaining Constant Composition*

Reactions that are pseudo-first order in analyte concentration will behave in this manner only as long as the concentrations of the other species involved in the rate law are kept constant. For example, many reaction rates are influenced by the pH of the solution and will be pseudo-first order in analyte concentration as long as the pH remains constant. For reactions involving the formation or consumption of hydrogen ions, the reaction rate may be determined potentiometrically by recording the change in pH as a function of time. To avoid possible interferences due to the change in pH as the reaction proceeds, a static approach may be employed. Rather than monitor the change in pH with time, the amount of acid or base required to maintain a constant pH over a finite time interval may be determined. The best example of this procedure involves the determination of urea according to reaction (15-28), which is catalyzed by the enzyme urease:

$$NH_2CONH_2 + 2H_2O + H_3O^+ \longrightarrow 2NH_4^+ + HCO_3^- + H_2O \tag{15-28}$$

Since the reaction consumes hydrogen ions, the amount of acid required to maintain a constant pH for a given time period, $\Delta t$, is directly proportional to the initial urea concentration as indicated by eq. (15-27).

## MULTICOMPONENT DETERMINATIONS

Several chemically similar components in a multicomponent mixture may be simultaneously determined under favorable circumstances without chemical separation by making use of differences in reaction rates. Although a variety of procedures have been developed to accomplish this, we will consider only two. Consider a sample containing two chemically similar substances, A and B, both of which react with reagent R, but at different rates. Let us assume for purposes of discussion that $k_A > k_B$ and that both reactions can be made pseudo-first order in analyte by making the concentration of R large enough. Let us also assume that the total concentration of A and B, $C_t$, can be determined as a function of time by some chemical or instrumental means. Figure 15-7 shows the decrease in the total concentration, $C_t$, with time due to reaction of A and B with R. Prior to the addition of R, the total initial concentration of A and B can be determined, and will be given by

$$C_0 = [A]_0 + [B]_0 \tag{15-29}$$

After the addition of R to the mixture, the total concentration $C_t$ at any time $t$ will be given by

$$C_t = [A]_t + [B]_t \tag{15-30}$$

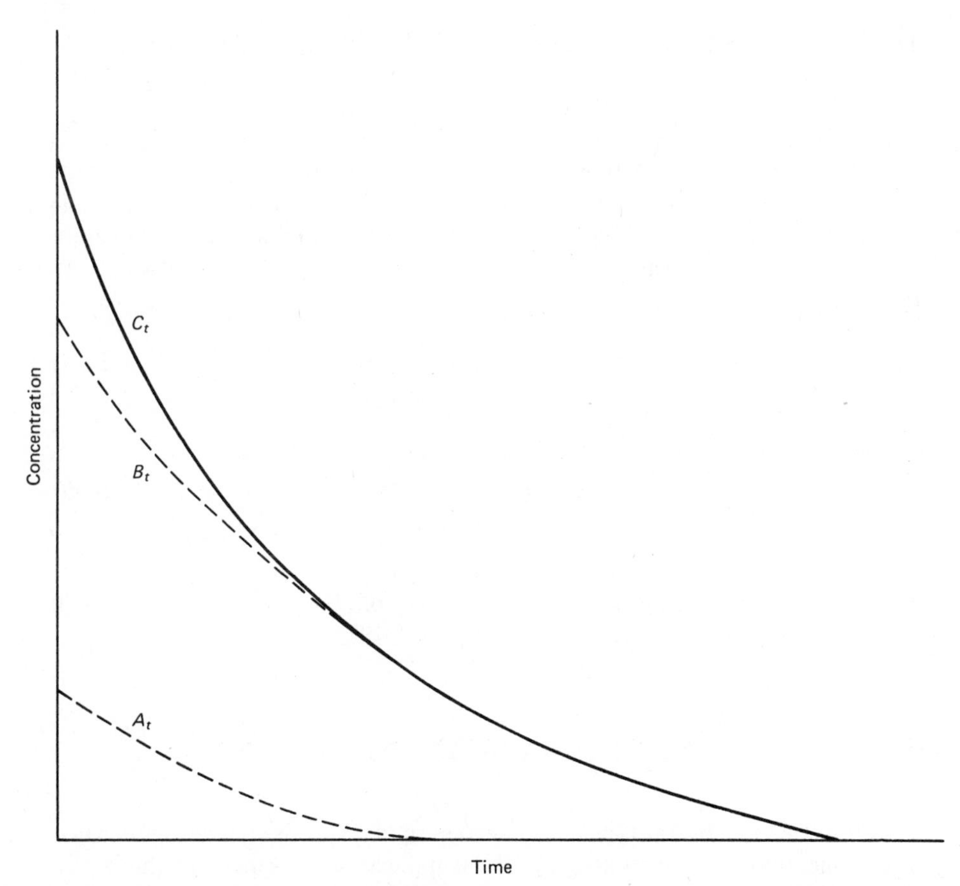


**Figure 15-7.** Change in total concentration, $C_t$, versus time for a mixture containing A and B. Dashed lines show individual changes in A and B with time.

If A and B both react independently with R and follow first- or pseudo-first-order kinetics,

$$C_t = [\mathrm{A}]_0 e^{-k_\mathrm{A} t} + [\mathrm{B}]_0 e^{-k_\mathrm{B} t} \tag{15-31}$$

This equation describes the time dependence of $C_t$.

At this point the initial concentrations, $[\mathrm{A}]_0$ and $[\mathrm{B}]_0$, can be determined in two ways. If $C_t$ is determined at two different times, $t_1$ and $t_2$, at the start of the reaction when A and B are both present, the initial concentrations may be determined by solving the following set of simultaneous equations for $[\mathrm{A}]_0$ and $[\mathrm{B}]_0$:

$$C_{t_1} = [\mathrm{A}]_0 e^{-k_\mathrm{A} t_1} + [\mathrm{B}]_0 e^{-k_\mathrm{B} t_1} \tag{15-32}$$

$$C_{t_2} = [\mathrm{A}]_0 e^{-k_\mathrm{A} t_2} + [\mathrm{B}]_0 e^{-k_\mathrm{B} t_2} \tag{15-33}$$

To solve eqs. (15-32) and (15-33) simultaneously for $[A]_0$ and $[B]_0$, the specific reaction rate constants $k_A$ and $k_B$ must be determined in a separate experiment with samples of pure A and B under similar reaction conditions.

If the initial concentrations of A and B and the rate constants $k_A$ and $k_B$ are such that the concentration of A decreases to zero before the concentration of B reaches zero, it is possible to determine the initial concentrations of A and B in another way. This is shown in Figure 15-7. After a period of approximately seven half-lives[1], the concentration of A is less than 1% of its initial value. At this point,

$$C_t = [A]_t + [B]_t \simeq [B]_t \quad \textbf{(15-34)}$$

and

$$C_t \simeq [B]_t = [B]_0 e^{-k_B t} \quad \textbf{(15-35)}$$

If log $C_t$ is plotted versus $t$ for the later stages of the reaction after A has completely reacted, a straight line with an intercept of $[B]_0$ will result. Once $[B]_0$ is known, $[A]_0$ may be determined from eq. (15-29).

## SOME TYPICAL DETERMINATIONS

In this section, several typical kinetic determinations of clinical significance will be mentioned to illustrate some of the applications of kinetic methods. We have already mentioned how urea may be determined using the enzyme urease.

Another typical clinical application of kinetics is the determination of protein-bound iodine. This determination makes use of the fact that the oxidation–reduction reaction,

$$2\,Ce(IV) + As(III) \longrightarrow 2\,Ce(III) + As(V) \quad \textbf{(15-36)}$$

is catalyzed by iodide ion. The proteins in a serum sample are first isolated and then ashed. In the process of ashing, the protein-bound iodine is converted to iodide ion. The reaction is initiated by the addition of fixed amounts of Ce(IV) and As(III) to the iodide-containing ash. The concentration of Ce(IV) is monitored spectrophotometrically at 420 nm at the start of the reaction and again

[1] One half-life is the time required for the concentration to drop to half its initial value. Thus, for a first-order reaction,

$$2.3 \log \frac{0.5[A]_0}{[A]_0} = -k_A t_{1/2}$$

and $t_{1/2} = 0.69/k_A$. After seven half-lives, the concentration of A will be $(\frac{1}{2})^7[A]_0 = (\frac{1}{128})[A]_0$.

after some fixed time. The decrease in the concentration of Ce(IV) is determined from the spectrophotometric measurements, and the concentration of iodine is determined by eq. (15-27).

The determination of glucose in blood serum is another example. In this case, the primary reaction is the conversion of glucose to gluconic acid,

$$C_6H_{12}O_6 + H_2O + O_2 \longrightarrow C_6H_{12}O_7 + H_2O_2 \qquad \textbf{(15-37)}$$

a reaction that is catalyzed by the enzyme glucose oxidase. Since the production of $H_2O_2$ is difficult to follow directly (because $H_2O_2$ does not have a conveniently measurable physical property), the production of $H_2O_2$ is determined indirectly by making use of a second "coupling" reaction. Several variations have been proposed. For example, the production of $H_2O_2$ from eq. (15-37) may be followed spectrophotometrically by using reaction (15-38):

$$\text{organic dye A} + H_2O_2 \xrightarrow[\text{peroxidase}]{\text{horseradish}} \text{colored product B} \qquad \textbf{(15-38)}$$

and monitoring the colored reaction product. Alternatively, the production of $H_2O_2$ may be followed by employing reaction (15-39):

$$H_2O_2 + 3I^- \xrightarrow[\text{catalyst}]{\text{molybdate}} I_3^- + 2OH^- \qquad \textbf{(15-39)}$$

The triiodide ion produced may then be conveniently determined potentiometrically, spectrophotometrically, or amperometrically. This scheme eliminates the need for the peroxidase enzyme required by reaction (15-38).

Finally, the enzyme lactic dehydrogenase may be determined by kinetic methods. Lactic dehydrogenase (LHD) is actually a mixture of isozymes, all of which exhibit activity toward nicotinamide adenine dinucleotide (NAD), by catalyzing the reaction between pyruvic acid and the reduced form of NAD,

$$\underset{\text{pyruvic acid}}{\begin{matrix} CH_3 \\ | \\ C{=}O \\ | \\ COOH \end{matrix}} + NADH_2 \xrightarrow[\text{dehydrogenase}]{\text{lactic}} NAD + \underset{\text{lactic acid}}{\begin{matrix} CH_3 \\ | \\ CHOH \\ | \\ COOH \end{matrix}} \qquad \textbf{(15-40)}$$

Since $NADH_2$ absorbs strongly at 320 nm, whereas NAD is nonabsorbing at this wavelength, the rate of decrease of $NADH_2$ may be monitored spectrophotometrically and related to the amount of LDH present.

## PROBLEMS

**1** Consider the reaction X + Y = Q + R for which the following initial rates have been experimentally determined:

| *Initial Concentration* X (*M*) | *Initial Concentration* Y (*M*) | *Initial Rate* |
|---|---|---|
| 1.0 | 1.0 | 0.030 |
| 0.5 | 1.0 | 0.015 |
| 0.5 | 0.5 | 0.00375 |

What is the differential rate law for the production of Q?

*Ans.* $\frac{dQ}{dt} = k[X][Y]^2$

**2** A particular reaction producing acid is followed by periodically withdrawing a sample and titrating with standard base. The volumes of base required to titrate a particular volume of sample for various times are

| *Time* (*min*) | 0 | 5 | 10 | 15 | 20 | 25 | 30 |
|---|---|---|---|---|---|---|---|
| *Volume of Base* (*mL*) | 0 | 5.0 | 9.0 | 12.2 | 15.0 | 17.5 | 19.4 |

What is the initial rate for the production of $H_3O^+$ if the concentration of the base is 0.1000 *M* and the samples withdrawn are 20.0 ml?

*Ans.* $1.1 \times 10^{-4}$ moles/liter sec.

**3** Consider the reaction A + B = C + D, where the progress of the reaction is determined by spectrophotometrically monitoring the disappearance of A. The concentration of A is measured at the following times:

| *Time* (*Min*) | 0 | 5 | 10 | 15 | 20 | 25 | 30 |
|---|---|---|---|---|---|---|---|
| *Mole/liter* A | 0.150 | 0.125 | 0.105 | 0.089 | 0.075 | 0.0625 | 0.053 |

(a) What is the order of the reaction with respect to A?
(b) What is the rate constant?
(c) What is the half-life for the reaction?
(d) How many half-lives are required to consume 99.9% of the initial amount of A?

*Ans.* (c) $1.21 \times 10^{-3}$ sec.

**4** The glucose level in a blood serum sample is to be determined by determining the amount of colored product B formed after 30 min of incubation at 37°C according to the following reactions:

$$C_6H_{12}O_6 + H_2O + O_2 \xrightarrow{\text{glucose oxidase}} C_6H_{12}O_7 + H_2O_2$$

$$\text{organic dye A} + H_2O_2 \xrightarrow{\text{horseradish peroxidase}} \text{colored product B}$$

A 0.5-mL blood serum sample is treated to precipitate the proteins, diluted to 10.0 mL, and then filtered. A 0.2-mL aliquot of the filtrate is added to a clean test tube, and standard enzyme containing organic dye A is added. After 30 min, $H_2SO_4$ is added to stop the reaction and the absorbance of colored product B is determined spectrophotometrically at 540 nm to be 0.124. Standards containing 1.00 mg/mL and 3.00 mg/mL are treated in an identical manner, giving absorbances of 0.100 and 0.300, respectively. Calculate the glucose level in the serum in mg% (1 mg% = 1 mg/100 mL of serum).

**5** P and Q are two chemically similar substances both of which react with R but with different rate constants. If the concentration of R is made suitably large, the reactions become pseudo-first order in P and Q. Since P and Q are chemically similar substances, they have similar absorption spectra. A mixture containing both P and Q is to be determined spectrophotometrically by monitoring the decrease in the total absorbance of P and Q with time. The absorbance is monitored at a wavelength where P has a molar absorptivity of 6200 cm-liter/mole and Q has a molar absorptivity of 5700 cm-liter/mole. At the start of the reaction, before the addition of R, the total absorbance due to the presence of P and Q is 0.375. The total absorbance is monitored as a function of time with the following results:

| *Time (sec)* | 0 | 15 | 30 | 60 | 90 | 120 | 180 |
|---|---|---|---|---|---|---|---|
| *Absorbance* | 0.375 | 0.320 | 0.278 | 0.218 | 0.177 | 0.146 | 0.103 |

Determine the concentrations of P and Q in the original sample, assuming measurements are made with a 1-cm cell and P has a larger rate constant than Q.

*Ans.* $4.8 \times 10^{-5}$ M Q

**6** The hydrolysis of ethyl acetate

$$CH_3CO_2C_2H_5 + H_2O \xrightarrow{OH^-} CH_3CO_2H + C_2H_5OH$$

has been shown to obey the differential rate law

$$-\frac{d[\text{ester}]}{dt} = k[OH^-][\text{ester}]$$

where the hydroxyl ion acts as a catalyst. As the reaction proceeds, however, the amount of hydroxyl ion decreases due to neutralization by the acid that is formed. If the amount of hydroxyl ion can be maintained at a constant level, the differential rate law becomes pseudo-first order,

$$-\frac{d[\text{ester}]}{dt} = k'[\text{ester}]$$

A sample containing ethyl acetate is determined on a specially designed instrument that continuously adds a fixed volume of standard base over a variable time period so as to maintain a constant pH. The instrument is calibrated by determining the time required to consume 5.0 mL of standard base. For a standard sample containing 3.2 mmoles of ester, 678 sec are required at the reaction temperature used. Calculate the number of millimoles of ester present in an unknown sample, if 530 sec are required to consume the standard base.

CHAPTER 16

# Absorption Methods in the Visible and Ultraviolet

Methods of analysis that measure the absorption of radiation constitute one of the most widely used analytical methods and have been developed for most elements and many organic compounds. All these methods use some means to measure the intensity of color of a solution or the absorbance, transmittance, or reflectance of radiant energy by a solid, liquid, or gas. The majority of the measurements concern the absorbance of a solution of a compound or element. If the measurement is made with the eye, the method is usually referred to as a **visual** colorimetric method, whereas methods that involve measurement using an instrument are known as **photometric** or **spectrometric** procedures. For a long time all such determinations were called *colorimetric*, since light in the visible region was utilized. However, wavelengths outside the visible region (such as the x-ray, ultraviolet, and infrared regions) are used in many analyses, so colorimetry is really a misnomer. In general, present practice is to call all methods photometric, spectrometric, or spectrophotometric, irrespective of the wavelength region, except when the eye is used to estimate intensity of color.

Photometric methods are used most often for minor constituents, which are present in quantities from trace amounts (as low as 0.00001%) up to about 2%. However, some methods have been proposed that can determine the amount of a major constituent or even the percent purity of a high-purity substance. Since such methods often call for special photometric measurements or extreme dilution, they are not as popular as the methods used for minor amounts. The relative error in most photometric measurements is between 0.1 and 1.0%, with 0.5% being accepted as the general average. As a consequence, ordinary titrimetric and gravimetric procedures are often preferable for major constituents. Whenever colorimetric or photometric methods can be used for repetitive samples, they are usually the method of choice because they tend to be more rapid and require less manipulative procedures.

## THE NATURE OF RADIANT ENERGY

Radiant energy may be defined as energy transmitted as electromagnetic radiation. It appears to have the characteristics or properties of both wave motion and particles. In diffraction and refraction, radiant energy shows the properties of waves, while in emission and absorption it has the properties of particles (called **photons**).

Radiant energy is characterized by the wavelength ($\lambda$) and the frequency ($\nu$) of the radiation. The **wavelength** is the distance between two points that are in phase on adjacent waves measured along the line of propagation. It is often thought of as being the shortest distance between the maxima or the minima of two adjacent waves. The **frequency** is the number of cycles in a unit of time or is the number of maxima that pass a given point in a unit of time. The frequency increases as the wavelength decreases.

The energy of a photon is given by

$$E = h\nu \qquad \textbf{(16-1)}$$

in which $E$ is the energy in ergs; $\nu$ is the frequency in hertz; and $h$ is Planck's constant, $6.62 \times 10^{-27}$ erg-sec. As a consequence, the energy is directly proportional to the frequency and inversely proportional to the wavelength. High energy is associated with short wavelengths and high frequencies, whereas long wavelengths and low frequencies have relatively lower energies.

| Energy change involved | Approximate wavelength | Region |
|---|---|---|
| Nuclear, inner-shell electrons | less than 10 nm | Gamma rays X rays |
| Ionization of atoms or molecules | 10 nm to 200 nm | Vacuum Ultraviolet |
| Valence electrons | 200 nm to 400 nm | Ultraviolet |
| Valence electrons | 400 nm to 800 nm | Visible |
| Molecular vibration | 800 nm to 25 μm | Infrared |
| Molecular rotation, Electro-magnetic spin | 25 μm to 25 cm | Far Infrared Micro Waves Radio Waves |

| Wavelength | Color absorbed | Color observed |
|---|---|---|
| 400 nm | | |
| | Violet | Yellow-green |
| 435 nm | | |
| | Blue | Yellow |
| 495 nm | | |
| | Green | Purple |
| 560 nm | | |
| | Yellow | Blue |
| 595 nm | | |
| | Orange | Greenish-blue |
| 650 nm | | |
| | Red | Bluish-green |
| 800 nm | | |

**Figure 16-1.** The electromagnetic spectrum. Designation of region boundaries are approximate. Wavelength units are related as follows: 1 m = 100 cm = 1000 mm = 1 million μm = 1 billion nm = 10 billion Å; or 1 Å = $10^{-1}$ nm = $10^{-4}$ μm = $10^{-7}$ mm = $10^{-8}$ cm = $10^{-10}$ m. The micrometer was formerly known as the micron (μ) and the nanometer as the millimicron (mμ).

The electromagnetic spectrum covers some 20 orders of magnitude and thus is difficult to show on a linear scale. Figure 16-1 shows the major regions of the spectrum, together with the approximate wavelengths of the region boundaries and the type of energy change that is caused in a substance by absorption of that wavelength of radiant energy. It should be noted that the visible region covers a very small portion of the spectrum.

## THE ABSORPTION PROCESS

When radiant energy impinges upon a substance, it can be absorbed, transmitted, reflected, or refracted by the substance. For the energy to be absorbed, it must correspond to the energy required to cause some change in an atom or molecule. As a consequence, some wavelengths of energy are absorbed and others are not absorbed by any given substance. The changes in the substance may be nuclear, electronic (changes in the energies of the electrons distributed about the atoms in the substance), vibrational (change in the average separation of the nuclei of two or more atoms), or rotational. The absorbed energy is usually dissipated as heat but may be emitted (as in fluorescence) or be utilized to initiate a chemical reaction (as in photochemistry).

As seen in Figure 16-1, the highest energies and shortest wavelengths are associated with the nuclear changes caused by gamma rays. These energies are measured in millions of electron volts (MeV). X-rays have less energy (in the range of several hundred to several thousand electron volts) and cause changes in the inner-shell electrons, which are firmly bound in the atom. All these changes are independent of the chemical combination of the atom or are the same irrespective of the chemical state of the atom involved.

Absorption in the ultraviolet and visible regions involves the transformation of the outer-shell or valence electrons from their ground or lowest energy state to an excited state and thus is dependent upon the state of chemical combination. As an example, hydrated manganese(II) ion is pale pink, permanganate ion is dark purple, and manganate ion is dark green.

In the infrared region, absorption of the shorter wavelengths causes vibrational changes such as bending or stretching, whereas molecular rotation is associated with absorption of the longer wavelengths. Absorption of the very low-frequency, long-wavelength radiation in the microwave and radio wave regions can cause changes in the electromagnetic spin characteristics.

When polychromatic light (light of many colors or of broad wavelength distribution) impinges upon a sample solution, light of some wavelengths is absorbed, whereas light of other wavelengths passes through the solution essentially unimpeded. With respect to the visible region, as shown in Figure 16-1, a solution has the color of light that is *not* absorbed or may be said to show the complementary color of the absorbed color. As an example, potassium dichromate solution is seen to be yellow-orange, since it absorbs at a blue wavelength of approximately 450 nm.

# NOMENCLATURE

Much confusion exists in the use of terminology in the science of photometry, and several different symbols and names for the same concept or property can be found in the literature. To obtain some consistency in nomenclature, a Joint Committee on Nomenclature in Applied Spectroscopy was established by the Society for Applied Spectroscopy (SAS) and the American Society for Testing Materials (ASTM) to recommend and define terms to be used by all writers. The definitions, symbols, and names used in this chapter are those recommended by this committee. A summary of the most common terms can be found in the December issue each year of *Analytical Chemistry*. As each new symbol or concept is introduced in the following discussion, some of the meanings or symbols previously used will be listed to aid the student in correlation between various sources of material. Some of the more common terms to be utilized are

**Absorbance,** $A$ [formerly known as optical density (OD), extinction ($E$), absorbancy]. The absorbance is the logarithm to the base 10 of the reciprocal of the transmittance, or, $A = \log 1/T$. Also defined as the logarithm to the base 10 of the ratio of the radiant power incident upon the sample to the radiant power transmitted by the sample, or $A = \log P_0/P$.

**Absorptivity,** $a$ (formerly known as the extinction coefficient, absorbancy index, specific extinction). The absorptivity equals the absorbance divided by the product of the concentration (grams per liter) and the sample path length (in centimeters), or, $a = A/bc$, where $b$ is path length and $c$ is concentration. The absorptivity is sometimes referred to as the **specific absorbance**, since it is the absorbance per unit optical path length and per unit concentration. Both the path length and the concentration units should be specified clearly if they are not in centimeters or grams per liter.

**Absorptivity, molar,** $\varepsilon$ (formerly known as molar extinction coefficient, molar absorbancy index, molar absorption coefficient). The molar absorptivity is the product of the absorptivity and the molecular weight, or $\varepsilon = a\text{MW}$. This is the value obtained for the absorptivity if the concentration is expressed as molar or moles per liter instead of grams per liter.

**Radiant power,** $P$ (formerly known as intensity, $I$). The radiant power is the rate at which energy is transported in a beam of radiant energy. It is the quantity measured by a detector such as the eye or a photocell. The radiant power incident upon a cell is given the symbol $P_0$, and that transmitted by a cell, the symbol $P$.

**Transmittance,** $T$ (formerly known as transmittancy). The transmittance is the ratio of the radiant power transmitted by a sample to the radiant power of the beam incident upon the sample, or $T = P/P_0$. It is assumed that the measurements are made at the same wavelength with parallel radiation which is incident on the sample at right angles to the plane parallel surfaces of the sample.

## FUNDAMENTAL LAWS OF ABSORBANCE

### THE BOUGUER OR LAMBERT LAW

Lambert, in 1760, expressed the relationship between the intensity of color of a solution and the depth of solution through which the light passed. Actually, the same law had been proposed by Bouguer in 1729, but this was not discovered until long after the law was known as the Lambert law. According to Lambert and Bouguer, the intensity transmitted by a solution diminishes in geometric (exponential, logarithmic) progression as the depth of solution increases linearly. That is, the relationship is not arithmetical as one might suppose. This can be expressed as

$$T = a^{-b} \qquad \textbf{(16-2)}$$

where $T$ is the transmittance; $a$ is a constant characteristic of the solution and the wavelength; and $b$ is the path length of the light through the solution. It may also be expressed logarithmically as

$$-\log T = -\log \frac{P}{P_0} = ab \qquad \textbf{(16-3)}$$

The relationship can be understood by reference to Figure 16-2, which shows the value of the incident light and the transmitted light for a series of cells of the same path length containing the same solution. For a transmittance of 0.5, the radiant power of the transmitted light after one cell is $0.5P_0$, after the second cell is $0.25P_0$, after the third is $0.125P_0$, and so on, so that the radiant power of the transmitted beam after $n$ cells will be $T^nP_0$, or $0.5^nP_0$ for the example in Figure 16-2.

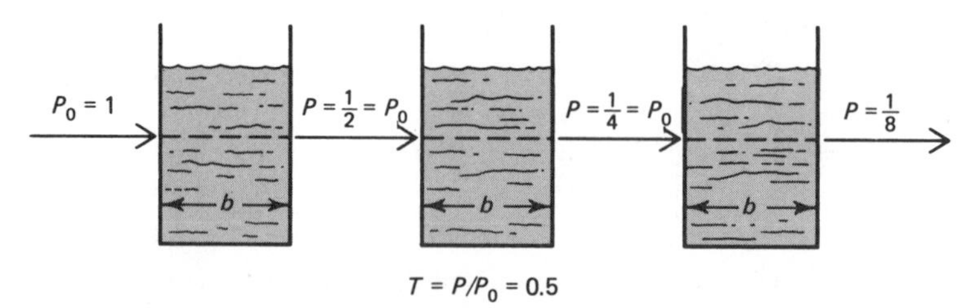


**Figure 16-2.** Diagrammatic representation of the Bouguer–Lambert law, showing the exponential effect of increasing the depth of the solution through which the light passes.

## THE BEER LAW

Beer enunciated the law relating transmittance to concentration in 1852. This law states that the intensity of light transmitted by a solution decreases geometrically (exponentially, logarithmically) as the concentration increases linearly, or

$$T = a^{-c} \tag{16-4}$$

where $T$ is the transmittance; $a$ is a constant depending upon the wavelength and the absorbing species; and $c$ is the concentration. This may also be expressed logarithmically as

$$-\log T = -\log \frac{P}{P_0} = ac \tag{16-5}$$

The relationship between concentration and transmittance can be understood by referring to Figure 16-3, which shows the value of the incident and transmitted light for a series of identical cells that contain solutions of different concentration. For a transmittance of 0.5, the radiant power of the transmitted beam for a solution of concentration $c$ is $0.5P_0$, for a concentration of $2c$ is $0.25P_0$, and for a concentration of $3c$ is $0.125P_0$, so that the radiant power transmitted through a cell containing a solution of concentration $nc$ would be $T^nP_0$, or, for $T = 0.5$, $0.5^nP_0$.

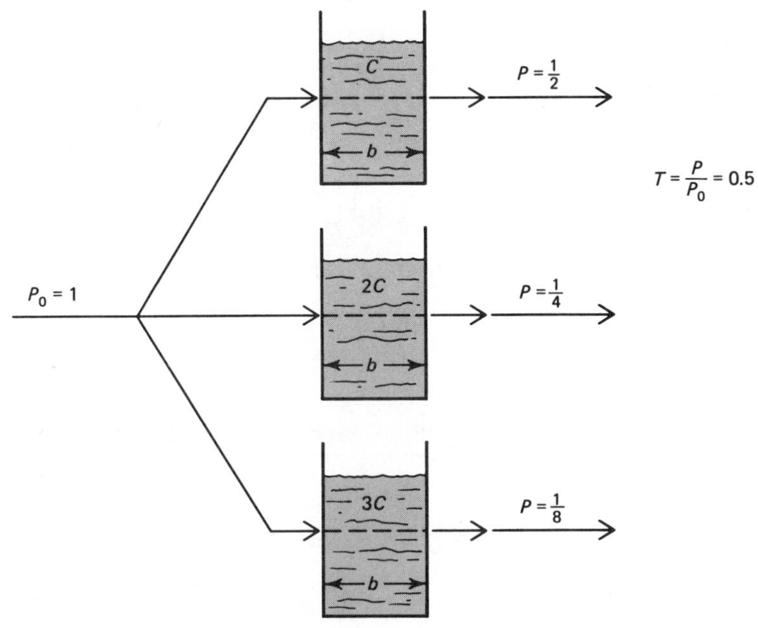


**Figure 16-3.** Diagrammatic representation of the Beer law, showing the exponential effect of increases in concentration.

## THE COMBINED BOUGUER–BEER LAW

The two laws just discussed are usually combined and used together in practically all absorption analysis. The combined law is often referred to simply as the Beer law, and is stated in the following ways:

$$A = abc \tag{16-6}$$

$$A = -\log T = -\log \frac{P}{P_0} = \log \frac{P_0}{P} = abc \tag{16-7}$$

where $A$ is the absorbance; $T$ is the transmittance; $P_0$ is the incident radiant power; $P$ is the transmitted radiant power; $a$ is the absorptivity; $b$ is the optical path length in centimeters; and $c$ is the concentration in grams per liter. The value of $a$ depends upon the units of $b$ and $c$. As a consequence, the path length and the concentration must be expressed clearly whenever they are not expressed in centimeters or grams per liter. Since concentrations in analytical procedures are often expressed as moles per liter (molar), the expression may be written

$$A = \varepsilon bc \tag{16-8}$$

where $\varepsilon$ is the molar absorptivity; $b$ is the depth in centimeters; and $c$ is the molar concentration.

Equations (16-7) and (16-8) are also used to calculate the absorptivity and/or the molar absorptivity under a given set of conditions.

• **Example 1.** Calculate the effective absorptivity and molar absorptivity for manganese if a potassium permanganate solution that contains 0.010 g of manganese in 100 mL has an absorbance of 0.434 using a 1.5-cm cell at a wavelength of 550 nm. Using eq. (16-7),

$$a = \frac{A}{bc} = \frac{0.434}{1.5 \times 0.1\ \text{(g/liter)}} = 2.89\ \text{liters/g-cm}$$

Using eq. (16-8), the molar concentration is 0.1/54.9 or $1.82 \times 10^{-3}$, and

$$\varepsilon = \frac{0.434}{1.5 \times 1.82 \times 10^{-3}} = 159\ \text{liters/mole-cm}$$

The absorptivities calculated in this example represent effective values for manganese, because the actual absorbing species is permanganate ion.

## DEVIATIONS FROM BEER'S LAW

There are no known exceptions to the Bouguer–Lambert law, which relates the absorbance at constant concentration to the depth of solution through which the energy passes, but many systems do not follow the Beer law concerning the relation of absorbance to concentration at constant optical path length. Deviations from Beer's law may be caused by chemical or instrumental factors.

If the absorbing species is in equilibrium with another species, dilution may not affect the concentration of each equally. A good example is the dichromate–chromate equilibrium shown in eq. (16-9).

$$Cr_2O_7^{2-} + H_2O \rightleftharpoons 2HCrO_4^- \rightleftharpoons 2H^+ + 2CrO_4^{2-} \quad \textbf{(16-9)}$$

This equilibrium is affected by the hydrogen ion concentration (pH), with dichromate predominating in acid solution and chromate in basic solution. Dichromate absorbs strongly at 350 nm, whereas chromate shows an absorption maximum at 375 nm. Variations of absorbance with dilution (concentration) for unbuffered solutions of dichromate and chromate (containing equal concentrations of Cr(VI)) at these two wavelengths are shown by the solid lines in Figure 16-4. The dashed lines represent the variation in absorbance of strongly acidic and basic solutions with dilution. In strongly acidic or basic solutions, the equilibrium is shifted in the direction of the absorbing species so that the variations of the hydrogen ion concentration caused by dilution do not affect the ratio of the concentrations of chromate and dichromate materially.

Since Beer's law is based on the use of monochromatic light, the use of polychromatic light will cause deviations. The extent of the deviation depends upon the range of wavelengths entering the solution and becomes greater as the bandwidth passed by the wavelength selection device becomes greater. The concentration of the absorbing species also plays a part, with deviations from the law becoming more apparent as the concentration increases. The effect is generally negligible for concentrations of 0.01 *M* or less.

The concentration range over which the Beer law is followed satisfactorily may be determined for a particular procedure on a specified instrument by the use of a **Ringbom plot**. In the Ringbom method, the percent *absorptance*, which

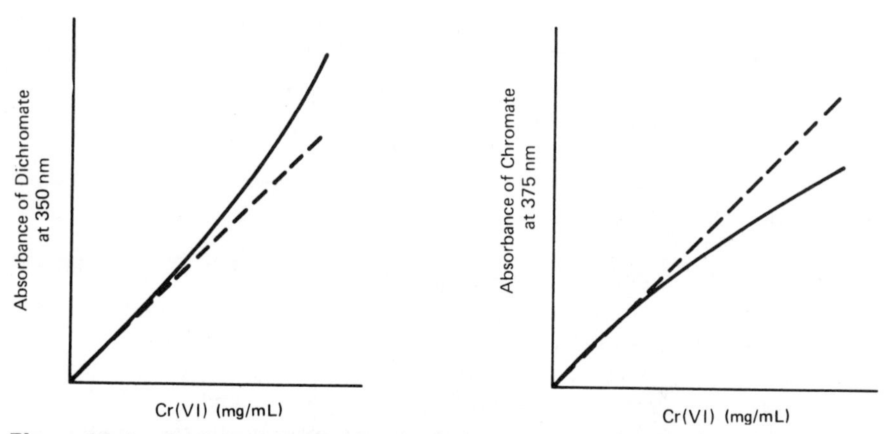


**Figure 16-4.** Qualitative relation of absorbance and concentration (dilution) for unbuffered solutions of potassium dichromate and potassium chromate. Dashed lines represent dilution of strongly acidic or strongly basic solutions.

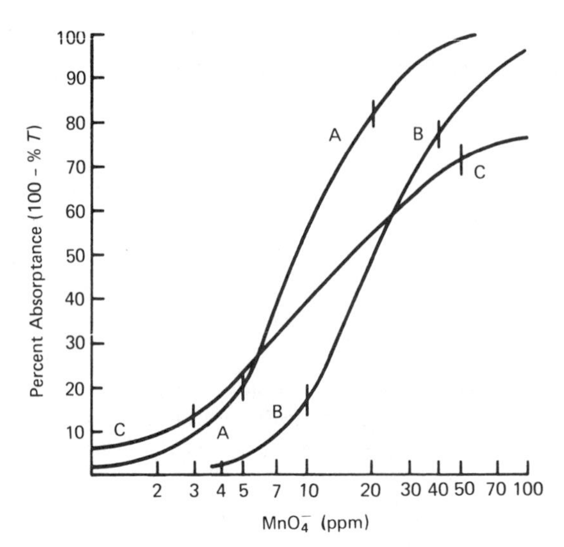


**Figure 16-5.** Simulated typical Ringbom plots. (A) Permanganate solutions measured at 525 nm in 1-cm cells on a spectrophotometer. (B) Permanganate solutions measured in the same cells on the same photometer at 550 nm. (C) Permanganate solutions measured on a filter photometer using a 530-nm filter and 1-cm cells. Satisfactory concentration ranges: A, 5–20 ppm; B, 10–40 ppm; C, 3–50 ppm.

is calculated by subtracting the percent transmittance from 100 (i.e., % absorptance = 100 − %$T$), is plotted against the logarithm of the concentration. When these functions are plotted in this manner, a sigmoid or S-shaped curve, the slope of which is dependent upon the absorptivity, is always obtained. Similar curves are produced using percent transmittance in place of percent absorptance if the vertical axis is plotted with 0% $T$ at the top and 100% $T$ at the bottom.

Typical simulated Ringbom plots for the measurement of permanganate solutions at different wavelengths and on different machines are shown in Figure 16-5. The steepest portion of the curve is the region of least error and can be used to estimate the range of concentration that should be determined by the method. Curve A in this figure represents measurements at 525 nm (the wavelength of maximum absorbance) and should be used for samples in the range 5–20 ppm of $MnO_4^-$ (2.3–9.2 ppm Mn). If most of the samples are in the range 10–40 ppm $MnO_4^-$ (4.6–18.5 ppm Mn), it would be better on this machine to make the measurements at 550 nm instead of 525 nm, even though sensitivity is greater at 525 nm. Curve C represents the decreased sensitivity (but increased range) to be expected using filter photometers.

## PHOTOMETRIC ERROR

For photometric measurements with a single-beam intrument, the smallest relative concentration error per unit photometric error occurs at 36.8% transmittance ($T$) or an absorbance ($A$) of 0.434, as shown in Figure 16-6. The error is small and does not vary much between 20% $T$ (0.700$A$) and 65% $T$ (0.190$A$) and increases only slightly between 65% $T$ and 80% $T$ (0.100$A$). Below 15% $T$ and above 80% $T$ the relative error increases rapidly. As a consequence, care should be taken in the design of photometric methods to ensure that conditions are specified so that the majority of the measurements will

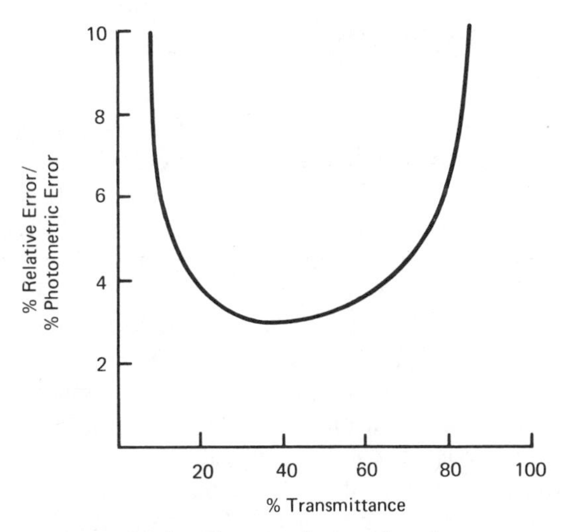


**Figure 16-6.** Photometric error-function curve.

be between 20% and 65% $T$ or between 0.190$A$ and 0.700$A$. In many cases, measurements between 15% $T$ and 80% $T$ can be used satisfactorily.

Even though the minimum relative concentration error can occur at other transmittance values for double-beam instruments, most analytical procedures are designed to measure transmittances in the range necessary for single-beam instruments, since this range is also satisfactory for many double-beam instruments.

Equations (16-7) and (16-8) are often used in the development of photometric methods to determine the sizes of cells that should be used with various concentration ranges. For example, what range of concentrations can be determined satisfactorily if the molar absorptivity is 750 liters/mole-cm and if a 10-cm cell is to be used? This involves calculation of the molar concentrations that will produce absorbances of 0.700 and of 0.190. Rearrangement of eq. (16-8) gives

$$c = \frac{A}{\varepsilon b}$$

and substitution of 0.190 and 0.700 gives

$$c = \frac{0.190}{750 \times 10} = 2.53 \times 10^{-5}\ M$$

$$c = \frac{0.700}{750 \times 10} = 9.33 \times 10^{-5}\ M$$

If larger concentrations need be measured, the use of 5-cm, 1-cm, or 0.1-cm cells or a different photometric system should be investigated.

## VISUAL COLORIMETRIC METHODS

All visual colorimetric methods use the eye to estimate the intensity of color of a solution compared to the intensity of color of a standard solution of the substance being determined. Because it is more practical to vary some factor until the two colors are exactly equal than it is to estimate the relative difference between two colors, all visual methods depend upon the matching of colors by the operator. There are several visual methods available and the selection of any one depends upon the accuracy with which the determination must be made and upon the number of tests that must be run over a given period of time. If a semiquantitative estimate is acceptable, the dilution method is the simplest to perform. If greater accuracy is needed or if a large number of samples are to be run, the standard series method is preferable, especially if a series of permanent standards is available. Both these methods utilize Nessler tubes—glass tubes with optically flat bottoms and uniform diameters—so that the solution can be viewed either horizontally or vertically.

### THE DILUTION METHOD

In the dilution method, the sample and a standard solution of almost equal concentration are placed in Nessler tubes and viewed horizontally. The darker solution is then diluted until the colors match and the volumes of the two solutions are measured. Since the absorbance, absorptivity, and path length for the two solutions are equal, the concentrations must be equal and the unknown can be calculated from the concentration of the standard and the volume ratio. This method is simple and is used when a rapid estimate of concentration is necessary; any uniform bore tubes may be used in the determination in place of Nessler tubes. Since the concentrations are equal and the volumes are known,

$$\frac{g_u}{V_u} = \frac{g_s}{V_s} \qquad \textbf{(16-10)}$$

where $g$ represents grams, $V$ represents volume, and the subscripts $u$ and $s$ refer to the unknown and standard, respectively.

### THE STANDARD SERIES METHOD

In this method, the color of the unknown is compared to the colors of a series of standards. The color of the unknown will match one of the standard colors exactly or else will fall between two of the colors. As a consequence, the concentration cannot be estimated closer than one-half the concentration increment between two standards. The procedure is slow due to the necessary preparation of several standard solutions with a constant concentration increment between successive standards. As a consequence, it is seldom used except when a large number of samples is to be run or when a series of permanent standards is available. Several comparators are on the market that furnish a series of permanent standards, either glass or solutions, so that comparisons can be made with little or no preparation. Such an instrument is beneficial in field work, where access to electrical current is limited.

The standard series method is the basis of all calibration curves used in photometry. These calibration procedures measure the absorbance or transmittance of a series of standard solutions of the proper concentration sequence and then graph the results so that future measurements of absorbance can be related directly to the concentration from the graph.

## PHOTOMETRIC INSTRUMENTS

To simplify measurements and to offset some of the errors associated with visual measurements, instruments have been designed that substitute a photosensitive device for the eye as the method of measurement. These instruments are called **photometers**, **spectrophotometers**, or **spectrometers** and are classified in several different ways. One classification is based upon the type of wavelength selector used (filter and spectrophotometers); another is based upon whether one or two light beams are used (single- or double-beam instruments); and a third is based upon whether the result is automatically plotted by a recorder or whether the reading must be made visually from a meter of some sort. A further classification is based upon the region of the spectra that is being measured (ultraviolet, visible, infrared). No matter what type of photometer is used, all have the same basic components, which are illustrated in the simplified diagrams of a single-beam and a double-beam photometer shown in Figure 16-7(a) and (b).

Radiant energy from some source passes through an entrance slit into the instrument and is focused by use of mirrors and lenses through a wavelength selector, the sample cell, and an exit slit onto a photosensitive device. The output is fed through the proper circuitry to a readout. Each of these main features will be discussed separately.

### RADIANT ENERGY SOURCES

Radiant energy sources vary with the region of the spectrum being utilized. In the ultraviolet region, the light source is usually a hydrogen discharge lamp; in the visible region, a tungsten lamp is used; and in the infrared, a globar or Nernst glower furnishes the radiant energy. The intensity of the energy emitted by the source is a function of the potential applied across the source. As a consequence, the potential must be controlled carefully between narrow limits for reproducible results. In the visual region, the tungsten lamp is operated at either 6 or 12 V dc, and most instruments incorporate a constant-voltage transformer in the system to help maintain a constant potential. A variation of

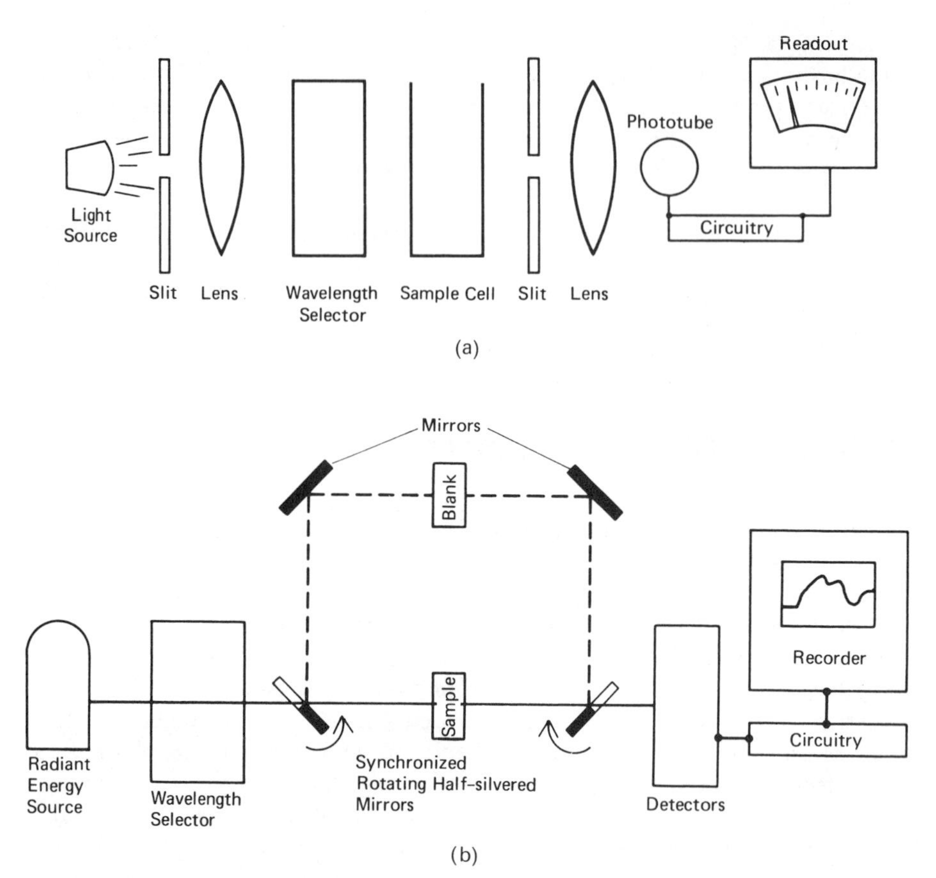


**Figure 16-7.** (a) Simplified schematic diagram of a single-beam photometer. (b) Simplified schematic diagram of a double-beam photometer employing a chopping device.

more than a few thousandths of a volt may cause a variation in the photocurrent from the detector of more than 0.2%.

The tungsten lamp is usually used in the visible and near-infrared regions between approximately 400 and 1200 nm, even though only about 15% of the energy is in the visible region. In the ultraviolet region below 400 nm, a hydrogen (or deuterium) discharge lamp is used down to approximately 200 nm (the spectral limit of the quartz envelope). The intensity of the deuterium lamp is almost three times as great as the intensity of the hydrogen lamp. The glowers and globars used in the infrared region are discussed in more detail in Chapter 17.

## THE OPTICAL SYSTEM

The optical system consists of the slits, lenses, mirrors, cell, and wavelength selector, but wavelength selectors will be discussed separately in the next section. The entrance slit on all photometers is variable in width and is used to control the amount of light entering the photometer. It is utilized to set the

100% transmittance or zero absorbance for the instrument at the beginning of a set of measurements.

The exit slit helps to govern the bandwidth of the light entering the detector. It is fixed in less expensive instruments and is variable in more expensive instruments. When fixed, it usually allows a bandwidth of 20–40 nm to pass; when variable, the slit may be controlled in many instruments to allow bandwidths as small as 5 nm or less.

The lenses are used to focus the light and mirrors may be used to extend the path length inside the instrument or to change the direction of the light beam. The sample cells, or **cuvettes**, are made of optical glass, plastic, or quartz. The path length through the cell must be controlled carefully for reproducible results between several cells. The outside surfaces must not reflect or diffract too much light and must always be kept scrupulously clean. Most cuvettes are square cylinders with a path length of 1.00 cm, but round or rectangular cuvettes can be used. In IR photometers, the optics and cells are often made of sodium chloride or other soluble salts.

## THE WAVELENGTH SELECTORS

The wavelength selectors are used to produce some degree of monochromatic light. Since the Beer and Lambert laws are based upon the use of monochromatic light, the use of polychromatic light will often decrease the concentration range in which the laws will hold. The efficiency of the wavelength selector is measured by the maximum percentage of incident light that will be transmitted and is based upon the half-width of the wavelength band that is allowed to pass. The **half-width** is the width of the band at one-half the height of the maximum transmittance. Wavelength selectors are classified as filters or monochromators.

### *Filters*

Filters may be made of colored glass, gelatine, paper, solutions, polyethylene, thin layers of metals, or other similar materials. A comparison of the efficiency of three types of filters is given in Figure 16-8. Filters made of colored glass usually have the lowest maximum percent transmittance and the widest bandpass and are consequently the poorest filters. Combination filters made of two or more interference filters show almost twice the efficiency of glass. An interference filter is one that goes from zero transmittance to maximum transmittance over a short wavelength span so that the combination of a filter that shows increasing transmittance with one that shows decreasing transmittance in the same wavelength region is more efficient than colored glass alone.

Multilayer metal interference filters are the most efficient filters, both as to peak transmittance and bandpass. The simplest example of this type of filter is shown in Figure 16-9 and consists of two highly reflecting but partially transmitting films of silver separated by a spacer film of nonabsorbing material.

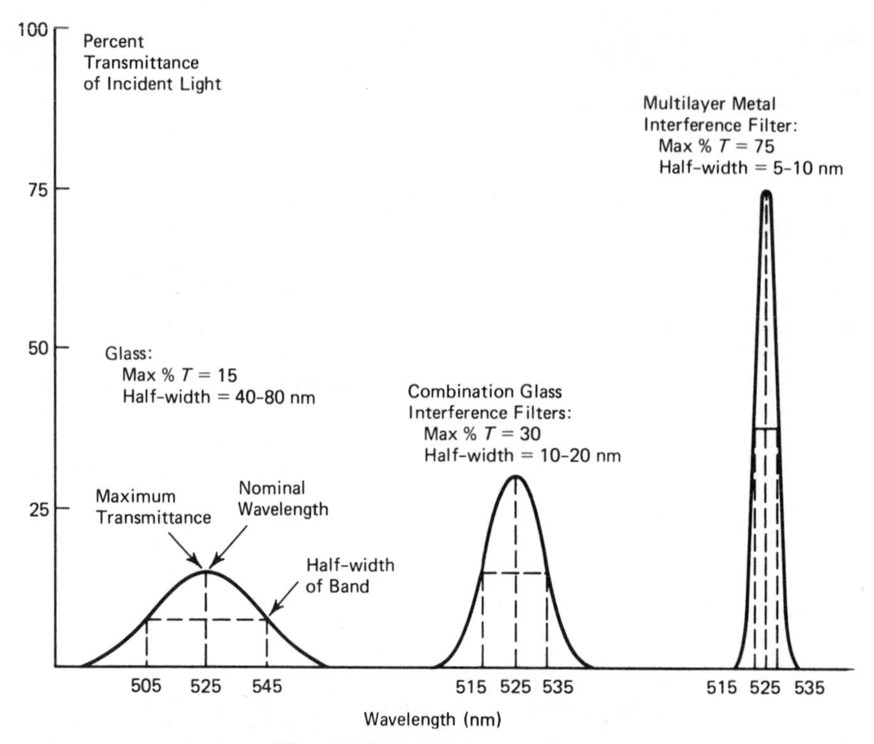


**Figure 16-8.** Efficiency of filters.

This combination is deposited on a glass plate by high-vacuum methods, and a cover plate is then cemented on for protection. Light incident upon the filter is reflected back and forth between the metal films so that only light that has a wavelength of twice the depth of the spacer will pass through the filter. More complex filters are made by alternating the spacer and metal films until as many as 25 layers are obtained. Such filters allow the passage of some second and

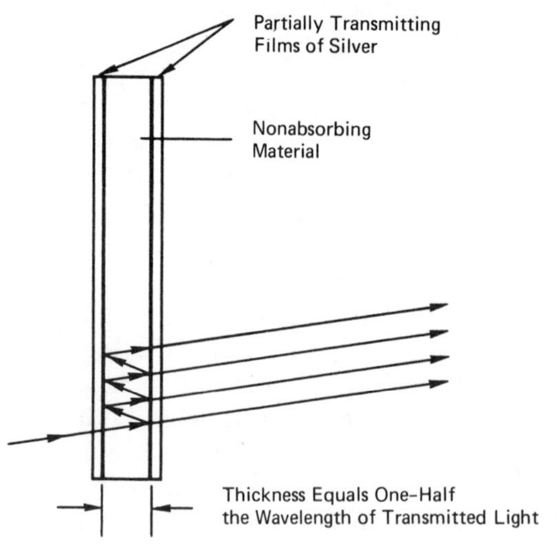


**Figure 16-9.** Simplified schematic diagram of a multilayer metal interference filter.

higher orders of wavelengths, but these may be screened out by glass blocking filters if necessary.

## *Monochromators*

The two types of monochromators used in photometers are prisms and diffraction gratings. **Prisms** may be made of glass, quartz, or other materials and may have various shapes. Most prisms are triangular in shape and are able to separate polychromatic light into individual wavelengths, since light is bent upon passing from one medium to another and the extent of bending is a function of the wavelength. Rotation of the prism in the light path allows the focusing of any particular wavelength through the sample and the exit slit. **Diffraction gratings** consist of a large number of closely spaced parallel lines ruled upon a surface. The surface can be flat or curved, and gratings are classified as *plane gratings* or *concave gratings*, depending upon the shape. When light strikes a

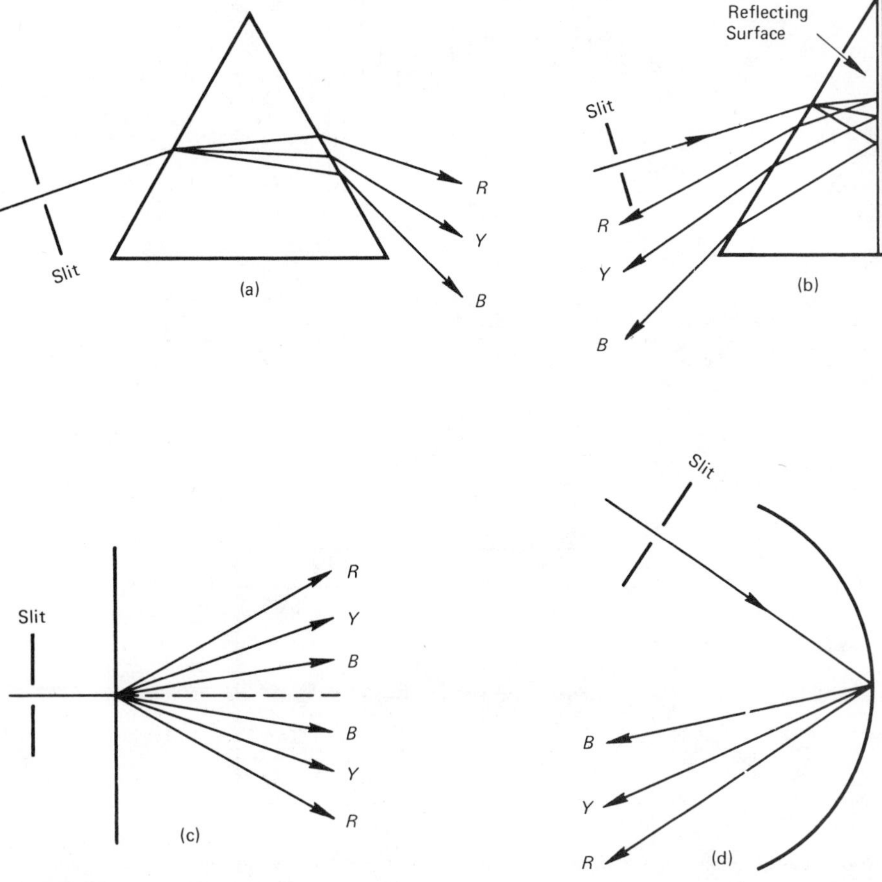


**Figure 16-10.** Monochromators: (a) Cornu prism; (b) Littrow prism; (c) plane transmission diffraction grating; (d) concave reflection diffraction grating. R, red; Y, yellow; B, blue.

diffraction grating, it is diffracted and the extent of diffraction depends upon the wavelength. As a consequence, rotation of a grating in the light path also allows light of a particular wavelength to be focused through the sample and the exit slit.

The effect of prisms and diffraction gratings in separating white light into its various wavelengths is shown in Figure 16-10. If the Cornu-type prism is made of quartz, two pieces must be cemented together to offset the double refraction or birefringence property of quartz.

Photometers that use filters are classified as filter photometers, and those using a prism or grating are classified as spectrophotometers. Filter photometers can operate only at the wavelengths of the filters available, but spectrophotometers operate at any wavelength simply by rotation of the prism or grating to

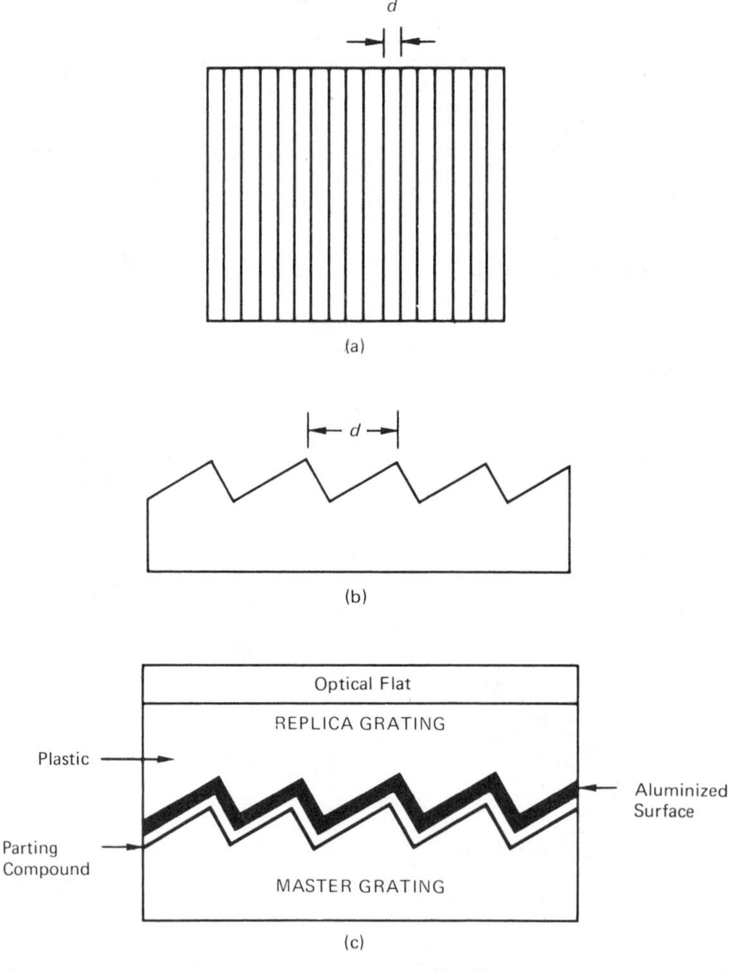


**Figure 16-11.** Echelette gratings, top and side view. Replica gratings are made by aluminizing a thin layer of a parting compound that has been applied to the master grating. The aluminized surface is covered with a soft plastic bonded to an optical flat, allowed to harden, and then removed from the master grating.

the proper position. Original gratings are very expensive and are not used except in the most expensive instruments. Several companies, however, satisfactorily use replica gratings in comparatively inexpensive spectrophotometers. Many of the moderately expensive instruments that formerly used prisms are now made with high-grade replica gratings. Figure 16-11 shows a highly magnified, simplified drawing of an echelette grating, together with a schematic diagram of the method by which replica gratings are made.

## LIGHT-SENSITIVE DEVICES

Light-sensitive devices that are used in the visible region usually may also be used in the ultraviolet but cannot ordinarily be used in the infrared owing to the lower energy associated with the infrared region. The devices used in the infrared instruments are discussed in Chapter 17.

### *Vacuum-Tube Photocells (Photoemissive Cells)*

The vacuum phototube contains a cylindrical cathode (negative electrode) and a wire anode sealed inside an evacuated glass envelope, as shown in Figure 16-12. The surface of the cathode is covered with a metal that will emit electrons when struck by light. The anode is maintained at a positive potential by means of an external circuit so that electrons emitted by the cathode will move to the anode and produce a flow of electrons or a current in the circuit. This current is amplified and fed to a readout device, which is often an ammeter calibrated in absorbance and/or percent transmittance.

The wavelength response and sensitivity of the vacuum phototube depends upon the cathode coating. Alkali metals are usually used, since they have low ionization energies and lose electrons readily. A sodium cathode tube responds

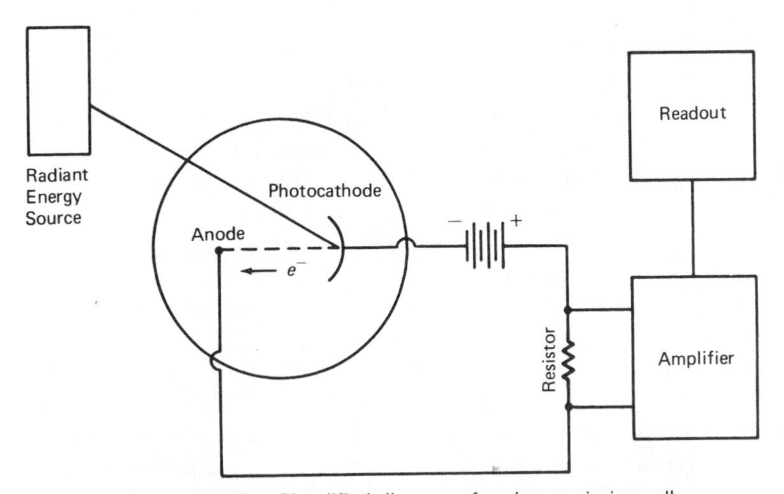


**Figure 16-12.** Simplified diagram of a photoemissive cell.

in the 300- to 500-nm wavelength region whereas a composite coating of silver, oxygen, and cesium responds to wavelengths from approximately 250 to 1200 nm, with maximum sensitivity in the region 700–900 nm. The composite coatings are as much as 1000 times more sensitive than the alkali metal alone.

The silver–oxygen–cesium composite cathode is made by plating a nickel or copper sheet with a thin layer of silver. This is oxidized carefully to form a coating of silver oxide, and then a layer of cesium is vacuum-distilled onto the surface and the entire cathode baked. This operation results in a layer of silver and a layer of cesium separated by a mixture of silver oxide and cesium oxide. Other composite coatings are prepared from silver–potassium, from antimony–cesium, and from bismuth–oxygen–silver–cesium to respond in different wavelength regions.

To function properly, the cathode must absorb light efficiently. To do this, it must approximate a semiconductor and must be neither transparent nor reflective. Also, the potential between the electrodes must be large enough that all the electrons emitted by the cathode will reach the anode but must not be so high that the leakage current is excessive. The necessary minimum potential, known as the **saturation photocurrent potential**, increases as the light intensity increases, since more electrons are emitted. The attainment of saturation photocurrent is necessary for the current produced by the photocell to be directly proportional to the light intensity.

### *Barrier-Layer Cells (Photovoltaic Cells)*

A barrier-layer cell is composed of a thin sheet of a conducting metal upon which a thin layer of a semiconductor has been deposited. The semiconductor is then covered with a grid or screen of conducting metal, such as silver, which is connected through the external circuit to the backing plate. The semiconductor may be selenium or cuprous oxide deposited on iron or copper as the backing plate. The grid acts as the collector electrode. External power is not required because the cell produces a potential.

When light passes through the grid and strikes the semiconductor, electrons flow from the semiconductor to the grid. The current thus produced flows through the external circuit to the backing plate. The term *barrier layer* comes from the fact that electrons will not flow in the opposite direction from the grid to the semiconductor. The current is small but can be measured with a galvanometer or microammeter if the external resistance is not too great. Owing to the low internal resistance, the current cannot be amplified readily. Under proper conditions, the amount of current is proportional to the intensity of the light striking the cell and thus can be used to measure the intensity of the incident light.

Barrier-layer cells suffer from fatigue and should not be irradiated for extended periods of time. Proper design of the circuit and selection of conditions for measurement can eliminate this difficulty in most cases. Barrier-layer cells are normally used only in the visible region in instruments that furnish fairly high levels of illumination. The response of the selenium cell is

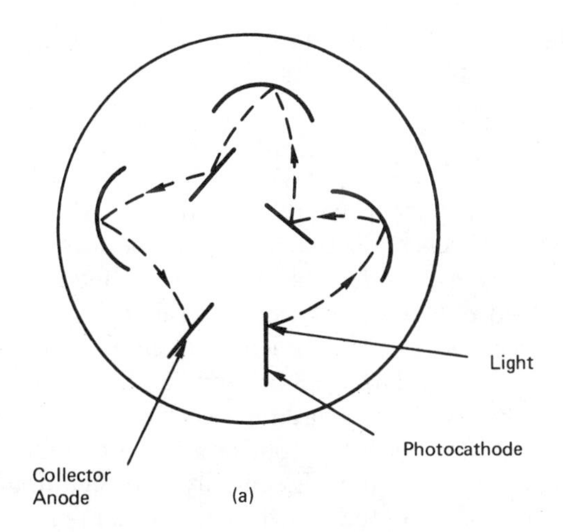


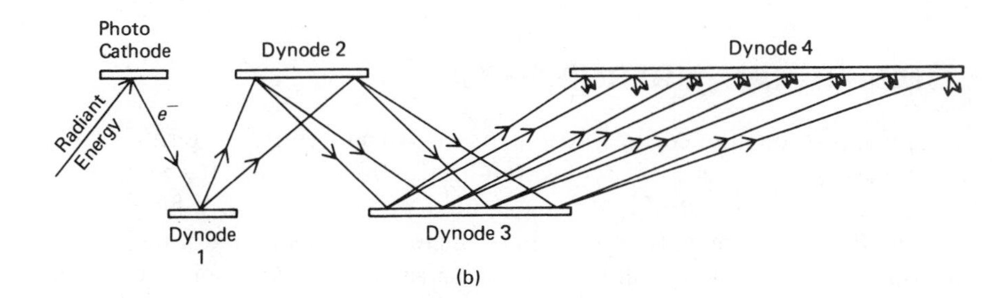


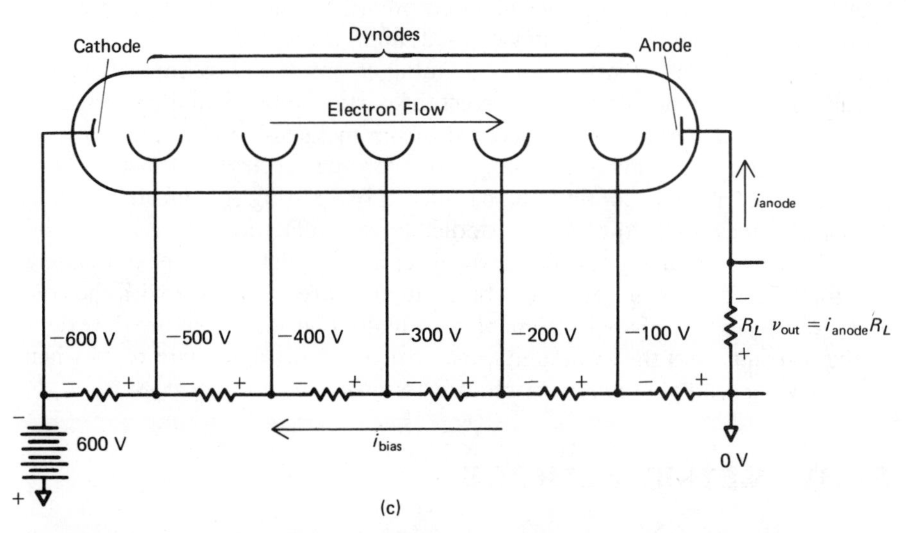


**Figure 16-13.** (a) Simplified diagram of a photomultiplier tube. (b) Simplified schematic of a photomultiplier tube, showing the increase in current per dynode if each dynode emits two electrons for each electron that strikes it. Most dynodes emit more than two electrons per strike. (c) Schematic diagram showing circuit used to make each successive dynode more positive than the previous one. The photoanode is held at ground potential while the photocathode is maintained at −600 V. Arrows indicate the direction of flow of conventional current.

approximately the same as the human eye, having the greatest sensitivity in the 500- to 600-nm region.

### *Photomultiplier Tubes*

Photomultiplier tubes have been devised that have several dynodes at increasing potential differences so that the electrons from the photocathode can be increased by each dynode. The dynodes are made of materials that, because of the applied potential, will emit several electrons for each electron that strikes the surface, and, as a consequence, the number of electrons striking the photoanode is increased. The amplification can be as great as $10^6$. Photomultiplier tubes can be used to increase the sensitivity for low light intensities. Since they are limited to a current output of a few milliamperes, they can be used only with very low intensity light. Essentially, photomultipliers are vacuum-tube photocells that provide their own internal amplification. Simplified diagrams are shown in Figure 16-13.

## COMMERCIAL INSTRUMENTS

As previously mentioned, the instruments used in photometric determinations may be classified in several different ways. One classification is on the basis of the type of wavelength selector used; thus, a filter photometer uses a filter as the wavelength selector whereas a spectrophotometer uses prisms or gratings. Other methods include single- or double-beam instruments; visible, ultraviolet, or infrared; manual, automatic recording (scanning), or digital readout; double or single monochromation (double monochromation uses both a prism and a grating); type of light-sensitive device used; and others.

In the selection of an instrument, the initial cost is often less important than the adaptability of the instrument to meet the particular needs of the purchasing laboratory. In general, the cost may vary from inexpensive filter photometers to very expensive automatic scanning and recording spectrophotometers. New and improved models continue to be offered by existing manufacturers, and new companies enter the field periodically. Manufacturers' brochures and representatives should be consulted whenever an instrument is to be purchased, and the various possibilities should be evaluated carefully in line with the cost and the expected use of the instrument. Two major considerations are the extent of the warranty and the availability and cost of repairs and spare parts when necessary, especially if the repairs require an out-of-town repair person.

## PHOTOMETRIC METHODS

### ABSORPTION SPECTRA

Each substance capable of absorbing light absorbs only those wavelengths of light that have energies that can cause some transformation in the atom or molecule absorbing the light. These changes can be nuclear, electronic, or

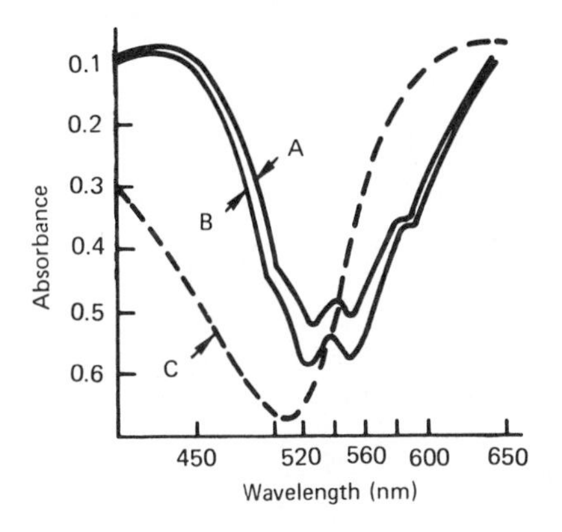


**Figure 16-14.** Absorbance curves. (A and B) Potassium permanganate of different concentrations; (C) iron(II) orthophenanthroline complex.

molecular, as discussed earlier. The wavelengths at which any give substance will absorb are determined by measuring the absorbance or transmittance of a solution as the wavelength is varied. The graph of absorbance versus wavelength is known as the **absorption curve** or the **absorption spectra** of the substance. Approximate absorption spectra for permanganate ion and the orthophenanthroline complex of ferrous iron are shown in Figure 16-14. Curves A and B represent the absorption curves of two solutions of potassium permanganate measured under the same conditions, differing only in concentration, with B being more concentrated than A. It should be noted that the shapes of the spectra are identical but that the greatest distance between the curves occurs at the peak, at 525 nm. This peak represents the maximum absorbance for permanganate and is known as the *wavelength of maximum absorbance or minimum transmittance.* There is a greater change in absorbance per unit change of concentration at this wavelength than at any other wavelength. As a consequence, the measurement of concentrations is more sensitive at the wavelength of maximum absorbance than at any other wavelength.

The greater sensitivity at the wavelength of minimum transmittance limits the concentrations that can be measured within the desired range of 20–80% transmittance (0.70–0.10 absorbance). If a wider range of concentrations needs to be determined, the measurements can be made at a wavelength of 550 nm, at which the sensitivity is less. Still wider ranges (with still less sensitivity) can be measured at wavelengths of 510 and 590 nm, at which "humps" appear on the spectra. Measurements should not be made at wavelengths on the steepest part of the curve, since minor errors in wavelength setting could cause major errors in the measured absorbance of the solution.

## DETERMINATION OF CONCENTRATION FROM ABSORBANCE

### *By Calculation*

The concentration for any given unknown solution can be calculated from the measured absorbance of the solution, and the absorbance of a standard solution determined under identical conditions

$$A = abc \tag{16-6}$$

so that

$$A_u = a_u b_u c_u$$

and

$$A_s = a_s b_s c_s$$

where the subscripts $u$ and $s$ refer to the unknown and standard solutions, respectively. Because the solutions are measured under the same conditions in cells of equal light path, $a_u = a_s$ and $b_u = b_s$, and the ratio of absorbances becomes

$$\frac{A_u}{A_s} = \frac{a_u b_u c_u}{a_s b_s c_s} = \frac{c_u}{c_s} \tag{16-11}$$

or

$$c_u = \frac{A_u c_s}{A_s} \tag{16-12}$$

Remember that eq. (16-12) does not hold for transmittance and that readings in $T$ or percent $T$ must be converted to $A$ ($-\log T = A$) before the calculation can be made.

## *Calibration Curves*

When a large number of repetitive samples are being run, it is preferable to construct a calibration or standard curve so that concentrations may be read directly from the graph. Standard curves are obtained by measuring the transmittance or absorbance of a series of standard solutions in the range of the solutions being measured in the determination. Each standard solution should be prepared in exactly the same way as the solutions being measured; that is, all the reagents added to the sample should be added to the standard and the standard should be taken through all the steps involved in the sample determination. The transmittance or absorbance values are then plotted against concentration. Absorbance values may be plotted on regular rectilinear graph paper, but transmittance or percent transmittance must be plotted on semilogarithm paper. The use of the logarithmic scale for transmittance obviates the necessity of converting transmittance to absorbance before using the graph. A representative calibration is shown in Figure 16-15, plotted on semilog paper

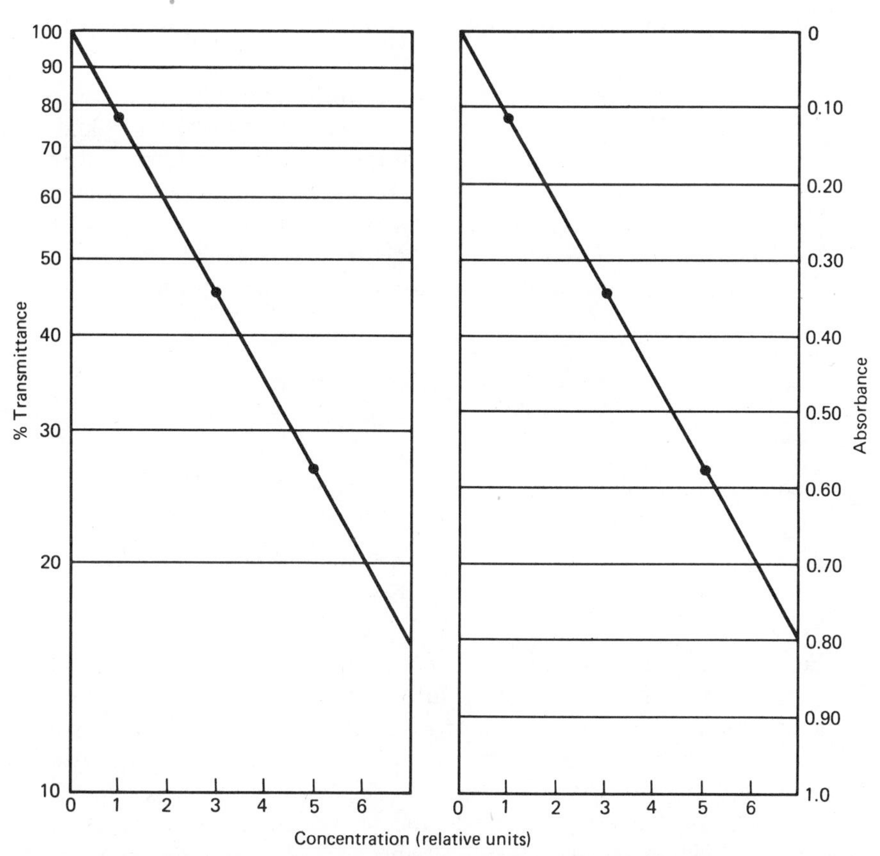


**Figure 16-15.** Calibration curves for the same measurement plotted as percent transmittance versus concentration on semilog paper and as absorbance versus concentration on regular graph paper. Points plotted are concentration (1, 3, 5); %$T$ (78, 46, 27) and $A$ (0.11, 0.34, 0.57). Note that $0.11 = -\log 0.78$, $0.34 = -\log 0.46$, and $0.57 = -\log 0.27$.

as percent $T$ versus concentration and on regular paper as $A$ versus concentration. It should be noted that the two calibrations will superimpose on each other when equivalent units are used for $A$, $T$, and concentration.

The meters or readout devices of most photometers are designed so that either percent transmittance or absorbance can be read. In most cases, fewer mistakes are made if the percent $T$ scale is used, since it reads from 0 on the left to 100 on the right in equal divisions. The absorbance scale reads from 2 on the left to 0 on the right, and the divisions vary in size throughout the scale. Since the photometers are set on 100% $T$ or 0 $A$ using the solvent or a blank solution, the zero concentration–zero absorbance (100% $T$) point should be considered as a measured point in constructing the graph. Ordinarily, the best straight line is drawn through all the points, including the zero–zero point. If the zero–zero point is obviously off the best straight line through the measured points, it usually indicates that an error has been made in the preparation of the standards or that there is an inherent positive or negative blank in the measurement that is not accounted for by the blank used to set the instrument.

## DETERMINATION OF pH AND p$K$

Most of the indicators used for acid–base work are weak organic acids or bases for which the molecular and ionic forms take on different colors and thus absorb at different wavelengths. The absorption curve in the intermediate color change region will show two absorbance peaks, one for the molecular form and one for the ionic form. Measurement of absorbances at either of these peaks will allow calculations of the pH if the p$K$ of the indicator is known.

The pH, p$K$, and concentrations of the colored forms are related by the expression

$$\text{pH} = \text{p}K_a - \log \frac{[\text{molecular form}]}{[\text{ionic form}]} \quad \textbf{(16-13)}$$

if the indicator is an acid and if a base by the expression

$$\text{pOH} = \text{p}K_b - \log \frac{[\text{molecular form}]}{[\text{ionic form}]} \quad \textbf{(16-14)}$$

in which the brackets indicate molar concentration. For very accurate calculations, the p$K$ values must be corrected for the ionic strength of the solution.

The molecular form of the indicator is the acid form for indicators that are weak acids and the ionic form is the basic form. The opposite holds true for indicators that are weak bases.

For indicators that are weak acids and for which the absorbance of the basic form is being measured, the absorbance at a pH above the color change range will be related to the total concentration of the indicator, since the indicator will be essentially completely converted to the basic form by the higher pH. This absorbance is also the maximum absorbance that can be attained in the solution at any pH value, so that

$$A_{\text{max}} = abc = ab[\text{In}]_{\text{tot}} \quad \textbf{(16-15)}$$

where $[\text{In}]_{\text{tot}}$ represents the total concentration of the indicator.

At pH values below the color-change region, the indicator will be essentially completely in the molecular form, so that minimum (or zero) absorbance will be obtained due to the very low concentration of the colored form being measured. Since $A_{\text{max}}$ represents the solution when the indicator is entirely in the basic form and $A_{\text{min}}$ represents the solution when the indicator is completely in the acid form, absorbance values intermediate between $A_{\text{max}}$ and $A_{\text{min}}$ will be obtained when the pH of the solution is in the color-change region of the indicator. Since concentrations are directly proportional to absorbance if the measurements are made under the same conditions, the concentration of the basic (ionic) form will be given by

$$[\text{basic form}] = A_{\text{obs}} - A_{\text{min}} \quad \textbf{(16-16)}$$

and the concentration of the acid (molecular) form will be

$$[\text{acid form}] = A_{max} - A_{obs} \tag{16-17}$$

where $A_{obs}$ represents the observed or measured absorbance of the solution. Substitution of these values in eq. (16-13) gives

$$\text{pH} = \text{p}K_a - \log \frac{A_{max} - A_{obs}}{A_{obs} - A_{min}} \tag{16-18}$$

If the absorbance of the molecular (acid form) is being measured, eq. (16-18) becomes

$$\text{pH} = \text{p}K_a - \log \frac{A_{obs} - A_{min}}{A_{max} - A_{obs}} \tag{16-19}$$

Similar expressions for pOH can be derived for indicators that are weak bases.

If small numbers of measurements are to be made, calculation by these equations represents a saving of time because only two standard solutions need be made, one of which is well above and the other well below the color-change region. If a large number of measurements are to be made, it is preferable to prepare a series of standards separated by 0.2 pH unit throughout the color-change region, together with two or more solutions outside the indicator range. These absorbance values can be plotted to give a standard curve similar to the one in Figure 16-16 for bromcresol green. The pH values can then be read directly from the curve without calculation.

• **Example 2.** Calculate the pH of a solution that contains bromthymol blue and has an absorbance of 0.365 at 617 nm if $A_{max}$ is 0.800 and $A_{min}$ 0.065 under

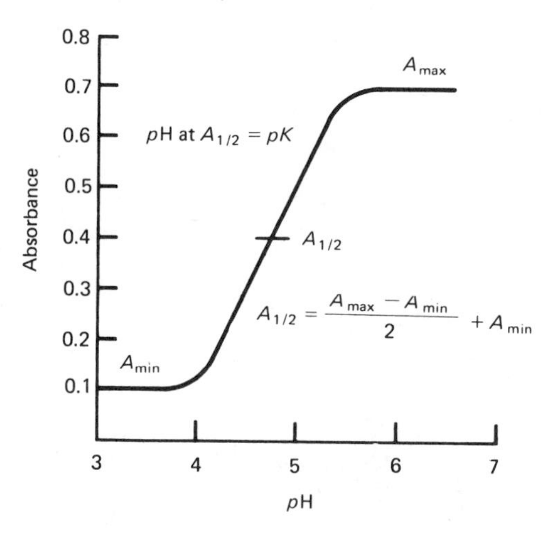


**Figure 16-16.** Absorbance curve for bromcresol green at 614 nm.

the conditions of measurement. The $pK_{In}$ for bromthymol blue is 7.10. Substitution in eq. (16-18) gives

$$pH = 7.10 - \log \frac{0.800 - 0.365}{0.365 - 0.065}$$

$$= 7.10 - \log 1.45 = 7.10 - 0.14 = 6.96$$

Equations (16-18) and (16-19) can also be used to calculate the $pK_a$ if $A_{max}$, $A_{min}$, $A_{obs}$, and the pH of a solution of an indicator are known. The $pK_a$ can also be obtained from the curve in Figure 16-16 because the pH at the inflection point equals $pK_a$ and can be calculated from the absorbance of any point on the curve together with the maximum and minimum absorbances.

• **Example 3.** Calculate the $pK_a$ of methyl red if a solution of pH 5.20 gave an absorbance of 0.48 at 530 nm, the wavelength of maximum absorbance of the acid form. Under the conditions of measurement, $A_{max}$ is 1.34 and $A_{min}$ is zero. Rearrangement of eq. (16-19) gives

$$pK_a = pH + \log \frac{A_{obs} - A_{min}}{A_{max} - A_{obs}}$$

and, by substitution of the proper values,

$$pK_a = 5.20 + \log \frac{0.48 - 0.00}{1.34 - 0.48}$$

$$= 5.20 + \log \frac{0.48}{0.86} = 5.20 + \log 0.56$$

$$= 5.20 - 0.25 = 4.95$$

These techniques can also be used to determine the $pK_a$ of many colorless acids by making the measurements in the ultraviolet region.

## DETERMINATION OF THE FORMULA OF A COMPLEX

### *The Mole-Ratio Method*

The general equation for the formation of a complex ion or compound is

$$M + nL \longrightarrow ML_n \qquad \textbf{(16-20)}$$

in which no charges are shown and $n$ represents the number of ligands attached to each metal atom. If the complex $ML_n$ absorbs at a wavelength at which M and L do not absorb (or do not absorb appreciably), the formula of the complex

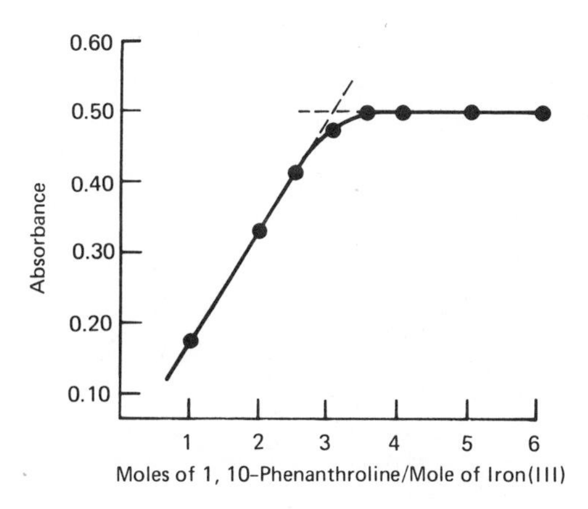


**Figure 16-17.** Mole-ratio plot for the blue tris(1,10-phenanthroline)iron(III) complex.

can be determined spectrophotometrically. A series of solutions is prepared in which the molar concentration of M is kept constant and the molar concentration of L is increased.

In a typical case, 1 mL of 0.01 *M* metal ion is added to each of several 50-mL volumetric flasks and then 1 mL of 0.01 *M* L is added to the first flask, 2 mL to the second flask, and so on until the necessary number of flasks is prepared. The color is then developed, each flask is diluted to volume, and the absorbance is measured at the proper wavelength. The absorbances are then plotted versus the mole ratio of L/M to give a graph similar to Figure 16-17 for the blue ferric–phenanthroline complex. The absorbance rises until the ratio of L/M attains that in the complex, since this region there will be excess M and as a consequence the amount of $ML_n$ that can be formed is controlled by the concentration of L. The absorbance remains constant for higher ratios of L/M than that in the complex, because in this region there is excess L and the amount of $ML_n$ that can be formed is controlled by the concentration of M, which is held constant. In the majority of cases, the dissociation of the complex causes a noticeable curvature in the region of the complex composition, and the correct ratio must be obtained by extrapolation of the straight-line portions of the graph.

### *The Method of Continuous Variations*

In this method for the determination of the formula of a complex, a series of solutions is prepared in which the total number of moles of M and L is kept constant while the number of moles of each component is varied. In a typical procedure, a total of 10 mL of 0.01 *M* metal ion and 0.01 *M* ligand is added to 11 different 50-mL flasks, but the ratio of mL of M to L is varied from 0 M, 10 L to 10 M, 0 L. After the development of color and dilution to volume, the absorbances are determined under the proper conditions and the values graphed to give a curve similar to that shown in Figure 16-18 for the red ferrous–phenanthroline complex. As the amount of iron is increased and the amount of 1,10-phenanthroline is decreased, the absorbances rise to a maximum and then decrease. In the region before the maximum is attained the amount of complex

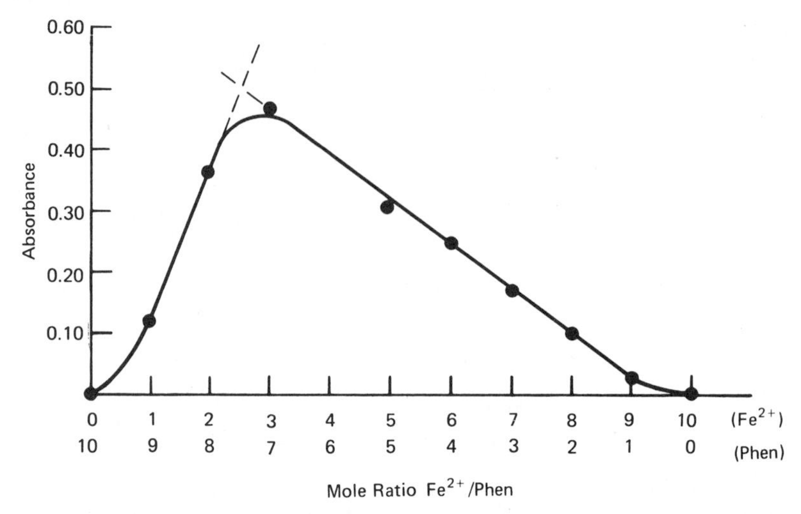


**Figure 16-18.** Continuous variation plot for the red tris(1,10-phenanthroline)iron(II) complex.

that can be formed is controlled by the amount of iron present and after the maximum by the amount of L present. The dissociation of the complex causes rounding of the curve in the region of maximum absorbance, so the correct ratio must usually be obtained by extrapolation of the straight-line portions of the graph.

## MULTICOMPONENT ANALYSIS

Under certain circumstances, it is possible to determine more than one component by spectrophotometric analysis using measurements at more than one wavelength and calculating the concentrations by the use of simultaneous equations. Since it is necessary to have $n$ equations relating absorbance and concentration for $n$ components, more than two components are seldom determined by this method, owing to the complexity of the calculations, unless a computer program and a computer are available.

If two components are being determined and neither of them absorbs at the wavelength of the other, no extended calculation is involved because this is the same as two separate single-component analyses. If, however, one or both absorb at the wavelength of measurement for the other, simultaneous equations involving measurements at two wavelengths are utilized.

In general two wavelengths ($\lambda_1$, $\lambda_2$) are selected at which the interference of the other component is at a minimum and the ratio of molar absorptivities is the greatest ($\varepsilon_1/\varepsilon_2$). For the calculation to be meaningful, the molar absorptivities must be determined accurately, Beer's law must be followed, the absorbances must be additive, and the measurements must be made under the same conditions at identical cell depths. The following equations assume that all the preceding conditions are met.

Since the absorbances are additive, the measured absorbance $A$ at a given wavelength will equal the sum of the separate absorbances at that wavelength or

$$A_{\lambda_1} = A_{1_{\lambda_1}} + A_{2_{\lambda_1}} \tag{16-21}$$

and

$$A_{\lambda_2} = A_{1_{\lambda_2}} + A_{2_{\lambda_2}} \tag{16-22}$$

Since $A = \varepsilon bC$ and since $b$ is constant, $\varepsilon C$ can be substituted for the individual absorbances in eqs. (16-21) and (16-22), to give

$$A_{\lambda_1} = \varepsilon_{1_{\lambda_1}} C_1 + \varepsilon_{2_{\lambda_1}} C_2 \tag{16-23}$$

$$A_{\lambda_2} = \varepsilon_{1_{\lambda_2}} C_1 + \varepsilon_{2_{\lambda_2}} C_2 \tag{16-24}$$

Simultaneous solution of eqs. (16-23) and (16-24) gives

$$C_1 = \frac{\varepsilon_{2_{\lambda_2}} A_{\lambda_1} - \varepsilon_{2_{\lambda_1}} A_{\lambda_2}}{\varepsilon_{1_{\lambda_1}} \varepsilon_{2_{\lambda_2}} - \varepsilon_{2_{\lambda_1}} \varepsilon_{1_{\lambda_2}}} \tag{16-25}$$

$$C_2 = \frac{\varepsilon_{1_{\lambda_1}} A_{\lambda_2} - \varepsilon_{1_{\lambda_2}} A_{\lambda_1}}{\varepsilon_{1_{\lambda_1}} \varepsilon_{2_{\lambda_2}} - \varepsilon_{2_{\lambda_1}} \varepsilon_{1_{\lambda_2}}} \tag{16-26}$$

The molar absorptivities are determined accurately at the two wavelengths on solutions of each component alone and may be used for similar measurements as long as the same cuvettes and spectrophotometer are utilized for the measurements on the mixtures. The absorptivities in the units of measurement should be substituted for molar absorptivities when concentrations are not specified as molar.

The simulated absorption curves shown in Figure 16-19 of the two components of a system used for simultaneous spectrophotometric analysis illustrate the factors that must be considered in the selection of the wavelengths to be used in the determination. The only suitable wavelength for the measurement of component H occurs at 495 nm, the wavelength of maximum absorption. There are three wavelengths, however, that are satisfactory for measurement of component G. The maximum at 570 nm should be selected for greatest sensitivity. If less sensitivity can be tolerated, measurement at the shoulder (at 630 nm) is preferable, since there is no interference from H at this wavelength. In case most of the samples to be measured contain relatively large amounts of G compared to H, measurement at 430 nm would probably be desirable, owing to the lower sensitivity at that wavelength.

• **Example 4.** As an example of the calculation of the composition of a two-component system, calculate the molar concentration of components G and H in a mixture if the absorbance is 0.30 at 495 nm and 0.40 at 570 nm using 1-cm cells. The absorbance of a 0.01 $M$ solution of G is 0.08 at 495 nm and 0.90 at

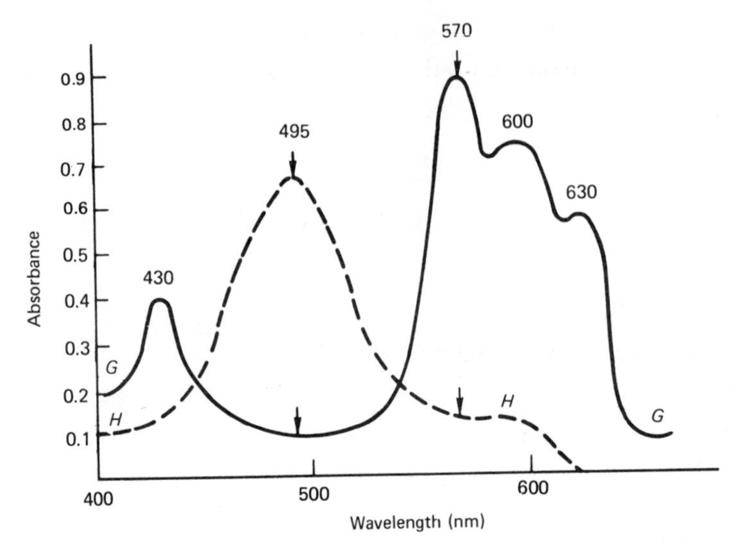


**Figure 16-19.** Simulated absorption curves for analysis of a two-component system.

570 nm and the absorbance of a 0.01 *M* solution of H is 0.67 at 495 nm and 0.12 at 570 nm using 1-cm cells.

The first step is to calculate the molar absorptivities of each component at the two wavelengths, or

$$\varepsilon_{G495} = \frac{0.08}{1 \times 0.01} = 8.0$$

$$\varepsilon_{G570} = \frac{0.90}{1 \times 0.01} = 90$$

$$\varepsilon_{H495} = \frac{0.67}{1 \times 0.01} = 67$$

$$\varepsilon_{H570} = \frac{0.12}{1 \times 0.01} = 12$$

Substitution in eqs. (16-25) and (16-26) gives

$$C_G = \frac{12 \times 0.30 - 67 \times 0.40}{8 \times 12 - 67 \times 90} = \frac{3.6 - 26.8}{96 - 6030}$$

$$= \frac{-23.2}{-5934} = 3.91 \times 10^{-3}\ M$$

$$C_H = \frac{8 \times 0.40 - 90 \times 0.30}{8 \times 12 - 67 \times 90} = \frac{3.2 - 27.0}{96 - 6030}$$

$$= \frac{-23.8}{-5934} = 4.0 \times 10^{-3}\ M$$

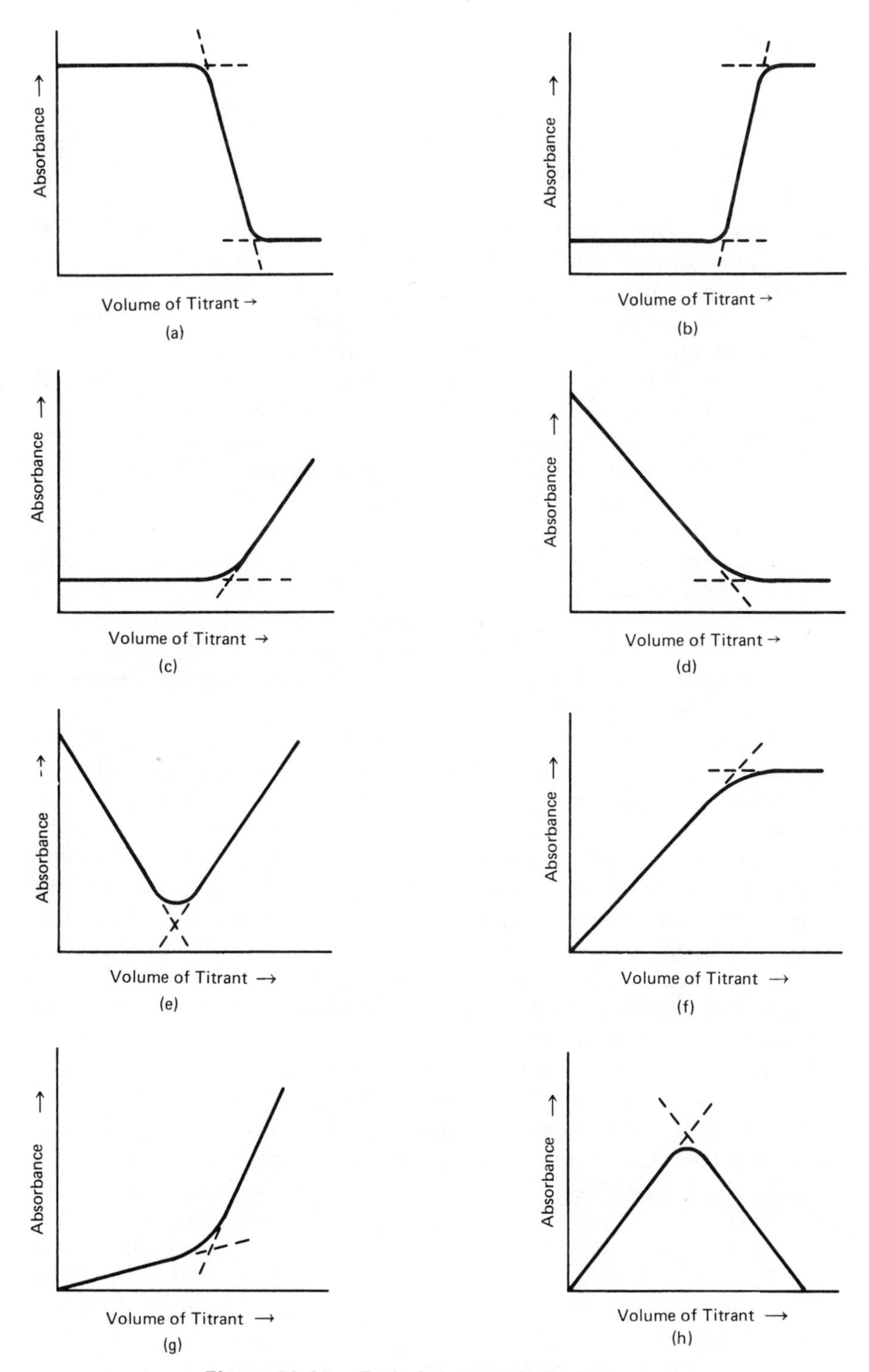


**Figure 16-20.** Typical photometric titration curves.

## PHOTOMETRIC TITRATIONS

Photometric titrations are possible for any titration in which the titrant, a reactant, or a product absorbs and has a sufficiently large molar absorptivity. In some cases in which none of the materials directly involved absorb satisfactorily, an indicator that absorbs can be added to determine the end point photometrically. During the titration, the absorbance of the solution is measured after each increment of titrant is added and the absorbance graphed versus the volume of titrant added. The possible shapes of titration curves are shown in Figure 16-20. Curve (a) represents the use of an indicator such as phenolphthalein in the titration of a base with an acid; (b) represents the reverse titration or the use of an indicator for which the absorbing species produced in the end point region is measured. Curve (c) represents a titration in which only the titrant absorbs; curve (d) a titration in which only the reactant absorbs; and (f) a titration in which only the product absorbs. Curves similar to (e) are obtained in titrations in which both the reactant and the titrant absorb, and (g) illustrates titrations in which the product and the titrant both absorb but have different molar absorptivities. Curve (h) is produced in titrations in which successive complexes are formed and in which only one complex absorbs, the maximum in this case representing complete conversion to the absorbing form. Curve (b) can also represent a titration in which two successive complexes are formed if the first complex does not absorb but the second does, as in the titration of bismuth and copper with EDTA.

Photometric titrations have the advantage that the presence of other absorbing species does not interfere unless they are present in relatively large quantities. Also, the measurement does not have to be made at the wavelength of maximum absorbance, since an important feature of photometric titrations is the sharp change in absorbance before and after the end point rather than the sensitivity of the measurement. Often a wavelength that is reliable but at which the molar absorptivity is small is used to allow measurements of higher concentrations.

The fact that only one of the reacting species or products need absorb extends photometric measurements to many substances that do not absorb. This fact also allows titration of substances that absorb in the ultraviolet as well as those that absorb in the visible region.

Absorbance titration curves usually show curvature in the end-point region, and the end point ordinarily is obtained by extrapolation of the straight-line portions of the graph well removed from the end-point region. Care must be exercised in the extrapolation, since there is a definite dilution effect during the titration. This can be offset in part by using a concentrated titrant (at least ten times the concentration of the solution being titrated); however, even with the more concentrated titrant, the absorbances should be corrected for the dilution effect. This can be accomplished by multiplying the measured absorbances by $(V + v)/V$, where $V$ is the initial volume of the titrated solution and $v$ is the volume of the titrant added up to that point in the titration.

The selection of the end point when an indicator is being used in acid–base titrations depends upon the pH of the end point and the pH range of the color change of the indicator. In some cases the first appearance of the measured form is desired; in some the disappearance of the measured form occurs at the end point; and in some the end point pH is indicated by the midpoint of the color-change region.

## TURBIDIMETRY AND NEPHELOMETRY

**Turbidimetry** involves the measurement of light transmitted by a suspension of particles (a turbid solution) and **nephelometry** is based on the measurement of the intensity of light scattered by a suspension. Turbidimetry thus measures the decrease in the intensity of a light beam as it passes through a solution and is similar to absorption photometry, even though the mechanism causing the decrease in intensity is completely different. Nephelometry, on the other hand, measures the light reflected from the particles and is similar to fluorometry even though the mechanisms for the production of the light are not related to each other in any way. Fluorometry is discussed in Chapter 19. Turbidimetry and nephelometry are usually discussed with absorption photometry because the instruments used and the techniques of measurement are similar. As there are relatively few analytical applications for these methods, they will not be discussed in detail in this text.

Nephelometry differs from turbidimetry in the arrangement of the photometer, since turbidity measurements are made in the same direction as the propagation of the light from the source, whereas the light scattered in nephelometry is usually measured at right angles to the direction of propagation of the light from the source. This difference is illustrated in Figure 16-21, which shows a single-beam photometer arranged both for photometry (turbidimetry) and nephelometry. Note that, in the nephelometric arrangement, the light source is at right angles to the light path to the detector.

The intensity of scattered monochromatic light varies with the wavelength of the incident light, even though there is no change in the wavelength of the light. As a result, red light is not scattered as strongly as blue light. The intensity also depends upon the number and the size of the particles, and, as a consequence,

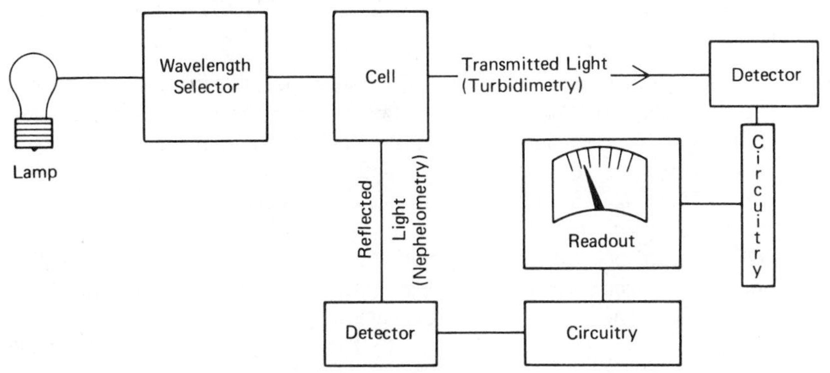


**Figure 16-21.** Simplified block diagram of a single-beam spectrophotometer showing the arrangements for nephelometry and turbidimetry.

care must be taken in procedures to ensure that conditions such as mixing, temperature, and agitation are as identical as possible in all the determinations run.

The intensity of the light also falls off as the number of particles is increased and thus with increase in concentration of the substance being measured. For this reason, only dilute solutions can be measured. Stray light is also a problem, since the particles reflect light from any source.

The simplicity of the measurement and the rapidity of the procedures are distinct advantages for turbidimetry and nephelometry, but these are offset in many cases by the lack of precision of the measurements. As a consequence, the methods are used primarily when the results need not be very accurate and when accurate procedures are tedious or require extremely careful technique. Sulfate in natural water is often estimated by a nephelometric procedure that measures the light scattered by a suspension of barium sulfate produced under reproducible conditions. The method is rapid, can be used for the concentrations normally found in natural waters, and is sufficiently accurate for most purposes.

## PROBLEMS

**1** Convert the following transmittance measurements to the corresponding absorbance: (a) 0.355*T*; (b) 35.5%*T*; (c) 0.368*T*; (d) 70%*T*; (e) 0.200*T*. *Ans.* (b) 0.450

**2** Convert the following absorbance readings to the corresponding percent transmittance values: (a) 0.699; (b) 0.109; (c) 0.500; (d) 0.434; (e) 0.222. *Ans.* (b) 77.8%

**3** Calculate the molar absorptivities of 0.0050 *M* solutions that were measured in 1-cm cells and had the absorbance values listed in problem 2. *Ans.* (b) 21.8

**4** Calculate the absorptivity in liters per gram centimeter, for 0.01% (w/v) solutions that were measured in 5.00-cm cells and showed the transmittance values listed in problem 1. *Ans.* (*b*) 0.900

**5** Calculate the range of molar concentrations that can be used to produce transmittance readings between 20.0 and 60.0% if 1.50-cm cells are to be used and the compounds have the molar absorptivities listed: (a) 11,100; (b) 5500; (c) 25; (d) 365; (e) 20,200. *Ans.* (b) $2.69 \times 10^{-5}$ to $8.47 \times 10^{-5}$ *M*

**6** (a) In the colorimetric determination of Mn in steel, the following data were obtained: weight of unknown, 0.65 g; weight of standard steel containing 0.40% Mn, 0.96 g. Both solutions were diluted to 250 mL before comparison in a colorimeter. The standard was set at 37.3 mm and the following readings were obtained for the unknown: 34.7, 34.5, 34.3, 34.7, 34.4, and 34.4. Calculate the percent Mn in the unknown.

(b) The standard and unknown solutions in part (a) were then measured photometrically in identical cells and the transmittance of the standard was 31.0% while that of the unknown was 28.1%. Calculate the percent Mn in the unknown. *Ans.* 0.639%

**7** Repeat problem 6(a), assuming that the standard contained 0.35% Mn and that 0.85 g was used.

**8** Repeat problem 6(a), assuming that the standard contained 0.54% Mn and that 0.59 g was used. Also assume that the standard was set at 25.0 and that the readings on the unknown were 26.2, 26.4, 25.8, 26.0, 26.6, and 26.2.

**9** After development of the $KMnO_4$ color and dilution to 250 mL, the absorbance of a 1.00-g sample of cast iron at 525 nm was 0.484 using a 1.00-cm cell. Using the same cell and wavelength, the absorbance of a solution prepared by diluting 5.00 mL of 0.0185 *M* $KMnO_4$ to 250 mL was 0.400. Calculate the percent Mn in the cast iron.

*Ans.* 0.614%

**10** Repeat problem 9, assuming that 5.00 mL of 0.0205 *M* $KMnO_4$ was used.

**11** Repeat problem 9, assuming that the absorbance of the sample was 0.434 and that of the standard was 0.415.

**12** A spectrophotometer was calibrated with standard solutions of glucose for which the concentrations were adjusted so that they had the same concentrations as protein-free filtrates from blood samples and were expressed as milligram percent in the original blood sample. The following data were obtained (milligram percent is equivalent to mg/100 mL):

| *Concentration (mg %)* | *% T* | *Concentration (mg %)* | *% T* |
|---|---|---|---|
| 50 | 84.0 | 150 | 59.0 |
| 75 | 76.8 | 200 | 49.0 |
| 100 | 70.2 | 250 | 41.0 |

A glucose tolerance test was run on a patient and the original blood sample before administration of glucose gave a reading of 68.4% *T*. A later sample, after dilution 1 : 1 with water before precipitation, gave a reading of 45.0% *T*. Determine the milligram percent glucose in each blood sample.

*Ans.* 108 and 448 mg%

**13** The transmittances of a series of $Cr(NO_3)_3$ solutions were determined using 1-cm cells at 550 nm and used to prepare a calibration curve for chromium solutions. The following data were recorded:

| *Concentration (M)* | *Transmittance* | *Concentration (M)* | *Transmittance* |
|---|---|---|---|
| Blank | 1.000 | 0.0300 | 0.387 |
| 0.0100 | 0.725 | 0.0400 | 0.281 |
| 0.0200 | 0.530 | 0.0500 | 0.207 |

(a) Graph the calibration curve as *T* versus *M* on semilog paper.
(b) Convert *T* to *A* and plot *A* versus *M* on rectilinear paper.
(c) A 1.234-g sample of chromite ore ($FeO{\cdot}Cr_2O_3$ + impurities) is dissolved and after proper treatment is diluted to 100.0 mL. The transmittance of this solution was 0.399. Read the concentration of chromium from each graph. Using the transmittance curve, calculate the percentage of chromite in the ore. Using the absorbance curve value, calculate the percentage of chromium in the ore.

**14** In the determination of acetone in biological fluids, the following calibration curve between absorbance and concentration was obtained:

| *Acetone Standards (mg/100 mL)* | *A* | *Acetone Standards (mg/100 mL)* | *A* |
|---|---|---|---|
| Reagent blank | 0.035 | 4.0 | 0.127 |
| 0.5 | 0.047 | 6.0 | 0.172 |
| 1.0 | 0.059 | 8.0 | 0.219 |
| 2.0 | 0.082 | 10.0 | 0.264 |

Samples of blood and urine from a normal subject and from a ketotic patient were analyzed with these results: normal blood, $A = 0.068$, ketotic blood, $A = 0.189$; normal urine, $A = 0.097$, ketotic urine, $A = 0.198$ (1 : 25 dilution). Calculate the acetone concentration in each sample.

**15** A standard chromium(III) solution was prepared by reduction of 5.00 mL of 1.200 *N* $K_2Cr_2O_7$ followed by dilution to 100 mL. This solution showed an absorbance of 0.310 when measured at 440 nm in a 1.00-cm cell. A 1.000-g sample of a chromium ore was dissolved, treated to ensure that all the chromium was in the trivalent state, and diluted to 100 mL. This solution showed an absorbance of 0.336 when measured in the same cell under the same conditions. Calculate the percent Cr in the ore.

**16** The absorbance of a 1-cm depth of a given concentration of chlorphenol red was 0.850 in 0.01 *M* NaOH, 0.000 in 0.01 *M* HCl, and 0.538 in pH 5.80 buffer solution. Calculate the $pK_{In}$ of chlorphenol red.

*Ans.* 6.04

**17** The transmittance of a 1-cm depth of a given concentration of bromcresol green was 0.158 in pH 7.0 buffer, 0.794 in pH 3.0 buffer, and 0.355 in pH 4.80 buffer. Calculate the $pK_{In}$ of bromcresol green.

**18** Determine the $pK_a$ and $K_a$ of each of the following acids from graphs of the absorbance data given:

| (*a*) | | (*b*) | | (*c*) | | (*d*) | |
|---|---|---|---|---|---|---|---|
| pH | *A* | pH | *A* | pH | *A* | pH | *A* |
| 1.00 | 0.987 | 4.20 | 0.023 | 1.50 | 0.028 | 6.00 | 0.300 |
| 2.00 | 0.974 | 4.40 | 0.041 | 2.00 | 0.055 | 6.50 | 0.327 |
| 3.00 | 0.960 | 4.60 | 0.092 | 2.50 | 0.080 | 7.00 | 0.354 |
| 4.00 | 0.920 | 4.80 | 0.365 | 3.00 | 0.110 | 7.50 | 0.580 |
| 5.00 | 0.800 | 5.00 | 0.720 | 3.50 | 0.205 | 8.00 | 1.020 |
| 6.00 | 0.650 | 5.20 | 0.920 | 4.00 | 0.560 | 8.50 | 1.239 |
| 7.00 | 0.500 | 5.40 | 0.963 | 4.50 | 1.000 | 9.00 | 1.275 |
| 8.00 | 0.350 | 5.60 | 0.983 | 5.00 | 1.440 | 9.50 | 1.300 |
| 9.00 | 0.200 | 5.80 | 1.004 | 5.50 | 1.785 | | |
| 10.00 | 0.080 | | | 6.00 | 1.880 | | |
| 11.00 | 0.040 | | | 6.50 | 1.923 | | |
| 12.00 | 0.027 | | | 7.00 | 1.950 | | |
| 13.00 | 0.013 | | | 7.50 | 1.975 | | |

*Ans.* (b) $pK_a = 4.88$

**19** The variation of absorbance at the wavelength of maximum absorbance with pH for several indicators is shown below. For each, (a) graph the results and determine the $pK_{In}$ from the graph; (b) calculate the $pK_{In}$ from the two pH readings closest to the $pK_{In}$; (c) read the pH of the unknown from the graph; (d) calculate the pH of the unknown using the value of $pK_{In}$ calculated in part (b)

| *Bromphenol Blue* | | *Methyl Red* | | *Bromthymol Blue* | | *Phenolphthalein* | |
|---|---|---|---|---|---|---|---|
| pH | *A* | pH | *A* | pH | *A* | pH | *A* |
| 2.60 | 0.100 | 3.80 | 1.60 | 5.60 | 0.000 | 7.60 | 0.000 |
| 2.80 | 0.100 | 4.00 | 1.60 | 5.80 | 0.000 | 7.80 | 0.000 |
| 3.00 | 0.100 | 4.20 | 1.60 | 6.00 | 0.000 | 8.00 | 0.000 |
| 3.20 | 0.142 | 4.40 | 1.54 | 6.20 | 0.042 | 8.20 | 0.035 |
| 3.40 | 0.235 | 4.80 | 1.20 | 6.60 | 0.258 | 8.40 | 0.100 |
| 3.60 | 0.359 | 5.20 | 0.80 | 7.00 | 0.500 | 8.80 | 0.298 |
| 3.80 | 0.479 | 5.60 | 0.40 | 7.40 | 0.734 | 9.20 | 0.500 |
| 4.00 | 0.600 | 6.00 | 0.08 | 7.60 | 0.800 | 9.60 | 0.700 |
| 4.20 | 0.720 | 6.20 | 0.00 | 7.80 | 0.800 | 9.80 | 0.768 |
| 4.40 | 0.835 | 6.40 | 0.00 | 8.00 | 0.800 | 10.00 | 0.800 |
| 4.60 | 0.900 | 6.80 | 0.00 | Unknown | 0.457 | 10.20 | 0.800 |
| 5.00 | 0.900 | Unknown | 0.88 | | | Unknown | 0.125 |
| Unknown | 0.525 | | | | | | |

**20** The *T*, using a 1-cm cell, of a $5.0 \times 10^{-4}$ *M* solution of substance K (MW 170) was 0.289 at 530 nm and 0.892 at 640 nm, and the *T*, using the same cell, of a $5.0 \times 10^{-4}$ *M* solution of substance L (MW 234) was 0.933 at 530 nm and 0.240 at 640 nm. Using the same cell, the *T* of an unknown solution was 0.661 at 530 nm and 0.381 at 640 nm. Calculate (a) the absorbances; (b) the molar absorptivities of K and L; (c) the molar concentrations of K and L in the unknown; (d) express these concentrations as ppm (mg/liter).

*Ans.* (c) K, $1.49 \times 10^{-4}$ *M*; L, $3.27 \times 10^{-4}$ *M*
(d) K, 25.3 ppm; L, 76.5 ppm

**21** Repeat problem 20 for the following solutions.

| | *Substance* | *Mol Wt* | *Concentration (M)* | *% T at $\lambda_1$* | *% T at $\lambda_2$* |
|---|---|---|---|---|---|
| (1) | S | 204 | $2.0 \times 10^{-5}$ | 26.9 | 95.5 |
| | T | 327 | $2.0 \times 10^{-5}$ | 91.2 | 22.9 |
| | Unknown | | | 63.1 | 44.7 |
| (2) | B | 100 | $3.0 \times 10^{-3}$ | 93.3 | 19.9 |
| | C | 150 | $3.0 \times 10^{-3}$ | 12.3 | 83.2 |
| | Unknown | | | 33.1 | 41.7 |

**22** In steel analysis, the manganese and chromium in a 1.000-g sample are often determined by a simultaneous photometric procedure that involves oxidation to permanganate and dichromate followed by dilution to 100 mL and photometric measurements at 420 nm and 530 nm. A 1.000-g sample that contained 0.50% manganese and no chromium showed absorbances of 0.050 at 420 nm and 0.70 at 530 nm. A 1.000-g sample that contained 1.00% chromium and no manganese showed absorbances of 0.65 at 420 nm and 0.10 at 530 nm. All measurements were made in 1.5-cm cuvettes. Samples of steel run by the method gave the following absorbances at the two wavelengths. Calculate the percentage of manganese and chromium in each.

| *Sample Number* | *A at 420 nm* | *A at 530 nm* |
|---|---|---|
| 1 | 0.45 | 0.30 |
| 2 | 0.30 | 0.45 |
| 3 | 0.18 | 0.63 |
| 4 | 0.63 | 0.18 |
| 5 | 0.35 | 0.35 |

*Ans.* (2) Mn, 0.44%; Cr, 0.63%

**23** The following data were obtained when a mole-ratio study was made for an iron(III) complex system. Plot $A$ versus mole ratio and determine the formula of the complex.

| *A* | *Mole Ratio Ligand to Metal* | *A* | *Mole Ratio Ligand to Metal* |
|---|---|---|---|
| 0.110 | 0.5 | 0.575 | 3.0 |
| 0.210 | 1.0 | 0.610 | 3.5 |
| 0.315 | 1.5 | 0.625 | 4.0 |
| 0.420 | 2.0 | 0.625 | 4.5 |
| 0.525 | 2.5 | 0.625 | 5.0 |

*Ans.* 2.95 or 3 to 1

**24** Repeat problem 23 for the following mole-ratio studies.

| *Mole Ratio Ligand to Metal* | *Absorbance Study Number* | | | |
|---|---|---|---|---|
| | (1) | (2) | (3) | (4) |
| 0.5 | 0.100 | 0.255 | 0.310 | 0.480 |
| 1.0 | 0.265 | 0.510 | 0.620 | 0.960 |
| 1.5 | 0.400 | 0.765 | 0.920 | 1.000 |
| 2.0 | 0.530 | 1.020 | 1.120 | 1.000 |
| 2.5 | 0.663 | 1.273 | 1.200 | 1.000 |
| 3.0 | 0.750 | 1.450 | 1.200 | 1.000 |
| 3.5 | 0.800 | 1.500 | 1.200 | 1.000 |
| 4.0 | 0.800 | 1.500 | 1.200 | 1.000 |
| 4.5 | 0.800 | 1.500 | 1.200 | 1.000 |

**25** The formula of a complex was determined by the continuous variation method using various volumes of $5 \times 10^{-4}$ $M$ solutions of the metal (M) and the ligand (L) followed by color development, dilution to 100 mL, and measurement at the wavelength of absorption of the complex. The following data were recorded. Plot the data and determine the formula of the complex.

| Metal (mL) | Ligand (mL) | A |
|---|---|---|
| 0 | 10 | 0.000 |
| 1 | 9 | 0.379 |
| 2 | 8 | 0.670 |
| 3 | 7 | 0.650 |
| 4 | 6 | 0.555 |
| 5 | 5 | 0.465 |
| 6 | 4 | 0.368 |
| 7 | 3 | 0.275 |
| 8 | 2 | 0.183 |
| 9 | 1 | 0.087 |
| 10 | 0 | 0.000 |

*Ans.* $ML_4$

**26** Repeat problem 25 for the following continuous variation studies:

| | | Absorbance Study Number | | |
|---|---|---|---|---|
| Metal (mL) | Ligand (mL) | (1) | (2) | (3) |
| 0 | 10 | 0.000 | 0.000 | 0.000 |
| 1 | 9 | 0.115 | 0.158 | 0.195 |
| 2 | 8 | 0.227 | 0.302 | 0.395 |
| 3 | 7 | 0.342 | 0.367 | 0.595 |
| 4 | 6 | 0.449 | 0.319 | 0.580 |
| 5 | 5 | 0.485 | 0.265 | 0.484 |
| 6 | 4 | 0.455 | 0.215 | 0.385 |
| 7 | 3 | 0.345 | 0.160 | 0.292 |
| 8 | 2 | 0.234 | 0.108 | 0.200 |
| 9 | 1 | 0.123 | 0.055 | 0.103 |
| 10 | 0 | 0.000 | 0.000 | 0.000 |

**27** The arsenate in a particular product is titrated photometrically with a bromate–bromide solution. Under the conditions chosen, only an inert impurity and the titrant absorb. A 0.200-g sample of the material was dissolved and diluted to 200 mL. A 10.00-mL aliquot was titrated with 0.00500 *N* bromate–bromide solution and the following data (corrected for dilution) were recorded:

| Added (mL) | A | Added (mL) | A | Added (mL) | A |
|---|---|---|---|---|---|
| 0.00 | 0.150 | 3.80 | 0.174 | 5.00 | 0.375 |
| 1.50 | 0.150 | 4.00 | 0.200 | 5.50 | 0.460 |
| 3.00 | 0.150 | 4.20 | 0.230 | 6.00 | 0.550 |
| 3.50 | 0.157 | 4.50 | 0.289 | 7.00 | 0.727 |

(a) Plot the data on rectilinear graph paper and determine the end point.
(b) Calculate the percent of As in the product. In the titration, the equivalent weight of arsenic is the atomic weight divided by 2 or 37.5.

**28** A 20.00-mL aliquot of HCl was diluted to 100 mL and titrated with 0.0100 *N* NaOH using phenolphthalein as indicator, with the following results:

| NaOH Added (*mL*) | *A* | NaOH Added (*mL*) | *A* | NaOH Added (*mL*) | *A* |
|---|---|---|---|---|---|
| 0.00 | 0.10 | 2.50 | 0.20 | 5.00 | 0.66 |
| 0.50 | 0.10 | 3.00 | 0.30 | 5.50 | 0.69 |
| 1.00 | 0.10 | 3.50 | 0.40 | 6.00 | 0.70 |
| 1.50 | 0.11 | 4.00 | 0.50 | 6.50 | 0.70 |
| 2.00 | 0.14 | 4.50 | 0.60 | 7.00 | 0.70 |

(a) Plot the data and determine the end point graphically, assuming that the absorbances have been corrected for dilution.
(b) Calculate the normality of the acid.

**29** A 10.00-mL sample of a weak acid was diluted to 100 mL and titrated with 0.025 *N* NaOH using bromthymol blue as the indicator, with the following results:

| NaOH Added (*mL*) | *A* | NaOH Added (*mL*) | *A* | NaOH Added (*mL*) | *A* |
|---|---|---|---|---|---|
| 0.00 | 0.200 | 3.00 | 0.245 | 5.00 | 0.800 |
| 1.00 | 0.200 | 3.50 | 0.500 | 5.50 | 0.800 |
| 2.00 | 0.200 | 4.00 | 0.775 | 6.00 | 0.800 |
| 2.50 | 0.210 | 4.50 | 0.795 | | |

(a) Graph the data and select the proper end point, assuming that the absorbances have been corrected for dilution.
(b) Calculate the normality of the acid.

**30** A 0.1027-g sample of the salt of a very weak monobasic acid is titrated with 0.1035 *N* HCl and the absorbance measured after each addition of HCl at a wavelength at which the weak acid absorbs. The data, after correction of the absorbances for dilution, are

| HCl (*mL*) | *A* | HCl (*mL*) | *A* | HCl (*mL*) | *A* |
|---|---|---|---|---|---|
| 0.00 | 0.000 | 2.50 | 0.501 | 4.00 | 0.597 |
| 1.00 | 0.198 | 3.00 | 0.552 | 4.50 | 0.600 |
| 2.00 | 0.395 | 3.50 | 0.585 | 5.00 | 0.600 |

(a) Plot the data and determine the end point.
(b) Calculate the equivalent weight of the acid.

**31** A 10.00-mL iron sample is diluted and titrated with 0.0505 *M* EDTA using salicylic acid as the indicator. The absorbance is measured after each addition of EDTA at the wavelength at which the iron salicylate complex absorbs. The corrected absorbances are:

| EDTA (*mL*) | *A* | EDTA (*mL*) | *A* | EDTA (*mL*) | *A* |
|---|---|---|---|---|---|
| 0.00 | 0.900 | 4.00 | 0.340 | 6.00 | 0.200 |
| 1.00 | 0.763 | 4.50 | 0.272 | 6.50 | 0.200 |
| 2.00 | 0.622 | 5.00 | 0.230 | 7.00 | 0.200 |
| 3.00 | 0.481 | 5.50 | 0.210 | | |

(a) Plot the data and determine the end point.
(b) Calculate the molar concentration of the iron solution, assuming that iron forms a 1 : 1 molar complex with EDTA.

CHAPTER

# 17 Infrared and Raman Spectrometry

The total energy of a molecule is simply the sum of all the possible forms of energy that the molecule may possess. In equation form, this may be represented as the following sum:

$$E_{tot} = E_{trans} + E_{elec} + E_{vib} + E_{rot} \qquad \textbf{(17-1)}$$

where $E_{trans} = E_{translational}$ and is the kinetic energy of the molecule; $E_{elec} = E_{electronic}$ and is the energy due to the arrangement of electrons in the molecule; $E_{vib} = E_{vibrational}$ and is the energy the molecule possesses due to vibrational motion; and $E_{rot} = E_{rotational}$ is the energy the molecule possesses due to rotational motion. $E_{trans}$ is dependent only on the kinetic energy of the molecule, and therefore does not provide information about molecular structure; however, $E_{elec}$, $E_{vib}$, and $E_{rot}$ are all quantized, and therefore depend on molecular structure. Since the vibrational states of a molecule are characteristic of the molecular structure, vibrational spectroscopy is invaluable in molecular structure determinations.

There are two techniques by which the vibrational states of a molecule may be observed. The first is by observing the absorption of infrared radiation, a technique known as **infrared spectrophotometry**. Because infrared absorption bands are characteristically sharp, their position in the spectrum is the basis for the identification of molecular species. The second technique involves observing frequency shifts produced when radiation is scattered from molecules, a technique known as **Raman spectroscopy**. Infrared absorption and Raman scattering follow different quantum mechanical rules, thus, the information obtained from Raman spectroscopy often complements that obtained from infrared spectrophotometry.

# INFRARED SPECTROPHOTOMETRY

## ABSORPTION OF INFRARED RADIATION

If the energy of a photon is equal to the energy difference between two quantized energy levels of the molecule, it is possible under certain circumstances for the molecule to absorb the photon. Since the energy of a photon is $h\nu$, where $h$ is Planck's constant and $\nu$ is the frequency of the radiation, photon absorption will be possible whenever

$$\Delta E = h\nu = \frac{hc}{\lambda} = hc\bar{\nu} \qquad \textbf{(17-2)}$$

where $\Delta E$ is the energy difference between the two quantized levels; $c$ is the speed of light; $\lambda$ is the wavelength of the radiation; and $\bar{\nu}$ is the wave number of the radiation, which is the reciprocal of the wavelength in centimeters.

Planck's relationship indicates that the energy of electromagnetic radiation depends on the wavelength of the radiation. By comparing the energies corresponding to different states of the molecule with the energies and wavelengths of electromagnetic radiation, it is possible to correlate particular molecular transitions with the wavelength of the radiation absorbed. For example, when a molecule absorbs a photon of visible or ultraviolet radiation, this corresponds to the energy difference between two electronic states in the molecule, resulting in a rearrangement of the valence electrons from the ground state configuration to some excited state configuration. Figure 17-1 shows the quantized energy states of a hypothetical diatomic molecule. It can be seen from the figure that $\Delta E_{\text{elec}} > \Delta E_{\text{vib}} > \Delta E_{\text{rot}}$. Using eq. (17-2) to calculate the wavelength of radiation corresponding to changes in the vibrational state of the molecule reveals that radiation in the infrared region of the spectrum is required. The infrared

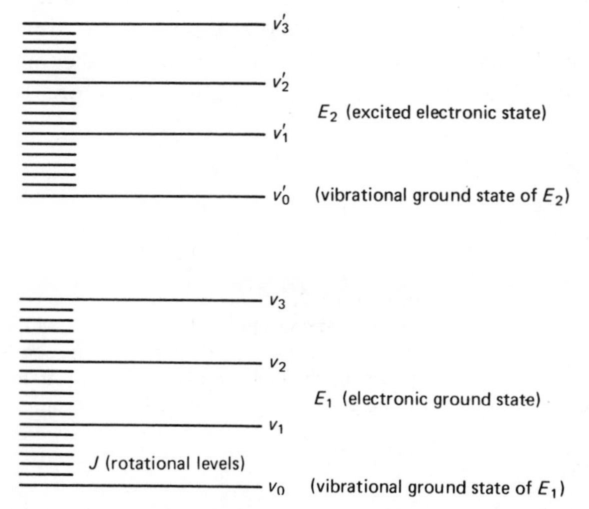


**Figure 17-1.** Quantized energy states for a hypothetical diatomic molecule.

region covers from 0.8 $\mu$m to 200 $\mu$m. Absorption of infrared radiation involves changes in the vibrational stretching or bending modes within a molecule or, at longer wavelengths, changes in molecular rotation.

In addition to the energy-matching requirement discussed previously, the electromagnetic radiation field must be capable of performing work on the molecule if absorption is to occur. For electrical work to be done on the molecule by the absorption of radiation, the transition must be accompanied by a change in the electrical center of the molecule. Thus, for molecules to absorb infrared radiation, there must be a change in the dipole moment of the molecule as it vibrates. The dipole moment of a molecule is determined by the positions of the positive and negative electrical centers of the molecule in relation to the molecular center of gravity. Any vibration that involves a change in the relative positions of the positive and negative centers with respect to the molecular center of gravity will cause a change in the dipole moment of the molecule as it vibrates. The molecule may therefore be thought of as an oscillating dipole, where the frequency of oscillation is the natural frequency of vibration of the molecule. From a classical viewpoint, absorption will occur when the frequency of the oscillating dipole due to molecular vibration is the same as the frequency of the electromagnetic radiation.

From a classical viewpoint, a vibrating molecule may be represented by a collection of balls and springs joined together as shown in Figure 17-2, where the atoms are represented by balls of different weight and the bonds by springs of different tension—each in proper relation to the atomic masses and bond strengths. If such a model is set in motion, the resulting motion may appear quite complex. However, even the most complex overall vibrational motion may be resolved into a number of fundamental or normal modes of vibration. If the vibrating model is viewed with light flashes of varying frequency, individual balls will appear stationary at certain frequencies. These particular frequencies are characteristic of the normal modes of vibration of the molecule.

It is reasonably simple to predict the number of normal modes of vibration for a molecule because the normal modes of vibration constitute the minimum amount of information required to describe the complex overall vibrational motion of the molecule. The total motion of the molecule is composed of translational motion, vibrational motion, and rotational motion. The amount of

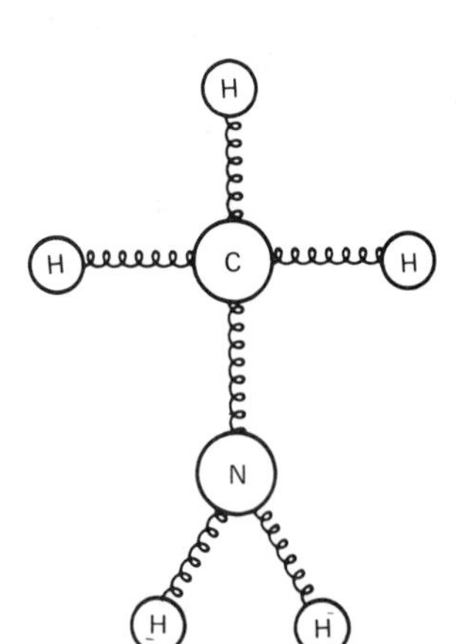


**Figure 17-2.** Schematic representation of methylamine by ball-and-spring model. This molecule will absorb at wavelengths characteristic of the bending modes of the methyl group and the amine group and of the stretching modes due to the C—H, C—C, C—N, and N—H groups.

information required to describe the motion of the molecule consists of the amount of information required to describe the translational motion of the molecule plus the amount of information required to describe the vibrational motion of the molecule plus the amount of information required to describe the rotational motion of the molecule. Since three coordinates are required to describe the position of the center of gravity of the molecule, three pieces of information must be specified to describe translation. For a nonlinear molecule, rotational motion may be described in terms of rotation about the $x$, $y$, and $z$ axes; hence three pieces of information are required to describe rotational motion. For a molecule consisting of $n$ atoms joined together, the maximum amount of information available consists of the $3n$ coordinates required to describe the position in space of each atom. Hence, the amount of information required to describe the vibrational motion must be

$$3n - \text{translational information} - \text{rotational information} \qquad \textbf{(17-3)}$$

or

$$3n - 6$$

For a linear molecule, there are $3n - 5$ modes of vibration, since only two pieces of information are required to describe the rotational motion of a linear molecule.

For a nonlinear molecule, we would therefore expect to observe $3n - 6$ absorption bands in the infrared spectrum at frequencies corresponding to the normal modes of vibration. It is not uncommon, however, for the observed number of absorption bands to be less than the calculated number predicted by eq. (17-3). Some factors that contribute to this discrepancy are as follows:

1 Some vibrations may be infrared-inactive (i.e., do not involve a change in the dipole moment).

2 One or more vibrations may have exactly the same energy (i.e., be degenerate).

3 Some vibrations may not be resolved by the instrument.

4 Some vibrations may be too weak to observe.

5 Some vibrations may lie outside the range of the instrument.

The two types of natural modes of vibration in molecules are the stretching and bending of the bonds between two atoms. Stretching is a rhythmical movement of the two atoms back and forth along the bond axis, and bending is a change in bond angles either between two atoms attached to a third atom or between a group of atoms and the rest of the molecule. The various types of bending and stretching modes are illustrated in Figure 17-3.

Stretching vibrations of two similar bonds joined by a common atom may be symmetrical or asymmetrical and thus cause two absorption bands, which are usually close together. An example is H—C—H in which the two hydrogens both move toward (or away from) the carbon at the same time or in which the one hydrogen moves toward while the other moves away from the carbon atom.

Four types of bending are possible for two atoms joined to a third atom or between a group of atoms and the molecule as a whole:

1 **Rocking**—the structural unit swings back and forth in the plane of the molecule. This also is known as **in-plane bending**.

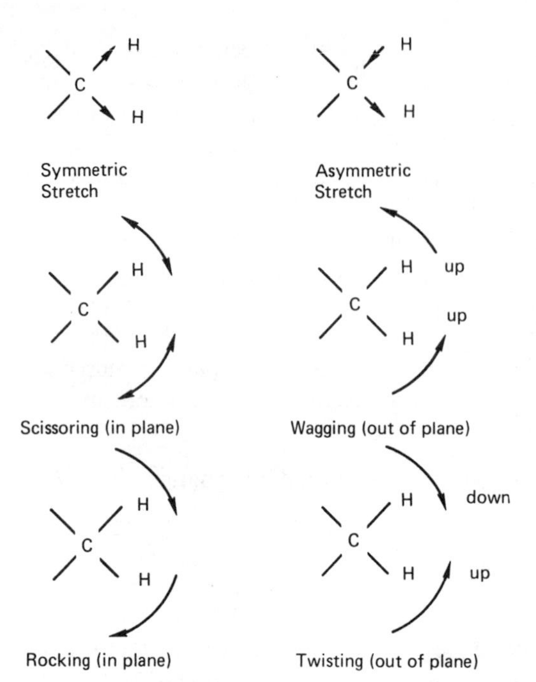


**Figure 17-3.** Vibrational modes of carbon and hydrogen. Up represents motion upward from the plane of the page; down represents motion downward from the plane of the page.

2 **Wagging**—the structural unit swings back and forth out of the plane of the molecule. This is also known as **out-of-plane bending**.

3 **Twisting**—rotation of the structural unit around the bond joining it to the molecule.

4 **Scissoring**—two atoms joined to a central atom move toward or away from each other, causing deformation of the bond angles.

The general regions of some common bending and stretching frequencies for carbon, hydrogen, and oxygen are shown in Figure 17-4 in bar graph form. Wavelengths are shown at the top and wave numbers at the bottom. Note that the wavelength scale is linear whereas the wave number scale is not.

The position of stretching absorption bands can be predicted fairly well by the use of **Hooke's law** for harmonic oscillators:

$$\bar{\nu}\ (\text{in cm}^{-1}) = \frac{1}{2\pi c}\sqrt{\frac{k(m_1 + m_2)}{m_1 m_2}} \qquad \textbf{(17-4)}$$

where $k$ is the force constant in dynes/cm; $c$ is the velocity of light in cm/sec; and $m_1$ and $m_2$ are the masses in grams of the individual adjacent atoms. The force constant $k$ can be estimated by **Gordy's rule**:

$$\frac{k}{10^5} = aN\left(\frac{\text{EN}_\text{A}\,\text{EN}_\text{B}}{d^2}\right)^{3/4} + b \qquad \textbf{(17-5)}$$

where $N$ is the bond order (number of bonds acting between the two atoms); $\text{EN}_\text{A}$ and $\text{EN}_\text{B}$ are the Pauling electronegativities of atoms A and B; $d$ is the

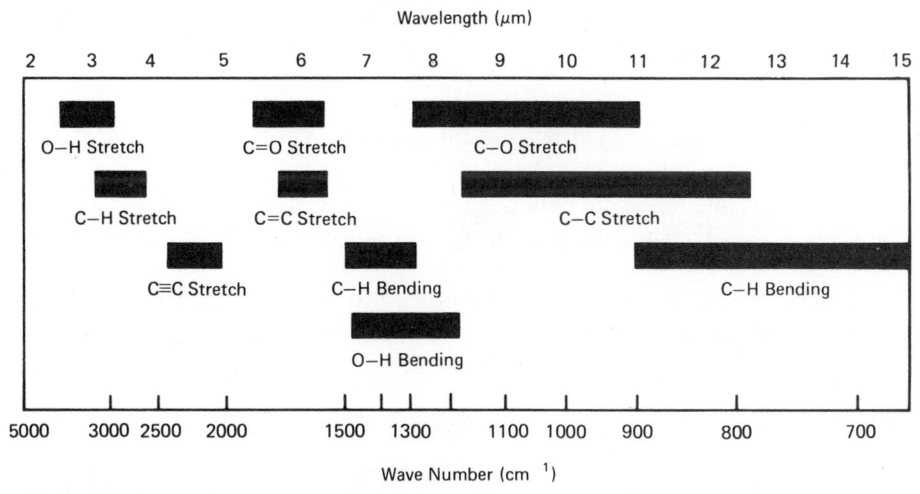


**Figure 17-4.** General regions for absorption in the infrared for some carbon, oxygen, and hydrogen group vibrations.

internuclear distance in angstroms; and the constants $a$ and $b$ are 1.67 and 0.30, respectively, for stable molecules exerting their normal covalency.

• **Example 1.** Calculate the force constant and the stretching frequency for the C—C bond, for which $EN_C = 2.5$, $d = 1.33$ Å, and $N = 1$.

$$\frac{k}{10^5} = 1.67 \times 1\left(\frac{2.5 \times 2.5}{(1.33)^2}\right)^{3/4} + 0.30$$

$$= 1.67\left(\frac{6.25}{1.77}\right)^{3/4} + 0.30$$

$$= 1.67 \times 2.58 + 0.30 = 4.30 + 0.30$$

$$= 4.60$$

This value can be substituted in eq. (17-4) to calculate the position of the absorption band in wavenumbers as follows:

$$\bar{\nu} = \frac{1}{2 \times 3.142 \times 3.0 \times 10^{10}} \times \sqrt{\frac{4.6 \times 10^5[(12/6.023 \times 10^{23}) + (12/6.023 \times 10^{23})]}{(12/6.023 \times 10^{23})(12/6.023 \times 10^{23})}}$$

$$= \frac{1}{1.885 \times 10^{11}} \sqrt{\frac{4.60 \times 10^5 \times 24 \times 6.023 \times 10^{23}}{144}}$$

$$= \frac{\sqrt{6.023 \times 10^{28}}}{1.885 \times 10^{11}} \times \sqrt{\frac{4.60 \times 24}{144}}$$

$$= 1302\sqrt{0.767} = 1302 \times 0.876 = 1140 \text{ cm}^{-1}$$

If a large number of calculations are to be made, note that, for single bonds, eq. (17-4) reduces to

$$\bar{\nu} = 1302\sqrt{\frac{k}{10^5} \times \frac{\text{sum of atomic weights}}{\text{product of atomic weights}}} \tag{17-6}$$

Force constants for single bonds are about 4–6 × 10⁵ dynes/cm, for double bonds 8–12 × 10⁵ dynes/cm, and for triple bonds approximately 12–18 × 10⁵ dynes/cm.

Bending vibrational modes usually absorb at lower frequencies than the stretching vibrational modes, as shown in Figure 17-4.

## PRINCIPLES OF INFRARED INSTRUMENT DESIGN

Although infrared spectrophotometers have the same basic components as instruments used in the visible and ultraviolet regions, there are certain noticeable differences. As with ultraviolet–visible spectrophotometers, conventional infrared instruments consist of a source, a wavelength-dispersing device, a sample cell, a detector, and a readout device. One difference between ultraviolet–visible and infrared spectrophotometers concerns the placement of the sample cell. For example, in most ultraviolet–visible spectrophotometers, the sample cell is placed after the monochromator to avoid exposing the sample to the full spectrum of the light source (which might cause photodecomposition of the sample). In contrast, virtually all infrared spectrophotometers place the sample before the monochromator in an effort to reduce the amount of stray light emitted by the sample cell that may strike the detector. Since most infrared instruments are used to record the absorption spectrum of compounds, most instruments are double-beam systems that permit automatic recording. Single-beam instruments are somewhat impractical in the infrared (except for certain

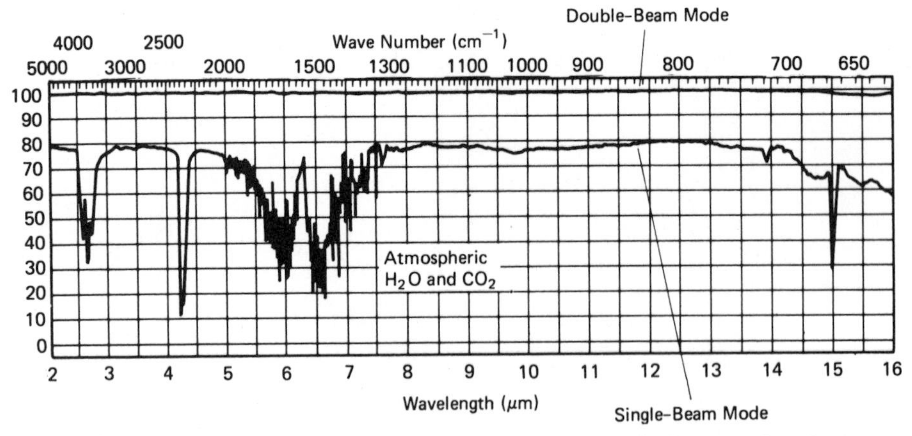


**Figure 17-5.** Effect of atmospheric water vapor and carbon dioxide on infrared spectra obtained in single- and double-beam modes. (Reprinted from E. Olsen, *Modern Optical Methods of Analysis*, New York: McGraw-Hill, 1975 with permission.)

applications) because of the absorption of infrared radiation by atmospheric water vapor and carbon dioxide present in the optical path. Figure 17-5 shows the effect of atmospheric water vapor and carbon dioxide on the infrared spectrum obtained in single- and double-beam modes with no sample present.

A final difference concerns the use of optical null detection. Most double-beam infrared spectrophotometers employ an optical null principle in which the radiant energy in the reference beam is matched with the radiant energy in the sample beam after the radiation has passed through the sample. In a region where the sample absorbs, the radiant energy of the sample beam will be reduced. To maintain equal energies in both the reference and sample beams, a wedge-shaped comb is inserted in the reference beam by an automatic mechanism, until both beams again have the same radiant energy. The use of an optical null system can result in a condition known as a "dead pen," if the sample absorbs strongly. That is, if the sample absorbs strongly, there will be little radiant energy in the sample beam and the reference beam will be almost completely blocked by the comb, so that the radiant energies in both beams are equal. In this situation, so little radiation strikes the detector that the system cannot find null balance and the pen fails to respond.

## *Sources*

The best sources of infrared radiant energy approximate black body radiation. Nernst glowers, Globars, and Nichrome heaters are used widely in commercial instruments. The **Nernst glower** is a mixture of zirconium and yttrium oxides formed in a small hollow rod, which, when maintained at a temperature of 1500–2000°C by an electrical current, furnishes maximum radiation at about 7100 $cm^{-1}$ (1.4 $\mu m$). The glower has a negative temperature coefficient of resistance and must be heated by a secondary heater before it will emit.

The **Globar** is a rod of sintered silicon carbide that furnishes maximum radiation at about 5200 $cm^{-1}$ (1.9 $\mu m$) when heated to temperatures of 1300–1700°C. It is self-starting and has a positive coefficient of resistance. The Globar is a less intense source than the glower but is preferable for wavelengths above 15 $\mu m$.

The **Nichrome heater** is made from coils or ribbons of Nichrome wire sealed in an inert atmosphere and heated to incandescence by passage of current. It operates in the range 800–900°C.

To offset stray light from short-wavelength (high-energy) radiation, the light beams must be chopped at low frequency (10 to 25 times per second), or double monochromation must be employed.

## *Detectors*

Thermal detectors are usually used, inasmuch as infrared radiation is in the low-energy region in which electromagnetic energy is readily converted into heat. In the higher energy regions of the near infrared, photoconductive cells respond better and are widely used in the special instruments designed primarily for the near-infrared region.

The **thermocouple** is the most popular thermal detector. A thermocouple works on the following principle: If two wires of different composition are joined at both ends and the ends are maintained at different temperatures, a potential will be developed. The magnitude of this potential is governed by the difference in temperature of the two junctions. In thermocouples used in the infrared region, the *hot* junction or receptor may consist of blackened gold foil attached to one weld of the two wires. The *cold* junction at the other end of the two wires is kept at constant temperature in the receiver housing. Light falling on the hot junction causes an increase in temperature, and the voltage produced is a measure of the intensity of the light.

Bolometers and thermistors are also used in Wheatstone bridge arrangements. One form of **bolometer** utilizes a thin film of a noble metal (usually platinum) evaporated onto a nonconducting backing material. Two such sensing devices, one exposed to the light, the other shielded from the light, act as two arms of a Wheatstone bridge circuit, as shown in Figure 17-6. Bolometers show a positive thermal coefficient of electrical resistance; that is, the resistance increases as the temperature increases. **Thermistors** are fused mixtures of metal oxides that have negative thermal coefficients of electrical resistance and may also be used as two arms of a Wheatstone bridge arrangement.

**Pneumatic cells**, of which the **Golay cell** is the best example, are filled with gases and can be either selective or nonselective. *Selective cells* are filled with an absorbing gas that sensitizes it to the corresponding gas in the gaseous mixture being determined. *Nonselective cells* are filled with gases that do not absorb appreciably and are heated by a solid absorbing receptor immersed in the gas. In the nonselective Golay cell, the radiation to be measured is absorbed by a blackened metal film located in the center of a small gas chamber filled

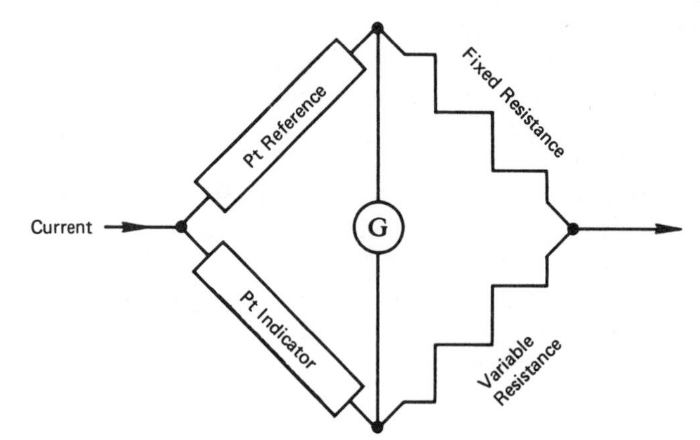


**Figure 17-6.** Schematic diagram of a bolometer using a Wheatstone bridge arrangement. The variable resistance is adjusted so that no current flows through the galvanometer (G) when the Pt resistance arm is not irradiated. Upon irradiation, the temperature and resistance increase, causing an imbalance, and current flows through the galvanometer. The increase in current can be measured or the resistance can be varied to restore balance. The resulting signal in either case is fed to a strip-chart recorder.

**TABLE 17-1.** Prism and Window Materials for the Infrared Region

| *Material* | *Approximate Range for Use as Windows* ($\mu m$) | *Optimum Range for Use as Prism* ($\mu m$) |
|---|---|---|
| Lithium fluoride | 0.1– 7.0 | 0.6– 6.0 |
| Barium fluoride | 0.2–13.5 | 0.3–13.0 |
| Sodium chloride | 0.2–17.0 | 0.2–15.4 |
| Potassium bromide | 0.2–26.0 | 10 –25 |
| Irtran (ZnS) | 0.2–13.0 | Not used |
| Silver chloride | 10 –25.0 | Not used |
| Cesium iodide | 1 –40 | 10 –38 |

with xenon; one side of the chamber is an expansible flexible mirror. The absorption of radiation warms the gas and the resulting pressure causes distortion of the flexible mirror. Light from an ordinary tungsten lamp is directed onto the mirror and reflected to an ordinary vacuum phototube. As the mirror is distorted, the amount of light falling on the photocell varies and the resulting difference in output is related to the intensity of the incident infrared radiation. The Golay cell can be used from the ultraviolet through the infrared regions if the proper window and construction materials are used.

### *Wavelength Selectors*

Both prisms and gratings are used as monochromators in the infrared region, but, since glass and quartz both absorb, other materials must be used except in the near infrared. No single material is suitable for windows and prisms over the entire infrared region, as shown in Table 17-1.

Sodium chloride is used more than any other material in commercial instruments, most of which cover from 1 or 2 $\mu m$ to 15 $\mu m$, since this is the region that is of prime importance in identification of functional groups in organic compounds.

All gratings used in infrared instruments are reflection gratings rather than transmission gratings owing to the low energies involved in this region. Front surface reflecting mirrors are used in place of lenses for the same reason.

## SAMPLE HANDLING

Beer's law holds in the infrared region the same as it does in the visible and ultraviolet regions, but the energies involved are much smaller and the absorption bands narrower in the infrared. As a consequence, the optical path lengths of the cells usually used are about 0.1 cm or less, except for gaseous samples. The narrow absorption bands combined with the short path lengths cause more instances of deviation from Beer's law in the infrared than in the visible or ultraviolet regions. Since both glass and quartz absorb in the infrared region, most cells are constructed with sodium chloride windows, which limits measurements to nonaqueous systems. Some insoluble materials such as silver chloride and Irtran are available for use as windows, but each has distinct disadvantages and is used only when absolutely necessary.

Liquid samples are usually run direct (neat) in cells with light paths of 0.01–1.00 mm. Cells with light paths in this range are prepared by sealing the windows with thin gaskets of copper and lead that have been treated with mercury. After the assembly is clamped together securely, the mercury penetrates the metals, causing expansion of the gasket and making a tight seal. Such cells are usually filled with a hypodermic needle. Both fixed light path and variable light path (demountable) cells are available. The small cell thicknesses (light paths) make it essentially impossible to repeat cell thicknesses exactly or to prepare a cell of an exact thickness. The length of the light path can be measured from the interference fringe patterns obtained by running the spectrum of the empty cell. For instruments linear in wave numbers, the length of the light path $b$ in centimeters is given by

$$b = \frac{n}{2(\bar{\nu}_1 - \bar{\nu}_2)} \tag{17-7}$$

where $n$ is the number of fringes between wave numbers $\bar{\nu}_1$ and $\bar{\nu}_2$. For instruments linear in wavelengths, the length of the light path in micrometers is given by

$$b = \frac{n\lambda_1\lambda_2}{2(\lambda_2 - \lambda_1)} \tag{17-8}$$

• **Example 2.** Calculate the lengths of the light path for a cell for which the interference pattern shows 17 fringes between 1000 and 1100 $cm^{-1}$ and for a cell for which there are 10 fringes between 9.0 $\mu m$ and 10.0 $\mu m$. Substitution in eqs. (17-7) and (17-8) gives

$$b = \frac{17}{2(1100 - 1000)} = \frac{17}{200} = 0.085 \text{ cm}$$

and

$$b = \frac{10 \times 9.0 \times 10.0}{2(10.0 - 9.0)} = \frac{900}{2 \times 1} = 450 \ \mu\text{m}$$

Solids are dissolved in nonaqueous solvents if possible. However, there is no single solvent that does not absorb in some portion of the infrared range, and care must be taken to ensure that there is no reaction of the solvent with the sample, such as hydrogen bonding. Substances that dissolve in both carbon tetrachloride and carbon disulfide can be run in these two solvents, since each is transparent in the regions in which the other absorbs. Selection of the proper solvent depends upon the solubility of the compound being determined and upon the region of interest. As an example, acetone could not be used as a solvent for a compound containing a ketone group, since it absorbs in this region (approximately 1600–1800 $cm^{-1}$).

Solids that can be ground to find powders may also be run as mulls or slurries with Nujol or white paraffin (mineral) oil. The suspension is encased in polyethylene film or pressed into the annular groove of a demountable cell. Solids may also be determined by the KBr pellet technique, in which the finely ground solid is mixed thoroughly with solid, high purity KBr and then compressed into a clear pellet by high pressure in an evacuable die.

## USE IN ANALYSIS

Measurements of infrared absorption are not used extensively for quantitative estimation, owing to the difficulties involved in obtaining reproducible measurements. In the first place, cell path lengths are often less than 0.1 mm, are hard to reproduce from cell to cell, and, unless the cell has plane parallel walls that allow measurement from interference patterns, difficult to estimate correctly. To offset this, measurements of the standard and the unknown are often made in the same cell. This presents its own set of problems.

The base line must be measured for each peak, since the base line does not remain at zero absorbance (100% transmittance) over the entire range. The problem of selecting the proper base line is illustrated in Figure 17-7, which shows three hypothetical curves and the possible base lines that may be drawn. Selection of the proper base line for curves (a) and (b) is relatively simple, the slanting base line chosen for (b) probably being used more than a horizontal base line such as is shown for curve (a). The complexity of curve (c) leads to the possibility of several base lines. Selection of the proper one is usually difficult unless a large number of measurements are made so that the reproducibility of the measurements using the various possible base lines can be studied to aid in the selection.

Since aqueous solutions cannot be used, the use of infrared for quantitative measurements is limited to those systems for which a nonaqueous solvent that does not absorb in the region of interest is available. The short optical path lengths used in the infrared combined with the use of very dilute solutions limits the use of infrared spectrometry in quantitative analytical procedures.

The main value of infrared measurements lies in the region of qualitative identification of substances, pure or in mixtures, and as a tool to aid in the determination of structure. This identification depends upon the fact that the infrared spectrum of a compound serves as a fingerprint of the compound and is unique to that compound. Because functional groups tend to absorb in definite regions, the spectra can also be used to aid in the determination of structure. Both of these aspects are covered thoroughly in several places in the literature and will not be considered in detail in this text. There are also several "libraries" or collections of infrared spectra, containing literally thousands of spectra that can be used to aid in the identification once the infrared spectrum of a compound has been determined.

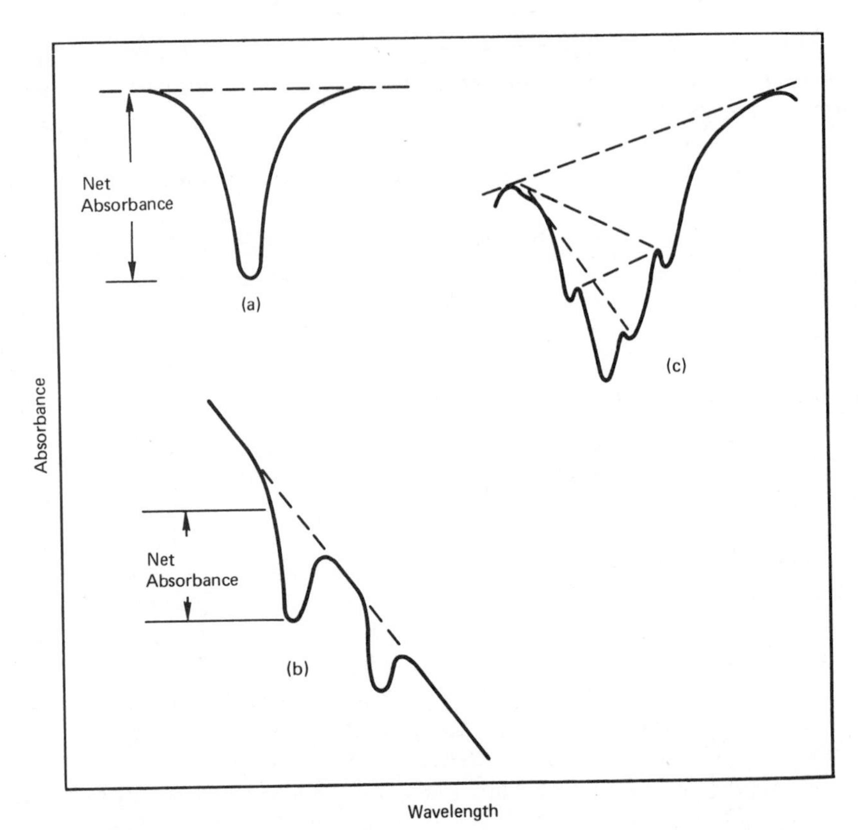


**Figure 17-7.** Base-line selection for infrared absorbance curves. The selection of the proper base-line is obvious in (a) and (b) but not in (c).

# RAMAN SPECTROSCOPY

## THEORY

Suppose that a collection of molecules is irradiated with a beam of photons whose energies are such that they cannot be absorbed by the given molecules. When a photon from this beam collides with a molecule, two phenomena are observed:

If the photon undergoes an elastic collision with the molecule, the photon will suffer a change in direction but otherwise will remain the same. This phenomenon is known as **Rayleigh scattering**, and is responsible for the sky appearing blue (shorter wavelength light scatters more readily than longer wavelength light).

If the photon undergoes an inelastic collision with a molecule, not only is the direction of travel of the photon altered (i.e., the photon is scattered) but the energy of the photon is changed as well. This effect is known as **Raman scattering**, after C. V. Raman who observed the phenomenon in 1928. Because elastic collisions occur more frequently than inelastic collisions, Rayleigh scattering is more intense than Raman scattering. If the energy of the photon is altered during Raman scattering, this means that the frequency of the radiation must also be altered.

To understand how the Raman effect is related to molecular vibrations, we must examine the scattering process in more detail. In any collection of molecules at room temperature, most of the molecules will be in the lowest vibrational energy level or the **vibrational ground state**. If a molecule in the vibrational ground state undergoes an inelastic collision with a photon, part of the photon energy will be transferred to the molecule, leaving it in a **vibrationally excited state**. Photons scattered in this manner will have a lower energy and therefore a lower frequency than the incident photons. The difference in energy between the incident and Raman scattered photons will be equal to the energy difference between the vibrational ground state and the particular vibrational excited state.

A few of the molecules that undergo inelastic collisions with photons may initially already be in a vibrationally excited state. Thus, during an inelastic collision, the molecule may transfer its vibrational energy to the photon, thereby increasing the energy and frequency of the photon. Photons scattered in this manner will have a higher frequency than the incident photons. Again, the energy difference between the incident and scattered photons will correspond to the difference between the initial vibrationally excited state and the ground vibrational state. These considerations are shown in Figure 17-8. Since most of

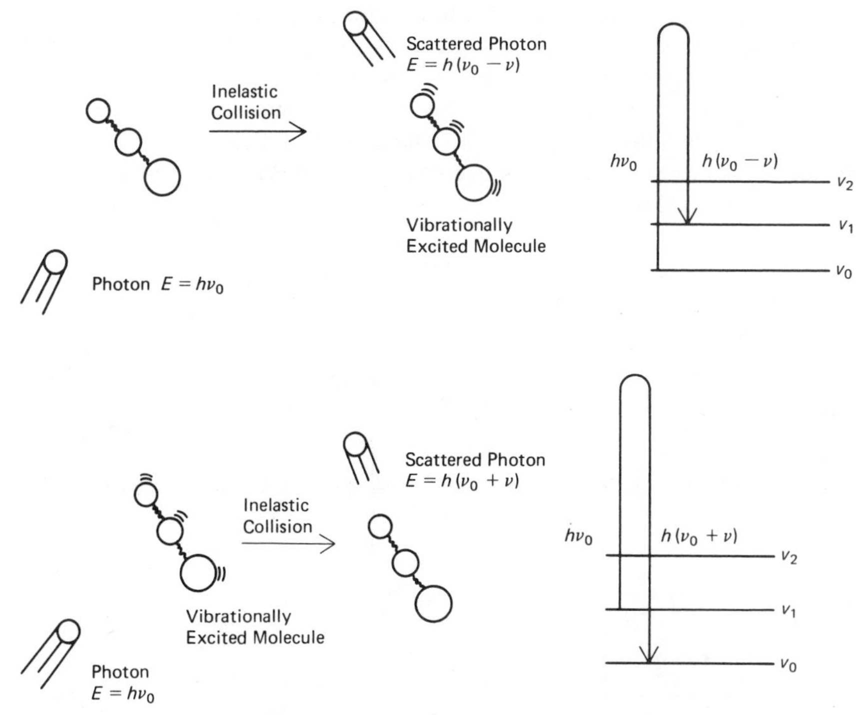


**Figure 17-8.** Schematic diagram of Raman effect.

the molecules are found in the ground vibrational state, most of the Raman-scattered photons have frequencies less than the incident photons.

Raman scattering will not necessarily be observed for all possible vibrational modes of the molecule. As with similar phenomena, the Raman scattering phenomenon is governed by quantum mechanical considerations. The criteria that determine whether a particular radiational interaction will occur are generally known as the **quantum mechanical selection rules**. The quantum mechanical selection rule governing Raman scattering states that Raman scattering corresponding to a particular vibrational mode will occur only if the polarizability of the molecule changes during the particular molecular vibration. The term *polarizability* refers to the ease with which the electron cloud surrounding the molecule may be repelled or attracted by the presence of an electric field.

## Principles of Raman Instrumentation

The Raman effect is reasonably weak, and sensitive instrumentation must be used to detect it. Figure 17-9 shows a schematic diagram of a typical Raman spectrometer, consisting of a monochromatic radiation source, a sample cell, a high-quality monochromator, and some means of photodetection.

In the early days of Raman spectroscopy, most instruments employed a mercury arc as a radiation source and photographic plates for photodetection. Since the intensity of the Raman scattering is directly proportional to the intensity of the radiation source, a high-intensity monochromatic source is required. Therefore, most instruments today employ a helium–neon laser as the radiation source, together with photoelectric photodetection. In addition to being intense, the helium–neon laser offers other advantages. For example, laser radiation is monochromatic. With the helium–neon laser, the wavelength of the emitted radiation is 632.8 nm, which is almost an ideal compromise excitation wavelength. It is important to remember in Raman spectroscopy that the molecule should not absorb the excitation wavelength to avoid interference

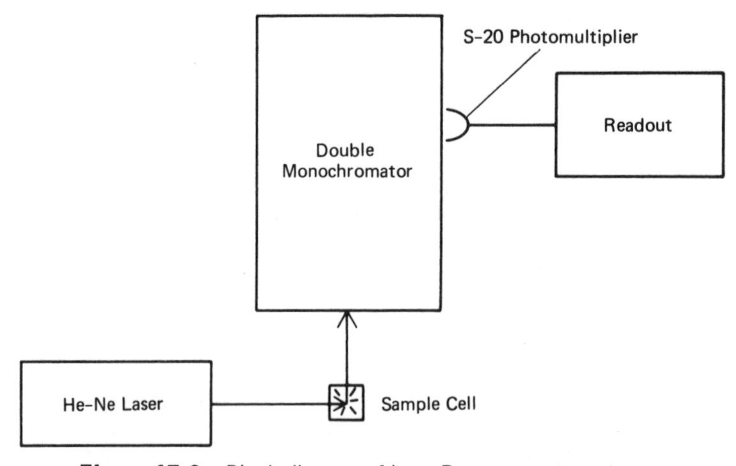


**Figure 17-9.** Block diagram of laser Raman spectrometer.

from molecular fluorescence. Because of its relatively long wavelength, this orange-red line is not absorbed, even by most colored substances. Nevertheless, the wavelength is not too long to prevent photodetection with a long-wavelength response photomultiplier. Furthermore, because of the highly collimated nature of laser radiation, the Raman spectrum of samples as small as 1 microliter ($\mu$L) may be determined.

Since Rayleigh scattering is about 1000 times more intense than Raman scattering, a high-quality monochromator with a very low stray light level is necessary to separate the weak Raman spectrum from the relatively intense Rayleigh line. To avoid interference from stray light, most instruments employ some form of double monochromation, such as two monochromators in series.

Even the best photomultipliers with an S-20 spectral response will not respond to wavelengths above 850 nm. Because $\bar{\nu} = 1/\lambda$ where $\lambda$ is the wavelength in centimeters, this corresponds to approximately 11,800 $cm^{-1}$. If a helium–neon laser is used as the excitation source, with an excitation line at 632.8 nm (15,800 $cm^{-1}$), the largest Raman shift that can be detected will be 4000 $cm^{-1}$, which corresponds to 2.5 $\mu$m. For this reason, most laser Raman spectrometers cover the spectral region from 100 to 4000 $cm^{-1}$, which corresponds to 2.5–100 $\mu$m.

## APPLICATIONS AND COMPARISONS

Table 17-2 compares Raman spectrometry with infrared spectrophotometry. Like its counterpart, infrared spectrophotometry, Raman spectrometry is most often used as a tool to obtain structural information on chemical substances and for qualitative analysis. Since the wavelengths employed in Raman spectrometry are in the visible region, it is unnecessary to resort to hygroscopic materials,

**TABLE 17-2.** Comparison of IR and Raman

| | *Infrared* | *Raman* |
|---|---|---|
| Selection rule | Change in dipole moment during vibration | Change in polarizability during vibration |
| Sample size | 1–10 mg solid, 0.8 $\mu$L–1 mL liquid | 1 $\mu$L–2.5 mL liquid |
| Sample type | Solids, liquids, gases | Liquids best, but solids and gases can be used |
| Aqueous samples | No | Yes |
| Wavelength range of typical instrument | 4000–625 $cm^{-1}$ | 4000–100 $cm^{-1}$ |

**TABLE 17-3.** Frequency Assignment Chart for Raman Spectra ($\phi$ = benzene ring; mono $\phi$ = mono substituted benzene; 1,2, $\phi$ = 1,2 disubstituted benzene; $\Delta$ = Raman shift in $cm^{-1}$)

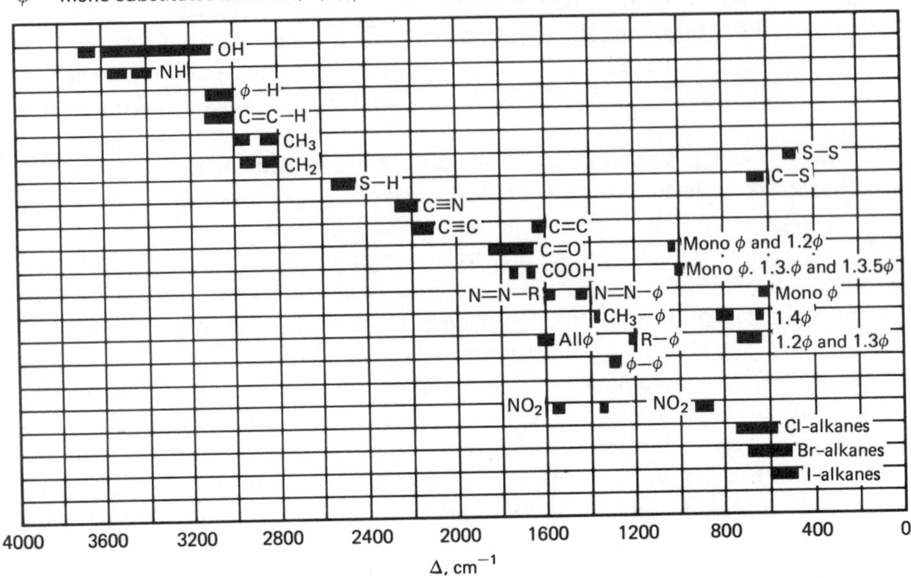


*Source:* E. Olsen, *Modern Optical Methods of Analysis*. New York: McGraw-Hill, 1975, with permission.

such as sodium chloride, for sample-cell windows. In contrast to infrared spectrophotometry, where water absorbs strongly, Raman spectrometry can be carried out in aqueous solutions and is therefore useful in biochemical studies. Table 17-3 shows a frequency assignment table for Raman spectra, similar to the group-frequency chart (Figure 17-4) given for infrared spectrophotometry. Although Raman spectra can be obtained on gaseous, liquid, and solid samples, liquid samples are the most convenient, with sample volumes from 1 $\mu$L to 2.5 mL commonly being used in laser Raman work. To obtain adequate spectra in Raman work, it is usually better to employ as concentrated a sample as possible.

## PROBLEMS

**1** Convert the following wavelengths to wave numbers in $cm^{-1}$: (a) 2.5 $\mu$m; (b) 4.0 $\mu$m; (c) 5.5 $\mu$m; (d) 6.0 $\mu$m; (e) 8.0 $\mu$m; (f) 20 $\mu$m.

*Ans.* (b) 2500 $cm^{-1}$

**2** Convert the following wave numbers to wavelengths in $\mu$m: (a) 3500 $cm^{-1}$, (b) 2300 $cm^{-1}$; (c) 1750 $cm^{-1}$; (d) 1400 $cm^{-1}$; (e) 1000 $cm^{-1}$; (f) 600 $cm^{-1}$.

*Ans.* (b) 4.35 $\mu$m

**3** Calculate the force constants for the following bonds using Gordy's rule. The numbers in parentheses are the bond lengths in angstroms and the Pauling electronegativities of the atoms, respectively: (a) C—H (1.05; 2.5, 2.1); (b) P=O (1.45; 2.1, 3.5); (c) C—Cl (1.76; 2.5, 3.0); (d) N—O (1.36; 3.0, 3.5); (e) N=O (1.15; 3.0, 3.5); (f) C≡C (1.22; 2.5, 2.5).

*Ans.* (b) $8.8 \times 10^5$ dynes/cm

**4** Calculate the wavenumbers of the absorption bands for the following bonds from the force constants given: (a) $k = 4.6 \times 10^5$ dynes/cm for C=C in $C_2H_4$; (b) $k = 3.4 \times 10^5$ dynes/cm for C—Cl in $CCl_4$; (c) $k = 5.6 \times 10^5$ dynes/cm for C—H in $C_2H_6$; (d) $k = 7.9 \times 10^5$ dynes/cm for C—C in $C_6H_6$; (e) $k = 12.1 \times 10^5$ dynes/cm for C=O in HCHO; (f) $k = 17.2 \times 10^5$ dynes/cm for C≡N in $CH_3CN$.

*Ans.* (b) 802 $cm^{-1}$

CHAPTER

# 18 Analytical Flame Spectrometry

Analytical flame spectrometry is extensively used in science, industry, agriculture, and medicine for the rapid, routine determination of over 60 metallic elements in a wide variety of sample types. The widespread use of these techniques is based on several factors: (1) the apparatus required is relatively inexpensive and easy to operate; (2) the technique is rapid, allowing large numbers of samples to be conveniently determined in a short time period; and (3) the technique is capable of determining trace quantities (less than 100 ppm) of metallic species, often without requiring prior chemical separation or pretreatment to avoid interference from other concomitants.

In this chapter, we will discuss the two most widely used analytical flame techniques, flame atomic emission and flame atomic absorption. Both techniques are based on the observation of electronic transitions of valence electrons in a free atomic vapor. Since the internal energy of a free uncombined atom is due solely to its electronic energy, and since the electronic energy states of an atom are quantized, electronic transitions in atomic species result in sharp line spectra, characteristic of the element. It is the extreme narrowness of the spectral lines produced by electronic transitions in atoms that accounts for the high specificity of analytical methods based on atomic spectroscopy.

Figure 18-1 shows an electronic energy level diagram for sodium, where the ordinate expresses the energy of a particular electronic state in electron volts (eV) or wave numbers. The most stable state, or the ground state, corresponds on the diagram to 0 eV. Lines drawn between various electronic states represent quantum mechanically allowed electronic transitions. The numbers associated with the various transitions represent the wavelength in angstroms corresponding to the energy difference between the two states.

In this chapter we will also consider the answers to three important questions: (1) How can we conveniently produce free atoms? (2) How can we observe these atoms? (3) How can we relate the observed atom population to the sample concentration?

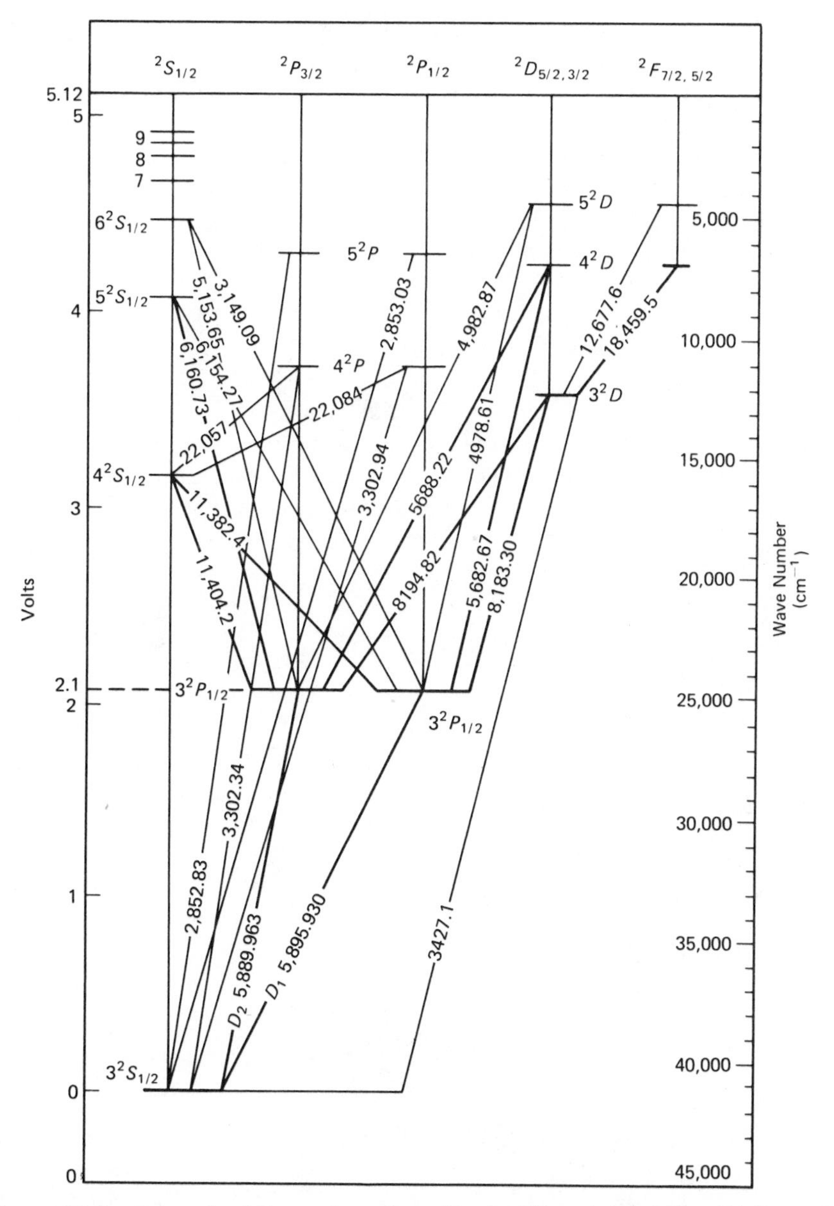


**Figure 18-1.** Energy-level diagram for sodium. (Reprinted from A. Mitchell and M. Zemansky, *Resonance Radiation and Excited Atoms*. New York: Macmillan, 1934, with permission.)

## THE PRODUCTION OF AN ATOMIC VAPOR

With the exception of the element mercury, free uncombined atoms exist only in the vapor state at high temperatures. A variety of means have been employed to generate the high temperatures required to vaporize a sample and to dissociate the molecular compounds thus formed into free atoms. Arcs and sparks, furnaces, and plasmas have all been used to generate atomic vapors in analytical atomic spectroscopy. Perhaps the most convenient and widely used atom reservoir for atomic spectroscopy is the ordinary combustion flame. Although a variety of fuel and oxidant combinations have been employed in analytical flame spectrometry, the most frequently used flames are the air–acetylene flame (2125–2400°C) and the nitrous oxide–acetylene flame (2600–2800°C). Because of its higher temperature, the nitrous oxide–acetylene flame is often used to determine elements that form refractory compounds in the flame.

## THE PREMIX BURNER

Figure 18-2 shows a schematic diagram of a premix slot burner. With a premixing-type burner, the combustion gases are mixed in a mixing chamber prior to issuance from the burner port. The sample, which must be in solution form, is aspirated pneumatically by the oxidant gas. The fine mist produced by

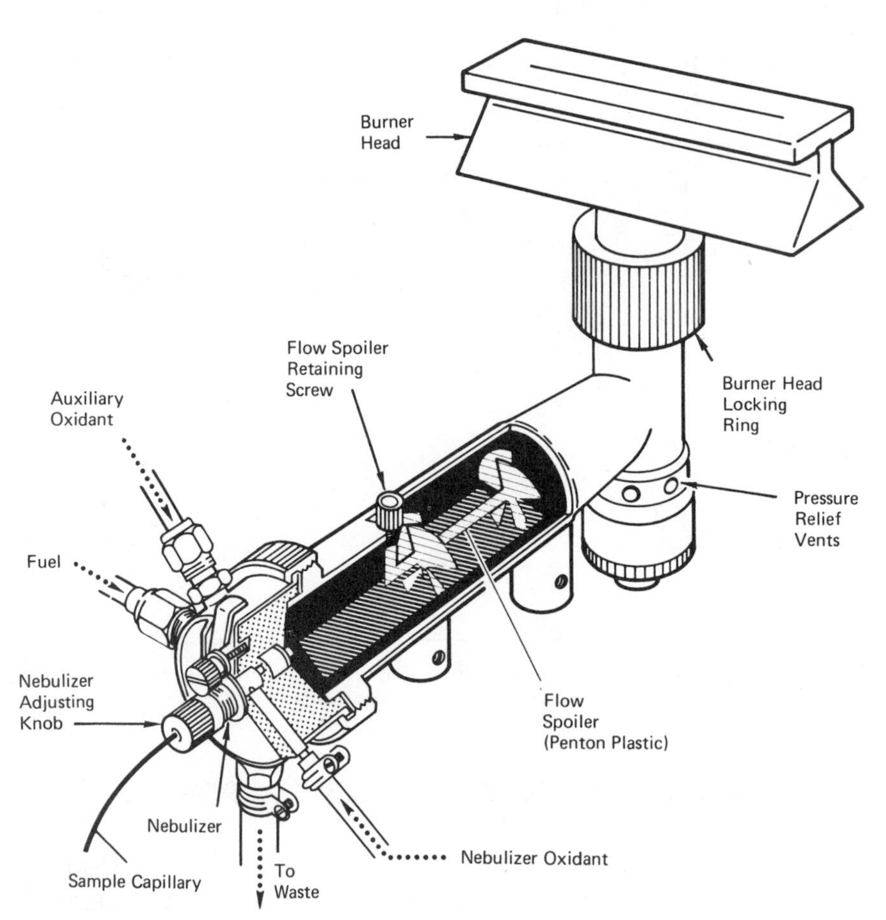


**Figure 18-2.** Spray chamber or premix type of nebulizer–burner. (Courtesy of Perkin–Elmer Corporation.)

the sprayer is mixed with the combustion gases by propellerlike "spoilers" within the mixing chamber. The larger droplets produced by the pneumatic aspirator are usually lost in collision with the spoiler blades. Only 5–12% of the nebulized solution actually reaches the flame with this arrangement, the remainder being drained away.

## THE ANALYTICAL FLAME

Since the combustion gases emerging from the burner port do so in a laminar manner, definite flame zones similar to those observed with an ordinary bunsen burner are formed. Figure 18-3 shows a schematic diagram of the flame structure formed with a premix burner. The preheating zone is a small, nonluminous region immediately above the burner port. In this region the emerging gases become heated to the ignition temperature. The bright region close to the burner head is the **primary reaction zone**. By the time the combustion gases reach this point, they have been preheated to the point where initial combustion is occurring. This region is not of much use analytically because of the intense flame background. The next zone encountered is the **interconal zone**. By the time the gases reach the interconal zone, partial combustion has occurred; however, further combustion is retarded because the available supply of oxidant is exhausted at this point. This region is important analytically because the flame background is low and also because the concentration of available oxygen is low at this point. The reducing nature of this region favors the formation of free atoms and is therefore useful with elements that have a tendency to form refractory oxides. The size of the interconal zone depends on the fuel/oxidant ratio. The bulk of the flame volume is made up of the **secondary reaction zone**. In this zone, final combustion of the partially combusted flame gases occurs as a result of diffusion of additional oxidant from the surroundings. The higher regions of the secondary reaction zone are seldom used analytically because the temperature begins to drop higher in the flame and the concentration of atomic vapor is diluted by diffusion and entrainment of atmospheric gases.

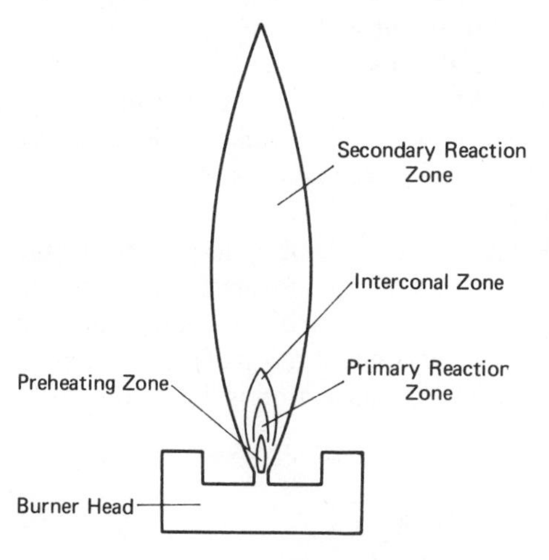


**Figure 18-3.** Flame structure obtained with premix burner. (Reprinted from E. Schweikert and M. Brand, eds., *Modern Techniques in Instrumental Analysis*. New York: Wiley, 1978, with permission.)

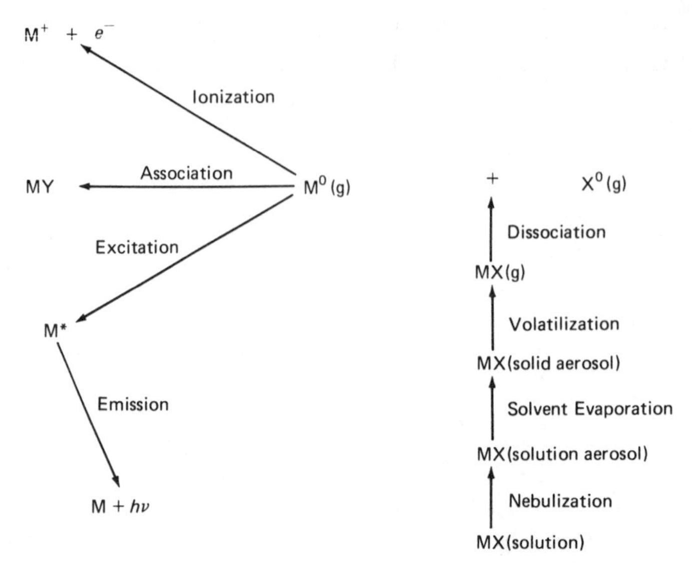


**Figure 18-4.** Events in free-atom formation. (Reprinted from E. Schweikert and M. Brand, eds., *Modern Techniques in Instrumental Analysis.* New York: Wiley, 1978, with permission.)

Analytical flames are often described as reducing or oxidizing, depending on the supply of available oxygen. In a flame where there is an excess of fuel, the supply of oxygen atoms is low in the interconal zone. Such a flame is referred to as a **reducing** flame. Conversely, a flame with an excess of oxidant will have a large amount of oxygen available, and therefore is referred to as an **oxidizing** flame.

## ATOMIZATION

The term *atomization* in flame spectrometry refers to the process of the production of free atoms. Figure 18-4 shows a schematic of the events believed to occur in free-atom production in a combustion flame. In the process of nebulization, the sample solution is converted into a fine mist. As the tiny droplets enter the flame, the solvent evaporates, producing an aerosol of tiny salt particles. As the aerosol is carried vertically in the flame, the salt particles begin to vaporize, forming a molecular vapor represented by MX. On absorbing additional energy from the flame, these molecular entities dissociate into free atoms. The free atoms thus formed may react with flame gas combustion products to form MY, or may undergo ionization at the flame temperature.

The extent of formation of free atoms in the flame is of great concern to the analytical atomic spectroscopist. The extent of free-atom formation is expressed in terms of the free-atom fraction, which is the ratio of the free-atom concentration of an element in the flame to the total concentration of that element in all possible forms found in the flame. The free-atom fraction is designated by $\beta$, where $\beta$ is given by

$$\beta = \frac{[M]}{[M] + [M^+] + [MX] + \cdots} \qquad \textbf{(18-1)}$$

where [M] is the concentration of free atoms; $[M^+]$ is the concentration of $M^+$ ions; and [MX] is the concentration of MX molecules. The free-atom fraction varies not only with the flame region observed but also with the flame stoichiometry.

For many elements the free-atom fraction in a given flame region depends heavily on the chemical environment in the particular flame region. For maximum sensitivity, the observation height and fuel/oxidant ratio should be optimized so that the largest free-atom fraction is observed.

## SPECTROSCOPIC OBSERVATION OF THE ATOMIC VAPOR

The atomic vapor produced by aspirating a sample solution may be observed in three ways. The emitted atomic line spectrum of the element may be monitored using a technique known as atomic emission spectrometry. Or the attenuation of appropriate radiation passing through the flame may be monitored using a technique known as atomic absorption. Finally, the fluorescent radiation emitted by a suitably irradiated flame may be observed using a technique known as atomic fluorescence. Our discussion will be limited to atomic emission and atomic absorption.

### ATOMIC EMISSION SPECTROMETRY

Of the free atoms produced in the flame, a certain percentage will occupy excited electronic states at the temperature of the flame. Since the electronic energy states of the atom are quantized, a definite amount of energy in the form of a photon will be emitted whenever an electronic transition from a higher excited state to a lower electronic state occurs. All photons emitted as a result of a particular transition in a given atom have close to the same energy; thus the emitted radiation is spectroscopically observed as a spectral line.

The intensity or, more correctly, the spectral radiance emitted by a collection of particular atoms for a given electronic transition will be given by

$$I = khvA_tN^*L \qquad \textbf{(18-2)}$$

where $I$ is the emitted spectral radiance; $k$ is a proportionality constant; $h$ is Planck's constant; $v$ is the frequency of the emitted radiation; $A_t$ is the number of transitions per second; $N^*$ is the number of excited atoms in the upper electronic state per cubic centimeter of flame gases; and $L$ is the path length of the flame. For a flame in thermal equilibrium at a given temperature, $N^*$ is directly proportional to $N$, the number of atoms in the ground electronic state per cubic centimeter of flame gases. Therefore,

$$I = k'hvA_tNL \qquad \textbf{(18-3)}$$

where $k'$ is a new constant of proportionality. Since the number of ground-state atoms per cubic centimeter of flame gases is directly proportional to the solution concentration $C$ of the given element in the solution that is aspirated,

$$I = k''C \tag{18-4}$$

where $k''$ is a new constant of proportionality. Since the spectral radiance directly proportional to the solution concentration of the element, quantitative analysis by flame atomic emission is possible.

## INSTRUMENTAL ASPECTS OF ATOMIC EMISSION SPECTROMETRY

Figure 18-5 shows a schematic diagram of a flame emission spectrometer, consisting of a flame excitation source with a gas-regulating system, external optics to focus the emitted light on the entrance slit of a monochromator, a monochromator to select a given wavelength, a detector, and finally an amplifier–readout system.

In atomic emission spectrometry, a high-quality monochromator is required to isolate the analytical spectral line from all other radiation emitted from the flame. Figure 18-6 shows a schematic diagram of a Czerny–Turner monochromator. Light enters the system through the entrance slit as a diverging bundle of rays. If such a bundle of rays were allowed to strike the dispersing device directly, a series of overlapping spectra would result because the angle with which radiation emerges from the dispersing device depends on *both* the wavelength of the radiation and the incident angle. To make sure that all the incident radiation strikes the dispersing device at the same angle of incidence, the diverging bundle of rays is first made parallel by a collimating mirror. Most instruments today employ a grating as a dispersing device. For our purposes it is not important to discuss the details of grating diffraction; it is only necessary to say that a bundle of radiation striking a grating with a particular angle of incidence is dispersed in

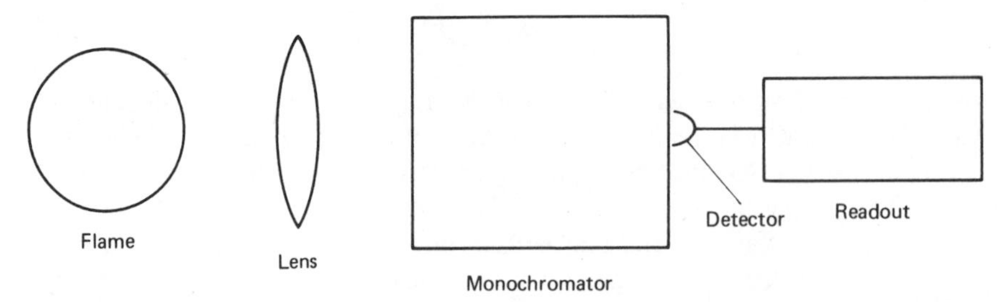


**Figure 18-5.** Schematic of flame emission spectrophotometer. (Reprinted from E. Schweikert and M. Brand, eds., *Modern Techniques in Instrumental Analysis*. New York: Wiley, 1978, with permission.)

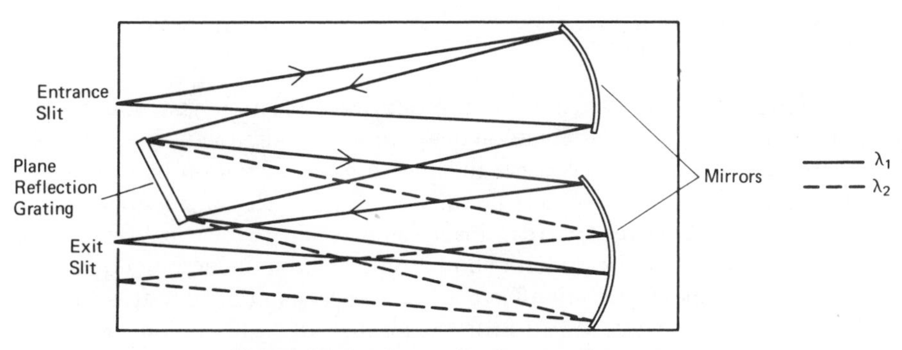


**Figure 18-6.** Czerny–Turner monochromator.

such a manner that radiation of different wavelengths will emerge at different angles.

A focusing mirror arranged to collect the series of parallel bundles of radiation (one parallel bundle for each wavelength, each parallel bundle diverging from other parallel bundles of different wavelength) will form an image of the entrance aperture in the focal plane of the mirror. The position of the image of the entrance aperture along the focal plane will depend on the angle with which that particular parallel bundle struck the focusing mirror (i.e., it will depend on the wavelength of the radiation). Since the entrance aperture of the monochromator generally takes the form of a slit, the image produced in the focal plane of the focusing mirror when the monochromator is illuminated with monochromatic light is a "spectral line" or image of the entrance slit. By placing another slit at the focal plane of the focusing mirror, only a narrow wavelength range is allowed to pass through to the detector at a time, and, by rotating the grating, any desired wavelength within the range of the instrument may be allowed to pass through to the detector.

Because the number of atoms in a particular excited state increases with the temperature of the atom reservoir, a high-temperature atom reservoir is desirable for atomic emission spectrometry. The nitrous oxide–acetylene flame (2600–2800°C) has frequently been used for atomic emission spectrometry. One advantage of this flame is that the burning characteristics are such that it may be used with an ordinary premix slot burner. One disadvantage to the nitrous oxide–acetylene flame is the high spectral flame background produced. The oxygen–acetylene flame (3060–3135°C) has also been used for atomic emission, but the burning characteristics of this flame are such that a specially designed burner must be used. In an effort to produce even higher temperatures than can be achieved with combustion flames, analytical spectroscopists are investigating the use of various forms of gas discharges. These plasma "flames" are capable of producing temperatures of about 8000°C. The sample is aspirated into the plasma in much the same manner as in conventional flame emission work. The high temperatures obtained with these systems not only increase the sensitivity of atomic emission but also tend to destroy many refractory compounds that often cause interferences in conventional flame atomic emission. In addition, because the plasma is formed in an inert gas such as argon, the spectral background is low.

## ATOMIC ABSORPTION

In atomic absorption, the free atoms produced in the atom reservoir are observed by monitoring the attenuation of appropriate radiation passing through the absorbing medium. To observe the absorption of radiation by the atomic vapor, several requirements must be satisfied. First, since the energy states of an atom are quantized, the atom can absorb only certain frequencies of radiation. During the process of absorption of a photon, the atom undergoes a quantum mechanically allowed transition from a lower energy initial state to a higher-energy, final electronic state. For strong absorption to occur, not only must the transition be quantum mechanically allowed but the energy of the photon, $h\nu$, must correspond exactly to the energy difference between the lower energy state and the higher energy state.

Furthermore, for significant absorption to occur, there must be an appreciable population of atoms present in the initial lower energy electronic state. For an atomic vapor in thermal equilibrium, the electronic state with the greatest population will be the lowest energy electronic state, the ground state. For this reason, atomic absorption will be observed to a significant extent only for transitions involving the ground state (i.e., **resonance transitions**). In Figure 18-1, the yellow sodium D lines at 5890 and 5896 Å are both resonance lines, whereas the line at 8195 Å is a nonresonance line.

Although it might seem initially that atomic lines should consist only of a single frequency and therefore be infinitely narrow, examination of a spectral line emitted by a free atom under very high resolution reveals that the emission line is not infinitely narrow but consists of a narrow spread of energies. The extent of this narrow spread of energies is specified by the half-width of the spectral line, which is simply the width of the energy distribution at half the maximum intensity. The broadening of a spectral line is due to two principal sources: Doppler broadening and collisional broadening.

**Doppler broadening** is caused by the random thermal motion of the emitting atoms in the flame. Thus, when a radiation source such as an emitting atom is moving toward an observer, the frequency emitted appears slightly higher. When the radiation source is moving away from the observer, the emitted frequency appears shifted to lower frequencies. Since an emitting atom in a flame can travel in any direction owing to random thermal motion, the contribution of all the Doppler shifts for each particular direction of motion combine to give a symmetrical profile similar to the Gaussian curve of error. **Collisional broadening** results from the perturbation of the energy levels of an excited radiating atom by collision with flame gas molecules.

According to **Kirchhoff's law**, a luminous object can absorb radiation only over the wavelength interval over which it also emits radiation; thus, an atom will be able to absorb radiation only over the wavelength interval covered by its emission line profile. This wavelength interval over which the atom may absorb radiation corresponds to the absorption profile for the atom. The absorption profile for an atom is also characterized by a half-width that is identical in magnitude to the emission half-width for the atom. Thus, an atom may absorb radiation corresponding to a resonance transition only over the absorption line width corresponding to the Doppler and collisional broadening conditions that exist *in the atom reservoir.*

Consider an atom reservoir containing a uniform concentration of absorbing atoms *n*, irradiated by a suitable radiation source emitting narrow spectral lines of the particular element. The greatest amount of absorption of resonance radiation for a given atom population will be observed if (1) the absorption line profile for the atoms in the atom reservoir and the emission line profile of the radiation emitted from the radiation source both peak at the same wavelength, and (2) the emission line profile of the radiation emitted from the radiation source lies entirely within the absorption line profile for the atoms in the atom reservoir. These considerations are shown in Figure 18-7, where the dashed line represents the absorption line profile for the atoms in the atom reservoir and the solid line represents the emission line profile for the radiation emitted by the light source. These two requirements provide the largest absorption, since the emission line maximum coincides with the point of maximum absorption on the absorption line profile, and also because the entire emission line profile lies within the absorption line profile. Thus, when these two criteria are fulfilled, the absorption measured is the absorption at the line center or the peak absorption.

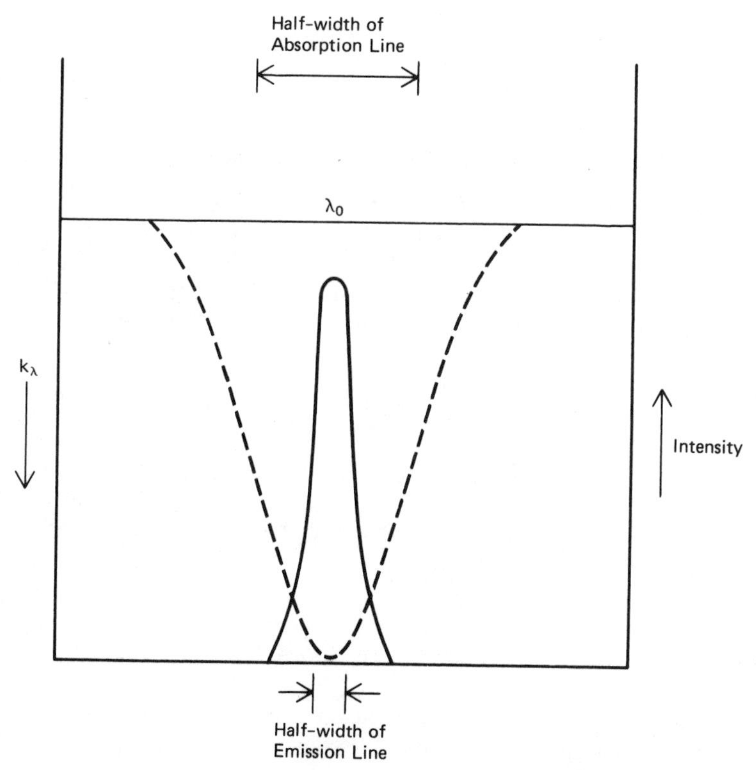


**Figure 18-7.** Line-width considerations in atomic absorption. (Reprinted from E. Schweikert and M. Brand, eds., *Modern Techniques in Instrumental Analysis.* New York: Wiley, 1978, with permission.)

Let us now turn our attention to the quantitative basis of atomic absorption. Consider an atom reservoir containing a uniform concentration of absorbing atoms, $n$, which is irradiated by a parallel beam of resonance radiation from a narrow emission line radiation source of spectral radiance, $I_0$ (such that the emission line width of the light from the source is less than the absorption line width for the atoms in the atom reservoir). Atoms raised to an excited state by absorption of resonance radiation will return to the ground state either by collisional deactivation or by reemision of a photon. Since reemission of a photon is equally probable to occur in any direction, the initial spectral radiance $I_0$ will be reduced after passage through the absorbing medium. If the path length of the parallel beam of radiation passing through the absorbing medium is $L$, the spectral radiance of the beam after traversing the medium will be given by

$$I = I_0 \exp(-\bar{k}L) \tag{18-5}$$

where $I$ is the spectral radiance of the beam emerging from the absorbing medium; $I_0$ is the spectral radiance of the beam prior to passing through the absorbing medium; and $\bar{k}$ is the average atomic absorption coefficient at the absorption line peak. Since the absorbance is defined by

$$A = -\log(I/I_0) \tag{18-6}$$

the absorbance produced by the atomic vapor will be

$$A = 0.43\bar{k}L \tag{18-7}$$

As long as the concentration of atomic vapor is not too high, and as long as the emission line width of the radiation from the light source is less than the absorption line width for the atoms in the atom reservoir, $\bar{k}$ will be directly proportional to the number of ground state absorbing atoms per cubic centimeter of flame gases. Since the number of ground state absorbing atoms per cubic centimeter of flame gases $n$ is directly proportional to the solution concentration $C$ of the solution aspirated into the flame,

$$A = k'n = k''C \tag{18-8}$$

where $k'$ and $k''$ are proportionality constants. Equation (18-8), which shows that the absorbance produced by an atomic vapor in a flame is directly proportional to the solution concentration aspirated into the flame, establishes the quantitative basis for atomic absorption.

## INSTRUMENTAL ASPECTS OF ATOMIC ABSORPTION

Figure 18-8(a) shows a schematic diagram of an atomic absorption spectrophotometer, consisting of a narrow-line radiation source, a chopper, focusing optics, a flame atom reservoir with associated gas regulation system, a monochromator, a detector, and an ac amplifier-readout system. Most atomic absorption spectrophotometers today employ a double-beam arrangement to

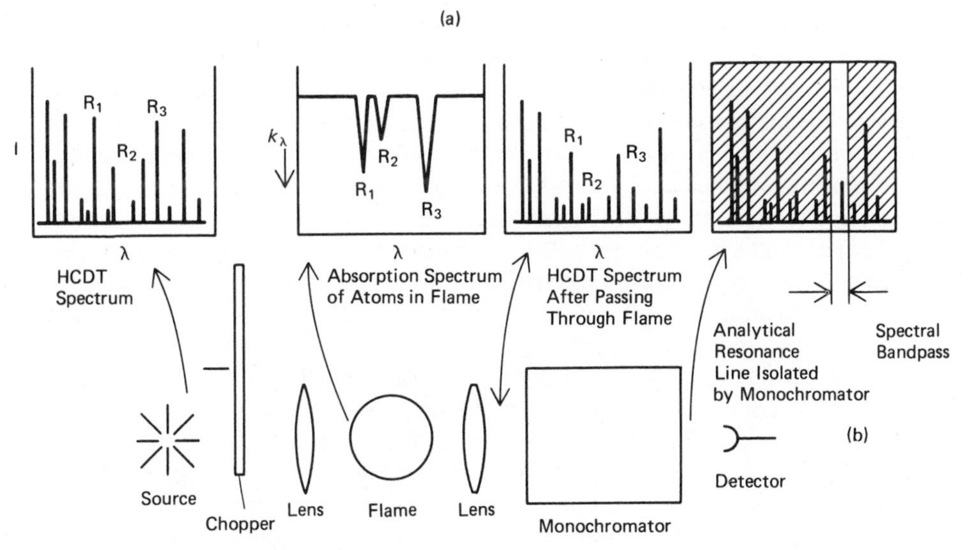


**Figure 18-8.** (a) Spectra showing role of monochromator in atomic absorption. (b) Atomic absorption spectrophotometer. (Reprinted from E. Schweikert and M. Brand. eds., *Modern Methods in Instrumental Analysis*. New York: Wiley, 1978, with permission.)

compensate for source-intensity fluctuations. Since the absorbance produced by a given atomic vapor is directly proportional to the path length of the absorbing medium for low atomic vapor concentrations, the long path length provided by the slot burner produces a convenient flame configuration for atomic absorption.

## *Hollow Cathode Discharge Lamp*

The hollow cathode discharge tube is the most commonly used light source in atomic absorption spectrophotometry. The lamp, shown in Figure 18-9, consists of an anode and a cathode, the cathode being in the form of an open cup

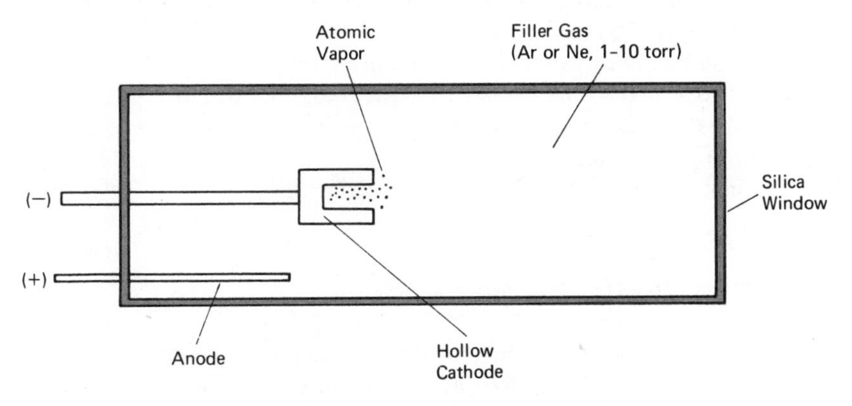


**Figure 18-9.** Hollow cathode discharge lamp.

containing the element of interest. The interior of the tube is filled with a few torr of an inert gas such as argon or neon. Application of several hundred volts across the electrodes causes an ionized gas or plasma to form. By proper design, the plasma may be confined to the interior of the cathode cup. Since the cathode is the negative electrode, positively charged inert gas ions present in the plasma will be accelerated toward the cathode, colliding with energies sufficient to eject atoms of the cathode material into the discharge, a process known as **sputtering**. The free sputtered atoms thus introduced into the plasma are subsequently excited by electronic collisions, emitting their characteristic spectrum. In addition to the spectrum of the cathode material, the spectrum of the inert gas will also be present.

The hollow cathode discharge lamp is almost an ideal light source for atomic absorption. By flowing a current on the order of 15 mA through the lamp, extremely narrow spectral lines of the cathode element are produced. The emission lines produced with a hollow cathode lamp are extremely narrow because the major sources of line broadening have been reduced. Because the interior of the lamp is maintained at low pressures, collisional broadening is negligible. In addition, since the spectral excitation of the sputtered atoms is due to electronic collisions in the plasma, the actual gas temperature of the emitting atomic vapor within the hollow cathode is relatively low, about 450°C. As a result, the amount of Doppler broadening is slight. Therefore, the emission linewidth for the radiation emitted from a hollow cathode discharge lamp will be less than the absorption line width for the atomic vapor in the flame, because significant Doppler and collisional broadening effects exist in the flame.

### *The Role of the Monochromator*

Figure 18-8(b) schematically illustrates the role of the monochromator in atomic absorption. The spectral lines labeled $R_1$, $R_2$, and $R_3$ represent three resonance lines of varying sensitivity emitted by the hollow cathode lamp. After passing through the flame, all three resonance lines are attenuated to different extents. The role of the monochromator is simply to isolate the particular analytical resonance line selected for the determination from the rest of the spectrum emitted by the hollow cathode lamp. Thus, as long as the analytical resonance line has been isolated from the other emitted lines, the absorbance measured is governed by the line width of the resonance line, not by the characteristics of the monochromator. For elements with simple spectra, the resolution required to isolate the resonance line is often quite low, and short focal length monochromators and wide slits are often satisfactory.

### *Elimination of Interfering Flame Emission*

When a sample containing the analyte (the substance being determined) is aspirated into the flame, the free atoms produced can be excited, emitting resonance radiation at exactly the same wavelength as the analytical resonance line emitted by the hollow cathode lamp. Thus, although the monochromator is capable of isolating the analytical resonance line from the remainder of the hollow cathode spectrum, it cannot discriminate between the resonance radiation emitted from the flame and that emitted from the hollow cathode tube because they both occur at precisely the same wavelength. To distinguish the signal produced by the hollow cathode radiation striking the detector from the signal arising from resonance radiation of the element emitted by the flame, an electronic approach is employed. By **chopping** (i.e., periodically interrupting the hollow cathode radiation with a spinning sectored disc), a pulsed signal consisting of a square-wave pulse train will be produced, since the light from the lamp will be periodically blocked from striking the detector. The pulsed signal produced at the detector from the intermittent hollow cathode radiation can be applied to the input of an ac amplifier, whose capacitive input will transmit only ac electrical signals, blocking those signals that are dc. The resonance radiation produced by flame emission is not periodically interrupted so that the signal produced by the detector from this component of the radiation passed by the monochromator is dc; consequently, it is not amplified by the ac amplifier.

## *Nonflame Devices*

In contrast to flame emission spectrometry, where the flame must serve the dual role of atomization and excitation, the only function of the flame in atomic absorption is the conversion of the nebulized sample into free atoms. Consequently, the flame may be easily replaced in atomic absorption by a miniature furnace. Figure 18-10 shows a schematic diagram of a heated graphite atomizer, which consists of an electrically heated graphite chamber that is capable of producing a high enough temperature to atomize the sample. In using the graphite furnace, the sample is usually introduced into the graphite chamber by means of an automatic pipette. The solvent is then evaporated at a low temperature. After the solvent has evaporated, any organic matter present in the sample

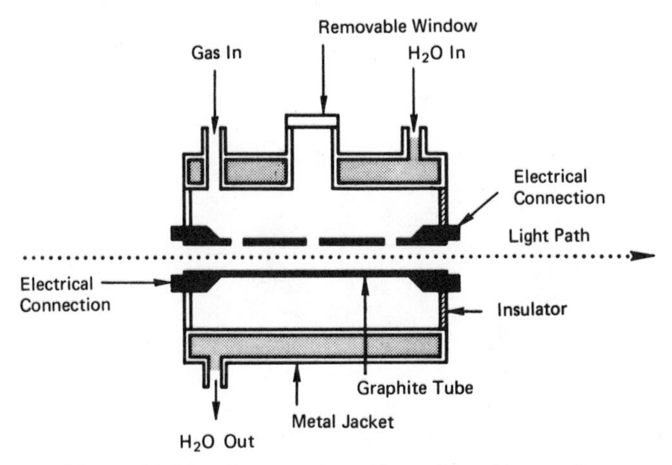


**Figure 18-10.** Cross section of heated graphite atomizer.

**TABLE 18-1.** Detection Limits in Flame Spectrometry

| | | *Detection Limits (ppm)* | | | | |
|---|---|---|---|---|---|---|
| | | | | *Atomic Absorption* | | |
| | | *Flame Emission** | | *Flame** | | |
| *Element* | *Wavelength (Å)* | $N_2O-C_2H_2$ | *Air–* $C_2H_2$ | $N_2O-C_2H_2$ | *Air–* $C_2H_2$ | *Nonflame*† |
| Ag | 3280.7 | 0.02 | | | 0.005 | 0.00004 |
| Al | 3961.5 | 0.005 | | | | |
| | 3092.8 | | | 0.1 | | 0.006 |
| Ba | 5535.5 | 0.001 | | 0.05 | | |
| Ca | 4226.7 | 0.0001 | 0.005 | | 0.002 | 0.00006 |
| Cd | 3261.1 | 2 | | | | |
| | 2288.0 | | | | 0.005 | 0.00002 |
| Co | 3453.5 | 0.05 | | | | |
| | 2407.2 | | | | 0.005 | 0.001 |
| Cr | 4254.4 | 0.005 | | | | |
| | 3578.7 | | | | 0.005 | 0.001 |
| Cu | 3274.0 | 0.01 | | | | |
| | 3247.5 | | | | 0.005 | 0.001 |
| Fe | 3719.9 | 0.05 | | | | |
| | 2484.3 | | | | 0.005 | |
| Hg | 2536.5 | | | | 0.5 | 0.0006 |
| K | 7664.9 | | 0.0005 | | 0.005 | 0.0002 |
| Li | 6707.8 | 0.00003 | | | 0.005 | 0.001 |
| Mg | 2852.1 | 0.005 | | | 0.0003 | 0.00001 |
| Mn | 4030.8 | 0.005 | | | | |
| | 2794.8 | | | | 0.002 | 0.0001 |
| Mo | 3903.0 | 0.1 | | | | |
| | 3132.6 | | | | 0.03 | 0.008 |
| Na | 5890.0 | | 0.0005 | | 0.002 | 0.00002 |
| Ni | 3414.8 | 0.03 | | | | |
| | 2320.0 | | | | 0.005 | 0.002 |
| Pb | 4057.8 | 0.2 | | | | |
| | 2833.1 | | | | 0.03 | 0.001 |
| Sn | 2840.0 | 0.5 | | | | |
| | 2246.0 | | | | 0.06 | 0.010 |
| Sr | 4607.3 | 0.0001 | | | 0.01 | 0.001 |

*Source:* E. Schweikert and M. Brand, eds., *Modern Techniques in Instrumental Analysis* (New York: Wiley, 1978), with permission.
* E. E. Pickett and S. R. Koirtyohann, *Anal. Chem.*, **41**, 28A (1969).
† A. Syty, *CRC Crit. Rev. Anal. Chem.*, **4**(2), 155 (1974).

is destroyed by ashing at a still higher temperature. Finally, full power is applied to the furnace, atomizing the sample. The atomization temperature is maintained until the signal returns to base line. Nonflame devices require only a few microliters of sample, and in many cases can detect as little as $10^{-11}$ g of analyte. Table 18-1 shows some typical detection limits obtained with nonflame graphite atomizers.

# QUANTITATIVE DETERMINATIONS BASED ON THE SPECTROSCOPIC OBSERVATION OF AN ATOMIC VAPOR

## SELECTION OF THE ANALYTICAL LINE

In general, more than one potential analytical line exists for any given element, so the first decision that an analyst must make is the selection of the appropriate analytical line for the given determination. One of the factors that must be considered in selecting an analytical line for a particular determination is the anticipated concentration range of the samples. If the samples to be determined are trace or ultratrace samples, it may be necessary to use the most sensitive analytical line available for the determination. On the other hand, if the expected concentration of the analyte in the sample is higher, a less sensitive line may avoid the need for excessive dilution of the sample. Ideally, the analytical line should be selected so as to match the linear range of the calibration plot, with the anticipated concentration range of the samples to be determined in such a way that various samples do not require different dilutions. For a given element, there is generally a wider selection of potential analytical lines for atomic emission spectrometry as compared with atomic absorption, because atomic emission is not limited only to resonance lines.

Another factor that should be considered in selecting an analytical line is the possibility of spectral interference. Spectral interference is caused by the presence of unwanted radiation within the spectral bandpass of the monochromator. In atomic absorption, spectral interference frequently occurs whenever a nonabsorbing spectral line is too close to the analytical resonance line, so that both lines are simultaneously passed by the monochromator. In atomic emission, spectral interference may be caused by the atomic emission of concomitants or by molecular band emission of concomitants.

## OPTIMIZATION

The instrumental variables that affect the instrument response in flame spectrometry include (1) the monochromator slit width, (2) the photomultiplier voltage, (3) the amplifier gain and filter time constant, (4) the fuel/oxidant ratio of the flame, and (5) the vertical portion of the flame observed by the

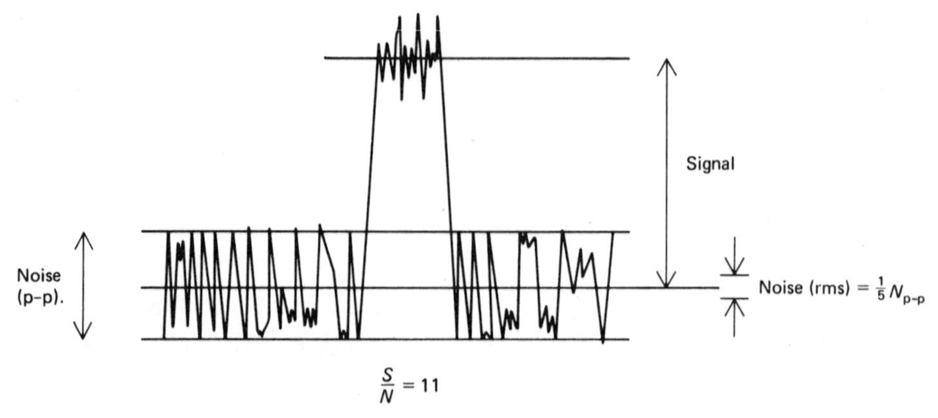


**Figure 18-11.** Determination of signal/noise ratio from recorder tracing. (Reprinted from E. Schweikert and M. Brand, eds., *Modern Techniques in Instrumental Analysis.* New York: Wiley, 1978, with permission.)

monochromator. The appropriate instrumental response to optimize with a flame spectrometer, as with any electronic device, is the signal/noise ratio. The **signal/noise ratio** (S/N) is the ratio of the average signal to the root-mean-square noise, as shown in Figure 18-11. The average signal is taken as the response difference between the average response with sample and the average response without sample. The root-mean-square noise is estimated as being about one-fifth the noise envelope drawn by the recorder pen. For a given analyte concentration, the optimum is that set of instrumental conditions giving the maximum S/N. By optimizing the S/N of the spectrometer output, the precision of the measurement is also maximized, and the detection limit, which is generally taken as that concentration producing a signal/noise ratio of 2, is minimized.

## INSTRUMENT CALIBRATION

In order to perform quantitative determinations by flame spectrometry, the instrument response must be calibrated in terms of the analyte concentration. This is commonly accomplished in two ways: (1) by using a calibration curve and (2) by the method of standard addition. With the **calibration curve** method, the instrument response is determined for a series of standard solutions of varying known analyte concentration. Calibration curves in flame spectrometry become nonlinear at high atomic vapor concentrations (corresponding to high solution concentrations), so that these regions are normally avoided when making the calibration curve.

It is a fundamental assumption in instrumental analysis that a constant proportionality exists between instrument response and solution concentration. Whenever some phenomenon occurs that causes the constant of proportionality to change between the calibration standards and the analytical samples, an interference is said to exist. In the analysis of a complex sample, the more closely

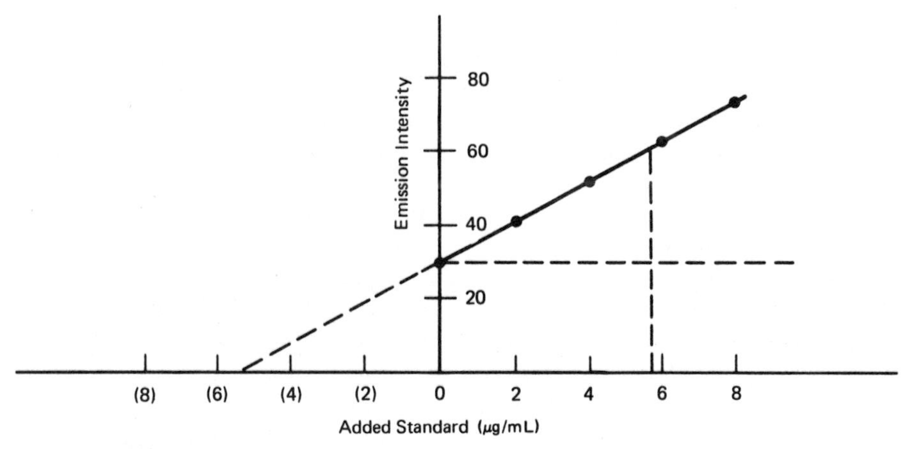


**Figure 18-12.** Typical graph for standard addition method. The sample contains 5.6 μg/mL of the element.

the calibration standards duplicate the other components present in the sample, the less likely an interference will be observed and the more likely a given determination will be accurate. In many cases, however, preparing calibration standards that duplicate the other components of the sample may be difficult if not impossible. In this situation, the method of **standard addition** may be useful. In this method, an aliquot of the sample is determined after dilution to a stated volume. A given volume of standard is then added to a second aliquot of the sample, which is then diluted to the same volume. If several aliquots of sample with different amounts of standard added are prepared in this manner and then determined with the flame spectrometer, the instrument response may be plotted versus the amount of standard added.

The procedure is illustrated in Figure 18-12 for the case of atomic emission spectrometry. The extrapolation of the calibration line intersects the zero emission line at the concentration of the unknown sample. Since the sample interferences are present in all the solutions determined, the method of standard additions should theoretically eliminate interferences. The method of standard additions should not, however, be considered as a panacea to be used without precautions. For example, accurate results will not be obtained unless the instrument response is zero when the analyte concentration is zero. In addition, the calibration curve must be linear. Finally, as with any extrapolation procedure, small variations in the instrument response can drastically shift the intersection point of the extrapolated line with the abscissa.

## INTERFERENCES

Perhaps the most effective procedure to ensure the accuracy of analytical determinations by flame spectrometry is to rely on a knowledge of the chemical and physical factors that can cause interferences. Since the instrument response obtained for a given solution concentration in atomic emission or atomic absorption depends on the free-atom concentration, anything that alters the extent of free-atom formation between the calibration standards and the samples will cause an interference. Interferences in flame spectrometry may be classified into three categories: transport interferences, evaporation interferences, and gas-phase interferences.

**Transport interferences** arise when the amount of sample reaching the flame is different for the standards and sample. Since the sample is pneumatically atomized, the amount of solution aspirated depends on the solution viscosity. To avoid this interference, the viscosity of the standards and samples should be the same. The surface tension of the sample is another factor that can alter the free-atom fraction observed by changing the drop size produced by the nebulizer, thereby changing the extent of desolvation of the droplets in a given region of the flame.

**Evaporation interferences** affect the rate of evaporation of the solid aerosol produced after desolvation of the droplets. The best known example of an evaporation interference is the effect of phosphate on the instrument response with a given concentration of calcium. The presence of phosphate severely depresses the signal observed for calcium compared with a similar concentration in the absence of phosphate. The effect is due to the formation of a refractory Ca—P—O compound upon desolvation. Since the compound formation reduces the free-atom concentration of calcium, the interference is observed regardless of whether the free-atom concentration is observed by atomic emission or atomic absorption. Evaporation interferences are minimized by three common procedures: (1) the addition of a substance known as a releasing agent which preferentially reacts with the interferent; (2) making all measurements higher in the flame, thereby allowing more time for the compound to break apart; and (3) the use of a high-temperature flame such as a nitrous oxide–acetylene flame.

The best example of a **gas-phase interference** results from shifts in ionization equilibria. Thus, if the analyte in the flame is ionized to a different extent in the unknown sample than in the calibration standards, the extent of free-atom formation will not be the same in both, and an interference will result. The ionization of an element in the flame obeys the equilibrium

$$M = M^+ + e^- \qquad \textbf{(18-9)}$$

For a system such as a flame in thermal equilibrium, the principle of mass action may be applied to the equilibrium. Thus, the ionization of the analyte may be easily suppressed by the addition of an easily ionizable substance known as an **ionization buffer**, which increases the electron concentration in the flame and drives the equilibrium shown in eq. (18-9) to the left. Therefore, for samples where ionization shifts may be a problem, it is good policy to add an ionization buffer to both the standards and the samples. This is easily accomplished by adding an easily ionizable cation such as cesium to both the samples and standards so that both contain approximately 1000 ppm of the buffer cation.

## COMPARISON OF ATOMIC EMISSION AND ATOMIC ABSORPTION

We have emphasized in this chapter that atomic emission and atomic absorption are two complementary techniques for observing an atomic vapor. Since the response obtained with both techniques ultimately depends on the concentration of free ground state atoms in the flame, both techniques are equally subject to the interferences previously described, which alter the free-atom concentration in the flame. As can be seen from Table 18-1, the detection limits obtained with the two techniques are comparable. It is a general rule of thumb that atomic absorption gives lower detection limits than atomic emission for elements whose most sensitive resonance lines are less than 300 nm. Atomic emission generally gives better detection limits compared to atomic absorption for elements whose most sensitive resonance lines are above 400 nm. For elements with resonance lines between 300 nm and 400 nm, both techniques have similar detection limits. The choice of which technique to use depends to a certain extent on the availability of the equipment.

## PROBLEMS

**1** The calibration of a flame photometer for potassium gave the following scale readings: 4.0 ppm, 100; 3.0 ppm, 74; 2.5 ppm, 62; 2.0 ppm, 49; 1.0 ppm, 25; 0.5 ppm, 12. A well-water sample, after dilution of 10.0 mL to 100 mL gave a scale reading of 56. Calculate the ppm of K in the water and also express the results as the molar concentration of KCl.

*Ans.* 23 ppm K, $5.9 \times 10^{-4}$ *M* KCl

**2** Repeat problem 1 for well waters that gave scale readings after similar dilution of (a) 92; (b) 45; (c) 20; (d) 70; (e) 65.

**3** A series of standards for the determination of $Na_2O$ in cement were prepared to contain various amounts of Na and the amounts of both Ca and K normally found in cements. In the procedure, a 1.000-g sample of the cement is dissolved and diluted to 100 mL. The results for the standards and several samples were as follows:

| *Concentration (ppm* Na) | *Scale Reading* |
|---|---|
| 75 | 100 |
| 60 | 84 |
| 45 | 67 |
| 30 | 48 |
| 15 | 26 |
| 0 | 0 |
| Cement 1 | 73 |
| Cement 2 | 40 |
| Cement 3 | 32 |
| Cement 4 | 88 |
| Cement 5 | 53 |

*Ans.* (2) 24 ppm Na, 0.32% $Na_2O$

**4** In the determination of calcium in antacid tablets by atomic absorption, a 0.5133-g sample was dissolved and diluted to 1000 mL. Three 5.00-mL aliquots were diluted to 50.0 mL each after addition of 0, 1.00, and 2.00 mL, respectively, of a calcium standard containing 0.500 mg Ca/mL. The absorbances of the three solutions were 0.310, 0.475, and 0.640, respectively. Calculate the number of mg Ca in the sample. If the average tablet weight was 0.7568 g and the sample was declared to contain 300 mg Ca per tablet, what was the percent of the declared value in the sample?

*Ans.* 180 mg, 88.5%

**5** Repeat problem 4 for the following tablets:

| | *Declared Amount (mg Ca)* | *Tablet Weight* | *Sample Weight* | *Absorbance* | | |
|---|---|---|---|---|---|---|
| | | | | *Sample* | *Sample +1 mL* | *Sample +2 mL* |
| (a) | 200 | 0.5000 | 0.5862 | 0.390 | 0.560 | 0.728 |
| (b) | 150 | 0.3826 | 0.2863 | 0.193 | 0.285 | 0.387 |
| (c) | 100 | 0.2500 | 0.2500 | 0.233 | 0.428 | 0.639 |

**6** A 0.20-mL sample of blood after dilution gave an emission reading of 31. Solutions B and C, containing the same amount of blood plus 50 and 100 μg/mL of K, respectively, gave emission readings of 47 and 64, respectively. Calculate the concentration of K in the blood in μg/mL.

CHAPTER

# 19 Miscellaneous Optical Methods

In this chapter we will discuss miscellaneous optical methods, including fluorescence, emission spectrography, and x ray. All of these involve emission of radiant energy in some way. Since the absorption of radiant energy is related to the energy levels of atoms and molecules and since emission is the reverse process, emission of radiant energy is a function of the electronic, vibrational, and rotational levels. In general, atoms and molecules are considered to be in their ground or lowest energy state under normal conditions and are raised to excited energy states by absorption of energy. After absorption, the atom or molecule returns to the ground state by conversion of the absorbed energy to heat energy (by transfer of the excess energy by collision) or by emission of the energy as radiant energy. In this chapter, we will consider the emission of radiant energy from excited atoms or molecules and the use of the emitted energy in analytical methods.

## ORIGIN OF SPECTRA

A complete discussion of the origin of atomic and molecular spectra is beyond the scope of this text, but a short review of atomic and molecular energy levels is given as a basis for a simplified explanation.

The electrons in an atom are arranged in **atomic orbitals** according to the rules of quantum mechanics. The electrons in the atomic orbitals are characterized by four quantum numbers—$n$, $l$, $m_l$, and $m_s$—which determine the energy of the electron. Different orbitals are designated by specifying the value for $n(1, 2, 3, 4, 5, 6)$ and the value for $l$. Values for $l$ are designated by small letters where $s$ means $l = 0$, $p$ means $l = 1$, $d$ means $l = 2$, and $f$ means $l = 3$. Finally, the number of electrons in an orbital or set of orbitals is represented by a small superscript on the letter. Thus the valence electron of sodium would be designated as $3s^1$ and the complete electronic configuration of sodium as $1s^2 2s^2 2p^6 3s^1$.

Under appropriate conditions an atom can be made to absorb energy by promoting an electron from a lower energy state to a higher energy state. The lifetime of this higher energy, or excited, state is very short, on the order of $10^{-8}$ sec. If the electron returns directly to the ground state in a single step, a definite amount of energy in the form of a photon will be emitted as a result of the transition because the electronic states of the atom are quantized. As an example, for sodium in the ground state, the valence electron occupies the 3*s* orbital. If sodium is placed in a Bunsen flame, however, it can absorb enough energy to raise the valence electron from the 3*s* orbital to a 3*p* orbital. When the valence electron returns to the 3*s* orbital, a photon of light at 589 nm will be emitted, producing the familiar yellow emission characteristic of sodium in flame tests. If enough energy is absorbed to raise the electron to a 3*d* orbital, it will fall back in two steps, the first transition being to a 3*p* orbital with the emission of a photon having a wavelength of 819 nm, the second being from the 3*p* orbital to the 3*s* orbital with the emission of a 589-nm photon. As a consequence of the fact that electrons can be excited to various energy levels and can return to the original level by more than one route, emission spectra of highly excited atoms can be very complex. In addition, because electronic transitions occur between quantized atomic orbitals, atomic spectra are line spectra.

The situation for molecules is somewhat more complex. Electrons in molecules are arranged in **molecular orbitals**. By analogy with atomic orbitals, the maximum number of electrons in any given molecular orbital is 2. Electrons in the molecule are added to the molecular orbitals in order of increasing energy following Hund's rule, which states that when orbitals of identical energy are available, electrons tend to occupy these singly rather than in pairs.

As in the case of atoms, electrons from lower-energy molecular orbitals may be raised to higher-energy molecular orbitals by absorption of appropriate energy photons. The existence of these quantum-mechanically allowed electronic transitions gives rise to the familiar molecular absorption spectra observed in the ultraviolet-visible region of the spectrum. In contrast to the electronic transitions in atoms, which result in line spectra, electronic transitions in molecules result in band spectra, owing to the various vibrational and rotational energy levels in which the molecule in any given electronic state can exist. This is illustrated in Figure 17-1 for a simple diatomic molecule. Under ordinary conditions, most molecules will exist in the lowest energy state that would correspond to the lowest energy vibrational state of the lowest energy electronic state. In Figure 17-1 this would be the $v_0$ state of $E_1$. Absorption of a photon of appropriate frequency can cause a transition from $E_{1,v_0}$ to any of the closely spaced vibrational-rotational levels associated with $E_2$. Because the molecule may be excited to any of the vibrational-rotational states associated with $E_2$, a range of frequencies may be absorbed, resulting in an absorption band.

As with the absorption of radiation in other regions of the electromagnetic spectrum, electronic transitions in molecules are governed by quantum-mechanical selection rules. Electronic states in a molecule may be classified according to whether the electrons in the molecular orbitals have paired or unpaired spins. If all the electrons have paired spins (the net spin of the entire system is zero), the state is referred to as a **singlet state**, whereas if two electrons with unpaired spins are present (the net spin is 1, since each electron has a spin

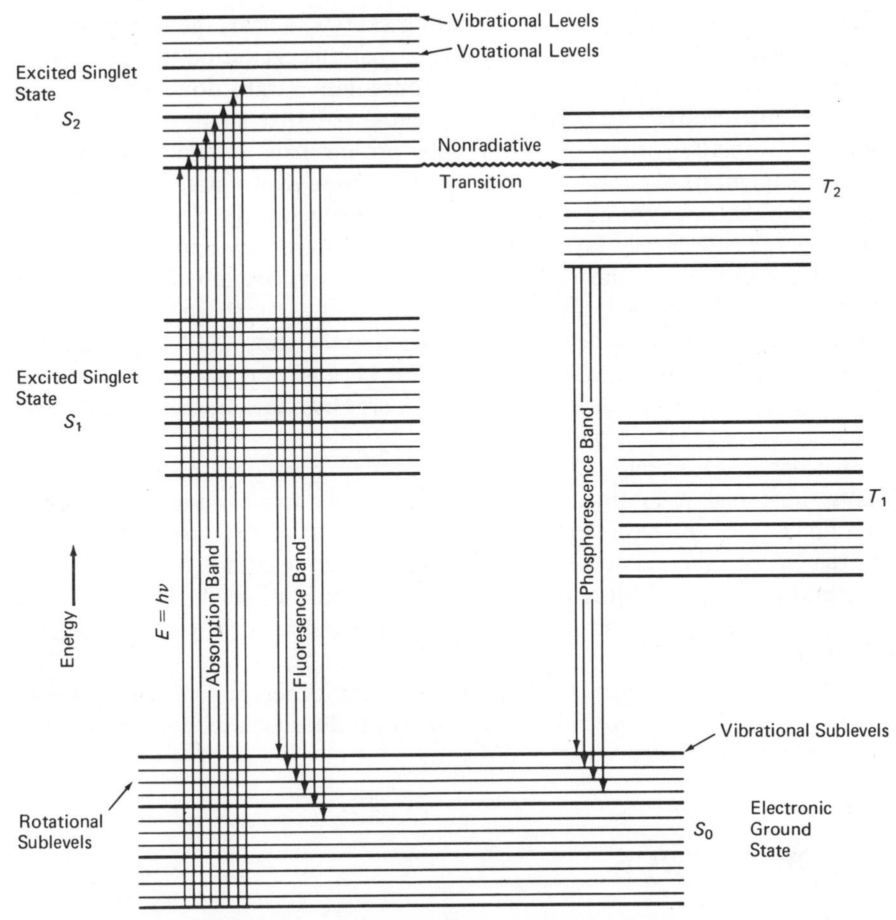


**Figure 19-1.** Simplified molecular energy level diagram.

of $\frac{1}{2}$), the state is referred to as a **triplet state**. The quantum-mechanical selection rule governing electronic transitions in molecules forbids singlet–triplet transitions.

A simplified energy diagram is shown in Figure 19-1 for a molecule that has a singlet ground state ($S_0$), two excited singlet states ($S_1$ and $S_2$), and two triplet states ($T_1$ and $T_2$). Associated with each electronic state are various vibrational-rotational sublevels.

When a photon of energy is absorbed by a molecule, an electron in the singlet ground state may be promoted to any vibrational-rotational sublevel of either singlet excited state. The lifetime of an excited singlet state is very short, on the order of $10^{-8}$ seconds. However, during the lifetime of the excited singlet state, it is extremely probable that the molecule will lose some of its excitation energy as heat, ending up in the vibrational ground state of the particular excited singlet state. An electronic transition from the vibrational ground state of the singlet excited state to the singlet ground state results in the emission of a photon. This singlet–singlet transition is known as **fluorescence**. Because the molecule loses some of its initial excitation energy as heat, the wavelength of the fluorescent radiation is generally longer than the excitation wavelength.

In some cases it is possible to have a nonradiative transition from a singlet excited state to a triplet excited state. Because radiative singlet–triplet transitions are formally forbidden by the quantum-mechanical selection rule, the molecule might be expected to remain in the excited triplet state forever owing to the fact that the ground state is a singlet. This is not the case, however, as the selection rule simply says that such transitions are very improbable; given enough time, therefore, the triplet–singlet transition will occur with the emission of a photon. This triplet–singlet transition is known as **phosphorescence** and is characterized by a time delay in the emission of the photon. Phosphorescence is generally observed only at low temperature, because at room temperature the triplet state will generally lose its excitation energy by nonradiative means before it has a chance to emit a photon.

In general, atomic and molecular transitions are related to the wavelength of energy absorbed or emitted, and the wavelength of energy emitted or absorbed is a function of the electron levels that are affected. X rays are produced by movement of inner-shell electrons, far ultraviolet by middle-shell electrons, near ultraviolet and visible by outer-shell or valence electrons, near and mid infrared by molecular vibrations, and far infrared and microwaves by molecular rotations.

We will discuss the emission methods in the order of increasing energy associated with the emission beginning with fluorescence and continuing through emission spectrography and x ray.

# FLUORESCENCE

## FLUORESCENCE SPECTRA

Both excitation (absorption) and emission fluorescence spectra are characteristic of compounds, with the excitation spectra normally showing peaks at lower wavelengths than the emission spectra. Because of the large number of possible excited states, both absorption and emission are characterized by band spectra rather than line spectra, and, expecially in solution, the bands tend to overlap or be fused together to give a structured spectrum similar to ultraviolet or visible absorption spectra. The long-wavelength portion of the excitation spectrum usually overlaps the short-wavelength portion of the emission spectrum, as shown in Figure 19-2.

The intensity of fluorescence is proportional to the concentration of the fluorescing species, but measurement is limited to very low concentrations. The fluorescent light is emitted in all directions and, as a consequence, some of the emitted light may be absorbed by other molecules before it leaves the cell. If the energy is emitted from the second molecule, it may not be in the direction of measurement and may have a longer wavelength, which is not within the wavelength band being measured. Also, the molecules at the back of the cell

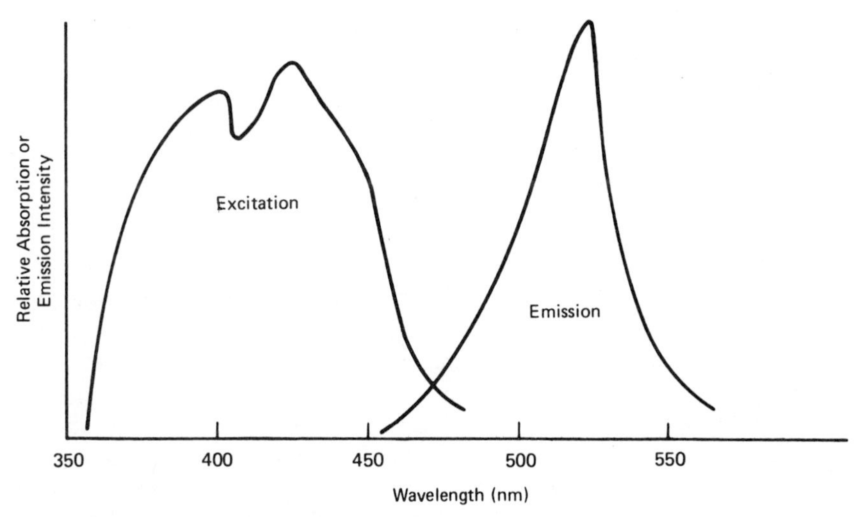


**Figure 19-2.** Typical excitation (absorption) and emission spectra for a fluorescent compound.

are being irradiated with light of less intensity, so there is a decrease in fluorescent intensity from the front to the back of the cell. As a consequence of this and the fact that the light is being emitted from all parts of the cell instead of from a single point, measurements are usually made of light emitted from the center of the cell and at right angles to the entering beam. The measurement of fluorescent intensity, however, is made against a small background; consequently, small differences in intensity can be measured with greater precision than an equivalent small decrease in intensity of a beam of light due to absorption.

If only a single absorbing and fluorescing species is present,

$$F = P_0 K(1 - 10^{-A}) \tag{19-1}$$

where $F$ is the intensity of fluorescent light reaching the detector, and $P_0$ is the intensity of the incident ultraviolet radiation. $K$ is a constant for a given system and instrument and corrects for the fluorescent light that does not reach the detector. $A$ is the absorbance of the solution at the nominal wavelength of the incident ultraviolet light. At low concentrations ($A = 10^{-2}$ or less) eq. (19-1) becomes

$$F = 2.30 P_0 KA = 2.30 K P_0 abc \tag{19-2}$$

where $a$ is the absorptivity; $b$ is the path length in the primary direction; and $c$ is the concentration. At higher values of $A$ and thus of $c$, $K$ is decreased as a result of reabsorption and decreased incident intensity at the position of measurement so that $F$ is no longer linearly proportional to $c$. This restricts measurements to concentrations of approximately 10 ppm or less.

Since fluorescence measurements are made on concentrations in the parts-per-billion and parts-per-million range, extreme care must be exercised to ensure that interferences are kept to a minimum. Errors may be caused by trace amounts of dirt left on poorly cleaned glassware, by soap films left on incompletely rinsed glassware, by grease from fingerprints, and by trace amounts of heavy metals present as impurities in reagents.

## INSTRUMENTATION

Photoelectric instruments for the measurement of fluorescence are known variously as fluorometers, fluorimeters, photofluorometers, spectrofluorometers, and fluorophotometers. These instruments have essentially the same components as found in the absorption photometers discussed in Chapter 16, but the arrangement is usually different. A typical arrangement is shown in Figure 19-3, in which the measurement is made at right angles to the incident radiation. Measurement at smaller angles has some advantages and is incorporated into some instruments. The mercury arc lamp creates a heat problem, so most instruments include some method of heat shielding or heat dissipation.

Single-beam filter fluorometers are simple and satisfactory for many measurements, but some of the more sophisticated instruments use gratings. Double-beam spectrofluorometers have certain advantages, but because of the expense of the double-beam construction are often restricted to research rather than routine procedures.

## APPLICATIONS

Visible fluorescence measurements are utilized for the determination of both inorganic and organic substances. The well-known fluorescence of certain uranium compounds is utilized in the determination of small quantities of

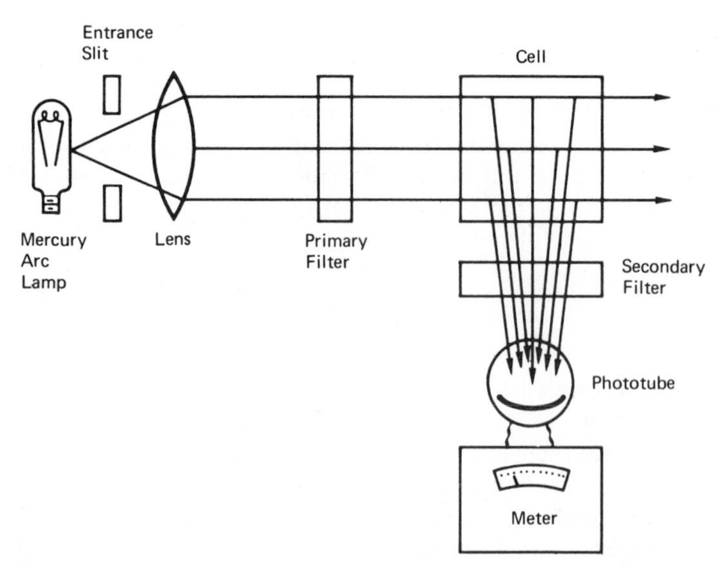


**Figure 19-3.** Simplified schematic diagram of a typical filter fluorometer.

uranium in various ores. Many metals that do not fluoresce in inorganic compounds form organometallic complexes that do fluoresce and thus can be determined by fluorometry.

There are more examples of fluorescence measurements in the organic field than in the inorganic. Fluorescence may be expected generally in molecules that contain multiple-conjugated double bonds capable of resonance, such as aromatic and heterocyclic compounds. Electron-donating substituents tend to favor or increase fluorescence, whereas electron-withdrawing substituents often diminish or destroy the fluorescence of a compound. Fluorescence of some organic compounds, such as fluorescein, is so affected by pH that they can be used as indicators in acid–base titrations.

Several vitamins, including thiamin and riboflavin, are determined regularly by fluorometry, and quinine and quinidine are used as standards in fluorescence procedures. Aflatoxins are also determined by fluorescence after extraction from foods that contain them.

Some fluorescent materials can be quenched by a change in conditions, such as pH or by addition of another compound. **Quenching** in this sense means the diminution or destruction of the intensity of fluorescence of a compound under a given set of conditions. Quenching is used in analytical procedures to offset various types of interferences and to determine mixtures of compounds. The fluorescence is measured, quenched, and then measured again. The difference is due to the substance quenched and thus is a measure of its concentration.

Fluorescence is also used as a method of locating colorless compounds in column, thin-layer, and paper chromatography. If the substance being separated fluoresces, after the separation the column, plate, or strip is irradiated with ultraviolet light, which causes the substance to fluoresce and reveal its location. If the substance does not fluoresce but quenches fluorescence, a fluorescent substance is incorporated with the column material. Again the column is irradiated with ultraviolet light after the separation, but this time the position is located as an area that does not fluoresce.

## EMISSION SPECTROSCOPY

The development of the spectrograph as an analytical tool actually began with the discovery by Bunsen and Kirchhoff that the spectra of colored flames were characteristic of the metallic salts that caused the colors. It was realized fairly early in the development that the more energetic the excitation the larger the number of lines that occur in the spectrum and the larger the number of elements that can be excited sufficiently to emit characteristic radiant energy. In the previous section we explored emission caused by the relatively low energies associated with absorption of visible and ultraviolet light (fluorescence). In this section, we will be concerned with higher energy excitation, which allows the determination of all metals and some nonmetals. With some of the metals, such as sodium and potassium, the spectra consist of only a few lines; with others, such as iron, literally thousands of distinct and reproducible lines are produced by the higher excitation methods.

Quantitative analysis with the spectrograph is based on the relation of the power or intensity of the emitted radiation to the amount of the element being excited in the sample under investigation. The intensity of the emitted light is a function of a large number of variables, which include the temperature of the sample caused by the excitation procedure, the shape and size of the electrodes, the distance of separation of the electrodes, and the electrode material, as well as the quantity of the element in the sample. As a consequence rigid standardization procedures must be utilized, and the spectra must be compared to the spectra of standard samples excited under the identical conditions.

In spite of these difficulties, modern automatic recording spectrographs are capable of giving the percentage of a number of elements directly in only a few minutes when the same type of samples are being investigated. Most metals can be detected in concentrations of 0.001% or less, and many of them can be quantitatively determined in these ranges. Spectrographs are useful in the qualitative determination of the composition of complex materials, expecially for the presence of minor or trace constituents, since the lines of all the metals present are excited and recorded.

## EXCITATION METHODS

Most spectrographs utilize either an ac arc, a dc arc, or an ac spark for excitation, but lasers have been used, especially when very small areas are being investigated.

The **dc arc**, produced by 50–250 V, is the most sensitive method and is widely used in qualitative identification procedures. Vaporization of the sample occurs from the heating caused by the passage of the current, the emission lines being primarily those of the neutral atoms. The arc, however, is unsteady in that it tends to flicker or wander and this causes uneven sampling of the surface. Since a relatively large amount of material is vaporized, the dc arc is very sensitive and can be used for the determination of materials present as very small traces.

The **ac arc** utilizes a potential of 1000 V or more and is steadier and more reproducible than the dc arc. For reproducible results, the electrode separation, the voltage, and the current must be carefully controlled. The ac arc is preferable for the analysis of residues of solutions that have been evaporated onto the surface of the electrodes.

The **ac spark** gives higher excitation energies and produces ionic spectra rather than atomic spectra in most cases. It is operated at high potentials (10,000–50,000 V) and is the preferred source whenever high precision rather than extreme sensitivity is required. The spark is more reproducible and stable than the arc and consumes less material, so higher concentrations may be determined. However, since the spark may strike to a particular small spot on the electrode in successive discharges, the results may not be representative.

There is less heating effect, so low-melting materials and solutions may be analyzed readily.

**Laser** beams, which can be focused on a small area of the sample surface to cause vaporization of even the most refractory materials, have been introduced as high-energy sources. If the vapor is not sufficiently excited thermally by the evaporation, an electric discharge is superimposed to cause satisfactory emission. No sample preparation is required, and the beam may be focused onto an area as small as 50 $\mu$m in diameter; thus, very small areas may be studied to determine whether or not localization of materials has occurred on surfaces during solidification. Because of the small area affected lasers are not satisfactory for the analysis of macro samples, unless the sample is completely homogeneous.

## ELECTRODES AND SAMPLE PREPARATION

The electrodes may be prepared from the material being investigated if there is a sufficient amount and the substance is a conductor that will withstand high temperatures. If the sample cannot be used as a **self-electrode**, it is usually placed in a small core in a spectrographic-grade graphite or carbon electrode. The **counterelectrode** (upper electrode) is also of graphite and is usually formed with a blunt point. The bottom or **sample electrode** may be formed in several different patterns but the most popular is the crater with a center post, shown in Figure 19-4(a). The center post helps to decrease the wandering of the arc. Steadier arcs may also be produced by mixing the sample with powdered graphite. In some cases, the counterelectrode may be made of a pure metal or an alloy of uniform composition and the spectrum used as the reference spectrum.

Solutions can be evaporated into the crater of the lower electrode or can be flowed over a rotating graphite disc. In the **porous cup technique**, a hole is drilled into the center of a carbon rod, leaving only about 1 mm of solid at the

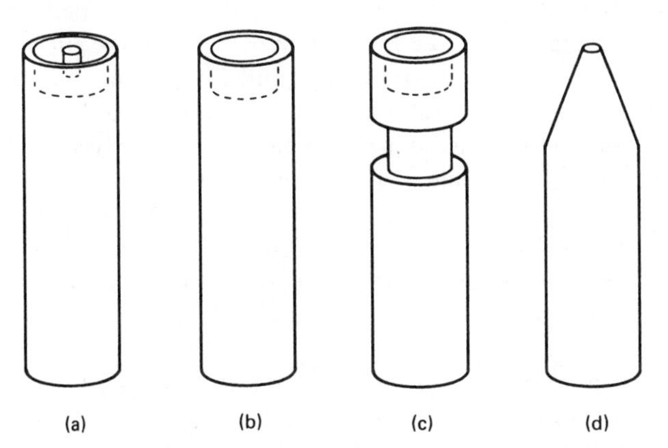


**Figure 19-4.** Typical electrode forms: (a) crater with center post; (b) crater; (c) narrow neck; (d) counter electrode.

bottom. The solution is placed in the cup, and the cup used as the upper electrode. During the analysis, the solution slowly seeps through the bottom of the cup and is vaporized in the arc or spark used.

## INSTRUMENTATION

Every spectrograph will have a slit, a dispersion device, and a recording device. The two main types of dispersion devices are prisms and gratings, and the two main types or recording devices are cameras and direct readers, which use photomultiplier tubes to eliminate the camera film and consequent development procedures. Since the spectral lines recorded are replica images of the slit, it is necessary that the edges be straight, parallel, and sharp. The slit should be continuously variable and both edges should move equally during adjustment in order to keep the center of the slit in the same position. Spectrographs are referred to as **prism** or **grating** instruments, since the usual method of classification is based on the dispersion device used.

Most of the prism instruments use the Cornu or Littrow mounting for the prism shown in Figure 19-5. The **Cornu mounting** involves the use of two pieces of quartz, one of which is right-handed and the other left-handed, to offset circular polarization. Quartz has the property of rotating the plane of polarized light and will also separate unpolarized light into two beams that are circularly polarized in opposite directions. The **Littrow mounting** utilizes only one piece of quartz with the back surface of the prism metallized that so it will reflect the light back through the prism and eliminate the polarization effects. The **Wadsworth mounting**, a modification of the Cornu, uses a mirror in the path of the emergent beam to produce a beam that is parallel to the entering beam.

Grating spectrographs usually use concave gratings in the Eagle, Wadsworth, or Rowland mountings, shown in Figure 19-6. All of these are based on the **Rowland circle**, a circle with a radius of curvature one-half the radius of the grating itself. If the slit and grating both lie on the Rowland circle, the images of the slit are focused somewhere on the circle. In the **Rowland mounting**, the film and grating are at right angles to the slit and all are on the Rowland circle. The **Wadsworth mounting** uses a concave mirror between the slit and grating so that the grating will be illuminated with parallel light. This mounting has the advantage that it is stigmatic rather than astigmatic, as are most of the other mountings. Stigmatic gratings focus both horizontal and vertical lines at the same place. The **Eagle mounting** uses a reflecting prism or mirror between the slit and the grating so that the slit does not have to lie on the Rowland circle. This is the most compact mounting and produces only slight astigmatism. The **Ebert** and **Czerny–Turner mountings** use plane gratings and concave mirrors instead of concave gratings. The Czerny–Turner is stigmatic and is, therefore, one of the better mountings.

Dispersion and resolving power are used to compare instruments. The **dispersion** is a measure of the ability of the spectrograph to separate two lines that have similar wavelengths ($\lambda$). It is often expressed as *linear dispersion*, $\Delta x/\Delta\lambda$, where $\Delta x$ is the distance in millimeters on the plate that separates two

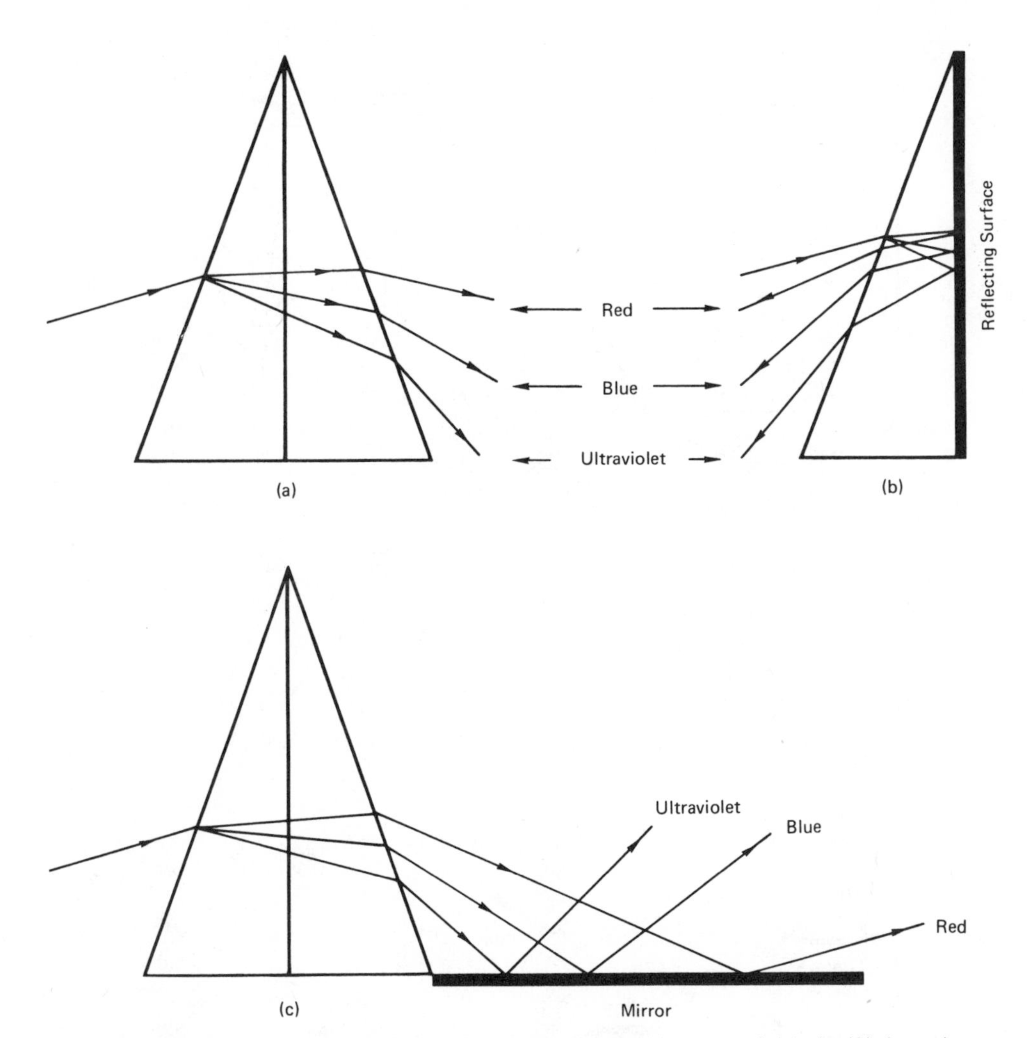


**Figure 19-5.** Prism mountings. (a) The Cornu (b) the Littrow, and (c) the Wadsworth. The Cornu and the Wadsworth utilize two pieces of quartz, one right-handed and the other left-handed. The Wadsworth also uses a mirror to render the light of minimum deviation parallel to the entering beam. Glass prisms for the Cornu and Wadsworth mountings are made of single pieces of glass.

lines $\Delta\lambda$ apart. The **resolving power** is a measure of the ability of the spectrograph to resolve two lines that have only slightly different wavelengths. The linear dispersion varies with wavelength for prism instruments but is essentially constant for grating instruments. The resolving power for prisms increases with the thickness of the prism and for gratings increases with the number of lines in the illuminated portion of the grating.

The majority of spectrographs utilize a camera and film as the recording device. The density of the spectral line recorded by the photographic emulsion is a function of the intensity of the emitted radiation under standardized conditions of exposure. Since the density of a line of given intensity varies from

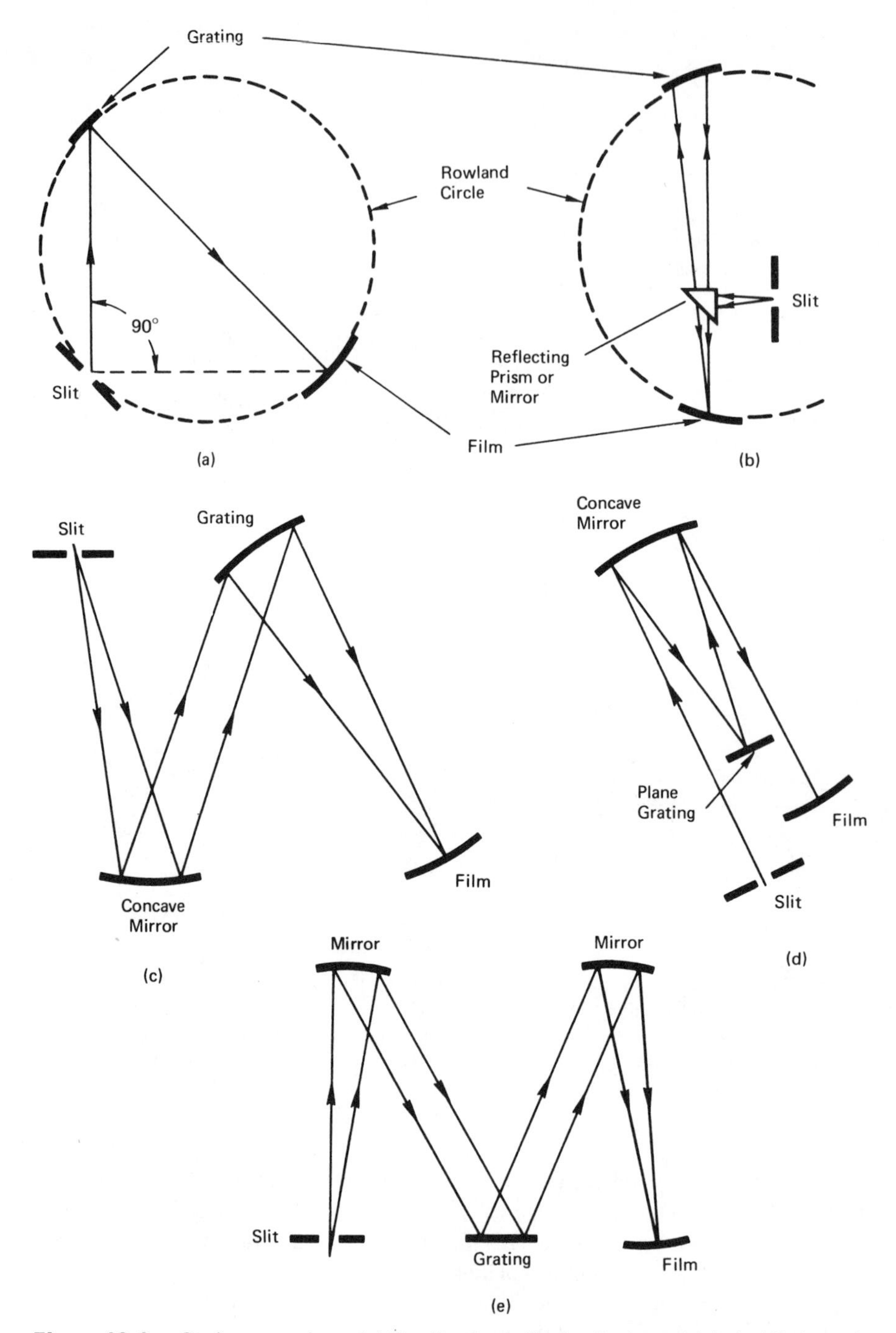


**Figure 19-6.** Grating mountings. (a) The Rowland, (b) the Eagle, and (c) the Wadsworth mountings use concave gratings, whereas (d) the Ebert and (e) the Czerny–Turner use plane gratings and concave mirrors.

emulsion to emulsion and is a function of the wavelength and conditions of development, each new batch of film purchased must be evaluated under the standardized conditions used for exposure and development.

## APPLICATIONS

### *Qualitative Identification*

One major use of the spectrograph is the identification of trace constituents as well as the major components of complex mixtures of materials, especially alloys and ores. The elements present in a sample can be determined by comparison of the spectrum of the unknown with the spectra of pure samples, or by measuring the wavelength of individual lines and using a table of wavelengths to ascertain which elements are present. Most of the work is done by comparison rather than calculation. The detection of trace constituents is usually based on the **raies ultimes** (RU), which are the last lines to disappear as the concentration of the element is continually decreased.

### *Quantitative Analysis*

Quantitative analysis involves the measurement of the density of the spectral line for an element on the photographic film. Since this density depends on a large number of factors, an internal standard is usually used. The intensity of the analytical line is then measured against a line of the internal standard.

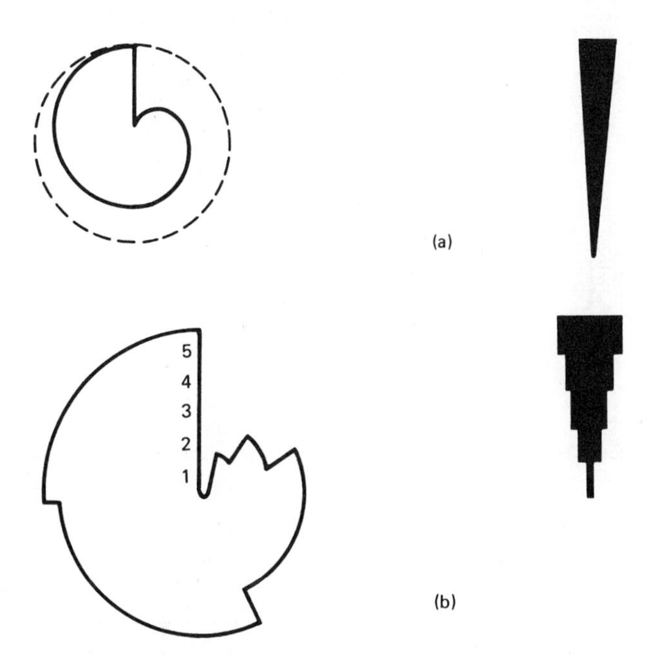


**Figure 19-7.** Rotating discs with types of line spectra obtained: (a) log sector; (b) step sector.

The lines elected for such comparison must be a **homologous pair**, a pair of lines that respond in the same way to variations in the excitation conditions.

The density of the spectral lines may be estimated visually, but better results are obtained by the use of a **densitometer**. One simple densitometer shines a light of given intensity through the film and records the amount of light transmitted by the film with an ordinary vacuum-tube photocell. The film is moved lengthwise over the light beam to give a continuous recording of the film density. Even better estimation can be obtained by using the rotating step or log sectors shown in Figure 19-7 in front of the slit. Since the width or density of the line decreases in each step (continuously with the log sector), the height of the line is a measure of the concentration. The use of densitometer estimation of the densities along a selected step in which the densities are not too great allows satisfactory accuracy of estimation for both major and minor components.

## X-RAY METHODS

X-ray methods in analytical chemistry may be classified as being absorption, emission, or diffraction methods. The absorption methods are similar to the absorption methods discussed in earlier chapters, except that the source of radiation is an x-ray tube instead of a tungsten lamp or other device. Since x rays involve electronic transitions in the inner electron shells associated with high energies, x radiation is characterized by short wavelengths and high frequencies. Absorption and fluorescence measurements are not affected by the chemical combination of the elements, except for the very light elements. Emission methods are mainly involved with the use of x rays to cause emission or fluorescence in other atoms, since the use of the substance being analyzed as

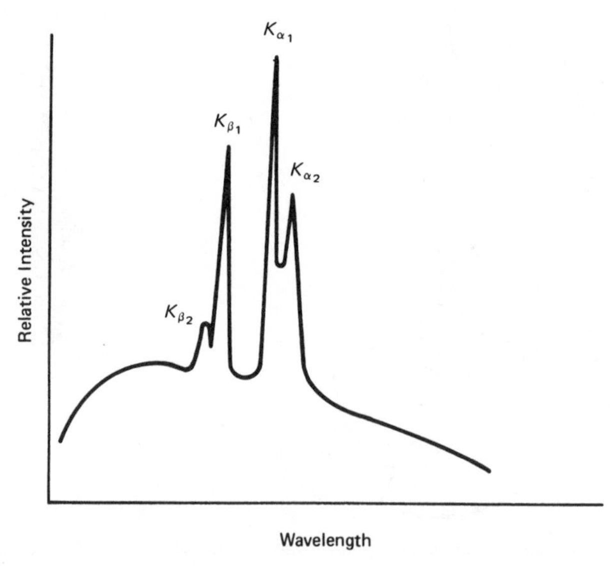


**Figure 19-8.** Typical spectral distribution of radiation obtained from an x-ray tube.

the target electrode in an x-ray tube with consequent determination of the wavelengths and intensities of the lines emitted is seldom used as an analytical method. X-ray diffraction methods are based on the diffraction of x rays by the crystal lattice of a solid and can be used to identify solid compounds or uncombined elements.

When an element is used as the anode or target in an x-ray tube and is bombarded with a stream of electrons from the cathode, it emits continuous background radiation known as **Bremsstrahlung** and sharp spectral lines characteristic of the element as shown in Figure 19-8. The continuum is caused by loss of energy by electrons as they strike the target, and the sharp lines are caused by the emission of photons of definite energy. The photons are emitted when the energy of the electron beam is sufficient to knock an electron from the inner or $K$ shell. An electron from an outer shell then falls into the hole caused by the electron lost and emits a photon of energy. The sharp lines are designated as $K_\alpha$, $K_\beta$, and so on, for transitions to the $K$ level. X rays due to transitions from the $L$ to $K$ shell are called $K_\alpha$, with the subscripts 1 and 2 being used to designate the sublevels of the $L$ shell from which the electron originated; x rays caused by transitions from the $M$ to the $K$ shell are known as $K_\beta$ and utilize the same type of numerical subscripts.

## ABSORPTION METHODS

X rays can be absorbed by matter in much the same way that other wavelengths of radiation, such as visible and ultraviolet, can be absorbed. The main difference is that the absorption of x rays is largely an atomic rather than a molecular phenomenon because electronic transitions in the inner shells of atoms are involved. The high energies associated with such electronic transitions are characteristic of the atoms and not of the chemical combination of the atoms. In general, the heavier atoms show greater absorption at lower wavelengths than the elements of lower atomic number. Each elements absorbs certain characteristic wavelengths better than other wavelengths.

The absorption of x rays follows the Beer law and is a function of concentration. The **linear absorption coefficient** $\mu$, which represents the fraction of energy absorbed per centimeter from an x-ray beam with a cross section of 1 $cm^2$ can be used, but the **mass absorption coefficient** $\mu_m$ is sometimes more convenient. The mass absorption coefficient is defined by eq. (19-3).

$$\mu_m = \frac{\mu}{\rho} \qquad \textbf{(19-3)}$$

where $\rho$ is the density of the absorbing material. The mass coefficient has been found empirically to be related to the wavelength and atomic properties of the absorbing substance according to eq. (19-4).

$$\mu_m = \frac{CNZ^4\lambda^n}{A} \qquad \textbf{(19-4)}$$

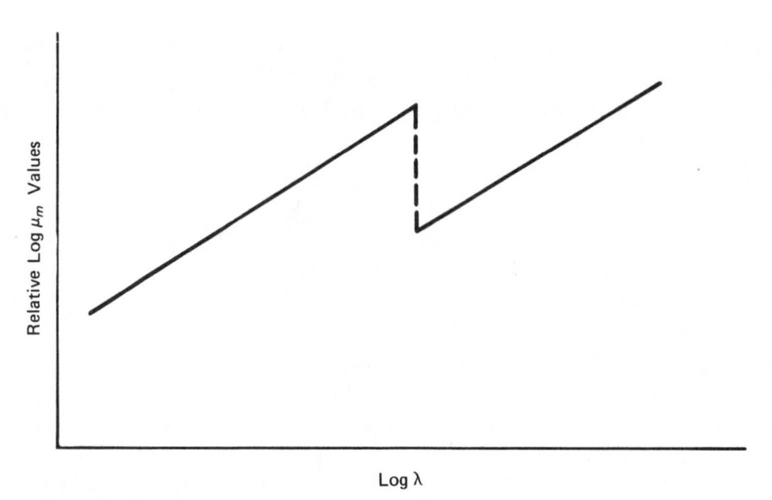


**Figure 19-9.** Typical x-ray absorption edge.

where $C$ is a constant within a given spectral range for all elements; $N$ is Avogadro's number; $Z$ is the atomic number; $A$ is the atomic weight; $n$ is an exponent with a value between 2.5 and 3.0; and $\lambda$ is the wavelength. The wavelength relationship is exponential so that, if logarithms are plotted, a straight line with a slope equal to the exponent of the wavelength should result. Figure 19-9 is a typical logarithmic plot of mass absorption coefficients versus wavelength. The sharp break is known as the ***K* critical absorption** or the ***K* absorption edge**. The absorption edges are caused by the fact that the radiation of longer wavelength (and thus lower energy) does not have sufficient energy to eject the $K$ electrons of the element and, consequently, is not absorbed as greatly as the shorter wavelengths with more energy. Heavy atoms also show similar absorption edges at longer wavelengths due to the ejection of $L$ and $M$ electrons. The value of $C$ in eq. (19-4) changes abruptly at each absorption edge.

The usefulness of x-ray absorption methods depends in part upon the measurement of one element that absorbs much more strongly than the other elements present. A good example is the determination of tetraethyllead (TEL) in gasoline. Lead is a heavy element that absorbs much more strongly than carbon, hydrogen, and the other elements present in gasoline. Even though the other elements do contribute to the absorption, the overall absorption of any sample is due primarily to the lead present, and lead concentrations in the range of 2 ppm can be determined readily with approximately 1% precision. Similarly, sulfur can be determined in gasoline and chloride in organic compounds that do not contain other heavy elements, such as sulfur, phosphorus, and iodine or bromine. The determination of chlorine in such compounds can often be completed in less than 5 min.

Absorption measurements at wavelengths just below and just above the absorption edge can be used to determine concentrations as well as to identify the element. A multichannel instrument is required, however, for satisfactory results.

Microradiography is an application that, utilizing the different adsorbing powers of various elements in an x-ray beam, is used to study the gross structure of small specimens under high magnification. The sample is placed in a compartment over a photographic film and irradiated. Upon development, positions where elements that absorb strongly are located appear as light spots, whereas the positions of those elements that absorb less strongly will appear dark.

## FLUORESCENT METHODS

When an unknown substance is caused to emit x radiation, the emission or fluorescent spectrum can be used as the basis of analysis, just the same as in ultraviolet and visible fluorescence methods. In practice, the sample is irradiated with x rays of shorter wavelength than the emission spectrum, and the intensity of the emitted radiation is measured. The intensity of the fluorescent x ray is approximately 1000 times smaller than the beam obtained by direct excitation with a stream of electrons. As a consequence, high-intensity x-ray tubes, very sensitive detectors, and suitable x-ray optics are necessary in x-ray fluorescence instruments.

For quantitative analysis, the instrument (Figure 19-10) is calibrated with standards of known composition and particle size. The packing of a finely divided sample will affect the apparent density and, because the emission is preceded by absorption, the sample density must be uniform. Internal standards may be incorporated into a sample to check and correct for nonuniformity of packing. The internal standard can also be used to determine and offset matrix effects that may be causing interference. Fluorescence x-ray analysis is inherently precise and rivals the accuracy of wet chemical methods for major components. It is not good for trace elements and is best adapted to the determination of

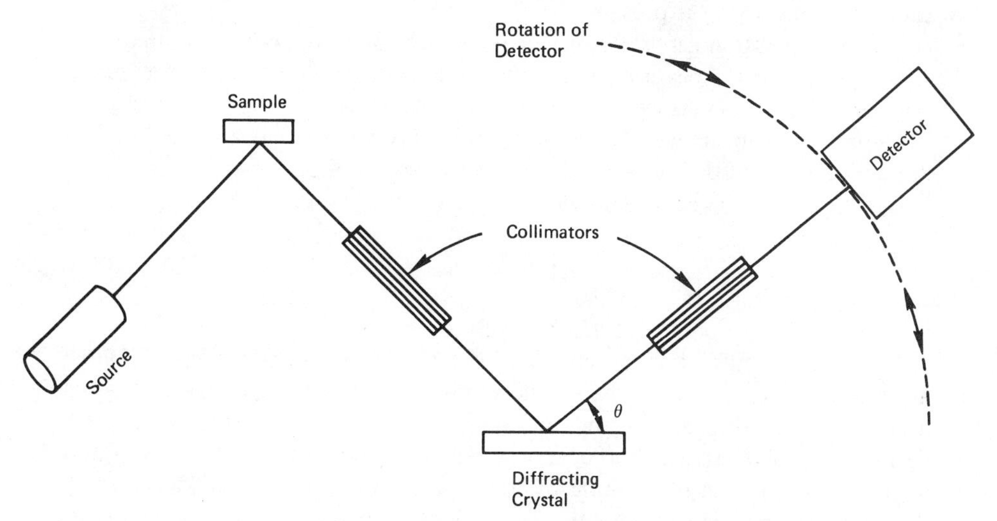


**Figure 19-10.** Simplified schematic diagram of an x-ray fluorescence spectrometer. The diffracting crystal or the detector and collimator may be rotated to give any desired value to the angle $\theta$.

elements for which the wet chemical methods are not reliable or require too long for completion.

X-ray fluorescence methods provide separate spectral information for each element in the sample and, consequently, are less subject to interference from the other elements in the sample than are x-ray absorption methods. For example, tetraethyllead in gasoline can be determined by fluorescence as well as absorption methods. In the absorption method, lead can be determined only because it absorbs much more strongly than any other element present. The measurement, however, is of the total absorption due to lead and the other elements. In the fluorescence method, the measurement involves the emission from lead atoms only and thus gives more accurate estimation of tetraethyllead than the absorption method.

X-ray fluorescence methods are very useful for the analysis of metal alloys, since the sample need not be dissolved or destroyed in any way. Simultaneous analysis of several elements is possible with the automatic equipment commercially available.

## X-RAY DIFFRACTION METHODS

If a monochromatic beam of x rays is passed through a single crystal or a powdered crystalline specimen, the rays will be diffracted and a **diffraction pattern** obtained that is characteristic of the crystalline substance. Because each crystalline substance produces a distinctive pattern, the patterns can be used for identification. Also, since the diffraction spectrum of a mixture is a simple additive combination of each component spectrum and the relative intensities of the patterns are a function of the relative concentrations of the components, the x-ray diffraction pattern of a mixture can be used to estimate the amount of each component that is present.

The equipment, in general, consists of a source of x rays, a filter or analyzing crystal to render the beam monochromatic, a specimen holder, and a detector, together with the necessary circuitry and recording device.

The relationship between the wavelength ($\lambda$) of the x-ray beam, the angle of diffraction ($\theta$), and the distance ($d$) between each set of planes shown in Figure 19-11 is given by the **Bragg equation,**

$$n\lambda = 2d \sin \theta \qquad \textbf{(19-5)}$$

in which $n$ represents the order of the diffraction.

To scan the diffraction spectrum of a crystalline material, the collimated x-ray beam from an x-ray tube is reflected from an **analyzing crystal** of known $d$ spacing onto the sample and the angle changed to select the wavelength desired. A **goniometer** is often used; this allows the detector to be moved in a circular path around the sample while the sample is being rotated at one-half the speed of the detector. A circular camera with the sample placed in the center may be used to record the diffraction pattern on x-ray film. In the photographic method, single crystal samples are usually rotated and powdered

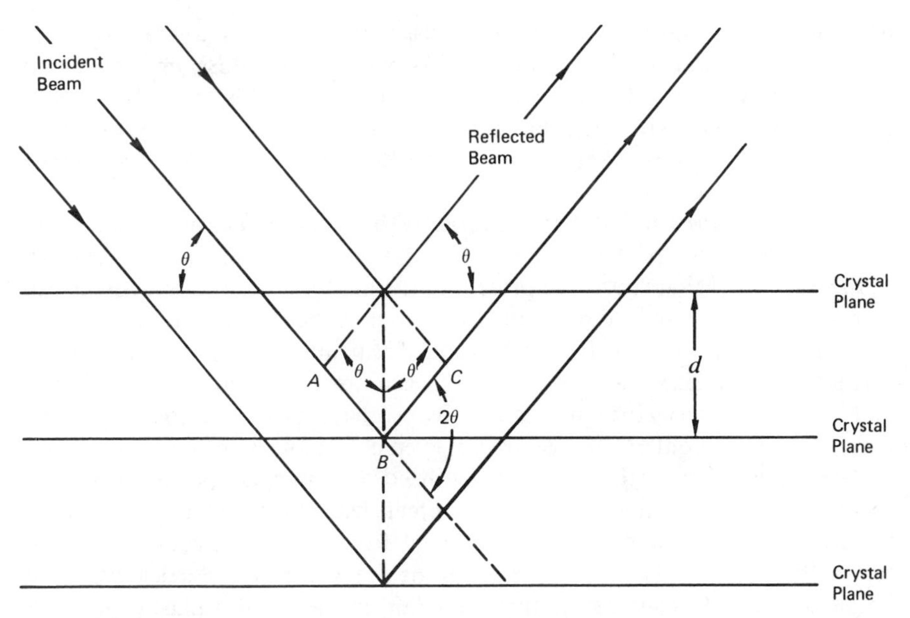


**Figure 19-11.** Diffraction of x rays by successive crystal planes. The distance *ABC* equals $n\lambda$.

samples held in a fixed position. The detector used with a goniometer can be a scintillation, proportional, or Geiger counter, with the intensity of the diffracted beam being recorded continuously on a strip-chart recorder or stepwise on a digital printer. The *d* spacings can then be determined from the chart or from the film. Better precision is obtained with the counters than with the photographic method. The principal advantage of the camera is that it provides a means of recordings many reflections at the same time.

X-ray diffraction is used primarily in the identification of crystalline compounds, even though each atom in the crystal has the power of scattering an x-ray beam incident upon it. The sum of all the scattered waves results in the diffraction pattern in which the beam, in effect, is diffracted from each allowed plane in the crystal. As a consequence, each crystalline substance produces a characteristic diffraction pattern that is a "fingerprint" of its atomic and molecular structure. Because different atoms have different scattering powers, two compounds having identical crystal lattices will produce different diffraction patterns. Also, different crystalline forms of the same substance will show different patterns. For example, there are several different crystalline iron(III) oxides, each of which can be identified by its x-ray diffraction pattern. There are several libraries of powder diffraction films and card files of *d* spacings available that include many thousands of compounds and can be utilized in the identification of unknown substances by comparing the films or the calculated *d* spacings.

One example of the use of x-ray diffraction patterns is in the determination of the compounds present in a mixture of soluble compounds, such as sodium nitrate and potassium chloride. Even though the amount of each ion can be determined quantitatively by wet methods, there is no way to state whether the mixture contains these two compounds or contains potassium nitrate and sodium chloride instead. The other possibility, of course, is that all four substances are present. The x-ray diffraction pattern, however, will indicate which of the possible substances are present and can also furnish an estimate of the relative amounts of each.

X-ray diffraction is adaptable to quantitative applications because the intensity of a given diffraction peak of an individual compound in a mixture is a function of the fraction of that substance present in the mixture. Calibration curves can be prepared from standard mixtures, but care must be exercised to take matrix effects into consideration in the comparison of peak intensities. As a consequence, quantitative determinations are usually accurate to within only 5 or 10% and are used primarily for semiquantitative estimations of the amounts present. Components of about 5% or less are not even detectable in most cases.

X-ray diffraction methods may be utilized for many purposes other than qualitative and quantitative analysis. A great deal of information concerning the structure of polymers can be obtained from x-ray diffraction patterns, especially as the to repeat distance along the polymer chain. Particle sizes can be measured and positions of stress or strain in metals and plastics located. Also, because the pattern is caused by the geometric arrangement in the crystal, it is possible with special techniques to arrive at ultimate structural information of crystalline substances. The structures of many important but complicated biochemical compounds have been determined in this manner.

## PROBLEMS

1 In an analysis of a sample of 2-grain quinidine sulfate tablets, 20 tablets taken from one bottle weighed 10.050 g. After grinding and mixing, a sample of 0.5035 g was weighed and agitated in 0.1 $N$ $H_2SO_4$. The undissolved filler was removed by filtration and the filtrate diluted to 500.0 mL with 0.1 $N$ $H_2SO_4$. A 1.00-mL aliquot of this solution was diluted to 1000 mL with 0.1 $N$ $H_2SO_4$.

A concentrated standard was prepared by transferring 0.1005 g of pure quinidine sulfate to a liter flask and diluting to volume with 0.1 $N$ $H_2SO_4$. A dilute stock standard was prepared by diluting 10.00 mL of the concentrated standard to 1 liter with 0.1 $N$ $H_2SO_4$. Working standards were prepared by diluting 0-, 10.00-, 25.00-, 35.00-, and 50.00-mL aliquots of the dilute standard to 100.0 mL with 0.1 $N$ $H_2SO_4$.

The working standards and the sample were measured on a Coleman Model 12C Photofluorometer, with the following results:

| *Solution* | *Meter Reading* |
|---|---|
| Blank | 0.0 |
| 10.00 mL | 20.1 |
| 25.00 mL | 50.0 |
| 35.00 mL | 69.9 |
| 50.00 mL | 100.0 |
| Unknown | 55.2 |

Calculate the concentrations of the working standards and prepare a calibration curve by plotting meter reading versus $\mu g$ of quinidine sulfate per mL for the working standards. Read the concentration of the unknown from the graph and calculate the number of mg of quinidine sulfate in an average tablet. Also, express the results as percentage of the declared amount.

*Ans.* 137.7 mg; 106.3%

**2** Repeat problem 1 using the following data:

| | *Weight of 20 Tablets (g)* | *Weight of Sample (g)* | *Meter Reading of Unknown* |
|---|---|---|---|
| (a) | 15.360 | 0.7880 | 59.3 |
| (b) | 6.124 | 0.3052 | 52.9 |
| (c) | 11.260 | 0.5590 | 49.8 |

CHAPTER 20

# Electrolytic Methods: Electrogravimetry, Polarography, and Coulometry

In Chapters 11 and 13, we discussed the generation of potential by galvanic cells and the applications of potentiometric measurements in analytical chemistry. In this chapter we will explore some of the electrolytic methods (often referred to as electrolysis) that are the reverse of potentiometric methods in that they require the use of an outside source of potential to force a current to flow through a system. In galvanic cells, the reactions that occur at the electrodes or half-cells would occur spontaneously if the electrode reactants were mixed in the same solution. In electrolysis the reactions that occur at the electrodes would not occur spontaneously if the electrode reactants were mixed in the same solution. In the case of electrolytic methods, the source of potential (either a battery, generator, or transformer) must produce sufficient energy to force a chemical reaction to take place which would not take place spontaneously upon mixing. This is known generally as **electrolysis** and the cell in which it takes place as an **electrolytic cell**.

Electrolytic methods include electrogravimetry (also referred to as electrodeposition and electroplating), polarography, and coulometry. In **electrogravimetry**, the passage of the current causes the deposition of a metal upon a previously weighed inert electrode. After deposition the electrode is again weighed to determine the weight of the metal deposited. In **polarography**, conditions are controlled so that the rate of deposition or reaction (and thus the amount of current) at a microelectrode is a function of the concentration of the electroactive ion in the solution. In **coulometry**, accurate measurement of the number of coulombs that pass through the system is used to calculate the amount of some substance present in the solution.

# ELECTROGRAVIMETRY

## DECOMPOSITION POTENTIAL

If a source of variable potential is applied to two inert platinum electrodes immersed in a solution of copper(II) sulfate and the potential is increased continuously, no appreciable current will flow through the system until some minimum potential is applied, after which the current will increase as the applied potential increases, as shown in Figure 20-1. The minimum potential necessary to cause a finite current to flow through an electrolytic cell is known as the **decomposition potential** and is composed of several factors, as indicated by eq. (20-1):

$$E_{a(\min)} = E_d = E_B + E_s + E_v \tag{20-1}$$

where $E_a$ = applied potential,
$E_d$ = decomposition potential,
$E_B$ = theoretical counter (or back) potential,
$E_s$ = system potential, and
$E_v$ = overvoltage.

Each of these terms will be discussed separately in relation to the copper(II) sulfate electrolytic cell shown in Figure 20-2. The **system potential** ($E_s$) is also known as the **resistance potential** and is the potential necessary to overcome the resistance of the system. $E_s$ can be calculated from Ohm's law, $E = IR$, in which $E$ equals the potential in volts, $I$ is the current (in amperes), and $R$ is the resistance in ohms. An increase of the applied potential above the decomposition potential causes an increase in this term because the amount of current flowing through the system increases with increase of potential in this region.

## THE THEORETICAL COUNTERPOTENTIAL

When a potential is applied to the electrolytic cell, the source of potential forces electrons to move through the connector to one of the electrodes and tends to pull electrons from the other electrode. The electrode receiving electrons thus becomes negatively charged, as a result of the excess of electrons,

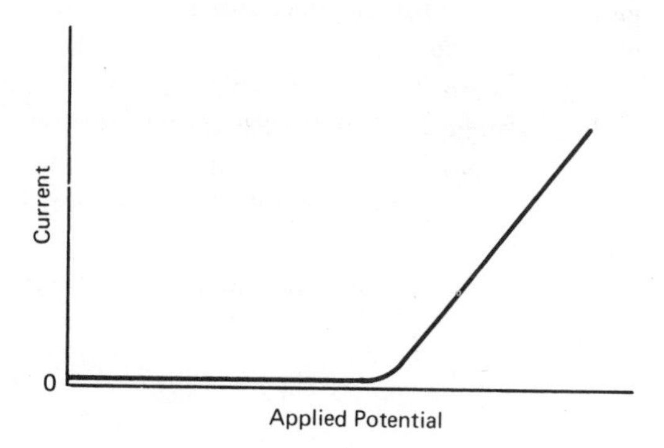


**Figure 20-1.** Variation of current with potential during electrodeposition.

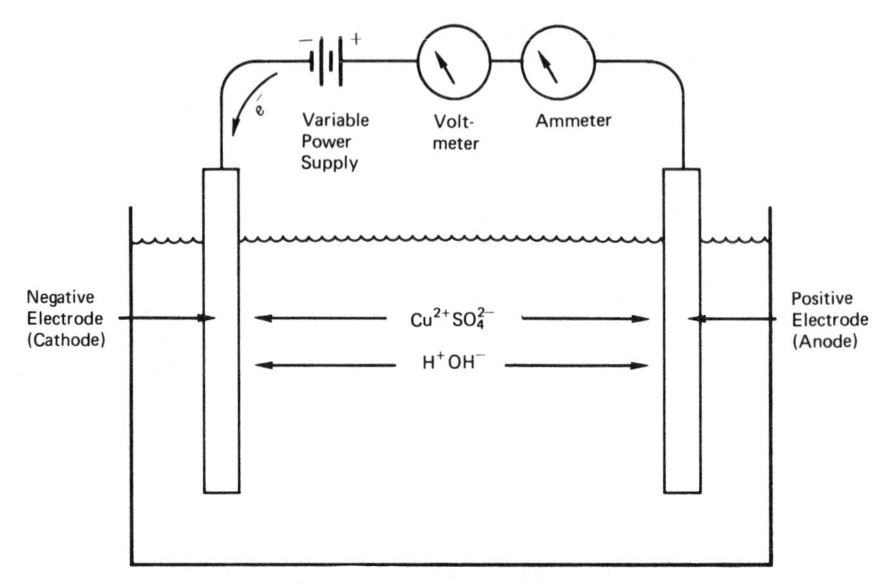


**Figure 20-2.** Simplified diagram of an electrolytic cell.

whereas the other becomes positively charged, as a result of a deficiency of electrons. As a consequence, the positively charged ions in the solution (copper and hydrogen) are attracted to and migrate toward the negative electrode (cathode), and the negative ions in the solution (sulfate and hydroxyl) migrate toward the positive pole (anode). As current flows through the system, copper ions are deposited on the negative electrode as copper atoms, since copper is discharged at lower potentials than hydrogen ion. At the positive electrode, hydroxyl ions are converted to oxygen molecules and hydrogen ions, since hydroxyl is easier to discharge than sulfate. As a general rule, metal ions or cations are discharged from aqueous solutions in the order of increasing reactivity, with the least reactive being plated first. Hydroxyl, in general, is discharged in preference to any other anion that contains oxygen, such as sulfate and nitrate.

As a result of the passage of the current, the cathode becomes a copper–copper ion half-cell and the anode becomes an oxygen half-cell. If the source of potential were suddenly removed, the electrolytic cell would operate as a galvanic cell, with the electrons moving in the opposite direction through the wire connecting the two half-cells. The potential that would be generated if the cell were operating as a galvanic cell is known as the **theoretical counterpotential** or **back potential** $E_B$). Because the electrons would flow in the opposite direction, the counterpotential operates in opposition to the applied potential and must be overcome before current will flow continuously through the electrolytic cell.

The Nernst equation for the copper half-cell is

$$E_{\mathrm{Cu}} = E^0_{\mathrm{Cu}} + \frac{0.059}{2} \log (\mathrm{Cu}^{2+}) \qquad \textbf{(20-2)}$$

and that for the oxygen half-cell is

$$E_{O_2} = E^0_{O_2} + 0.059 \log (H^+) \qquad \textbf{(20-3)}$$

because the equation for the discharge of hydroxyl is

$$4OH^- \longrightarrow O_2 + 2HOH + 4e^- \qquad \textbf{(20-4)}$$

and the $4OH^-$ can be assumed to be produced by the ionization of water,

$$4HOH \longrightarrow 4H^+ + 4OH^- \qquad \textbf{(20-5)}$$

which can be rearranged as

$$4OH^- \longrightarrow 4HOH - 4H^+$$

and substituted into eq. (20-4) to give

$$4HOH - 4H^+ \longrightarrow O_2 + 2HOH + 4e^-$$

or

$$2HOH \longrightarrow O_2 + 4H^+ + 4e^- \qquad \textbf{(20-6)}$$

for which the logarithm term of the Nernst equation is

$$E_{O_2} = E^0_{O_2} + \frac{0.059}{4} \log \frac{(H^+)^4(O_2)}{(HOH)^2}$$

which is equivalent to eq. (20-3), since the solution is saturated with oxygen and the water is in the standard state.

In practically all calculations involving electrolytic cells, molar concentrations can be substituted for active concentrations in the Nernst equation without major error. The theoretical counterpotential can be calculated as the potential the system would show if operating as a galvanic cell by

$$E_B = E_{\text{anode}} - E_{\text{cathode}} \qquad \textbf{(20-7)}$$

or, in the case of the copper sulfate electrolytic cell,

$$\begin{aligned} E_{B(O_2, Cu)} &= E_{O_2} - E_{Cu} \\ &= E^0_{O_2} + 0.059 \log [H^+] - E^0_{Cu} - \frac{0.059}{2} \log [Cu^{2+}] \end{aligned}$$

Note that both the hydrogen ion concentration and the copper ion concentration affect the value of the theoretical counterpotential. As the electrolysis proceeds, the hydrogen ion concentration increases and the copper ion concentration decreases. As a consequence, the theoretical counterpotential increases during the electrodeposition.

## THE OVERVOLTAGE

If the decomposition potential is calculated only from $E_B$ and $E_s$ and this potential is applied to the system, a continuous current will not flow through the cell. In other words, the experimentally determined $E_d$ is larger than the value calculated from the theoretical counter potential and the system potential. The difference between these calculated and experimental values is called the **overvoltage** ($E_v$). The mechanism that causes overvoltage is not completely understood, although several theories have been proposed to account for it. The simplest explanation is to consider the overvoltage to be the voltage necessary to overcome the opposition to the liberation of products at a finite rate or the potential necessary to overcome the electrical "inertia" of the system.

The magnitude of the overvoltage depends on several factors. Usually, overvoltage is low for deposition of metals and high for liberation of gases. The physical condition of the electrode surface influences the overvoltage for the liberation of gases because the overvoltage is usually lower from a rough-textured surface than from a smooth-polished surface. The current density and the temperature are also involved with the overvoltage, in general increasing with increase in current density and decreasing with increase in temperature.

## SEPARATION AND DETERMINATION OF METALS BY ELECTRODEPOSITION

One important analytical feature of electrodeposition is the ability to determine one metal quantitatively in the presence of one or more other metal ions. Since the decomposition potential ($E_d$) increases as the metal ion is plated, the decomposition potential for small concentrations is larger than the decomposition potential for larger concentrations. The conditions for quantitative deposition of a metal ion from the solution is that the applied potential ($E_a$) must be greater than the decomposition potential that occurs when the concentration of the ion being plated has been reduced to a negligible value. A negligible value in this case is considered to be 0.1 mg/100 mL or in the range $10^{-6}$ to $10^{-7}$ $M$ for most metals.

For quantitative separation of two metals, the applied potential must be greater than the decomposition potential at negligible concentration for the first ion to be deposited, and must also be smaller than the decomposition potential of any other metal ion in the solution.

It is possible to predict whether or not a given separation is theoretically possible by calculating the values of the decomposition potentials of all the ions in the solution to see which will be deposited first and then to calculate the decomposition potentials of all the other ions in the solution after the first metal has been quantitatively deposited. Comparison of these two potentials will indicate whether or not the separation is theoretically possible. However, since we are interested only in differences in potential, any factor that is constant (or has an equal value for all similar calculations) may be neglected for the comparison.

Because all the calculations for decomposition potentials involve the same electrolytic system, the $E_s$ will be the same and may be neglected. Also, since only one ion (usually hydroxyl) is being discharged at the anode, both the anode potential and the anode overvoltage will be the same and need not be calculated, since they would cancel each other when the decomposition potentials are subtracted to obtain the difference for comparison. This leaves only the cathode potential, which must be calculated since the overvoltage at the cathode is negligibly small in most cases. Since the cathode potential is subtracted [has a negative sign in eq. (20-7)], the smaller (or more negative) the cathode potential, the larger will be the decomposition potential.

• **Example.** Calculate whether or not the quantitative separation of copper and zinc is theoretically possible from a solution containing 10 mmoles of copper(II) nitrate, 20 mmoles of zinc nitrate, and 10 mmoles of nitric acid in 100 mL. First calculate the molar concentrations.

$$M_{\mathrm{Cu}} = \frac{10 \text{ mmoles}}{100 \text{ mL}} = 0.10\,M$$

$$M_{\mathrm{Zn}} = \frac{20 \text{ mmoles}}{100 \text{ ml}} = 0.20\,M$$

The concentration of hydrogen ion need not be calculated because the anode potential will not be utilized.

Next, calculate the cathode potential for each of these ions. The $E^0$ value for Cu is 337 mV and for Zn is $-763$ mV.

$$\begin{aligned} E_{\mathrm{Cu}} &= 0.337 + \frac{0.059}{2} \log 0.10 \\ &= 0.337 + 0.030\,(-1) \\ &= 0.337 - 0.030 = 0.307 \text{ V} \end{aligned}$$

and

$$\begin{aligned} E_{\mathrm{Zn}} &= -0.763 + \frac{0.059}{2} \log 0.20 \\ &= -0.763 + 0.030(-0.70) \\ &= -0.763 - 0.021 = -0.784 \text{ V} \end{aligned}$$

Since the decomposition potential for zinc is larger (Zn has a more negative cathode potential) than the decomposition potential for copper, copper will be deposited first. The next step is to calculate the cathode potential for copper when the copper has been reduced to a negligible value, or $10^{-7}$ *M*.

$$E_{Cu}(10^{-7}) = 0.337 + \frac{0.059}{2} \log 10^{-7}$$

$$= 0.337 + 0.030(-7)$$

$$= 0.337 - 0.210 = 0.127 \text{ V}$$

Since this value is smaller (more positive) than the zinc cathode potential, the copper should be quantitatively deposited before zinc begins to plate.

Even though such calculations show that quantitative separation is theoretically possible, they do not indicate the applied potential that should be used. In the separation of copper from zinc in brass analysis, the applied potential is usually set between 1.5 and 2.0 V to produce a current of between 1 and 5 A. At this potential and current, the copper in a 1-g brass sample can be deposited quantitatively in less than 1 hr. Experimentally it has been found that, with the usual type of constant-voltage or constant-current electrodeposition equipment in which the potential is set to give rather large constant currents, the $E^0$ values should differ by at least 0.5 V in order for the separation to be quantitative. As an example, theoretical calculations show that tin should not plate with copper from a nitrate solution, whereas in actual practice tin always plates with the copper. This is the reason tin must be separated from the solution prior to electrodeposition of copper in brass analysis.

## CONTROLLED CATHODE POTENTIAL

The main reason for the necessity of the larger than theoretical difference is that it is the actual cathode potential that governs electrodeposition. In deposition at constant applied potential, sudden fluctuations in the overvoltage of the anode cause fluctuations in the cathode potential, which in turn may cause incomplete deposition of the metal being plated or partial deposition of a second metal. If the cathode potential, instead of the applied potential, is held constant, separations of metals with closer $E^0$ values may be achieved satisfactorily.

The cathode potential can be controlled during electrolysis by inserting a reference electrode into the solution near the cathode and measuring the potential between the cathode and the reference electrode, usually a saturated calomel electrode. Variations in the potential between the cathode and reference are used to activate a servomechanism that increases or decreases the applied potential by the amount necessary to return the cathode potential to the preset value. In this system, the applied potential is varied to take care of sudden fluctuations in anode overvoltage rather than allowing the cathode potential to vary. Using this method, copper can be plated quantitatively in the presence of tin without partial deposition of the tin with the copper.

## FACTORS THAT AFFECT ELECTROLYTIC DEPOSITS

The metal deposit must be pure, dense, firmly adherent, lustrous, and easily washed and dried if electrodeposition is used as an analytical method. Factors that affect the physical characteristics of the deposit include the formation of complex ions, the current density, agitation, and evolution of gas. Each of these will be examined separately.

### *Formation of Complex Ions*

Quite often, deposition from a solution containing a complexing agent improves the physical characteristics of the plated metal. As an example, silver can be quantitatively deposited from a nitrate solution, but the deposit is large-grained and nonadherent. Deposition of silver from an ammoniacal solution or a cyanide solution results in a hard, adherent and lustrous deposit. The presence of complexing agents can also affect the order of deposition of metals. For instance, copper is plated quantitatively before cadmium from nitric acid solution, but cadmium is plated quantitatively before copper from cyanide solutions.

### *Current Density*

Current density, which governs the rate of deposition, is probably the most important variable in the control of the physical characteristics of the plated metal. Deposition at high current densities can cause nonadherent deposits as well as "treeing" and "noduling," especially at sharp corners. Low current densities, on the other hand, may require an exorbitant length of time for quantitative deposition. Ordinarily, a compromise is attained by starting the deposition at low current densities and increasing the current density after a satisfactory "base plate" has been deposited.

The amount of time necessary to deposit any given weight of metal at a definite current strength may be calculated from the fact that 96,500 (actually 96,487) coulombs are required to deposit 1 eq wt of any metal. This is known as **Faraday's law**, and is the basis of coulometry, which is discussed later in this chapter. Assuming 100% current efficiency, the amount of time necessary to deposit a given amount of a metal at a specified current strength is given by

$$\frac{\text{wt deposited}}{\text{eq wt}} = \frac{It}{96{,}500} \qquad \textbf{(20-8)}$$

in which $I$ is the current in amperes and $t$ is the time in seconds.

### *Agitation*

If electrodeposition is carried out without agitation, the metal ions in the immediate vicinity of the electrode are depleted and are replenished only by diffusion, which is a slow process. As a consequence, an extremely long time may be necessary for complete deposition, and very low current densities must be used. With adequate stirring, however, the cation concentration is maintained homogeneously throughout the solution because the deposited cations are replaced as the agitation brings fresh solution to the electrode area. Agitation also offsets the high overvoltages that may be caused by depletion of the electroactive ion in the immediate vicinity of the electrode.

### *Evolution of Gas*

Evolution of gas during the deposition of a metal leads to spongy, nonadherent deposits, especially at the beginning of the deposition. The use of an electrode depolarizer in the solution offsets in part the increase in decomposition potential as the metal concentration is lowered by plating. A **depolarizer** is a substance that is more readily reduced at the cathode than hydrogen ion. In the deposition of copper from nitric acid solution, the nitrate ion acts as depolarizer, since it can be reduced to ammonia or nitrous acid at the cathode.

# POLAROGRAPHY

Polarography derives its analytical importance from the unique current–voltage relations that occur when an easily polarized electrode, such as a dropping mercury electrode (DME), is used together with a large depolarized electrode, such as a mercury pool or a saturated calomel electrode. The position of the **polarographic wave** along the potential axis may be used to identify the substance being electrolyzed under certain circumstances, but, more important, the **limiting current** that can be obtained under these conditions of electrolysis is governed by the concentration of the electroactive ion in the solution. As a consequence, polarography can be used for qualitative identification of a substance, as well as a measure of the concentration of that substance in the solution. The current–voltage curve is called a **polarogram**, and the method is often referred to as **voltammetry**, since it depends upon measurement of current as potential is varied.

## BASIC PRINCIPLES

In ordinary constant-potential or constant-current electrolysis using two macroelectrodes and relatively large currents, the current increases continuously with increase in potential after the decomposition potential is attained as shown

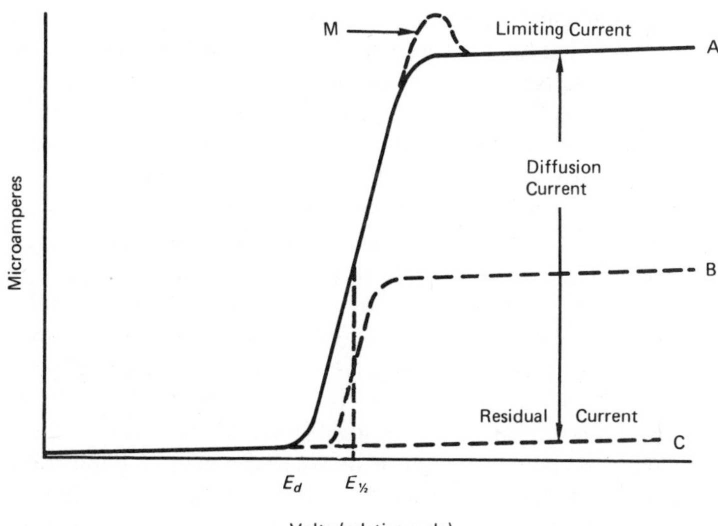


**Figure 20-3.** A typical polarogram. Curve B represents a solution with a lower concentration than curve A. Curve C is the extrapolated residual current. $E_d$ is the decomposition potential and $E_{1/2}$ is the half-wave potential. The dashed portion of curve A at *M* represents the distortion of the polarographic wave due to adsorption currents.

in Figure 20-1. If the conditions of electrolysis are changed to incorporate a microelectrode as the cathode, a macroelectrode as the anode, and allow only small currents to flow with no agitation, the current–voltage curve attains a plateau or limiting value at some potential greater than the decomposition potential. A typical polarogram is shown in Figure 20-3. A very small current known as the **residual current** flows before the decomposition potential is reached and is also present when no electroactive ion is present in the solution. The difference between the limiting current and the residual current is known as the **diffusion current**.

Under these conditions the macro electrode shows an almost constant potential, because it is essentially depolarized, and the micro electrode, which is polarized, assumes the potential applied to it over a certain range. If the microelectrode is made the cathode and an external potential applied, no current will flow until the decomposition potential is reached. Further increase in potential causes current to flow at a finite rate. If the solution is stirred, the current will increase as the potential is increased at a rate determined by Ohm's law, since the slope of the line will be inversely proportional to the resistance of the system. If the solution is not stirred and the applied potential is held constant at a value slightly exceeding the decomposition potential, the current will flow but will decrease to some constant value after a given time. If, however, the solution is not stirred and the applied potential is increased continuously, a potential will be reached at which the current continues to flow but does not increase with increase in applied potential.

The limiting current so obtained is caused by the fact that the region immediately surrounding the electrode (known as the **diffusion layer**) has become depleted in electroactive ions, and new ions can enter this region only by diffusion from the body of the solution. The plating of ions from the diffusion layer causes a decrease in concentration and sets up a concentration gradient between the electrode and the body of the solution. Because of the concentration gradient, electroactive ions diffuse into the layer at a rate that is dependent upon the difference in the concentration in the body of the solution and the concentration in the diffusion layer. Since the rate of deposition and thus the current is a function of the applied potential and increases with applied potential, it can be assumed that at some potential ions will be deposited as soon as they enter the diffusion layer, that is, that the concentration in the diffusion layer has been reduced essentially to zero. When this condition is reached, the diffusion into the area depends solely upon the concentration in the body of the solution. Further increase in potential cannot increase the rate of deposition, so the current becomes constant or reaches its limiting value. The limiting value obtained is thus dependent upon the concentration and can be related to it. Curves A and B in Figure 20-3 represent polarograms obtained with solutions of different concentrations.

It should be noted that only very small currents flow, so the amount of electroactive material discharged at the electrode is negligible compared to the total amount of electroactive ion in the solution. This means that the concentration in the body of the solution is not affected measurably by the deposition, and the polarogram can be repeated over and over again with essentially the same results.

## ELECTRODES

The **indicator** or **microelectrodes** can be stationary solid, rotating solid, or dropping mercury. The *solid microelectrode* is usually a platinum or gold wire and has the disadvantages of poor reproducibility, high temperature coefficients, and a time lag before the current becomes constant. As a consequence, it is not used except for special cases. The *rotating microelectrode* gives diffusion currents with no time lag, but the diffusion current depends in part upon the speed of rotation which makes the reproducibility poor.

The most popular indicating microelectrode is the **dropping mercury electrode**. This electrode consists of a reservoir of mercury attached to a capillary (usually 0.05 mm ID) that is immersed in the solution being tested. The mercury flows through the capillary, forming in drops that fall of their own weight. The drop size depends primarily upon the internal diameter of the capillary, and the rate of flow depends primarily upon the pressure of the mercury or upon the difference in height between the top of the mercury reservoir and the tip of the capillary. The average area of the drops is directly proportional to $m^{2/3}t^{1/6}$, where $m$ is the number of milligrams of mercury per second and $t$ is the drop time in seconds. The diffusion current increases during the life of the drop and falls precipitously to zero when the drop falls so that a fluctuating current is obtained.

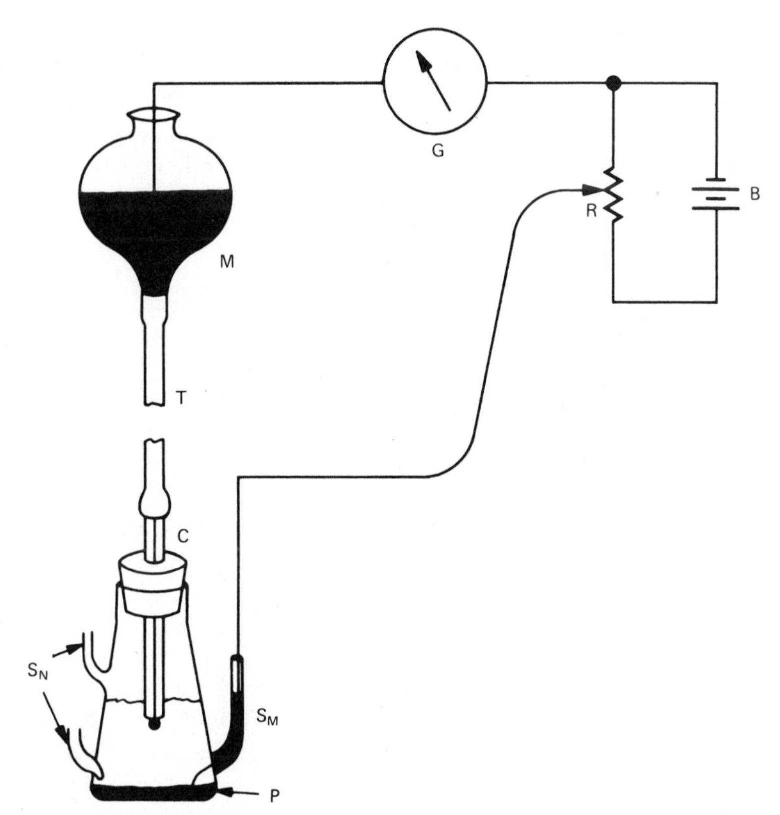


**Figure 20-4.** Simple polarographic circuit. M, mercury reservoir; T, Tygon tube; C, capillary; $S_N$, side arms for deaeration by nitrogen gas; $S_M$, side arm filled with mercury; P, mercury pool; R, variable resistance; B, battery or other source of potential; G, galvanometer.

The fluctuation is reproducible, so the average area of the drops and the average diffusion current is readily determinable.

The most popular reference electrode is the mercury pool at the bottom of the polarographic cell, even though it has the disadvantage of the exact potential not being known unless determined independently when needed. A saturated calomel electrode is often used as the reference electrode and is connected to the system through a salt bridge. The potential of a saturated calomel electrode is not affected by the small currents that flow during polarographic assay.

The essential components of a polarographic circuit using a dropping mercury electrode are shown in Figure 20-4.

## THE LIMITING CURRENT

Although it is the diffusion current of the electroactive ion that causes the polarographic wave, other contributing factors may affect the value of the limiting current. These include the residual current, the migration current, and adsorption currents.

### The Residual Current

The reason for the small residual current that flows in the absence of an electroactive ion, or before the decomposition potential is reached in the presence of an electroactive ion, is not clearly understood. It is probably related to some sort of migration current and can be thought of as a leakage current due to the applied potential between the electrodes.

### The Migration Current

Implicit in the explanation given earlier for the limiting current is the understanding that ions can reach the electrode only by diffusion from the body of the solution. However, because of the electrical gradient between the electrodes, ions in the solution also migrate to the electrodes as a result of the attraction of the electrodes for the oppositely charged ions. This migration current must be eliminated or corrected for if the true diffusion current is to be obtained. The current is carried through the solution by all the ions that are present, not just the electroactive ions. The effect of the migration of the electroactive ion can thus be offset by the addition of a large concentration of an inert compound or supporting electrolyte. In practice, potassium chloride or a similar salt is added so that the concentration is approximately 100 times the concentration of the electroactive ion. The inert ions carried to the cathode by migration diffuse back into the body of the solution as a result of the concentration gradient that is set up.

### Adsorption Currents

Polarograms obtained with a DME are often distorted by maxima, as shown at point $M$ on curve A of Figure 20-3. These **polarographic maxima** may be offset by adding a surface active material to the solution in small concentration and are consequently thought to be due to adsorption at the mercury interface. Gelatin is often used, but the amount must be carefully controlled, because gelatin also affects the drop time. Ordinarily, concentrations between 0.002 and 0.01 % may be utilized satisfactorily.

### The Diffusion Current

The main component of the limiting current is of course the diffusion current, which is concentration dependent and the basis for quantitative measurements. The relationship of the various factors that affect the diffusion current is given by the **Ilkovic equation**:

$$i_d = 607nD^{1/2}Cm^{2/3}t^{1/6} \tag{20-9}$$

where $i_d$ = average current in microamperes during the life of the drop
$n$ = number of faradays required per mole of electrode reaction
$D$ = diffusion coefficient of the electroactive ion in square centimeters per second
$C$ = concentration in millimoles per liter
$m$ = flow rate of mercury in milligrams per second
$t$ = drop time in seconds.

The Ilkovic equation is seldom used to calculate diffusion current because of the lack of precise values for the diffusion coefficients under the conditions used in polarography. The equation is very important, however, because it shows that the diffusion current is directly proportional to the concentration and also indicates the factors that must be controlled for reproducible results. Any factor that affects the diffusion of ions in the solution, such as temperature, viscosity, ionic strength, and so on, must be regulated between rather narrow imits if the results are to be reproducible. The characteristics of the capillary, which control the drop size and flow rate, are also important and must be held constant.

## THE HALF-WAVE POTENTIAL

The potential at one-half the wave height is known as the half-wave potential ($E_{1/2}$) and is constant for a given electroactive material under a given set of conditions, irrespective of the concentration of the electroactive ion, as shown in Figure 20-3. As a consequence, $E_{1/2}$ values can be used for qualitative identification. The half-wave potential is related to the standard reduction potential, $E^0$, of the oxidation–reduction system involved in the electrode reaction.

## ANALYTICAL APPLICATIONS

A polarogram gives both the qualitative composition (by the half-wave potential) and the quantitative composition (by the diffusion current) of the solution being measured. To determine the concentration of an electroactive ion, the diffusion current is determined from the polarogram by subtraction of the residual current from the limiting current and the concentration is either calculated or read from a calibration curve. Two or more ions can be determined in the same solution if the half-wave potentials are sufficiently different so that there is no interference or overlap between the two polarographic waves. Most metal ions can be determined by cathodic reduction, and many negative inorganic ions can be measured by anodic oxidation if not by cathodic reduction. Polarography also has many applications in the determination of organic compounds.

Oxygen interferes in many determinations and must be removed from the solution before the polarogram is run. The most popular method for deoxygenation is to deaerate by bubbling nitrogen through the system for several minutes. In some cases oxygen can be removed by the addition of sodium sulfite if the solution is neutral or alkaline.

Best results are obtained with $10^{-3}$ or $10^{-4}$ $M$ solutions of samples, but concentrations from $10^{-2}$ to $10^{-6}$ can be measured satisfactorily.

## EVALUATION METHODS

### Direct Comparison

When a large number of samples are to be run, construction of a calibration curve of diffusion current versus concentration for a series of standards is the preferable method. Individual samples are then read from the curve. Care must be taken to reproduce the conditions if a standard series is not run with each series of samples.

### Standard Addition

In this method, the polarogram of a definite volume of the unknown solution is run and the polarogram repeated after the addition of a known volume of a standard solution to the unknown in the polarographic cell. The concentration of the unknown can be calculated from the expression

$$C_u = \frac{C_s V_s i_{d_1}}{(V_u + V_s) i_{d_2} - V_u i_{d_1}} \qquad \textbf{(20-10)}$$

in which the subscripts $u$ and $s$ refer to unknown and standard, respectively; $V$ represents volume; $C$ represents concentration; $i_{d_1}$ is the diffusion current of the unknown; and $i_{d_2}$ is the diffusion current of the unknown plus standard.

### Internal Standard

In this method, the polarogram of the unknown is run after dilution of an aliquot to a known volume. A second sample is prepared by adding a known volume of a standard solution of another ion to an equal-size aliquot of the unknown, followed by dilution to the same volume utilized for the sample polarogram. The second ion is known as the **pilot** or **reference ion**. The concentration of unknown can then be calculated from the expression

$$\frac{i_{d_s}}{i_{d_u}} = \frac{I_s C_s}{I_u C_u} \qquad \textbf{(20-11)}$$

where the $s$ and $u$ subscripts refer to the internal standard ion and the unknown ion, respectively; $i_d$ represents the diffusion current; $C$ the concentration; and $I$ the diffusion current constant ($I = 607nD^{1/2}$). The ratio $I_s/I_u$ can be calculated from known values of the diffusion coefficients or can be evaluated from the polarograms of standard mixtures.

### *Absolute Method*

This method involves calculation of the concentrations from measured diffusion currents using eq. (20-9). This method is limited by the lack of precise values for the diffusion coefficients for a large number of systems and is used primarily to substantiate the applicability of the Ilkovic equation to the relationships that govern the diffusion current.

## COULOMETRY

Coulometry is based on the exact measurement of the quantity of electricity that passes through a solution during the occurrence of an electrochemical reaction. When an electrode reaction takes place with 100% efficiency, the quantity of substance reacting can be related to the quantity of electricity passed through the system by application of Faraday's law, which states that 1 faraday of electricity causes the electrochemical change of 1 g eq wt of the electrode reactant. The substance actually being determined may be oxidized or reduced at one of the electrodes or may react with one of the electrode products.

One faraday of electricity equals 96,500 coulombs (96,487 if greater precision is required). A coulomb is the amount of electricity that passes a given point when a current of 1 A flows for 1 sec (or $Q = It$). Since 1 faraday causes the change of 1 g eq wt at the electrode, 1 faraday of electricity is equivalent to one Avogadro's number of electrons ($6.02 \times 10^{23}$).

The essential requirements for coulometric measurements are that a single reaction must take place at the *working* electrode, that the electrode must be 100% efficient, and that the amount of current must be measured accurately. The measurement of the amount of current requires an accurate estimation of the volume of a gas liberated or the weight of a solid deposited when the current flows through a device known as a coulometer; alternatively a constant current, together with a precisely determined time interval, may be used.

Coulometric methods eliminate the necessity of preparation and storage of standard solutions, because the electron becomes the primary standard in all coulometric measurements, and the reactant is generated in controlled amounts within the electrolytic cell. The method is capable of great accuracy, since the accuracy depends only upon the measurement of the current and upon the sensitivity with which the end point can be determined. Coulometry is more versatile than electrogravimetry or polarography because reactions involving unstable reactants or gaseous or volatile materials can be utilized with ease. The methods may be carried out at constant potential or at constant current. At constant potential, the current decreases as a logarithmic function of time and can be determined by integrating with respect to time or measuring the area under the current–time curve. These constant potential methods are useful in the determination of microgram quantities of metals by cathodic deposition or anodic oxidation (**stripping**). In many cases, the trace metal is deposited and then stripped in the same determination to attain greater accuracy and sensitivity.

## COULOMETERS

A second electrolytic cell that shows 100% current efficiency may be inserted into the circuit to measure the amount of electricity that passes through the working cell. Electrolytic cells used in this way are known as coulometers. Most coulometers either cause deposition of a solid that can be weighed accurately or generation of a gas the volume of which can be measured accurately.

The most popular metal coulometers are the silver and copper coulometers. The silver cell can be used with smaller amounts of current because the equivalent weight of silver is 107.87 whereas that of copper is 31.77. Gas coulometers usually involve the electrolysis of potassium or sodium sulfate solutions with platinum electrodes to generate hydrogen and oxygen at the two electrodes. The volume of the individual gases or a mixture of the two gases can be measured with a gas burette after adjustment to atmospheric pressure and determining the temperature. The volumes must be corrected to standard conditions before calculation of the amount of current. The electrolysis of hydrazine sulfate to produce nitrogen and hydrogen at the electrodes is preferable for low current densities.

Mechanical or electronic integrators may be used with the constant-potential methods to determine the amount of electricity. The integrators are based on the logarithmic relation between the time and the current and are used to estimate the area under the time–current curve.

For constant-current methods, we need a constant source of current and an accurate time measurement. The simplest measurement device is the use of a precision stopwatch together with several batteries connected in series with a current-limiting resitance. Since the voltage of the battery tends to decay with use and since the resistance of the resistor changes with temperature, such systems are not used when high accuracy is desired. Electronic instruments are available commercially that allow current regulation to within $\pm 0.01\%$ and permit selection between several different values for the current before the measurement is begun. The time interval is usually measured with an electric stopclock powered by a synchronous motor, which is actuated by the same switch that is used to start the current. This switch may be turned on and off for short intervals several times as the end point is approached in a titration. Consequently, any lag time in starting the motor or any coasting time when the motor is cut off can cause an appreciable error in the time interval. Total titration times of several hundred seconds are used in most cases to offset small errors in time measurements.

## COULOMETRIC TITRATIONS

Many of the uses of coulometry are in the field of coulometric titrations in which the substance being titrated is oxidized or reduced at one of the electrodes or the titrant is generated at one of the electrodes by an electrochemical reaction. Since a titrant can be generated in the system, coulometric titrations are not limited to the titration of substances that are electrochemically reactive under the conditions encountered in the electrolytic cell. A simplified schematic diagram of an apparatus for coulometric titration with internal generation of titrant is shown in Figure 20-5.

Coulometric titrations are very sensitive and can be adapted to automatic operation and remote control. No standard solutions are required because the electron is the standard. This obviates the necessity of preparation and storage of unstable titrants or of the preparation of small amounts of a standard solution when only a few determinations must be run.

The accuracy and precision of coulometric titrations is limited only by the

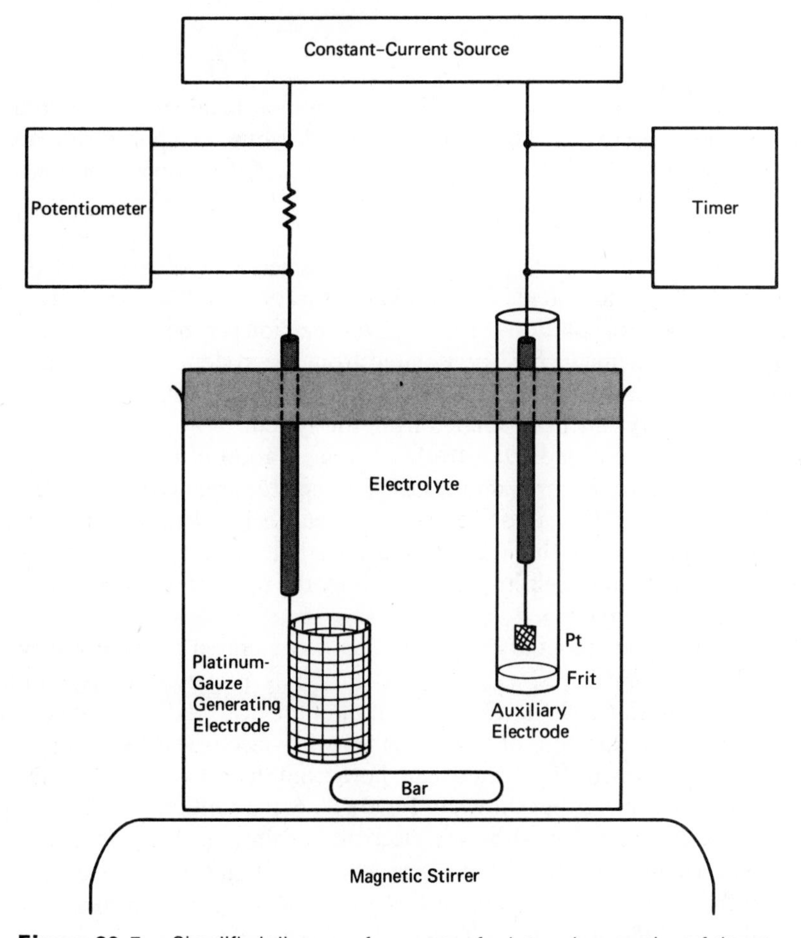


**Figure 20-5.** Simplified diagram of apparatus for internal generation of titrant.

accuracy and precision involved in the measurement of the current and the time and by the precision with which the end point can be located. In many cases involving small quantities the sensitivity of end-point selection is the governing factor. The end point can be determined with a visual indicator or by photometric or potentiometric methods.

Both primary and secondary coulometric titrations are possible. **Primary** or **direct** titrations at constant current include those in which the substance being determined reacts directly at the electrode. **Secondary** or **indirect** titrations at constant current involve the generation of an active intermediate at the electrode, followed by the reaction of the intermediate with the substance being determined.

Examples of direct coulometric titration include the titration of halide ions by electrogenerated silver ion. Mercaptans in petroleum products can also be determined by electrogeneration of silver ion after dissolving the sample in a mixture of methanol and benzene to which ammonium hydroxide and ammonium nitrate are added to decrease the electrical resistance of the solution.

A good example of a secondary or indirect method is the titration of iron(II) (ferrous) ions with cerium(IV) (ceric) ions in sulfuric acid solution by the reaction

$$Fe^{2+} + Ce^{4+} \longrightarrow Fe^{3+} + Ce^{3+} \qquad \textbf{(20-12)}$$

In this case, an excess of cerium(III) (cerous) ions is added to the solution. The $Ce^{3+}$ ions are oxidized at the anode to $Ce^{4+}$ ions, which in turn react immediately with the $Fe^{2+}$ ions by reaction (20-12). The end point occurs when the first excess of $Ce^{4+}$ is generated and can be detected photometrically or potentiometrically.

Secondary methods include the cathodic generation of hydroxyl for the titration of very dilute acids or for the preparation of dilute carbonate-free strong base standard solutions. Because hydrogen ion is generated at the anode, the anode compartment must be isolated from the system in a separate compartment. Dilute bases can also be titrated by electrogenerated hydrogen ion if the cathode compartment is isolated from the system.

Another example of indirect titration is the titration of metals with EDTA. In this case the mercury or cadmium EDTA chelate is reduced at the electrode and the liberated EDTA anion reacts with metal being determined. There are also many applications in organic analysis that involve the generation of bromine or iodine at the electrode followed by the reaction of the halogen with the organic compound.

The titrant ion in secondary titrations can be generated externally as well as internally. In some cases, external generation is preferred to isolation of the nonworking electrode. In the titration of dilute acids with electrogenerated hydroxyl or the titration of dilute bases with electrogenerated hydrogen, external generation of the titrant can be accomplished using a double-arm electrolytic cell such as that shown in Figure 20-6. If a soution of sodium sulfate or potassium sulfate is used as the electrolyte solution, the electrogenerated hydroxyl ions flow out the cathode arm while electrogenerated hydrogen ions flow out the anode arm. The gaseous hydrogen and oxygen generated are swept out of the electrode arms by the flow of the electrolyte solution.

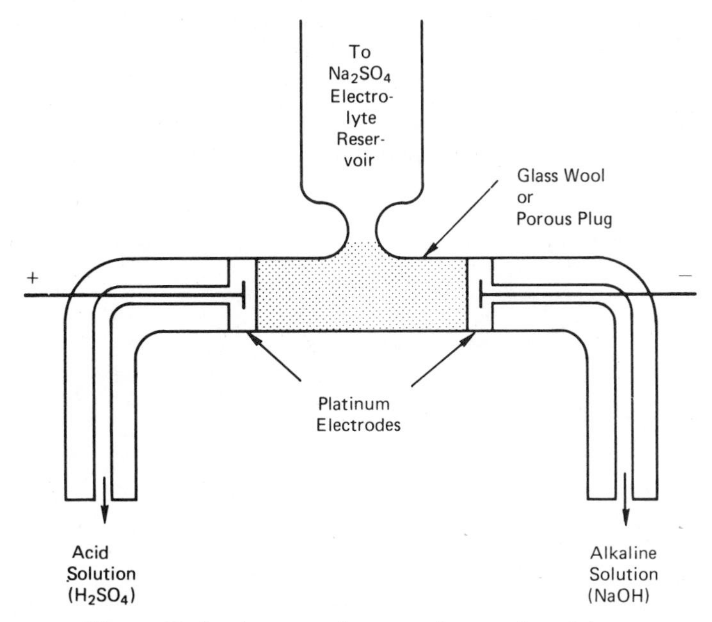


**Figure 20-6.** Apparatus for external generation of titrant.

# PROBLEMS

**1** A sulfuric acid solution that contains 12.0 g of $CuSO_4 \cdot 5H_2O$ in 200 mL was electrolyzed. When the pH reached 0.40, five-sixths of the copper had been deposited. Calculate the theoretical counter potential at this point.

*Ans.* 909 mV

**2** A solution that is 0.20 *M* in cupric nitrate and 0.10 *M* in nitric acid is electrolyzed in a cell with a copper cathode and a smooth platinum anode. When a potential of 2.08 V is applied, the resistance is 2.0 Ω and a current of 0.10 A flows. Calculate the overvoltage of the cell.

**3** A solution that contains 20 mmoles of $Cu(NO_3)_2$, 50 mmoles of $Zn(NO_3)_2$, and 100 mmoles of $HNO_3$ in 100 mL is electrolyzed. Calculate (a) the original anode potential and the original counter potential for each metal ion; (b) state which ion is deposited first as the potential is increased continuously from zero; (c) calculate the hydrogen ion concentration in the solution and the counter potential after the deposition of this metal ion is quantitative; (d) calculate the minimum concentration of the second ion that would have to be present to cause deposition of some of the metal with the first metal to plate; (e) state whether or not the separation is quantitative.

*Ans.* (d) $10^{30}$ *M*

**4** A solution that contains 5 mmoles of $AgNO_3$ and 20 mmoles of $Ni(NO_3)_2$ is 100 mL and which is buffered to a pH of 1.00 is electrolyzed. Is quantitative separation of the two metal ions possible?

**5** A solution containing 1.5 g of $CuSO_4 \cdot 5H_2O$ is electrolyzed using a current of 0.700 A. If no cation other than copper is reduced at the cathode, calculate the length of time required for complete deposition. Explain why the actual time necessary for complete deposition often differs from the theoretically calculated time.

6 A sample of brass weighing 0.9300 g was electrolyzed after solution in $HNO_3$ and filtration of the metastannic acid. The cathode deposit weighed 0.8210 g and the anode deposit weighed 0.0153 g. Calculate the percentage of copper and lead in the sample.

7 A solution containing the electroactive substance A in 0.1 *M* KCl gave the following diffusion currents during polarographic analysis. Draw and label the polarogram and determine the half-wave potential.

| V | $\mu$A | V | $\mu$A | V | $\mu$A |
|---|---|---|---|---|---|
| 0.00 | 0.0 | 0.35 | 3.0 | 0.90 | 38.8 |
| 0.10 | 0.2 | 0.40 | 6.5 | 1.0 | 39.6 |
| 0.20 | 0.4 | 0.60 | 20.0 | 1.1 | 39.8 |
| 0.30 | 1.0 | 0.80 | 33.5 | 1.2 | 40.0 |
| | | 0.85 | 37.0 | | |

*Ans.* 0.6 V

8 Solutions of cadmium sulfate containing 0.01% gelatin and 0.10 *M* $K_2SO_4$ gave the following limiting currents when run at a potential of −1.00 V. Draw the calibration curve and calculate the concentration of a similar cadmium solution which shows a diffusion current of 10.5 $\mu$A.

| *Concentration* (m*M*) | *Limiting Current* ($\mu$A) |
|---|---|
| 0.00 | 2.0 |
| 0.50 | 8.5 |
| 1.00 | 15.0 |
| 1.50 | 21.5 |
| 2.00 | 28.0 |

*Ans.* 0.80 mM

9 Repeat problem 8 for the following unknowns: (a) diffusion current ($\mu$A) equals (1) 22.3, (2) 17.6, (3) 8.0; (b) limiting current ($\mu$A) equals (1) 22.3, (2) 15.6, (3) 27.0.

10 Utilizing the calibration data given in problem 8, calculate the concentrations of solutions prepared as follows:

(a) 5.0 mL of 0.10 m*M* standard cadmium solution is added to 20.00 mL of an unknown cadmium solution. Before the addition the diffusion current was 14.0 $\mu$A and after the addition the diffusion current was 21.0 $\mu$A. Calculate the concentration of the unknown solution.

(b) A solution containing 50.0 mL of an unknown cadmium solution and 10.0 mL of 2.00 m*M* zinc standard solution in 100 mL was prepared and gave a diffusion current (corrected for residual current) of 16.0 $\mu$A for cadmium and a diffusion current (corrected for residual current) of 12.0 $\mu$A for zinc. Calculate the concentration of the unknown solution.

11 In coulometric titrations, a milliampere-second is equivalent to how many milligrams of (a) iron(II) to (III); (b) hydroxyl; (c) arsenic(III) oxide to (V); (d) copper(II) to (0); (e) chloride ions; (f) hydrogen ions; (g) silver ions.

*Ans.* (a) $5.8 \times 10^{-4}$ mg

**12** A gas coulometer containing $Na_2SO_4$ produced 22.4 mL (corrected to STP) of hydrogen during an electrolysis. How many coulombs passed through the system? How many coulombs would have been involved if the gas measured had been oxygen instead of hydrogen?

**13** The cathode in a silver coulometer weighed 1.0363 g before use and 1.0970 g after use. How many coulombs passed through the coulometer?

*Ans.* 54.3

**14** Repeat problem 13 for a copper coulometer for which the cathode increased in weight from 5.6482 g to 6.0721 g.

**15** Calculate the normal concentration of a solution of hydrochloric acid if a 10.00-mL aliquot required 235 sec to attain the pink color of phenolphthalein. The voltage drop across the 100-$\Omega$ resistor was 0.729 V.

**16** In the coulometric determination of arsenious acid by generation of iodine from potassium iodide, a 25.00-mL aliquot of the acid required 8 min and 40 sec using a constant current of 3.00 mA. Calculate the normal and the molar concentration of the arsenious acid.

CHAPTER

# 21 Solvent Extraction

Solvent extraction involves the partition or distribution of a solute between two immiscible liquids in contact with each other. The process has become increasingly important in the analysis of metals, because it furnishes clean separations in a short span of time and has the further advantages of simplicity of technique and equipment. In the analysis of metals, we are primarily interested in separations in or from aqueous solutions and will limit our discussion to partition between aqueous and organic liquids. In order to cross the phase boundary, the metal must be incorporated into a neutral species. Since metal salts, even though neutral, are usually not especially soluble in organic solvents and since solubility in the organic phase is a necessity for extraction, we will first discuss the types of metal extraction systems that can convert the metal ion to a neutral species for extraction.

## TYPES OF METAL EXTRACTION SYSTEMS

The two main classes of metal extraction systems are the **chelate extraction** system and the ion association extraction system. In general, the chelate systems utilize nonpolar organic liquids whereas the ion association systems require polar organic solvents. The two systems may be classified on the basis of type of organic solvent used rather than type of neutral species formed.

In the chelate extraction system, the metal ion reacts with a **chelating agent** to form a neutral chelate compound that can cross the boundary. A chelate compound is a metal–organic compound in which the metal is incorporated into a ring structure with the organic compound. Although a metal–organic chelate can be charged or uncharged, only the uncharged chelates will extract into the organic liquid. Some of the characteristics of chelates are discussed in Chapter 14.

In the ion association extraction system, which is also known as **ion pairing**, the extractable species is formed by the association of ions, one of which must be very large compared to the other. In the two most important types, the metal ion is incorporated into a large bulky ion in which it is essentially surrounded by the organic portion and thus "shielded" from the organic liquid so that it becomes soluble in the organic phase. The organic extracting solvent also acts as the complexing agent in one type but not in the other. An example of the type in which the organic complexing agent is not the extraction solvent is the reaction of copper(I) with 2,9-dimethyl-1,10-phenanthroline to form a large bulky ion with a unit positive charge that can be extracted into chloroform by association with a small anion. An example of the type in which the complexing agent and the extracting solvent are the same organic compound is the extraction by ethyl ether of iron(III) from a solution 6 *M* in hydrochloric acid. In this case, the ether displaces all or part of the water that is coordinated to the ferric ion, and this ion associates with one or more chloride ions to cross the boundary. In a few cases, a small negatively charged metal complex ion will associate with a bulky organic ion to cross the boundary. An example is the association of $ZnCl_4^{2-}$ with two tribenzylammonium ions $(C_6H_5CH_2)_3NH^+$ for extraction into trichloroethylene. The type in which the solvent itself displaces the coordinated water before extraction is also called the **oxonium system**, since all these solvents must contain an oxygen that has a pair of electrons available for bonding. Common solvents in this group are ethers, alcohols, ketones, and esters.

## PRINCIPLES OF SOLVENT EXTRACTION

### THE DISTRIBUTION LAW

Whenever a solute is soluble in both of two immiscible liquids in contact with each other, the solute will distribute itself between the two liquids so that the ratio of concentrations is a constant. Mathematically,

$$K_d = \frac{C_o}{C_w} \tag{21-1}$$

where $K_d$ is the distribution constant or coefficient; $C_o$ is the concentration of the extractable (distributable, partitionable) species in the organic phase; and $C_w$ is the concentration of the extractable species in the water layer. The **distribution constant** is also known as the **partition coefficient** and given the symbol *P*. Note that it is the concentration of the distributable species that is specified in eq. (21-1). If there is any type of interaction between the distributable species and any other component in either of the two phases, the distribution coefficient does not show the total amount extracted or the total concentration of the metal ion in each of the two phases. As a consequence, analytical chemists prefer to use the distribution ratio, *D*, which is defined by

$$D = \frac{M_{t_o}}{M_{t_w}} \tag{21-2}$$

where $M_{t_o}$ is the total molar concentration of all species containing the metal ion in the organic phase and $M_{t_w}$ is the total molar concentration of all species containing the metal ion in the water phase.

$K_d$ equals $D$ only if all the metal ion in each of the two phases is present as the partitionable species, that is, there are no species in either phase that contain the metal other than the extractable species. For systems in which $D$ is equal to $K_d$, the ratio of concentrations is the same as the ratio of the solubilities of the extractable species in the two phases. This holds true for many extraction systems in which there is little interaction, but care must be exercised in assuming that the concentration ratio and solubility ratio are the same (or that $D$ and $K_d$ are equal), especially if the system is affected by the pH. As an example, consider the distribution of 8-hydroxyquinoline (also called 8-quinolinol and oxine) between water and chloroform. The $K_d$ is 720, but the value of $D$ varies from 0.006 at pH 0 to 60 at pH 4 and 720 at pH 7. Above pH 9 the values of $D$ decrease continuously to 0.05 at pH 14.

## EFFICIENCY OF EXTRACTION

Because the primary interest in most extractions is the efficiency of the extraction under a given set of conditions, the use of the percent of extraction ($\%E$) is more meaningful than either $D$ or $K_d$. The $\%E$ and $D$ are related by the expression

$$\%E = \frac{100D}{D + V_w/V_o} \tag{21-3}$$

or

$$\%E = \frac{100DV_o}{DV_o + V_w} \tag{21-4}$$

in which $V_o$ and $V_w$ represent the volumes of the organic and water phases, respectively. If these two volumes are equal, the denominator in eq. (21-3) reduces to $D + 1$. As the extraction efficiency approaches 100%, the distribution ratio approaches infinity, as shown in Figure 21-1 for systems using equal volumes of the two phases.

The amount of solute left unextracted after a given series of batch extractions is even more meaningful to the analyst than $\%E$. The amount left unextracted after one extraction can be calculated from $D$ by

$$D = \frac{(w - w_1)/V_o}{w_1/V_w} = \frac{(w - w_1)V_w}{w_1 V_o} \tag{21-5}$$

in which $w$ is the amount of solute originally present in an aqueous solution with a volume of $V_w$, and $w_1$ is the amount left unextracted after equilibrium

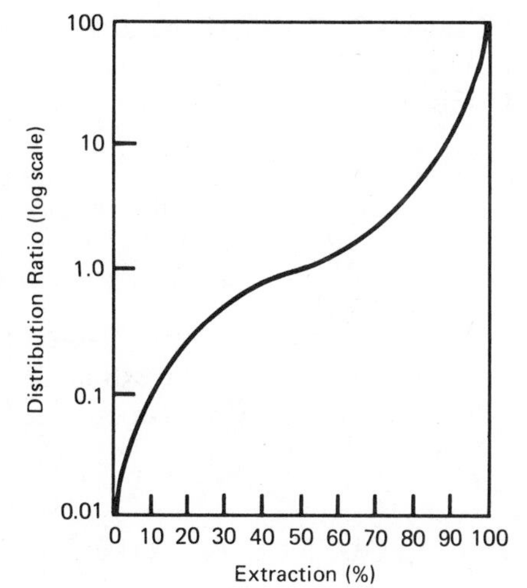


**Figure 21-1.** Efficiency of extraction as a function of the distribution ratio for systems using equal volumes of extractant and solution.

with a volume $V_o$ of organic extractant. Sequential calculations with eq. (21-5) for a series of extractions leads to the expression

$$w_n = w\left(\frac{V_w}{DV_o + V_w}\right)^n \qquad \textbf{(21-6)}$$

where $w_n$ is the amount left unextracted after $n$ extractions, using the volume $V_o$ of the organic liquid for each extraction. Equation (21-6) can be used to calculate the amount unextracted after a given series of extractions or can be used to estimate the number of extractions that will be needed for quantitative removal of a solute.

• **Example 1.** Calculate the mg of iron(III) left unextracted from 100 mL of a solution that contains 200 mg of $Fe^{3+}$ and is 6 *M* in HCl after three extractions with 25-mL portions of ethyl ether. The *D* for this extraction is approximately 150.

Substitution in eq. (21-6) gives

$$w_n = 200\left(\frac{100}{150 \times 25 + 100}\right)^3 = 200\left(\frac{100}{3850}\right)^3$$

$$= 200 \times (0.026)^3 = 3.5 \times 10^{-3} \text{ mg}$$

Equation (21-6) may also be used to prove that several extractions using small volumes of extractant are more efficient than one extraction with a large volume.

• **Example 2.** Calculate the amount of iron left in the iron–ether system of Example 1 if only one extraction with a 75-mL portion of ether had been used. In this case,

$$w_n = 200\left(\frac{100}{150 \times 75 + 100}\right) = 200\left(\frac{100}{11350}\right)$$
$$= 200 \times 8.8 \times 10^{-3} = 1.8 \text{ mg}$$

As a consequence, the three extractions with 25-mL portions are approximately 500 times more effective than one extraction with 75 mL of ether.

## THE SEQUENCE OF THE EXTRACTION PROCESS

Even though the actual reactions may differ from system to system, all extraction processes involve three basic steps:

### *Formation of a Distributable Species*

This step involves the reactions of the metal ion in the water phase with the complexing agent to form either a neutral molecular species or an extractable ion pair. Assuming that a neutral molecular species is formed, the reaction can be represented by

$$M + nL \longrightarrow ML_n \qquad \textbf{(21-7)}$$

where M represents a metal ion and L a ligand. Charges are omitted for simplicity. Since complexes are formed in steps when more than one ligand is involved, there may be appreciable quantities of intermediate species present. Also, the metal may react with other complexing agents present to form other complexes, such as $MA_n$, and may also hydrolyze to form $M(OH)_n$. Each of these interfering reactions also occurs in steps.

If an ion pair is formed in which the metal is incorporated into a large positive organic ion, the reactions may be represented as

$$M^{n+} + xL \longrightarrow ML_x^{n+} \qquad \textbf{(21-8)}$$

$$ML_x^{n+} + nA^- \longrightarrow (ML_x^{n+} \cdot nA^-) \qquad \textbf{(21-9)}$$

Similar equations can be written for the formation of a negative metal complex ion and ion pair. Again intermediates may be present, and the metal ion can react with other complexing agents or undergo hydrolysis.

### *Distribution of the Distributable Species*

After formation, the distributable species moves across the boundary until equilibrium is established and the equilibrium concentrations satisfy the distribution law. This may be represented by

$$(ML_n)_o \rightleftharpoons (ML_n)_w \qquad \textbf{(21-10)}$$

where the parentheses denote active concentrations and the subscripts $o$ and $w$ refer to the organic and water phases, respectively. Molar concentrations are usually substituted for activities in most extraction calculations.

### *Interactions in the Organic Phase*

After moving across the phase boundary, the extractable complex may polymerize or dissociate or may interact with other components in the organic phase. All these possible interferences will affect the distribution and may cause significant differences between $D$ and $K_d$.

## EFFECT OF pH ON CHELATE EXTRACTION SYSTEMS

As seen in the previous section, a large number of interferences may be involved in the extraction of metal chelates from water solution. In this discussion of the effect of pH, we will consider only the reactions that occur in the aqueous phase during the extraction of a metal ion with a weakly acidic chelating agent that is only slightly soluble in water. The following factors must be considered:

1 The ionization of the reagent (HL) to form the anion ($L^-$), which reacts with the metal to form the chelate ($ML_n$):

$$HL \rightleftharpoons H^+ + L^- \qquad K_{aR} = \frac{[H^+][L^-]}{[HL]} \qquad \textbf{(21-11)}$$

2 The stepwise formation of the chelate:

$$\begin{array}{llll} M^{n+} + L^- & \rightleftharpoons & ML^{(n-1)+} & k_1 = \dfrac{[ML^{(n-1)+}]}{[M^{n+}][L^-]} \\ \vdots & & \vdots & \vdots \\ ML_{n-1}^+ + L^- & \rightleftharpoons & ML_n & k_n = \dfrac{[ML_n]}{[ML_{n-1}][L^-]} \end{array}$$

for which the overall formation constant ($K_{stab}$) is

$$K_{stab} = k_1 \cdot k_2 \cdots k_n = \frac{[ML_n]}{[M^{n+}][L^-]^n} \quad \textbf{(21-12)}$$

3 The competing reactions for the metal ion such as hydrolysis and the formation of anion complexes:

$$\begin{aligned} M^{n+} + yOH^- &\rightleftharpoons M(OH)_y^{n-y} \\ M^{n+} + zA^- &\rightleftharpoons MA_z^{n-z} \end{aligned} \quad \textbf{(21-13)}$$

4 The distribution of the reagent itself between the two phases for which $K_{d_R}$ is

$$K_{d_R} = \frac{[HL]_o}{[HL]_w} \quad \textbf{(21-14)}$$

5 The distribution of the chelate between the two phases for which $K_{d_C}$ is

$$K_{d_C} = \frac{[ML_n]_o}{[ML_n]_w} \quad \textbf{(21-15)}$$

If the total concentration of all hydrolysis products plus the total concentration of all anion complexes is considered to be negligible, the distribution ratio is given by

$$D = \frac{M_{t_o}}{M_{t_w}} = \frac{[ML_n]_o}{[M^{n+}] + [M^{(n-1)+}] + \cdots + [ML_{n-1}^+] + [ML_n]_w} \quad \textbf{(21-16)}$$

By substituting the appropriate values from the constants for eqs. (21-11), (21-12), (21-14), and (21-15) into eq. (21-16) and assuming that the concentration of intermediate species is also negligible, we obtain[1]

$$\begin{aligned} D &= \frac{K_{stab} K_{d_C} K_{a_R}^n}{K_{d_R}^n} \left[\frac{[HL]_o}{[H^+]}\right]^n \\ &= K_{comb} \left[\frac{[HL]_o}{[H^+]}\right]^n \end{aligned} \quad \textbf{(21-17)}$$

in which $K_{comb}$ represents the combination of all the constants.

Equation (21-17) shows that for chelate extraction systems the extent of extraction is affected by the concentration of the extracting agent in the organic phase and the hydrogen ion concentration in the water phase as well as by the

[1] The steps in the conversion of eq. (21-16) to eq. (21-17) should be performed by the student.

stability of the complex, the acidic strength of the chelating agent, and the distribution coefficients of both the reagent and the complex. Since the last three of these are controlled by constants, and inasmuch as the concentration of the chelating agent is also a constant in most procedures, eq. (21-17) shows primarily the effect of variation in pH upon the extractability of a metal ion for a given chelate extraction system.

Another important aspect of eq. (21-17) is that it shows the distribution ratio to be independent of the concentration of the metal ion. If the pH is such that the metal ion undergoes significant hydrolysis, eq. (21-17) may be modified as follows:

$$D = \frac{K_{\text{comb}}}{K_h} \frac{[\text{HL}]_o^n}{[\text{H}^+]^{n-y}} \qquad \textbf{(21-18)}$$

where $K_h$ is the hydrolysis constant, and $y$ is the average number of hydroxyls attached to each metal ion. A similar correction for the effect of the formation of anion complexes may be utilized when necessary.

Whenever the concentration of the chelating agent in the organic phase is held constant, eq. (21-17) may be rewritten as

$$D = K^*_{\text{comb}}[\text{H}^+]^{-n} \qquad \textbf{(21-19)}$$

where $K^*_{\text{comb}}$ equals the product of $K_{\text{comb}}$ and $[\text{HL}]_o^n$. Because $D$ and $\%E$ are related by the equation

$$D = \frac{\%E}{100 - \%E} \qquad \textbf{(21-20)}$$

substitution of eq. (21-20) into eq. (21-19) followed by conversion to the logarithmic form gives

$$\log \%E - \log (100 - \%E) = \log K^*_{\text{comb}} + n\text{pH} \qquad \textbf{(21-21)}$$

which indicates that curves of percent extraction versus pH depend upon the value of $K^*_{\text{comb}}$ and the value of $n$. A set of typical curves for different divalent metals with a given chelate extractant is shown in Figure 21-2. The position of the extraction curve on the pH axis depends on the value of $K^*_{\text{comb}}$ and the slope depends upon the value of $n$ as shown in Figure 21-3.

The pH value at 50% extraction is known as the $\text{pH}_{1/2}$ and is an important characteristic of metal extraction systems because it can be used together with the pH range in which extraction occurs to estimate whether or not a given separation is possible. For the system shown in Figure 21-2, $M_1$ and $M_2$ can be separated by a single extraction by setting the pH of the water phase at 3.5, at which $M_1$ will completely extract and $M_2$ will not extract appreciably. Similarly, $M_2$ and $M_3$ could not be separated by a single extraction because the ranges overlap. The $\text{pH}_{1/2}$ values shown in Figure 21-2 are 2.2 for $M_1$, 4.8 for $M_2$, and 5.7 for $M_3$. The pH range over which extraction occurs is approximately

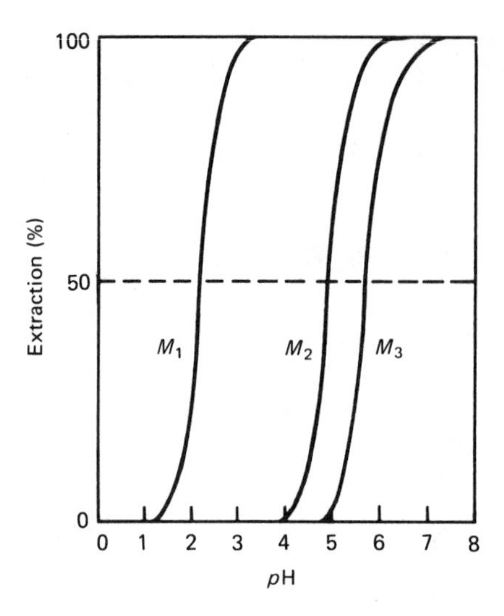


**Figure 21-2.** Typical theoretical extraction curves for different divalent metals ($M_1$, $M_2$, $M_3$) with a given chelate extraction system. The pH value at 60% extraction is known as $pH_{1/2}$. The same type curves are obtained for the use of different chelates with the same metal.

2.3 pH units for each metal. Note that, if the pH of the water phase is adjusted to 6.5, both $M_1$ and $M_2$ will be completely extracted and $M_3$ will be partially extracted by a single extraction. For those metals such as $M_2$ and $M_3$, which have overlapping ranges, separation is sometimes possible if the pH is set for maximum extraction and the extraction is repeated several times.

Equation (21-17) can also be used to estimate the separability of two metals with the same extraction system at a given pH. Using the subscripts 1 and 2 to represent the different metals, the separation ratio can be calculated as

$$R_{sep} = \frac{D_1}{D_2} = \frac{K_{stab_1} K_{dc_1}}{K_{stab_2} K_{dc_2}} \tag{21-22}$$

since all the other factors in eq. (21-17) would be identical and cancel in the ratio. If at least 99% extraction of one metal with no more than 1% extraction of the other is taken as a quantitative separation, the separation ratio equals $10^4$.

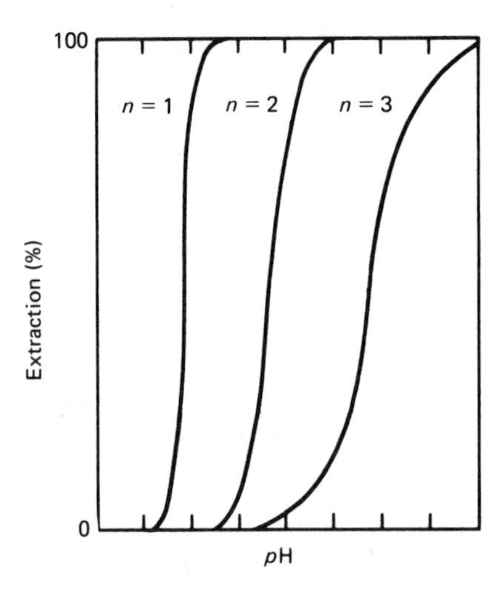


**Figure 21-3.** Typical theoretical extraction curves showing the effect of the metal ion charge on the slope.

This means that the product of the stability constant and distribution constant for one metal must be at least $10^4$ times this product for the second metal for quantitative separation with one extraction. If the separation ratio is less than $10^4$, separations may still be accomplished by several extractions or by a continuous-extraction system.

# EXTRACTION TECHNIQUES

## BATCH EXTRACTIONS

The most common type of extraction technique is the batch extraction, in which the organic liquid is added to the solution to be extracted in a separatory funnel (Figure 21-4). After agitation for a sufficient length of time, the layers are allowed to separate, the top opened, and the lower or heavier layer allowed to drain through the stopcock. If further extractions are necessary and the organic layer is the lower layer, a second portion of the organic liquid is added and the process repeated. If the organic liquid is the upper layer, the aqueous layer is transferred to a second separatory funnel and a second portion of the organic liquid added. The process is repeated as often as in necessary to ensure satisfactory separation. The method for calculating the efficiency of batch separations was shown in Examples 1 and 2.

## STRIPPING OR BACK EXTRACTION

After extraction into the organic layer, it is often necessary to return the extracted material to water for the analytical procedure to be used in the determination. This back extraction is known as stripping. For those systems

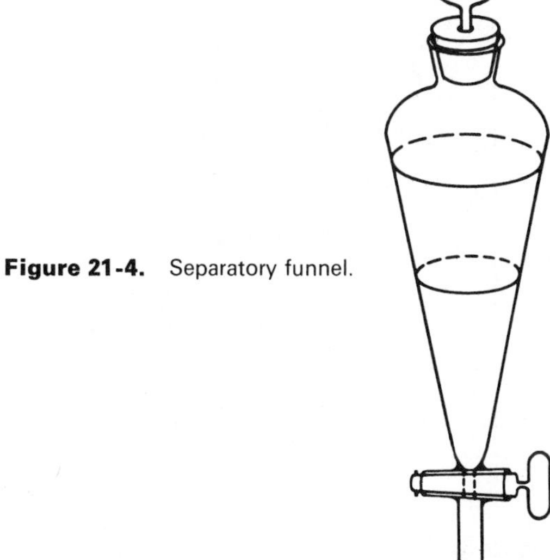

**Figure 21-4.** Separatory funnel.

that are pH sensitive, the organic solution is agitated with an aqueous solution that has been adjusted to a pH below the pH range in which the metal is extracted into the organic phase. Sometimes, the organic liquid containing the solute can be agitated with an aqueous solution containing a complexing agent which forms complexes with the metal that are more soluble in water than in the organic liquid used in the extraction.

## CONTINUOUS EXTRACTION

If a large number of multiple batch extractions is necessary to obtain a satisfactory separation, it is often preferable to use a continuous-extraction procedure. Extraction equipment has been devised for continuous extraction using organic solvents which are lighter (less dense) than water or heavier (have a greater density) than water. Examples are shown in Figure 21-5. The apparatus shown in Figure 21-5(a) is used when the organic solvent is lighter than the solution to be extracted, whereas that in 21-5(b) is used for organic liquids heavier than the solution to be extracted. The solutions may be agitated during the extraction, or the entry tube for the solvent may be rotated to ensure more uniform distribution and more efficient extraction. Agitation should be minimal for systems that have a tendency to produce emulsions.

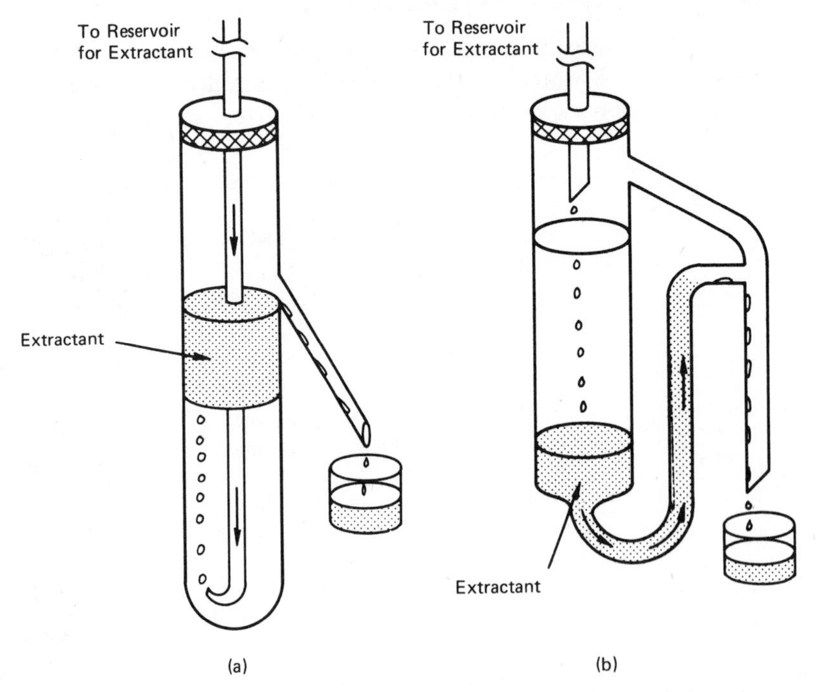


**Figure 21-5.** Continuous-flow solvent extractors: (a) for extractants lighter than water; (b) for extractants heavier than water.

## COUNTERCURRENT EXTRACTION

In a true countercurrent extraction, the two immiscible solvents contact each other as they flow through one another in opposite directions. Such extraction procedures can be very efficient because fresh extractant is brough into contact with the solute-depleted phase, and the solute-enriched extractant is brought into contact with fresh aqueous solution. Most applications, however, involve a pseudo-countercurrent procedure in which successive batches of extractant moving in one direction are brought into contact with successive batches of solution moving in the opposite direction. The amount of solute in either phase after any given number of contact steps can be calculated from a binomial expansion.

Various devices have been proposed for countercurrent extractions, but, since most of the applications have been in the field of organic separations rather than metal ion separations, they will not be discussed in detail in this text.

## EXTRACTION OF SOLIDS

Even though the extraction or leaching of solids is not a true liquid–liquid partitioning process, it is an important example of an extraction phenomenon. In the simplest procedure, the solid is covered with the liquid, and, after agitation for a suitable period, the liquid and solid are separated by decantation, centrifugation, or filtration.

Most continuous-extraction procedures for solids utilize the Soxhlet extractor shown in Figure 21-6. A volatile solvent is introduced into the lower

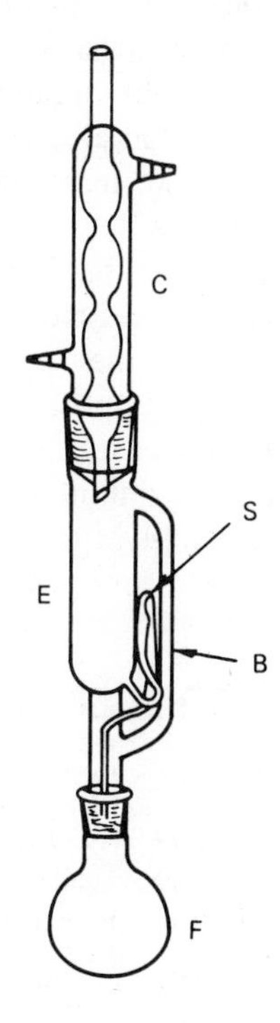


**Figure 21-6.** Soxhlet extractor. C, condenser; E, extractor; S, siphon; B, boiling vapors; F, flask.

flask F and the solid is placed in tube E encased in paper, cloth, or a porous cup. Upon heating, the solvent boils and the vapors bypass tube E by going through tube B to the condenser C. The condensed liquid falls onto the solid in tube E. When sufficient liquid has been collected, it syphons back into F by tube S. As a consequence, the volatile extractant is continuously recycled, the extracted components concentrate in the flask solution, and the remaining solid is left in tube E. Soxhlet extractions may be set up and allowed to run unobserved for long periods of time and thus save analyst time. Soxhlet extractors are standard equipment in laboratories that analyze biological and other samples for fats and oils, but they are seldom used for inorganic extraction procedures.

## ADVANTAGES AND DISADVANTAGES

The main advantages of the extraction technique are the rapidity, simplicity, and versatility of extraction methods. When necessary, separations can be effected at low or high temperatures and in inert atmospheres. Both micro- and macroscale procedures are available, depending on the size of the sample.

There are also some disadvantages associated with extraction procedures. For one, methods are still empirical, even though the basic theory has been developed satisfactorily, owing to the lack of distribution ratio and pH data for many systems. In many extraction procedures, emulsions may form when the extractant and sample are mixed, as a result of the presence of constituents not being determined. In the simplest of these cases, the length of time required for the separation of phases into two distinct layers is increased; and in those procedures in which very stable emulsions are formed, the method may not be utilized at all unless prior removal of the emulsifying component is possible.

Another disadvantage lies in the mutual solubility of the organic reagent and water. In many extractions, it is necessary to saturate the organic liquid with water or with other components of the solution to be extracted. As an example, the ether used in the extraction of iron(III) from steel solutions in hydrochloric acid must be saturated before use by equilibration with 6 $M$ HCl. Backwashing may be necessary in some multiple batch extraction procedures to remove impurities that have been extracted along with the desired component. In this case, the combined extractants are equilibrated with a fresh aqueous phase containing the optimum concentrations of extraction or other agents. Hopefully, the major portions of the impurity and negligible quantities of the desired constituent will back-extract into the aqueous phase.

## PROBLEMS

1 The partition coefficient for the extraction of iron from a 5 $M$ HCl solution into ether is 143. Assuming that 1.00 g of iron is present in 100 mL of acid and is extracted with ether, calculate the mg of iron left unextracted after (a) one extraction with 100 mL of ether; (b) two extractions with 50 mL of ether; (c) four extractions with 25 mL of ether.

*Ans.* (a) 6.9 mg

**2** Repeat problem 1 for the extraction of 100 mL of $OsO_4$ (1 mg/mL) with chloroform for which $D = 6$.

**3** Repeat problem 1 for the extraction of 50 mL of $RuO_4$ (3 mg/mL) by chloroform for which $D = 58$.

**4** For a given system, $D = 5$ and the volume of extractant to be used in each step equals 10% of the volume of the solution to be extracted. Calculate the number of steps necessary to extract 99.9% of the solute.

*Ans.* 17

**5** Repeat problem 4 for systems in which $D$ equals (a) 75, (b) 150, (c) 720, (d) 900.

**6** For oxine (8-hydroxyquinoline), $K_1 = 8 \times 10^{-6}$, $K_2 = 1.4 \times 10^{-10}$, and $K_d$ for chloroform–water is 720. Calculate (a) the value of the distribution ratio $D$ for oxine at pH 12; (b) the percentage extraction at this pH.

*Ans.* (a) 5.1; (b) 83.6%

**7** Repeat problem 6 for (a) pH 2, (b) pH 5, (c) pH 7, (d) pH 9.

**8** Show by calculations that four extractions with 24-mL portions of organic liquids are better than one extraction with 99 mL. Assume that $K_d$ is 100 and that the original solution contained 200 mg in 100 mL.

**9** Explain the difference between $K_d$ and $D$ as applied to extraction showing why $K_d$ remains constant even though $D$ can vary greatly with pH for some systems.

**10** Explain how you would separate three divalent metals for which the $pH_{1/2}$ values are 2.5, 7.5, and 10.5 for extraction into the organic phase using a particular chelate system. Assume that the total extraction range for each metal is 3.0 pH units. End up with a single metal ion in each of three water solutions.

CHAPTER

# 22 Column and Plane Chromatography

Under favorable circumstances, a solution containing a mixture of two or more components can be poured through a column of a finely divided solid and the substances will be adsorbed on the solid at the very top of the column. Continued washing of the column with a liquid may cause the components to move down the column at different rates and to be separated into individual zones if the column is long enough. Such separations are called **chromatographic** separations and the process is known as **chromatography**. Complete and partial separation of components on a column is illustrated in Figure 22-1. Column 1 of Figure 22-1 represents the adsorption of the components of the mixture at the top of the column; column 2 shows the partial separation of the components as they move down the column; and column 3 indicates the complete separation into individual zones moving at their own characteristic rate down the column as more liquid is added.

Chromatographic separations are based on physical properties such as adsorption and partition rather than on chemical changes such as precipitation, complex formation, neutralization, or oxidation–reduction; as a consequence, the original substances are available for testing after the separation. All chromatographic methods involve a mobile phase and an immobile phase and are therefore differential migration methods. The mobile phase usually provides the driving force that moves the components to be separated through the medium, and the immobile phase usually provides the selective retarding force, which causes each component to move through the medium at its own rate. The mobile phase can be liquid or gas, and the immobile phase is usually a finely divided solid or a finely divided solid coated with a thin layer of a liquid. The driving force is the flow of the liquid or gas, and the retarding forces are due either to adsorption on a solid or to partition between two fluids (gas–liquid or liquid–liquid).

Chromatographic methods can be classified as to the mobile phase (gas or liquid chromatography), as to the retardation mechanism (adsorption or partition), and as to the equipment used (column, thin layer, paper).

The first reported chromatographic separation was the separation of plant pigments by the Russian botanist, Tswett, in 1906. Tswett allowed a petroleum

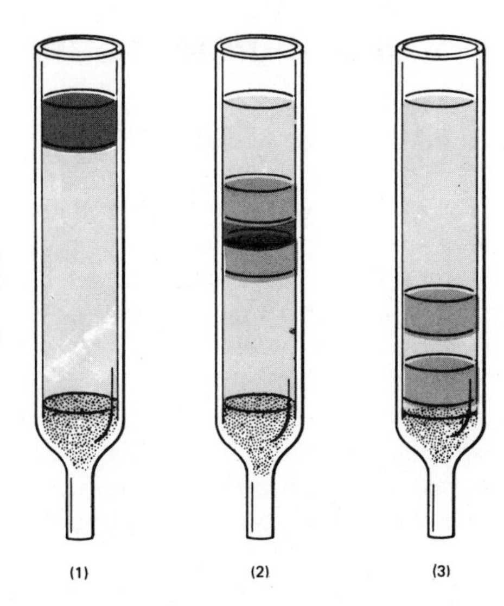


**Figure 22-1.** Column chromatography. Column 1 shows adsorption of mixture at top of column; column 2 shows partial resolution of components; and column 3 shows complete resolution of components.

ether extract of plant leaves to percolate through a column of finely divided calcium carbonate and noticed that the pigments were strongly adsorbed near the top of the column and were separated into yellow and green bands. Continued washing of the column with the petroleum ether solvent caused movement of the bands down the column, with still further separation or resolution.

Tswett called the process chromatography, since he considered it to be the separation of colored materials by selective adsorption. It is now known that colorless materials can also be separated by the process; in fact, chromatography is used to separate more colorless materials than colored. Several other names have been suggested from time to time, but none has replaced the term *chromatography*, and it is now firmly fixed in the literature.

Chromatography was not really developed as a separation method until after the work of Kuhn and Lederer on the separation of plant carotenes in 1931. There was gradual acceptance of the method between 1931 and 1943, and since 1943 the number of publications in the field has increased rapidly with constant acceleration. Now literally thousands of papers that include some type of chromatography appear annually.

There is probably no completely satisfactory definition of chromatography. One that probably covers the majority of chromatographic methods is that chromatography is the resolution of mixtures by differential migration from a narrow zone in porous media, the migration being caused by flow of liquid or gas. This definition does not include electrophoresis, in which the migration is caused by electrical potential. However, it is generally applicable because there is no mention of the retarding mechanism that causes the differential movement.

A complete discussion of all phases of chromatography is beyond the scope of this book. The student is referred to various advanced analytical texts and to reference works on the general subject and on particular aspects of the various types of chromatographic separation. We shall consider here only the simpler aspects of column chromatography and paper chromatography (gas chromatography will be discussed in Chapter 23).

# COLUMN CHROMATOGRAPHY

## SEPARATION MECHANISM

The following discussion is based upon the premises that the retarding force is selective adsorption upon a solid, that the driving force is the flow of a liquid, and that the column can be considered to be composed of a series of "theoretical plates" in which equilibrium is attained between the adsorbed substance and the dissolved substance.

In any adsorption process, where equilibrium is attained between the dissolved and adsorbed solute, the relation between the two is given by the typical adsorption isotherm, shown in Figure 22-2. From the isotherm it is apparent that the extent of adsorption increases with concentration and that there is a greater rate of change of extent of adsorption per unit change of concentration at low concentrations.

If a solution of concentration $c$ is equilibrated with a given amount of solid adsorbent, the concentration in the solution will be decreased to $c_1$ as a result of adsorption of $m_1$ of the solute upon the adsorbent. If the solution is removed by filtration and again equilibrated with fresh solid adsorbent, the concentration will be reduced to $c_2$ as a result of adsorption of $m_2$ of solute on the adsorbent. If this process is continued, the concentration in the solution will be reduced to a negligible value after a sufficient number of equilibrations have been made.

Assume that the solution is poured down a column of solid adsorbent instead of being equilibrated successively with different batches of solid adsorbent.

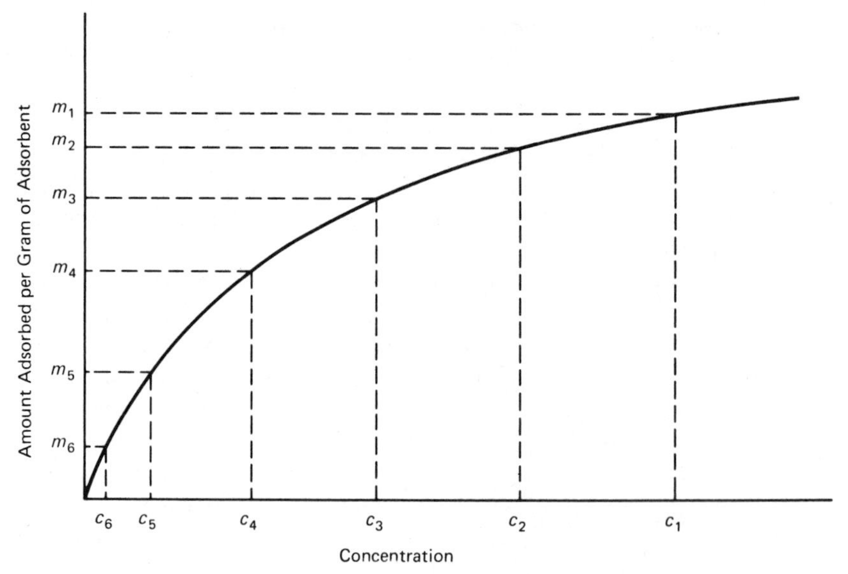


**Figure 22-2.** Removal of solute from solution by a series of batch adsorptions using fresh adsorbent for each step. The solution and adsorbent are separated by filtration after each stage. The concentration of the solution is decreased in each step until only a negligible amount of solute remains in the solution.

Also, assume that the column is composed of a number of thin layers or "theoretical" plates, each of which acts as one batch in successive batch separations. That is, assume that equilibrium is attained between the dissolved solute and the adsorbed solute in each layer. As the solution moves through each layer, some solute is removed by adsorption until a negligible amount remains, as indicated in Figure 22-3A.

If a slug of pure solvent equal to the volume of one theoretical plate is now poured through the column, some of the solute will be desorbed from the

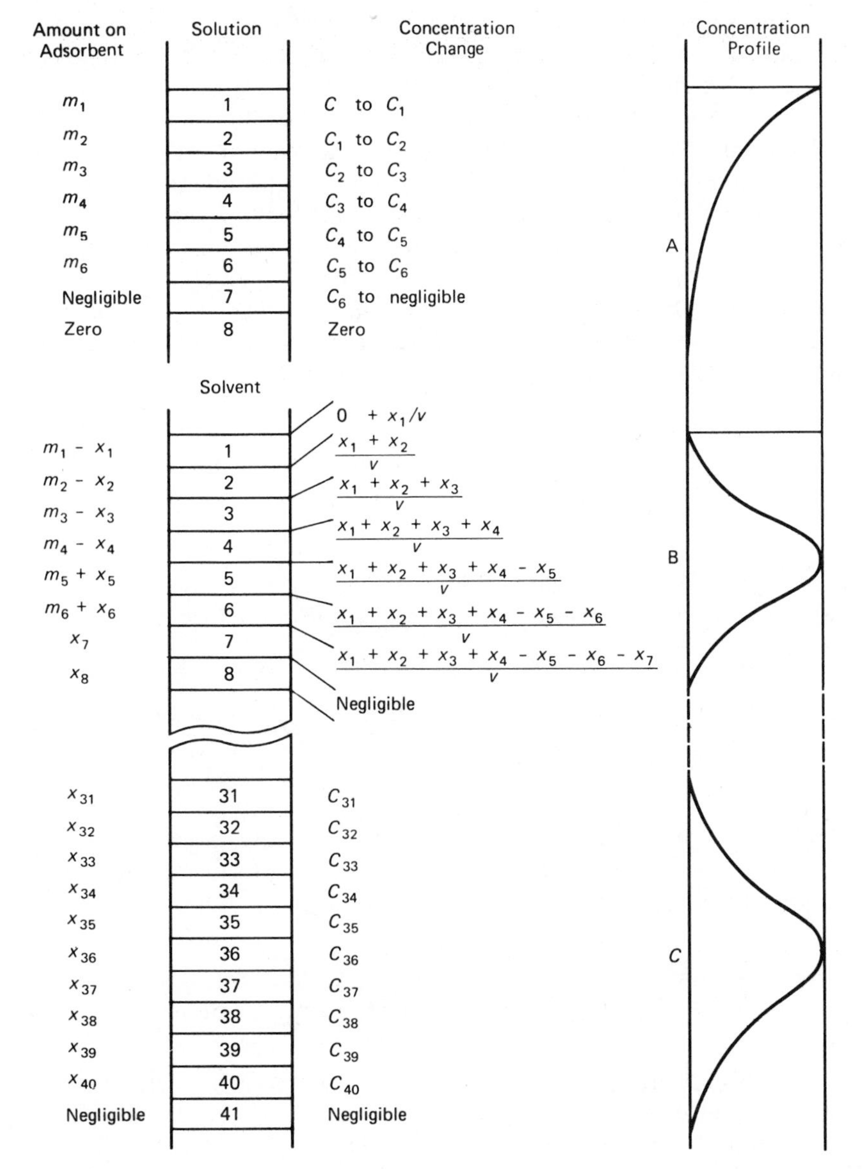


**Figure 22-3.** (A) Adsorption on the column. (B) Effect of solvent. (C) Distribution as zone moves down the column.

first layer to establish equilibrium between the dissolved solute and the adsorbed solute. This will continue as the solvent (now a solution) passes through the successive plates until the concentration leaving one plate exceeds the equilibrium concentration in the next plate (note plate 5 of Figure 22-3B). In this plate some of the dissolved solute will be adsorbed, and the concentration will be decreased to the proper equilibrium value. As a consequence, the solute will be moved down the column by decreasing the amount adsorbed at the top of the zone and increasing the amount adsorbed toward the bottom of the zone, as indicated by Figure 22-3B.

As more solvent is poured through the column, the substance is moved down the column and is distributed between the solid and the liquid, as shown in Figure 22-3C. Note that the concentration distribution in the zone does not follow the normal distribution curve, since the distribution curves in Figure 22-3 indicate definite tailing. The tailing is caused primarily by the lateral diffusion in the solution and by failure to attain equilibrium in each plate.

For separations to be effective, the rate of movement of the individual components must be sufficiently different and the column must be long enough. The rate of movement of a substance down a column may be stated in several different ways. One method is the ***R* factor**, which is defined as the ratio between the rate of movement of a zone and the rate of movement of the solvent. To be uniform, the leading edges are usually measured and stated as the $\boldsymbol{R_f}$

$$R_f = \frac{\text{rate of movement of leading edge of solute zone}}{\text{rate of movement of leading edge of solvent}} \qquad \textbf{(22-1)}$$

Since rate equals distance divided by time and since the time is the same for both the zone and the solvent, distance can be substituted for rate in eq. (22-1) to obtain

$$R_f = \frac{\text{distance moved by leading edge of solute zone}}{\text{distance moved by leading edge of solvent}} \qquad \textbf{(22-2)}$$

The $R_f$ factors are characteristic of the solutes and the conditions under which the separation is performed and can be used to identify substances if conditions are reproduced.

## METHODS OF OPERATING CHROMATOGRAPHIC COLUMNS

Methods of operation of chromatographic columns are usually based upon the liquid poured through the column after the mixture to be separated has been adsorbed on the column. Use of the original solution is known as **frontal analysis**, use of a different solvent is referred to as **displacement analysis**, and use of the same solvent is called **elution analysis**.

## Elution Analysis

Elution analysis is the type of chromatography introduced by Tswett. Pure solvent is poured through the column after the sample is adsorbed in a narrow zone at the top of the column. Each component moves down the column at a rate that is a measure of the strength with which it is adsorbed. Elution analysis can be used to resolve very complex mixtures if the column is sufficiently long and the flow rate is slow enough that equilibrium can be established in each theoretical plate. Each solute is distributed in a zone that moves down the column

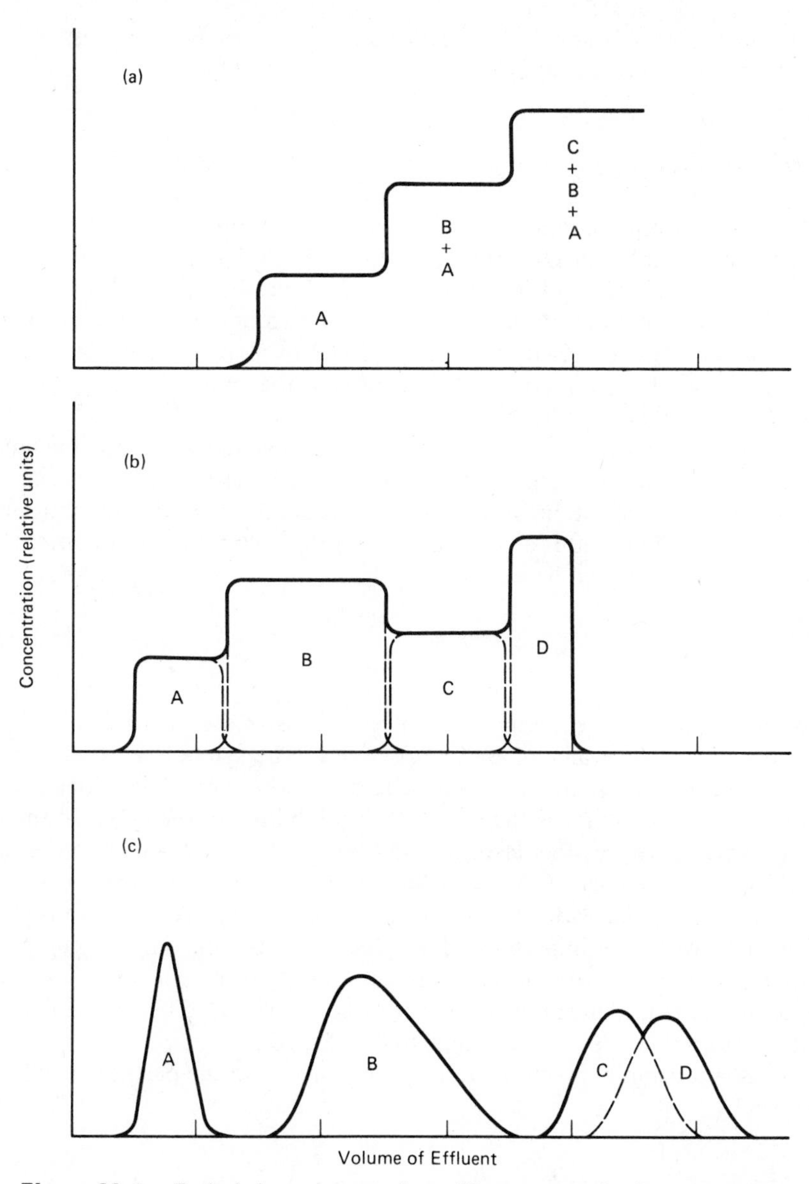


**Figure 22-4.** Typical chromatograms from different types of column operation.

at a characteristic rate and can be determined in the eluate as it comes out the bottom of the column. A typical chromatogram is shown in Figure 22-4(c), which shows two components appearing as completely separated zones and two as incompletely separated zones. The longer the column or the smaller the $R_f$ factor, the more diffuse is the zone.

Elution analysis can be used to separate and determine the components of a mixture because the area under the curve in the chromatogram is a measure of the amount of the substance present in the mixture. Also, because the $R_f$ factors are characteristic and reproducible under controlled conditions, each component can be determined in the eluate by analyzing that portion of the eluate in which it is contained. Elution analysis is by far the most popular type of column chromatography.

### *Partition Analysis*

Partition chromatography is similar to elution chromatography but differs in the mechanism of selective retardation. In adsorption, the retarding force is the selective adsorption of the solutes on the solid; in partition, the retarding force is the distribution of the solutes between two liquids. As a consequence, partition uses two immiscible solvents instead of a solid and a solvent. One solvent is coated on an inert solid and is used as the immobile phase, while the other is used as the mobile phase.

As discussed in Chapter 21, a solute that is soluble in two immiscible liquids will be distributed between the two liquids when the liquids are in contact. The ratio of concentrations in the two liquids is known as the distribution or partition coefficient and is a constant at a given temperature. For convenience the equation is repeated here.

$$K_d = \frac{C_o}{C_w} \qquad \textbf{(22-3)}$$

in which $C_o$ and $C_w$ represent the concentrations of the distributable species in the organic and water phases, respectively. The symbols $P$, $D$, and $\alpha$ are often used for the distribution coefficient. The ratio of concentrations is often approximately the same as the ratio of the solubilities of the solute in the two solvents, but there are other factors in addition to solubility that affect the value of the distribution coefficient. (See Chapter 21.)

As the solutes are washed through the column, they are continuously partitioned between the immobile liquid phase and the moving solvent phase; consequently, they move through the column at their own individual rates, which are controlled in part by the partition coefficients. A differential migration is thus set up that allows separations to be effected.

The relationship between the rate of movement and the partition coefficient is given by

$$R_f = \frac{A_m}{A_m + PA_s} \qquad \textbf{(22-4)}$$

$$\text{where } R_f = \frac{\text{distance solute zone migrates}}{\text{distance solvent front migrates}}$$

$$A_m = \text{cross-sectional area of mobile phase}$$

$$A_s = \text{cross-sectional area of immobile phase}$$

$$P = \frac{\text{solute concentration in immobile phase}}{\text{solute concentration in mobile phase}}$$

This relationship can be used to calculate $R_f$ factors from known partition coefficients, and the calculated values agree well with measured values.

For column partition separations to be effective, the two solvents must be essentially insoluble in each other, the flowing solvent must be saturated with the immobile liquid, and the substances to be separated must be soluble in both liquids and must have different distribution ratios.

Calculation of $R_f$ Factors from Partition Coefficients. Equation (22-4) can be used to calculate $R_f$ factors from the column characteristic and previously determined partition coefficients.

• **Example 1.** Calculate the $R_f$ factors for two components (A and B) for which the partition coefficients are 0.200 and 2.50, respectively. The column used is 2.00 cm in diameter, 30.0 cm long, and contains 80.0 g of alumina [density ($d$) = 3.60 g/mL], 10.0 mL of water, and 62.0 mL of the eluting solvent.

The cross-sectional area ($A$) of the column can be calculated from the diameter and is also the sum of the individual cross sections:

$$A = \pi r^2 = A_i + A_m + A_s \qquad \textbf{(22-5)}$$

where $A_i$ is the cross-sectional area of the inert phase. Each of the individual cross sections may be calculated from the volume divided by the length of the column:

$$A_i = \frac{V_i}{L} = \frac{\text{g inert phase}}{dL} = \frac{80.0}{3.6 \times 30} = 0.74 \text{ cm}^2 \qquad \textbf{(22-6)}$$

$$A_m = \frac{V_m}{L} = \frac{62.0}{30.0} = 2.07 \text{ cm}^2 \qquad \textbf{(22-7)}$$

$$A_s = \frac{V_s}{L} = \frac{10.0}{30.0} = 0.333 \text{ cm}^2 \qquad \textbf{(22-8)}$$

where $L$ represents column length in centimeters; $V$ represents volume; and the subscripts $i$, $m$, and $s$ represent the inert, mobile, and stationary phases, respectively.

It is advisable to check the column cross-section area obtained by summing the individual cross sections against that calculated from the diameter. In this case,

$$A = 3.14 \times 1^2 = 3.14 = 2.07 + 0.74 + 0.333$$

and the check is satisfactory.

Substitution in eq. (22-4) gives

$$R_{f_A} = \frac{2.07}{2.07 + 0.200 \times 0.333} = \frac{2.07}{2.14} = 0.967$$

$$R_{f_B} = \frac{2.07}{2.07 + 2.50 \times 0.333} = \frac{2.07}{2.90} = 0.714$$

The reverse calculation of partition coefficients from $R_f$ factors can also be made. Solving eq. (22-4) for $P$ gives

$$P = \frac{A_m(1 - R_f)}{A_s R_f} \quad \textbf{(22-9)}$$

• **Example 2.** Using the same column as in Example 1, calculate the partition coefficients of two components (C and D) that have $R_f$ factors of 0.400 and 0.800 respectively. Substitution in eq. (22-9) gives

$$P_C = \frac{2.07(1.00 - 0.400)}{0.33 \times 0.400} = \frac{1.242}{0.132} = 9.41$$

$$P_D = \frac{2.07(1.00 - 0.800)}{0.33 \times 0.800} = \frac{0.414}{0.266} = 15.6$$

Efficiency of Separation. The efficiency of a given column separation can be stated in terms of the fraction of the most rapidly moving component left on the column when the eluate is cut midway between the peaks of the two elution bands. Figure 22-5 illustrates the effect of incomplete separation.

• **Example 3.** Assume the column in Example 1 is used to separate two components (B and E) that have partition coefficients of 2.50 and 2.83, respectively. What fraction of E would be left on the column if the eluant is cut midway between the peaks? To solve this problem we must be able to calculate the area of the darkened portion in Figure 22-5. To do this, we need the ratio of the two $R_f$ factors and the volume of eluant required to wash component B off the column and component E to the bottom of the column. The ratio of $R_f$ values is given by

$$\frac{R_{f_B}}{R_{f_E}} = \frac{A_m + P_E A_s}{A_m + P_B A_s} = \frac{2.07 + 2.83 \times 0.333}{2.07 + 2.50 \times 0.333}$$

$$= \frac{3.01}{2.90} = 1.04 \quad \textbf{(22-10)}$$

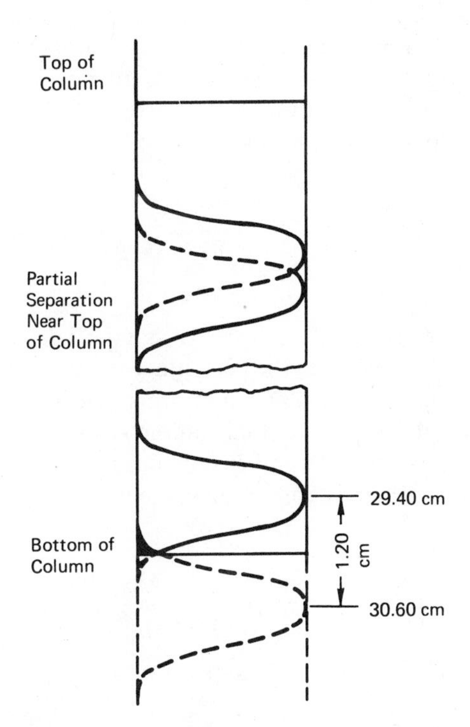


**Figure 22-5.** Partial separation in partition chromatography. The dimensions shown are for Example 3 in the text. (Not drawn to scale.)

This means that component B moves down the column 1.04 times as fast as component E. When the peak of component E reaches the bottom of the column, the peak for B would be at 30.0 × 1.04, or 31.2 cm, if the column were long enough. Since the eluate is to be cut midway between the peaks, and the peaks are essentially 1.20 cm apart, the E peak would be at 29.4 cm and the B peak at the equivalent of 30.6 cm when the cut was made.

Next we must find the volume of the eluant required to move the peak of component B an equivalent of 30.6 cm down the column. To do this, we need the number of theoretical plates in the column and the effective plate volume for component B. The effective plate volume ($V_h$) is given by

$$V_h = h(A_m + PA_s) \tag{22-11}$$

in which $h$ is the height equivalent of a theoretical plate (HETP). The height equivalent of a theoretical plate for the column in this example is 0.00300 cm, so the effective plate volume is

$$V_{h_B} = 0.00300(2.07 + 2.50 \times 0.333) = 0.00871 \text{ cm}^3$$

The number of theoretical plates ($N$) for a column distance of 30.6 cm is 30.6/0.00300, or 10,200, so the volume of the eluant needed to wash component B 30.6 cm down the column is

$$V_B = V_h N = 0.00871 \times 10{,}200 = 88.8 \text{ mL} \tag{22-12}$$

For systems with a large number of theoretical plates, the shapes of the peaks are the same as the Gaussian distribution curves. For a Gaussian distribution, the area under any portion of the curve is related to the number of theoretical plates and the elution volume by

$$t = \frac{V}{V_h\sqrt{N}} - \sqrt{N} \tag{22-13}$$

where $t$ is the number of standard deviations away from the center of the curve. Values of $t$ as the percentage of the total area have been calculated for an infinite number of plates and are given in Appendix 8. Even though the number of plates in any actual column is finite, the table values can be used without major error if the number of plates is large.

Substitution of the value from eq. (22-12) in eq. (22-13) gives

$$\begin{aligned} t &= \frac{88.8}{0.00871\sqrt{10{,}000}} - \sqrt{10{,}000} = \frac{88.8}{0.871} - 100 \\ &= 102.0 - 100 = 2.0 \end{aligned}$$

For $t = 2.0$, the percentage of the total area under the curve is 95.4. The darkened area in Figure 22-5 represents one-half the difference between the table value and 100 or 4.6/2, or 2.3. Consequently, 2.3% of the original amount of B would be left on the column if the eluant is cut midway between the peaks. Note that the number of theoretical plates used in the calculation of $t$ is based on the actual column length, since the value of $t$ required is the value at the end of the column—not at the peak maximum.

## Gel Permeation Chromatography

Gel permeation chromatography involves the use of a porous material with definite-size pores to separate molecules according to molecular size. Since the larger molecules also have higher molecular weights, the separation is often considered to be based on molecular weights. The technique is also known as **gel filtration** and can be considered a type of partition in which the molecules distribute themselves between the liquid inside and outside the gel bead. Because of the limited size of the pores and the different sizes of molecules, some of the larger molecules will not be able to enter the porous material, which usually is a three-dimensional network capable of acting as a "molecular sieve." By controlling the extent of crosslinkage and thus the amount of swelling, the pore size can be controlled so that several different size ranges are available.

The mechanism of separation by exclusion is shown in Figure 22-6. The large open circles represent the gel beads, whereas the solid circles represent large

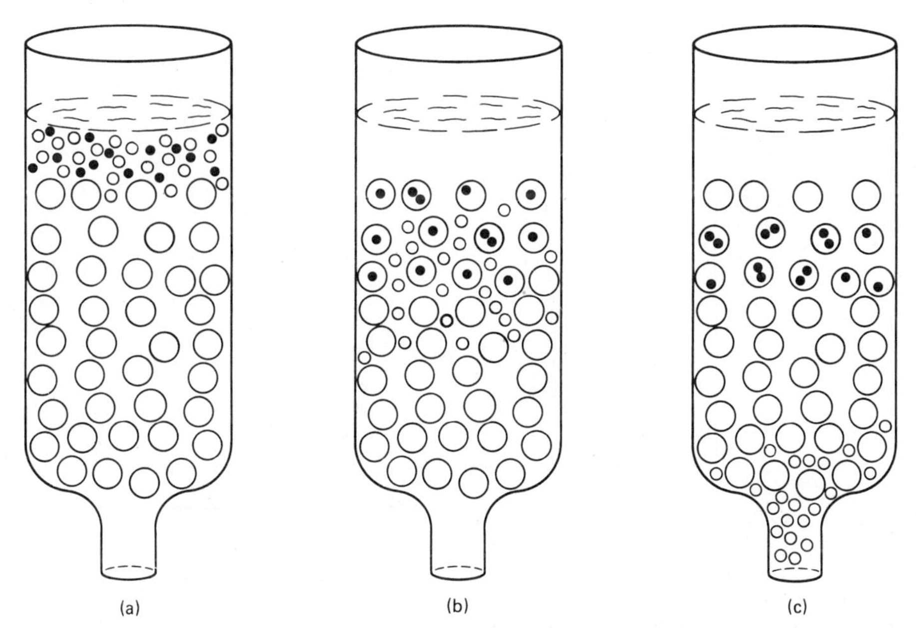


**Figure 22-6.** Gel permeation chromatography: (a) sample added to solvent in column; (b) addition of more solvent causes movement down column and small molecules diffuse into gel beads; (c) large molecules are washed off the column while the small molecules are still inside the gel beads. Large open circles, gel beads; small open circles, large molecules; small closed circles, small molecules.

molecules and the small dots represent small molecules. As the mixture moves down the column, the smaller molecules distribute themselves between the liquid inside and outside the particle, and thus are retarded, whereas the large molecules move at the rate of the solvent and emerge first from the bottom of the column. A separation is thus achieved. For the smaller molecules the distribution between the inner and outer liquid is controlled by a distribution or partition coefficient similar to the distribution coefficient for a solute between two immiscible liquids. Since different molecules will have different distribution coefficients, separations of molecules that are small enough to enter the resin may also be achieved if the column is long enough and there is a sufficient difference between the coefficients.

Gel permeation chromatography is employed in the separation, fractionation, and purification of a large number of compounds, including proteins, carbohydrates, lipids, addition polymers such as polyethylene, and condensation polymers such as nylon. Included in the list of biologically important compounds that have been separated are hormones, vitamins, enzymes, antigens, nucleic acids, and fatty acid esters. One advantage of this technique in the separation of these materials is that usually it does not affect the biological activity of the compounds. It can be performed on a small or large scale by modifying the size of the column and, consequently, is used in commercial as well as research desalting operations. Another use is the determination of the molecular-weight distribution in polymers, since there is a linear relationship between the

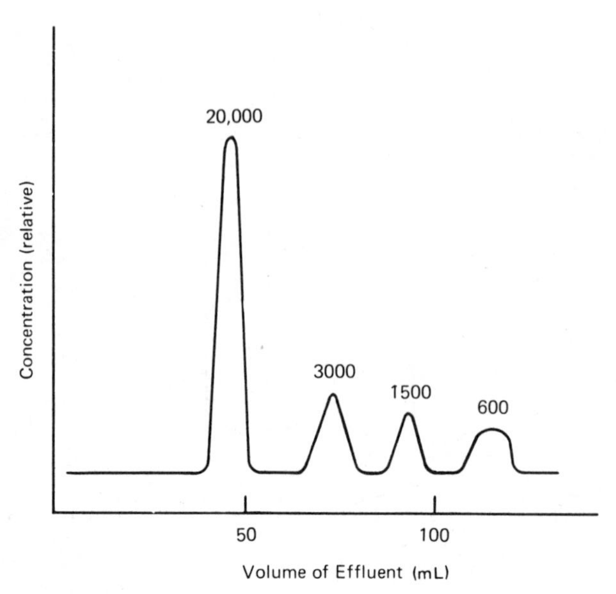


**Figure 22-7.** Typical molecular-weight distribution elution curve for polymers. The numbers represent the average molecular weight of the components being eluted.

logarithm of the molecular weight and the elution volume for a series of compounds of similar shape and density. A typical elution curve for a molecular-weight distribution experiment is shown in Figure 22-7.

## HIGH-PRESSURE LIQUID CHROMATOGRAPHY

High-pressure liquid chromatography (HPLC) also known as high-performance chromatography and high-speed chromatography (HSLC), is a recent development in which solvents are forced through a chromatographic column at pressures up to 3000 pounds per square inch (psi) at room or elevated temperatures. High-speed selective separations of many compounds not readily determinable by gas chromatography or ordinary column chromatography may be achieved by adsorption, partition, gel permeation, ion pairing, or ion exchange mechanisms using the high-pressure technique. Ultraviolet, fluorescent, and refractive index detectors are the most popular at present, and a large amount of research is being directed to the development of new and improved detectors, so better detectors should be available in the near future.

Some of the advantages attributed to high-pressure liquid chromatography are that many nonvolatile compounds can be determined without derivitization, that sample sizes may vary from the microgram range up to several grams, and that high resolution is attainable. The technique is coming into widespread use in the separation and determination of pesticides, aromatic and other hydrocarbons, and biologically important compounds such as vitamins, sulfa drugs, narcotics, steroids, and nucleic acids. The fact that relatively large samples can be used accounts for the popularity of the technique for preparative purposes.

The chromatogram for high-pressure liquid chromatography is similar to the elution curve shown in Figure 22-4(c).

### *Instrumentation*

A block diagram of a typical high-pressure liquid chromatograph is shown in Figure 22-8. Most instruments are operated at a constant flow rate but can be operated at constant pressure. At constant pressure the flow rate may vary with time, whereas maintenance of a constant flow rate often requires variation in pressure during the run. Although pressures up to 10,000 psi may be used, most instruments do not operate at greater than 3000 psi. The use of a polar solid and nonpolar solvent is known as **normal-phase chromatography** and the reverse (polar solvent and nonpolar solid) is called **reverse-phase chromatography**.

Two or more solvents may be used at any desired ratio, and the ratio may be varied with time for gradient elution. The solvents are moved through the column by pulse-free pumping systems. Typical flow rates vary from 0.5 to 3.0 mL/min.

The injection port is placed as close to the top of the column as possible and is designed to withstand the maximum pressures that can be attained by the instrument. Normally, a microliter syringe is used to inject a small sample into the flowing solvent through a self-sealing elastic diaphragm. Larger samples and viscous samples are introduced into the system from a special valve fitted with one or more loops of tubing that contain the sample.

Most columns are prepared from stainless steel with fittings that can withstand the necessary pressures. Diameters (ID) vary from 2 to 8 mm and lengths from 15 cm to more than 1 m. The dead volume between the injection port and the column and in the column itself must be as small as possible to offset lateral diffusion and peak broadening. Tightly wound metal capillary columns of much greater length have been utilized. This type of column must be tightly wound to cause turbulent flow over the packing.

The most popular detector is the ultraviolet absorption detector, which can be operated at fixed wavelength or at variable wavelengths. Many detectors measure the absorption at 254 nm, at which most organic compounds containing double bonds or aromatic structures absorb. Other wavelengths may be used for compounds that show greater absorbance at different wavelengths.

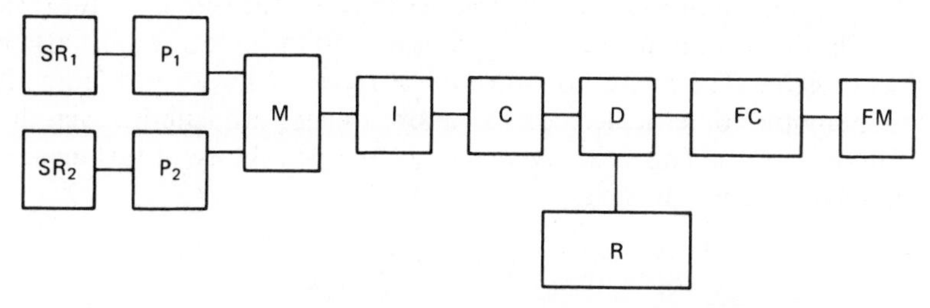


P—pumps; SR—solvent reservoirs; M—variable ratio solvent mixer; I—injection port; C—column; D—detector; R—recorder; FC—fraction collector; FM—flow meter.

**Figure 22-8.** Block diagram of a high-pressure liquid chromatograph.

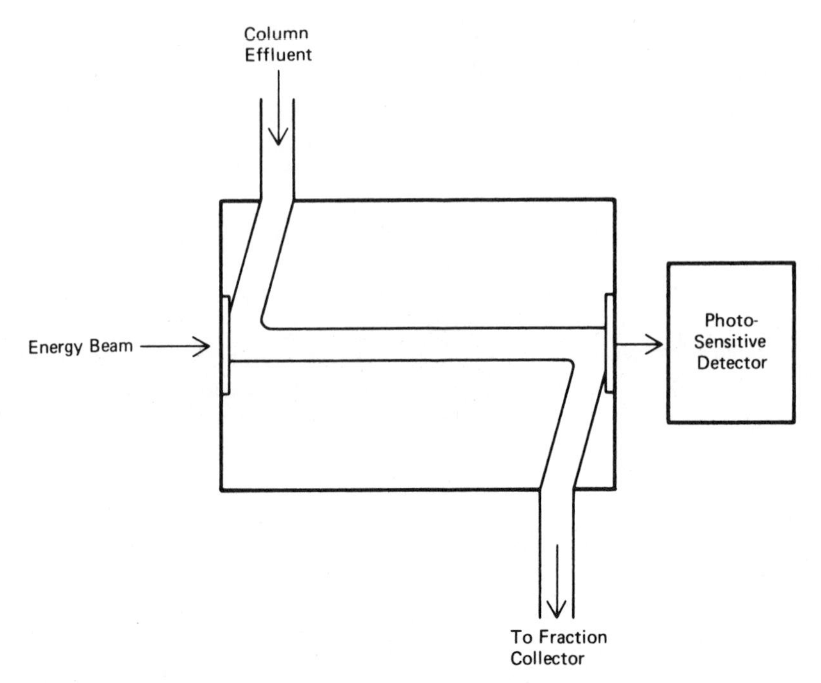


**Figure 22-9.** Simplified diagram of a Z-type microcell.

Fluorescent detectors that increase the sensitivity greatly for compounds capable of fluorescence have recently been introduced. The differential refractive index detector, which measures slight changes in the refractive index due to changes in the composition of the column effluent, was one of the first used but is not popular, owing to its low sensitivity. Other detectors that have been introduced include a flame ionization detector, an electron capture detector, an electrochemical detector, and a visual colorimetric detector using ninhydrin for amino acids. Each of these has special uses in separations for which it can increase the sensitivity or specificity.

Another intriguing development is the combination of a mass spectrometer with a high-pressure liquid chromatograph to allow positive identification of column effluents. The output of the detector is fed to a recorder, which draws the chromatogram. Many instruments are equipped with dual (or more) pen recorders so that more than one variable may be measured in the column effluent. The dead volume between the column and the detector must also be as small as possible. The micro flow-through Z-type cell shown in Figure 22-9 is used in many photometric detectors to allow longer path lengths, even though the volume is small. This cell also decreases the possibility of turbulence and mixing in the column effluent.

### *Packing Materials*

Packing materials vary according to the type of separation involved: solid/liquid (SLC), liquid/liquid (LLC), gel permeation (GPC), and ion exchange

(IEC). The packing supports may be porous solids of uniform diameter approximately 10 μm or solid glass beads of 30–50 μm coated with a thin layer (1–2 μm) of porous material. These coated glass supports are known as **pellicular beads**, and their larger size allows the use of lower pressures for any given flow rate. The porous solids or pellicular layers may be coated with or bonded to a liquid for LLC separations. The particles for gel permeation methods are made by crosslinking styrene and divinylbenzene to form small, rigid gel particles with carefully controlled pore sizes. Polymer beads with pore sizes ranging from 10 to $10^5$ nm allow a wide range of molecular-weight separations. For instance, the $10^5$-nm size can be used for separations of compounds with molecular weights ranging from 1,000,000 to 20,000,000, whereas the 10-nm size is effective for compounds with molecular weights from 0 to 700.

Open-pore polyurethane (OPP) is also used for gel permeation work. This material is composed of agglomerated spherical particles (1- to 10-μm diameters) bonded to each other in a rigid, highly permeable structure. It can also be prepared with controlled pore sizes similar to the polystyrene polymer. Other packings that separate compounds by a size-exclusion principle are prepared by bonding an ether to porous silica particles. Ion exchange packing materials are prepared from polystyrene–divinylbenzene copolymer that has been treated to attach ionic functional groups to the polymer skeleton (see Chapter 24). Strong or weak cation exchangers and strong or weak anion exchangers are available as permeable micro beads or as coated pellicular particles. The pellicular beads show much lower capacities than the solids.

Other packings are prepared by chemically bonding various compounds to the porous solids or pellicular beads. A widely used material for reverse-phase chromatography has a $C_{18}$ group bonded chemically to porous silica particles to furnish a nonpolar surface layer. Compounds terminating in cyano, amine, phenyl, or other functional groups can be bonded to a porous solid and used for specific types of separations.

### Solvent–Packing Combinations

The selection of the particular packing and solvent system used depends upon the composition of the sample and the type of compounds that must be separated. Many factors are involved, but the most important are usually the polarity of the solvent, the sample, and the packing surface. Figure 22-10 shows some typical solvent–packing combinations. Success in affinity separations is achieved by establishing the proper balance between the attraction of the solvent and the packing for the sample constituents, owing to the fact that both are in competition for the components. Poor retention occurs when the polarities of the sample and solvent are more similar than the polarities of the sample and packing. Better retention can be obtained by changing the polarity of the solvent or by changing to a packing that is more like the sample. Most good separations are obtained by matching the polarity of the sample and packing and using a solvent that has a markedly different polarity. Improved separations can often be achieved by using reverse-phase chromatography when changes in solvent

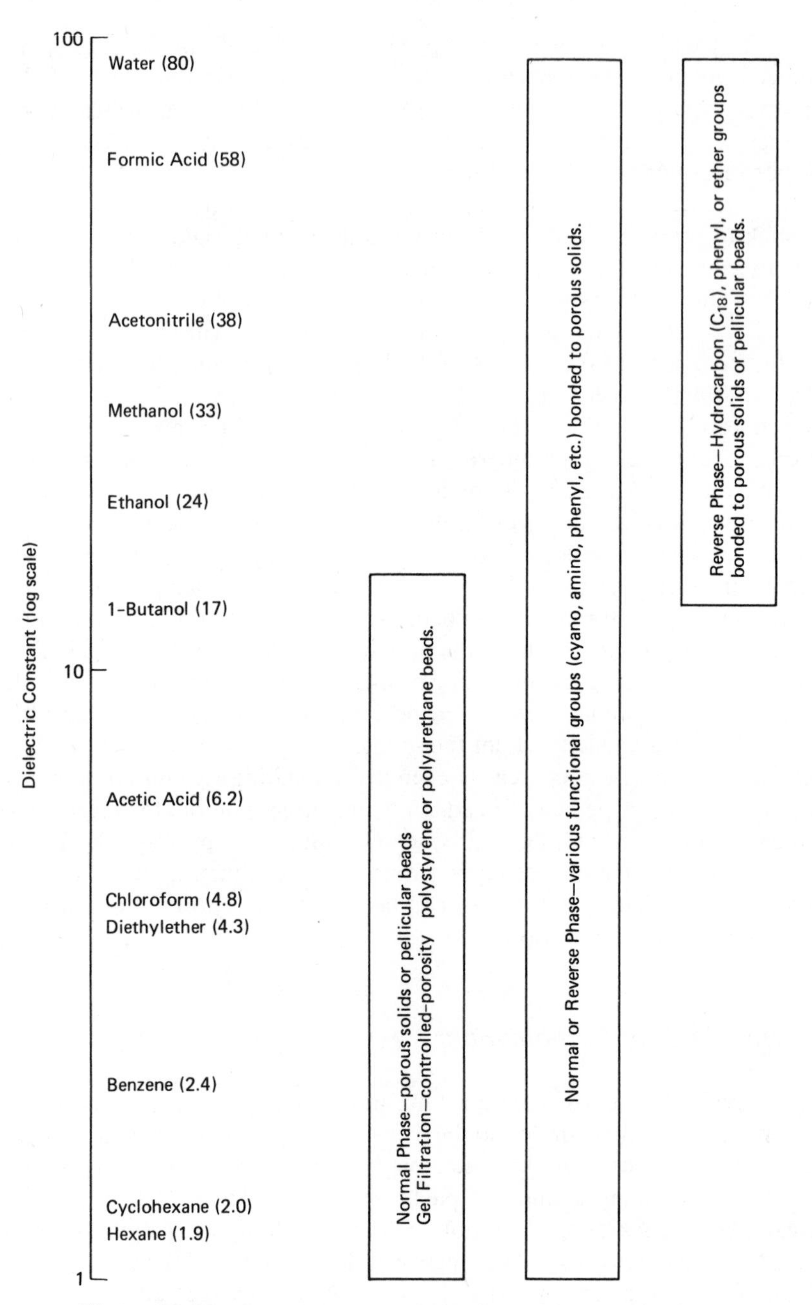


**Figure 22-10.** Typical solvent and column packing combinations.

polarity fail to cause changes in elution times that improve the resolution of the components.

With ion exchange packings, variations in pH, strength of resin, common and competing ion concentrations, temperature, flow rate and column length will affect the separation; for gel filtration, variation of the pore size, solvent strength, column length, and flow rate are the main variables to be considered.

Recently, **ion pair partition chromatography** (IPC) has been developed, by which organic acids, esters, and bases can be separated using a porous silica support coated with a film of solution containing a large hydrophobic anion or cation. If an aqueous film is used, the mobile phase is a polar organic solvent, but if the film consists of a long-chain amine or similar functional group bonded to the silica, the mobile phase is usually an aqueous solution containing a miscible polar organic compound or is a mixture of two miscible polar organic solvents. The $C_{18}$ type of packing is often used with methanol–water, acetonitrile–water, acetonitrile–methanol, and similar mixtures, with the solvents being programmed to vary the ratio continuously from a set mixture by increasing the concentration of one of the solvents at a definite rate.

## *Applications*

The use of HPLC in basic and applied research and in industrial, governmental, and environmental analyses and control is growing rapidly. Many mixtures of organic compounds that defy separation by older techniques can be separated easily and rapidly by HPLC. It is used to support basic synthetic research, by allowing rapid identification and estimation of the products under various reaction conditions or to isolate a particular product. It is used by industry to control production quality, to ensure the purity of purchased materials, and to follow shelf stability of materials subject to decomposition with time. In medical and biological research, it is used to detect and determine drugs, foodstuffs, metabolic products, and other materials in body fluids under various conditions and diets. Governmental laboratories utilize HPLC in the analysis of various pharmaceuticals, foodstuffs, and other products to ensure that they meet specifications and contain no harmful amounts of pesticides or carcinogenic compounds. In environmental work, HPLC helps to speed up the analysis of trace and major contaminants in air and water and to monitor less desirable contaminants in areas where their concentration may vary as a result of industrial or other activities. Mixtures may be separated by the mode (SLC, LLC, GPC, IPC, IEC) most satisfactory for that sample, and many separations can be improved by the use of solvent programming (gradient elution). Samples containing hormones, drugs, vitamins, nucleic acids, pesticides, carcinogens, metabolic products, synthetic organic chemicals, polymers, organic acids, and many other mixtures can be analyzed by HPLC.

An illustration of the separations that can be accomplished using HPLC is shown in Figure 22-11, which is a reverse-phase chromatogram of a mixture of 13 polynuclear aromatic hydrocarbon (PNA) standards; the structural formulas of the compounds are shown in Figure 22-12. This separation was obtained from

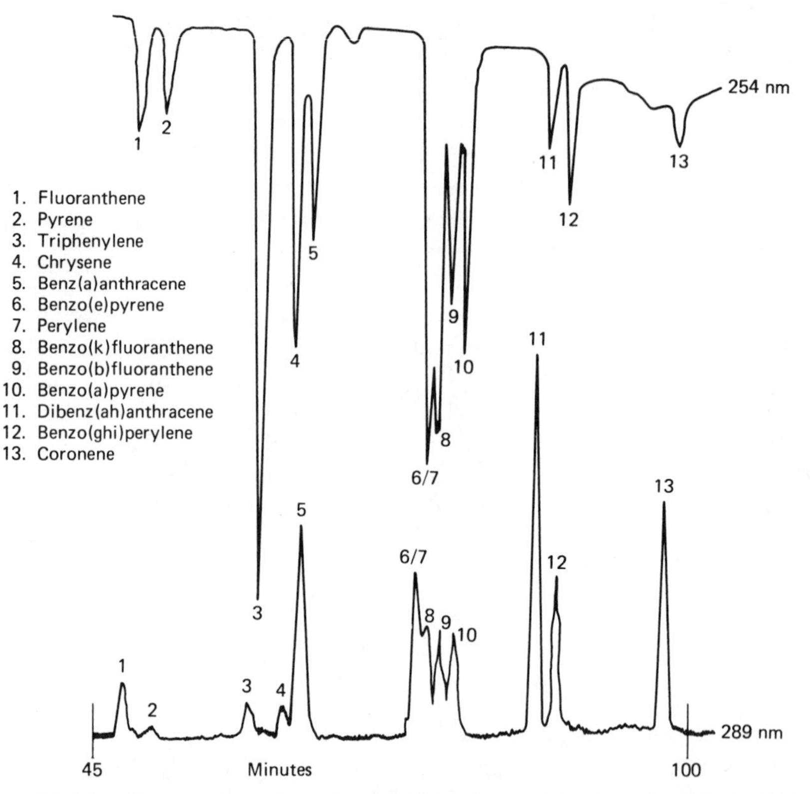


**Figure 22-11.** Reverse-phase separation of polynuclear hydrocarbons by HPLC. Column: 3 × 30 cm × 4 mm; packing: 10-$\mu$m $\mu$Bondapak $C_{18}$ (hydrocarbon bonded to porous silica); mobile phase: linear program from 40 to 80% acetonitrile in methanol–water (1 + 1) during 90 min; flow rate: 1 ml/min; detectors: ultraviolet absorption set at 254 nm and at 289 nm. (Courtesy J. P. Hanus and E. R. Biehl, Dallas District, Food and Drug Administration.)

the injection of 20 $\mu$L of a solution containing 2 ng/$\mu$L of each component during an investigation of polynuclear aromatics present in oil-contaminated shellfish by the Dallas District of the Food and Drug Administration. In the actual analysis, a homogenated sample of the contaminated shellfish is extracted with acetonitrile and the extracted PNAs partitioned into petroleum ether, which is then evaporated. The residue is saponified before a second partition into petroleum ether. The sample extract is passed through a silica gel column and the PNAs are eluted with petroleum ether and a petroleum ether–acetonitrile mixture (99 + 1). The solvent is removed by evaporation and the residue made to a small volume with toluene. This solution is then separated into fractions by gel permeation HPLC using toluene as the mobile phase. The fractions are evaporated, then the residues are dissolved in a small volume of acetonitrile and injected separately into the chromatograph. Each of the four fractions retained contains from one to seven of the individual compounds, which allows better detection and quantitation. The separation of the four-ring isomers labeled 2, 3, 4, and 5 in Figure 22-11 is indicative of the possibilities of this technique. Also, the use of dual detectors usually improves the quantitation of

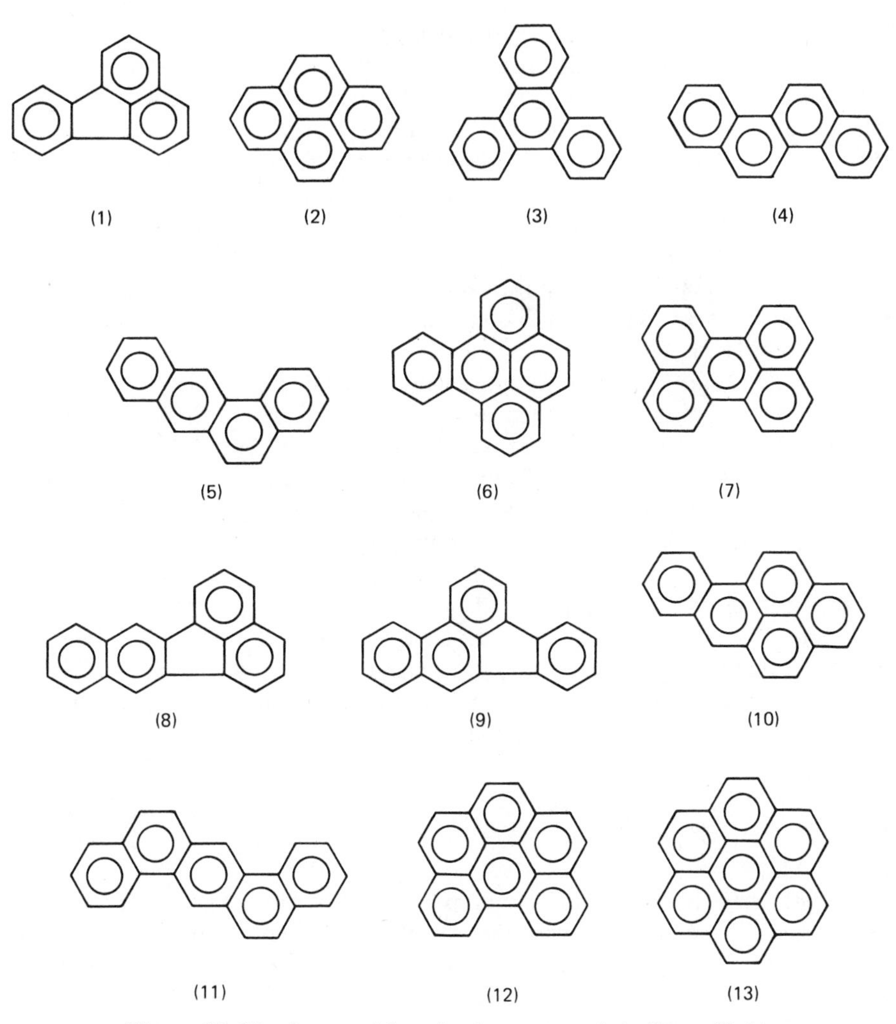


**Figure 22-12.** Structural formulas for compounds in Figure 22-11.

one or more of the components of a complex mixture. The variation of retention times between compounds 2, 7, 10, 12, and 13 shows the effect of increasing molecular weight.

## PLANE CHROMATOGRAPHY

### THIN-LAYER CHROMATOGRAPHY

Thin-layer chromatography (TLC) is similar to column adsorption chromatography except that it is performed using a thin layer of adsorbent (0.025 mm or less) spread evenly over a glass plate. The sample is deposited as a small spot, and the solvent is allowed to move upward over the spot, usually by capillary action. Since the thin layer is used, sample sizes must be much smaller than with columns, and the time required for separations is much shorter. The general principles that apply to separations on a column also apply to thin-layer chromatography, so thin-layer experiments are often used to determine proper column conditions before the column separation is run. Thin-layer separations are especially beneficial when the sample size is limited. The size of the spot made by a component can be used to estimate the amount present if a series of standards is run on the same plate. Replication from plate to plate is not precise, since it is difficult to prepare plates that have identical thicknesses of adsorbent layer.

Plates are usually prepared by spreading a thin layer of a slurry over a glass plate. The layer is then smoothed out by some device that ensures approximately equal thickness of the desired depth over the entire plate, and the plate is dried at room or elevated temperatures. Various devices on the market are of aid in the preparation of uniform plates, and several companies market prepared plates of constant thickness and composition.

The chromatograms must be run in a closed system to prevent evaporation of the solvent as it flows through the adsorbent. The closed system can be a simple wide-mouth jar with a screw top; a device that allows a cover plate to be used above the thin layer in a sandwich arrangement, with a small air space between the adsorbent layer and the cover; or a thermostated closed container, which may cost up to several thousand dollars.

## PAPER CHROMATOGRAPHY

The principles of paper chromatography are similar to the general principles of column chromatography, with the exception that paper is used as the solid immobile phase in place of a finely divided solid adsorbent. The sample is spotted on the paper as with thin-layer chromatography, and the solvent or mobile phase is allowed to flow over the spot to cause the separation.

The exact retarding mechanism that causes the differential migration is not known. In some cases it appears that the main mechanism is adsorption on the surface of the cellulose fibers; in others, it is thought to be due primarily to partition between the flowing solvent and the water adsorbed on the paper; in still others, ion exchange seems to play an important role. Since it is difficult to determine which of these possible retarding mechanisms is the actual cause of the differential migration, it is usually assumed that all three play a part in all separations.

The partition viewpoint is substantiated by the fact that some $R_f$ factors can be predicted satisfactorily from known partition coefficients. However, this is not true for a large number of cases. Carboxyl groups are formed on the cellulose during processing or manufacture, and the number can be increased by certain treatments. These carboxyl groups can furnish sites for ion exchange of metal or organic ions and thus play a part in the retardation mechanism. The flow of solvent through the paper is by capillary action through the interspaces between fibers and in the capillary openings in the fibers themselves. Adsorption probably occurs in the fiber capillaries and also upon the surface of long fibers.

## *General Methods*

Chromatographic paper is usually available in large sheets (45 × 55 cm.) or in 2.5 cm.-wide strips in rolls of several hundred feet. Since the "grain" of the paper has some effect upon the flow rates, the long direction of the paper, the "machine" or "grain" direction, should be used.

**Ascending Chromatography.** The sample is spotted about 1.3 cm from the end of the paper strip, and the strip is allowed to hang down with the lower 0.6 cm immersed in a liquid. The liquid flows upward by capillary action. The chromatographic chamber must be closed to ensure equilibrium between the liquid and vapor in the chamber, to prevent evaporation of the solvent during the development of the chromatogram. The chamber can be a test tube, a milk jar, a specimen flask, or a commercially available chromatographic chamber, which can be very expensive. The effectiveness of the separation is limited by the fact that the solvents will rise only to a limited height by capillary action. The procedure is also known as **countergravity separation**.

**Descending Chromatography.** The sample is spotted some distance from one end of the strip, which is folded between the spot and the short end of the strip. The strip is suspended at the fold, and the short end is allowed to dip into a tank of the solvent being used. The solvent rises to the fold and then moves down the paper over the spot by capillary action. The strip must be protected from siphon action by a dry rod or container edge. This technique permits longer strips, which is the equivalent of lengthening the column in column chromatography. The chromatogram must be developed in a closed chamber to prevent evaporation of the solvent. Simple equipment, such as a graduated cylinder covered by a beaker, up to commercially available equipment, which can run into several thousand dollars, can be used.

**Two-Dimensional Chromatography.** In two-dimensional chromatography a large sheet of chromatographic paper is used together with two solvents. The sample is spotted in one corner and run in one direction by one solvent. The chromatogram is then dried, turned 90°, and developed with the second solvent, as shown in Figure 22-13. The advantage of two-dimensional over single-dimension chromatography is that several substances that are not separated by the first solvent may be separated by the second, so the efficiency of separation is increased tremendously. Two-dimensional paper chromatography is used to separate complex mixtures of amino acids produced by hydrolysis of protein. Many complex mixtures that cannot be separated by any other method can be resolved effectively by two-dimensional chromatography with proper selection of the two solvents used. As in the other methods, a closed chamber must be used.

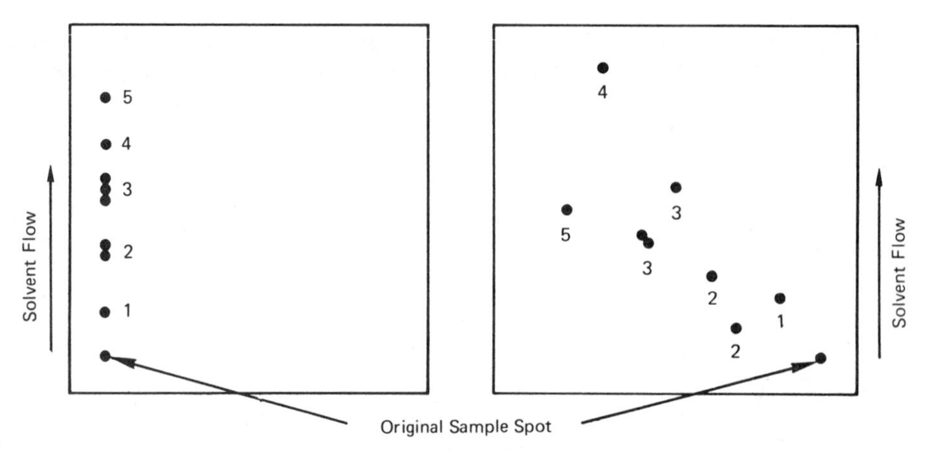


**Figure 22-13.** Two-dimensional paper chromatography. The sample is spotted in a corner and run vertically. The paper is then dried, rotated 90°, and run with a second solvent. In this example, spot 2 was resolved into two components by the second solvent. Spot 3 appears to have three components, two of which were not resolved by the second solvent.

## *Evaluation of Chromatograms*

The location of the various zones or spots that indicate the positions of the components of the mixture can be determined readily if all the components are colored. For colorless components, the separated components must be converted into colored compounds before location is attempted. This can be done by spraying the chromatogram with a color-producing reagent, such as ninhydrin for amino acids. Other procedures involve streaking or dipping the chromatogram. In some cases radioactive tracers may be utilized, and the locations may be determined by covering the chromatogram with photographic paper followed by later development of the exposed photographic film.

The amount of the various components can be evaluated by the size of the spot, the intensity of the color, or the amount of darkening of the photographic film. Qualitative identification is made by the $R_f$ factor.

The $R_f$ factor for paper chromatography is defined as the ratio of the rate of movement of the solute zone to the rate of movement of the solvent. The leading edge, the center of the spot, or the trailing edge can be used to make such ratio measurements. For uniformity, and since it is usually easier to measure and is more reproducible, the leading edge is used and the ratio as previously mentioned is called the $R_f$ factor. The $R_f$ factor is re-defined here for convenience:

$$R_f = \frac{\text{rate of movement of leading edge of zone}}{\text{rate of movement of leading edge of solvent}} \tag{22-1}$$

and since the time involved for both solute and solvent zones is the same, distances may be substituted for rates to give

$$R_f = \frac{\text{distance moved by leading edge of solute zone}}{\text{distance moved by leading edge of solvent}} \tag{22-2}$$

The $R_f$ factors are reproducible as long as all conditions that affect them are held constant or are reproduced. They are also characteristic of the individual compound and thus can be used to identify substances as long as conditions are reproduced. Conditions that affect $R_f$ factors include the solvent used, the paper used, the method of development, vapor equilibrium with the solvent and the temperature.

Because two substances may have almost equal or identical $R_f$ factors with a given solvent under a given set of conditions, reproduction of an $R_f$ factor under one set of conditions is not considered absolute identification. However, if the substance has the same $R_f$ factor as a known standard under several sets of conditions using different solvents, different types of paper, different temperatures, and different methods, that is essentially sure proof the substances are identical.

## ELECTROPHORESIS AND ELECTROCHROMATOGRAPHY

**Electrophoresis** was developed for the separation of charged colloidal particles in an electric field. The term was originally applied to the migration of charged colloidal particles only. As the techniques were developed, it became apparent that electrophoretic systems could be applied to almost any charged particle or any substance about which an electrical double layer exists. A large number of names have been suggested to supplant electrophoresis, among them ionography, cataphoresis, electromigration, and electrochromatography. Electrophoresis is well established in the literature and will probably be retained as the name of the general process, although the other names may be applied to particular types of electrophoresis. For instance, the term **electrochromatography** is generally used to describe a combination electrophoretic–paper chromatography system in which the electrical field is applied at right angles to the flow of the chromatographic mobile phase.

Electrophoresis is often defined as the migration of large molecules and small aggregates of molecules under the influence of an electrical field applied to the medium in which the particles are suspended. Proteins, viruses, and colloidal particles are examples of this type of substance. However, electrophoresis can also be applied to inorganic ions and smaller molecules such as amino acids, carbohydrates, and lipids, and the definition should include these types of substances.

Electrophoresis systems are classified according to the medium in which the particles are suspended and upon the material used as support. **Free** or **boundary electrophoresis** refers to the method developed originally by Tiselius in 1937, which involves the vertical movement of a boundary between a buffered protein solution and a less dense supernatant solution of the buffer alone. These separations are usually performed in a U-tube, and the boundary movements are followed by a complex optical system.

**Paper electrophoresis**, which utilizes paper strips as the support, is probably the most popular type of electrophoresis. The paper is saturated with a buffered liquid that acts as the conduction medium. Various types of gels such as starch, agar, polyacrylamide, and cellulose acetate are also used as supports and in some cases can be cast into sheets or strips. Our discussion will be limited to paper electrophoresis, since the principles, techniques, apparatus, and terminology used in paper electrophoresis are also applicable to the other systems.

### *Paper Electrophoresis*

**Theory.** The migration of charged particles in paper depends upon the magnitude and sign of the charge on the particle, the size of the particles, the adsorption properties of the paper, the applied voltage, and the current flow. The charge governs the direction of movement of the particle because positive particles migrate toward the negative pole, and vice versa. Also, the greater the charge, the greater the **mobility** of the particle. The mobility is also a function of the particle size, since larger particles move more slowly.

The adsorption properties of the paper can either favor or hinder the separation of two substances. If the faster moving particle is the most strongly adsorbed, the separation is not as effective; if the opposite is true, the difference in mobilities will be increased. The applied voltage and the current flow not only affect the rate of migration but also affect the temperature, since significant amounts of heat are generated by the passage of the current through the medium. The increase in temperature increases the evaporation of the solvent, with a resultant increase in the concentration of the electrolytes in the buffer medium. As a consequence, a cooling system is usually incorporated into the apparatus to dissipate the heat generated. Ordinarily, it is desirable to use the highest potential and current flow that will produce a separation in the shortest possible time. High voltages also increase the heat and other problems, so compromise conditions that will produce adequate separation in a practical length of time are utilized.

The relationship between the mobility of the particle and the applied voltage is given by

$$u = \frac{p_m L_m}{Vt} \qquad \textbf{(22-16)}$$

in which $u$ is the mobility of the particle in square centimeters per volt per second; $p_m$ is the measured distance the particle moves in centimeters; $L_m$ is the distance between the electrodes in centimeters; $V$ is the applied voltage; and $t$ is the time in seconds.

**Apparatus.** The simplest technique for paper electrophoresis consists of a strip of paper moistened with buffer prepared in a sandwich arrangement between two glass plates, with the ends dipping into two electrode vessels, as shown in Figure 22-14. The electrode tanks usually contain a porous divider. This entire apparatus is enclosed in a covered container to offset evaporation of the solvent. Various modifications of this simple apparatus are available. The sample can be spotted in the middle of the strip or at either end, but most are spotted in the middle.

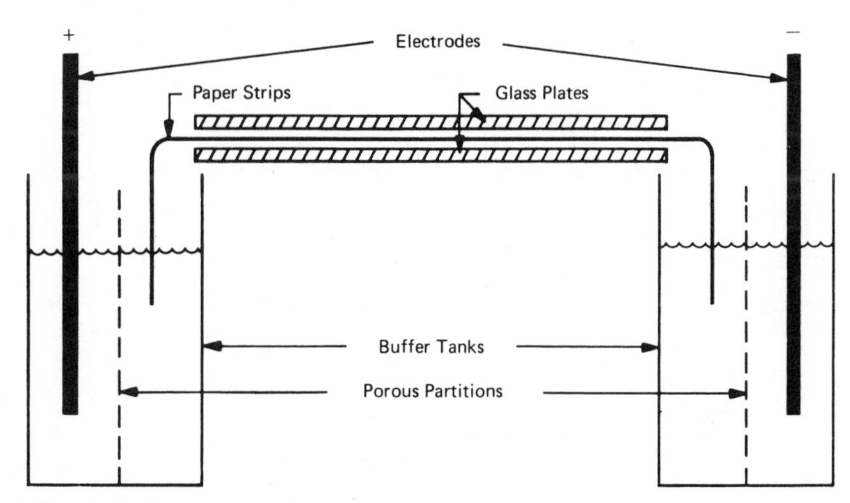


**Figure 22-14.** Simplified diagram of apparatus for sandwich-type electrophoresis.

**Applications.** Serum electrophoresis is widely used as a diagnostic tool in the medical profession. The plasma proteins in normal blood give a unique electrophoretic pattern in which the albumins are separated from the various types of gamma globulins. As the percentages of these change in certain diseases, the electrophoretic pattern changes and the change is usually typical of the disease. Thus, certain types of disease are evident simply by looking at the electrophoretic patterns. Electrophoresis of enzyme fractions, lipids, and carbohydrates can also be utilized as diagnostic aids.

Organic substances separated by electrophoresis include aliphatic amines, dyes, amino acids, dipeptides, polypeptides, organic acids, proteins, and the hydrolysis products of nucleic acids. Electrophoretic separations of metal ions include arsenic, antimony, and tin; silver, mercury, and lead; lithium, sodium, and rubidium; silver, copper, nickel, and iron; and many others.

Typical electrophoretic separations of some amino acids are shown in Figure 22-15. Amino acids are unique in that they show an isoelectric point or pH at which they are electrically neutral or exist in the zwitterion condition. At this particular pH, they do not migrate during electrophoresis. At pH values below the isoelectric point, the amino acid will become positively charged by the protonation of the amine nitrogen as a result of the larger hydrogen ion concentration. At pH values above the isoelectric point, the amino acid becomes negatively charged as a result of anion formation by the reaction of the larger number of hydroxyl ions with the carboxyl group. The isoelectric points for amino acids such as alanine ($CH_3CHNH_2COOH$), which have one carboxyl and one amine group, are close to pH 7, whereas those for amino acids such as aspartic acid ($HOOCCH_2CHNH_2COOH$), with two carboxyls and one amine group, are two or more units below 7. Amino acids that have two amine groups and one carboxyl group, such as lysine ($H_2NCH_2CH_2CH_2CH_2CHNH_2COOH$), have isoelectric points at pH values two or more units above 7. By adjustment of the pH of the buffer, any given amino acid can be made positive, negative,

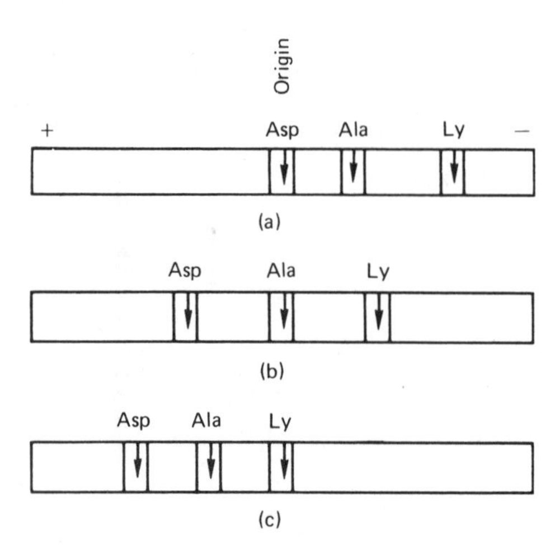


**Figure 22-15.** Electrophoresis of a mixture of alanine (Ala), aspartic acid (Asp), and lysine (Ly) at various pH values: (a) pH equals the isoelectric point of aspartic acid; (b) pH equals the isoelectric point of alanine; and (c) pH equals the isoelectric point of lysine. All strips were spotted in the center.

or neutral. In a mixture of all three of these acids at the pH of the isoelectric point of alanine, the aspartic acid will be negative, the alanine neutral, and the lysine positive; thus electrophoresis will produce the pattern shown in Figure 22-15(b). In Figure 22-15(a) and (c), the two components, which are not at their isoelectric point, move in the same direction because the isoelectric points of both are either above or below the pH being used.

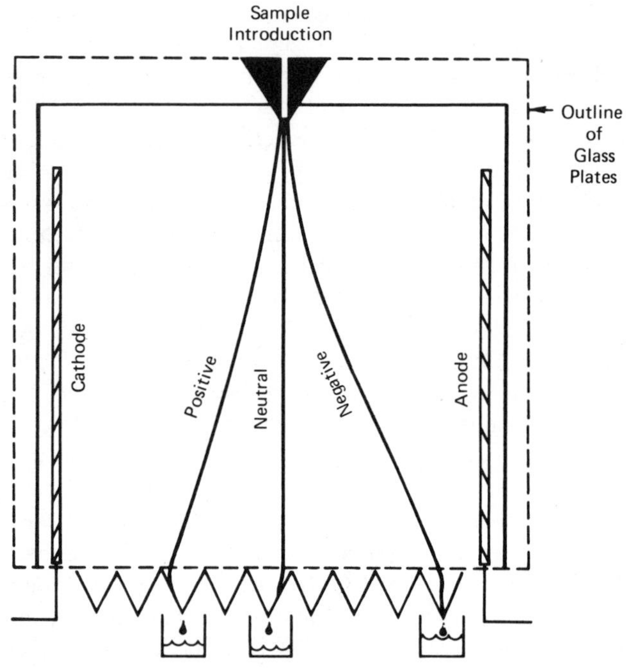


**Figure 22-16.** Continuous paper electrochromatography. The paths of the charged components are diverted by the electric field.

### *Paper Electrochromatography*

In paper electrochromatography a thick sheet of paper moistened with an electrolyte is clamped between two glass plates held in a vertical position with the electrodes along the side edges, as shown in Figure 22-16. The sample is added continuously at the top center of the sheet, and the electrolyte is allowed to flow continuously down the full width of the paper. The bottom edge is serrated to allow the separated components to be caught as they emerge from the paper. The vertical edges of the paper are waxed to seal the sides of the unit. In some equipment, the paper is hung inside a closed container instead of being sandwiched between glass plates.

Paper electrochromatography is similar to two-dimensional paper chromatography in that two effects are present. One is the chromatographic effect of selective retardation; the other is the electrophoretic effect of the electrical field. Since the sample is being added continuously as in frontal analysis, there would be no separation if the paths of the charged substances were not affected by the electrical field. In practice, each charged material present follows its own characteristic path to the bottom and can be collected as it drips off the serrated edge. The apparatus can also be operated in a discontinuous manner by placing a discrete sample at the top of the paper and allowing the chromatographic developing solvent to flow down the paper by capillary action. The separation is stopped before the sample reaches the bottom, and the paper is removed. Colorless materials are located by spray techniques.

## PROBLEMS

**1** Determine the $R_f$ values for phenol and salicylic acid if a column 2 cm in diameter and 30 cm long contains 83 g of alumina ($d = 3.6\ g/cm^3$), 9.6 mL of water, and 61 mL of a butanol–pyridine mixture. The partition coefficients are 0.194 and 3.42, respectively.
*Ans.* 0.77, 0.65

**2** Using the column described in problem 1, what are the partition coefficients of gallic acid and protocatechuic acid if their $R_f$s are 0.43 and 0.74, respectively?
*Ans.* 8.4; 2.2

**3** A column having an inside diameter of 14 mm is filled with silica gel ($d = 0.61$) containing 10% water by weight, 15 g being required and filling the tube to the 25-cm level. Calculate $A_m$. The $R_f$ values for pyrogallol and phenol obtained with this column were 0.51 and 0.97, respectively. What are the partition coefficients?

**4** Using the column of problem 3, calculate the $R_f$ values if the partition coefficients of citric acid and lactic acid are 8.3 and 3.9, respectively.

**5** A 30-cm column has an $A_m$ of 0.44, an $A_s$ of 0.06, and an $A_i$ of 1.03. A mixture of succinic and fumaric acid is placed at the top of the column. What is the percent fumaric acid that is left on the column if the eluant is cut midway between the peaks of the bands? $h = 0.0025$ cm; $P_{succinic} = 8.7$; $P_{fumaric} = 8.3$.
*Ans.* 7.5 ± 1%

**6** A 20-cm column was used to separate glutamic acid from monomethyl succinic acid. A butanol–formic acid–water mixture was used as the mobile phase and it was found that the respective $R_f$ values were 0.84 and 0.86. If $A_m$ was 0.40 and $A_s$ was 0.097, what are the partition coefficients? If the plate height is 0.002 cm, what will be the percent of monomethyl succinic acid contamination in the glutaric acid if the eluant is cut midway between the peaks of the bands?

**7** 2,4-Lutidine has an $R_f$ of 0.75 using a butanol–formic acid–water (75 : 15 : 10) eluant. Under the same conditions, 2,6-lutidine has an $R_f$ of 0.79. How long a column would it take to separate these two compounds if a purity of 99.9% is desired? $h$ = 0.0028; $A_m$ = 0.53; $A_s$ = 0.173; $V$ = 18.7 ml.

*Ans.* 28 cm

**8** What HETP would be required if the compounds used in problem 7 had been separated with a 60-cm column?

*Ans.* 0.0060

**9** How many milliliters of eluant would be required to remove 95% of the 2,6-lutidine described in problem 7?

*Ans.* 19 mL

**10** Prolinol has an $R_f$ = 0.26 using *n*-butanol saturated with 0.1% aqueous $NH_3$ as an eluant. Aspartidiol has an $R_f$ of 0.23 under the same conditions. How long a column would it take to separate these two compounds if a contamination of not more than 3% is desired? $h$ = 0.0032 cm; $A_s$ = 0.27; $A_m$ = 0.64; $V$ = 26.4 mL.

**11** If an 80-cm column had been used in problem 10, what HETP would have been required?

**12** How many milliliters of the saturated *n*-butanol would it take to ensure a minimum of 90% purity in the system employed in problem 10?

**13** An alumina column having an $A_s$ of 0.16 and an $A_m$ of 0.57 was found to contain 6200 plates in an 18-cm section during the separation of some sulfur compounds. If we assume that sulfanilic acid ($R_f$ = 0.51) and sulfamethazine ($R_f$ = 0.55) behave similarly, what degree of separation can be expected?

**14** If the system employed in problem 13 is used, how long a column would be required to have a separation that would be 99% complete?

**15** A *t*-amyl alcohol–glacial acetic acid–water mixture (4 : 1 : 1) has been used as an eluting agent to separate mixtures of dyes. How long a column would it take to separate bromothymol blue ($R_f$ = 0.88) from metacresol purple ($R_f$ = 0.66)? $h$ = 0.0023 cm; $A_s$ = 0.13; $A_m$ = 0.38; $V$ = 22.7 mL.

CHAPTER

# 23 Gas Chromatography

Gas chromatography is similar to column chromatography, except that a gas is used as the mobile phase in place of a liquid. Gas chromatographic separations are based upon selective adsorption on a solid (called **gas–solid chromatography**, GSC) or upon partition between the gas and an immobile liquid phase (called **gas–liquid chromatography**, GLC). The latter is by far the most popular and has many more applications than gas–solid chromatography. Credit for discovery of column partition chromatography is usually given to A. J. P. Martin. In a paper published in 1941, Martin described the possibilities of GLC, but essentially no work was done until after Martin published the first successful gas–liquid chromatographic separation in 1951. Since that time, the field has grown by leaps and bounds, and today there are several journals devoted solely to gas chromatography, as well as literally hundreds of papers each month in other periodicals which deal at least in part with GSC or GLC. In the early days, progress was hindered by slow development of the necessary instrumentation, but as the great possibilities of GLC as a separation method were realized and more and more applications were proposed, tremendous strides were made in the development of immobile phase liquids, sensitive detectors, fast recorders, and proper circuitry for stable operation.

In 1952, Martin and Synge were awarded the Nobel Prize in chemistry for their work on the development of partition chromatography.

## INSTRUMENTATION

All gas chromatographs, whether GSC or GLC, have the same basic components, shown in Figure 23-1. The carrier gas from a tank is fed through a reducing gauge to the inlet or reference side of the detector. The sample is injected into the gas stream after it leaves the reference side of the detector and before it enters the column. After passing through the column, the gas, with the separated components of the sample, passes through the exit or sample side of

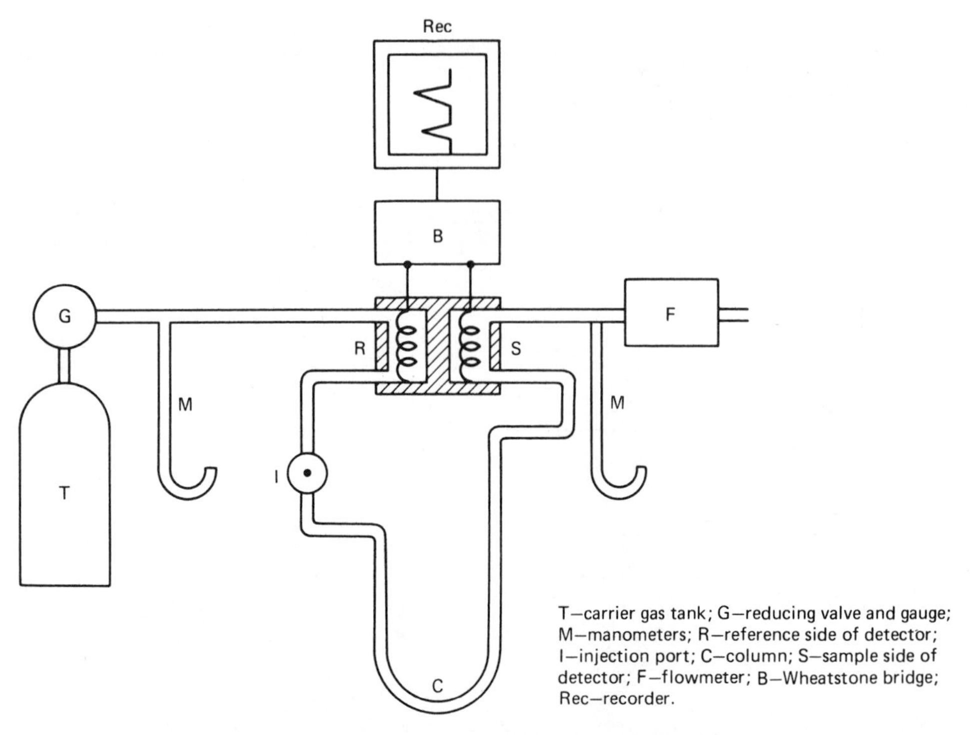


**Figure 23-1.** Typical components of a gas chromatograph.

the detector and then through a flow meter and exit vent. The detector is connected to a recorder. Manometers may be used to measure inlet and outlet pressure.

The gas is the driving force for the movement of the zones through the column, and the solid (for GSC) or liquid (for GLC) furnishes the selective retarding force. The detector may be considered as the brain of the chromatograph and the column as the heart.

## CARRIER GAS

Hydrogen, helium, nitrogen, and air are the most widely used gases. Hydrogen probably has more advantages than the others but is dangerous to use; helium is second best but is expensive; nitrogen is inexpensive but gives reduced sensitivity; and air is used only when the oxygen in the air is beneficial to the detector or separation. Helium is used more than the others in spite of the expense, since it has excellent thermal conductivity, is chemically inert, and has a low density, which allows greater flow rates. Hydrogen has a better thermal conductivity and lower density but has the disadvantages that it may react with unsaturated compounds and that it creates a fire and explosion hazard.

## SAMPLE INJECTION

The sample injection system is important because one of the outstanding features of gas chromatography is the utilization of very small samples. The injection system must introduce the sample in a reproducible manner and must vaporize it instantaneously so that the sample will enter the column as a single **slug**.

Liquid samples are usually injected with a hypodermic syringe through a self-sealing silicone rubber septum into a small inlet chamber, which may be heated to cause flash evaporation. The temperature must not be so high that it decomposes the sample. The manipulation of the syringe to obtain reproducible results is virtually an art, attained only by repeated practice. Solid samples may be dissolved in volatile liquids for injection, or may be injected directly if they can be liquefied. Gas samples require special gas-sampling valves for introduction into the carrier gas stream.

## DETECTORS

### *Differential Thermal Conductivity Detector*

One of the first detectors used was the differential thermal conductivity detector; the operation of this detector is based upon the fact that the temperature and thus the resistance of a wire through which a current is flowing is dependent upon the thermal conductivity of the gas in which it is immersed. The thermal conductivity of a gas is a function of its composition. The detector has two sides (known as the reference and sample sides) in which identical wires are used. These wires are connected to a Wheatstone bridge arrangement and heated by the passage of a current. The carrier gas flows through the reference side before the sample is injected; thus, the reference side always contains gas of the same composition. The gas flows through the sample side as it exits from the column and thus contains the separated components as well as the carrier gas. When there is no component exiting from the column, the gas in the two sides is identical, and the Wheatstone bridge remains in balance because the wires have the same temperature. As soon as a sample component enters the sample side, the composition of the gas in the sample side differs from the reference side. This causes a change in temperature of the wire and an imbalance in the bridge. The circuit is designed to measure the extent of the imbalance and to feed this signal to a recorder. Because the degree of imbalance is a function of the concentration of the component in the carrier gas, the recorder draws the elution curve for the chromatographic separation.

### *Ionization Detectors*

All ionization detectors are based on the electrical conductivity of gases. At normal temperatures and pressures, gases act as insulators but will become conductive if ions or electrons are present. If conditions are such that the gas molecules themselves do not ionize, the change in conductivity due to the presence of a very small number of ions can be detected. The most popular ionization detectors are the flame ionization and the electron capture detectors.

**The Flame Ionization Detector.** Very few ions are produced by a hydrogen flame burning in oxygen or air; however, the introduction of a volatile carbon compound greatly increases the number of ions produced. This effect is the basis for the flame ionization detector, which has proved to be one of the simplest and most satisfactory of the sensitive ionization detectors. (See also the discussion of ionization detectors in radioactivity measurements in Chapter 26.)

A typical flame ionization detector is shown in Figure 23-2. Hydrogen is added before the detector if it was not used as the carrier gas. The mixture is burned in air (or oxygen) in the detector. The platinum jet is one electrode of the cell and the collector is the other. A sufficient potential to collect all ions is used.

The exact mechanism by which the ions are produced is still somewhat obscure. The flame temperature is too low to explain the high ionization observed by thermal ionization, so **chemi-ionization** must be responsible. In chemi-ionization, sample fragments react with energetic flame intermediates in exothermic reactions. The energy released is retained in the product molecule and leads to ionization before thermal randomization of the energy occurs.

This detector is remarkably insensitive to the presence of water vapor or air in the carrier gas, and the background current is low so that small quantities can be measured with proper amplification. The main drawbacks are the different response to different substances and to the fact that the sample is destroyed. The detector responds to all organic compounds except formic acid, with the response being greatest for hydrocarbons.

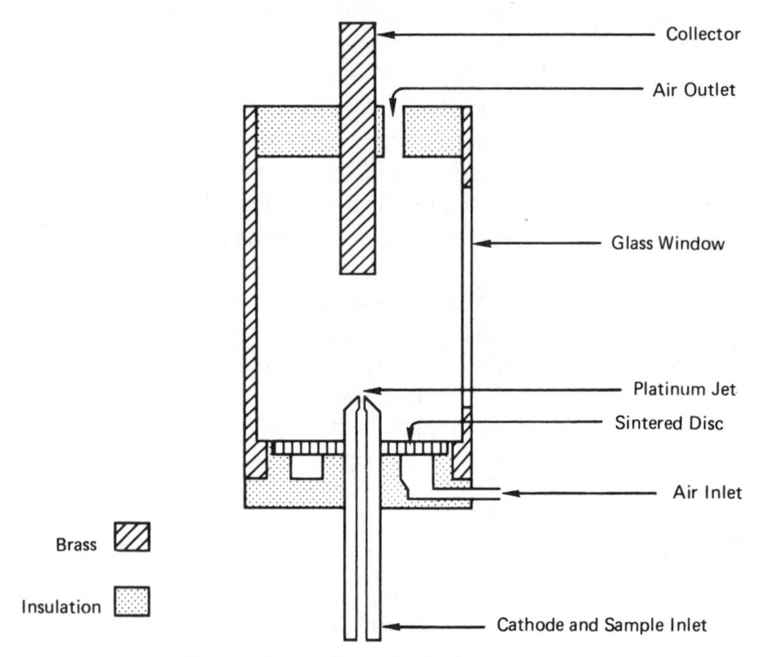


**Figure 23-2.** Flame ionization detector.

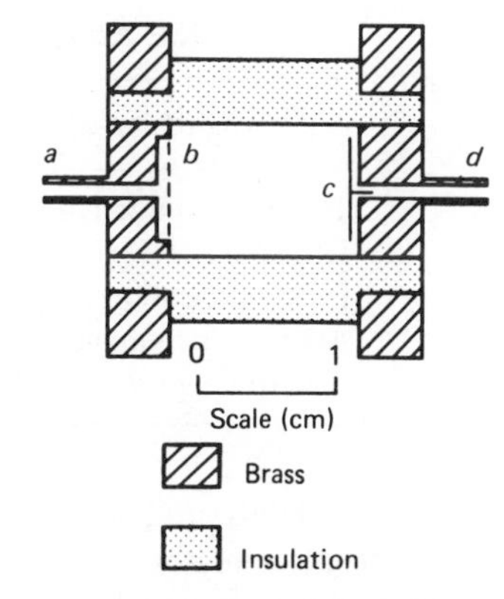


**Figure 23-3.** Electron capture detector: (a) anode and inlet; (b) diffuser to offset turbulence; (c) tritium foil; (d) cathode and outlet.

The Electron Capture Detector. The electron affinity of different substances can be used as the basis for an ionization detector known as the electron capture detector. With this detector, a radioactive source that is a $\beta$-emitter is used to produce an electric current between a pair of electrodes. This current is known as the **standing current**. The introduction of a substance with an affinity for electrons will cause absorption by "electron capture," resulting in a decrease in the standing current. The magnitude of the decrease is a function of the concentration of the absorbing species, since the mathematics of the absorption are similar to those in optical absorption methods. Figure 23-3 is a diagram of an electron capture detector in which a metal foil coated with a tritium-containing compound is used as a steady source of slow electrons.

The electron capture detector is especially sensitive to halogens, oxygen, nitrogen, phosphorus, and other electron-capturing atoms. The presence of the absorbing group is usually detected by comparison of the peak heights of the electron capture chromatogram, with the peak heights of a flame ionization or thermal conductivity chromatogram run simultaneously on the effluent.

Nitrogen is used for the majority of electron capture determinations, but hydrogen and helium can also be used. The primary uses are in the quantitative determination of halogen-containing pesticides and the qualitative identification of the peaks of compounds that contain halogens and other electron-capturing groups. Quantitative determination is usually performed with other detectors for the majority of organic compounds. Since carbon does not have a great affinity for electrons, the members of homologous series with a common electrophore will have similar absorption coefficients.

### *Other Detectors*

Many other detectors have been proposed and used for the detection and determination of the separated constituents in the column effluent. Among these are the **gas density balance**, which responds to variations in molecular weight; the **integral coulometric detector**, used for halogen-containing compounds, which titrates the halides with electrolytically generated silver ion; the highly selective **thermionic detector**, which incorporates a screen or porous

**TABLE 23-1.** Relative Sensitivities of Selected Detectors

| *Type of Detector* | *Minimum Weight Detectable (g/sec)* |
|---|---|
| Thermal conductivity | $10^{-5}$ |
| Ultrasonic | $10^{-7}$ |
| Flame ionization | $10^{-11}$ |
| Argon or helium activation | $10^{-12}$ |
| Electron capture | $10^{-14}$ |

block coated with an alkali halide over the flame in a flame ionization detector; the **ultrasonic detector**, which measures the velocity of sound in the effluent gas; the **argon detector**, which uses argon as the carrier gas; and the **flame photometric detector**, which uses the flame in flame ionization detectors as the radiation source for a flame photometer and is selective and sensitive for phosphorus- and sulfur-containing compounds.

### *Relative Sensitivity*

The relative sensitivity of these detectors is shown in Table 23-1. Other detectors, not discussed in this text, are included for comparison.

## COLUMNS

The column can be constructed of glass or metal tubing and, for analytical work, is usually 4–8 mm ID. It can be any length from a few centimeters to over a hundred meters and can be coiled, bent, or straight. Capillary columns are also used for GLC separations. The column is packed uniformly with the immobile phase—a solid adsorbent for GSC or a solid coated with a high-boiling liquid for GLC. In capillary columns for GLC, the walls are coated with a thin layer of the substrate, since it would be impossible to pack such a column with particles of solid. Adsorbents for GSC include silica, alumina, and various grades of diatomaceous earth. The solid supports for GLC include firebrick and various grades of Chromosorb. Mesh sizes from 30/50 to 80/100 are used, with 30/60 and 60/80 being the most popular.

### *Substrates*

The solid support is coated with a high-boiling liquid called the substrate, which acts as the immobile phase in GLC. There are literally hundreds of substrates commercially available. A few of the most popular are listed in Table 23-2. The general requirements for the liquid phase are (1) good solvent properties for the components, (2) differential partitioning of sample components, (3) high thermal stability, and (4) low vapor pressure at the column temperature.

**TABLE 23-2.** Selected Typical Substrates

| *Substrate* | *Solute Type for Which Used* | *Temperature* (°C) |
|---|---|---|
| Polyglycols (Ucon, Carbowaxes) | Amines, ethers, alcohols, esters, ketones, aromatics | 100–200 |
| Dinonyl phthalate | Polar compounds in general | 130 |
| Squalane | Low-molecular-weight paraffins | 140 |
| Paraffin oil (Nujol) | Paraffins, olefins, and halides | 150 |
| Didecyl phthalate | Polar compounds in general | 170 |
| Silicone oils (DC200) | Paraffins, olefins, esters, ethers | 200 |
| Apiezon L grease | Polar compounds in general | 300 |
| Silicone gums | Polar compounds in general | 350 |

The amount of the liquid utilized is specified in terms of percent **loading** by weight. The usual coating procedure is to dissolve the substrate in a volatile liquid, cover the support with the solution, and then allow the solvent to evaporate slowly. Loading varies from 2 to 30% with 5–10% being the most common. The column is usually **conditioned** at the maximum temperature of use for several hours to remove all the substances that can be eluted from the substrate at the temperature involved.

## TEMPERATURE CONTROL

The temperature at which a chromatogram is run will affect the retention times or retention volumes of the components, and, as a consequence, precise control of the column temperature is required for adequate replication in repetitive measurements. In some cases the column temperature is held constant at room or other temperatures, but many instruments are designed so that the temperature can be increased at a uniform rate during the determination. Continuous increase of the temperature is known as **temperature programming**.

Temperature programming is especially helpful for complex mixtures that contain both low and high retention-time components. The components with small retention times will come off at the lower temperatures, and the retention times for the components with higher values will be reduced as the temperature rises.

The thermal regulation system should be sensitive to variations of 0.01°C and maintain control to ±0.1°C. The thermal compartment should be arranged so that it may be cooled down rapidly after a run at elevated temperature so that downtime between determinations will be as short as possible.

During temperature programming the base line may rise as a result of variation in the gas flow and increased **bleeding** of the substrate at the higher temperatures. This effect can be minimized by using two identical columns with identical detectors and splitting the gas flow so that half goes through each column. The sample is then injected into only one of the carrier streams and the detectors connected so that the temperature effect from the nonsample column can be used to correct the readout from the sample column.

## EVALUATION

As for other chromatographic methods, the efficiency of a column is expressed by the number ($N$) of theoretical plates in the column or by the height equivalent of a theoretical plate (HETP). The larger the number of theoretical plates or the smaller the HETP, the more efficient the column is for separations. A theoretical plate is that distance on the column in which equilibrium is attained between the solute in the gas phase and the solute in the liquid phase and is equivalent to one equilibrium stage in a distillation.

The number of theoretical plates in any column can be calculated from the elution curve, as shown in Figure 23-4. The distance $d$ from the point of injection to the peak maximum can be measured in units of length, time, or volume and is usually called the **retention time** $t$ or **retention volume** $V_r$. In Figure 23-4, $H$ is the height of the peak maximum; $W$ is the width of the peak measured as the distance between the intersections of the tangents to the inflection points with the base line; $\beta$ is the width of the peak at a height of $H/e$; $t$ is the time lapse between the injection and the emergence of the peak maximum; $t_i$ and $t_f$ are the time lapses between injection and the initial rise of the peak and the return of the peak to the base line, respectively; and the area is calculated as the height multiplied by the width of the peak at one-half the height.

The Glueckauf equation is probably the simplest to use since the width is measured at $H/e$ instead of the intersections of the tangents with the base line. The values by the Glueckauf equation and the Van Deemter or Martin and Synge equation agree satisfactorily. The other equations shown are utilized when the peak is not symmetrical.

The number of theoretical plates is a good measure of the efficiency of a

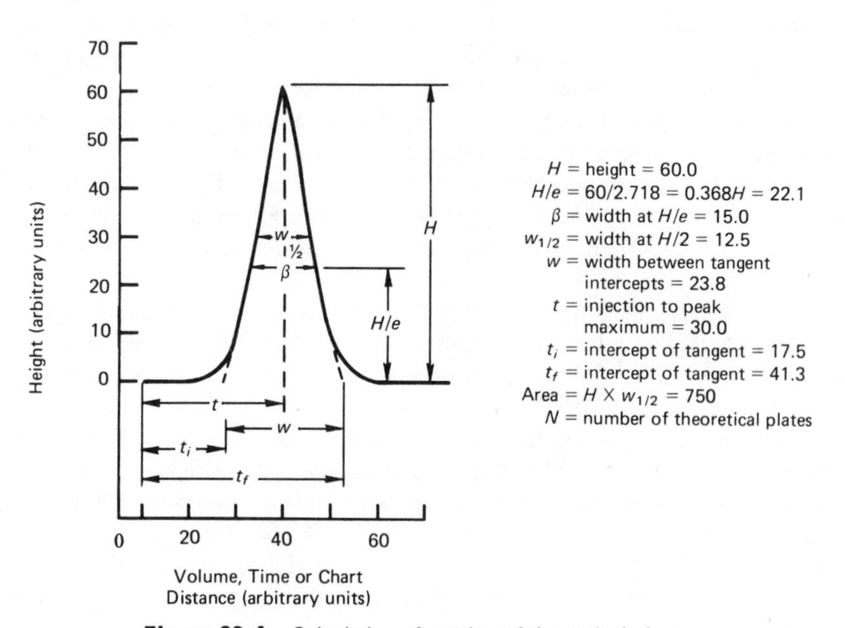


**Figure 23-4.** Calculation of number of theoretical plates.

column but does not give any information concerning the effect of various parameters upon the efficiency. The Van Deemter equation, which is based upon the rate theory, gives the relationship of the HETP to the flow rate and includes the effect of many operational parameters in the various terms of the equation. The **Van Deemter equation** may be written in simplified form as

$$\text{HETP} = A + \frac{B}{u} + Cu \qquad \textbf{(23-1)}$$

where $u$ is the gas velocity or flow rate; $A$ is a constant that involves the packing effect in the column and the particle diameter and is known as the **eddy diffusion term**; $B$ is a constant that includes the effect of diffusion in the gas phase and the correction for tortuosity of path and is called the **molecular diffusion term**. In the nonequilibrium term, $C$ is a constant that reflects the resistance to mass transfer between the gas and liquid and includes such factors as the thickness of the film, diffusion in the liquid phase, and the fraction of the solute in each phase.

The graph of a typical Van Deemter study, in Figure 23-5, shows the effects of each term of the equation on the relationship between the flow rate and the

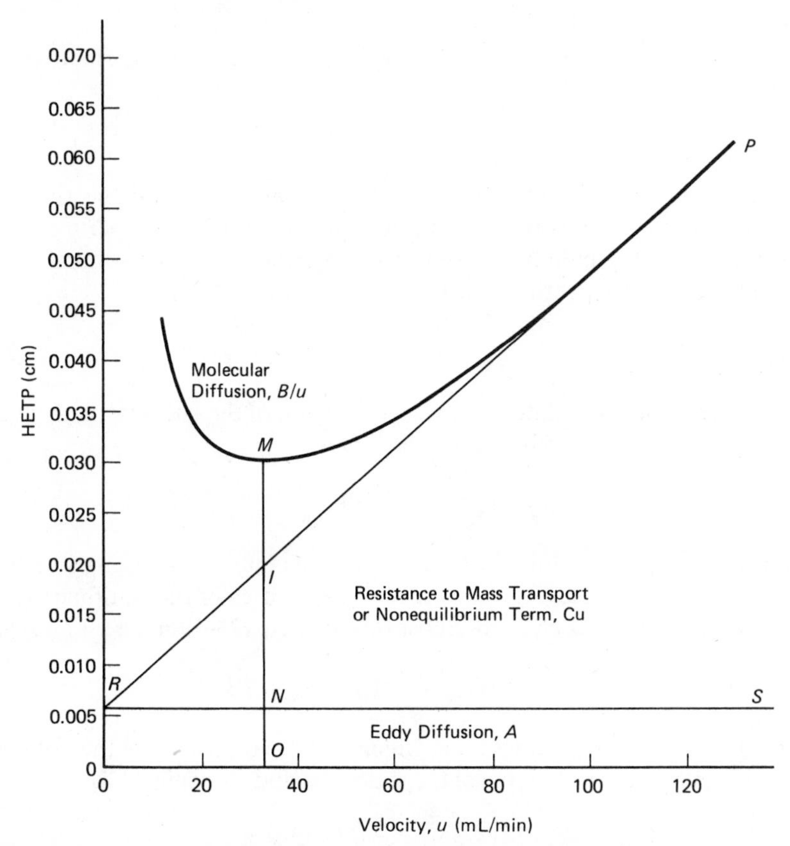


**Figure 23-5.** Typical Van Deemter graph. Optimum flow, 33; minimum HETP, 0.030; values of constants as read from graph: A, 0.006; B, 0.363; C, 0.0004.

HETP. Since $A$ is constant and does not include the flow rate, it contributes the same amount to the HETP, irrespective of the flow rate. The second term increases as flow rate decreases and contributes the largest effect at low flow rates; the third term contributes its greatest effect at high flow rates because the effect increases as the flow rate increases. Owing to the opposing effects of the second and third terms, there is an optimum flow rate at which there will be a minimum HETP. In Figure 23-5, the optimum flow rate is 33 mL/min, which gives a minimum HETP of 0.030. Since the HETP increases sharply at flow rates below the optimum, it is preferable to operate at the optimum or slightly higher flow rates. The values of the various constants can be determined from the graph by calculating the contribution of each term to the optimum flow rate. The value of $A$ is obtained by extrapolating the tangent to the curve at high flow rates to zero flow rate (line $PR$ in Figure 23-5). The contribution of the second term is given by line $MI$, and the contribution of the third term by the line $IN$. $B$ is calculated by multiplying the value of $MI$ by the optimum flow rate, and $C$ is found by dividing the value of $IN$ by the optimum flow rate.

## THE RETENTION VOLUME

The uncorrected or experimental retention volume for a chromatogram is given by

$$V_r = t_r F_c \tag{23-2}$$

where $t_r$ is the time in minutes on the time axis from the point of injection to the peak maximum and $F_c$ is the volumetric flow rate in milliliters per minute. As seen previously for column chromatography, the retention volume is also related to the number of theoretical plates ($N$) by eq. (23-3)

$$V_r = NV_h \tag{23-3}$$

where $V_h$ is the effective plate volume. The relation of the effective plate volume to the partition coefficient is given by eq. (23-4)

$$V_h = h(A_m + PA_s) \tag{23-4}$$

in which $h$ is the height equivalent of a theoretical plate, $A_m$ is the cross-sectional area of the mobile phase, $A_s$ is the cross-sectional area of the stationary phase, and $P$ is the partition coefficient. Substitution of eq. (23-4) into eq. (23-3) gives

$$V_r = Nh(A_m + PA_s) = V_m + PV_s \tag{23-5}$$

since $NhA$ equals height times area, which equals volume. $V_m$ is the volume of the mobile phase, $V_s$ the volume of the substrate, and $P$ is defined as

$$P = \frac{\text{wt solute per mL substrate}}{\text{wt solute per mL gas}} \tag{23-6}$$

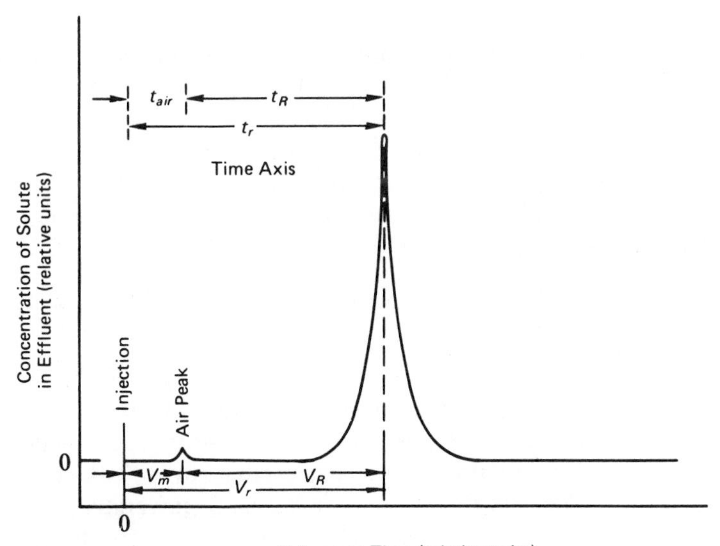


**Figure 23-6.** Typical chromatogram showing air peak. $V_r$ is the uncorrected or experimental retention volume; $V_m$ is the interstitial or dead volume, or the volume of the mobile phase; $V_R$ is the adjusted retention volume. On the time axis, $t_{air}$ is the time between injection and the air peak, $t_r$ is the uncorrected retention time, and $t_R$ is the adjusted retention time.

The uncorrected retention volume must be corrected for the dead or interstitial volume, which is also equal to the volume of the mobile phase. In each chromatogram a small amount of air is injected with each sample and, since air is not retarded in its flow through the column, appears as an **air peak** soon after the injection. Figure 23-6 shows a typical chromatogram with an air peak. The retention volume represented by the air peak is essentially the interstitial volume of the system and can be used to correct the retention volume. Rearrangement of eq. (23-5) gives

$$V_R = V_r - V_m = PV_s \tag{23-7}$$

in which $V_R$ is known as the **adjusted retention volume**. In some texts the adjusted retention volume is given the symbol $V'_R$.

The adjusted retention volume must also be corrected for the change in the volume of the gas due to the pressure gradient along the column. The pressure gradient correction is given by

$$f = \frac{3[(p_i/p_o)^2 - 1]}{2[(p_i/p_o)^3 - 1]} \tag{23-8}$$

in which $p_i$ is the inlet pressure and $p_o$ is the outlet pressure. The adjusted retention volume corrected for the pressure gradient is known as the **net retention volume**

$$V_N = f\,V_R \tag{23-9}$$

The same correction applied to the experimental retention volume provides the **corrected retention volume** (also known as the **limiting retention volume**):

$$V_R^0 = f V_r \tag{23-10}$$

Columns that have high liquid-phase loadings will exhibit large retention values. In order to include this factor in the specification of retention volumes, the **specific retention volume** $V_g$ is defined as

$$V_g = \frac{273 V_R^0}{T_c W_s} = \frac{273 P}{T_c \rho_s} \tag{23-11}$$

where $T_c$ is the column temperature on the absolute scale; $W_s$ is the weight of substrate in the column; $\rho_s$ is the density of the substrate and $P$ is defined by eq. (23-6). The specific retention volume corresponds to the volume of carrier gas required to remove half the solute at a specified temperature (0°C unless otherwise specifically stated) from a hypothetical column that contains 1 g of the liquid phase and has no pressure drop or apparatus dead space.

A common method for reporting retention information is in the form of the **relative retention ratio** by utilizing some solute as a standard and expressing the ratio of the retention volumes or retention times of the other components to the retention volume or time of the standard. Equation (23-12) shows the relationships:

$$R_{V_g} = \frac{V_g}{V_{g_s}} = \frac{P}{P_s} = \frac{V_R}{V_{R_s}} \tag{23-12}$$

where the subscript $s$ refers to the standard; $n$-heptane is often used as the standard.

For uniformity of presentation, the values of $V_g$ should be tabulated or plotted as $V_g$ versus $1/T$. In many articles the log of the relative retention volumes is plotted against $1/T$ and data are given for at least two temperatures, together with the partition coefficient and $V_g$ of the standard.

The procedure used to calculate relative retention ratios is to analyze the sample, the standard, and a gas (usually air) that is not retarded by the column. The time for the air peak is subtracted from all the other retention times to obtain the adjusted retention times. Because the ratio of the adjusted retention times is the same as the ratio of the adjusted or specific retention times, the times can be used directly to calculate the relative retention ratio.

## RESOLUTION

The efficiency of the separation of the components of a mixture is probably the most important aspect in all types of chromatography and is often expressed as the **separation factor** or the **resolution** between two peaks. Examples of good and poor separations for columns with a small number of plates are shown in

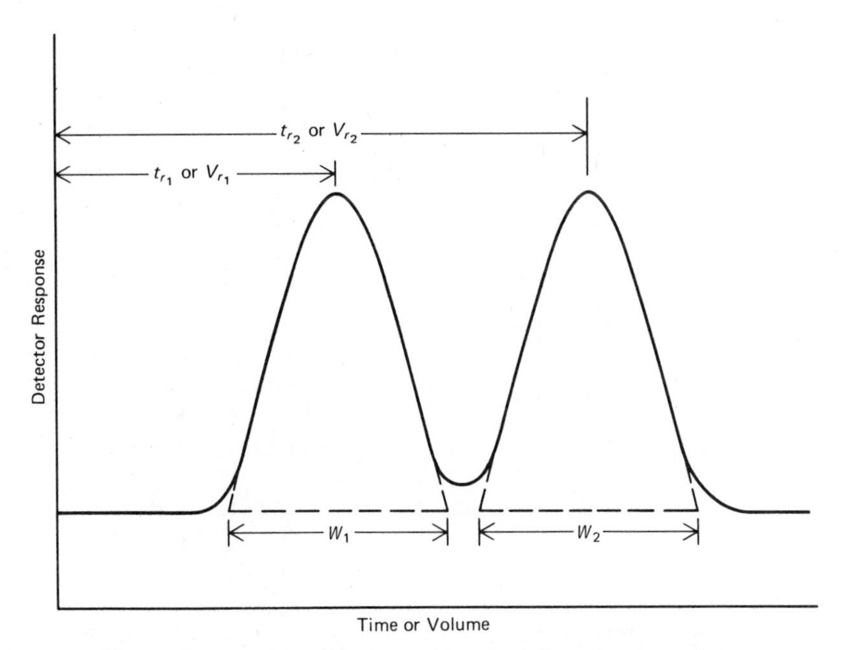


**Figure 23-7.** Measurements used in calculation of peak resolution.

Figure 22-4(c) and also in Figure 24-2. The resolution is often expressed in terms of the distance of separation of the peak maxima and the width of the peaks using the expression

$$R = \frac{2(V_{r2} - V_{r1})}{W_2 + W_1} = \frac{2(t_{r2} - t_{r1})}{W_2 + W_1} \quad \textbf{(23-13)}$$

in which $V_r$ and $t_r$ represent the retention volume and retention time, as shown in Figure 23-7. The value of $R$ in this figure is approximately 1.14. Values of $R$ equal to or greater than 1.5 indicate base-line or essentially complete resolution.

## APPLICATIONS

### QUALITATIVE

If the conditions of flow rate and temperature are reproduced, the retention time (or volume) for a given compound on a given column will always be the same and consequently can be used to identify the compound. Care must be taken to ensure that none of the other possible components of the unknown mixture will have the same retention time.

For a given column and a given set of conditions, the retention times show a linear relationship to the boiling points of the components. It is sometimes possible to utilize a retention time-boiling point graph prepared from one series

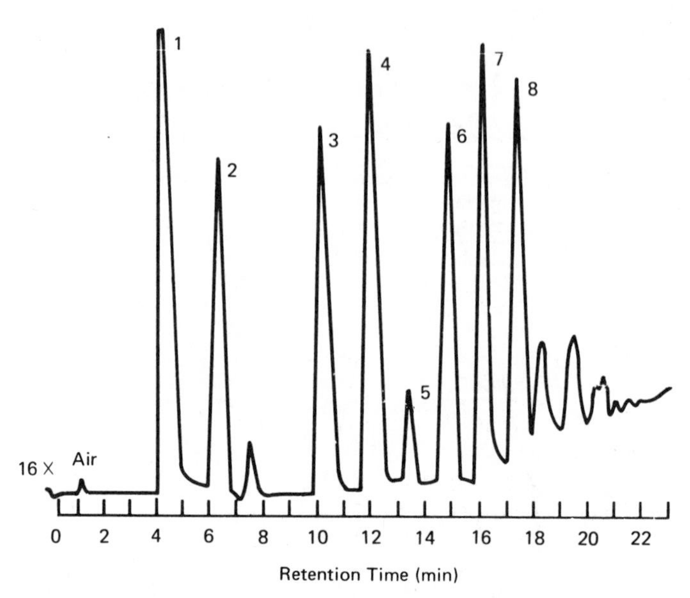


**Figure 23-8.** Chromatogram of trichlorosilane (1) containing: (2) silicon tetrachloride; (3) 1,1,3,3,-tetrachlorodisiloxane; (4) 1,1,1,3,3-pentachlorodisiloxane; (5) hexachlorodisiloxane; (6) 1,1,2,3,3-pentachlorotrisiloxane; (7) 1,1,1,2,3,3-hexachlorotrisiloxane; and (8) 1,1,1,2,3,3,3-heptachlorotrisiloxane on 10% SF-96. (From Burson, K. R., Kenner, C. T., Anal. Chem., *41*, 870 (1969), with permission)

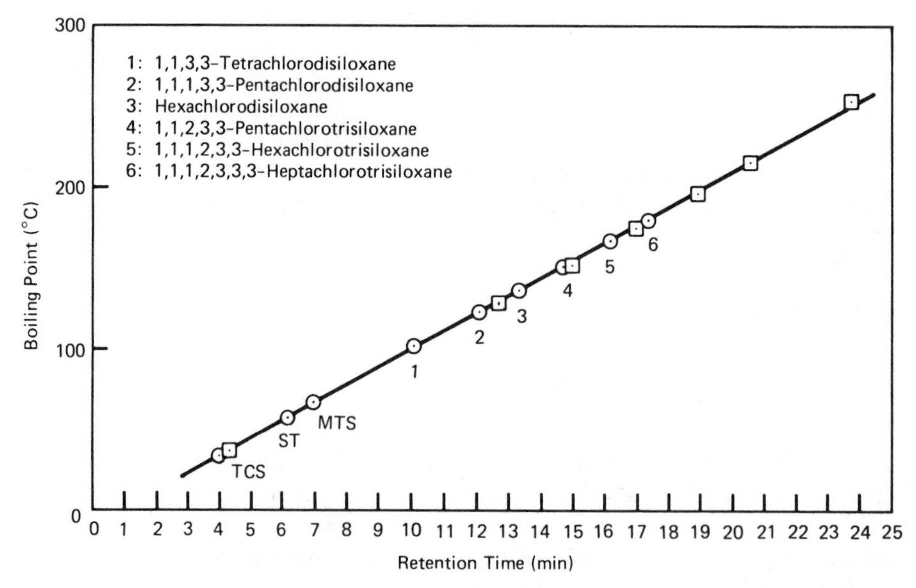


**Figure 23-9.** Boiling point–retention time relationships. Square with dot, known *n*-alkanes; circle with dot, silanes and siloxanes; TCS is trichlorosilane; ST is silicon tetrachloride; MTS is methyltrichlorosilane. (From K. R. Burson, Master's Thesis, Southern Methodist University, 1968.)

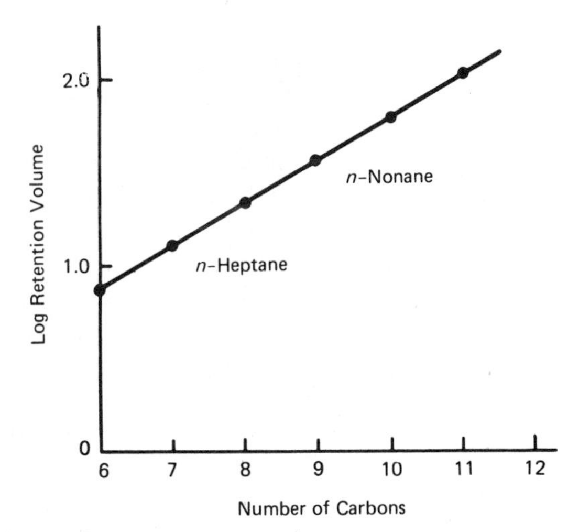


**Figure 23-10.** Relationship between log of retention volume and the number of carbon atoms in *n*-alkanes.

of pure compounds to determine the boiling points of a second series for which pure samples are not available in sufficient quantity. As an example, Figure 23-8 is a programmed temperature chromatogram of a sample of trichlorosilane that had become contaminated with siloxanes by hydrolysis. Figure 23-9 is a graph of retention times versus boiling points determined under the same conditions for a group of *n*-alkanes. The boiling-point values of the various components of the mixture as read from this curve agree satisfactorily with the literature values obtained by other methods. In Figure 23-9 the points enclosed in squares are the *n*-alkane values and those in circles are the values for the components of the trichlorosilane sample.

For homologous series, the log of the retention volume is often a linear function of the number of carbon or other atoms or $CH_2$ groups. Figure 23-10 is a graph of some *n*-alkanes on a particular column under a given set of conditions. The equation of this line may be derived from any two points and used to estimate the retention volumes of larger *n*-alkanes.

## QUANTITATIVE

There are literally thousands of examples in the literature of the use of GSC and GLC for quantitative analysis. All quantitative data are based on the fact that the area under the curve is a function of the quantity of the component present. For this to be true, (1) the ouput of the detector–recorder system must be linear with concentration and (2) the flow rate must be constant so that the distance on the time axis can be converted to the volume of carrier gas. The areas can be measured by triangulation or by hand with a planimeter, but some chromatographs are equipped with disc integrators that calculate the area directly. Peak height can also be related to weight of component for a given apparatus and column under a given set of conditions by careful calibration. Peak height is usually not as satisfactory as area, since the relationship between height and weight is not linear owing to peak broadening at the longer retention times.

Combination of a gas chromatograph with a rapid-scan mass spectrometer allows analysis and positive identification of the components as they are eluted. Most combined operations are controlled by a computer that calculates the areas under the chromatographic peaks and normalizes them. The eluted components are automatically diverted to the mass spectrometer for mass spectrometric analysis and positive identification. Other combinations include rapid-scan infrared spectrometers connected with the effluent from the chromatograph, but some rapid-scan infrared instruments do not have good resolution. In some cases, the effluent peaks can be trapped as they emerge from the exit tube by various types of cold traps for later analysis by mass spectrometry or infrared absorption.

One of the main benefits of the gas chromatography–mass spectrometer (GC/MS) and gas chromatography–infrared (GC/IR) instruments is that the mass or infrared spectrum aids in the positive identification of unknown components of mixtures. In a typical case, the unknown may be separated into fractions by various extraction procedures. These fractions are then run by ordinary GLC to obtain the retention times and best resolutions of the peaks present under different conditions. After a satisfactory set of GLC conditions is found, the sample is injected into the GC/MS and the mass spectra of the peaks is determined. Evaluation of the mass spectra and the retention times often suggests possible compounds for which known pure samples are obtained and injected under identical conditions. If the known and unknown do not agree, other compounds may be tried until a match is obtained.

An example of this type of identification is shown in Figure 23-11. In this case an unmarked, unknown white tablet was received by the Dallas District of the Food and Drug Administration. It was extracted, injected, and the GLC chromatogram indicated the presence of three main components, one of which

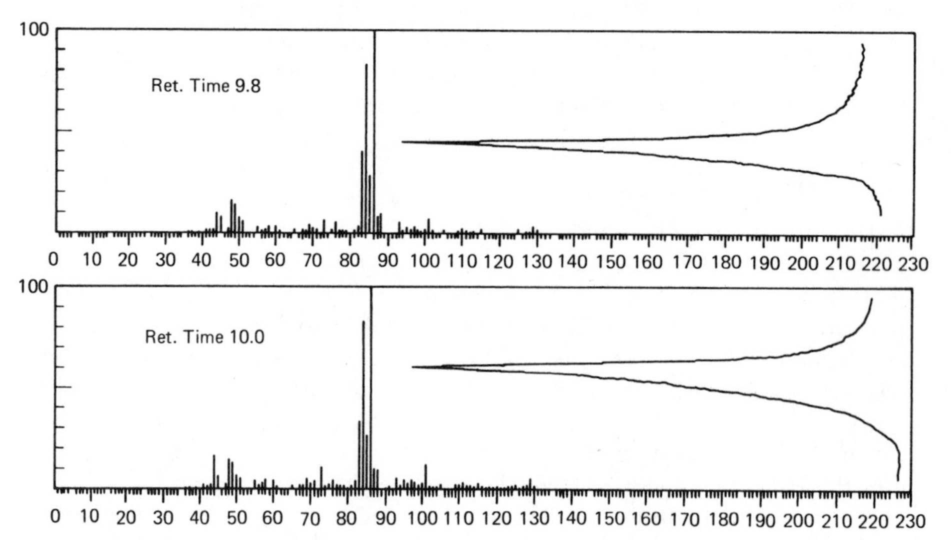


**Figure 23-11.** GC/MS spectra of diisopropylamine dichloroacetate (upper scan) and of one of an unknown tablet. (Courtesy of G. D. Castillo and H. Guerrero, Dallas District, Food and Drug Administration.)

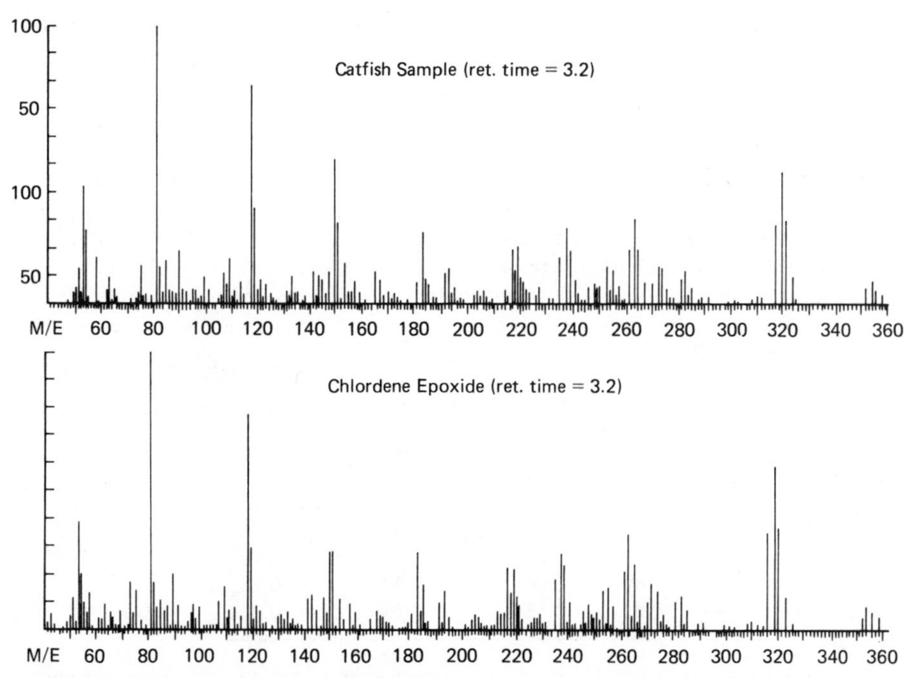


**Figure 23-12.** Mass spectra of chlordene epoxide and of an unknown peak in a fresh catfish sample. (Courtesy of G. D. Castillo, M. T. Jeffus, and C. T. Kenner, Dallas District, Food and Drug Administration.)

was tentatively identified from the mass spectrum as diisopropylamine dichloroacetate. A know pure sample of this compound was obtained and injected under identical conditions, with the results shown in Figure 23-11. Comparison of the two chromatograms and the mass spectra confirmed that one of the components of the tablet was actually diisopropylamine dichloroacetate. The other components were identified and confirmed by the same procedure.

Another example of the positive identification of an unknown component that was reported[1] from the Dallas District of the Food and Drug Administration is shown in Figure 23-12. During the routine analysis for pesticide residues in 33 samples of fresh fish collected for a particular survey, three fresh catfish samples showed an extraneous peak when run by the standard method normally used. These three samples were run by the CG/MS and the extraneous peak fed into the same spectrometer. Preliminary identification indicated a chlorinated hydrocarbon, even though none of the more than 25 pesticides usually determined have this particular retention time. Further study suggested that the compound might be chlordene epoxide and, since no known sample was available, some was synthesized in the laboratory. The retention time and the mass spectrum of the synthesized material matched the mass spectrum of the unknown material and further matched a mass spectrum of chlordene epoxide that had not been reported previously but was made available. As a consequence,

[1] G. D. Castillo, M. T. Jeffus, C. T. Kenner, *J. Assoc. Official Anal. Chemists,* **61**, 1 (1978).

the substance was proved to be chlordene epoxide and was reported as the first evidence of chlordene epoxide being found in the edible portion of fresh catfish.

## PROBLEMS

**1** For a particular gas chromatographic column under a given set of conditions, the terms of the Van Deemter equation were: $A = 0.06$ cm; $B = 0.38$ cm$^2$/sec; $C = 0.05$ sec. (a) Calculate the value of $h$ at several flow rates and graph the results; (b) show the contribution of each term on the graph; (c) read the optimum flow rate and the corresponding minimum HETP from the graph.

*Ans.* (c) 0.33 cm at 2.5 cm/sec

**2** Repeat problem 1 for the following gas chromatographic columns:

| | *A (cm)* | *B (cm²/sec)* | *C (sec)* |
|---|---|---|---|
| (a) | 0.10 | 0.30 | 0.05 |
| (b) | 0.25 | 0.38 | 0.10 |
| (c) | 0.06 | 0.40 | 0.12 |
| (d) | 0.15 | 0.25 | 0.08 |

**3** The following values for HETP and flow rate were obtained on a particular gas chromatographic column under a given set of conditions. (a) Graph the data; (b) what is the optimum flow rate and the corresponding minimum HETP? (c) determine the values of $A$, $B$, and $C$ from the graph at the optimum flow rate; (d) calculate the values of $A$, $B$, and $C$ using the data of the three points closest to the optimum flow rate and three simultaneous equations.

| HETP | 0.78 | 0.66 | 0.62 | 0.62 | 0.65 | 0.79 | 0.92 | 1.18 | 1.33 |
|---|---|---|---|---|---|---|---|---|---|
| *u (cm/sec)* | 2.0 | 3.0 | 4.0 | 5.0 | 6.0 | 8.0 | 10.0 | 14.0 | 16.0 |

*Ans.* (b) 0.61 cm at 4.5 cm/sec

**4** Repeat problem 3 for the following sets of conditions:

| *(a)* | | *(b)* | | *(c)* | | *(d)* | |
|---|---|---|---|---|---|---|---|
| HETP (cm) | *u* (cm/sec) | HETP (cm) | *u* (cm/sec) | HETP (cm) | *u* (cm/sec) | HETP (cm) | *u* (cm/sec) |
| 0.67 | 2.0 | 0.48 | 1.0 | 0.189 | 0.71 | 0.350 | 0.80 |
| 0.56 | 3.5 | 0.39 | 1.5 | 0.135 | 1.38 | 0.275 | 1.30 |
| 0.55 | 4.5 | 0.35 | 2.0 | 0.125 | 2.72 | 0.250 | 1.95 |
| 0.57 | 6.0 | 0.33 | 2.8 | 0.135 | 3.50 | 0.259 | 3.00 |
| 0.64 | 8.0 | 0.35 | 3.8 | 0.164 | 5.00 | 0.290 | 4.00 |
| 0.71 | 10.0 | 0.38 | 5.0 | 0.187 | 6.12 | 0.350 | 5.55 |
| 0.80 | 12.0 | 0.45 | 7.0 | 0.205 | 7.00 | 0.418 | 6.80 |
| 0.90 | 14.3 | 0.61 | 10.2 | | | | |
| | | 0.70 | 12.3 | | | | |
| | | 0.83 | 15.0 | | | | |
| | | 0.98 | 18.0 | | | | |

**5** Adjusted retention times (in minutes) for straight-chain saturated paraffins on 60/80 mesh Chromosorb containing 20% by weight of silicone oil (DC 200) are tabulated at 50°C (carrier flow = 100 mL/min) and at 100°C ($F$ = 82 mL/min):

| *Temp.* (°C) | $C_1$ | $C_2$ | $C_3$ | $C_4$ | $C_5$ | $C_6$ | $C_7$ | $C_8$ |
|---|---|---|---|---|---|---|---|---|
| 50 | 0.00 | 0.30 | 0.90 | 2.30 | 6.50 | 16.0 | 38.7 | 92.6 |
| 100 | 0.00 | 0.20 | 0.50 | 1.20 | 2.40 | 4.90 | 9.70 | 19.0 |

(a) Calculate the adjusted retention volumes ($V_R$) of each component.
(b) At each column temperature, calculate the retention ratio of each compound relative to $n$-pentane.
(c) Plot log $V_R$ versus the carbon number.

**6** A gas chromatographic analysis of a number of components of white gasoline was conducted using a thermal conductivity cell as the detector. The components, with their respective integrated areas, are given in the following table. Calculate the weight percent composition of the original gasoline sample.

| *Component* | *Area* | *Component* | *Area* |
|---|---|---|---|
| A | 4 | F | 17 |
| B | 8 | G | 15 |
| C | 10 | H | 26 |
| D | 14 | I | 20 |
| E | 8 | J | 12 |

*Ans.* A 3.0%

**7** Consider the following data obtained using a silicone 702 column:

| | *Adjusted Retention Volume* (*mL*) | | | |
|---|---|---|---|---|
| *Temperature* (°C) | 55 | 75 | 95 | 135 |
| Methyl acetate | 50 | 30.0 | 18.5 | 8.3 |
| Ethyl acetate | 135 | 64 | 37.0 | 13.2 |
| $n$-Propyl acetate | 316 | 141 | 66.0 | 21.5 |
| $n$-Butyl acetate | 1000 | 316 | 135.0 | 35.5 |

(a) Plot these data as log $V_R$ versus reciprocal temperature.
(b) Determine the retention volume for ethyl acetate at 38°C and at 120°C.
(c) Also determine the retention volume for $n$-amyl acetate at 80° and 150°C.

*Ans.* (b) 18 mL at 120°C; (c) 33 mL at 150°C

**8** Consider the following data obtained using a di-*n*-decyl phthalate column.

| | *Adjusted Retention Volume* (*mL*) | | |
|---|---|---|---|
| *Temperature* (°C) | 50 | 100 | 150 |
| Acetone | 21.5 | 19.3 | 16.0 |
| Butanone | 58 | 44.2 | 33.5 |
| 3-Pentanone | 145 | 96 | 65 |
| 2,4-Pentanedione | 400 | 229 | 142 |

(a) Plot these data as log $V_R$ versus reciprocal temperature.
(b) Determine the retention volume for methyl ethyl ketone at 25°C and 120°C.
(c) Also determine the retention volume for 3-hexanone at 75°C and 125°C.

**9** Consider the following data obtained using a silicone DC-200 column:

| | *Adjusted Retention Volume* (*mL*) | | |
|---|---|---|---|
| *Temperture* (°C) | 50 | 100 | 150 |
| Acetone | 8.2 | 9.6 | 9.7 |
| Butanone | 22.0 | 20.7 | 18.2 |
| 3-Pentanone | 54.4 | 42.9 | 33.2 |
| 2,4-Pentanedione | 119 | 82 | 61 |

Using these data, together with those given in problem 8
(a) Prepare plots of the retention volume on silicon DC-200 versus the retention volume on di-*n*-decyl phthalate for each of the three temperatures, including your calculated data for 3-hexanone.
(b) From your results, which compound is not properly a member of this series?
(c) Predict an approximate retention volume for the 2,4-pentanedione at 25°C.

CHAPTER

# 24 Ion Exchange

Ion exchange is the exchange of ions of like sign between a solution and a solid-insoluble body in contact with it. For such an exchange to take place, the solid must be permeable to the solution and must contain ions of its own. Many naturally occurring substances, such as clays, the humic acids in the soil, and some man-made resins, have this property of exchanging ions. The most popular exchange materials in use today are the synthetic organic ion exchangers, the majority of which are derived from crosslinked polystyrene. These polymers are prepared by copolymerization of styrene and divinylbenzene and are relatively simple. They are more stable toward strong acids and bases than the naturally occurring siliceous materials and have high capacities.

## SYNTHETIC ION EXCHANGE MATERIALS

### STRUCTURE

Typical structures of polystyrene resins are shown in Figure 24-1. The molecular weight is very high, and the material is therefore quite insoluble but permeable to water and solutions. The solid granules swell when wet, but the amount of expansion is limited or controlled by the crosslinking that joins the long chains together, so that the greater the crosslinkage, the less the swelling. The structure is similar to a sponge or a three-dimensional fish net, with the sulfonic acid groups attached firmly to the skeletal framework. Since the sulfonic acid group is a strong acid and thus essentially completely ionized, the result is a sequence of negatively charged ionic groups that are fixed in position surrounded by hydrogen or other positive ions that are free to move. Even though the cations can move freely about inside the resin, they cannot leave the resin particle unless replaced by another ion of equal charge because electrical neutrality must be maintained. The inside of the particle is very similar to a concentrated solution of a strong acid, except that the negative charges are held firmly in position instead of being free to move as they are in solution.

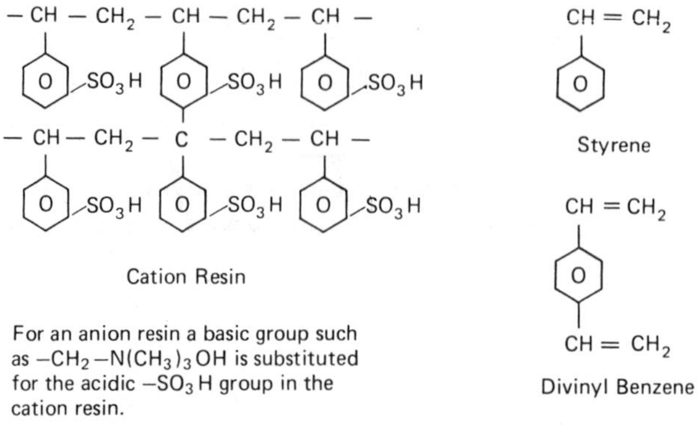

$-CH-CH_2-CH-CH_2-CH-$ with $SO_3H$ groups on each ring; $-CH-CH_2-C-CH_2-CH-$ with $SO_3H$ groups on each ring

Cation Resin

$CH=CH_2$

Styrene

$CH=CH_2$ / $CH=CH_2$

Divinyl Benzene

For an anion resin a basic group such as $-CH_2-N(CH_3)_3OH$ is substituted for the acidic $-SO_3H$ group in the cation resin.

**Figure 24-1.** Structure of polystyrene ion exchange resins.

## MECHANISM OF EXCHANGE

When a synthetic or naturally occurring ion exchanger is placed in a solution, ions may enter the solid particle and be exchanged for the ions of like charge that are free to move. Ions of the same charge as the fixed exchange positions will be excluded from the particle. For a cation exchanger in the hydrogen form in contact with a solution of sodium chloride, the exchange may be represented as

$$Na^+ + H^+R^- \rightleftharpoons Na^+R^- + H^+ \tag{24-1}$$

in which $R^-$ represents the fixed negative sites on the polymer structure. The exchange is a reversible reaction that will attain equilibrium and follows the law of mass action to a first approximation. If a solution containing a relatively small amount of sodium ion is percolated through a relatively large amount of resin, the sodium will be quantitatively removed from the solution and replaced with an equivalent amount of hydrogen ions. Similarly, if a concentrated solution of an acid is passed through a cation resin in the sodium form, the sodium will be essentially completely removed from the resin, which will be converted to the hydrogen form. Such regeneration of the resin may be repeated many times without apparent damage to the resin.

## SELECTIVITY RULES

The usual synthetic organic ion exchangers exhibit only slight selectivity for different ions and are not specific for any ion. There are, however, some general rules that may be used to predict the behavior of resins in contact with ionic solutions. For the strong cation exchangers such as sulfonated polystyrene, the higher the charge on an ion, the more firmly it is held by the resin. A trivalent cation such as aluminum will be more firmly bound to the resin than a divalent ion such as calcium, and calcium more than a monovalent ion such as sodium. In exchange reactions, trivalent ions will tend to displace divalent and monovalent ions from the resins, and divalent ions will displace monovalent ions.

There is also some selectivity among ions of the same valence group. In these cases, apparently the size of the hydrated ion is the principal factor. The smaller the hydrated ion, the more strongly the ion is held by the resin. Note that it is the diameter of the hydrated ion and not the bare ion as it exists in crystalline solids that is effective. In any periodic table group, the elements with lower atomic number bind more water molecules than the elements with higher atomic number and thus actually are larger in the hydrated state than those with higher atomic number. In the alkali metals (group IA), the ${}_{3}Li(H_2O)_x^+$ ion is the largest and the ${}_{55}Cs(H_2O)_x^+$ ion is the smallest, and ${}_{19}K(H_2O)_x^+$ would be held more firmly than ${}_{11}Na(H_2O)_x^+$.

The extent of crosslinkage also affects the selectivity of a resin to some extent, especially with large ions. The greater the amount of crosslinkage, the less the swelling and the smaller the size of the pores through which water and ions enter the resin particle. In some cases, highly crosslinked resins will actually exclude large organic ions.

Although the preceding generalizations apply to the selectivity of all forms of strong cation exchangers and to all forms of strong anion exchangers, other considerations will affect other types of exchange resins. The weak acid or carboxyl type of cation exchanger binds hydrogen much more strongly than other cations, and weak base anion exchangers bind hydroxyl more strongly than other anions. The newer chelating resins, which have iminodiacetate or other chelating groups for the functional groups attached to the resin structure, show greater specificity than the ordinary cation and anion resins. The iminodiacetate resins are essentially specific for divalent ions, especially the transition elements, which tend to form coordination complexes readily.

## DISTRIBUTION COEFFICIENTS

One measure of the strength with which a metal ion is held by a resin is the distribution coefficient, which is defined as

$$D = \frac{\text{mmoles metal/g dry resin}}{\text{mmoles metal/ml solution}} \qquad \textbf{(24-2)}$$

The larger the distribution coefficient, the more strongly the metal is held by the resin. Distribution coefficients are measured by a batch process in which a given weight of resin is equilibrated with a known volume of a solution of known concentration. The concentration in the solution is determined after the equilibration, and the amount of metal in the resin is calculated from the values before and after equilibration.

# APPLICATIONS

## DEIONIZATION OF WATER

If natural or domestic water is passed first through a cation resin in the hydrogen form and then through an anion resin in the hydroxide form, the resulting water is said to be demineralized or deionized. Deionized water can be used as a substitute for distilled water in many applications and is often much less expensive than distilled water. When the resins become loaded with ions after continued use, they can be regenerated into the hydrogen and hydroxyl forms by use of inexpensive, commercial grades of acid and base.

Cation exchangers (either naturally occurring or synthetic) in the sodium form are used to soften water by exchanging sodium ions for calcium and magnesium ions and can be regenerated with a concentrated sodium chloride solution. Mixed-bed deionization (using a cation and anion resin mixed together instead of in separate columns) is now being used in many deionization installations. The resins are separated by a flotation process before regeneration and mixed again after regeneration.

Because the exchange of ions follows the laws of equivalency, a cation exchanger in the hydrogen form can be used to determine the total salt content of natural or domestic waters. The pH is adjusted to some value before passing the water through a column of the exchanger and is then titrated back to the same pH after passage through the column. This procedure can also be utilized with an anion exchanger, but the results are not as satisfactory.

## SEPARATION OF IONS OF OPPOSITE CHARGE

In many analytical methods, an ion interferes with the determination of an ion of opposite charge, and often these interfering ions may be removed by passage through a column of an appropriate exchanger. As an example, phosphate ores, which usually contain calcium phosphate or calcium apatites, are dissolved in acid for analysis. The phosphate can be titrated from the $H_2PO_4^-$ form to the $HPO_4^{2-}$ form readily, but calcium causes precipitation as the pH is raised. Passage of the acid solution through a cation exchanger in the hydrogen form prior to titration removes the calcium ions and allows the titration to be completed satisfactorily. There are many hundreds of examples of this type of separation in the literature using either anion or cation resins.

## CONCENTRATION OF TRACE CONSTITUENTS

One very important use of ion exchange resins is the concentration of trace constituents. Large volumes of liquid containing small amounts of ionic materials can be passed through a column of an appropriate exchanger. The ionic materials that have been held by the column can then be eluted with a small volume of solution for analysis. An excellent example of this type of use is the development by the Public Health Service of a simple field method for identifying samples of milk that may be contaminated with radioactive fallout. Radioactive iodine-131 ($^{131}I$) is the ion of greatest public health interest and is separated from raw milk by passing a gallon or more through a plastic cartridge containing the appropriate resin. The cartridge is then sent to the Public Health Service for determination of the amount of radioactivity. Similar methods are available for determination of radioactive cations such as strontium-90 ($^{90}Sr$) and cesium-137 ($^{137}Cs$).

## CHROMATOGRAPHIC SEPARATIONS OF SIMILAR IONS

Ion exchange resins may also be used in the chromatographic separation of similar ions that are difficult to separate by chemical means. If a mixture of relatively small amounts of two cations (or anions) is poured onto a column of exchanger, both ions will be removed from the solution in the top portion of the resin. If one cation (A) is held more strongly by the resin than the other cation (B), use of a satisfactory eluting agent will cause B to move down the column faster than A, and under favorable circumstances all of B will flow out the

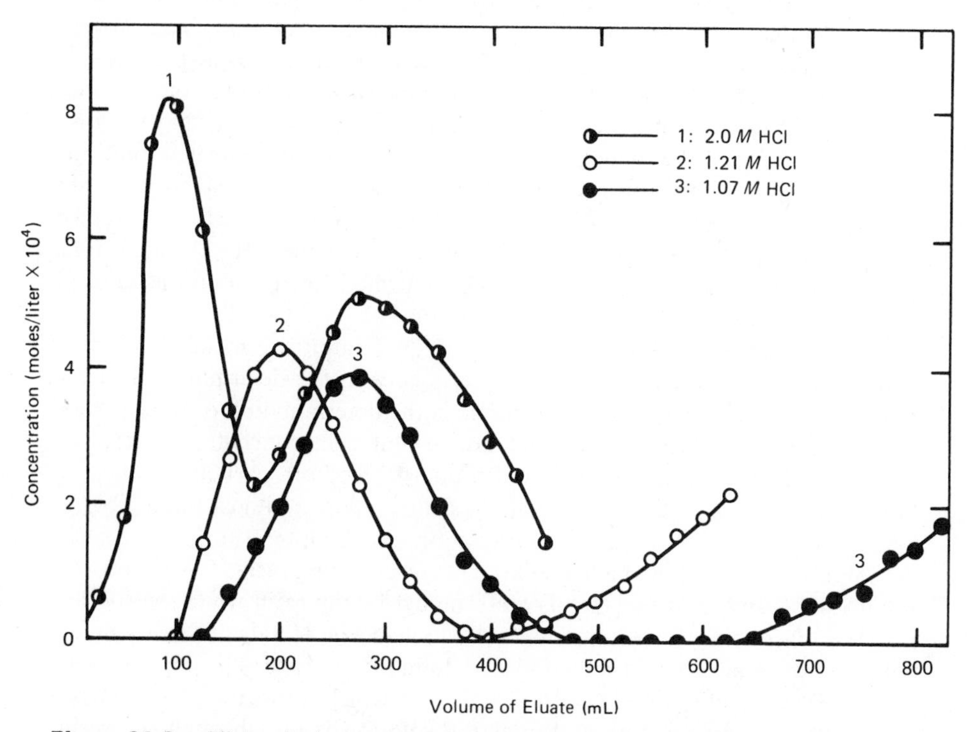


**Figure 24-2.** Effect of eluant concentration upon the separation of magnesium from calcium using a strong acid cation exchanger. Flow rate, 8 mL/min; resin height, 6.3 cm; 1.5 mg of magnesium plus 4.5 mg of calcium. (From D. N. Campbell, Master's Thesis, Southern Methodist University, 1953.)

bottom of the column before A appears in the effluent. By proper choice of eluting agents, ion exchange resin, column length, flow rate, particle size, and other variables, separations of complex mixtures may be effected by this procedure. Examples are the separation of sodium from potassium, calcium from magnesium, and chloride from bromide. Perhaps one of the most spectacular successes has been the separation of the rare earth elements, which has allowed the preparation of very pure samples of many of these metals.

Figure 24-2 shows the effect of the variation in the concentration of the hydrochloric acid used as eluant upon the separation of 1.5-mg magnesium from 4.5-mg calcium using a strong acid cation resin. A partial separation is obtained with 2.0 *M* HCl and a complete separation with 1.21 *M* HCl, but the volume of eluate that contains neither ion is very small. A satisfactory separation, with a large enough volume of eluate that contains neither calcium nor magnesium ions, is effected by the use of 1.07 *M* HCl. In an actual analysis, the first 100 ml of the eluate would be discarded and the magnesium determined in the next 425 mL by atomic absorption or titration.

## USE OF CHELATING RESINS

Ordinary strong acid or weak acid cation exchange resins cannot be used to remove divalent metals from brine solutions such as ocean water and oil-well brines because the high concentration of sodium ions effectively displaces them from the resin. This is true for the removal of any trace ion from a concentrated solution. Use of a chelating resin such as an iminodiacetate exchanger allows separation of trace divalent metals from brines and other concentrated solutions of monovalent cations, since monovalent cations are not bound to the resin and thus will not displace the divalent metals. Examples include the separation of manganese and other divalent ions from oil-well brines, of polyvalent ions from sugar solutions, of calcium from concentrated caustic solutions, and of copper from water.

The iminodiacetate resin is also used to concetrate trace metals from high ionic strength solutions to eliminate interference in the determination of the metals by atomic absorption. An example is the determination of trace metals in ores after basic fusion. Ores that are insoluble in concentrated acids or bases are often rendered soluble by fusion with sodium carbonate, and the high concentration of sodium ions that results makes it almost impossible to determine trace metals by atomic absorption (see Chapter 18). In these cases the solution is poured through a column of the chelating resin after neutralization to pH 7 and the trace metals are retained by the resin while the sodium ions pass through. After elution with dilute acid, the metals are easily determinable by atomic absorption without sodium interference. The improvement in the measurement is illustrated by Figure 24-3, which shows the recorder trace for 3 ppm of copper before and after the solution is put through the resin. Figure 24-4 shows the effect of pH upon the recovery of various metals from solutions using this procedure and emphasizes the necessity for proper pH adjustment before the solution is poured over the resin. The recovery of

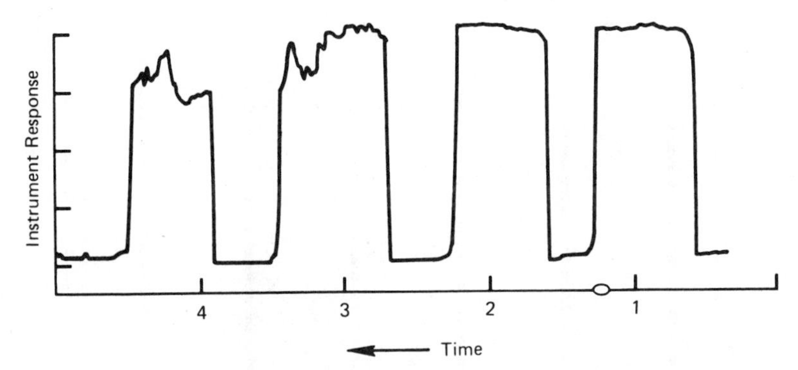


**Figure 24-3.** Typical recorder responses for 3 ppm copper: (1) standard copper solution; (2–4) fused standard oxide. (2, after passage through resin column; 3, 4, before passage through resin column). (From T. W. Freudiger, Master's Thesis, Southern Methodist University, 1971.)

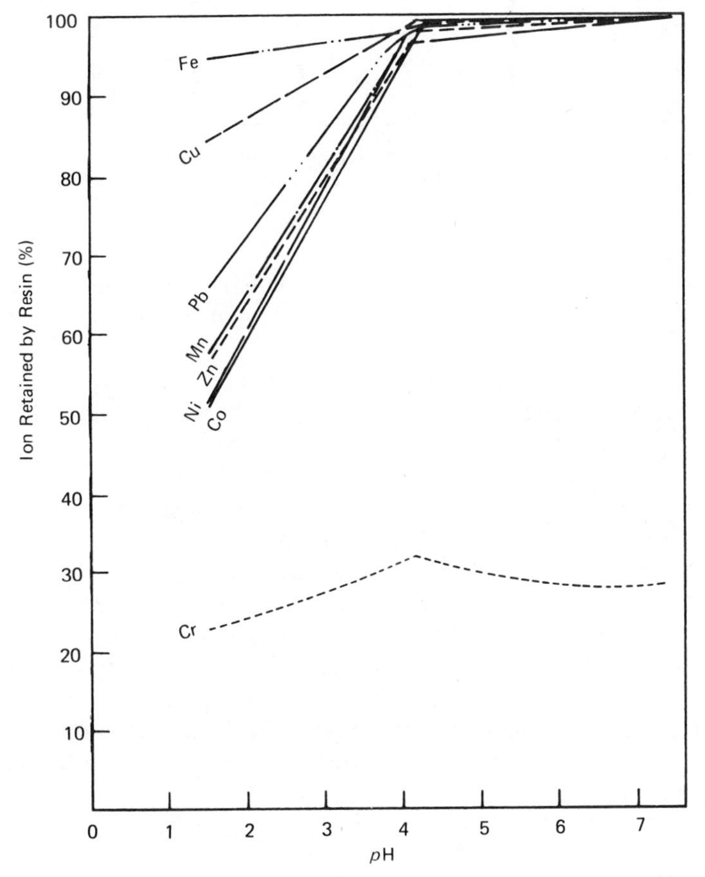


**Figure 24-4.** Recovery of various metals from a chelating resin as a function of the pH of the solution of the metal ion. Chromium recovery is low, owing to formation of chromate or dichromate during fusion. (From T. W. Freudiger, Master's Thesis, Southern Methodist University, 1971.)

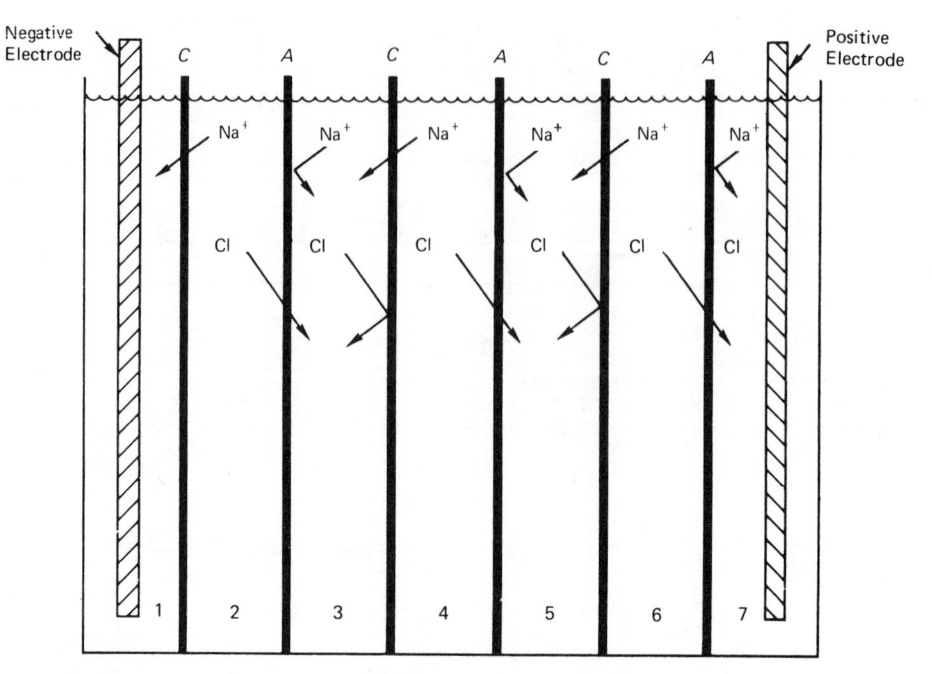


**Figure 24-5.** Electrolytic desalting using ion exchange membranes. C, cation exchange membrane; A, anion exchange membrane. Concentration occurs in cells 1, 3, 5, and 7 and desalting occurs in cells 2, 4, and 6.

chromium is very poor after fusion, since only trivalent chromium is retained by the resin and any chromium present is usually oxidized to chromate during fusion.

The use of the chelating resin for concentration of trace metals can also be applied in food analysis. The foodstuff is digested in a nitric acid–sulfuric acid–hydrogen peroxide medium to remove the carbonaceous material, and the resulting solution is not suitable for atomic absorption measurement without extensive dilution. Adjustment of the strongly acid medium to pH 7 followed by passage through the resin bed and elution with dilute acid allows determination of many metals in the parts per million or smaller range. The procedure is being used to determine cadmium, lead, manganese, zinc, nickel, and cobalt in various types of foods, including vegetables, fish, milk, grain cereals, and fruits. When lead is present, a strontium salt must be added during the digestion to coprecipitate the lead. After filtration and washing, the precipitated sulfates are converted to carbonates and dissolved before determination. The filtrate and washings are then diluted and neutralized to pH 7 before passage through the column.

## ION EXCHANGE MEMBRANES

Ion exchange membranes are now prepared by casting the resin as a thin film or by incorporating finely ground resin particles in a plastic but inert matrix. Membranes prepared from cation exchange resins are permeable to cations but not to anions, since the high concentration of fixed negative charges in the resin matrix offers a high resistance to the passage of anions. Similarly, membranes of anion exchange materials readily pass anions but almost completely exclude cations. These membranes are now used in one method of desalinization of water by electrodialysis, increasing the efficiency from about 18 to 80% over ordinary nonselective membranes. In a multicompartment electrodialysis cell formed by alternating anion and cation membranes such as that shown in Figure 24-5, the salt tends to concentrate in every other cell; thus, desalting occurs in every other cell.

CHAPTER

# 25 Automated Analysis

As we move further and further into a technological age, the demands for more and better knowledge, for more reliable products, and for more efficient services have increased astronomically. Even though all areas of chemical endeavor are experiencing increased pressure for better output, the chemists and technicians who deal with this problem most often are those involved in analytical chemistry either as a primary or secondary interest. The analyst finds himself needing new and faster techniques that need to be more precise but should cost less. In some cases no solutions to these problems exist at this time. However, many, if not most of them, may be solved by converting slow manual techniques of chemical analysis and data handling to faster automated or semiautomated analyses with computer manipulation and presentation of the data.

Many chemistry students, upon graduation and subsequent employment, do not have sufficient knowledge and experience in even the simplest types of automated techniques, primarily because of the relative newness of these techniques in analytical chemistry and also because clinical and industrial laboratories have been the avant garde for analytical automation. The purpose of this chapter is to introduce some of the basic terms, theories, and applications of this new field.

Strict definition of **automated chemistry** is difficult, since automatic procedures can vary from a simple self-filling burette to control of complicated production processes through automatic sampling, analysis, computer calculation, and consequent necessary changes in production parameters to keep the process working properly. A workable definition is that automation includes any process, procedure, or technique that can be accomplished without physical manipulation by a chemist or technician and/or does not require the physical presence of an operator. This definition covers the applications in almost all research, industrial, control, clinical, or government laboratories.

The obvious advantages of automated chemical analysis include speed, better reproducibility, and lower cost. Other advantages are the partial or total elimination of many analytical hobgoblins, such as time- or temperature-dependent reactions, nonstoichiometry, and unfavorable equilibrium effects.

However, there are also many problems and disadvantages involved in the transformation from manual to automated chemical procedures.

One disadvantage is the usually large initial cost of the equipment, especially if computers are involved. There is also the cost in time and money of the development and setting up of the actual automated procedures and of training personnel in the use of the equipment. Originally, it was believed that any manual procedure could be automated, but it is now apparent that often whole new chemical approaches have to be developed before automation can be accomplished for some analyses. Automation can result in a large amount of very precise but also very inaccurate data, owing to poor system design and inadequate knowledge of the chemistry or kinetics of the test being run.

Finally, an automated laboratory may find itself with too much data, simply because of the ease with which extra data can be generated by the automated system. Most analysts try to obtain as much data as the system can generate, because data not believed important when the system is designed can often help in the interpretation of deviant results when the system is in operation. The augmented efficiency due to automation is lost if the analysis of the extra data consumes most of the time saved by automating. The time involved, however, can often be reclaimed by further automation.

The use of electronic standardization, compilation, and printout devices, coupled with high-speed digital computers, may alleviate the problem of data analysis. Indeed, the role played by computers and computing devices in the overall view of analytical automation probably will become as important as the various analytical devices themselves. Computers not only serve to collect, reduce, interpret, and present data in finalized form, but they also can be used to control the generation of the data itself. This task is accomplished by having the computer react directly with the analytical device through continuous monitoring of various parameters, changing them if need be for best results. But perhaps the most interesting role of the computer in analytical chemistry is its capacity to simulate chemical processes mathematically and thus generate new methods of chemical analyses. Computers will be discussed in more detail in the latter part of this chapter.

The types of automated techniques and devices now in use in chemical analysis can be classified under two main groups:

1. Nondiscrete or continuous-flow techniques.
2. Discrete techniques:
   a. Devices that permit only one analytical function.
   b. Devices that permit more than one analytical function.

## CONTINUOUS-FLOW OR NONDISCRETE TECHNIQUES

The **continuous-flow system** was probably the first true example of a successful automated technique for analytical chemistry. It was introduced almost 10 years ago and is now the most widely used automatic procedure. Continuous-flow analysis is defined as the continuous detection and monitoring of some chemical species in a continuously flowing stream of liquid.

Although continuous-flow systems have proved their worth in research (especially in biochemistry), they really came into their own in industrial quality control and medical/clinical laboratories. It is the nature of these laboratories to have a large number of analyses of the same type. A recent and very significant application of continuous-flow systems is in the area of air and water pollution, where continuous, hour-by-hour monitoring of concentrations of various pollutants is essential.

## NATURE AND FUNCTION OF BASIC EQUIPMENT

Basically the system is modular in form and consists of:

1 A sampling device.

2 A pump.

3 An area where separations based on extractions, filtrations, dialyses, or distillations may take place.

4 An area where reaction chemistries take place, generally coiled glass tubing.

5 A detection device for direct or indirect quantitation of the desired ingredient.

6 A readout device, usually but not necessarily a strip-chart recorder.

Figure 25-1 shows a schematic diagram of the modules of an instrument system. The modular nature of the system readily facilitates the incorporation of other devices and components when a higher level of sophistication is needed.

The pumping device consists of a series of flexible plastic or rubber tubes, called **pump tubes**, that are stretched across a flat surface. A series of rollers attached to a continuously rotating chain presses against the tubing as shown in Figure 25-2. As the chain moves, the rollers roll over the tubing, producing a constant squeezing action, so the liquid or air may be pumped through the tubing. Some of the roller bars are always in contact with the tubes, so an even nonpulsating flow of liquid is realized. The newer pump systems can hold up to 28 different tubes; thus, very sophisticated analyses may be accomplished.

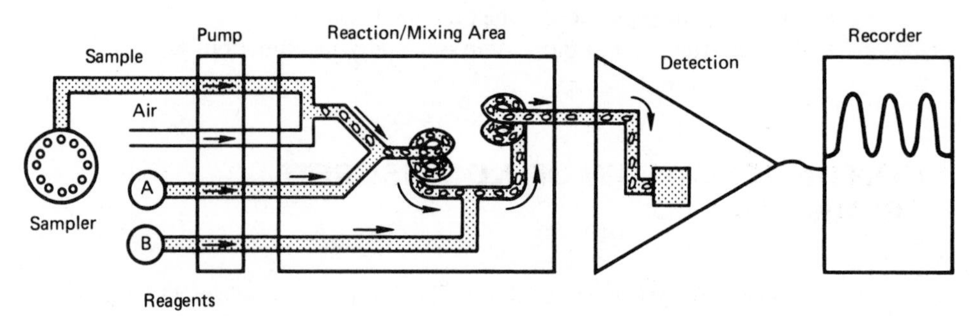


**Figure 25-1.** Block diagram of a continuous-flow automated system.

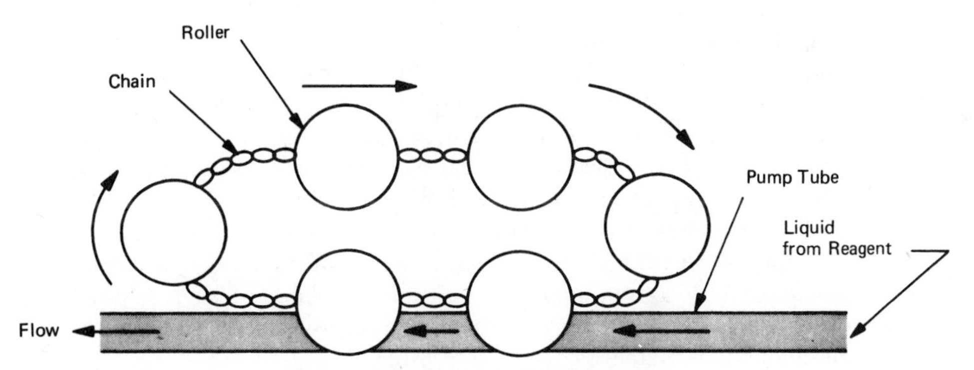


**Figure 25-2.** Proportionating pump.

One end of the pump tube is immersed in the reagent and the other end connects with different devices for the next step in the analysis. The pump tubes can be made of various materials, so almost any type of solvent may be used. Different volumes of liquid can be pumped by utilizing tubes with different diameters; thus, different amounts of reagents can be mixed together in the exact proportions called for by the chemistry of the situation. For example, if the manual method called for mixing reagent A with reagent B in a ratio of 2 : 1, this can be accomplished in a continuous-flow system by having the pump tube delivering reagent A have an internal diameter 1.4 times that of the B tube. The chain rollers, along with the flexible tubing, make up what is called a **proportionating pump** and is the heart of any continuous-flow system.

The sampler consists of a circular plate that holds small plastic cups containing the samples to be analyzed. The sample plate is rotated in a timed sequence by means of a cam. At the same time the cam controls the movement of a small metal probe in and out of the sample cups as they pass by. When the probe is not aspirating a sample, it is automatically placed in a wash bath to aspirate the system with wash. The probe's residence time in both the sample and the wash is controlled by the cam.

The actual chemical reactions take place after the sampling and pumping steps. Several streams of liquids (reagents) may be combined and mixed in small helical glass coils. The combined reagents are then merged with the sample stream and mixed to cause the chemical reaction used for quantitation to occur. Most often, but not necessarily, this reaction results in the formation of a colored solution.

The solution now flows to a detector for measurement of some chemical parameter that is a direct function of the concentration of the desired component in the original sample. The most widely used detection device is a simple filter photometer although more complicated detectors are sometimes used. The filter photometer uses multilayer interference filters to furnish monochromatic light of the desired wavelength. Recently, spectrophotometers have been utilized that have the most modern optical and electronic systems made especially for continuous-flow analyses. These newer spectrophotometers permit selection of any wavelength throughout the normal ultraviolet–visible range, with either single- or double-beam capability. In any case, the solution absorbs light which is monitored by a photosensitive device, and the changes in absorbance are usually recorded graphically by a strip chart recorder.

Although spectrophotometric or filter photometric detection is the most common form of detection in continuous-flow systems, many others have been suggested and are in use. Electrochemical methods have become prominent, especially since an ever-increasing number of ion selective electrodes have been adapted to these systems. Since little or no sample preparation is needed, several electrodes may be placed serially to determine several chemical species simultaneously. Even very complicated electrochemical techniques, such as polarography using a dropping mercury electrode, have been successfully incorporated into continuous-flow systems.

Other detection techniques include atomic absorption, atomic and molecular fluorescence, and scintillation counting. One of the more interesting of these newer techniques is thermometric detection, in which changes in temperature caused by a chemical reaction are used to determine analyte concentration. Thermometry usually requires relatively larger amounts of the analyte than photometry and electrochemical methods.

No matter what detection method is used, a series of peaks and valleys appear on the recorder chart as a series of samples, standards, and washes pass through the flow cell. The heights of these peaks are generally directly proportional to the concentration of the desired constituent in the samples.

## THEORETICAL CONSIDERATIONS AND PROBLEMS OF OPERATIONS

There are two primary prerequisites that must be satisfied to permit the use of continuous-flow analysis systems. The first of these is that all the samples picked up by the probe must be treated in the same manner; that is, the unknown sample and the known standard must both undergo the same reaction steps, at the same temperature, and in the same time sequence. Many problems of chemical interference are eliminated in this manner; however, the partial and sometimes total elimination of unfavorable kinetics and equilibrium factors is often more important than elimination of interferences. Since all samples experience the same timed sequence of events, the chemical reactions employed need not go to completion. Even in the case of incomplete reaction, the samples and standards can be quantitated because they should have reached the same percent of reaction completion.

The other prerequisite is the characteristic of continuous-flow processes known as the steady-state condition. This condition is attained when there is no change in concentration of the detected species with respect to time in the flow cells. A typical recorder tracing for the steady-state condition is shown in Figure 25-3. When the pump tubes are placed in the reagents and the sampling probe is placed in the wash, the system becomes stabilized. This initial state results in a straight base line being drawn on the recorder. Placing the sample probe into a sample cup, aspirating for a time, and then replacing it into the wash, would result in a rising curve, a plateau section,

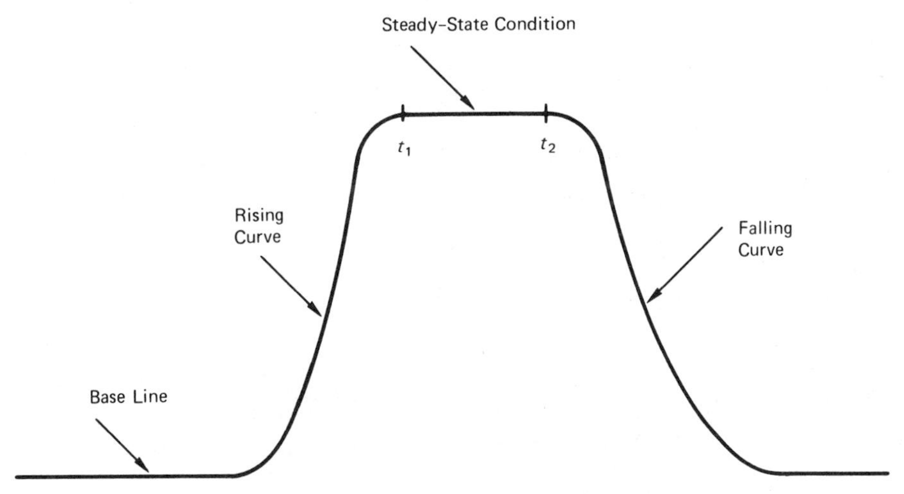


**Figure 25-3.** The steady-state condition.

and finally a falling curve. The plateau section of the curve represents the steady state condition and indicates that conditions have stabilized. In practice, the full steady state condition may not be attained, because the percent of the steady state condition reached is a function only of the sampling time and not of the concentration. A decrease in sampling time allows more samples to be assayed per unit time. Best accuracy and precision are obtained, however, when the steady state is attained and used for the measurement.

There are, however, limits to the rate at which multiple samples can be analyzed. These limits have been shown to be functions of the half-wash time and a lag time. The effect of these two parameters is manifested in the transient state of quantitation (i.e., in the rising and falling portions of the graph). As a result of narrow peaks, short wash times, and/or differences in sample concentrations, the curves will tend to overlap, so that the falling portion of one curve will be added to the rising portion of the next, resulting in erroneous data. The discreteness or integrity of the sample will not have been preserved. This phenomenon is the result of sample-to-sample interaction caused by the thin film of liquid that adheres to the tubing, as illustrated in Figure 25-4. The sample stream is segmented by the introduction of small bubbles of air into the stream. However, the air does not fit snugly against the inner wall of the tubing and a small amount of the first sample will become mixed with the second sample, and so on. This sample dispersion is one of the fundamental problems of continuous-flow analysis.

Figure 25-5 indicates that the theoretical output of a series of aspirated samples should be in the form of a square wave, but that in practice the square wave is seldom obtained. The square-wave form is the theoretical output, since the sample probe is either in the sample or in the wash solution. The deformation of this square wave is a measure of the amount of interaction or dispersion that the sample slug has undergone. The amount of dispersion has been shown to be both theoretically and practically related to the flow rate of the solution, the bubble segmentation rate, the residence time in the system,

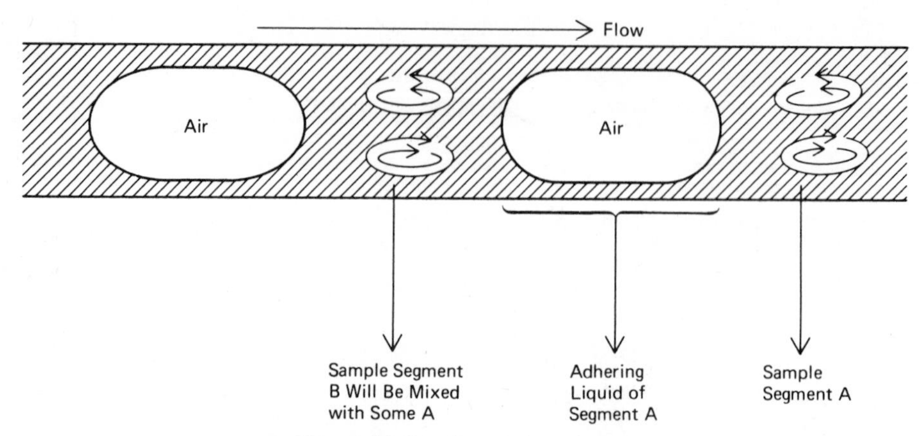


**Figure 25-4.** Sample interaction.

the surface tension of the solution, and the viscosity of the solution. In general, dispersion may be reduced by using small-bore tubing with small bubble size and high bubble frequency. Short reaction times, which reduce total system time, are also advantageous. These observations indicate that the best results are obtained when nonsegmented bubble flow is kept to a minimum. Such conditions occur in extraction steps and in the flow cell.

In general, it may be said that sample dispersions can be minimized by several methods: changing the flow rate, the sampling and wash times, or the original sample concentration. There are methods other than precise system design that can aid in solving the problem of dispersion. Since the rise and fall times of the sampling curves are indicative of the amount of dispersion, they may be regenerated by various computer methods, thus reducing appreciably the amount of apparent interaction. However, the simplest method to preserve sample integrity and decrease sample interaction is the periodic introduction of small amounts of air into the system, as stated before.

Air is introduced into the system by means of a pump tube in such a way that the flow stream will be broken up into a series of small segments of liquid and air. When no air is present, the flow of liquid is not uniform throughout the cross section of the tubing, since the flow next to the tube wall is slower because of friction. This causes some of the liquid to be held back and to contaminate the next sample. The surface tension at the interfaces of the air segments significantly

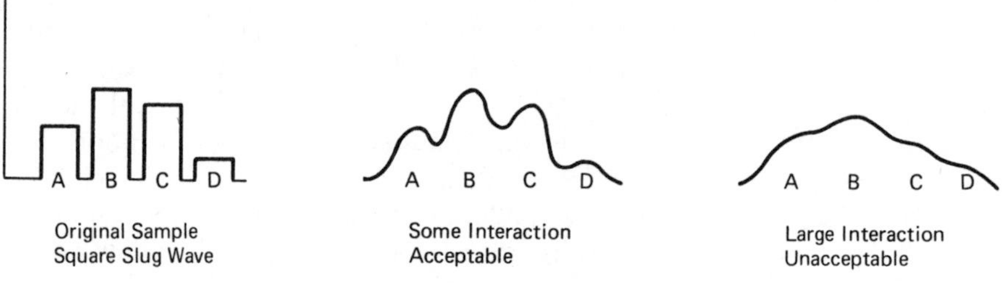


**Figure 25-5.** Dispersion effects.

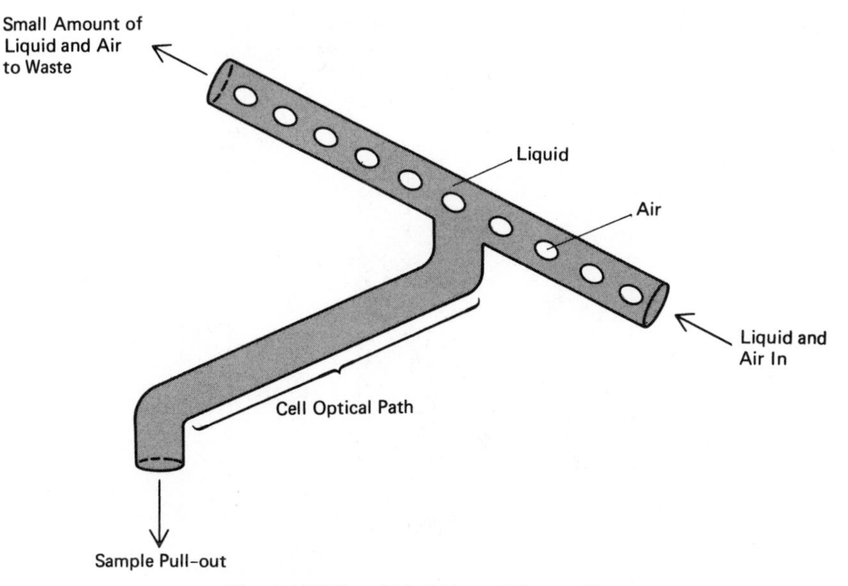


**Figure 25-6.** Debubble and flow cell.

reduces the interaction of the wall with the liquid. The air also serves to facilitate the mixing of solutions in the mixing coils. As the small volumes of liquid, segmented by air, go through a mixing coil, they experience a tumbling effect; the more dense part of the liquid falls through the less dense part. The final result is a homogeneous liquid segment. The air, however, must be removed before the sample goes through the flow cell for quantitation. Air removal is accomplished by a debubbling system, which consists of a T-shaped piece of glass (Figure 25-6). The air and liquid segments enter from the middle section. Most of the liquid falls down toward the flow cell, whereas the air and a small amount of liquid go up through the top to waste. There are some continuous-flow systems in which the bubbles of air are not removed so that dispersion is further limited in the flow cell itself. In these cases, electronic devices have been incorporated that permit the detector to disregard the signal interruption caused by the bubbles.

There are problems other than maintaining sample integrity associated with continuous-flow systems. During the continuous use of the instrument, the base line may begin to drift because of a buildup of foreign materials along the sides of the tubing and flow cells. Use of different volumes of sample in the sample cups may cause variable results. System bias may also exist. These and other problems, except perhaps the bias, may be solved by careful attention to system on the part of the analyst.

## SPECIAL TECHNIQUES AND DEVICES

Thus far, we have discussed only the basic continuous-flow system and operation. Many chemical analyses, however, require more than simply mixing a reagent and sample together and measuring the absorbance of the color produced

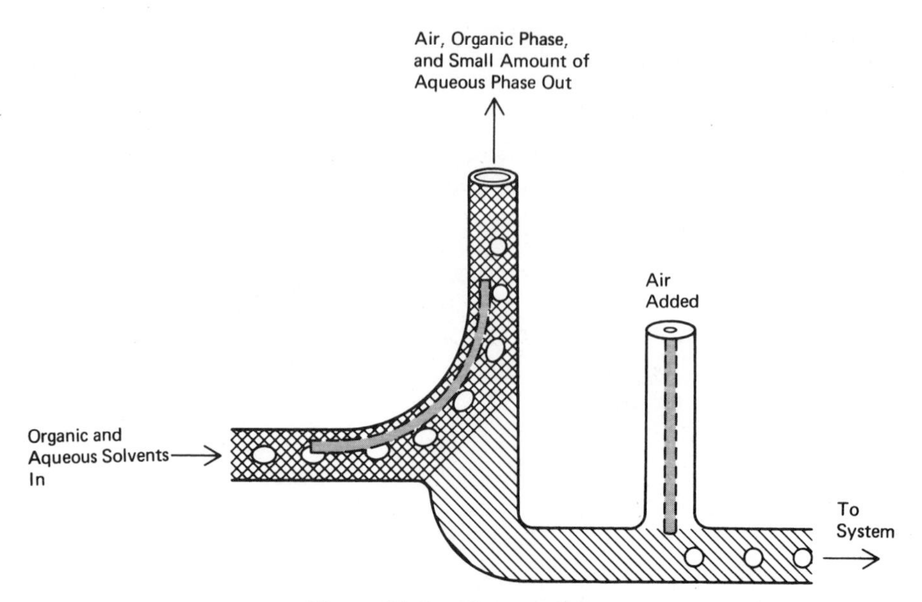


**Figure 25-7.** Phase separation.

with a colorimeter. In many cases the more sophisticated manual techniques can also be included in continuous-flow processes.

Many reactions require heating for a specified length of time. An automated continuous-flow system meets this requirement by passing the flow stream through a large coil of glass tubing placed into a constant-temperature oil bath. The length of the coil needed may be calculated easily from the flow rate, the inside diameter of the tubing, and the length of time needed for heating.

Extraction procedures may also be incorporated into the system because adequate mixing of two immiscible liquids is accomplished in the small coils used for mixing. The efficiency of the extraction is greatly increased if small glass beads are added to the coils so that the liquid/liquid interface area is increased. The two immiscible liquids then flow together into a separator (Figure 25-7). The separator contains a small piece of a fluorinated polymer that has hydrophyllic properties. It is wetted preferentially by the more polar (aqueous) solvent and thus serves to direct the air, organic phase, and a small amount of the aqueous phase to waste. At the same time the main portion of the aqueous phase is directed downward, air is introduced, and the sample is allowed to continue to the next step. Note that, since every sample and standard will be treated in the same way, it is not necessary that the chemical specie of interest be 100% extracted or even that all the aqueous phase be recovered. Other separators have been designed to retain the organic phase and discard the aqueous phase.

Other techniques, such as filtration, centrifugation, and dialysis, can be performed by the system using a dialyzer (Figure 25-8). This module consists of two flat pieces of material with a groove on their surfaces. A membrane is placed on one of the surfaces, and then the other piece is placed on the first with the grooves aligned. The setup resembles a sandwich. Two streams of liquid that are separated physically only by the membrane are passed simul-

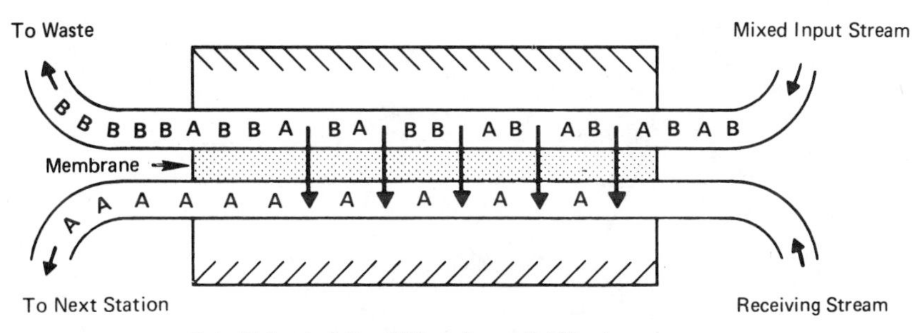


**Figure 25-8.** Dialysis.

taneously through the grooves. The module simulates a filter or centrifuge by keeping all solids in the upper sample stream while allowing small ions and molecules to diffuse through the membrane into the recipient stream. If dialysis is desired, the membrane must also be impermeable to certain macromolecules and ions.

Often, the sample is not in a form suitable for the liquid sampler described earlier. Such samples could be aspirin tablets, food, soil, and so on. A module has been designed for continuous-flow systems that allows the use of solid samples. This module consists of a carousel that contains small plastic cups into which the solid samples may be placed. As the carousel moves, the cups are tipped serially to spill the contents into a funnel that leads to a device similar to a common kitchen blender (Figure 25-9). Solvent is added through a valve and the blades of the blender disrupt and homogenize the solid samples so that the analyte sought may be dissolved. During this time the rest of the continuous-flow analytical train is aspirating wash through a three-way valve. After the sample has undergone homogenation, it is aspirated for the required length of time. After the sampling valve switches back to the wash reservoir, the homogenization chamber is washed several times with solvent, which is discarded. This is all accomplished through several electronically controlled valves. The system is now ready to accept a new sample.

Sometimes the solvent used to homogenize the sample is not the best one for the subsequent analytical steps. Recently, a module has been designed that allows changing solvents on stream. This module is known as the evaporation-to-dryness module and is illustrated in Figure 25-10. The module works by adding one solvent containing the analyte sought to a thin, moving solid band resembling a metal wire. The band moves continuously through an area of partial vacuum that is heated. The first solvent evaporates, leaving the solute on the surface of the band. As the band moves out of the evaporation area, the second solvent is added to redissolve the analyte and reintroduce it into the analytical system. Note that an effective increase or decrease in analyte concentration can be obtained by simply varying the volume ratio of the two solvents.

In addition to the techniques discussed, others, such as distillation, column chromatography, and digestion, have been successfully incorporated into continuous-flow automated systems.

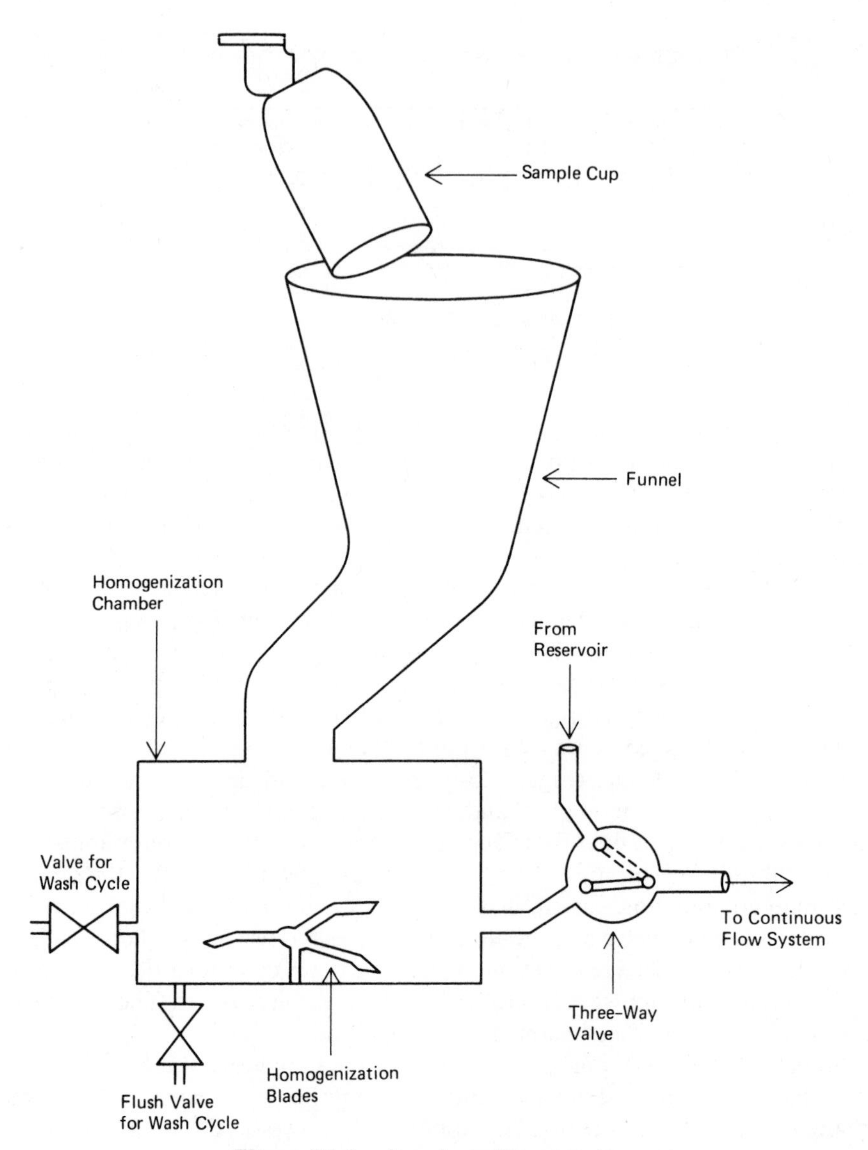


**Figure 25-9.** Sampler and homogenizer.

## APPLICATIONS

Continuous-flow automatic procedures have been adapted to many analyses that must be run many times in a single day. Some examples are blood analyses in hospitals and clinical laboratories and single tablet or vial analyses in pharmaceutical control and testing laboratories. As many as 20 different determinations can be made from a single blood sample by use of automated analytical equipment at a rate of 150 samples per hour. A typical example of an automated analysis in the area of environmental controls is the monitoring of

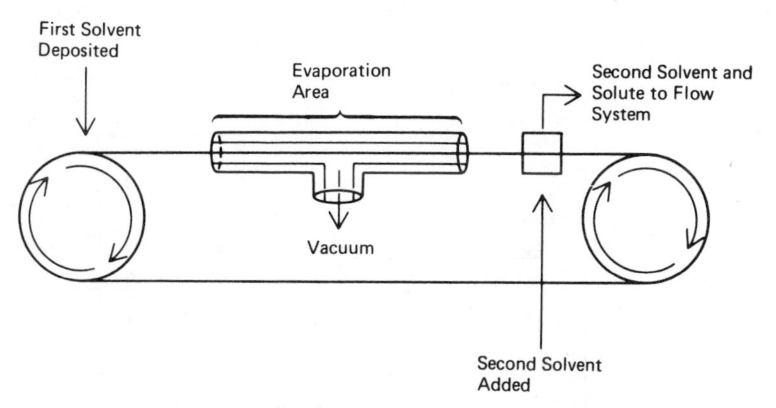


**Figure 25-10.** Evaporation-to-dryness module.

natural waters for their ammonia content. Trace ammonia may be determined colorimetrically by forming a chlorinated ammonia compound and allowing it to react with an excess of phenol to form an indophenol. This last compound is blue in alkaline solution; thus, its concentration, which is directly proportional to the amount of ammonia originally present, may be determined with a colorimeter. The equations for the sequence of reactions are

$$\mathrm{NH_3 + OCl^- \xrightarrow{OH^-} \underset{\text{chloramine}}{NH_2Cl} + OH^-}$$

$$\mathrm{C_6H_5{-}OH + NH_2Cl + 2OCl^- \xrightarrow{OH^-} Cl{-}N{=}C_6H_4{=}O}$$

$$\mathrm{Cl{-}N{=}C_6H_4{=}O + C_6H_5{-}OH \xrightarrow{OH^-} \underset{\text{indophenol gives blue color}}{{}^-O{-}C_6H_4{-}N{=}C_6H_4{=}O}}$$

The schematic diagram in Figure 25-11 indicates a possible configuration of pump tubes and modules to accomplish the analysis. Follow through the system, noting that the air is added to the sample stream first to minimize dispersion. Next, sodium hydroxide is added to the sample, then it and the air are mixed in a mixing coil. Phenol is added and mixed and then the hypochlorite salt. The reaction mixture is passed through a jacketed mixing coil heated to 90°C, mixed again, and finally debubbled and passed through the flow cell of the colorimeter. The absorbance is read at 630 nm and the values of ammonia may be determined by comparing the peak heights of the samples to those of the standards. In this way 60 samples per hour may be determined unattended. Manually, an experienced analyst would be hard-pressed to do ten samples in an hour. The advantages of automating the method are instantly obvious, especially when it is realized that an environmental laboratory may have hundreds of ammonia assays to do every day.

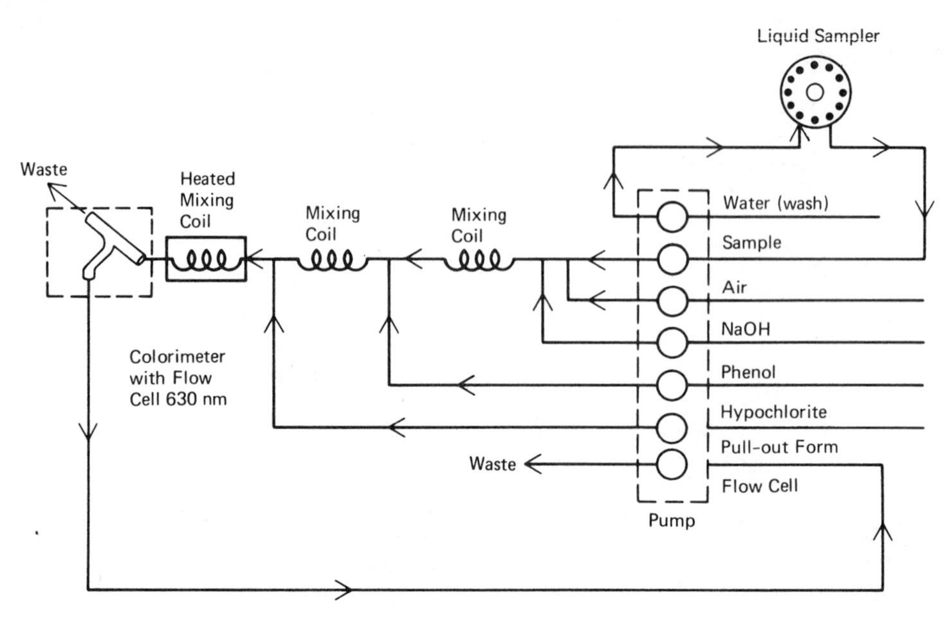


**Figure 25-11.** Ammonia system.

The preceding illustration shows the value of continuous-flow analysis for a simple analytical situation, but how will it serve for a much more complicated problem? To illustrate, imagine the following situation. You wish to use a continuous-flow system to provide the dissolution, separations, and sample cleanup steps for an analysis of some natural product by high-performance liquid chromatography (HPLC; see Chapter 22 for details). Although many chromatographic procedures themselves are relatively straightforward, the sample preparation is often tedious and time-consuming.

If the sample, which contains a small amount of natural product, is the leaf of a plant, it would be in the form of a solid incorporating many chemical species that would probably interfere in the analysis. These must be removed. In addition, assume that the natural product molecule is not an ultraviolet chromaphore. Since most HPLC instruments utilize an ultraviolet detector, the natural product molecule must be reacted with another compound to form an ultraviolet-absorbing species. As a further complication, assume that the solvent used to dissolve the natural product molecule, the solvent in which the chromophore is formed, and the solvent used to introduce it to the LC column are all different. In summary, the necessary analytical steps are

1 Homogenize the sample and extract the natural product molecule into aqueous buffer solution.

2 Filter the solution to remove bits and pieces of insoluble plant parts.

3 Dialyze the solution to remove proteinaceous material that could cause emulsion problems.

4 By pH control, extract the natural product molecule into organic solvent A to prepare for reaction to form the chromophore.

5 Reaction to form chromophore.

6 By use of the evaporation-to-dryness module, change from organic solvent "A" to methanol in preparation for injection to the HPLC.

7 Inject into the LC.

8 Quantitate the chromatogram.

All of the steps may be incorporated into a single continuous-flow system. Indeed, a system almost as complicated as the one just described is commercially available for the HPLC analysis of fat-soluble vitamins. It will analyze about six samples per hour. This is probably more than an experienced analyst could do in a whole day. There are, however, many significant design problems in such a system. The system must be very stable with respect to time. The HPLC can only take a sample from the flowing stream at the steady-state part of a peak (see Figure 25-3). This means that the initial sampling apparatus controls admission of the sample stream to the HPLC column. The admission of sample may be accomplished through the use of a microprocessor (discussed later in this chapter) that acts as a timing device. During the time that the HPLC is not sampling the flowing stream, a valve is switched so that only mobile phase from a reservoir is introduced into the HPLC column. The whole example serves to illustrate that it is not only the chromatograph that deserves the name "high performance."

## NONCONTINUOUS FLOW OR DISCRETE ANALYSIS

Noncontinuous automatic methods and techniques of analysis are designed for specific uses and generally are much more complicated than continuous-flow systems. They are inherently different from the previous system in that the integrity of the sample is always preserved, since there is little or no possibility of interaction between the samples. Discrete automatic analysis systems may be divided into two major categories:

1 Those that perform only one analytical function with the sample.

2 Those that perform several analytical techniques on the sample.

The first category of automated discrete systems are those used to perform only one analytical function. Examples include automated gas chromatographs, atomic absorption spectrometers, and titrators. Such systems are considered automated because a sampler and a recording unit have been attached to a unit basically made for manual use. The ubiquitous circular sample holder and probe are again used for timed reference and sample introduction. These systems are very useful when a great many samples of the same type are analyzed every day.

Those discrete systems in the second category are usually modular in form, since some type of operator intervention is often required when changing from one type of analysis to another. Although discrete systems are capable of performing multiple types of analyses in many cases as well as or better than continuous systems, they have not gained enthusiastic reception because of their very high initial cost. Most of these multiple analysis instruments are found in clinical or medically oriented research laboratories where they have been designed to perform a sequence of specific analyses.

Basically discrete systems consist of the following modular devices:

1 A sampling device.
2 Some type of conveyor belt or its equivalent to move the sample to and from the different processing stations.
3 A detection and quantitation device (usually a colorimeter or spectrophotometer).
4 A recorder, digital readout, or other tabulating device.

The systems commercially available generally resemble each other except for the manner in which the sample is moved and the way the wet chemistry functions are performed. In a typical system, the samples are placed on a rotation rack that turns periodically with time. A sampling probe aspirates a predetermined amount of sample and transfers it to its own discrete reaction vessel. The transference of the sample and other liquids may be accomplished by the in-and-out motion of a mechanical plunger or by pneumatic means from vacuum pumps. The sample now moves along what is essentially a conveyor belt from station to station in a timed sequence. At the various stations the sample can have various reagents added to it, be stirred, heated, or simply allowed to stand. Attempts to incorporate extraction and dialysis functions to the systems have met with little success. Ultimately, at the last station, the sample is placed into a colorimeter for quantitation. In effect, the entire system merely duplicates mechanically the pre-existing manual procedures. Unfortunately, most mechanical devices are prone to mechanical failures.

## COMPUTERS IN ANALYTICAL CHEMISTRY

Automated methods for chemical analysis by discrete or continuous-flow methods represent only a partial solution to the total automation process. After automating the analytical procedure, the analyst is usually faced with the unpleasant prospect of manipulating voluminous amounts of raw data in order to turn it into useful information. In a typical case, the data must first be inspected to determine whether or not the automated system was functioning properly. The raw data must then be reduced to a form that can be used to represent the results of the analysis and, finally, the analyst must tabulate the results in formal report form. This entire time-consuming data processing and evaluation can often be offset or eliminated by the use of a computer. The computer, in this sense, can be considered as another analytical instrument that can be used together with other instruments to reduce the time involved in presenting the results of the analyses.

## COMPUTER SYSTEMS

The physical parts of a computing system are called **hardware**. In the simplest case the system consists of three distinct units: a device for input of information; a **central processing unit** (CPU) where the actual computation is performed; and a device for collecting the output information.

Input devices vary among different brands of systems and with the intended use of the computer. The most popular type are machines that gather information from holes or slots punched in paper cards or in paper tapes. Access to the computer in many other cases is by the use of a machine similar to a typewriter, known as a **teletype**. Whatever the system, the input device conveys the instructions and data to the CPU, where the actual computing takes place. After computation, the results are presented through the output device. The most popular of these are very rapid line printers and the typewriter-like devices used for the input.

A computer must of course, be told what to do, how to do it, and where and in what form to deliver the results. This task is accomplished by the input of a logically sequenced set of instructions called a **program**. The program may be written in any one of several **computer languages**, of which FORTRAN and BASIC are the most widely used by chemists. One can become proficient in their use by completion of a one-semester computer language course or by a 6-week self-study program. It is preferable for a chemist or technician to write his own computer programs for the solution of his own particular problem rather than allowing a professional programmer, unfamiliar with the chemical aspects of the problem, to write it.

There are many different digital computer systems on the market today, from massive systems costing $100,000 a month to lease to very small computer–calculator combinations costing less than $300.00 to buy outright. However, most often analytical laboratories will own a small programmable minicomputer or will utilize a time-sharing system.

The minicomputer is a small stand-alone system that will fit on a desktop or lab bench. Moderately priced (about the same as a good spectrophotometer), they may be programmed in any one of several languages. Teletypes are generally used for input and output. Most systems are modular in form so that more devices, memory, and so on, may be added as more sophistication is required. The main disadvantage of such a system is that it will handle only small to moderate-size programs and problems. The implication is, of course, that the programmer must be both efficient and imaginative.

The other system often used in the laboratory is the time-sharing system. In this case, the only equipment actually in the laboratory is an input–output device, usually a teletype, called a **terminal**. Access to the computer is by mean of a telephone line, since it may be in another part of the building or 1,000 miles away in another city. These computers are of the massive variety, with giant memories, and are able to handle simultaneously several hundred different programs from several hundred different users. The cost of using such a system is comparatively low, because payment is shared by many people. The principal advantage is the ability to handle large numbers of relatively large programs.

With the recent advances in the technology of making large-scale integrated circuits, a very small computer called a **microcomputer** has come into being. The microcomputer consists of a microprocessor in conjunction with memory chips and interfaces for input and output capability. The microprocessor is the heart of the microcomputing system. It functions in the same way as the CPU does in larger computing systems. A **microprocessor** is made by having negative and positive regions on a silicon crystal. These regions constitute combinations of diodes, resistors, capacitors, transistors, and amplifiers that can perform arithmetical and logical operations. A photoengraved film of evaporated metal serves to connect the electronic elements on the crystal. Such a microprocessor costs less then $300. Microcomputer systems are used primarily in analytical chemistry for closed-loop control systems and optimization of instrumental operating conditions. Microcomputers are able to perform these functions by comparing various experimental conditions in terms of voltages with preprogrammed standards within the microcomputer system's memory. The microcomputer then becomes an active part of the analytical process. Because microcomputers can be programmed for different situations, they exhibit much more flexibility than control units that must be attached directly to analytical instrumentation. A microcomputing system can perform arithmetic and logical operations in the execution of its program just as larger computers do to execute statistical analysis, data presentation, and so on.

## APPLICATIONS

### *Batch Processes*

The two principal ways in which an analyst uses a computer are batch processes and real-time processes. In a batch process the analyst must still gather the raw data from his experiment, check it for errors or abnormalities, and enter it into the computer. The computer then sorts the data, calculates the results, and generates a formal presentation or report of results. Actually only one-half of the total data manipulative process has been automated at this point.

Batch processes may also be used for statistical analysis of chemical data and for curve fitting. The curve-fitting methods, such as the least-squares method, are of particular interest because many calibration curves are not linear. Such calculations are very long and tedious by hand but require only microseconds for computer calculation. Batch methods are also used by chemists for numerical solutions to complex chemical equilibrium systems that may require several simultaneous mathematical equations to describe the competing equilibria accurately. Mathematical solutions of such problems without a computer become almost impossible as the complexities increase.

### *Real-Time Processes*

The next step toward complete automation is the use of real-time processes. A real-time process is one in which the computer and the analytical instrument are actually connected to each other by an **interface**. The interface is an instrument that converts the signals from the automatic chemical analyzer to an electrical code that the computer can understand. As the data are generated, the computer instantaneously gathers them, performs calculations, and prints the final answers.

• **Example.** At this point, an example of an extremely simple system that illustrates the total concept is in order. Let us assume that we wish to interface to a microcomputer the equipment for the ammonia analysis mentioned in an earlier section. After physically connecting the instrument, the problem must be defined exactly for the computer program. The output from the colorimeter in this case results in a continuous series of peaks on a recorder, which represent the absorbances of a series of known standards and unknown samples. Since the Beer–Lambert law is obeyed, the computer must calculate the ratio of each sample absorbance to one or more of the standard absorbances.

A set of instructions (i.e., a computer program) must be written to do the following things:

1 Take electrical signals from the colorimeter at a predetermined time interval (as one per millisecond).
2 Find and use only the largest in each peak and disregard the rest.
3 Differentiate between standard and sample peaks.
4 Perform the actual chemical calculations.
5 Print out the results.

The second instruction is the most difficult to program. A simple method is to have the computer compare a datum point to the one before it; if the second value is larger, forget the first and take another. The new datum point is compared to the remaining one in the same way. This process is continued until the first point is larger than the second (i.e., the highest point in the peak). The computer is told to keep this value in memory and to start the process all over for the second peak; and so on. If the standard and sample solutions were always placed on the sampling tray in the same order (e.g., one standard then ten samples, one standard, ten samples, etc.), the computer would then know to compare the ten unknown peaks in the sequence to the first and last standard peaks in the sequence. Multiple standards are used to compensate for base-line drift.

The total system will now perform the sampling, the chemical reactions, the data reduction and calculations, and finally present the results as a typed formal report. All the analyst has to do is load the sample tray, call up his program, push the "go" button, and finally remove his results from the teletype. However, a very high level of total system reliability must be achieved in order for this level of automation to be feasible.

Other systems, of course, may be interfaced to the computer. The computer can, in real time, integrate the areas of gas chromatograph peaks, reduce the data from a mass spectrograph, or perform a myriad of other duties. The highest level of sophistication is reached when the computer is allowed to interact with the analytical device itself, controlling the various parameters of the entire experiment.

This, then, is the direction in which analytical chemistry is moving. The rate of progress in the design and utilization of totally automated systems is limited only by our imagination and should accelerate rapidly in the near future.

CHAPTER

# 26 Miscellaneous Methods

## NUCLEAR MAGNETIC RESONANCE

Nuclear magnetic resonance (NMR) is used primarily to elucidate the detailed structure and configuration of organic molecules rather than for the determination of the concentration of a particular species. Although NMR can be used to determine whether or not certain nuclei are present in a sample, as well as the bonding or configuration in which these nuclei occur, it does not detect the presence of atomic nuclei that have even numbers of both protons and neutrons, such as $^{12}C$ and $^{16}O$.

### MAGNETIC PROPERTIES OF NUCLEI

NMR is based upon the fact that most nuclei appear to be spinning around an axis. This spin results in a magnetic field along the axis of rotation due to the motion of the charged nucleus around the axis, as shown in Figure 26-1. If the nucleus is placed in an external magnetic field, the nuclear field will tend to align itself with the external field. The torque exerted by the external field will then cause the nuclear field to tilt and rotate or **precess** around the direction of the external field. The precession frequency is a function of the strength of the external field; the greater that strength, the larger the rate of precession. Precession frequencies for most magnetic nuclei are between 0.1 and 60 megahertz (MHz) for fields of from 1000 to 14,000 gauss.

The rotation and precession of a nucleus in the external field can be described as being similar to the motion of a child's top, which rotates about its own axis and also precesses around the tip when the rotational axis is not vertical.

If the precessing nuclei are placed in an alternating radio-frequency field that is at right angles to the external magnetic field, the nuclei will accept energy from the radio-frequency field when the radio frequency equals the precession frequency. The effect of the radio-frequency field, in this case, is similar to the effect of striking one of two tuning forks of the same pitch that

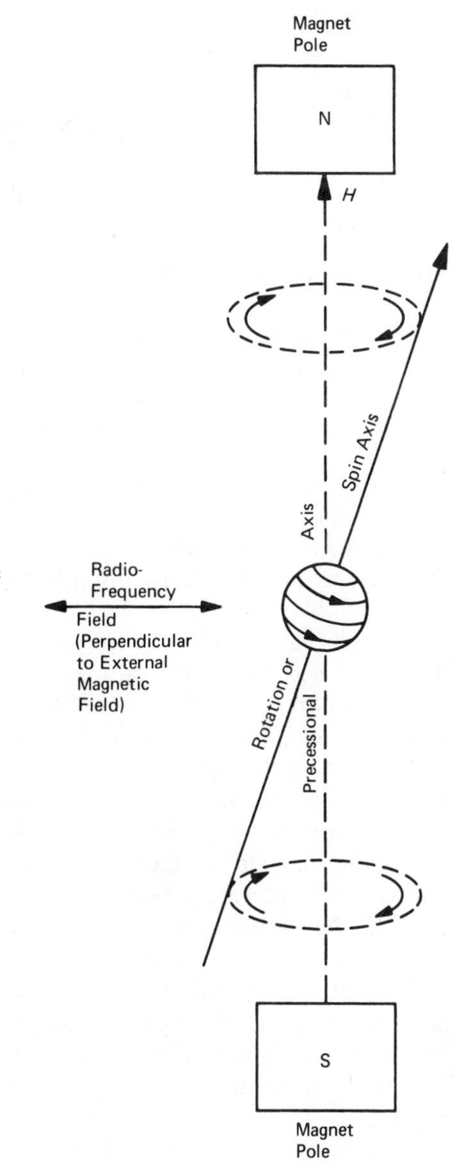


**Figure 26-1.** Effect of external magnetic field on a spinning nucleus.

are close together—when one is struck and starts vibrating, the other will also begin to vibrate. The additional rhythmic force added by the radio-frequency field will cause the nucleus to **wobble** or **resonate**. This wobbling or resonance can be detected by a receiving coil whose axis is perpendicular to both the external magnetic field and the radio-frequency field.

## NUCLEAR SPIN ENERGY LEVELS

The spin levels of each nucleus are quantized, and each nucleus is characterized by a nuclear spin number $I$, which can have values of 1/2, 1, 3/2, 2 . . . (in units of $h/2\pi$, where $h$ is Planck's constant.). For a given nucleus with spin number $I$, there are $2I + 1$ possible orientations of the nucleus in the magnetic field. Thus, for $I = 1/2$, there are two possible orientations; for $I = 1$, there are three. The energy difference ($\Delta E$) between the spin levels is given by

$$\Delta E = h\nu = \frac{\mu H}{I} \quad \textbf{(26-1)}$$

where $H$ is the strength of the magnetic field; $\nu$ is the radio frequency; and $\mu$ is the magnetic moment of the spinning nucleus. This equation can be rewritten in terms of the angular frequency ($\omega$) as

$$\frac{\omega}{H} = \frac{2\pi\nu}{H} = \frac{2\pi\mu}{hI} \quad \textbf{(26-2)}$$

The value of $\omega/H$ is a constant that is characteristic of any nuclear species having a finite value of $I$, and is known as the **gyromagnetic ratio** (sometimes symbolized as $\gamma$). The value of $\Delta E$ is very small ($10^{-2}$ cal/mole at 60 MHz), so the radio-frequency field need not be very powerful, but the detector must be quite sensitive.

The spin number of an atom may be calculated by vector addition of the individual neutron and proton spins with the restriction that protons can cancel only protons and neutrons can cancel only neutrons. Atoms such as $^{12}C$ with 6 protons and 6 neutrons, $^{16}O$ with 8 protons and 8 neutrons, and $^{32}S$ with 16 protons and 16 neutrons have $I = 0$, and consequently do not produce an NMR signal since they have only one spin state. Each of these atoms has an even number of protons and an even number of neutrons. Nuclei that have either an odd number of protons or an odd number of neutrons have $I = 1/2$ and are very sensitive to NMR. Examples are $^{1}H$, $^{15}N$, $^{17}O$, $^{19}F$, and $^{31}P$. Nuclei for which both the number of protons and the number of neutrons are odd have a spin number of 1. Examples are $^{2}H$ and $^{14}N$. Most NMR instruments are arranged to detect $^{1}H$ magnetic resonance, since hydrogen gives a strong NMR response; the other main elements present in organic compounds, carbon and oxygen, do not respond and also do not interfere with the hydrogen signal.

## CHEMICAL SHIFT

Information concerning the types of protons in organic compounds is obtained from chemical shift measurements. Since the nucleus undergoing resonance is surrounded by its own electrons and those of adjacent atoms, it will be **shielded** somewhat from the magnetic field. The shielding effect is in the range of 0.01 gauss in a field of 10,000 gauss and consequently is only a few parts per million of the external field strength. Shielding, although small, is different for different bonding conditions and will cause slight differences in the proton resonance frequency. As a result, each different type of bonding for hydrogen in the molecule will result in a slightly different resonance frequency (or a slightly different external magnetic field strength for the same frequency). The different responses can be used to determine the number of different types of protons in the molecule. The difference between the resonance frequency of an atom in two different chemical environments is called the **chemical shift** and is given the symbol $\delta$. For significant measurements, the chemical shift must be measured against a standard. The almost universal standard used today is $(CH_3)_4Si$ (tetramethylsilane or TMS), but water and benzene may also be used. The shift is usually calculated from the expression

$$\delta\ (\text{ppm}) = \frac{H_s - H_r}{H_r} \times 10^6 \qquad \textbf{(26-3)}$$

because most NMR instruments vary the external field rather than the frequency. In eq. (26-3), $H$ represents the external magnetic field strength and the subscripts $s$ and $r$ refer to the sample and reference, respectively. Compared with TMS, most proton resonances in organic compounds occur at field strengths less than $H_r$; consequently most $\delta$-values are actually negative although they are commonly written as if they were positive. Protons that are highly shielded require a higher field strength for resonance and are said to occur **upfield** with respect to less highly shielded protons; conversely, less highly shielded protons give resonances that are said to be **downfield**. Thus, the direction of the chemical shift gives information concerning the relative shielding of a nucleus in the molecule. Approximate $\delta$ values using TMS as the reference can be assigned for various molecular environments. For example, the value for $CH_3$ in $CH_3$—C is in the range 1–2, for $CH_3$ in $CH_3$—N is approximately 2–3, and for $CH_3$ in $CH_3$—O is between 3.5 and 4.5. The low-resolution chemical shift spectrum for ethyl alcohol ($CH_3CH_2OH$) is shown in Figure 26-2. The areas under the peaks have

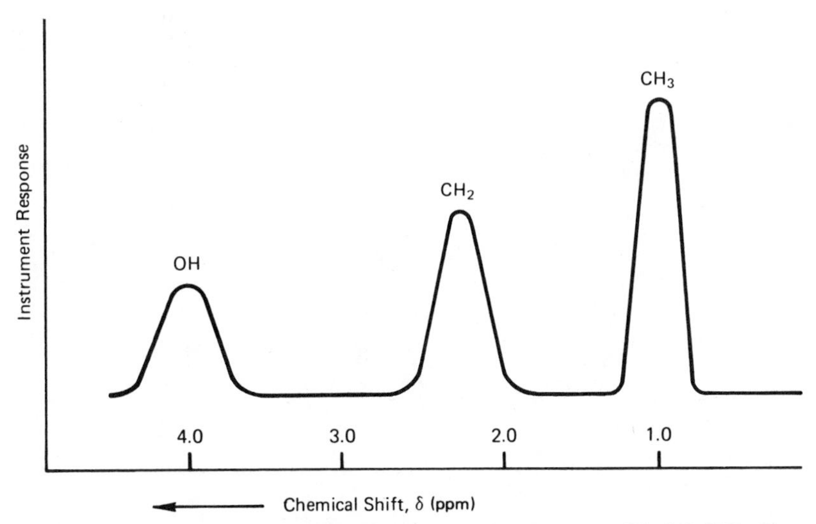


**Figure 26-2.** Low-resolution NMR spectrum of ethyl alcohol ($CH_3CH_2OH$). The area under the peaks is in the approximate ratio 1:2:3 in the order OH, $CH_2$, $CH_3$. (Not drawn to scale.)

the same ratio as the number of hydrogens in the group, so that relative areas give information concerning the number of hydrogens that have the same environment in the molecule. Note that the hydrogens in $CH_3$ are the most shielded, since the $\delta$ value taken as a positive number is the lowest, whereas the hydrogen in the OH group is less shielded, owing to the higher electronegativity of the oxygen compared to carbon.

## SPIN–SPIN SPLITTING

With the advent of high-resolution NMR spectrometers, more information concerning structure became available from NMR measurements because fine structure of the absorption bands could be observed. It was found that each of the individual peaks, as shown by low-resolution instruments, actually were split into several peaks, shown only by the higher resolution instruments.

Protons and other spinning nuclei interact with each other to cause mutual splitting of their absorption bands into multiplet structures if they are on the same or adjacent carbon atoms. This effect, known as **spin–spin splitting** or **spin–spin coupling**, seldom extends beyond protons on adjacent carbons unless there are intervening multiple bonds, and is negligible if the protons are in equivalent bonding environments. The separation between adjacent multiplet peaks is called the **coupling constant** and is usually given the symbol $J$. The value of $J$, in hertz, is independent of the frequency used to make the measurements, but is very dependent upon bond type and bond angles. An increase in bond angles causes a decrease in the coupling constant. Approximate values for the coupling constants can be assigned for various molecular environments, but rarely exceed 20 Hz.

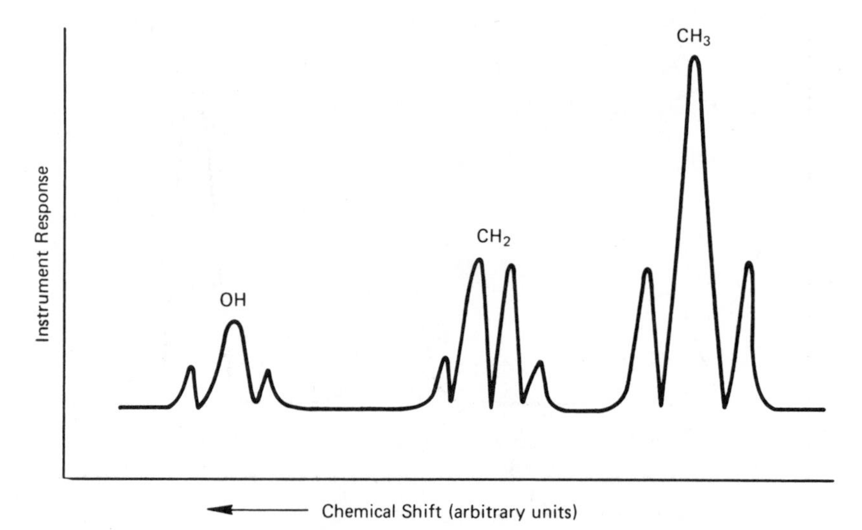


**Figure 26-3.** High-resolution NMR spectrum of ethyl alcohol. Note that OH and $CH_3$ peaks are triplets in the approximate ratio of 1 : 2 : 1 and that the $CH_2$ peak is a quartet in the approximate ratio of 1 : 3 : 3 : 1. (Not drawn to scale.)

The number of peaks into which a given signal will be split is a function of the number of hydrogen atoms causing the splitting and can be calculated as $2nI + 1$, where $n$ is the number of hydrogens causing the splitting and $I$ is the quantum spin number. A high-resolution scan of ethyl alcohol is shown in Figure 26-3. Note that the single bands shown in Figure 26-2 are split into several multiplet bands in the high-resolution spectrum. The OH and the $CH_3$ peaks are split into three peaks as a result of coupling with the $CH_2$ group, and the $CH_2$ peak divides into four peaks as a result of coupling with the $CH_3$ group. Each of the four peaks in the $CH_2$ group is further split into two peaks because of the COH group, but this effect can only be seen on high-resolution instruments. The ratio of intensities (areas under the bands) for doublets is 1 : 1, for triplets is 1 : 2 : 1, and for quartets is 1 : 3 : 3 : 1. When additional lines appear as a result of second-order splitting, complicated proton couplings may be simplified in some cases by spin-decoupling, which involves superimposing a second and relatively intense radio frequency field on the sample. Proper adjustment of the two radio-frequency fields will cause the system to be saturated for the interfering protons so that they do not interfere with the protons being studied.

## INSTRUMENTATION

The NMR spectrum can be scanned either by changing the frequency of the radio-frequency oscillator or by making small changes in the strength of the external magnetic field. Most instruments utilize a fixed radio-frequency source and a variable magnetic field. A simplified diagram of a crossed-coil type of spectrometer is shown in Figure 26-4.

The sample is usually contained in a small glass tube, which can be rotated on some instruments. The sample holder is placed between the poles of a dc electromagnet and inside the two crossed coils. The magnetic field can be varied from 0 to as high as 23,000 gauss in some instruments. The oscillating radio-frequency field is supplied by a low-power crystal oscillator and is applied at right angles or in a direction that would be perpendicular to the plane of the paper in Figure 26-4.

The resonance frequency is picked up by a second coil around the sample holder, which is perpendicular to both the external magnetic field and the radio-frequency transmitter coil. When nuclear transitions are induced in the sample, energy is absorbed, causing a decrease in the signal received by the receiver coil. This change is detected and amplified by a high-gain detector tuned to the same frequency as the transmitter. The resultant signal is fed to the readout, which is usually a strip-chart recorder or an oscillograph.

The magnetic field is usually set to a value close to that needed and the field is varied by the use of coils wound around the poles of the magnet so that a narrow sweep may be made through that portion of the magnetic field which produces resonance frequencies in the range of the anticipated precession frequencies.

Many instruments are equipped with an electronic integrator that will trace a second curve composed of a series of steps whose height is proportional

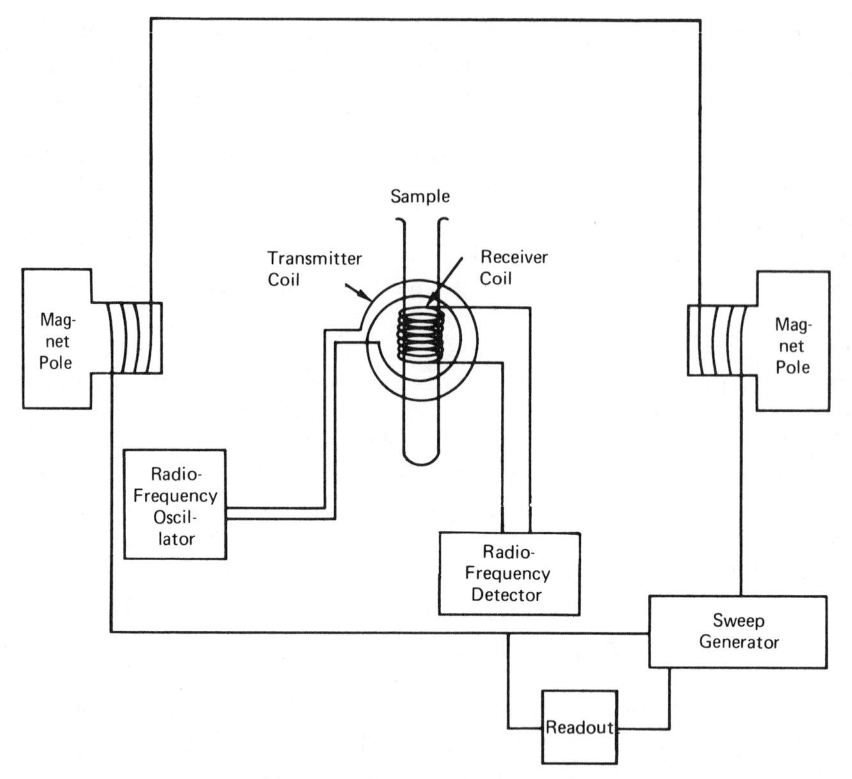


**Figure 26-4.** Simplified diagram of an NMR spectrometer.

to the number of nuclei under the respective peaks. The difference in height between any two steps is related to the number of protons causing the increase in elevation and can be used for quantitative estimation if the empirical formula is known.

The design requirements for high-resolution NMR spectrometers are difficult to attain, a factor that contributes to the relatively high cost of high-resolution instruments. The magnetic field must be uniform over a volume large enough to contain the sample, placing rigorous limitations on the construction and size of the magnet faces. Also, the homogeneity of the field in the critical area must be maintained to within approximately 1 part in 60 million for adequate high resolution. To maintain this degree of homogeneity, specially shaped windings that are powered by an adjustable dc current are usually required.

## APPLICATIONS

The primary applications of NMR are in the field of structure determination and delineation. There are libraries of NMR spectra comparable to those for optical absorption spectra and x-ray diffraction spectra. Comparison of the NMR spectrum of an unknown compound with those in the library may lead directly to identification of the unknown. Even if this is not the case, the NMR spectrum can often help distinguish between two possible isomeric structures and in many cases can furnish the critical information that the analyst needs to delineate the exact structure of a complicated compound.

Integration of the NMR spectrum furnishes information that can be used for quantitative estimation of one or more components present in a mixture. It can also be used to determine the percentage of hydrogen in a pure compound with a relative precision of $\pm 0.5\%$.

# MASS SPECTROMETRY

Mass spectrometry is based on the principles (1) that the velocity of a charged particle is affected by an electrical field, (2) that the direction of a charged particle is affected by a magnetic field, and (3) that both effects are functions of the **mass/charge ratio** ($m/e$) of the particle. The **mass spectrometer** is an instrument in which molecules of a substance are broken down into ions by an **ionizing beam** of electrons, and the ions produced are sorted according to their mass or mass/charge ratio. The resulting record is known as a **mass spectrum** and is characteristic of the compound and the conditions of ionization and measurement. This technique provides information concerning the molecular weight of the compound and the weights of various portions of the compound; thus, it is a useful tool in the determination of structure. It can also be used as the basis of qualitative identification of pure compounds and of quantitative determination of the composition of mixtures.

Several types of mass spectrometers are available for use in analysis and identification, but the three most popular are the electromagnetic mass spectrometer, also known as the magnetic deflection mass spectrometer, which can be arranged for single, double, or cycloidal focusing; the time-of-flight mass spectrometer; and the quadrupole mass spectrometer. In all of these there are essentially three separate and distinct components: an **electron source**, together with an **ion accelerator**; an **analyzer** or **flight tube**; and an **ion detector**. The principles of operation differ; the magnetic deflection mass spectrometers depend on deflection of ions in a magnetic field, the time-of-flight instruments separate ions on the basis of ionic velocities, and the quadrupole method is based on the effect of direct current and radio-frequency fields upon the lateral movement of the ions. Most spectrometers measure positive rather than negative ions because collisions between electrons and molecules usually produce more positive than negative ions. This is due to the fact that many collisions result in the loss of an electron by a molecule or neutral fragment rather than production of two oppositely charged ions.

## INSTRUMENTATION

### *Magnetic-Deflection Mass Spectrometers*

A simplified schematic diagram of a typical 180° mass spectrometer is shown in Figure 26-5. Although gaseous samples may be introduced directly into the

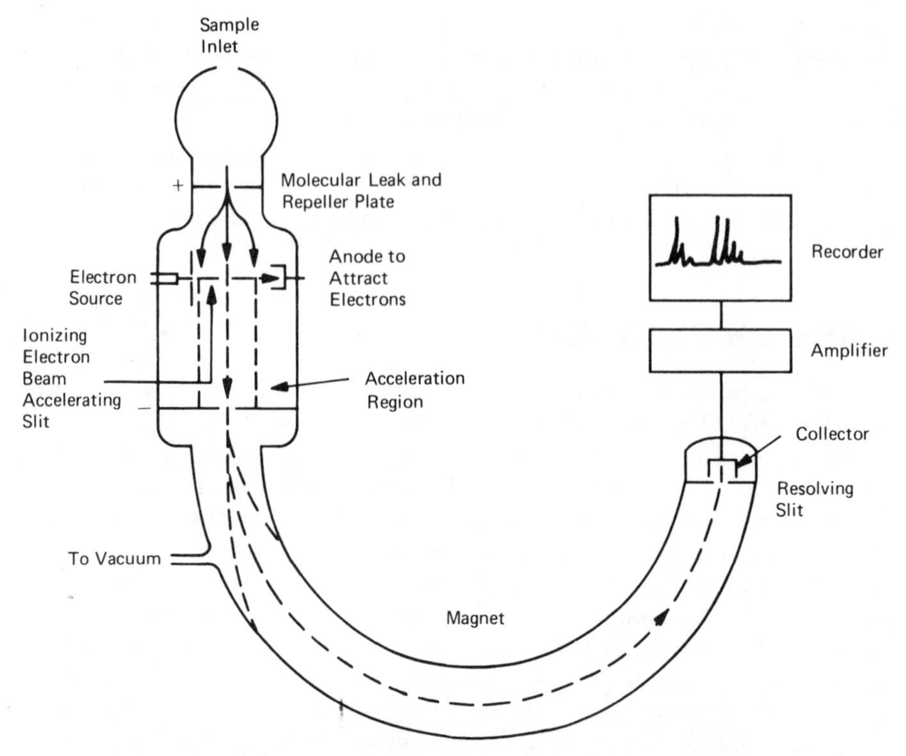


**Figure 26-5.** Simplified diagram of a 180° mass spectrometer.

mass spectrometer through a small aperture or **molecular leak**, solids and liquids (with a sufficiently high vapor pressure to vaporize without decomposition) are usually introduced from a heated sample inlet. Molecules entering through the molecular leak are broken into fragments and/or ionized by the stream of electrons in the ionizing beam. The charged particles thus formed are accelerated by the accelerating potential of 6–100 V applied between the positive repeller plate (molecular leak) and the accelerating slit, but move in straight lines so that only those ions formed directly in line with the slit will pass into the magnetic deflection tube. Ions formed in other regions of the ionizing beam will be neutralized by the accelerating slit plate upon contact.

Ions passing through the slit will have kinetic energy given by eq. (26-4)

$$\tfrac{1}{2}mv^2 = eV \qquad \textbf{(26-4)}$$

where $m$ is the mass of the ion; $v$ is the velocity of the ion; $e$ is the charge on the ion; and $V$ is the accelerating potential. The ions leaving the electrical field through the slit enter the magnetic field and experience a force that causes them to move in a curved path. The radius of this curvature may be obtained by equating the expressions for the centripetal and centrifugal forces exerted on the ion:

$$\frac{mv^2}{r} = Hev \qquad \textbf{(26-5)}$$

where $H$ is the strength of the magnetic field and $r$ is the radius of curvature of the ion. Substitution of the value of $v$ from eq. (26-4) into eq. (26-5) gives

$$\frac{m}{e} = \frac{H^2r^2}{2V} \tag{26-6}$$

or

$$r = \sqrt{\frac{2V}{H^2} \times \frac{m}{e}}$$

which shows that the radius of curvature depends upon the accelerating potential, upon the strength of the magnetic field, and upon the ratio $m/e$. In practically all magnetic-deflection spectrometers, $r$ is fixed so that only particles with a given $m/e$ ratio will pass through the resolving slit for given values of $H$ and $V$. Either $H$ or $V$ can be varied continuously to scan the ions of different $m/e$ ratio formed, but in most spectrometers the magnetic field is kept constant and the voltage is varied. Ions that do not have the $m/e$ ratio focused on the resolving slit at a given potential will strike the walls of the magnetic field tube and be neutralized and pumped away by the vacuum.

The current produced by the ion beam striking the collector is a measure of the number of those ions produced by the ionizing beam and gives information as to the relative ratios of each type of ion produced. Owing to the large number of sample molecules introduced into the ionizing beam, the laws of probability hold, so the dissociation fragments will always occur in the same relative abundance for a given compound under a given set of conditions. This is the basis for the use of mass spectrometry in the quantitative analysis of mixtures and in the qualitative identification of unknown compounds.

The current produced in the collector is amplified and fed to the recorder, which records the mass spectrum. Most instruments record with five different sensitivities simultaneously in the ratio of 1, 3, 10, 30, and 100, so that all the spectrum peaks (which may vary in intensity by a factor of 50,000) can be estimated more accurately.

The resolution of a mass spectrometer is a measure of its ability to distinguish between two different mass numbers and is defined as the mean mass of the two numbers divided by the difference between them. Thus, separation of particles with $m/e$ of 200 from particles with $m/e$ of 201 requires a resolution of 200, whereas separations of particles with $m/e$ of 999 and 1001 requires a resolution of 500. Modern instruments can handle compounds with molecular weights up to 2000 with resolution in the range 1000–2000. Arrangements that allow double focusing of the ion beam greatly improve both the $m/e$ range and the resolution.

### *Time-of-Flight Mass Spectrometers*

Time-of-flight mass spectrometers do not utilize a magnetic field to change the direction of charged particles but depend upon the measurement of the time it takes an ion to move a fixed distance after acceleration by an applied electrical field. The kinetic energy of the particles is the same as in the magnetic-deflection mass spectrometer, or as given by eq. (26-4). Since all the ions fall through the same potential drop, they will all have the same kinetic energy, and eq. (26-4) may be rewritten

$$\frac{m}{e} = \frac{2V}{v^2} \tag{26-7}$$

$$= \frac{1}{v^2} \times 2V = \frac{1}{v^2} k_1 \tag{26-8}$$

where $k_1$ is twice the value of the accelerating potential. As a consequence, the velocity of the ion is related to the $m/e$ ratio at constant potential and since

$$v = \frac{d}{t} \tag{26-9}$$

where $d$ is distance and $t$ is time, eq. (26-8) may be shown as

$$\frac{m}{e} = \frac{t^2}{d^2} k_1 \tag{26-10}$$

Time-of-flight spectrometers utilize a fixed distance for ion travel, so that $d$ in eq. (26-10) is a constant. Rearrangement of eq. (26-10) gives

$$t = k_2 \sqrt{\frac{m}{e}} \tag{26-11}$$

where $k_2$ equals $d^2/2V$ or $d^2/k_1$. This shows that the time of flight for fixed distances under a given accelerating potential is proportional to the square root of the $m/e$ ratio, which is the basis of the mass spectra produced by time-of-flight instruments.

Figure 26-6 is a simplified schematic for a time-of-flight mass spectrometer.

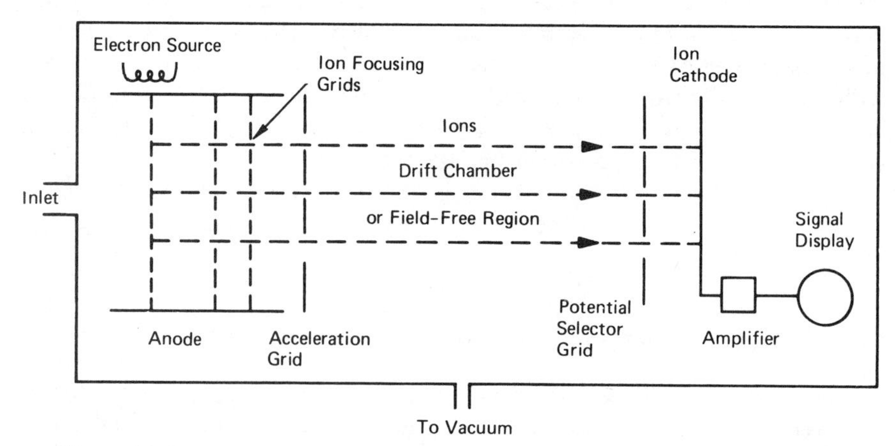


**Figure 26-6.** Simplified schematic diagram for a time-of-flight mass spectrometer.

The sample molecules enter the inlet and are ionized by a pulsed electron stream. The charged particles pass through focusing grids and an accelerating grid into the field-free evacuated drift chamber. After traversing the drift chamber they strike the ion cathode receiver. The current produced is amplified and fed to a signal display, which is usually an oscilloscope. The oscilloscope traces are photographed on Polaroid film, which can be developed rapidly to obtain the complete mass spectrum in a very short length of time.

The ionizing beam from the electron source is pulsed by energizing the beam for 0.25 $\mu$sec every 100 $\mu$sec, and the charged particles, as a result of the accelerating potential, attain velocities that are a function of their $m/e$ ratios, with the lightest ions attaining the highest velocities. Owing to the pulsing of the ionizing beam, bursts of particles enter the drift chamber and separate into **bunches**, since the faster moving particles pull ahead of the slower-moving particles. The arrival of each bunch of ions produces a current in the detector; this is fed to an oscilloscope whose time base is synchronized with, and triggered by, the original electron beam pulses.

Since a complete mass spectrum can be obtained in a few microseconds, the time-of-flight spectrometer is excellent for kinetic studies in fast reactions and can be used to analyze individual effluent peaks in gas chromatography. Resolution is usually in the neighborhood of 200.

### *Quadrupole Mass Spectrometers*

The quadrupole mass spectrometer, shown in Figure 26-7, is another instrument by which ions can be separated according to their $m/e$ ratio without the

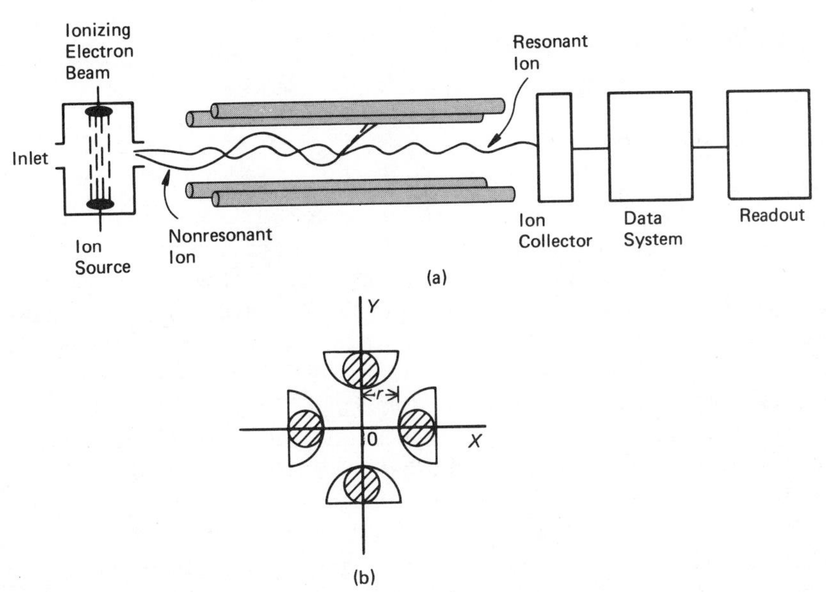


**Figure 26-7.** (a) Schematic diagram of a quadrupole mass spectrometer. (b) Geometry of rods in the quadrupole mass spectrometer.

need of a magnet. This technique uses four parallel rods so positioned that their axes are at the corners of a square and the ion beam shoots down the center of the system. The rods can be either hyperbolic or cylindrical but must be precisely straight and parallel. Diagonally opposite rods are connected together electrically, the two pairs to the opposite poles of a dc source and simultaneously to a radio-frequency oscillator.

Since neither the dc nor the ac field has a component parallel to the rods, the fields have no effect on the speed but do have an effect on the lateral movement of the ions. Most of the fragments introduced into the quadrupole will oscillate with increasing amplitude and strike one of the electrodes, but one particular $m/e$ ratio can pass completely through the system and be detected. Mass scanning can be accomplished by varying the radio frequency or the magnitude of the radio-frequency and dc voltages. Since no detector slit is necessary, a relatively large number of ions can pass through to the ion collector. The absence of a magnet decreases the size and cost, which allows the quadrupole to be used in many gas chromatograph–mass spectrometer combination instruments. The solution of a series of rather complicated equations shows that, for singly charged ions, the $m/e$ can be calculated from

$$\frac{m}{e} = \frac{0.136V}{r_0^2 f^2} \qquad \textbf{(26-12)}$$

where $V$ is the accelerating voltage, $r_0$ is the rod separation in centimeters, and $f$ is the radio frequency in megahertz. Resolution is usually in the range of 500.

## MASS SPECTRA

### *Electron Impact Ionization*

The fragments that are formed when molecules of a compound are bombarded with high-speed electrons is a function of the molecular structure and the relative strength of the bonds between the atoms. Molecular or **parent ions** are caused by the loss of an electron from the molecule and can be produced with relatively low energy electron bombardment; however, as the energy of the electron beam is increased, the parent ion will be formed with large amounts of excess energy. This excess energy is rapidly and evenly distributed throughout the molecule, and, as soon as the energy is sufficient to exceed the dissociation energy of some bond in the molecule, that bond will be ruptured and fragment ions will be produced.

The greater the excitation energy, the greater the number of fragments produced. At electron energies on the order of 50–70 eV, fragment ions appear in abundance. The mass spectrum of a molecule shows all the fragment ions and the relative abundance of each. Since sufficient molecules are present for the laws

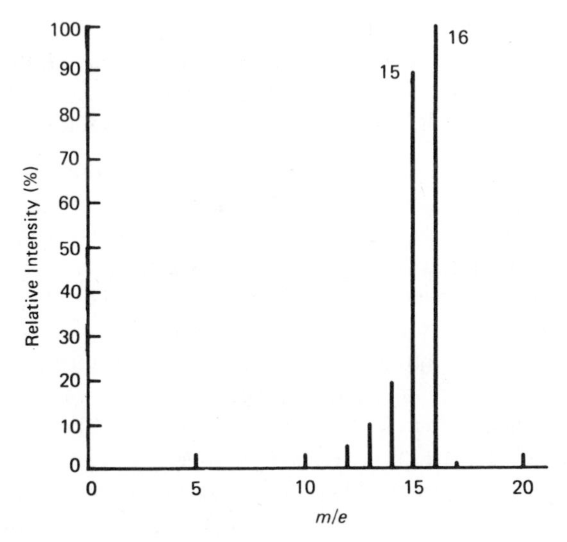


**Figure 26-8.** Mass spectrum of methane.

of probability to hold, the **fragmentation** or **cracking pattern** is characteristic of the compound and serves as a "fingerprint" for its identification. Mass spectra obtained on different instruments under different energy conditions may show slightly different fragmentation patterns, but the pattern will always be the same on the same instrument when the conditions are reproduced.

The mass spectrum of methane ($CH_4$) is shown in Figure 26-8. This very simple spectrum illustrates several important characteristics of mass spectra:

1 The intensities of the peaks are usually expressed as relative intensity; that is, the most intense peak in the entire spectrum is assigned a value of 100 and the other peaks are expressed as a percentage of this value.

2 the *m/e* ratio is often referred to as the **mass number** rather than the *m/e* ratio value, since most peaks are for singly charged ions. Doubly charged ions may occur and cannot be distinguished from the singly charged ions unless the mass of the fragment is odd or an instrument with very high sensitivity is used.

3 The peaks in mass spectra usually appear in **clusters** or groups with different intensity lines separated by one or two mass numbers. The most intense line in any cluster is called the **major** peak.

4 The major peak in the cluster at the largest mass number is usually the parent peak or **parent mass**. However, for all organic compounds and for some inorganic compounds, there will be one or more small **isotope peaks** at higher mass numbers than the parent mass owing to normal isotopic distribution of the elements present. The parent peak is seldom the most intense peak in the entire spectrum, except for simple compounds, and may not appear at all in the spectra of high-molecular-weight compounds or of compounds that contain an easily ruptured bond.

5 The relative intensities of the major peaks in the various clusters may be used as an indication of the bond strength (actually the probability of

rupture) of the bond or bonds ruptured to form the fragment. This must be confirmed by other measurements, such as **appearance potentials** for accurate prediction of bond strength. The appearance potential is the potential at which the peak appears as the voltage on the electron source is increased.

In the methane spectrum, the small peak at mass 17 is the isotropic peak due to $^{13}C$; the most intense peak at mass 16 is the parent peak; and the fragment peaks at 12, 13, 14, and 15 represent the loss of 4, 3, 2, and 1 hydrogens, respectively. Note that the relative intensities decrease from mass 15 to mass 12 because of the greater energy required and the fact that the probability of rupture decreases with each additional hydrogen lost.

The isotopic peaks can often give considerable information concerning the number of a particular type of atom in the compound or fragment because the intensities in the cluster due to isotopic distribution will be in the same ratio as the probability of the various isotopes occurring in the parent ion or fragment. For example, consider the peak at *m/e* 17 in the methane spectrum. This peak is approximately 1.1 % of the peak at *m/e* 16. Since the $^{13}C$ in natural compounds always represents 1.1 % of the $^{12}C$ present, the number of carbons may be found by dividing the relative intensity of the isotope peak due to $^{13}C$ by 1.1. In the case of methane, the peak indicates that only one carbon is present.

Similarly, the number of atoms of chlorine in a molecule or fragment can be determined from the relative intensities of the peaks in the clusters, since the natural isotopic ratio of $^{35}Cl$ to $^{37}Cl$ is 100 : 33. The calculated relative intensities of the peaks for parent ions or fragments containing from one to seven chlorine atoms is shown in Figure 26-9. The spectra of trichlorosilane and silicon tetrachloride in Figure 26-10 and of the methyl chlorosilanes in Figure 26-11 are examples. Note that the peaks in each fragment show the correct pattern for chlorine isotopes and that there are smaller peaks between

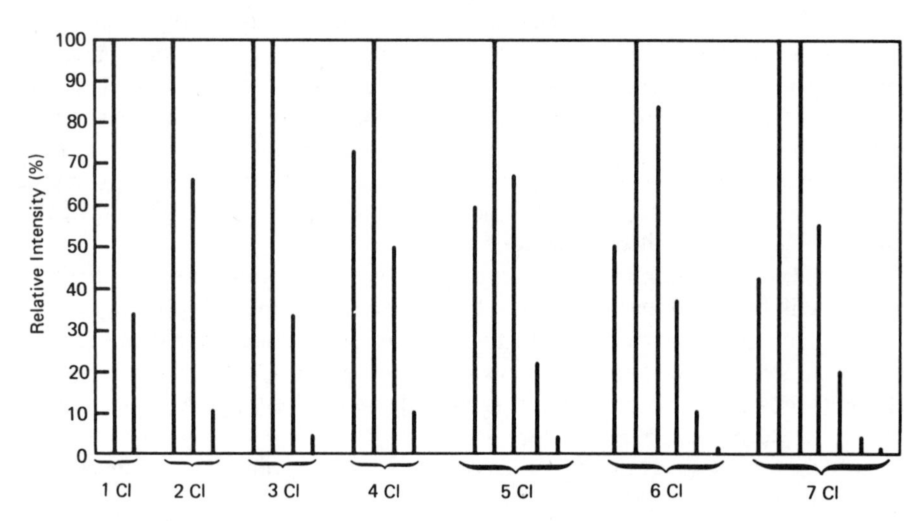


**Figure 26-9.** Multiple chlorine atom peak intensities. (From K. R. Burson, Master's Thesis, Southern Methodist University, 1968.)

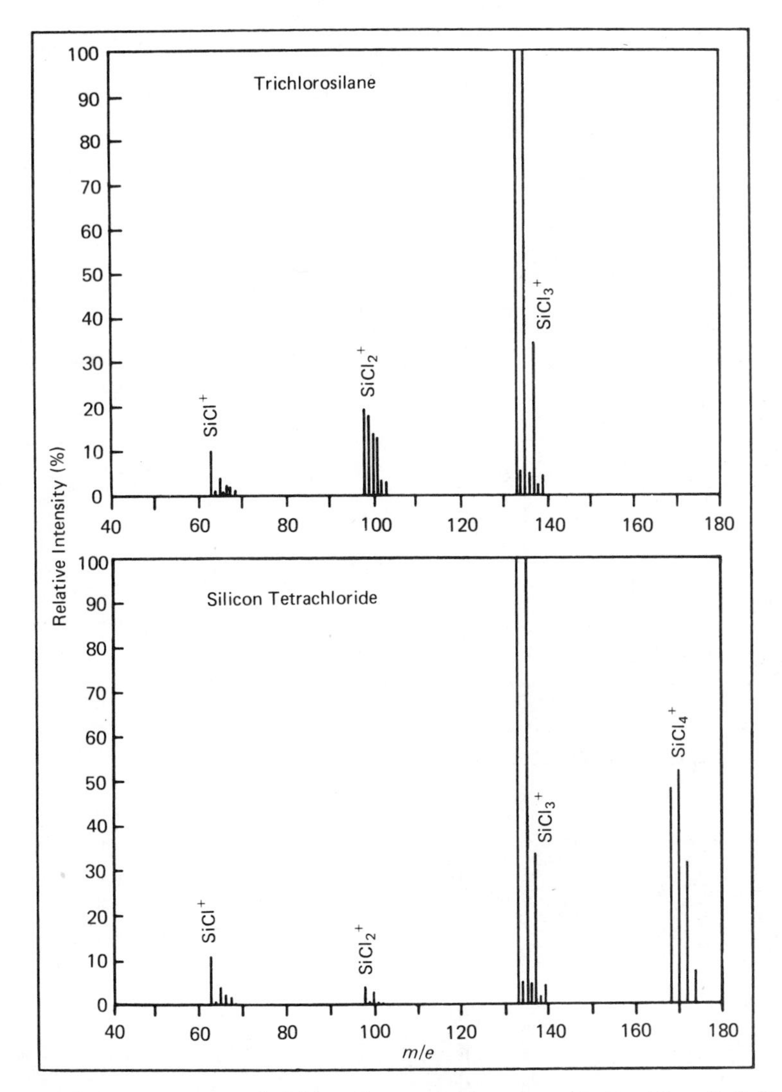


**Figure 26-10.** Mass spectra of trichlorosilane and silicon tetrachloride. (From K. R. Burson, Master's Thesis, Southern Methodist University, 1968.)

these "pattern" peaks. These smaller peaks are due to the presence of the $^{29}Si$ isotope or the $^{13}C$ isotope.

## *Chemical Ionization*

If the source pressure is maintained at a much higher pressure than the rest of the mass spectrometer and methane gas is introduced into the electron beam of the source, ion–molecule reactions will occur to form heavier ions, such as $CH_5^+$, $C_2H_5^+$, and $C_3H_5^+$. These ions can then be used to bombard the sample

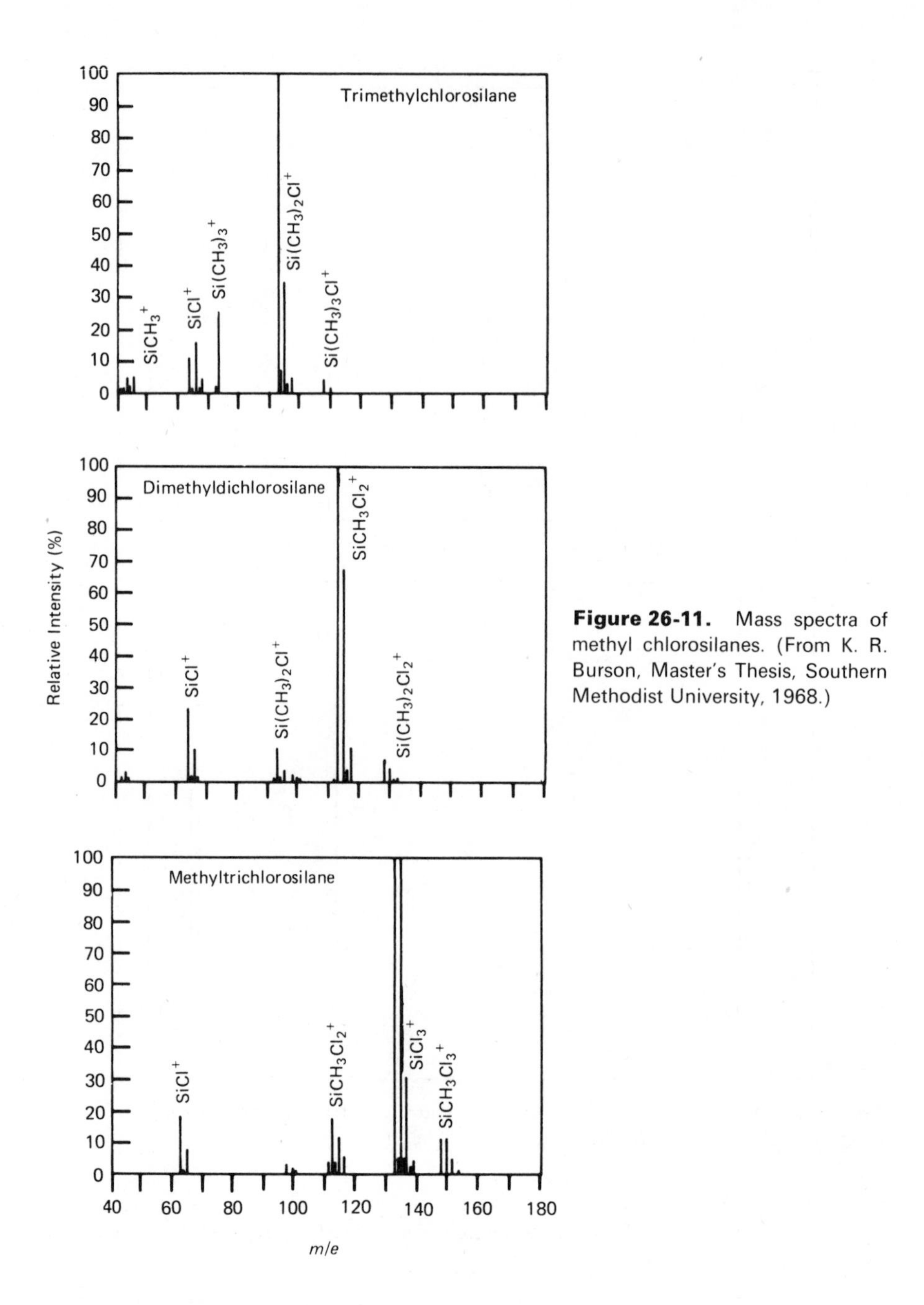


**Figure 26-11.** Mass spectra of methyl chlorosilanes. (From K. R. Burson, Master's Thesis, Southern Methodist University, 1968.)

and cause ionization by means of ion–molecule interactions. Since $CH_5^+$ is the most abundant, most interactions will form $(M + 1)^+$ ions, but $(M + 29)^+$ and $(M + 41)^+$ ions are also formed. The mass spectrum produced by chemical ionization is always simpler than that produced by **electron impact** (EI) methods. The appearance of the $(M + 1)^+$ peak simplifies determination of the molecular weight of the compound, especially in those cases where there is no parent peak in the EI spectrum.

## CORRELATION OF MASS SPECTRA WITH STRUCTURE

One big advantage of mass spectrometry is that it not only gives information concerning the molecular weight but also offers methods of differentiating between the structural isomers of organic compounds. As an example, both butane and isobutane have the same molecular weight and show the same parent peak. They also break down into fragments that have the same *m/e* ratios but occur in different relative percentages. Both produce mass 15 peaks ($CH_3^+$), mass 29 peaks ($CH_3CH_2^+$), mass 43 peaks [$CH_3CH_2CH_2^+$ or $(CH_3)_2CH^+$] and the parent peak at mass 58. One of the distinguishing features is that the $CH_3{}^+$ peak is larger in isobutane than in normal butane, since there are more $CH_3$ groups that can be ruptured. The $C_3H_7^+$ peak is also larger, for the same reason.

Peaks at certain mass numbers furnish indications as to structure. For example, mass 43 ($C_3H_7^+$) is large for linear hydrocarbons containing more than four carbons, whereas masses 31, 45, and 59, representing various numbers of $CH_2$ groups attached to OH, indicate the presence of oxygen as an alcohol or an ether. Mass 77 is present when the compound contains a benzene ring. A few general rules for prediction of fragmentation patterns include:

1 Rupture is favored at branched carbon atoms, with the positive charge remaining with the branched fragment.

2 Rupture at the beta position is probable for double bonds and hetero atoms.

3 Ring compounds show peaks at mass numbers characteristic of the ring, such as the peak at 77 for the benzene ring.

4 Saturated rings lose side chains at the alpha position.

5 Compounds containing a carbonyl group tend to rupture at the carbonyl.

6 For compounds containing Cl, Br, Si, and so on, the characteristic patterns due to isotropic distribution aid in determining the number of such atoms in each fragment, and this gives information as to possible structural units.

In any case, the proper procedure after obtaining the mass spectrum is to compare the information obtainable from it with information from other sources. It is always a good plan to write possible structures for each fragment and possible structures of the compound from this information. Such procedures often lead to deduction of the correct structure by eliminating various other possible structures as not being capable of forming the fragments actually found. General rules for organic structures that are often helpful include the fact that compounds of even molecular weight, which contain only C, H, O, and N, contain even numbers of hydrogen atoms; if the molecular weight is divisible by 4, the number of hydrogens is also divisible by 4.

## ANALYSIS OF MIXTURES

Probably the most important aspect of mass spectrometry from the analyst's point of view is the capability of quantitative determination of the composition of complex mixtures, especially of organic compounds. The system employed in mass spectrometry is essentially the same as that used in infrared or ultraviolet absorption photometry, in that the spectra of the pure compounds must be determined first. An inspection of the mass spectra of all the compounds present in the mixture leads to the selection of peaks that have adequate intensity and show maximum freedom from interference from other components present. If possible, monocomponent peaks (parent ion peaks if low-voltage electron beams are used) are selected. Because this is seldom possible for all the components, simultaneous equations are set up and solved for the correct composition.

As an example of peak selection, consider the analysis of a mixture of the three chlorosilanes shown in Figure 26-11. The parent peak at 148 for methyltrichlorosilane and the $Si(CH_3)_3^+$ peak at 73 for trimethylchlorosilane are both monocomponent, but both have relatively low sensitivity. The $SiCl_3^+$ peak at 133 in the methyltrichlorosilane spectrum, the $SiCH_3Cl_2^+$ peak at 113 for dimethyldichlorosilane, and the $Si(CH_3)_2Cl^+$ peak at 93 in the spectrum of trimethylchlorosilane are all intense peaks and probably would be selected, even though there would be some interference in the two peaks at 93 and 113. The interference could be calculated fairly easily because the 133 peak is a monocomponent peak.

For complex mixtures, the solution of the many simultaneous equations necessary is a long and tedious task unless a high-speed digital computer and a satisfactory program are available. As a consequence, most mass spectrometers used for analysis feed the peak information directly into a computer, and the computer makes the necesaary calculations and then presents the answer on a high-speed typwriter or linear printer in a matter of only a few minutes. There are many examples of this type of calculation in the literature, so no specific examples will be given here.

The use of on-line mass spectrometers for continuous monitoring of liquid or gaseous streams in oil refineries is a good example of the usefulness of mass spectral analysis. In this application, the mass spectrometer is set next to the line in which a particular mixture is being carried from one step in the distillation or manufacturing process to the next step. In most cases, the composition is critical and must be held within certain narrow limits for optimum production of the desired product. Periodically, the mixture is sampled and led through the mass spectrometer, which is adjusted to the correct parameters. The data are fed into a small computer, which calculates the result. If the composition is outside of specification, the computer will either ring a bell or light a light on a control panel to indicate that something is wrong. In some cases the computer is programmed to correct the condition producing the out-of-specification material automatically without operator intervention.

## OTHER APPLICATIONS

Mass spectrometers are useful in many ways other than the estimation of molecular weights, the determination of structure, the qualitative identification of compounds, and the analysis of complex mixtures. One of these uses is as a

**leak detector** for high-vacuum systems. The instrument is set to determine helium only and is then connected to the vacuum system. A fine jet of helium is sprayed around the parts suspected of leaking; the mass spectrometer then detects the entrance of the helium into the system. Very small leaks can be detected in this way, and several leaks can be located without shutting off the instrument to repair each as it is found.

The original use of mass spectrometers was in the determination of the atomic weights and isotopic distribution of elements. The information obtained by Thompson and Aston on a mass spectrometer between 1910 and 1920 was very important in the early days of the development of the atomic theory, especially in the determination of accurate atomic weights. The use of the mass spectrometer to separate fissionable $^{235}U$ from ordinary $^{238}U$ during the early research on nuclear reactors and atomic bombs is well known to most school children, even though the instrument is not usually referred to as a mass spectrometer is most elementary books.

By analogy with the use of radioactive isotopes in the study of metabolic reactions (which led to many fundamental discoveries in chemistry), mass spectrometry permits the use of stable isotopes as tracers in biochemical and other studies. It can also help delineate the course of many organic reactions by determining the composition of intermediates. This is especially true today, since mass spectrometers can now be hooked directly to the effluent tube of a gas chromatograph to analyze each effluent peak as it emerges from the column. The combination of the mass spectrometer and the gas chromatograph has been possible only in the recent past with the advent of rapid-scan instruments. Most of the early instruments required from 20 min to 1 hr to scan the mass spectrum from 0 to 200, but some of the present spectrometers can scan the entire spectrum from 0 to 2000 in less than 1 sec.

The development and characterization of ultra-high-vacuum systems requires a knowledge of the total pressure of the residual gases and of the partial pressures of each of the individual gases present. Special mass spectrometers have been developed which permit such measurements in ultra-high-vacuum systems down to as little as $10^{-16}$ torr, a value that corresponds to approximately one molecule/cm$^3$. Portable instruments that are compatible with ultra-high-vacuum systems and can be connected directly to the system are being developed in the solid-state electronics industry where extremely high vacuums are necessary, both in research and in production.

## RADIOACTIVE METHODS

After the accidental discovery of radioactivity by Becquerel in 1896, practically all work with radioisotopes involved the use of the heavy naturally occurring radioactive elements such as uranium and radium until the advent of the cyclotron in 1932 and the atomic pile in 1942. The atomic pile or nuclear reactor is now used to produce a large number of radionuclides of the lighter elements in sufficient quantities and purity so that radioactive methods today

**TABLE 26-1.** Selected Beta-Emitting Radioisotopes

| Radio-isotope | Half-life | Energy of Radio Products *(MeV) | |
|---|---|---|---|
| | | Beta | Gamma |
| $^{3}H$ | 12.3 yr | 0.02 | none |
| $^{14}C$ | 5720 yr | 0.16 | none |
| $^{24}Na$ | 15.0 hr | 1.39 | 1.37 |
| $^{32}P$ | 14.2 days | 1.71 | none |
| $^{35}S$ | 87 days | 0.17 | none |
| $^{42}K$ | 12.4 hr | 3.55 | 1.52 |
| $^{56}Mn$ | 2.6 hr | 2.81 | 2.06 |
| $^{59}Fe$ | 45.1 days | 0.46 | 1.30 |
| $^{60}Co$ | 5.3 yr | 0.32 | 1.33 |
| $^{131}I$ | 8.1 days | 0.61 | 0.36 |
| $^{203}Hg$ | 47 days | 0.21 | 0.28 |

*Only the most energetic and/or the most abundant are listed for those elements that produce more than one beta or gamma.

use more artificial than naturally occurring radioactive isotopes. A few of the more important man-made radioisotopes widely used in analytical procedures or as sources of radiation are shown in Table 26-1.

## RADIOACTIVE DISINTEGRATIONS

The nuclei of naturally occurrring radioisotopes decompose by emission of either an **alpha** or a **beta particle** and the emission may or may not be followed by **gamma radiation**. All these disintegration products are different, and consequently involve different methods of detection and determination. The beta particle emitted by the naturally occurring radionuclides is a high-speed, energetic electron but some of the man-made isotopes emit a **positron** or positive electron, which is also known as **beta radiation**. The alpha particle is the nucleus of a helium atom, whereas gamma radiation is electromagnetic energy with a high frequency and short wavelength. The negative beta particle is sometimes referred to as a **negatron** to distinguish it from the positron. Analytically, the negatron is much more important than the positron, since the majority of radionuclides utilized in analytical procedures are negatron emitters.

Alpha particles have a nuclear charge of 2 positive and a mass of 4 **atomic mass units (amu)** and consequently have high energy and high ionizing power but poor penetration power. An alpha particle of 5 MeV will be stopped by 3.5 cm of air but will produce 25,000 ion pairs/cm of travel. The beta particle has a nuclear charge of either 1 positive or 1 negative and essentially negligible mass (approximately 0.0005 amu). As a result, beta particles have less energy and less ionizing power than alpha particles but much greater penetrating power. A negatron of 0.5 MeV has a penetrating power of approximately 1 m in air and will produce approximately 60 ion pairs/cm. Gamma radiation, since it is electromagnetic energy, has the greatest penetrating power and the lowest ionizing power. Certain nuclei can emit **neutrons**, which have a mass of 1 amu but no charge and consequently have high penetrating power but essentially no ionizing power.

The rate of radioactive decay is an important property that must be taken into consideration in analytical procedures. The rate of decay depends upon the decay constant $\lambda$ and is usually expressed in terms of the half-life $t_{1/2}$, which is defined as the length of time necessary for one-half of any given sample to decompose. The half-life is related to the decay constant by the expression

$$t_{1/2} = \frac{0.693}{\lambda} \qquad \textbf{(26-13)}$$

Half-lives range from microseconds for some nuclei to billions of years for others. The half-lives at either extreme are not satisfactory for analytical purposes, since the elements with very short half-life would be completely disintegrated before the experiment could be completed whereas those with an excessively long half-life do not produce sufficient disintegrations for adequate measurement over short periods of time. The effective life for most analytical purposes is approximately 10 half-lives, so most radionuclides used in analysis have half-lives from a few hours to a few thousand years. For the shorter-lived elements, corrections are often applied to account for disintegration that occurs during the course of the experiment.

## INTERACTION OF RADIATIONS WITH MATTER

Radioactive disintegrations are detected by the interaction of the disintegration products with matter, and each type of disintegration product interacts differently. Alpha and beta particles interact primarily to cause ionization of the molecules or atoms through which they traverse, even though some of the molecules may be dissociated into neutral fragments or simply raised to an excited energy state. Gamma rays can lose energy by the photoelectric effect, the Compton effect, and pair production. Each of these mechanisms will be discussed separately.

The **ionization of molecules** by alpha and beta particles is utilized in the detection and measurement of these types of decomposition. The absorption of beta particles by matter follows an exponential relationship up to a certain distance, at which the absorption becomes infinite. This distance is known as the **range** of the beta radiation and, for energies above 0.6 MeV, the range/maximum energy relationship is given by

$$\text{range} = 543E_{\text{max}} - 160 \qquad \textbf{(26-14)}$$

where the range is given in milligrams per square centimeter of absorber. Since the distance of travel is a function of the density of the absorbing medium, the range is often expressed in terms of weight per unit of surface area because density equals weight per unit volume and volume equals area times thickness. This allows comparison between absorbers of widely different densities. The range is usually determined by inserting various thicknesses of aluminum between the sample and the detector.

Alpha particles lose energy by the same mechanisms as beta particles but produce a much larger number of ion pairs per unit distance of travel. Owing to their relatively large mass and charge, alpha particles lose only a small fraction of their energy with each ion pair produced, so that essentially all the particles produced by nuclei of a given isotope will travel the same distance in an absorber.

The **photoelectric effect** is an important method for loss of energy by low-energy gamma rays and by higher energy gammas in heavy-element absorbers. To produce the photoelectric effect, the entire energy of a gamma ray is given to an electron in an inner shell of the absorbing atom, with the resultant ejection of the electron from the atom. Emission of x radiation also occurs after the ejection when an electron from a higher energy shell falls into the vacant orbit. In the **Compton effect**, the gamma utilizes only a portion of its energy to eject an electron, and the remaining energy appears as a new photon or gamma moving in a different direction. This effect is important for light-absorbing elements and for gammas with less than 3 MeV. When a high-energy gamma penetrates the electron cloud and strikes the nucleus, the gamma is annihilated with the consequent production of a positron and an electron. This is known as **pair production** and is important when the absorber is a heavy element and the gamma energy is greater than 1.02 MeV. This energy value is twice the energy corresponding to the rest mass of the electron, which is 0.51 MeV. Incidentally, whenever a positron and a negatron meet, they are annihilated and produce two gamma rays, each of which has an energy of 0.51 MeV.

The ionization caused by the passage of gamma radiation through air is almost entirely due to the interaction of the photoelectrons, the Compton electrons, and the positrons and electrons caused by pair production with the molecules of the air. The production of a positive ion and negative ion (an **ion pair**) from a gaseous molecule requires approximately 35 eV, and the number of ion pairs produced per centimeter of travel in air is the **specific ionization**. The specific ionization for gamma radiation is small.

The neutron is electrically neutral and does not cause ionization on passage through air. Neutrons, however, are readily captured by certain nuclei, which then undergo radioactive disintegration. The secondary radioactivity thus produced causes ionization and can be used to detect neutrons. For example, $^{10}B$ captures a neutron and decomposes to form $^{7}Li$ and an alpha particle.

## DETECTION OF RADIOACTIVITY

The most important methods for detection of radioactivity in analytical applications involve either **scintillation counters**, which measure the light emitted by a substance when struck by radiations, or devices that measure the extent of ionization caused by the radiations. Such devices are known as **ionization**

**chambers, proportional counters**, and **Geiger counters**. Photography is also used in some cases since the radiations will darken a photographic plate. This is known as **autoradiography** or **radioautography** and is used primarily in the location of radioactive elements in complex structures and in the **film badges** worn by personnel to detect and measure the extent of exposure to radiation.

## *Scintillation Counters*

The fact that radiations from radioactive nuclei will cause scintillation or flashes of light when they strike the surface of certain substances has been known for a long time and was used as the basis for the early **spinthariscopes**, which were used to count visually the number of radiations being emitted by a given sample. Present-day scintillation counters work on the same principle but use photomultiplier tubes to detect the flashes and measure the energy of the incident radiation, since the energy in each flash of light is a function of the energy of the particle or ray that caused it. A simplified diagram of a well-type scintillation counter is shown in Figure 26-12.

Solid scintillators used include anthracene, *p*-terphenyl, naphthalene, and stilbene for beta particles and sodium iodide activated with 1% thallium(I)

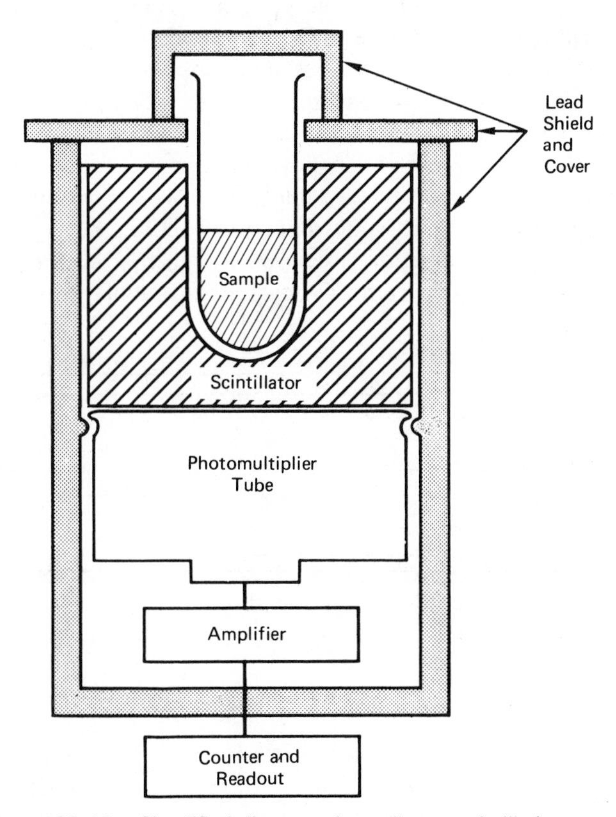


**Figure 26-12.** Simplified diagram of a well-type scintillation counter.

iodide for gamma radiation. Liquid scintillators are also used for low-level betas such as those from $^3H$ and $^{14}C$. The compound containing the isotope is dissolved in toluene or xylene and a scintillator added in the range 1–10 g/liter. Often a secondary scintillator is added to increase the pulse height. Combinations often used include *p*-terphenyl plus 1,6-diphenyl-1,3,5-hexatriene(DPH) and 2,5-diphenyloxazole(PPO) plus 1,4-bis-[2(5-phenyloxazolyl)]benzene (POPOP). For liquid scintillators, the photomultiplier tube is dipped directly into the solution.

When working with low-level activity, special precautions must be taken to discriminate against spurious pulses in the photomultiplier caused by shot-effect circuit noise or cosmic radiation. This may be accomplished by using two identical photomultiplier tubes in a circuit that includes an AND gate (see Chapter 27), which will transmit a signal to the counter only if it receives *simultaneous* signals from both photomultipliers.

## *Ionization Detectors*

All ionization detectors utilize a gas-ionization tube, which contains two electrodes enclosed in a glass envelope. The cathode is a cylindrical metal tube the diameter of which is essentially the same as the internal diameter of the envelope. The anode is a thin wire that runs down the center or axis of the cylindrical metal tube. The two electrodes are connected to an external variable potential source and a current-measuring device, which also has the ability to count pulses of current. A simplified diagram is shown in Figure 26-13.

Assume that the potential is raised continuously from zero in a gas-ionization tube, which is filled with an easily ionizable gas such as argon, and constructed as shown in Figure 26-13. Also assume that a radioactive emitter is placed close to the tube and that the emissions are slow enough that each can produce a distinct and separate pulse of current. The current output of the tube under these

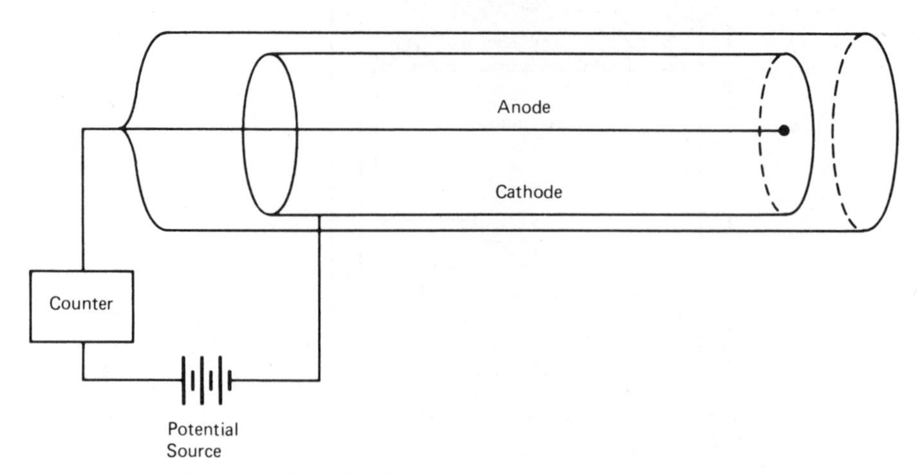


**Figure 26-13.** Simplified diagram of a Geiger–Müller tube.

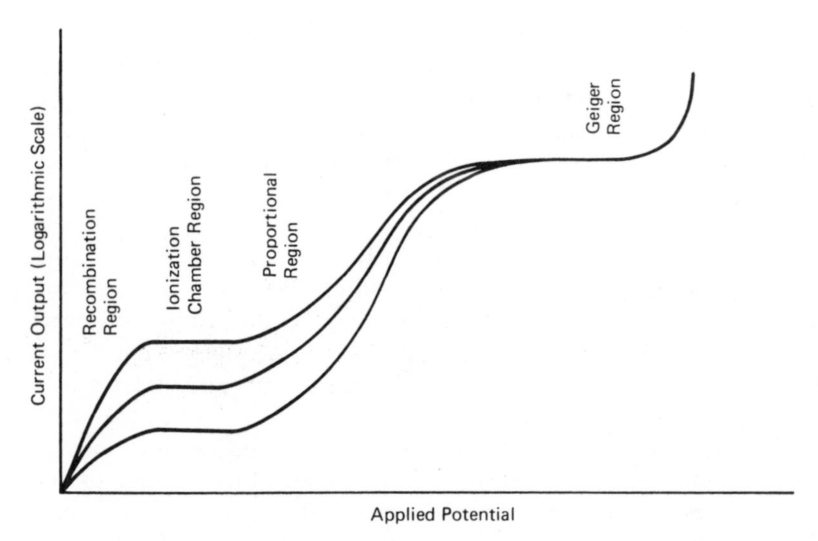


**Figure 26-14.** Current–potential diagram for a gas-ionization tube.

conditions is shown schematically in Figure 26-14. Several distinct regions should be noted. At low potentials, many of the ion pairs recombine before reaching the electrode, even though the fraction recombining decreases as the potential increases. This region is not satisfactory for measurements.

At slightly higher potential, all the ions pairs produced by the ionizing event do reach the electrodes, but they are not energetic enough to cause further ionization on the path to the electrodes. This is the ionization-chamber region and can be used to make measurements, since a saturation current is obtained that is related to the specific ionization of the radiation. In this region the tube measures the integrated effect of a large number of ionizing events. In practice the ionization chamber is used primarily as a pulse counter for alpha particles, which have a high specific ionization. It is also used for beta and gamma radiation, but the output is a continuous rather than a pulsed current.

As the potential is raised beyond the ionization chamber region, the increased energy of the ion pairs allows them to cause further ionization on their way to the electrodes. In this region the size of the pulse is proportional to the number of ion pairs produced by the primary particle, but the current is amplified as much as 10,000 times as a result of the formation of ions during the flight to the electrodes. This is known as the **proportional counter region** because the pulse height is proportional to the energy of the radiation. The **dead time** (or the length of time after an ionizing event has begun before the counter will respond to another ionizing event) is in the range 1–2 $\mu$sec and consequently the tube can be used for very high counting rates (as much as 50,000 to 200,000 per second). As the potential is raised further, the current increases but the amplification is no longer linear, so measurements are not satisfactory. If the potential is raised high enough, however, a second plateau in the current–voltage curve is attained. This is known as the **Geiger region**.

In the Geiger region, the accelerating potential of the electrons formed in the ionizing events is so great that they are able to cause further ionization on their way to the anode. The electrons produced by these secondary ionizations also have enough energy to cause further ionization, so a veritable avalanche of electrons strikes the anode. Photons are emitted when the electrons strike the anode and spread the ionization throughout the tube to produce a continuous discharge in less than 1 $\mu$sec after the original ionizing event. Each discharge builds up to a constant pulse that is large enough to be sent to the counter without further amplification.

The positive ions produced by the ionization do not move as fast as the electrons and, consequently, do not reach the cathode when the electron avalanche is collected on the anode. The travel time of the positive ions is usually about 200 $\mu$sec, during which time they act as a virtual positive shield about the anode, rendering the tube insensitive to further ionizing particles. If only argon is present in the tube, the positive ions will emit photons when they strike the cathode and again spread ionization throughout the tube. To offset this, a second gas, such as chlorine, methane, or ethanol, is added to the tube to act as a **quencher**. These substances have a lower ionization potential than argon, so that essentially all the positive ions that reach the cathode come from the quencher molecules. Upon neutralization, these ions do not emit photons but decompose into neutral fragments. As a consequence, the quenching agent is exhausted with use, so the useful life of most tubes is approximately $10^8$ counts.

Counting rates are limited to about 15,000 per minute, owing to the dead time of 200–270 $\mu$sec. The big disadvantage of the Geiger–Müller tube is the fact that there is no way to determine the energy of the ionizing particle, since each pulse builds up to a constant value irrespective of the energy of the entering radiation. Geiger tubes are used primarily for beta emitters because alpha particles are stopped by the glass envelope and most gamma radiation passes through the tube without triggering an ionization event. Most tubes are fitted with a thin mica end window or side window to reduce as much as possible the absorption of the beta particles by the glass envelope.

### *Auxiliary Equipment*

Since scintillation, proportional, and Geiger counters all deliver their signals as pulses of current, equipment must be provided to register or count the pulses. As a consequence, each counter requires some sort of **counting chamber** to hold the sample, together with a scaler and count registering unit. A simplified diagram of the interior of a counting chamber is shown in Figure 26-15. The shelf support is made of Lucite with 1-cm spacings between the grooves. The sample is supported on a Lucite plate that can be inserted into the grooves for the shelf position desired. The entire unit, together with the ionization tube, is encased in a heavy metal sheathing chamber to decrease background radiation and to protect personnel.

The scalar is an electronic counting device. For binary scalers, the output is arranged so that a single pulse is obtained for each 2, 4, 8, . . . , $2^n$ incident particles. For decade scalers, each stage passes on each tenth count so that decimal readings are produced. Most scalers can be set for a predetermined counting time or for a predetermined number of counts.

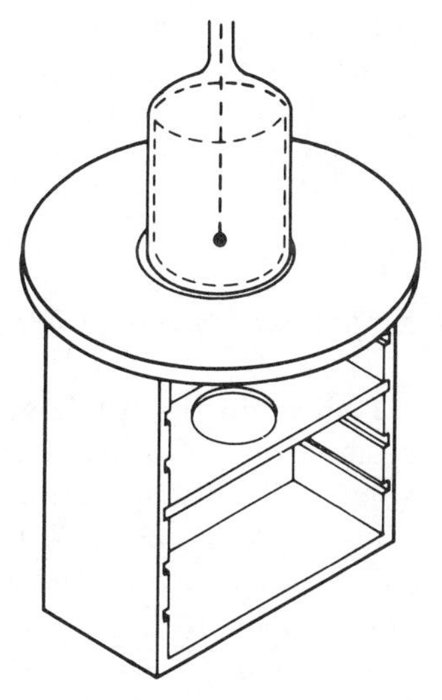

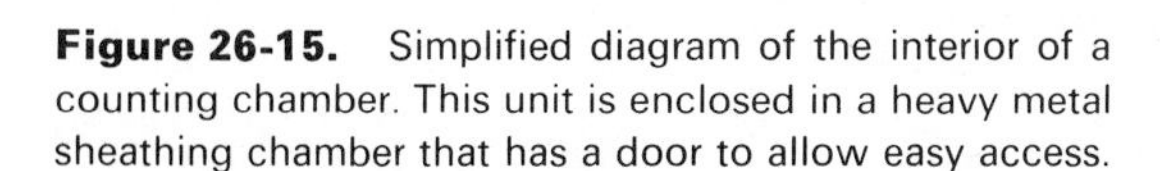

**Figure 26-15.** Simplified diagram of the interior of a counting chamber. This unit is enclosed in a heavy metal sheathing chamber that has a door to allow easy access.

Whenever the pulse height is proportional to the energy of the particle or radiation, as in proportional and scintillation counters, the measurement of pulse height can be used to distinguish or discriminate between particles and rays of different energies. **Pulse-height discriminators** are designed with a large number of narrow **energy channels**, each of which responds only to particles that have a specific energy. This allows both measurement of the energy of the particles and the number of disintegrations. The energy of the particles can be used to identify the isotope that produced the particles, and the number of disintegrations is a measure of the amount of that particular isotope present.

## ERRORS IN THE MEASUREMENT OF RADIOACTIVITY

### *Background*

There is a significant amount of radiation in the atmosphere at any location due to cosmic rays and the presence of radioactive isotopes in the immediate surroundings. Even though the shielding of the counting chamber will decrease this **background radiation**, it never eliminates it completely. As a consequence, the amount of the background must always be determined and subtracted from the sample counts. Background counts are normally run several times longer than sample counts, to assure a better statistical value.

### *Counter Geometry*

The arrangement of the sample and the counter should always be reproducible and the actual geometry correction of each shelf determined by counting a known weight of a standard on each shelf. A radioactivity standard is a sample with a known number of disintegrations per second per gram. Since the total number of disintegrations during the counting period is known, division of the total number of disintegrations by the corrected number counted gives the multiplier that should be used for any measurement made on that particular shelf.

### *Statistical Nature of Measurements*

Radioactive decay follows the laws of probability, since a very large number of atoms are involved in even the smallest samples. That is, the number of atoms that will disintegrate in a given length of time is a function of the decay constant. Although we cannot predict the exact time when any given atom will decompose, we can predict the fraction of any total number of atoms that will disintegrate in a given time. The most convenient statistical measure to employ is the **standard deviation** ($\sigma$). One standard deviation or $1\sigma$ is the maximum deviation from the "true" value that should be expected in 68% of a large series of measurements. In radioactivity measurements, it can be shown that, when the half-life is long compared to the duration of the experiment, the standard deviation is simply the square root of $n$, the total number of counts, and the reliability of the results can be expressed as $n \pm \sqrt{n}$.

### *Coincidence Errors*

Coincidence errors are due to the dead time of the counter and occur whenever two disintegrations are so close together that the counter is not able to count both. The recovery time varies greatly between the various types of counters and must be calculated for each counter used. If the dead time of the counter is designated by $\gamma$, the observed counting rate by $r_o$, and the true rate of disintegration by $r_t$, the relationship is given by

$$r_t = r_o + \gamma r_o r_t \qquad \textbf{(26-15)}$$

or

$$r_t = \frac{r_o}{1 - \gamma r_o}$$

## APPLICATIONS

### *Tracer Techniques*

The tracer technique (**tagging**) has proved to be a valuable tool for many kinds of research. In this procedure, a radioactive element is incorporated into a compound that is then allowed to react, and the radioisotope is followed during the course of the reaction. For example, the complicated photosynthesis reaction has been almost completely elucidated by incorporating $^{14}C$ into $CO_2$ and allowing plants to use it. The pathways of metabolism in plants and animals are being studied by tagging simple sugars with $^{14}C$; amino acids with radioactive carbon, nitrogen, or sulfur; and lipid material with $^{14}C$, $^{32}P$, or $^{3}H$.

Many projects initiated as medical research soon fall into the category of diagnosis and/or treatment. Thus, $^{24}Na$ in NaCl was injected into the bloodstream to study the speed and course of normal distribution and circulation of the blood. This technique is now used to determine the extent of circulation in crushed arms and legs. The information aids the physician in determining whether amputation is necessary and where to operate if it is necessary.

$^{131}I$ is probably one of the best examples of a radioactive isotope used for research that led to diagnostic procedures and treatment. Iodine is taken up by the thyroid gland for inclusion in the hormone thyroxin. Feeding radioactive iodine and then measuring the amount of radioactivity in the thyroid gland can be used to diagnose hypo- and hyperthyroidism. In case of hyperthyroidism, larger amounts can then be used to reduce the size of the gland by irradiation from within the gland. This element is also used in kidney function tests and in the location of brain tumors.

$^{32}P$ has been used to study the formation of bones and teeth, certain types of anemias, and to aid in the location of breast cancers and brain tumors. $^{35}S$ has been used to study various aspects of protein metabolism and the turnover rate of plasma proteins in the blood. $^{59}Fe$ has proved helpful in the study of the production and life of red blood cells and hemoglobin.

Typical industrial uses of the tagging procedure include the incorporation of radioactive sulfur into vulcanized rubber to study the wear characteristics of tires. Automobile engine pistons have been made radioactive by irradiation in an atomic pile in order to determine the rate and positions of wear of the pistons in actual use. Radioactive isotopes have been intimately mixed with dirt to learn the effectiveness of different detergents in washing clothes and dishes. The addition of $^{203}Hg$ to the mercury in the gauges and valves in oil fields that are automatically pumped, helped catch the thieves who stole the mercury, when they tried to sell it back to the oil company.

## *Isotope Dilution*

Isotope dilution analysis can be used to determine the yield or amount of material when nonquantitative analytical or isolation methods must be used. A known amount of the compound being determined, which has been tagged with a radioisotope, is added to the unknown mixture containing the compound being determined. The mixture is then put through the nonquantitative isolation procedure, and some portion of final product is separated, weighed, and the activity determined. Comparison of the final activity with the known activity of the added radionuclide allows calculation of the result. The relationship used is

$$W_u = W_s\left(\frac{A_s}{A_p} - 1\right)$$

where $W_u$ is the weight of the inactive material being determined; $W_s$ is the weight of the tagged compound added; $A_p$ is the corrected activity of the product separated; and $A_s$ is the known activity of the tagged compound added. The reverse procedure can be used to determine the weight of a compound containing a very active radioisotope which must be diluted before it can be safely measured.

### *Activation Analysis*

Many nuclei become radioactive when bombarded by slow **(thermal)** neutrons by capture of the neutrons to form an isotope that is heavier by one mass unit. These heavier isotopes have their own decay schemes, which allow both qualitative identification and quantitative estimation. Irradiation is accomplished in nuclear reactor or by a radionuclide–beryllium source for the length of time necessary for the activity to be induced. This can range from a few minutes to several months. Upon removal from the neutron flux, the activity of the sample is analyzed with a multichannel analyzer, which indicates not only the type and energy of the particles produced but also the relative number of disintegrations in a given time. The radionuclide is identified by the type and energy of the particle or radiation produced, and the amount can be estimated from the number of disintegrations. In practice, standards are irradiated at the same time as the samples, and the analysis is carried out by simple comparison of the results. The method is very sensitive and can detect very minute amounts of elements. It is especially advantageous when only small amounts of sample are available. The calculations require knowledge of the neutron flux in the irradiation chamber, the number of nuclei available, and the capture cross section of the nuclei, as well as other factors. Many examples of the necessary calculations may be found in the literature.

### *Miscellaneous Analytical Uses*

Radioisotopes have proved to be extremely helpful to the analytical chemist, not only in the development of new procedures but in the improvement of existing procedures. Included in such a list are the use of radioisotopes to determine the effect of variables on the elution curves from chromatographic and ion exchange columns, the determination of the solubility of slightly soluble compounds, the effectiveness of washing a precipitate, the efficiency of washing laboratory equipment, the surface area of solids, the type and extent of coprecipitation and adsorption, reaction mechanisms, the aging mechanisms of precipitates, the location of end points in titrations, and many others.

# PROBLEMS

**1** The following mass spectral data were obtained for the four isomeric butanols. Identify each.

| | Relative Abundance | | | |
|---|---|---|---|---|
| *m/e* | A | B | C | D |
| 27 | 51 | 10 | 16 | 42 |
| 31 | 100 | 36 | 20 | 63 |
| 41 | 61 | 21 | 10 | 56 |
| 43 | 61 | 14 | 10 | 100 |
| 45 | 7 | 1 | 100 | 5 |
| 56 | 91 | 2 | 1 | 2 |
| 59 | 1 | 100 | 18 | 5 |

*Ans.* A is 1-butanol

**2** Which fragments are represented by each of the mass numbers in problem 1?

*Ans.* 31 is $—CH_2OH^+$

**3** Indicate the weakest bond in each molecule in problem 1.

*Ans.* A, the C—C bond in $R—CH_2CH_2OH$

**4** Indicate the points of rupture in the following molecules that would lead to the formation of each mass fragment listed: (a) 2-methyl-2-butanol: 73, 59; (b) diisopropyl ether: 87, 59, 45, 43; (c) acetophenone: 120, 105, 77.

*Ans.* 73 represents loss of $CH_3$

**5** The following mass spectral data were obtained for the isomeric propanols. Identify each.

| | Relative Abundance | | | Relative Abundance | |
|---|---|---|---|---|---|
| *m/e* | A | B | *m/e* | A | B |
| 15 | 3.8 | 10.7 | 39 | 4.0 | 5.5 |
| 19 | 1.0 | 6.5 | 43 | 3.2 | 16.8 |
| 27 | 15.2 | 15.5 | 45 | 4.4 | 100.0 |
| 29 | 14.1 | 9.5 | 60 | 6.4 | 0.5 |
| 31 | 100.0 | 5.8 | | | |

**6** Which fragments are represented by each of the mass numbers in problem 5?

**7** Indicate the weakest bond in each molecule in problem 5.

**8** Indicate the points of rupture in each of the following molecules that would lead to each of the mass fragments shown: (a) cyclobutane: 27, 28, 41, 56; (b) 1-butanol: 27, 31, 41, 56; (c) 2-butanone: 27, 29, 43, 72; (d) 3-methyl-1-butyne: 27, 39, 53, 67.

**9** Identify the compound that has the following mass spectrum:

| *m/e* | 1 | 16 | 17 | 18 | 19 | 20 |
|---|---|---|---|---|---|---|
| *Relative Abundance* | $<1$ | 1 | 21 | 100(P) | $<1$ | $<1$ |

**10** Identify the compounds that contain two carbons and have the following mass spectra:

| | *Relative Abundance* | | | | | | | | | |
|---|---|---|---|---|---|---|---|---|---|---|
| *m/e* | 15 | 27 | 28 | 29 | 30 | 31 | 43 | 44 | 45 | 46 |
| (a) | — | 23 | — | 24 | — | 100 | — | — | 35 | 16(P) |
| (b) | — | 5 | — | 100 | — | — | 27 | 46(P) | — | — |
| (c) | 5 | 33 | 100 | 22 | 26(P) | — | — | — | — | — |

**11** Using the method of simultaneous equations, determine the mole % composition for the unknown mixture from the mass spectral data given.

| | *Relative Abundance* | | | | |
|---|---|---|---|---|---|
| *m/e* | *Benzene* | *3-Hexyne* | *Isopropyl-cyclopropane* | *4-Methyl-1-pentyne* | *Unknown* |
| 41 | — | 92.5 | 73.2 | 90.4 | 100.0 |
| 43 | — | 2.7 | 24.2 | 78.4 | 35.0 |
| 51 | 20.5 | 16.6 | 2.0 | 6.4 | 33.2 |
| 52 | 19.6 | 7.8 | 0.8 | 2.2 | 10.3 |
| 56 | — | 1.6 | 100.0 | 1.3 | 45.4 |
| 63 | 3.4 | 9.0 | 0.5 | 4.1 | 51.4 |
| 67 | — | 100.0 | 2.0 | 100.0 | 74.7 |
| 69 | — | 0.2 | 12.4 | 0.1 | 5.6 |
| 78 | 100.0 | 1.2 | — | 0.3 | 30.2 |
| 81 | — | 12.3 | 0.1 | 9.6 | 35.3 |
| 82 | — | 93.5 | — | 2.8 | 42.2 |
| 84 | — | 0.2 | 1.7 | — | 0.8 |

**12** Repeat problem 11 for the following mixture from the mass spectral data given.

| | *Relative Abundance* | | | | |
|---|---|---|---|---|---|
| *m/e* | *1-Propanol* | *2-Propanol* | *Propanol* | *Ethanethiol* | *Unknown* |
| 28 | 6.1 | 1.8 | 68.9 | 41.5 | 36.1 |
| 29 | 15.9 | 11.3 | 100.0 | 93.5 | 75.1 |
| 31 | 100.0 | 6.4 | 3.2 | 0.1 | 100.0 |
| 34 | — | — | — | 24.1 | 9.3 |
| 45 | 4.2 | 100.0 | 0.1 | 22.4 | 51.4 |
| 47 | — | 0.2 | — | 78.6 | 30.5 |
| 58 | 0.5 | 0.2 | 38.9 | 10.8 | 12.2 |
| 59 | 9.2 | 3.5 | 1.4 | 7.7 | 13.5 |
| 60 | 6.4 | 0.5 | — | 1.5 | 7.0 |
| 62 | — | — | — | 100.0 | 39.7 |

**13** A 50.0-mg sample of $U_3O_8$ showed a net activity of 1272 cpm (counts per minute) on shelf 1 and 255 cpm on shelf 2. Calculate the geometry correction of each shelf. The activity of 1 mg of U is 748 disintegrations per minute (dpm).

*Ans.* shelf 1, 0.040

**14** Repeat problem 13 for the following counting chambers.

| | | *Net Activity* | |
|---|---|---|---|
| | $U_3O_8$ *(mg)* | *Shelf 1* | *Shelf 2* |
| (a) | 28.3 | 1063 | 201 |
| (b) | 56.4 | 9942 | 2137 |
| (c) | 51.3 | 5636 | 1216 |
| (d) | 49.3 | 1344 | 278 |

**15** A sample of compound Y is known to have compound X as an impurity. Compound X is also available in radioactive form (X*) with an activity of 250,000 dps/g. A 10.0-g sample of the impure compound is mixed intimately with a 10.0-mg sample of X* and 15.0 mg of X + X* is separated from the mixture. The activity of the separated sample is 250 dps. Calculate the percentage of X present as impurity.

*Ans.* 1.35%

**16** Repeat problem 15 for the following isotopic dilutions.

| | *Radioisotope* | | *Sample Separated* | |
|---|---|---|---|---|
| *Sample Taken (g)* | *Weight Added (mg)* | *Activity (dps/g)* | *Weight (mg)* | *Activity (dps)* |
| (a) 25.0 | 5.0 | 225,000 | 50.0 | 425 |
| (b) 5.0 | 7.5 | 300,000 | 15.0 | 225 |
| (c) 10.0 | 10.0 | 150,000 | 10.0 | 125 |
| (d) 20.0 | 10.0 | 275,000 | 12.5 | 250 |

**17** A 100-$\mu$L sample of radioactive ethanol (250,000 cpm/mL) was added to 100.0 mL of an aqueous solution of ethanol. After thorough mixing, a 1.00-mL sample of ethanol was separated from the mixture. A 50-$\mu$L aliquot of the separated ethanol had an activity of 200 cpm. Calculate the volume percentage of ethanol in the original solution.

*Ans.* 6.15% (v/v)

**18** Repeat problem 17 for the following isotopic dilutions.

| | *Radioisotope* | | *Sample Separated* | | |
|---|---|---|---|---|---|
| *Sample Taken (mL)* | *Volume Added ($\mu$L)* | *Activity (cpm/mL)* | *Volume (mL)* | *Aliquot Counted ($\mu$L)* | *Activity (cpm)* |
| (a) 500.0 | 50.0 | 100,000 | 1.00 | 100.0 | 250 |
| (b) 50.0 | 25.0 | 300,000 | 1.00 | 50.0 | 300 |
| (c) 250.0 | 10.0 | 200,000 | 1.00 | 25.0 | 150 |
| (d) 100.0 | 100.0 | 250,000 | 1.00 | 25.0 | 175 |

CHAPTER

# 27 Solid State Microelectronics

The advent of inexpensive semiconductor devices has caused the use of tubes in electronic circuits to be severely curtailed and essentially all instrumental circuits are now designed to utilize solid state or semiconductor devices. These devices are **bipolar** because they are constructed with *p* and *n* "doped" semiconductor materials, which form **junctions** where *p* and *n* materials are interfaced. Other types of semiconductor devices include the **field-effect transistor** (FET) and the **metal oxide semiconductor** (MOS).

Definitions of terms used in this chapter are as follows:

**Ampere:** The practical unit of the flow of electric current. One ampere is the current that will flow with a potential of 1 V through a conductor with a resistance of 1 Ω. Symbol, A.

**Boolean algebra:** The algebra of classes: the symbolical study of the significance of relations between classes.

**Capacitance:** That property of a system of conductors and dielectrics that permits the storage of electricity when potential differences exist between the conductors. It is expressed as the ratio of charge (quantity of electricity) to the potential difference. The standard unit is the farad.

**Capacitor:** A device that introduces capacitance into an electric circuit. It permits electricity to be stored temporarily.

**Coulomb:** The unit used for measuring the quantity of an electrical charge. It is equal to the charge on $6.24 \times 10^{18}$ electrons.

**Diffusion length:** The average distance that a minority carrier travels before combining with a majority carrier in a semiconductor material.

**Farad:** One farad equals the amount of capacitance present when 1 V can store 1 coulomb of electricity. Symbol, f

**Henry:** The electromagnetic unit of inductance, defined as the inductance present that will cause 1 V to be induced if the current changes at the rate of 1 A/sec. Symbol H.

**Inductance:** The resistance offered to a change in electromotive force in a circuit due to the self-induced electromotive force caused by variation of the magnetic field. The direction of the induced electromotive force is that which counteracts changes in the current. Symbol, *L*.

**Inductor:** A device that introduces inductance into an electrical circuit.

**Ohm:** The practical unit of electrical resistance. An ohm is the resistance that allows 1 A of current to flow through a conductor at 1 V potential. Symbol, Ω.

**Schematic:** The graphical representation of an electrical circuit.

**Semiconductor:** A solid substance that is intermediate between metallic conductors and nonconducting insulators. It has a negative thermal coefficient of resistance and conducts better at high temperatures than at low.

**Volt:** The practical unit of electric potential, potential difference, electromotive force, or electrical pressure. One volt is the difference of potential that causes a current of 1 A to flow through a resistance of 1 Ω. Symbol, V.

**Watt:** A unit of measure of electrical power: One watt equals the power required to keep a current of 1 A flowing under a potential drop of 1 V. Symbol, W.

## SEMICONDUCTOR MATERIALS

The two most commonly used semiconductor materials are silicon and germanium. In the crystalline state these elements form a diamond lattice structure in which each atom forms a covalent bond with each of its four adjacent neighbors. The shared electrons in the covalent bond (sometimes referred to as the **valence energy band**) are not available for conduction of the electric current because each electron is bound to the two adjacent atoms. As a result, the crystal structure is neutral, but the neutrality can be upset either by thermal agitation or by imperfections in the crystal structure that cause the rupture of some of the covalent bonds. When a bond is broken, the free electron makes a quantum jump to a higher energy band known as the **conduction band**. The electrons in the conduction band are available for the conduction of current; consequently, some conduction does take place in semiconductors when a voltage potential is placed across the material. Under the same applied potential, the amount of current conducted by a semiconductor is much less than that which flows through metallic conductors and much greater than that which passes through an insulator. After an electron is raised to the conduction band, its absence in the valence band is termed a **hole** and can be regarded as a fictitous positive charge. In the neutral structure this hole–electron pair generation is balanced by the recombination of holes and electrons at the same rate.

Impurities that act as imperfections in the crystal lattice may be introduced during crystal growth or by diffusion after growth is complete. If the impurity introduced is an element with five electrons in the outer or valence shell, it is known as a **donor** and electrons are generated in the conduction band of the crystal lattice by its introduction. This type of semiconductor is known as *n type*. If, however, the impurity introduced is an element with three electrons in the outer or valence shell, it is known as an **acceptor** and holes are generated in the crystal lattice by its introduction. This type semiconductor is known as *p type*. In the *n*-type semiconductor the electrons are referred to as **majority**

**carriers**, whereas the holes are **minority carriers**. The opposite holds true for the *p*-type material, in which the hole is the majority carrier and the electron is the minority carrier.

# SEMICONDUCTOR DEVICES

## *p-n* JUNCTIONS

The junction between *n*-type and *p*-type semiconductors has important properties that are the primary basis of semiconductor devices. The *p-n* junction is also known as a **diode junction** or a **semiconductor diode**. Thermal agitation will cause electrons to move from the *n*-type material across the junction to the *p*-type material; holes will move from the *p*-type material to the *n*-type. Since *n* material and *p* material are both neutral before they are brought into contact with each other, the movement of the electrons across the junction leaves the *n* material with a net positive charge. Similarly, the movement of holes across the junction leaves the *p* material with a net negative charge. As the net charge on the two materials increases, the movement of holes and electrons across the junction decreases. With time, the net charge will attain a value at which point further migration across the boundary ceases and equilibrium is established. If an external electrical force is now applied across the junction in a direction that helps in the transfer of electrons from *n*-type to *p*-type material, the equilibrium will be upset and current will flow. If the electrical force is applied in the opposite direction, the equilibrium will be upset but current will not flow because holes would be required to move into the *p*-type material and electrons into the *n* type. The structure and the graphical representation or symbol of a typical diode is shown in Figure 27-1(a). The terminal connected to the *p* material is called the **anode** and the terminal connected to the *n*-type material is called the **cathode**. For current to flow, a positive electrical potential must be applied from the anode to the cathode.

## TRANSISTORS

The junction **transistor** is composed of two *p-n* junctions connected in either an *n-p-n* or a *p-n-p* configuration. These two structures and the symbols for them are shown in Figure 27-1(b) and (c). The center section is known as the **base** region, whereas the region that is normally forward-biased with respect to the base is known as the **emitter**. The region that is normally reverse-biased with respect to the base is called the **collector**. Bias in this case refers to the electrical potential between the two points. Forward bias is that potential which causes current to flow across the base–emitter diode junction.

The operation of the transistor is controlled through the very narrow base region. Majority carriers will flow from the emitter to the base when a forward bias potential is placed across the base–emitter junction. If a strong reverse bias

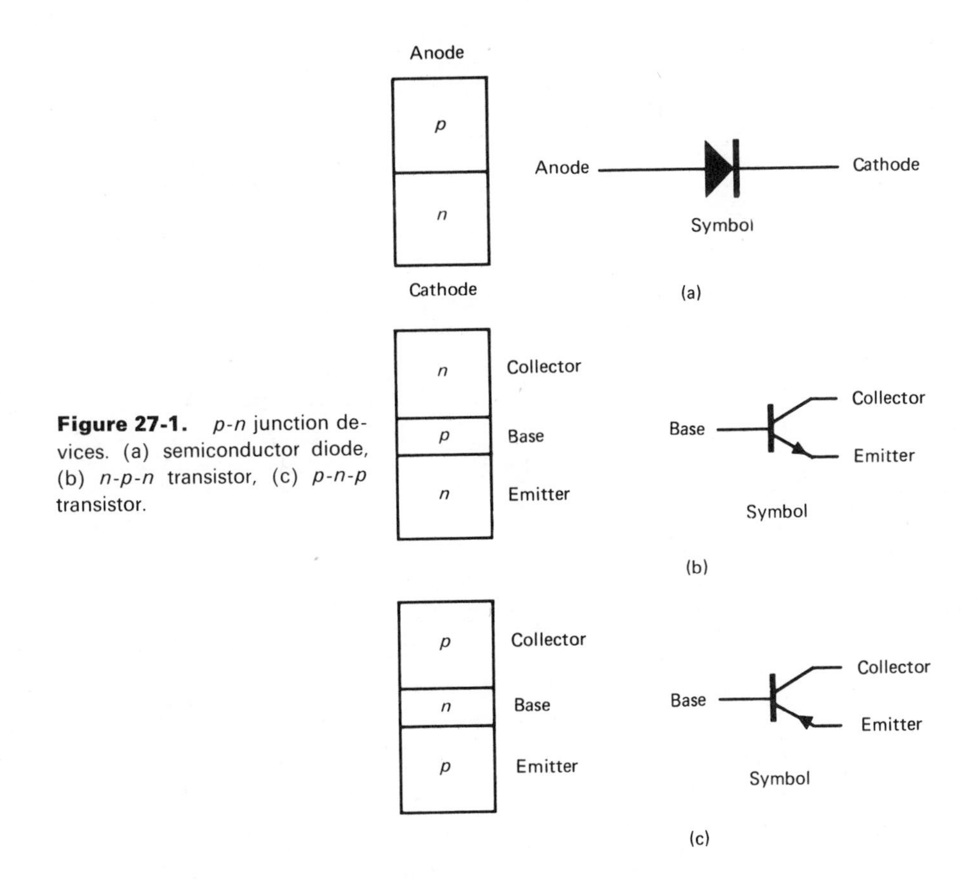


**Figure 27-1.** *p-n* junction devices. (a) semiconductor diode, (b) *n-p-n* transistor, (c) *p-n-p* transistor.

is now placed on the collector, most of the majority carriers flowing from the emitter will be swept across the base, which is narrower than the average diffusion length. Even though only a small fraction of the majority carriers actually flow into the base, control of the base current allows control of the much larger collector current. This property is the basis of the current gain from base to collector that allows the use of transistors as amplifiers.

## SEMICONDUCTOR CIRCUITS

### DIODE RECTIFIERS

A rectifier is a device used to convert ac current to dc current and utilizes the circuit shown in Figure 27-2(a). When a potential is applied to the load resistor $R$ through the diode $D$, the diode conducts current only during the positive half-cycle, as shown in Figure 27-2(b), since the diode can conduct current in the forward bias only. If a capacitor is added across the resistor as shown in Figure 27-2(c), the capacitor will be charged to the input potential level during

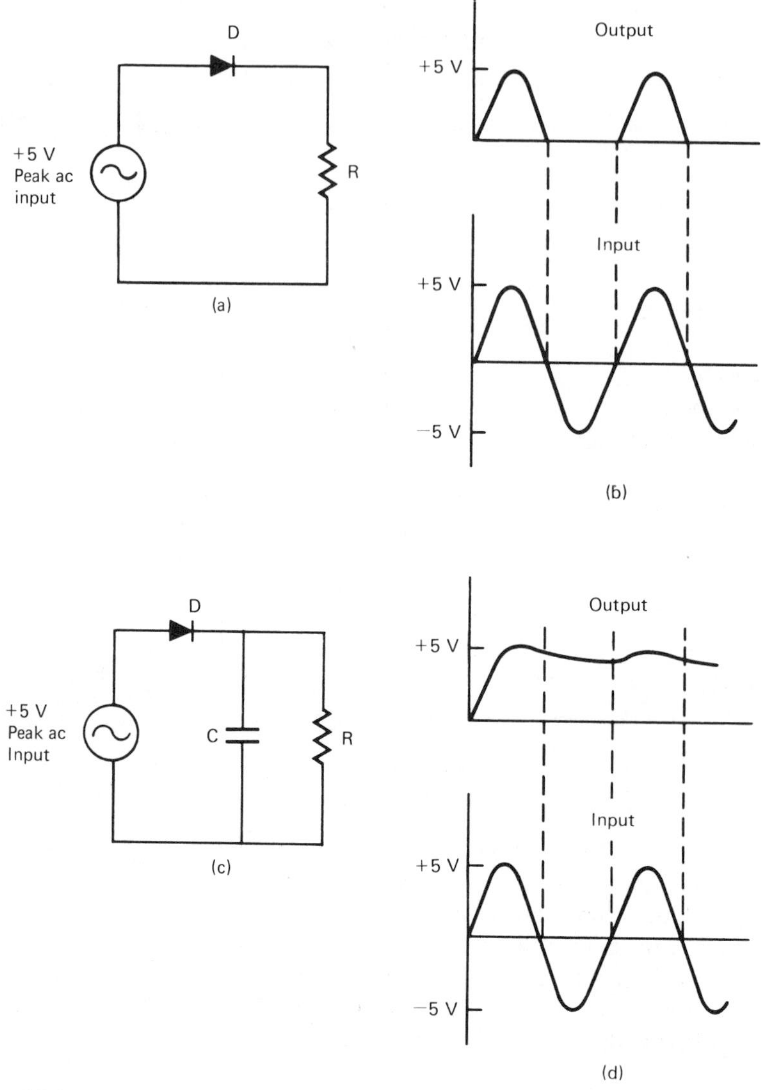


**Figure 27-2.** Diode rectifiers. (a) Diode rectifier, (b) and (d) voltage diagram, (c) diode rectifier with capacitor.

the positive half-cycle. When the diode ceases to conduct in the forward direction as a result of the decrease in input voltage, the energy stored in the capacitor will begin to be dissipated as power in the resistor at a rate equal to

$$P_{\text{dis}} = \text{power dissipated in resistor} = \frac{V_c^2}{R} \qquad \textbf{(27-1)}$$

where $V_c$ is the voltage across the capacitor at any instant of time. If the capacitor can store a large amount of energy, the amount dissipated as power in the resistor will not reduce the total amount of energy stored in the capacitor appreciably before the next positive cycle of the ac source recharges the capacitor. As a consequence, the output voltage across $R$ has a high dc component with a small ac ripple voltage on top of it, as shown in Figure 27-2(d). This ripple voltage has a frequency equal to the frequency of the input ac source.

## TRANSISTOR AMPLIFIERS

The current flowing in the collector of a transistor is a function of the base current and can be determined approximately by

$$I_c = \beta I_b \tag{27-2}$$

where $I_c$ is the collector current; $I_b$ is the base current; and $\beta$ is the gain of the transistor and is given by

$$\beta = \frac{\Delta I_c}{\Delta I_b} \tag{27-3}$$

The emitter current $I_e$ is the sum of the collector current and base current, or

$$I_e = I_b + I_c \tag{27-4}$$

Equations (27-2) and (27-3) indicate that any variation in the base current causes a correspondingly larger variation in the collector current, or, that a current applied to the base will be amplified in the collector because the change in collector current equals the change in the base current multiplied by the gain of the transistor:

$$\Delta I_c = \Delta I_b \beta \tag{27-5}$$

The $\beta$ of a typical transistor is about 50 but will vary from device to device. The equivalent circuit of an *n-p-n* transistor is shown in Figure 27-3, in which $R_b$ is the input base resistance of the transistor. The current will flow only in the direction indicated by the arrow. The application of a positive voltage from base to emitter will cause a current to flow in the base equal to

$$I_b = \frac{V_{be}}{R_b} \tag{27-6}$$

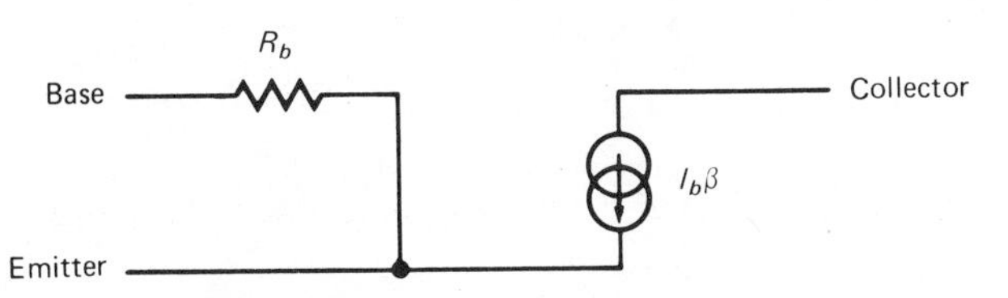


**Figure 27-3.** Transistor equivalent circuit.

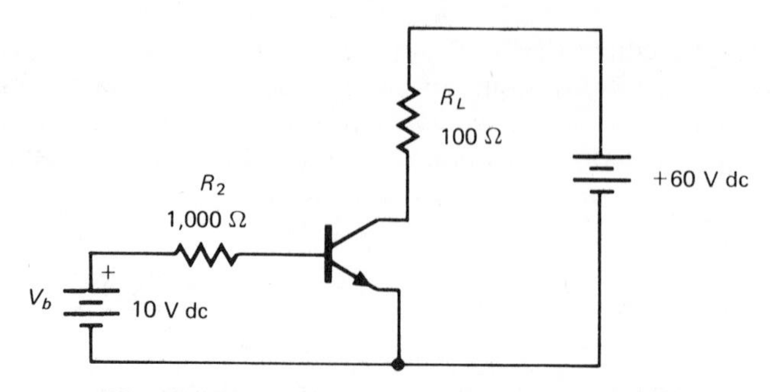


**Figure 27-4.** Typical transistor amplifier circuit.

where $V_{be}$ is the voltage applied from the base to the emitter. This base current flow causes a current to flow from the collector to the emitter, as indicated in Figure 27-3.

A typical transistor amplifier circuit is shown in Figure 27-4. The 100-Ω resistor $R_L$ represents the load and is placed in series with the *n-p-n* transistor collector and the collector voltage supply. The 1000-Ω resistor $R_2$ is placed in the base to allow more precise control of the base current than would be possible if the very small base resistance of the device itself is used. $R_2$ must be large enough that the device base resistance is negligible in comparison. If the base voltage source $V_b$ is adjusted to 10 V, the base current will be

$$I_b = \frac{V_b}{R_2} = \frac{10\text{ V}}{1000} = 10^{-2}\text{ A} = 10\text{ mA}$$

and the power delivered from the base voltage source is

$$\begin{aligned} P_{\text{base source}} &= V_b I_b = 10\text{ V} \times 10\text{ mA} \\ &= 100\text{ mW} = 0.1\text{ W} \end{aligned} \tag{27-7}$$

Assuming a transistor $\beta$ of 50, the collector current will be

$$I_c = I_b \beta = 10\text{ mA} \times 50 = 0.5\text{ A}$$

With this collector current the voltage drop across the load resistor $R_L$ is

$$V_{R_L} = I_c R_L = 0.5\text{ A} \times 100 = 50\text{ V}$$

and the power in the load is

$$P_{\text{load}} = V_{R_L} I_c = 0.5\text{ A} \times 50\text{ V} = 25\text{ W}$$

while the total power delivered from the collector voltage source is

$$P_{V_c} = 60\text{ V} \times 0.5\text{ A} = 30\text{ W}$$

The difference in power delivered from the collector voltage source and the power dissipated in the load resistor must be dissipated in the transistor. As a consequence, the power dissipated in the transistor should be minimized in order that maximum amplification be obtained from the circuit. The power gain $G$ of the circuit in Figure 27-4 from input base source power to the output load power is

$$G = \frac{P_{\text{load}}}{P_{\text{base source}}} = \frac{25}{0.1} = 250 \qquad \textbf{(27-8)}$$

A power gain of 250 is not uncommon in transistor circuits. If all the power delivered from the collector voltage source could be dissipated in the load, the gain for this circuit would be 300 instead of 250.

## INTEGRATED CIRCUITS

An integrated circuit is composed of all the resistors, transistors, and diodes necessary to perform a complete circuit function and is usually constructed on a single piece of semiconductor material. Figure 27.5(a) is a photomicrograph of a large-scale integrated circuit (LSI) constructed on a single small chip of a metal oxide semiconductor (MOS) material such as those used in programmable hand-held calculators and for control of microwave ovens and ignition systems. Figure 27.5(b) is a picture of the packaged chip showing the dimensions of the package and the number of connection pins in the completed package. Note that the chip itself is much smaller than the package. Manufacturers also produce medium-scale integrated circuits (MSI) and small-scale integrated circuits (SSI), which have fewer components than the LSI circuit shown. The primary advantages of integrated circuits compared to circuits using tubes or transistors include the greatly reduced size and weight, the increased reliability, and the decreased power consumption.

Integrated circuits are classified in two broad categories: linear circuits and digital or logic circuits. Linear circuits are those in which the output is a continuous function of the applied input; digital or logic circuits are those which function according to the principles of Boolean algebra.

A discussion of all the available types of integrated circuits is beyond the scope of this text since manufacturers are producing a seemingly unlimited quantity of variations and configurations. Examples of simple but typical circuits which are widely used in instrument design will be presented.

### LINEAR INTEGRATED CIRCUITS

The most popular linear circuit is probably the **operational amplifier**. These devices are widely available and have essentially ideal characteristics for all practical purposes. The symbol of an ideal amplifier is shown in Figure 27-6(a).

**Figure 27-5.** (a) Integrated circuit such as that used in programmable hand-held calculators, microwave ovens, and automotive aplications. ROM, read-only memory with 8192-bit capacity; RAM, random-access memory with 256-bit capacity; ALU, arithmetic logic unit that can add or subtract and perform logical and arithmetic comparisons. (Courtesy of Texas Instruments, Incorporated, Dallas, Texas.)

In an ideal amplifier the voltages at the inverting (INV) and noninverting (NI) terminals must be equal, and the input resistance of both must be infinite. The theoretical gain of such an amplifier is infinite, and gains of 100,000 or more are attained in actual practice. The output of the operational amplifier constitutes a voltage source that will supply whatever current is necessary to maintain the required voltage.

An operational amplifier may be connected for inverting operation as shown in Figure 27-6(b) or for noninverting operation as shown in Figure 27-6(c). In the inverting connection, the NI terminal voltage is set at zero and a potential $V_{in}$ applied to the INV terminal. The resulting current $I_{R_{in}}$ through $R_{in}$ can be found from

$$I_{R_{in}} = \frac{V_{in} - V_{INV}}{R_{in}} \tag{27-9}$$

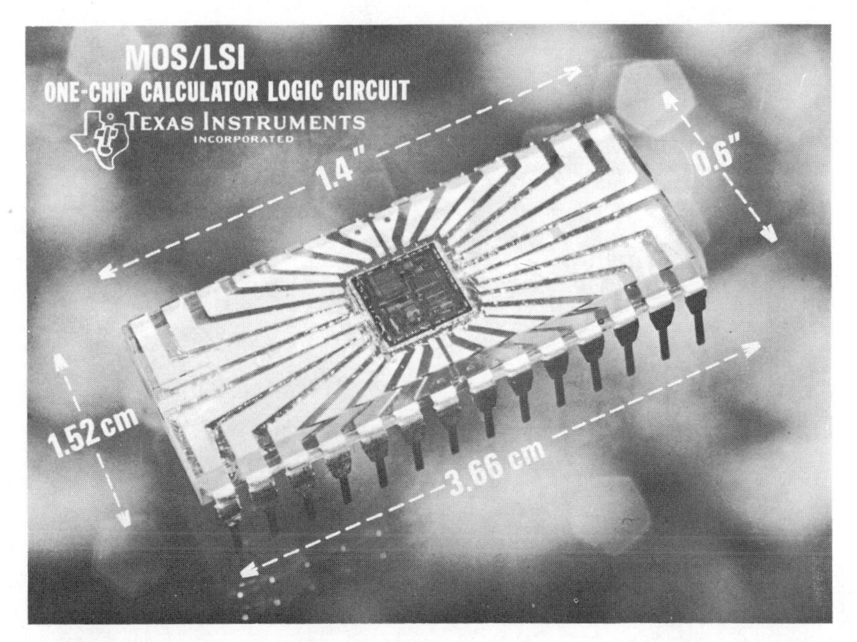


**Figure 27-5.** (b) Packaged integrated-circuit chip such as shown in Figure 27-5(a) with cover removed. MOS/LSI, metal oxide semiconductor/large-scale integrated circuit. (Courtesy Texas Instruments, Incorporated, Dallas.)

where $V_{\mathrm{INV}}$ is the voltage at the inverting terminal and also equals the voltage at the noninverting terminal, $V_{\mathrm{NI}}$. Because $V_{\mathrm{NI}}$ equals zero, eq. (27-9) becomes

$$I_{R_{\mathrm{in}}} = \frac{V_{\mathrm{in}}}{R_{\mathrm{in}}} \tag{27-10}$$

No current can flow into the amplifier itself due to the infinite resistance and because the current at any node must equal zero, thus

$$I_{\mathrm{in}} = I_{\mathrm{out}} \tag{27-11}$$

where $I_{\mathrm{out}}$ is the current flowing through the feedback resistor $R_{FB}$ in the output circuit; however,

$$I_{\mathrm{out}} = \frac{V_{\mathrm{INV}} - V_{\mathrm{out}}}{R_{RB}} = \frac{-V_{\mathrm{out}}}{R_{FB}} \tag{27-12}$$

since $V_{\mathrm{INV}}$ equals zero. Substitution in eq. (27-11) gives

$$\frac{V_{\mathrm{in}}}{R_{\mathrm{in}}} = \frac{-V_{\mathrm{out}}}{R_{FB}}$$

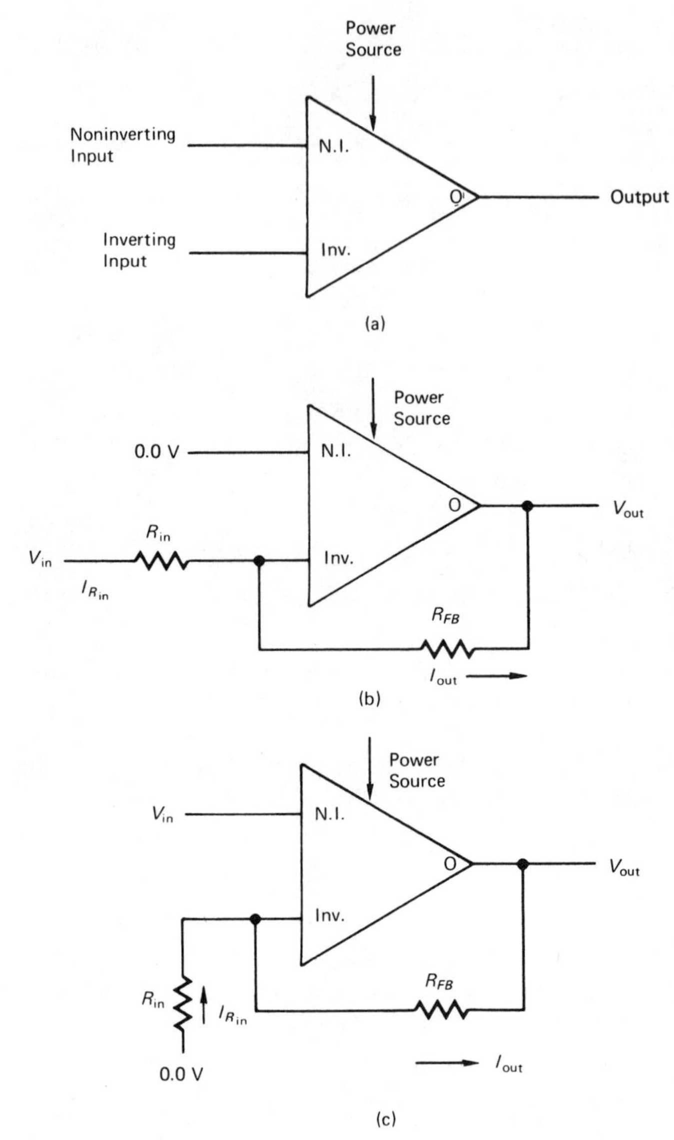


**Figure 27-6.** Operational amplifier circuits.

which, upon rearrangement, becomes

$$\frac{V_{out}}{V_{in}} = \frac{-R_{FB}}{R_{in}} = \text{gain} \qquad \textbf{(27-13)}$$

Equation (27-13) shows that the amount of gain can be controlled precisely by proper adjustment of the values of $R_{in}$ and $R_{FB}$.

The noninverting connection shown in Figure 27-6(c) can be analyzed as follows. Because $V_{INV}$ equals $V_{NI}$ and because no current can flow into the INV terminal,

$$I_{R_{in}} = \frac{V_{INV} - 0.0\ V}{R_{in}} = \frac{V_{NI}}{R_{in}} \tag{27-14}$$

Also,

$$I_{R_{in}} = I_{out} = \frac{V_{out} - V_{INV}}{R_{FB}} \tag{27-15}$$

and, since

$$V_{NI} = V_{in} = V_{INV}$$

substitution in eqs. (27-14) and (27-15) gives

$$\frac{V_{in}}{R_{in}} = \frac{V_{out} - V_{in}}{R_{FB}} \tag{27-16}$$

which, upon sequential rearrangement, becomes

$$\frac{V_{in} R_{FB}}{R_{in}} = V_{out} - V_{in}$$

$$V_{out} = \frac{V_{in} R_{FB}}{R_{in}} + V_{in}$$

$$\frac{V_{out}}{V_{in}} = \frac{R_{FB}}{R_{in}} + 1$$

$$\frac{V_{out}}{V_{in}} = \frac{R_{FB} + R_{in}}{R_{in}} = \text{gain} \tag{27-17}$$

According to eq. (27-17), the gain of a noninverting circuit can be controlled by selection of the proper values for $R_{in}$ and $R_{FB}$.

These basic inverting and noninverting operational amplifiers are often found in testing and measurement instruments and similar devices.

## DIGITAL INTEGRATED CIRCUITS

Digital integrated circuits are those circuits that are widely used in computers and whose outputs are a function of their inputs in a way that can be described by Boolean algebra. The inputs and outputs of digital devices are binary in value; that is, they are either in the logic 0 state or they are in the logic 1 state. For most high-speed logic circuits, including transistor–transistor logic devices

(TTL), which we will discuss, the logic 1 state equals +5 V dc and logic 0 is equal to 0 V dc. Other types of logic circuits include resistor–transistor logic (RTL), diode–transistor logic (DTL), and emitter-coupled logic (ECL). Each of these types of logic has certain advantages, but TTL is probably used more than any other at present.

The two basic functions that can be considered to be the building blocks for all other functions are the AND function and the OR function. The AND function follows the truth table given in Figure 27-7(b) and is symbolized as shown in Figure 27-7(a). A **truth table** defines the relationship between the inputs and the outputs of a device or circuit. Thus, the output of the AND circuit will be in the logic 1 state (+5 V) only when *all* inputs are at the logic 1 state and will be at logic 0 when *any* of the inputs are in logic 0 state. The OR function follows the truth table in Figure 27-7(d) and is symbolized as shown in Figure 27-7(c) The output of the OR circuit is in the logic 1 state when *any or all* of the inputs are in the logic 1 state and will be in logic 0 state only when *all* the inputs are in the logic 0 state.

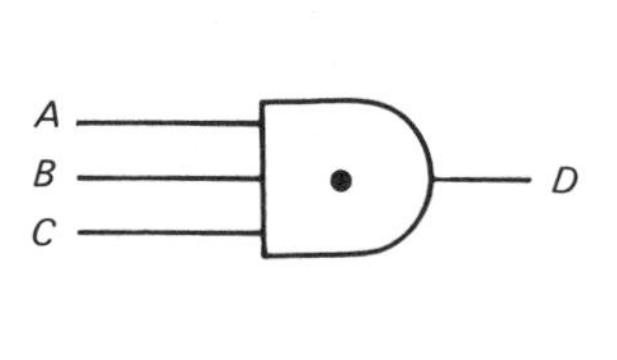


| Input | | | Output |
|---|---|---|---|
| *A* | *B* | *C* | *D* |
| 0 | 0 | 0 | 0 |
| 0 | 0 | 1 | 0 |
| 0 | 1 | 0 | 0 |
| 0 | 1 | 1 | 0 |
| 1 | 0 | 0 | 0 |
| 1 | 0 | 1 | 0 |
| 1 | 1 | 0 | 0 |
| 1 | 1 | 1 | 1 |

AND Function

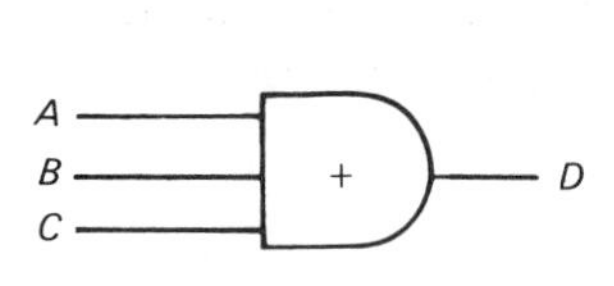


| Input | | | Output |
|---|---|---|---|
| *A* | *B* | *C* | *D* |
| 0 | 0 | 0 | 0 |
| 0 | 0 | 1 | 1 |
| 0 | 1 | 0 | 1 |
| 0 | 1 | 1 | 1 |
| 1 | 0 | 0 | 1 |
| 1 | 0 | 1 | 1 |
| 1 | 1 | 0 | 1 |
| 1 | 1 | 1 | 1 |

OR Function

**Figure 27-7.** Symbols and truth tables for AND and OR functions.

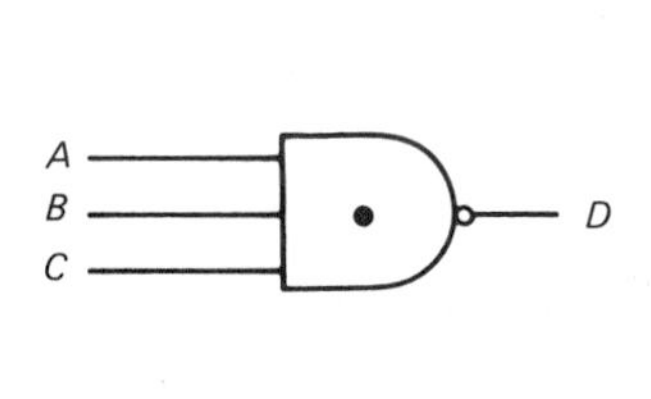


| Inputs | | | Output |
|---|---|---|---|
| A | B | C | D |
| 0 | 0 | 0 | 1 |
| 0 | 0 | 1 | 1 |
| 0 | 1 | 0 | 1 |
| 0 | 1 | 1 | 1 |
| 1 | 0 | 0 | 1 |
| 1 | 0 | 1 | 1 |
| 1 | 1 | 0 | 1 |
| 1 | 1 | 1 | 0 |

NAND Function

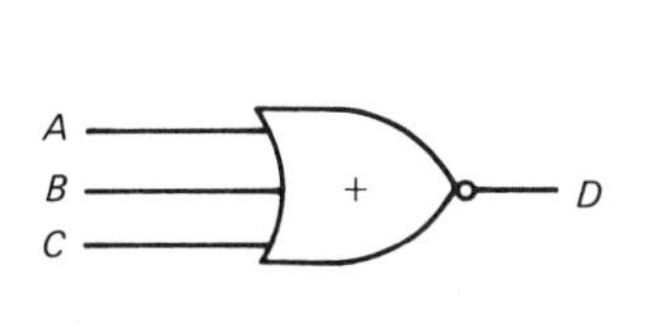


| Inputs | | | Output |
|---|---|---|---|
| A | B | C | D |
| 0 | 0 | 0 | 1 |
| 0 | 0 | 1 | 0 |
| 0 | 1 | 0 | 0 |
| 0 | 1 | 1 | 0 |
| 1 | 0 | 0 | 0 |
| 1 | 0 | 1 | 0 |
| 1 | 1 | 0 | 0 |
| 1 | 1 | 1 | 0 |

NOR Function

**Figure 27-8.** Symbols and truth tables for NAND and NOR functions.

The actual techniques used to adapt these basic logic functions to solid-state integrated circuits make it easier to generate NAND and NOR functions than it is to generate AND and OR functions. The NAND and NOR functions are the inverse of the AND and OR functions, respectively. The truth tables and symbols for the NAND and NOR functions are shown in Figure 27-8.

The operation of a very simple circuit may be examined using an everyday occurrence that all of us have experienced. Assume that you are walking down a street and are approaching an intersection that has a traffic control signal in operation. The decision as to whether to cross the street when you arrive at the intersection will actually be based on several prior decisions, depending upon the factual or logical input to your brain. The first decision to make is whether you want to cross the street or turn the corner. If you do want to cross the street, the next decision is based upon whether the light is green, orange, or red when you arrive at the intersection. If the light is red or orange, you would wait, but if the light is green, you would then check to see if any cars are entering the intersection at right angles to your path or turning the corner into your path. If no cars are interfering with your progress, you would finally check to see if

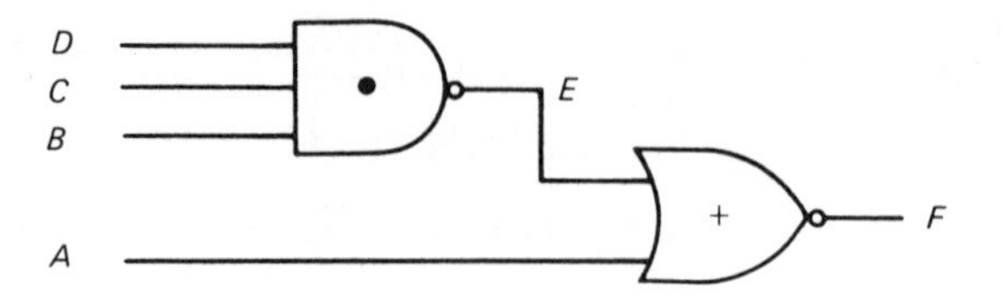


*A* input = 0, if we want to cross street; 1, if not
*B* input = 0, if signal is against us; 1, if for us
*C* input = 0, if cars are entering intersection; 1, if not
*D* input = 0, if there are obstructions in our path; 1, if not
*E* output = 0, if *B, C, D* are all equal to 1 state; 1, if one or more is 0.
*F* output = 1, if we are to cross intersection; 0, if we are not to cross.

(a)

| Inputs | | | | Outputs | |
|---|---|---|---|---|---|
| *A* | *B* | *C* | *D* | *E* | *F* |
| 0 | 0 | 0 | 0 | 1 | 0 |
| 0 | 0 | 0 | 1 | 1 | 0 |
| 0 | 0 | 1 | 0 | 1 | 0 |
| 0 | 0 | 1 | 1 | 1 | 0 |
| 0 | 1 | 0 | 0 | 1 | 0 |
| 0 | 1 | 0 | 1 | 1 | 0 |
| 0 | 1 | 1 | 0 | 1 | 0 |
| 0 | 1 | 1 | 1 | 0 | 1 |
| 1 | 0 | 0 | 0 | 1 | 0 |
| 1 | 0 | 0 | 1 | 1 | 0 |
| 1 | 0 | 1 | 0 | 1 | 0 |
| 1 | 0 | 1 | 1 | 1 | 0 |
| 1 | 1 | 0 | 0 | 1 | 0 |
| 1 | 1 | 0 | 1 | 1 | 0 |
| 1 | 1 | 1 | 0 | 1 | 0 |
| 1 | 1 | 1 | 1 | 0 | 0 |

(b)

**Figure 27-9.** Schematic and truth table for decision to cross intersection.

any other obstructions (pedestrians, building, or road construction, etc.) are in your path. You will decide to cross the street with the green light only if all these decisions are in your favor.

The circuit in Figure 27-9(a) represents the logic used in making the decision. Let input A be 0 if you want to cross the street and 1 if you do not. Input B could then be set for logic 1 if the light is green and 0 if red or orange; input C for logic 1 if no cars are interfering and 0 if they are; and input D set for logic 1 if there are no obstructions and 0 if there are. The truth table shown in Figure 27-9(b) represents the logic in the decision on whether to cross the street.

You will decide to cross the street only if you want to cross (logic 0), if the light is green (logic 1), and there are no cars (logic 1) or other obstructions in your path (logic 1). Output E in the circuit is logic 0 only if B, C, and D are all logic 1 and is logic 1 otherwise. As a consequence, the decision to cross (logic 1 for output F) will be made only when the sequence is 0, 1, 1, 1, 0, 1, which is shown in line 8 of the truth table.

There are many other logic functions available to the designer in addition to the NAND and NOR gates. However, if the designer has a good understanding of these two basic functions, other functions can be understood and utilized because the others use essentially the same fundamental building blocks and logic.

CHAPTER

# 28 Laboratory Techniques

The importance of learning and using good laboratory technique cannot be overemphasized because good technique is the prime prerequisite for accurate analysis. Most students are able to develop the manual dexterity necessary for the proper handling of analytical equipment, but constant practice and care are necessary until the correct operations become automatic through repeated performance. In learning techniques, students should first understand the steps required for correct use of the equipment and should then practice the actual measurements using movements that are comfortable. Repeated practice teaches the manual dexterity necessary, and going slowly at the beginning will help in learning the movements. After the muscles have been trained to perform the technique, students can increase the speed of the operation so that they become both accurate and quick. One of the main aids in learning and using good laboratory practice is to remember to be relaxed when making measurements because tension usually causes poor muscular control and consequent errors.

The techniques given here are generally used in most of the analyses performed by an analyst. The more refined or special techniques that are not used repeatedly are usually described in the specific procedures in which they are needed.

## TRANSFER OF LIQUIDS FROM BEAKERS

In most cases, transfer of a liquid from a beaker is performed with the equipment at room temperature, but there are occasions when hot solutions must be transferred. As a consequence, students should learn how to use beaker tongs to hold the beaker in the proper manner so that the transfer is easy to control. The hands and arms should be in comfortable and normal positions, which allow relaxed and efficient handling during the pouring of the liquid. The best way to do this is to have the tongs encircle the sides of the beaker with the handles at right angles to the lip of the beaker and held so that the wrist and palm of the

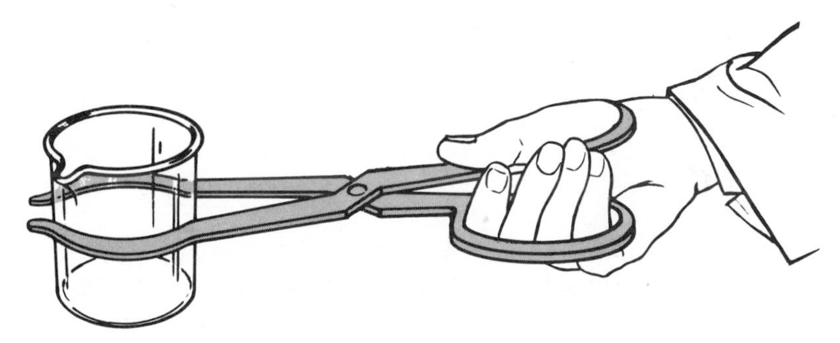

**Figure 28-1.** Use of beaker tongs.

hand is upward, as shown in Figure 28-1. This position allows proper flexing of the wrist for adequate control of the beaker during the pouring operation.

A stirring rod should always be used when pouring from a beaker. The rod should be placed on the lip of the beaker and the liquid poured down the rod, as shown in Figure 28-2. This ensures that all the liquid goes down the rod into the receiver. When the rod is not used, some of the liquid will flow down

**Figure 28-2.** Technique for pouring from a beaker.

the outside of the beaker and be lost to the transfer. Remember that for a transfer to be quantitative, all the liquid must end up in the receiving vessel, not on the outside of the original beaker.

## FILTRATION

The purpose of a filtration is to remove a solid from contact with a solution. The solid is usually a precipitate that has been formed in the liquid, and the liquid is often referred to as the **mother liquor**. The liquid that passes through the filtration device is known as the **filtrate**. Various devices are used in the filtration process, the most common being paper, fritted glass crucibles, and Gooch crucibles. The crucibles are ordinarily used for suction filtration and paper for gravity filtration.

### GRAVITY FILTRATION

Filter-paper circles should be folded into half- and then quarter-circles, as shown in Figure 28-3. One corner at the open end of the fold should be torn

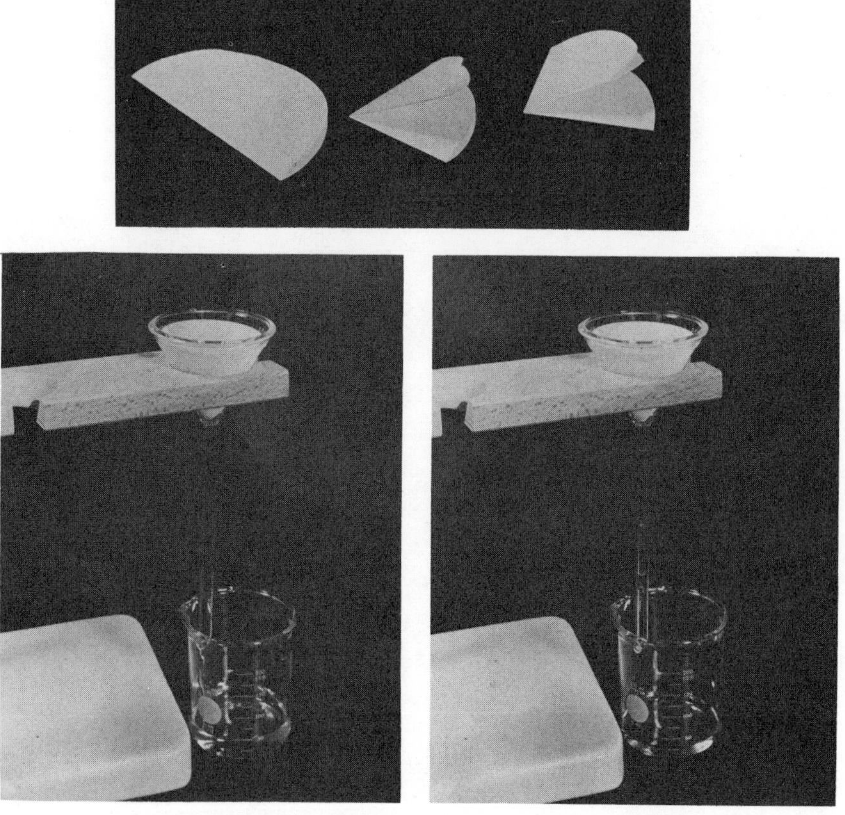

**Figure 28-3.** Preparation of funnel for gravity filtration.

off and the paper should be seated in the funnel with the torn corner to the outside next to the glass. While holding the paper in the funnel, fill it with pure solvent (usually water) until the liquid level is above the top of the paper. Press down the folds and the top of the paper while the liquid is running out to be sure that no air can leak through the paper into the stem of the funnel. This will ensure a full column of water in the stem, which is necessary for adequate and rapid filtration. Discard the solvent used to wet and seal the filter. Figure 28-3 shows a good funnel (stem is full of water) and a poor funnel (stem is filled partly with air). During actual filtration, never fill the paper more than three-fourths full.

## SUCTION FILTRATION

Suction filtration is preferred when possible because it is much faster than ordinary gravity filtration. Gooch crucibles and fritted glass crucibles are used for suction filtration, since the power of the suction will break most ordinary filter paper. Paper can be used if the bottom of the paper is supported by a platinum or porcelain filter cone, but this is not a popular procedure. The crucible is supported on a rubber ring in a funnel in a suction flask, as shown in Figure 28-4. Some of the other methods of support that may be used

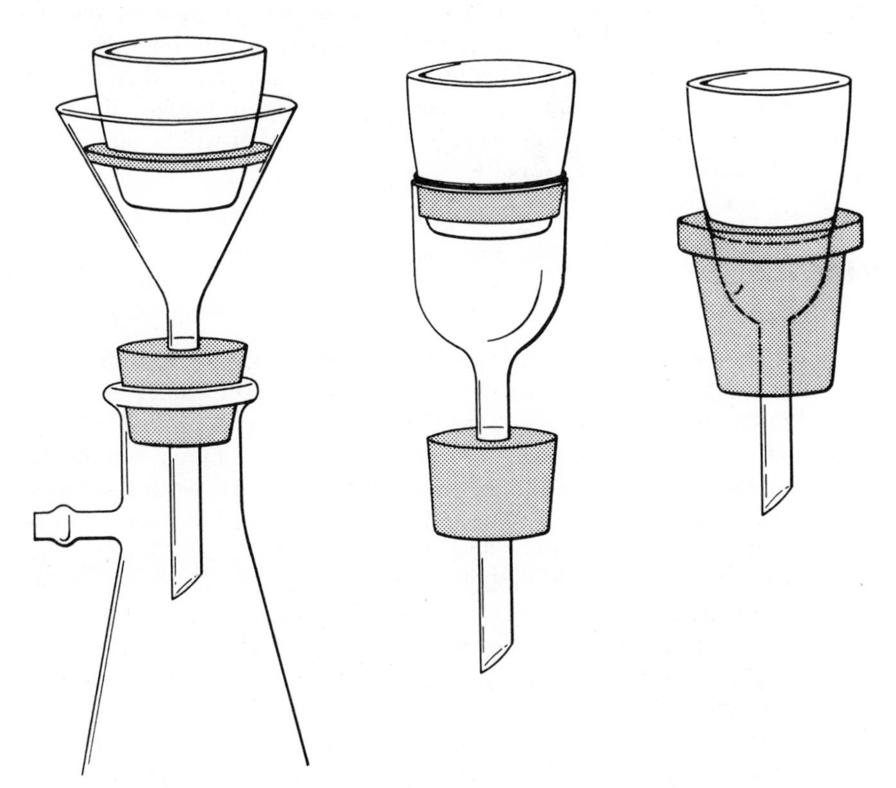

**Figure 28-4.** Methods of support for crucibles during suction filtration.

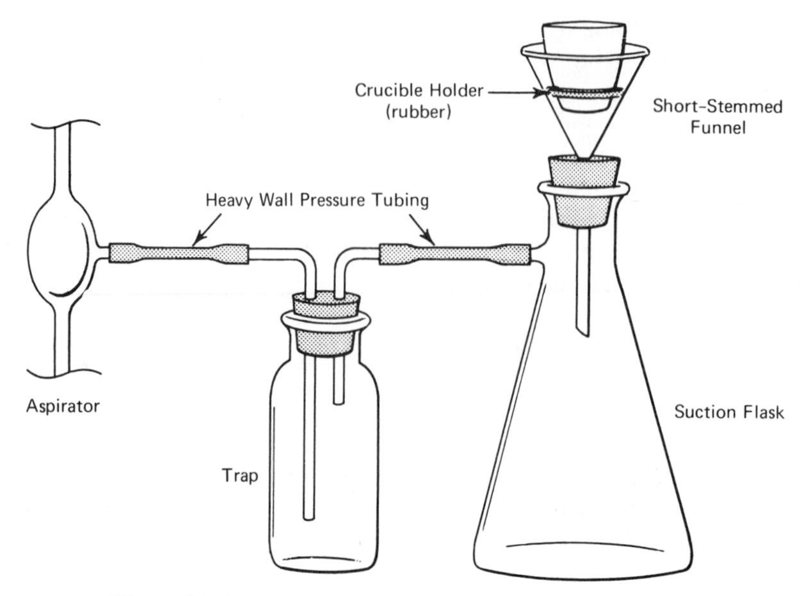


**Figure 28-5.** Arrangement of apparatus for suction filtration.

are shown in the same figure. The suction flask is always separated from the aspirator with a trap to protect the liquid filtrate from contamination with tap water, as shown in Figure 28-5. Whenever there is a change in water pressure on the water line, the suction pulled by the aspirator changes, and, when pressure falls, it is possible for the water to flow from the aspirator to the flask if the trap is not in the line. When water pressure increases, the water will be pulled from the trap if placed in the line properly.

## WASHING

A precipitate is often washed by decantation before being physically transferred to the paper or crucible. In this procedure, which is shown in Figure 28-6, the precipitate is left undisturbed and the mother liquor is poured from the beaker into the filtration device until no more can be poured without the precipitate also leaving the beaker but the filtration device is never filled more than three-fourths full. The precipitate in the beaker is covered with some volume of wash solution, agitated, and allowed to settle. This wash liquid is then decanted in the same way as the mother solution, and the operation is continued as many times as desired. Washing by decantation is desirable when possible, since it is more efficient in removing impurities and is much faster than washing in the filtration device after transfer of the precipitate. The rate of filtration decreases significantly after the precipitate has been transferred.

During the washing by decantation, the liquid should be poured down a rod, as shown in Figure 28-6(a). When the precipitate itself is to be transferred, it must be washed from the beaker with a jet of water from the plastic wash

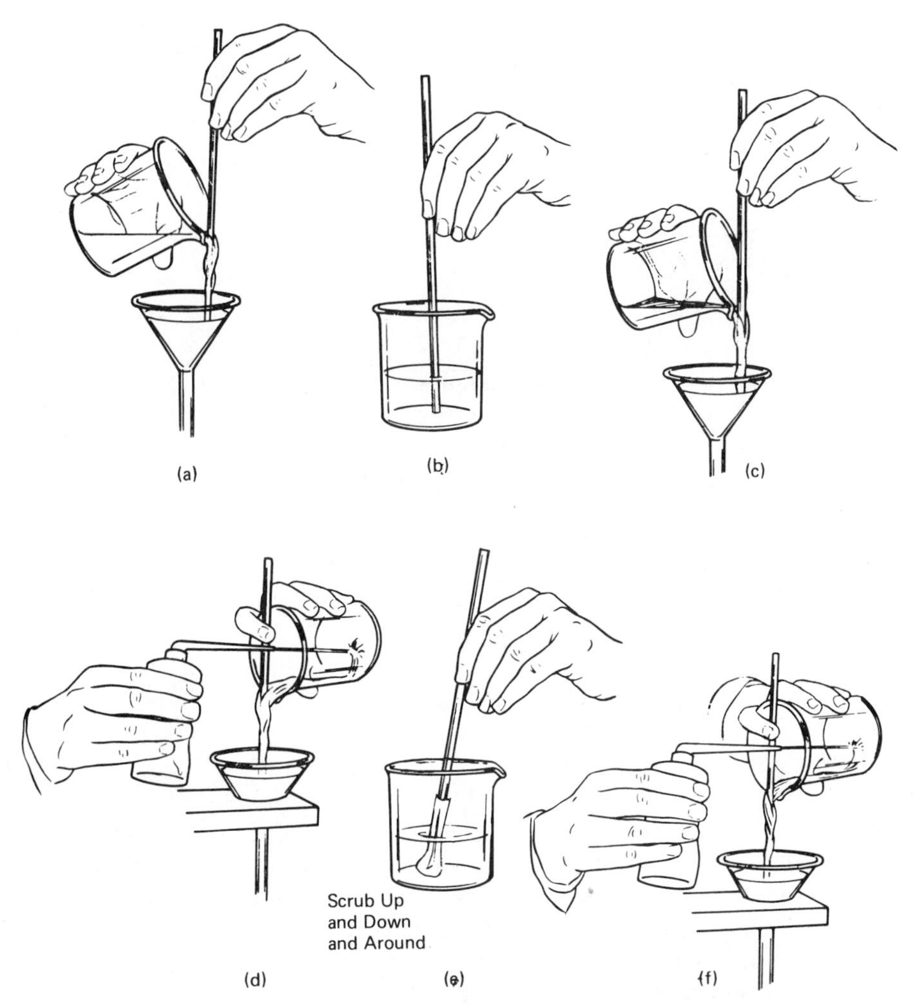


**Figure 28-6.** Transfer of precipitate after washing by decantation.

bottle and allowed to flow down the rod in the same manner, as shown in Figure 28-6(d). Be careful during this part of the transfer to keep the rod centered over the filtration device so that the solid will not be lost outside the device.

After as much of the solid has been transferred as possible, wash down the rod and return it to the beaker. Then scrub the beaker and rod with a policeman to loosen any solid adhering to the walls, as shown in Figure 28-6(e). Wash any loosened solid into the filtration device and continue washing the precipitate as directed for the particular procedure being run.

There are several shapes of rubber policemen on the market for attachment to the end of a stirring rod. Regardless of shape, a policeman must be used only for scrubbing glass containers to remove adherent particles. The rubber portion of the policeman rod should never be used as a stirring rod because the rubber will deteriorate and there is grave danger of contamination of the solution from material on the policeman.

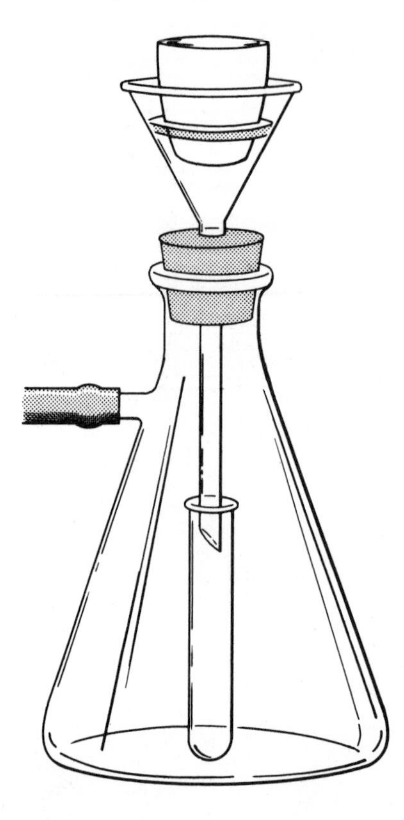

**Figure 28-7.** Use of test tube to catch filtrate to check for completeness of washing.

After washing is completed, it is usually necessary to check the filtrate for the presence or absence of some ion, which will indicate whether or not the washing is satisfactorily complete. For gravity filtration a few drops of the filtrate are caught on a watch glass. The proper reagents are then added to the watch glass to complete the test. For suction filtration, a few drops or milliliters of the filtrate are caught in a test tube inserted in the flask, as shown in Figure 28-7. The test is either made in the test tube, or a small amount is transferred to a watch glass for the test.

## IGNITION OF PRECIPITATES

The paper must be removed from precipitates that have been caught in paper before the precipitate can be dried or ignited for weighing, because it is not possible to dry the paper itself reproducibly or with sufficient accuracy for most quantitative procedures. The precipitate in the paper is transferred to an ignition crucible, and the paper is removed by drying and charring at low temperature, followed by high-temperature ignition to convert the carbon to carbon dioxide.

The proper method of removing and folding the wet filter paper with the precipitate is shown in Figure 28-8. After flattening, the paper is folded sequentially down from the top, up from the bottom, and finally in from the sides

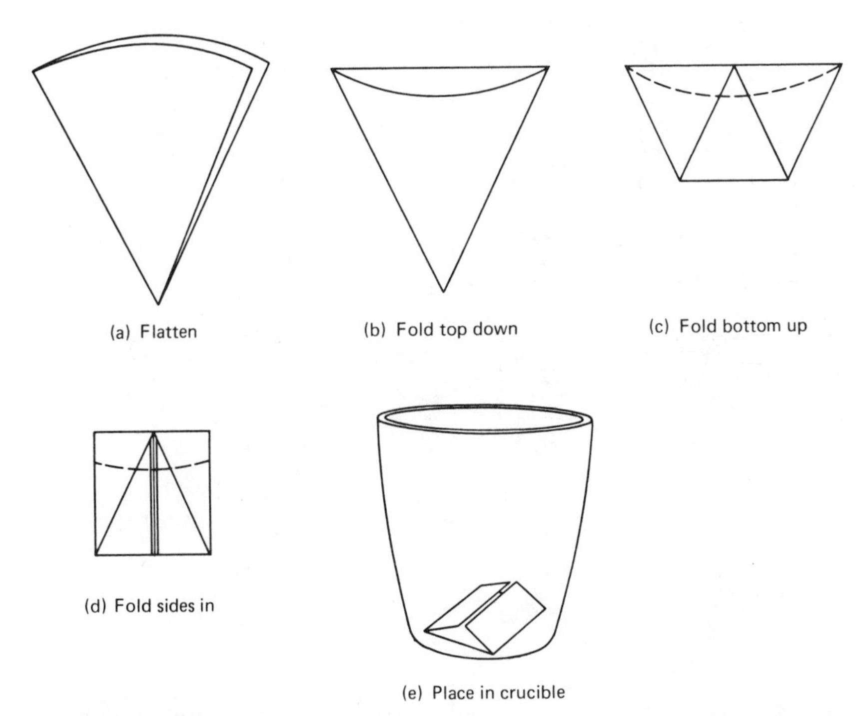


**Figure 28-8.** Procedure for folding filter paper containing a precipitate prior to ignition.

before insertion in the crucible. The purpose of the folding steps is to ensure that the precipitate is separated from the crucible by several thicknesses of paper.

The paper is dried and charred at low temperature, keeping the crucible in an upright position with the top at a slight angle to allow a small opening, as shown in Figure 28-9(a). During this part of the procedure, the temperature should be low enough that the bottom of the crucible never becomes red hot. High temperatures during drying often cause spattering of the precipitate onto the sides and top, where it is easily lost in handling. After the paper is dried, the temperature should be raised slightly to cause charring to start. As the smoke from the charring decreases in volume, the temperature is raised again, and this is continued until no more smoke appears even though a relatively hot flame is being used. *Do not allow the smoke or the paper to catch on fire during this part of the procedure.* After the paper is completely charred, the crucible is placed at an angle, and the top is supported on the clay triangle as shown in Figure 28-9(b). The flame from the burner should be increased to its maximum and placed at the bottom of the crucible so that the flame gases do not enter the top of the crucible. This position allows air to flow over the precipitate and paper and to burn the carbon more rapidly. Flame gases are reducing and often cause reduction of the precipitate to an undesirable form if allowed to enter the crucible while the precipitate is hot. The remainder of the carbon may be removed by heating in an ignition furnace at a specified temperature if one is available. Use of the furnace allows reproducibility of temperatures and

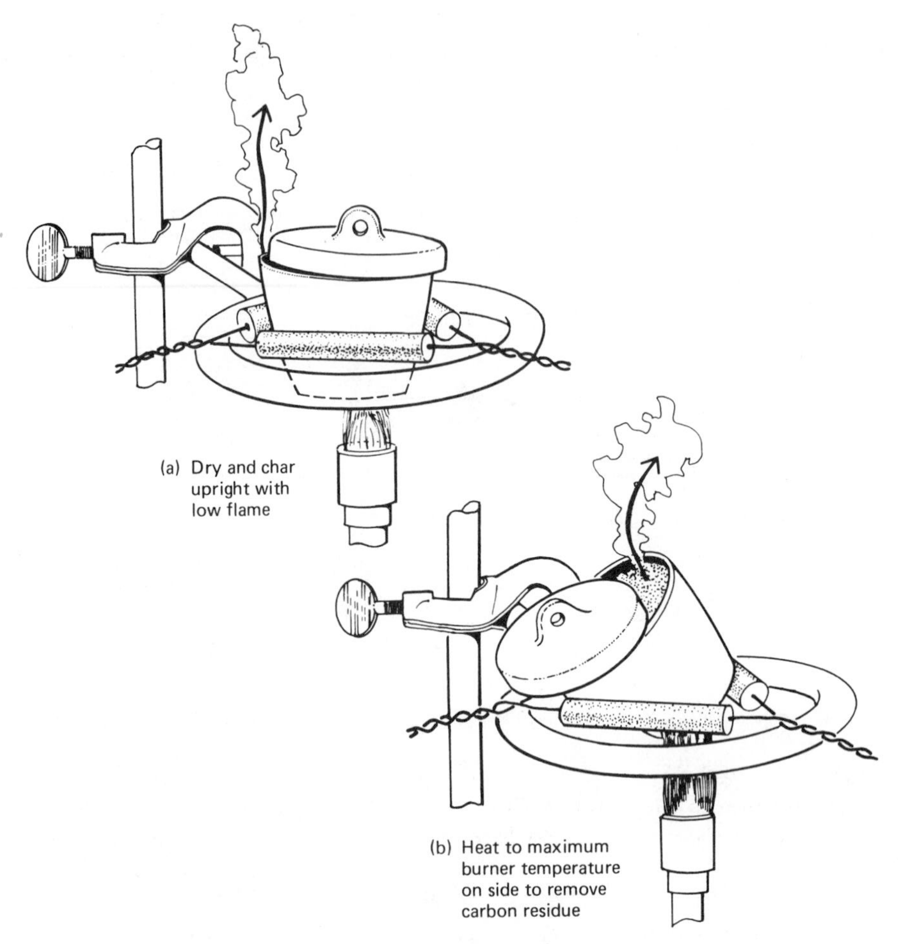


**Figure 28-9.** Drying, charring, and ignition of filter paper containing a precipitate.

conditions that are not possible for an open flame. If the top of the crucible is to be weighed, the carbon must be burned off before weighing. This can be done by heating to red heat over an open flame.

After removal of all paper and carbon, the crucible should be cooled to a little above room temperature and then stored in a desiccator before weighing. Hygroscopic solids should be stored in the desiccator while still above 100°C.

## DRYING AND THE USE OF DESICCATORS

Solids such as unknowns and standards must be dried before weighing to rid them of adsorbed moisture. This is usually performed in a drying oven kept between 100 and 115°C. Heating for 1 hr is usually sufficient, but, when time is available, solids should be reheated and reweighed until a constant weight is attained. The solid to be dried is placed in a weighing bottle, and the weighing

**Figure 28-10.** Arrangement for drying solids in drying oven.

bottle is placed in a beaker covered with a supported watch glass. The contents of the weighing bottle should be marked clearly on the bottle, and the beaker should be marked clearly so that mistakes will not be made in removing the beakers from the oven. The bottle top should be removed from the weighing bottle but should be dried at the same time, as shown in Figure 28-10. Crucibles that do not have to be ignited are dried in the same way. The general rule to be followed in any drying or heating operation is that the equipment should be given the same heat treatment before use as it will be given during the procedure being followed.

The desiccator shown in Figure 28-11 is used primarily to store dry materials to keep them dry. It consists of a lower chamber filled with a liquid or solid desiccant or drying agent and an upper chamber in which the objects stored are placed on a perforated porcelain plate. It is provided with an airtight ground-glass cover lubricated with stopcock grease, and the cover is removed by sliding it sidewise. Red-hot objects are always allowed to cool to approximately 100°C or less before being placed in the desiccator. When a very hot object is placed inside, the air becomes heated and expands before the top is replaced; upon cooling, a vacuum is created inside the desiccator. This makes the top difficult to remove, and the rush of air when the top is removed may cause a fine solid to be blown from its container. Since the capacity of the desiccant to maintain a dry atmosphere is limited, the desiccator must be kept closed as much as possible. Leaving the top off for extended periods of time will saturate the desiccant with moisture so that it is no longer effective. Indicating desiccants are available that change color as the ability of the solid to remove moisture is diminished.

**Figure 28-11.** Storage of dried materials in desiccator.

The most popular solid desiccants are anhydrous magnesium perchlorate, activated alumina, and silica gel, all of which can be purchased in the indicating form for admixture with the ordinary product. All of these may also be regenerated at comparatively low temperatures and used over and over again. Silica gel requires 120°C, alumina 175°C, and magnesium perchlorate 240°C for regeneration.

## EVAPORATION AND DIGESTION

In many procedures it is necessary to reduce the volume of a solution before the next step in the procedure can be performed. Evaporations are usually carried out on a steam hot plate or over a steam bath in order to maintain as high a temperature as possible without causing the solution to boil. Whenever a solution is allowed to boil, there is loss of the solution itself because some of the microdroplets of spray caused by the ebullition escape from the container. Any solute in these droplets would not be determined in the determination step, and this causes a negative error in the analysis.

During evaporation, the cover should be supported by evaporation hooks, as shown in Figure 28-10. The elevated cover allows the vapors to leave the beaker freely and thus speeds the evaporation. Even though it is undesirable to do so, at times it is necessary to reduce the volume of solution more rapidly than

**Figure 28-12.** Storage of liquids in beakers. Cover glass should rest on beaker and should be in this position during digestion of precipitates or for any situation in which the liquid is boiling.

possible by regular evaporation. In this case, the solution can be boiled gently with the cover resting on the beaker top, as shown in Figure 28-12. Vigorous boiling and frothing must be avoided during the evaporation, to ensure the smallest possible loss of solute.

It is often necessary to keep a solution at elevated temperature with minimum loss of solvent. Examples are the digestion of a precipitate to increase particle size and the slow dissolution of a solid. In these cases the solution is kept below the boiling point with the beaker covered, as shown in Figure 28-12.

## USE OF VOLUMETRIC APPARATUS

It is necessary for the student to distinguish carefully between the volume that a vessel holds and the volume that it delivers. If exactly 1 liter of a liquid is poured into a clean dry flask, the amount of liquid that can be poured from the flask will be less than 1 liter because an appreciable quantity of the liquid will cling to the sides of the flask. The quantity of liquid delivered by the flask will depend upon the shape of the flask, the cleanliness of the inside surface, the time allowed for drainage, the viscosity of the liquid, the surface tension of the liquid, and the angle at which the flask is held, as well as other factors. If it is desired that the flask deliver exactly 1 liter, the neck of the flask must be marked at points that vary depending upon the characteristics of the fluid in the flask. The neck, however, may be marked at a point that shows where the flask contains exactly 1 liter, irrespective of the liquid in the flask. Volumetric flasks are marked in this way—to show the amount contained by the flask, not the amount the flask will deliver. Burettes and transfer (volumetric) pipettes, on the other hand, are marked so that they will deliver a definite volume of liquid.

### BURETTES

Usually two 50-mL burettes are furnished each student in a quantitative analysis class. One of these will be a Geissler burette with a glass or Teflon stopcock; the other will be a Mohr burette fitted with a glass bead in rubber tubing

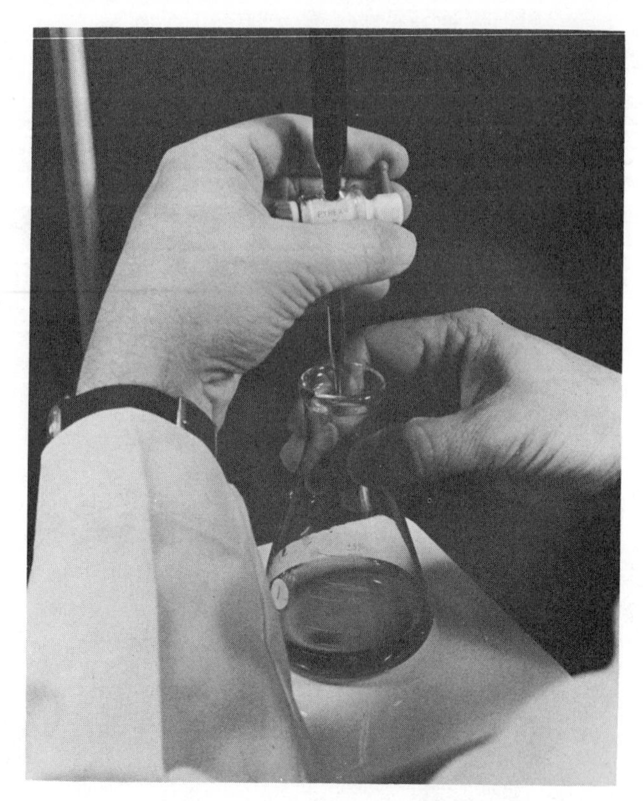

**Figure 28-13.** Operation of stopcock. Always use left hand for stopcock and right hand to agitate the solution.

to control the flow of liquid. The Mohr burette is used for basic titrants such as sodium hydroxide, and the Geissler is used for practically all other titrants.

During titrations, the burette stopcock is normally operated with the left hand, as shown in Figure 28-13. This leaves the right hand free for stirring the liquid or rotating the titration flask. Left-handed students should reverse this and use the right hand for operating the stopcock. Proper control of the stopcock to ensure delivery of small or large volumes accurately requires relaxed muscular control and extensive practice. Muscular tension often causes the student to pull the stopcock out of the burette and thus ruin the titration.

Many errors can be involved in the use of burettes, especially by beginning students:

1 Dirty burettes can cause an error because the volume of a liquid that clings to a dirty glass surface is different from the volume that clings to a clean surface. Water and dilute solutions form a smooth unbroken surface on clean glassware but will form droplets on a dirty surface. Burettes should be cleaned periodically and also whenever it appears that they are not draining evenly. Usually, cleaning with a detergent and a burette brush is satisfactory, but at times it is necessary to use dichromate cleaning solution (concentrated sulfuric acid saturated with potassium dichromate) to get the burette really clean.

The student should practice the utmost caution in the use of dichromate cleaning solution, because it is very corrosive and will cause serious burns if allowed to remain on the skin for more than a few seconds. It can also cause almost immediate deterioration of certain types of cloth, which may lead to embarrassing situations if other clothing is not available. Clothing that has been heavily contaminated with cleaning solution should be removed immediately irrespective of any possible embarrassment.

2 If the liquid is allowed to flow out of the burette rapidly, the student should wait at least 1 min before estimating the position of the meniscus to allow for proper drainage. An error of as much as 0.20 mL can result from reading the position of the meniscus too quickly.

3 In filling the burette, care must be taken to ensure that no bubble of air remains in the tip. If a bubble is entrapped there and dislodged when the liquid is delivered, the reading will be in error by the volume of the bubble.

4 If a solution of known concentration is poured into a clean burette that is wet with water, the solution is diluted, and the concentration is no longer known exactly. The burette should either be dried before it is filled with the solution or it should be rinsed at least three times with small volumes of the solution before filling. This rinsing procedure is usually necessary, since the burette is seldom dry just before use.

5 An error due to parallax is encountered whenever the eye is not at the exact level of the meniscus, as shown in Figure 28-14. To offset this error, students should position their eyes so that the back of the milliliter mark above the meniscus appears to be below the front portion and the back of the milliliter mark below the meniscus appears to be above the front portion.

6 Actually, any part of the meniscus may be utilized as long as it is reproducible, but the same portion of the meniscus must be read every time the position of the meniscus is estimated or an error will be involved. A piece

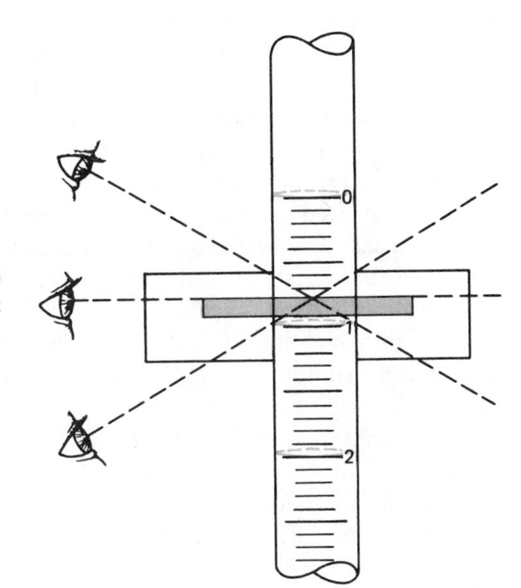


**Figure 28-14.** Parallax. Always keep eye on level of meniscus to offset parallax during estimation of meniscus position.

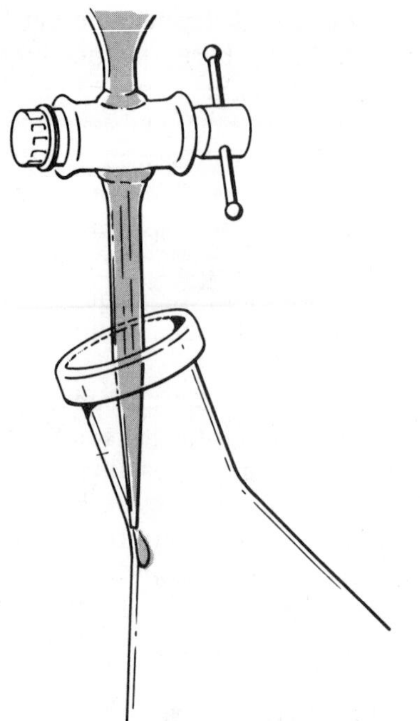

**Figure 28-15.** Partial drops. Touch burette to side of flask to remove partial drops. If titration vessel is a beaker, use the stirring rod to remove partial drops.

of white paper or a piece of white paper upon which a large black mark (with smooth upper edge) is made may be placed behind the burette to sharpen the bottom of the meniscus and aid in the estimation of the correct reading. With intensely colored solutions such as potassium permanganate, the upper part of the meniscus is read.

7 The relative error in measuring the volume of liquid delivered by a burette decreases as the volume increases. Since an error of 0.02 mL is possible in each estimation of the meniscus, an error of 0.04 mL is possible. This represents a relative error of 1 ppt for 40.00 mL, an error of 2 ppt for 20.00 mL, and an error of 40 ppt for a volume of 1.00 mL. The delivery of more than one burette-full for any one titration is to be avoided, since this requires refilling the burette and consequently two additional readings.

8 It is often necessary when approaching the end point of a titration to add a partial drop of titrant. To do this, allow the partial drop to form on the tip of the burette and then touch the tip to the side of the titration flask as shown in Figure 28-15. The flask can then be tilted so that the added amount is mixed with the solution, or the titrant can be washed down with water from the wash bottle. The burette tip should never be rinsed, since this may dilute the titrant in the tip and cause an error.

9 Burettes will often leak slowly through the stopcock if the stopcock is not greased and inserted properly. The use of a large quantity of grease to offset a leak is not advisable, because it can plug the bore of the stopcock or the tip of the burette. The correct way to clean and grease a stopcock is shown in Figure 28-16. The burette is usually wet when the stopcock is cleaned and

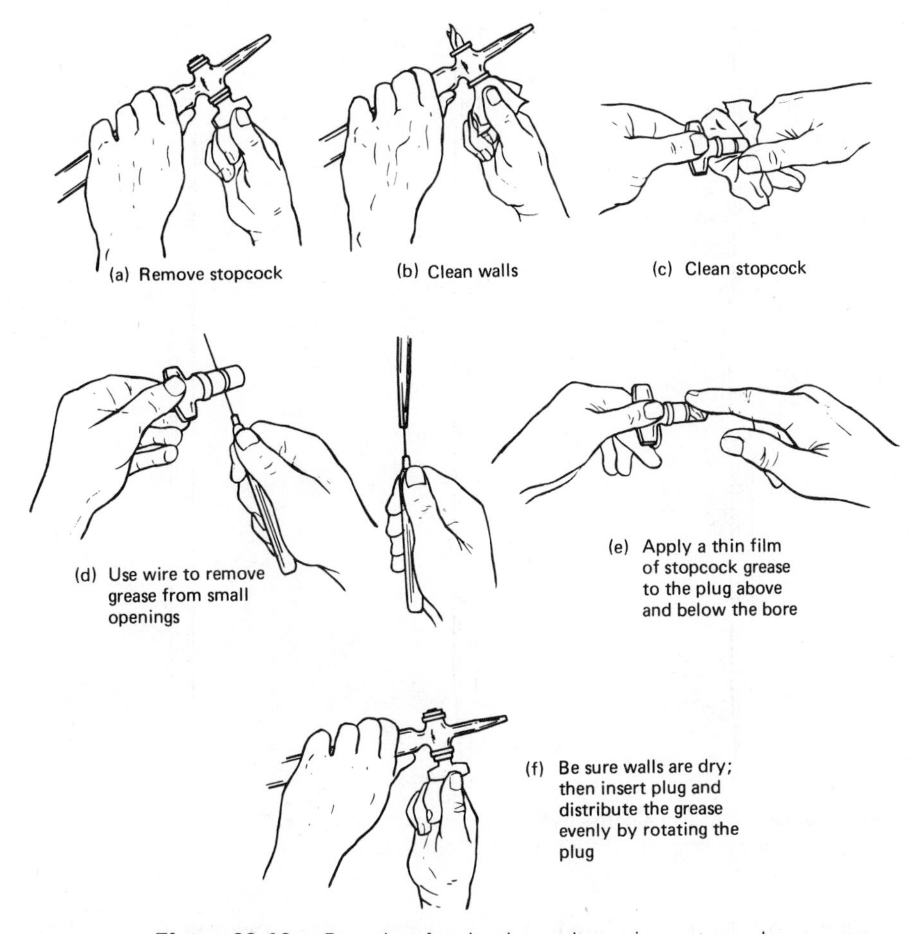


**Figure 28-16.** Procedure for cleaning and greasing a stopcock.

replaced, but the stopcock plug and walls must be dry to ensure a satisfactory seal. As a consequence, the burette is always held in a horizontal position during the removal, cleaning, regreasing, and reinsertion of the stopcock plug to ensure that the walls remain dry. The most common error other than getting water between the plug and walls is to use too much stopcock grease, with consequent clogging of the burette during later operation. A satisfactory seal will be transparent, will have no grease in the bore of the stopcock or tip of the burette, and will allow the plug to be rotated easily. A cloudy seal ordinarily indicates that the walls or plug were not dry or else that too much grease was put on the plug.

## PIPETTES

The two main types of pipettes used in the quantitative analysis laboratory are shown in Figure 28-17. The volumetric or transfer pipette is used to deliver accurately a definite quantity of liquid such as 10.00 or 25.00 mL. The Mohr or

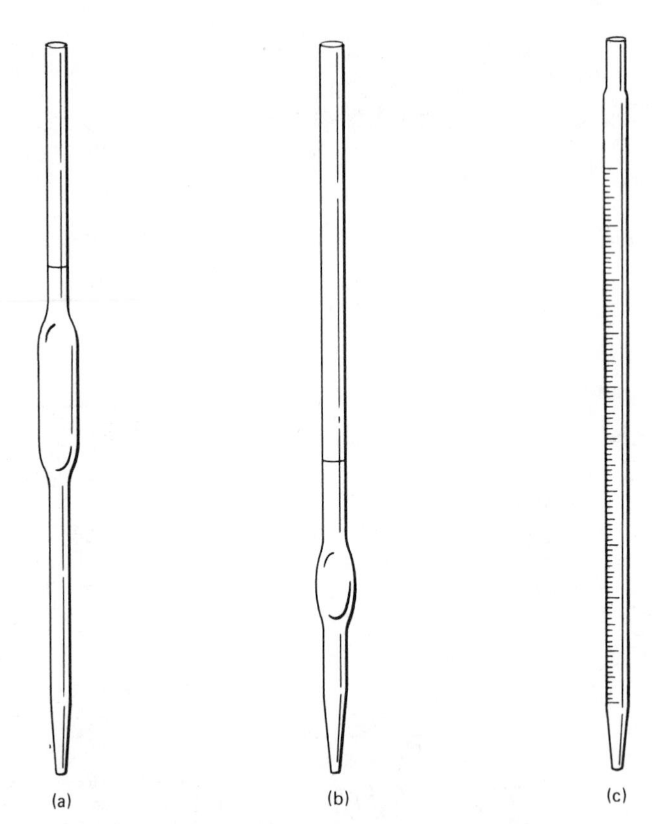


**Figure 28-17.** Types of pipettes used in quantitative analysis. (a) Transfer pipette, drainage type; (b) Transfer pipette, blow-out type; (c) Mohr pipette.

graduated pipette is graduated so that different volumes can be delivered as desired. The transfer pipette is considered to be more accurate, because only one setting of the meniscus is necessary; whereas, for the Mohr pipette, the volume is obtained from the difference between two estimations of the meniscus. Most pipettes are calibrated to deliver a definite volume after a specified drainage time; thus, some liquid must be left in the tip of the pipette after delivery. Some pipettes are calibrated, however, so that the last portion must be blown out rather than left in the tip. The blow-out-type pipette is characterized by one or two etched lines around the top of the pipette stem, and the bulb or expanded portion is closer to the tip than in the drainage type. It is seldom that the difference causes an error, since most laboratories will have only one of these types on hand, but analysts should be aware that there is a difference and should also be able to identify the different types easily. Some pipettes are marked TC (to contain) rathern than TD (to deliver). The contents of the TC pipettes must be washed out of the pipette for accurate measurement.

To pipette accurately and reproducibly, the student must learn how to regulate the flow from the pipette. Since the control of the index finger is usually better than the control of the other fingers, the index finger should be used for this purpose.

The mouth should never be used to produce the necessary suction to pull the liquid into the pipette. A rubber bulb or other suction device should be used for this purpose. Many of the liquids that must be pipetted are corrosive to tissue or poisonous, and thus should not be allowed to get into the mouth. No matter how expert an analyst becomes in mouth pipetting, there are times when inadequate control occurs and a harmful substance is allowed to get in the mouth. One of the authors, even though he knew better, once was unable to

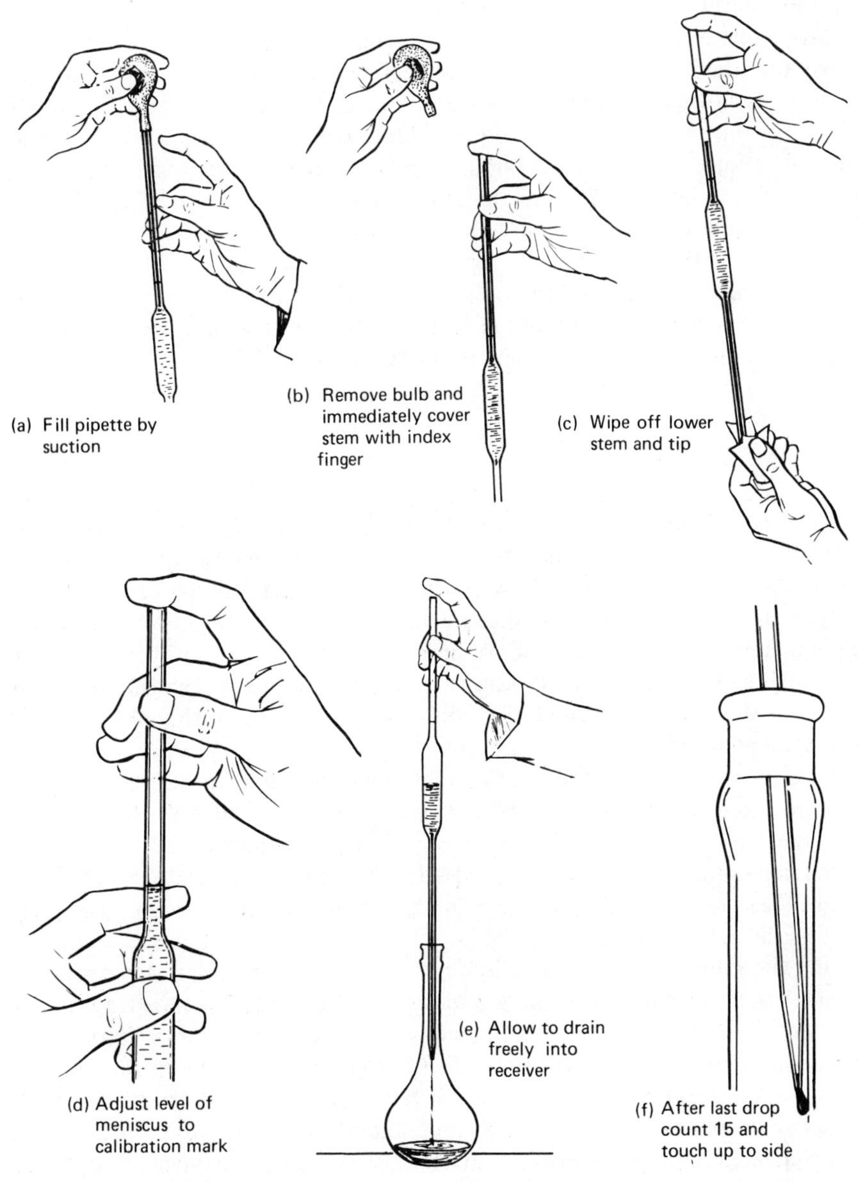


**Figure 28-18.** Correct procedure for use of pipette.

taste food for over a week because he used mouth suction for a concentrated sodium hydroxide solution.

The correct sequence to be used in pipetting is shown in Figure 28-18. The rubber bulb is used to pull the liquid into the stem above the calibration mark, and the right index finger is placed over the top of the stem as the bulb is removed. The pipette is rotated to ensure adequate contact with the finger. The bulb is operated with the left hand. This operation is difficult but can be learned with repeated practice. The thumb should never be used to control the flow. The tip and lower stem of the pipette must be wiped dry with absorbent paper before the liquid level is allowed to fall to the calibration mark, since the paper will remove some of the liquid from the tip. Consequently, if the stem and tip are wiped after the liquid level has been adjusted to the calibration mark, too small a volume will be delivered from the pipette. The meniscus is adjusted to the calibration mark by allowing the liquid to flow slowly until the meniscus reaches the calibration mark. Any drop hanging to the tip must be removed by touching the side of the container. The pipette is then moved slowly to a position over or in the receiving container, and the liquid is allowed to drain freely until the last drop falls. For the drainage-type pipettes, the pipette is allowed to drain for a specified length of time, and the tip is then touched to the side of the container to remove any partial drop that has formed. Usually, it is sufficient to count to 15 after the last drop falls and then touch the side.

## VOLUMETRIC FLASKS

Volumetric flasks are used to prepare solutions of known concentration and are calibrated to contain rather than to deliver. They are designed, as shown in Figure 28-19, with a long neck, which has a much smaller diameter than the body of the flask in order to reduce the error in the estimation of the position of the meniscus. The two methods of preparing a solution of known concentration are dilution of an accurately known volume of a more concentrated solution and dissolving a known weight of a solid in sufficient solvent to make a definite volume of solution.

Solids may be dissolved in another container and then transferred quantitatively to the flask, or they may be dissolved directly in the flask. In the latter case, the flask should be filled approximately three-fourths full and rotated periodically until the solid is dissolved. More solvent can be added if necessary, but the flask should never be filled so full that the liquid cannot be rotated before all the solid is dissolved.

After the solid is dissolved or a liquid pipetted into the flask, it is necessary to fill the flask to the calibration mark and then mix completely before transferring the solution to a storage vessel. Inadequate mixing after dilution is probably one of the most common errors made by beginning students in quantitative analysis. Adequate mixing is attained if the flask is inverted, turned upright, rotated one-quarter turn, inverted, and again turned upright, this process being repeated through at least four inversions. The proper sequence is shown in Figure 28-19.

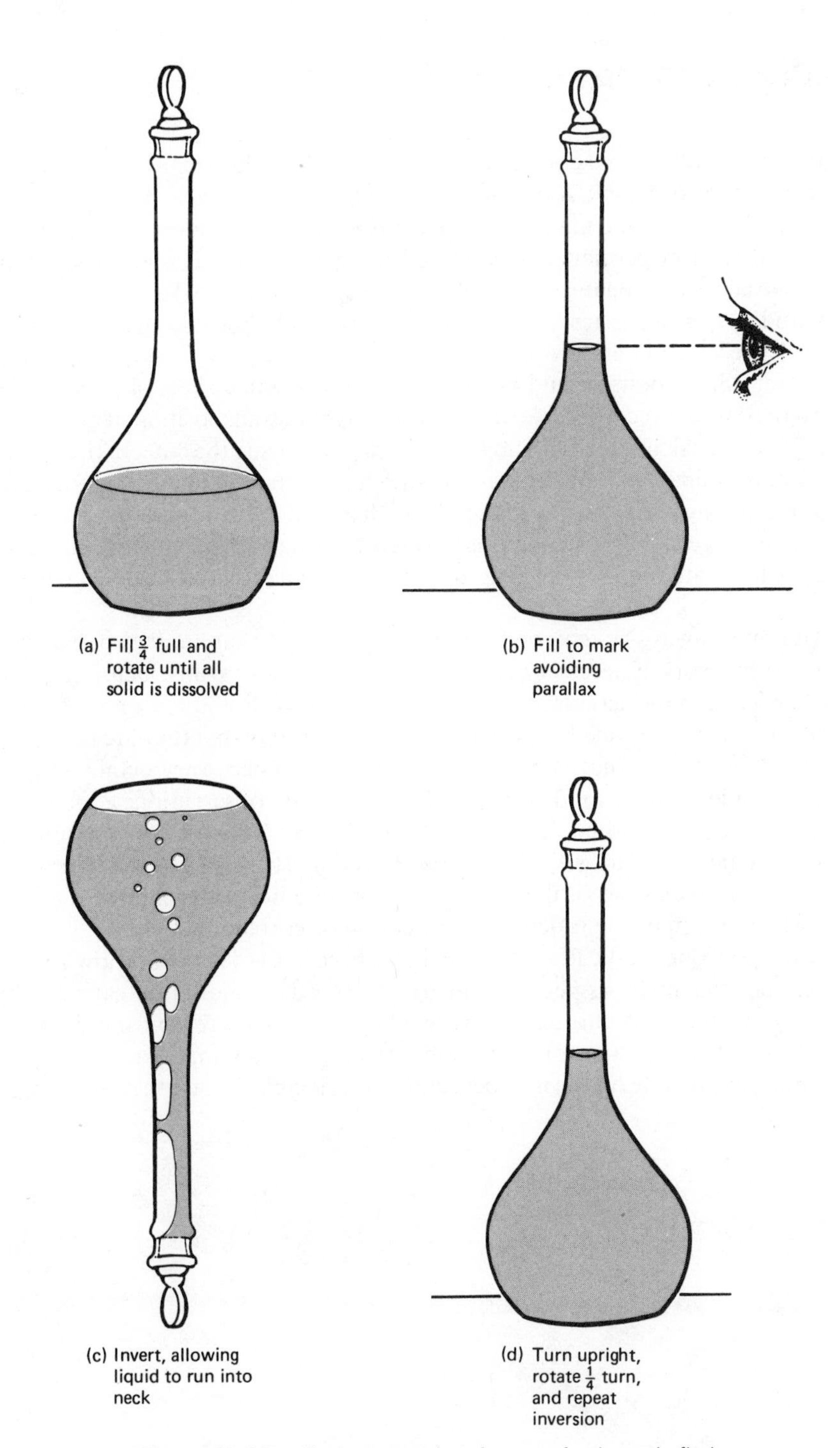


**Figure 28-19.** Correct procedure for use of volumetric flask.

## RECORDING DATA

All data must be taken correctly and should be recorded on the data sheets in a permanent record book. *Never take data on loose sheets of paper.* Loose sheets of paper are lost easily and many errors of transposition are made in later transfers of the data to more permanent locations. Taking data properly is very important in industrial, government, clinical, and research labs as well as in academic labs, and students should learn to practice the correct methods so that they automatically record all data properly.

Related calculations should be performed on the same sheet of paper upon which the data are recorded in order to offset errors of transposition. If not on the same page, the calculations should be on the page opposite the data. In the bound data books utilized in industrial and research labs, the calculations are usually made on the page opposite the data page so that data are always recorded on the right-hand page and the calculations are on the left-hand page. In some manuals, data are listed at the top of the page and related calculations at the bottom of the page.

Data must always be recorded in sufficient detail so that another person can continue the work if necessary and so that a supervisor or instructor can find any particular data necessary to check calculations. For these reasons, data should always be identified, dated, and signed on the day that they are recorded. Short statements of what is being recorded and any necessary details of procedure should be used to identify the data. Also, any observations concerning possible errors, variations from expected colors or precipitates, or unusual results should be noted with the figures. It is almost impossible to remember the details of an analysis or the meaning of a set of unidentified figures at a later date when they may be needed for clarification or correction.

Never erase or eradicate data that have been recorded. If a known error causes the student to suspect certain results, the data should be canceled by drawing crossed lines through it. An explanation of the reason the data are suspect should be written next to or under the crossed-out material. In crossing out data, be sure to leave the numbers and words legible, in case they are needed later.

APPENDIX

# 1 Literature of Analytical Chemistry

The literature of analytical chemistry is constantly growing as new methods are developed and new principles are discovered that can form the basis of an analytical procedure. Most of the new information is in research articles published in various journals or periodicals. Because it is impossible to read all the analytical research papers, even those dealing with relatively narrow fields, many analytical chemists read only a few articles and keep abreast of the literature by reading abstracts and reviews. The best sources of abstracts are *Chemical Abstracts*, *Analytical Abstracts*, *Chemisches Zentralblatt*, and *Gas Chromatography Abstracts*. Each issue of some of the analytical periodicals have rather complete abstracts of the articles in that issue as well as abstracts of papers in other journals.

Review articles are probably the easiest and quickest way to keep up with the progress in any particular field. *Analytical Chemistry* devotes one issue each year to very complete biennial reviews in various fields of pure and applied analytical chemistry. There are also annual "reviews" and "advances" in the various aspects of analysis and instrumentation, which can be beneficial in obtaining information on recent developments.

It is possible to list only a few of the periodicals and books devoted to analytical chemistry to furnish the student with an introduction to the literature available and to suggest sources for further reading and study. Many excellent sources of information must be omitted in any such list.

## PERIODICALS

*Analytical Chemistry* (U.S.A.)
*Analytica Chimica Acta* (International)
*Analyst, The* (British)
*Chemist-Analyst* (U.S.A.)
*Journal of Agricultural and Food Chemistry* (U.S.A.)
*Journal of the Association of Official Analytical Chemists* (U.S.A.)

*Journal of Chemical Education* (U.S.A.)
*Journal of Chromatographic Science* (International)
*Journal of Gas Chromatography* (International)
*Journal of Pharmaceutical Sciences* (U.S.A.)
*Separation Science* (International)
*Talanta* (International)
*Zeitschrift für Analytische Chemie* (German)
*Zhurnal Analiticheskoi Khimii* (Russian; English translation availabile)

## MULTIVOLUME COLLECTIONS OF PRINCIPLES AND METHODS[1]

Bard, A. J., ed. *Electroanalytical Chemistry*. New York: Marcel Dekker, 1967.
Belcher, R., and L. Gordon. *International Series of Monographs on Analytical Chemistry*. New York: Pergamon Press, 1961.
Elving, P. J., and I. M. Kolthoff. *Chemical Analysis*. New York: Wiley–Interscience, 1941.
Furman, N. H., and F. J. Welcher, eds. *Standard Methods of Chemical Analysis*, 6th ed. New York: Van Nostrand Reinhold, 1963.
Kolthoff, I. M., and P. J. Elving, eds. *Treatise on Analytical Chemistry*. New York: Wiley–Interscience, 1959.
Welcher, F. J. *Organic Analytical Reagents*. New York: Van Nostrand Reinhold, 1947.
Wilson, C. L., and D. W. Wilson, eds. *Comprehensive Analytical Chemistry*. New York: Van Nostrand Reinhold, 1959.

Also, many series entitled *Advances in . . .*, *Recent Developments in . . .*, *Progress in . . .*, *Annual Review of . . .*, *Topics in . . .*, etc.

## OFFICIAL METHODS[2]

*A.S.T.M. Methods for Chemical Analysis of Metals*. Philadelphia: American Society for Testing Materials.
Hanson, N. W. *Official, Standardized and Recommended Methods of Analysis*. London: Society for Analytical Chemistry.
Horwitz, W. *Official Methods of Analysis*. Washington, D.C.: Association of Official Analytical Chemists.
Kline, E. K., et al. *Standard Methods for the Examination of Water, Sewage, and Industrial Wastes*. New York: American Public Health Association.
Wichers, E., et al. *Reagent Chemicals*. Washington, D.C.: American Chemical Society.

[1] Dates are for latest edition or first volume.

[2] Use latest edition.

# TEXTS AND REFERENCE BOOKS

## ELEMENTARY AND ADVANCED TEXTS

Ayres, G. H. *Quantitative Chemical Analysis*, 2nd ed. New York: Harper and Row, 1968.

Christian, G. D. *Analytical Chemistry*, 2nd ed. New York: Wiley, 1977.

Fischer, R. B., and D. G. Peters. *Quantitative Chemical Analysis*, 3rd ed. Philadelphia: Saunders, 1968.

Fritz, J. S., and G. H. Schenk. *Quantitative Analytical Chemistry*, 3rd ed. Boston: Allyn and Bacon, 1974.

Hamilton, L. E., and S. G. Simpson. *Quantitative Chemical Analysis*, 12th ed. New York: Macmillan, 1964.

Kenner, C. T. *Analytical Separations and Determinations*. New York: Macmillan, 1971.

Kolthoff, I. M., E. B. Sandell, E. J. Meehan, and S. Bruckenstein. *Quantitative Chemical Analysis*, 4th ed. New York: Macmillan, 1969.

Laitinen, H. A., and W. E. Harris. *Chemical Analysis*, 2nd ed. New York: McGraw-Hill, 1975.

Meites, L., and H. C. Thomas. *Advanced Analytical Chemistry*. New York: McGraw-Hill, 1958.

Peters, D. G., J. M. Hayes, and G. M. Hieftje. *Chemical Separations and Measurements*. Philadelphia: Saunders, 1974.

Schenk, G. H., R. B. Hahn, and A. V. Hartkopf. *Quantitative Analytical Chemistry*. Boston: Allyn and Bacon, 1977.

Skoog, D. A., and D. M. West. *Fundamentals of Analytical Chemistry*, 3rd ed. New York: Holt, Rinehart and Winston, 1976.

Vogel, A. I. *Quantitative Inorganic Analysis*, 3rd ed. New York: Wiley, 1961.

# INSTRUMENTAL ANALYSIS (GENERAL TEXTS AND MANUALS)

Ewing, G. *Instrumental Methods of Chemical Analysis*, 4th ed. New York: McGraw-Hill, 1975.

Guilbault, G. G., and L. G. Hargis. *Instrumental Analysis Manual*. New York: Marcel Dekker, 1970.

Kenner, C. T., *Instrumental and Separation Analysis*. Columbus, Ohio: Merrill, 1973.

Mann, C. K., T. J. Vickers, and W. M. Gulik. *Instrumental Analysis*. New York: Harper & Row, 1974.

Pecsok, R. L., L. D. Shields, T. Cairns, and I. G. McWilliam. *Modern Methods of Chemical Analysis*, 2nd ed. New York: Wiley, 1976.

Reilley, C. N., and D. T. Sawyer. *Experiments for Instrumental Methods*. New York: McGraw-Hill, 1961.

Robinson, J. W., *Undergraduate Instrumental Analysis*. New York: Marcel Dekker, 1970.

Skoog, D. A., and D. M. West. *Principles of Instrumental Analysis*. New York: Holt, Rinehart and Winston, 1971.

Willard, H. H., L. L. Merritt, Jr., and J. A. Dean. *Instrumental Methods of Analysis*, 5th ed. New York: Van Nostrand Reinhold, 1975.

## THEORY AND SPECIAL TOPICS

Altgelt, K. H., and L. Segal, eds. *Gel Permeation Chromatography*. New York: Marcel Dekker, 1971.

Bates, R. G. *Determination of* pH. New York: Wiley, 1964.

Berg, E. W. *Physical and Chemical Methods of Separation*. New York: McGraw-Hill, 1963.

Berl, W. G., ed. *Physical Methods in Chemical Analysis*. New York: Academic Press, Inc., 1960.

Bishop, E., ed. *Indicators*. New York: Pergamon Press, 1972.

Boltz, D. F. *Colorimetric Determination of Nonmetals*. New York: Wiley–Interscience, 1958.

Christian, G. D., and F. J. Feldman. *Atomic Absorption Spectroscopy*. New York: Wiley, 1970.

Dal Nogare, S., and R. S. Juvet. *Gas–Liquid Chromatography*. New York: Wiley–Interscience, 1962.

Dean, J. A. *Chemical Separation Methods*. New York: Van Nostrand Reinhold, 1969.

Dean, J. A. *Flame Photometry*. New York: McGraw-Hill, 1960.

DeGalan, L. *Analytical Spectrometry*. London: Adam Hilger, 1971.

Duval, C. *Inorganic Thermogravimetric Analysis*, 2nd ed. Amsterdam: Elsevier, 1962.

Eisenman, G., ed. *Glass Electrodes for Hydrogen and Other Cations*. New York: Marcel Dekker, 1967.

Eisenman, G., R. G. Bates, and G. Mattock. *The Glass Electrode*. New York: Wiley–Interscience, 1966.

Feigl, G., *Spot Tests*, 5th ed. Amsterdam: Elsevier, 1962.

Friedlander, G., J. W. Kennedy, and J. M. Miller. *Nuclear and Radiochemistry*. New York: Wiley, 1964.

Fritz, J. S. *Acid–Base Titrations in Non-aqueous Solvents*. Boston: Allyn and Bacon, 1973.

Gordon, L., M. L. Salutsky, and H. H. Willard. *Precipitation from Homogeneous Solution*. New York: Wiley, 1959.

Hamilton, L. F., S. G. Simpson, and D. W. Ellis. *Calculations of Analytical Chemistry*, 7th ed. New York: McGraw-Hill, 1969.

Heftman, E. *Chromatography*, 2nd ed. New York: Reinhold, 1967.

Helfferich, F. *Ion Exchange*. New York: McGraw-Hill, 1962.

Heyrovsky, J., and J. Kuta. *Principles of Polarography*. New York: Academic Press, 1966.

Jones, R. A. *An Introduction to Gas–Liquid Chromatography*. New York: Academic Press, 1970.

Kirkland, J. J. ed. *Modern Practice of Liquid Chromatography*. New York: Wiley–Interscience, 1971.

Lundell, G. E. F., J. I. Hoffman, and H. A. Bright. *Chemical Analysis of Iron and Steel*. New York: Wiley, 1931.
Lynden-Bell, R., and R. K. Harris. *Nuclear Magnetic Resonance Spectroscopy*. New York: Appleton-Century-Crofts, 1971.
Malmstadt, H. V., and C. G. Enke. *Digital Electronics for Scientists*. New York: Benjamin, 1969.
Meites, L. *Polarographic Techniques*, 2nd ed. New York: Wiley–Interscience, 1965.
Mellon, M. G. *Analytical Absorption Spectroscopy*. New York: Wiley, 1950.
Milne, G. W. A., ed. *Mass Spectrometry: Techniques and Applications*. New York: Wiley–Interscience, 1971.
Moody, G. J., and J. D. R. Thomas. *Selective Ion Sensitive Electrodes*. Watford, England: Merrow Publishing Co. Ltd., 1971.
Morrison, G. H., and H. Freiser. *Solvent Extraction in Analytical Chemistry*. New York: Wiley, 1963.
Parker, F. S. *Applications of Infrared Spectroscopy in Biochemistry, Biology, and Medicine*. New York: Plenum Press, 1971.
Perrin, D. D. *Masking and Demasking of Chemical Reactions*. New York: Wiley, 1970.
Price, W. J. *Analytical Absorption Spectroscopy*. London: Heyden, 1972.
Rieman, W., and H. F. Walton. *Ion Exchange in Analytical Chemistry*. Oxford: Pergamon, 1970.
Ringbom, A. *Complexation in Analytical Chemistry*. New York: Wiley–Interscience, 1963.
Samuelson, O. *Ion Exchange Separations in Analytical Chemistry*. New York: Wiley, 1963.
Sandell, E. B. *Colorimetric Determination of Traces of Metals*, 3rd ed. New York: Wiley–Interscience, 1959.
Schenk, G. H. *Absorption of Light and Ultraviolet Radiation*. Boston: Allyn and Bacon, 1973.
Schwartzenbach, G. *Complexometric Titrations*, transl. by H. Irving. New York: Wiley–Interscience, 1957.
Schwartzenbach, G., and H. Flaschka. *Complexometric Titrations*, 2nd English ed., translated by H. M. N. H. Irving. London: Methuen, 1969.
Siggia, S., ed. *Instrumental Methods of Organic Functional Group Analysis*. New York: Wiley, 1972.
Siggia, S. *Quantitative Organic Analysis via Functional Groups*, 3rd ed. New York: Wiley, 1963.
Sillen, L. G., and A. E. Martel. *Stability Constant of Metal-Ion Complexes. I, Inorganic Ligands; II, Organic Ligands*. London: The Chemical Society, 1964.
Silverstein, R. M., and G. C. Bassler. *Spectrometric Identification of Organic Compounds*, 2nd ed. New York: Wiley, 1967.
Stewart, J. E. *Infrared Spectroscopy: Experimental Methods and Techniques*. New York: Marcel Dekker, 1970.
Wagner, W., and C. J. Hull. *Inorganic Titrimetric Analysis*. New York: Marcel Dekker, 1971.
Walton, H. F. *Principles and Methods of Chemical Analysis*, 2nd ed. Englewood Cliffs, N.J.: Prentice-Hall, 1964.
Welcher, F. J. *The Analytical Uses of Ethylenediaminetetraacetic Acid*. New York: Van Nostrand Reinhold, 1958.

Wilson, A. J. C. *Elements of X-Ray Crystallography*. Reading, Mass.: Addison-Wesley, 1969.
Winefordner, J. D., ed. *Spectrochemical Methods of Analysis*. New York: Wiley, 1970.
Winefordner, J. D., S. G. Schulman, and T. O'Haver. *Luminescence Spectrometry in Analytical Chemistry*. New York: Wiley–Interscience, 1972.
Winton, A. L., and K. B. Winton. *The Analysis of Foods*. New York: Wiley, 1945.

# 2 Inorganic Nomenclature

Several methods of naming inorganic compounds have been utilized in the past, but the method used to the greatest extent today is the system proposed by a committee of the International Union of Pure and Applied Chemistry. This system is described in this appendix and is correlated with some of the older systems still in use. The naming of binary compounds is discussed before ternary compounds.

## BINARY COMPOUNDS

Binary compounds contain two different elements, and the most electropositive element is named first because it is written first in the formula. To name a binary compound, give the name of the most electropositive element followed by the name of the most electronegative element ending in "ide." The majority of such compounds are those formed by reaction of a metal with a nonmetal and thus are classified as being salts. For metals with variable valence, the oxidation number is shown by use of a roman numeral in parentheses following the name of the metal.

| | | | |
|---|---|---|---|
| $NaCl$ | sodium chloride | $FeCl_2$ | iron(II) chloride |
| $Al_2O_3$ | aluminum oxide | $Fe_2O_3$ | iron(III) oxide |
| $Ca_3P_2$ | calcium phosphide | $SnBr_4$ | tin(IV) bromide |
| $CdS$ | cadmium sulfide | $Cu_2O$ | copper(I) oxide |

In the older system, variable valence of metals was indicated by use of the suffixes *-ous* and *-ic* on the name of the metal, with -ous indicating the lower and -ic the higher valence. Thus, $FeCl_2$ is ferrous chloride and $FeCl_3$ is ferric chloride.

Binary compounds may also be formed between two nonmetals with the most electropositive being written first and named first. For elements of variable valence that can form more than one compound, prefixes are used with each nonmetal to indicate the number of atoms of each element present. Prefixes in common use are *mono-* for one, *di-* for two, *tri-* for three, *tetra-* for four, *penta-* for five, *hexa-* for six, *hepta-* for seven, and *octa-* for eight. The prefix "mono" is understood if no prefix is used. In the following examples, older methods of naming are given in parentheses when there is a difference.

| | |
|---|---|
| $N_2O$ | dinitrogen monoxide (nitrous oxide) |
| $NO$ | nitrogen oxide or mononitrogen monoxide (nitric oxide) |
| $NO_2$ | nitrogen dioxide |
| $N_2O_3$ | dinitrogen trioxide (nitrogen trioxide) |
| $SO_2$ | sulfur dioxide |
| $SO_3$ | sulfur trioxide |
| $IF_7$ | iodine heptafluoride |
| $BrF_5$ | bromine pentafluoride |

A few polyatomic ions have special names and are treated as if they were single atoms or elements in naming their compounds. Examples are $OH^-$ (hydroxide), $CN^-$ (cyanide), and $NH_4^+$ (ammonium). Thus, NaOH is sodium hydroxide, KCN is potassium cyanide, and $NH_4Cl$ is ammonium chloride.

If a binary hydrogen compound acts as an acid when it is dissolved in water, it is usually named as an acid with the prefix *hydro-* and the suffix *-ic*. Thus, HCl is hydrochloric acid, $H_2S$ is hydrosulfuric acid, HCN is hydrocyanic acid, HI is hydroiodic acid, and HBr is hydrobromic acid. These compounds also may be named as hydrogen chloride, hydrogen sulfide, hydrogen cyanide, and so on. This latter type of name is generally used when referring to the actual compound, whereas the acid name is used to refer to the water solution of the compound. Thus, we speak of hydrogen chloride gas, the water solution of which is hydrochloric acid.

## TERNARY COMPOUNDS

Ternary compounds are those containing three different elements. It has already been noted that some ternary compounds, such as $NH_4Cl$, KOH, and HCN, are named as if they were binary compounds. Chlorine, nitrogen, sulfur, phosphorus, and several other elements each form oxyacids (ternary compounds with hydrogen and oxygen), which differ from each other in their oxygen content. Usually, the most common acid of a series bears the name of the acid-forming element, ending with the suffix *-ic*. This may be noted in the names chloric acid ($HClO_3$), sulfuric acid ($H_2SO_4$), nitric acid ($HNO_3$), and phosphoric acid ($H_3PO_4$). The names of acids containing one oxygen atom more than the -ic acid retain the suffix -ic and have the prefix *per-* added. The name perchloric acid for $HClO_4$ illustrates this rule. An acid that contains one less oxygen atom than the -ic acid is named with the suffix *-ous*. Examples are chlorous acid ($HClO_2$), sulfurous acid ($H_2SO_3$), nitrous acid ($HNO_2$), and phosphorous acid ($H_3PO_3$). Acids with one less oxygen atom than the -ous acid are named by adding the prefix *hypo-* and retaining the ending -ous. Thus, HClO is hypochlorous acid, and $H_2SO_2$ is hyposulfurous acid.

Metal salts of the oxyacids (compounds in which a metal replaces the hydrogen of the acid) are named by naming the metal and then the acid radical. The ending -ic of the oxyacid name is changed to *-ate* and the ending -ous of the acid is changed to *-ite* for the salt. The salts of perchloric acid are perchlorates, those of sulfuric acid are sulfates, those of nitrous acid are nitrites, and those of hyposulfurous acid are hyposulfites. This system of naming applies to all inorganic oxyacids and their salts. Examples of the names of the oxyacids of chlorine and the corresponding sodium salts are as follows:

| *Acids* | | *Salts* | |
|---|---|---|---|
| HClO | hypochlorous acid | NaClO | sodium hypochlorite |
| $HClO_2$ | chlorous acid | $NaClO_2$ | sodium chlorite |
| $HClO_3$ | chloric acid | $NaClO_3$ | sodium chlorate |
| $HClO_4$ | perchloric acid | $NaClO_4$ | sodium perchlorate |

If an acid contains more than one hydrogen, it may react partially with a base to form acid salts, which still contain some hydrogen. Examples are NaHS, $NaHCO_3$, $KH_2PO_4$, and $CaHPO_4$. Such salts can be named by stating the electropositive elements present followed by the usual name of the acid radical; thus, NaHS is sodium hydrogen sulfide, $NaHCO_3$ is sodium hydrogen carbonate, $KH_2PO_4$ is potassium dihydrogen phosphate, and $CaHPO_4$ is calcium hydrogen phosphate. If the acid that produces the salt contains two hydrogens (is diprotic), the salt may be named by using the prefix "bi" in front of the name of the acid radical; thus, NaHS is sodium bisulfide, $NaHCO_3$ is sodium bicarbonate, $NaHSO_4$ is sodium bisulfate, and $Mg(HSO_3)_2$ is magnesium bisulfite. If the oxyacid contains three hydrogens (is triprotic), the acid salts are often called the primary and secondary salts. Thus, $NaH_2PO_4$ is primary sodium phosphate, and $Li_2HPO_4$ is secondary lithium phosphate.

## COORDINATION COMPLEXES

Coordination complexes include those compounds or ions in which an atom or a group of atoms is attached to a metal ion by coordinate bonding in which the atom or group of atoms furnishes the electron pair for the bond. The metal ion is called the **central atom** and the atoms or groups of atoms attached are known as **ligands**. The total number of electron pair bonds formed by the metal is the **coordination number** of the metal. The complex formed can be positive, negative, or neutral. In naming coordination complex compounds, the following rules are followed:

1 Name the cation first followed by the anion.

2 In naming the coordination sphere, whether it be a cation, an anion, or a neutral molecule, the ligands are named first, followed by the name of the metal.

3 In the coordination sphere, name the negative ligands first, giving them the suffix *-o* as in "chloro." The neutral ligands are named next, followed by those which are positive.

4 The number of each ligand is indicated by using the prefixes *di-*, *tri-*, *tetra-*, *penta-*, *hexa-*, and so on. The *bis-*, *tris-*, and so on, Greek prefixes are sometimes used for large ligands.

5 The valence or oxidation number of the central atom is shown as a roman numeral in parentheses directly following the name.

6 If the complex is a cation or a neutral molecule. the name of the metal is used without change, but if the complex is an anion, the suffix *-ate* is appended to the name of the metal.

7 The name of the coordination sphere is written as one word, without separating the names of the individual ligands or metal.

8 In writing the formula, enclose the coordination sphere in brackets and show neutral ligands first, followed by charged ligands. Examples, using typical complexes, are as follows:

*Cations*

| | |
|---|---|
| $[Cu(NH_3)_4]SO_4$ | tetramminecopper(II) sulfate |
| $[Co(NH_3)_4Cl_2]Cl$ | dichlorotetramminecobalt(III) chloride |

*Neutral molecules*

| | |
|---|---|
| $[Pt(NH_3)_2Cl_4]$ | tetrachlorodiammineplatinum(IV) |

*Anions*

| | |
|---|---|
| $K_3[Co(NO_2)_6]$ | potassium hexanitrocobaltate(III) |
| $Na_3[Mn(C_2O_4)_3]$ | sodium trioxalatomanganate(III) |
| $K_3[FeF_6]$ | potassium hexafluoroferrate(III) |

The names used for some of the more common ligands not shown in the examples include: $Br^-$, bromo; $I^-$, iodo; $PO_4^{3-}$, phosphato; $CN^-$, cyano; $NO_3^-$, nitrato; $OH^-$, hydroxo; $H_2O$, aquo.

# 3 General Solubility Rules

It should be realized that the generalizations listed apply to the simple compounds of the more common metals and acids. There are many exceptions among the less common metals and complex compounds.

1 Practically all the ammonium and alkali metal salts are soluble. The alkali metals are the group I metals: lithium, sodium, potassium, rubidium, and cesium.

2 Most nitrates, chlorates, and acetates are soluble.

3 All chlorides are soluble except silver, mercurous, and lead chlorides. Lead chloride is soluble in hot water.

4 All sulfates except barium, strontium, and lead sulfates are soluble. Calcium sulfate is only moderately soluble.

5 All carbonates, phosphates, arsenates, and sulfides are insoluble with the exception of the ammonium and alkali salts. The alkaline earth sulfides are hydrolyzed in water.

6 All hydroxides except sodium, potassium, ammonium, barium, and strontium are insoluble. Calcium hydroxide is only moderately soluble.

APPENDIX

# 4 Solubility Product Constants of Some Salts Arranged by Alphabetical Order of the Name of the Metal

| Salt | $K_{sp}$ | $pK_{sp}$ |
|---|---|---|
| Aluminum | | |
| hydroxide, $Al(OH)_3$ | $4.6 \times 10^{-23}$ | 32.34 |
| phosphate, $AlPO_4$ | $6.2 \times 10^{-19}$ | 18.20 |
| Barium | | |
| carbonate, $BaCO_3$ | $5.1 \times 10^{-9}$ | 8.29 |
| chromate, $BaCrO_4$ | $1.2 \times 10^{-10}$ | 9.93 |
| Fluoride, $BaF_2$ | $1.0 \times 10^{-6}$ | 6.00 |
| iodate, $Ba(IO_3)_2$ | $6.0 \times 10^{-10}$ | 9.22 |
| oxalate, $BaC_2O_4$ | $1.5 \times 10^{-8}$ | 7.82 |
| sulfate, $BaSO_4$ | $1.0 \times 10^{-10}$ | 10.00 |
| Bismuth | | |
| phosphate, $BiPO_4$ | $1.3 \times 10^{-23}$ | 22.89 |
| sulfide, $Bi_2S_3$ | $1.0 \times 10^{-96}$ | 96.00 |
| Cadmium | | |
| hydroxide, $Cd(OH)_2$ | $1.2 \times 10^{-14}$ | 13.93 |
| sulfide, CdS | $7.1 \times 10^{-17}$ | 16.15 |
| Calcium | | |
| carbonate, $CaCO_3$ | $1.2 \times 10^{-8}$ | 7.92 |
| fluoride, $CaF_2$ | $3.5 \times 10^{-11}$ | 10.46 |
| iodate, $Ca(IO_3)_2$ | $6.5 \times 10^{-7}$ | 6.19 |
| oxalate, $CaC_2O_4$ | $2.3 \times 10^{-9}$ | 8.64 |
| phosphate, $Ca_3(PO_4)_2$ | $2.0 \times 10^{-29}$ | 28.70 |
| Cerium(III) (cerous) | | |
| hydroxide, $Ce(OH)_3$ | $6.3 \times 10^{-21}$ | 20.2 |
| Chromium(III) (chromic) | | |
| hydroxide, $Cr(OH)_3$ | $6.0 \times 10^{-31}$ | 30.22 |
| phosphate, $CrPO_4$ | $2.4 \times 10^{-23}$ | 22.62 |
| Cobalt(II) (cobaltous) | | |
| hydroxide, $Co(OH)_2$ | $2.5 \times 10^{-16}$ | 15.60 |
| sulfide, CoS | $3 \times 10^{-26}$ | 25.52 |

| *Salt* | $K_{sp}$ | $pK_{sp}$ |
|---|---|---|
| Copper(I) (cuprous) | | |
| iodide, $Cu_2I_2$ | $1.1 \times 10^{-12}$ | 11.93 |
| sulfide, $Cu_2S$ | $1 \times 10^{-49}$ | 49.00 |
| thiocyanate, $Cu_2(SCN)_2$ | $1.9 \times 10^{-13}$ | 12.73 |
| Copper(II) (cupric) | | |
| carbonate, $CuCO_3$ | $2.5 \times 10^{-10}$ | 9.60 |
| hydroxide, $Cu(OH)_2$ | $2.2 \times 10^{-20}$ | 19.66 |
| oxalate, $CuC_2O_4$ | $2.9 \times 10^{-8}$ | 7.54 |
| sulfide, CuS | $8 \times 10^{-36}$ | 35.1 |
| Iron(II) (ferrous) | | |
| hydroxide, $Fe(OH)_2$ | $3.2 \times 10^{-14}$ | 13.50 |
| sulfide, FeS | $5.0 \times 10^{-18}$ | 17.30 |
| Iron(III) (ferric) | | |
| hydroxide, $Fe(OH)_3$ | $2.5 \times 10^{-39}$ | 38.60 |
| phosphate, $FePO_4$ | $1.4 \times 10^{-22}$ | 21.87 |
| Lead | | |
| carbonate, $PbCO_3$ | $3.4 \times 10^{-14}$ | 13.48 |
| chromate, $PbCrO_4$ | $1.8 \times 10^{-14}$ | 13.75 |
| fluoride, $PbF_2$ | $2.7 \times 10^{-8}$ | 7.57 |
| hydroxide, $Pb(OH)_2$ | $1.2 \times 10^{-15}$ | 14.93 |
| iodide, $PbI_2$ | $6.5 \times 10^{-9}$ | 8.19 |
| oxalate, $PbC_2O_4$ | $8.3 \times 10^{-12}$ | 11.08 |
| phosphate, $Pb_3(PO_4)_2$ | $8.0 \times 10^{-43}$ | 42.10 |
| sulfate, $PbSO_4$ | $1.0 \times 10^{-8}$ | 8.00 |
| sulfide, PbS | $7.1 \times 10^{-29}$ | 28.15 |
| Magnesium | | |
| ammonium phosphate, $MgNH_4PO_4$ | $2.5 \times 10^{-13}$ | 12.60 |
| carbonate, $MgCO_3$ | $1 \times 10^{-5}$ | 5.00 |
| fluoride, $MgF_2$ | $6.4 \times 10^{-9}$ | 8.19 |
| hydroxide, $Mg(OH)_2$ | $1.2 \times 10^{-11}$ | 10.92 |
| oxalate, $MgC_2O_4$ | $8.6 \times 10^{-5}$ | 4.07 |
| Manganese(II) (manganous) | | |
| hydroxide, $Mn(OH)_2$ | $1.9 \times 10^{-13}$ | 12.72 |
| oxalate, $MnC_2O_4$ | $1.1 \times 10^{-15}$ | 14.96 |
| sulfide, MnS | $1.1 \times 10^{-15}$ | 14.96 |
| Mercury(I) (mercurous) | | |
| bromide, $Hg_2Br_2$ | $1.3 \times 10^{-21}$ | 20.89 |
| chloride, $Hg_2Cl_2$ | $1.3 \times 10^{-18}$ | 17.88 |
| iodide, $Hg_2I_2$ | $1.2 \times 10^{-28}$ | 27.92 |
| sulfide, $Hg_2S$ | $1.0 \times 10^{-48}$ | 48.00 |
| Mercury(II) (mercuric) | | |
| hydroxide, $Hg(OH)_2$ | $3.0 \times 10^{-26}$ | 25.52 |
| iodide, $HgI_2$ | $8.8 \times 10^{-12}$ | 11.06 |
| sulfide, HgS | $3.0 \times 10^{-52}$ | 51.52 |
| Nickel(II) (nickelous) | | |
| hydroxide, $Ni(OH)_2$ | $6.2 \times 10^{-16}$ | 15.21 |
| sulfide, NiS | $3.0 \times 10^{-21}$ | 20.52 |

| *Salt* | $K_{sp}$ | $pK_{sp}$ |
|---|---|---|
| Silver | | |
| bromide, $AgBr$ | $4.9 \times 10^{-13}$ | 12.31 |
| carbonate, $Ag_2CO_3$ | $8.1 \times 10^{-12}$ | 11.09 |
| chloride, $AgCl$ | $1.56 \times 10^{-10}$ | 9.81 |
| chromate, $Ag_2CrO_4$ | $1.3 \times 10^{-12}$ | 11.89 |
| cyanide, $AgCN$ | $1.6 \times 10^{-14}$ | 13.80 |
| hydroxide, $AgOH$ | $2.6 \times 10^{-8}$ | 7.59 |
| iodate, $AgIO_3$ | $3.1 \times 10^{-8}$ | 7.51 |
| iodide, $AgI$ | $8.3 \times 10^{-17}$ | 16.08 |
| nitrite, $AgNO_2$ | $1.6 \times 10^{-4}$ | 3.80 |
| oxalate, $Ag_2C_2O_4$ | $1.1 \times 10^{-11}$ | 10.96 |
| phosphate, $Ag_3PO_4$ | $1.3 \times 10^{-20}$ | 19.89 |
| sulfide, $Ag_2S$ | $3.3 \times 10^{-52}$ | 51.48 |
| thiocyanate, $AgSCN$ | $1.4 \times 10^{-12}$ | 11.85 |
| Strontium | | |
| carbonate, $SrCO_3$ | $1.1 \times 10^{-10}$ | 9.96 |
| chromate, $SrCrO_4$ | $3.6 \times 10^{-5}$ | 4.44 |
| fluoride, $SrF_2$ | $2.5 \times 10^{-9}$ | 8.60 |
| iodate, $Sr(IO_3)_2$ | $3.3 \times 10^{-7}$ | 6.48 |
| oxalate, $SrC_2O_4$ | $5.0 \times 10^{-8}$ | 7.30 |
| phosphate, $Sr_3(PO_4)_2$ | $1.0 \times 10^{-31}$ | 31.00 |
| sulfate, $SrSO_4$ | $2.8 \times 10^{-7}$ | 6.56 |
| Thallium(I) (thallous) | | |
| bromide, $TlBr$ | $2.0 \times 10^{-6}$ | 5.70 |
| chloride, $TlCl$ | $1.5 \times 10^{-4}$ | 3.82 |
| iodide, $TlI$ | $2.8 \times 10^{-8}$ | 7.55 |
| sulfide, $Tl_2S$ | $1.0 \times 10^{-21}$ | 21.00 |
| Tin(II) (stannous) | | |
| hydroxide, $Sn(OH)_2$ | $3.2 \times 10^{-26}$ | 25.50 |
| sulfide, $SnS$ | $1.0 \times 10^{-26}$ | 26.00 |
| Zinc | | |
| carbonate, $ZnCO_3$ | $2.1 \times 10^{-11}$ | 10.68 |
| hydroxide, $Zn(OH)_2$ | $3.3 \times 10^{-17}$ | 16.48 |
| phosphate, $Zn_3(PO_4)_2$ | $1.0 \times 10^{-32}$ | 32.00 |
| sulfide, $ZnS$ | $1.6 \times 10^{-24}$ | 23.80 |

APPENDIX

# 5 Ionization Constants of Weak Acids and Bases

In general, the value of the first ionization constant for inorganic acids that contain oxygen depends upon the ratio of oxygen to hydrogen. For acids with the general formula $H_nXO_n$, $K_1$ is approximately $10^{-7}$ or less, and the acid is very weak; for $H_nXO_{n+1}$ acids, $K_1$ is in the range $10^{-2}$ to $10^{-5}$ and $K_2$ is often approximately $10^{-7}$, so the acid is considered to be weak or moderately strong. For acids with the formula $H_nXO_{n+2}$, $K_1$ is usually about $10^3$ and $K_2$ is approximately $10^{-2}$, so the acid is classified as a strong acid. $K_1$ for acids with the formula $H_nXO_{n+3}$ is usually infinitely large, and the acids are known as very strong acids. Also, the difference in successive ionization constants for the inorganic oxygen acids generally is approximately $10^{-4}$. It should be realized that there are many exceptions to these general statements, and that the statements do not apply to organic acids. Actually, the only acids in common laboratory use considered to be strong acids are hydrochloric, nitric, perchloric, and sulfuric.

## ACIDS

| *Acid* | *Equilibrium* | $K_a$ | $pK_a$ |
|---|---|---|---|
| Acetic | $HC_2H_3O_2 \rightleftharpoons H^+ + C_2H_3O_2^-$ | $1.8 \times 10^{-5}$ | 4.76 |
| Arsenic | (1) $H_3AsO_4 \rightleftharpoons H^+ + H_2AsO_4^-$ | $6.0 \times 10^{-3}$ | 2.22 |
| | (2) $H_2AsO_4^- \rightleftharpoons H^+ + HAsO_4^{2-}$ | $1.1 \times 10^{-7}$ | 6.98 |
| | (3) $HAsO_4^{2-} \rightleftharpoons H^+ + AsO_4^{3-}$ | $3.0 \times 10^{-12}$ | 11.53 |
| Arsenious | $HAsO_2 \rightleftharpoons H^+ + AsO_2^-$ | $6 \times 10^{-10}$ | 9.2 |
| Benzoic | $HC_7H_5O_2 \rightleftharpoons H^+ + C_7H_5O_2^-$ | $6.31 \times 10^{-5}$ | 4.20 |
| Boric | $HBO_2 \rightleftharpoons H^+ + BO_2^-$ | $6.0 \times 10^{-10}$ | 9.22 |
| Butyric | $HC_4H_7O_2 \rightleftharpoons H^+ + C_4H_7O_2^-$ | $1.51 \times 10^{-5}$ | 4.82 |
| Caproic | $HC_6H_{11}O_2 \rightleftharpoons H^+ + C_6H_{11}O_2^-$ | $1.32 \times 10^{-5}$ | 4.88 |
| Carbonic | (1) $H_2CO_3 \rightleftharpoons H^+ + HCO_3^-$ | $4.3 \times 10^{-7}$ | 6.37 |
| | (2) $HCO_3^- \rightleftharpoons H^+ + CO_3^{2-}$ | $5.6 \times 10^{-11}$ | 10.25 |

| *Acid* | *Equilibrium* | $K_a$ | $pK_a$ |
|---|---|---|---|
| Chromic | (2) $HCrO_4^- \rightleftharpoons H^+ + CrO_4^{2-}$ | $3.2 \times 10^{-7}$ | 6.50 |
| Citric | (1) $H_3C_6H_5O_7 \rightleftharpoons H^+ + H_2C_6H_5O_7^-$ | $8.3 \times 10^{-4}$ | 3.08 |
| | (2) $H_2C_6H_5O_7^- \rightleftharpoons H^+ + HC_6H_5O_7^{2-}$ | $2.2 \times 10^{-5}$ | 4.66 |
| | (3) $HC_6H_5O_7^{2-} \rightleftharpoons H^+ + C_6H_5O_7^{3-}$ | $4.0 \times 10^{-7}$ | 6.40 |
| Cyanic | $HCNO \rightleftharpoons H^+ + CNO^-$ | $2.2 \times 10^{-4}$ | 3.66 |
| Cyanuric | $HC_3H_2N_3O_3 \rightleftharpoons H^+ + C_3H_2N_3O_3^-$ | $1.8 \times 10^{-7}$ | 6.74 |
| EDTA | (1) $H_4Y \rightleftharpoons H^+ + H_3Y^-$ | $1 \times 10^{-2}$ | 2.0 |
| | (2) $H_3Y^- \rightleftharpoons H^+ + H_2Y^{2-}$ | $2.1 \times 10^{-3}$ | 2.67 |
| | (3) $H_2Y^{2-} \rightleftharpoons H^+ + HY^{3-}$ | $6.9 \times 10^{-7}$ | 6.16 |
| | (4) $HY^{3-} \rightleftharpoons H^+ + Y^{4-}$ | $5.5 \times 10^{-11}$ | 10.26 |
| Formic | $HCOOH \rightleftharpoons H^+ + HCOO^-$ | $1.7 \times 10^{-4}$ | 3.77 |
| Fumaric | (1) $H_2C_4H_2O_4 \rightleftharpoons H^+ + HC_4H_2O_4^-$ | $9.6 \times 10^{-4}$ | 3.02 |
| | (2) $HC_4H_2O_4^- \rightleftharpoons H^+ + C_4H_2O_4^{2-}$ | $4.1 \times 10^{-5}$ | 4.39 |
| Hydrazoic | $HN_3 \rightleftharpoons H^+ + N_3^-$ | $1.9 \times 10^{-5}$ | 4.72 |
| Hydrocyanic | $HCN \rightleftharpoons H^+ + CN^-$ | $4.9 \times 10^{-10}$ | 9.31 |
| Hydrofluoric | $HF \rightleftharpoons H^+ + F^-$ | $2.4 \times 10^{-4}$ | 3.62 |
| Hydrosulfuric | (1) $H_2S \rightleftharpoons H^+ + HS^-$ | $1.0 \times 10^{-7}$ | 7.00 |
| | (2) $HS^- \rightleftharpoons H^+ + S^{2-}$ | $1.2 \times 10^{-13}$ | 12.92 |
| Hypobromous | $HBrO \rightleftharpoons H^+ + BrO^-$ | $2.5 \times 10^{-9}$ | 8.60 |
| Hypochlorous | $HClO \rightleftharpoons H^+ + ClO^-$ | $3.0 \times 10^{-8}$ | 7.53 |
| Hypoiodous | $HIO \rightleftharpoons H^+ + IO^-$ | $5 \times 10^{-13}$ | 12.3 |
| Lactic | $HC_3H_5O_3 \rightleftharpoons H^+ + C_3H_5O_3^-$ | $1.39 \times 10^{-4}$ | 3.86 |
| Maleic | (1) $H_2C_4H_2O_4 \rightleftharpoons H^+ + HC_4H_2O_4^-$ | $1.0 \times 10^{-2}$ | 2.00 |
| | (2) $HC_4H_2O_4^- \rightleftharpoons H^+ + C_4H_2O_4^{2-}$ | $5.5 \times 10^{-7}$ | 6.26 |
| Malic | (1) $H_2C_4H_4O_5 \rightleftharpoons H^+ + HC_4H_4O_5^-$ | $4.0 \times 10^{-4}$ | 3.40 |
| | (2) $HC_4H_4O_5^- \rightleftharpoons H^+ + C_4H_4O_5^{2-}$ | $8.9 \times 10^{-6}$ | 5.05 |
| Malonic | (1) $H_2C_3H_2O_4 \rightleftharpoons H^+ + HC_3H_2O_4^-$ | $1.40 \times 10^{-3}$ | 2.85 |
| | (2) $HC_3H_2O_4^- \rightleftharpoons H^+ + C_3H_2O_4^{2-}$ | $8.0 \times 10^{-7}$ | 6.10 |
| Nitrous | $HNO_2 \rightleftharpoons H^+ + NO_2^-$ | $5.1 \times 10^{-4}$ | 3.29 |
| Oxalic | (1) $H_2C_2O_4 \rightleftharpoons H^+ + HC_2O_4^-$ | $5.6 \times 10^{-2}$ | 1.25 |
| | (2) $HC_2O_4^- \rightleftharpoons H^+ + C_2O_4^{2-}$ | $5.2 \times 10^{-5}$ | 4.28 |
| Phenol | $C_6H_5OH \rightleftharpoons H^+ + C_6H_5O^-$ | $1.3 \times 10^{-10}$ | 9.89 |
| Phosphoric | (1) $H_3PO_4 \rightleftharpoons H^+ + H_2PO^-$ | $7.5 \times 10^{-3}$ | 2.12 |
| | (2) $H_2PO_4^- \rightleftharpoons H^+ + HPO_4^{2-}$ | $6.2 \times 10^{-8}$ | 7.21 |
| | (3) $HPO_4^{2-} \rightleftharpoons H^+ + PO_4^{3-}$ | $4.7 \times 10^{-13}$ | 12.33 |
| Phthalic | (1) $H_2C_8H_4O_4 \rightleftharpoons H^+ + HC_8H_4O_4^-$ | $8.0 \times 10^{-4}$ | 3.10 |
| | (2) $HC_8H_4O_4^- \rightleftharpoons H^+ + C_8H_4O_4^{2-}$ | $4.0 \times 10^{-6}$ | 5.40 |
| Propionic | $HC_3H_5O_2 \rightleftharpoons H^+ + C_3H_5O_2^-$ | $1.34 \times 10^{-5}$ | 4.87 |
| Succinic | (1) $H_2C_4H_4O_4 \rightleftharpoons H^+ + HC_4H_4O_4^-$ | $6.5 \times 10^{-5}$ | 4.19 |
| | (2) $HC_4H_4O_4^- \rightleftharpoons H^+ + C_4H_4O_4^{2-}$ | $3.3 \times 10^{-6}$ | 5.48 |
| Sulfuric | (2) $HSO_4^- \rightleftharpoons H^+ + SO_4^{2-}$ | $1.0 \times 10^{-2}$ | 2.00 |
| Sulfurous | (1) $H_2SO_3 \rightleftharpoons H^+ + HSO_3^-$ | $1.3 \times 10^{-2}$ | 1.89 |
| | (2) $HSO_3^- \rightleftharpoons H^+ + SO_3^{2-}$ | $6.3 \times 10^{-8}$ | 7.20 |
| Tartaric | (1) $H_2C_4H_4O_6 \rightleftharpoons H^+ + HC_4H_4O_6^-$ | $3.0 \times 10^{-3}$ | 2.52 |
| | (2) $HC_4H_4O_6^- \rightleftharpoons H^+ + C_4H_4O_6^{2-}$ | $6.9 \times 10^{-5}$ | 4.16 |

# BASES

| *Base* | *Equilibrium* | $K_b$ | $pK_b$ |
|---|---|---|---|
| Ammonia | $NH_3 + H_2O \rightleftharpoons NH_4OH \rightleftharpoons NH_4^+ + OH^-$ | $1.8 \times 10^{-5}$ | 4.74 |
| Aniline | $C_6H_5NH_2 + H_2O \rightleftharpoons C_6H_5NH_3^+ + OH^-$ | $4.6 \times 10^{-10}$ | 9.34 |
| Benzylamine | $C_7H_7NH_2 + H_2O \rightleftharpoons C_7H_7NH_3^+ + OH^-$ | $2.0 \times 10^{-5}$ | 4.70 |
| Ethylamine | $C_2H_5NH_2 + H_2O \rightleftharpoons C_2H_5NH_3^+ + OH^-$ | $5.6 \times 10^{-4}$ | 3.25 |
| Hydrazine | $N_2H_4 + H_2O \rightleftharpoons N_2H_5^+ + OH^-$ | $3 \times 10^{-6}$ | 5.5 |
| Hydroxylamine | $NH_2OH \rightleftharpoons NH_2^+ + OH^-$ | $6.6 \times 10^{-9}$ | 8.18 |
| Methylamine | $CH_3NH_2 + H_2O \rightleftharpoons CH_3NH_3^+ + OH^-$ | $5 \times 10^{-4}$ | 3.3 |
| α-Naphthylamine | $C_{10}H_7NH_2 + H_2O \rightleftharpoons C_{10}H_7NH_3^+ + OH^-$ | $9.9 \times 10^{-11}$ | 10.01 |
| β-Naphthylamine | $C_{10}H_7NH_2 + H_2O \rightleftharpoons C_{10}H_7NH_3^+ + OH^-$ | $2 \times 10^{-10}$ | 9.70 |
| Phenylhydrazine | $C_6H_5N_2H_3 + H_2O \rightleftharpoons C_6H_5N_2H_4^+ + OH^-$ | $1.6 \times 10^{-9}$ | 8.80 |
| Pyridine | $C_5H_5N + H_2O \rightleftharpoons C_5H_5NH^+ + OH^-$ | $2.3 \times 10^{-9}$ | 8.64 |
| Quinoline | $C_9H_7N + H_2O \rightleftharpoons C_9H_7NH^+ + OH^-$ | $1 \times 10^{-9}$ | 9.0 |
| Silver hydroxide | $AgOH \rightleftharpoons Ag^+ + OH^-$ | $1.1 \times 10^{-4}$ | 3.96 |
| Triethylamine | $(C_2H_5)_3N + H_2O \rightleftharpoons (C_2H_5)_3NH^+ + OH^-$ | $2.6 \times 10^{-4}$ | 3.58 |

# 6 Stability (Formation) Constants of Complexes

The table value is the logarithm of the stability (formation) constant. Stepwise constants are designated by $k_1, k_2, \ldots, k_n$. The cumulative constant is the product of the stepwise constants (i.e., $K_{stab} = k_1 k_2 \ldots k_n$). Hence, $\log K_{stab} = \log k_1 + \log k_2 + \cdots + \log k_n$. Where values of the stepwise constants are not given, the parenthetical number before $\log K_{stab}$ indicates the value of $n$ for the overall reaction $M + nL \rightleftharpoons ML_n$.

| Ligand | Cation | $\log k_1$ | $\log k_2$ | $\log k_3$ | $\log k_4$ | $\log K_{stab}$ |
|---|---|---|---|---|---|---|
| $NH_3$ | $Ag^+$ | 3.34 | 3.83 | | | 7.17 |
| | $Cd^{2+}$ | 2.51 | 1.96 | 1.30 | 0.79 | 6.56 |
| | $Co^{2+}$ | 2.07 | 1.60 | 0.93 | 0.64 | 5.24 |
| | $Cu^{2+}$ | 3.99 | 3.34 | 2.73 | 1.97 | 12.03 |
| | $Ni^{2+}$ | 2.67 | 2.12 | 1.61 | 1.07 | 7.47 |
| | $Zn^{2+}$ | 2.18 | 2.25 | 2.31 | 1.96 | 8.70 |
| $CN^-$ | $Ag^+$ | | | | | (2) 21.1 |
| | $Cd^{2+}$ | 5.18 | 4.42 | 4.32 | 3.19 | 17.11 |
| | $Co^{2+}$ | | | | | (6) 19.09 |
| | $Cu^+$ | $\log(k_1 k_2) = 24$ | | 4.59 | 1.70 | 30.3 |
| | $Fe^{2+}$ | | | | | (6) 24 |
| | $Fe^{3+}$ | | | | | (6) 31 |
| | $Ni^{2+}$ | | | | | (4) 31.3 |
| | $Zn^{2+}$ | | | | | (4) 16.7 |
| $C_2O_4^{2-}$ | $Fe^{3+}$ | 9.40 | 6.78 | 4.00 | | 20.18 |
| | $Mg^{2+}$ | 3.82 | 1.43 | | | 5.25 |
| | $Pb^{2+}$ | | | | | (2) 6.54 |
| $F^-$ | $Al^{3+}$ | 6.13 | 5.02 | 3.85 | 2.74 | (6) 19.84 |
| | $Fe^{3+}$ | 5.17 | 3.92 | 2.91 | | (6) 15.3 |
| | $Th^{4+}$ | 7.70 | 5.78 | 4.51 | | (3) 17.99 |
| | $Zr^{4+}$ | 8.73 | 7.34 | 5.83 | | (3) 21.90 |
| $Cl^-$ | $Ag^+$ | | | | | (2) 5.28 |
| | $Cd^{2+}$ | 2.00 | 0.60 | 0.10 | 0.30 | 3.00 |
| | $Cu^+$ | | | | | (2) 5.54 |

| *Ligand* | *Cation* | log $k_1$ | log $k_2$ | log $k_3$ | log $k_4$ | log $K_{stab}$ |
|---|---|---|---|---|---|---|
| | $Hg^{2+}$ | 6.74 | 6.48 | 0.85 | 1.00 | 15.07 |
| | $Pd^{2+}$ | 6.1 | 4.6 | 2.4 | 2.6 | 15.7 |
| | $Zn^{2+}$ | | | | | (4) 0.20 |
| $I^-$ | $Ag^+$ | | | | | (2) 13.7 |
| | $Cd^{2+}$ | 2.28 | 1.64 | 1.08 | 1.10 | 6.10 |
| | $Hg^{2+}$ | 12.87 | 10.95 | 3.78 | 2.23 | 29.83 |
| $SCN^-$ | $Ag^+$ | | | | | (2) 8.78 |
| | $Fe^{3+}$ | 3.03 | | | | |
| | $Hg^{2+}$ | 9.48 | 10.30 | 1.70 | | (4) 21.9 |
| $S_2O_3^{2-}$ | $Ag^+$ | | | | | (2) 13.22 |
| EDTA* | $Ag^+$ | 7.30 | | | | |
| | $Ba^{2+}$ | 7.76 | | | | |
| | $Ca^{2+}$ | 11.00 | | | | |
| | $Cd^{2+}$ | 16.46 | | | | |
| | $Co^{2+}$ | 16.31 | | | | |
| | $Cu^{2+}$ | 18.79 | | | | |
| | $Fe^{2+}$ | 14.33 | | | | |
| | $Fe^{3+}$ | 24.23 | | | | |
| | $Mg^{2+}$ | 8.69 | | | | |
| | $Mn^{2+}$ | 13.79 | | | | |
| | $Ni^{2+}$ | 18.56 | | | | |
| | $Pb^{2+}$ | 18.04 | | | | |
| | $Sr^{2+}$ | 8.63 | | | | |
| | $Zn^{2+}$ | 16.26 | | | | |

| *Ligand* | *Cation* | *Equilibrium*† | log $K_{stab}$ |
|---|---|---|---|
| $OH^-$ | $Al^{3+}$ | $Al(OH)_3 + OH^- \rightleftharpoons Al(OH)_4^-$ | 2.60 |
| | $Cr^{3+}$ | $Cr(OH)_3 + OH^- \rightleftharpoons Cr(OH)_4^-$ | −2.0 |
| | $Pb^{2+}$ | $Pb(OH)_2 + OH^- \rightleftharpoons Pb(OH)_3^-$ | −1.70 |
| | $Sn^{2+}$ | $Sn(OH)_2 + OH^- \rightleftharpoons Sn(OH)_3^-$ | −3.30 |
| | $Zn^{2+}$ | $Zn(OH)_2 + 2OH^- \rightleftharpoons Zn(OH)_4^{2-}$ | −1.0 |

* EDTA forms only 1 : 1 complexes with the cations.

† The values given for the various cations with $OH^-$ as the ligand are for the equilibrium between the solid amphoteric hydroxide and the anionic complex.

APPENDIX

# 7 Selected Standard Electrode (Half-Cell) Reduction Potentials

The direction of the half-reaction equation and the sign of the potential conform to the IUPAC (Stockholm) conventions.

| *Half-reaction* | *E* (V) |
|---|---|
| $F_2 + 2e^- \rightleftharpoons 2F^-$ | +2.65 |
| $O_3 + 2H^+ + 2e^- \rightleftharpoons O_2 + H_2O$ | +2.07 |
| $S_2O_8^{2-} + 2e^- \rightleftharpoons 2SO_4^{2-}$ | +2.01 |
| $H_2O_2 + 2H^+ + 2e^- \rightleftharpoons 2H_2O$ | +1.77 |
| $Ce^{4+} + e^- \rightleftharpoons Ce^{3+}$ (perchlorate solution) | +1.70 |
| $MnO_4^- + 4H^+ + 3e^- \rightleftharpoons MnO_2 + 2H_2O$ | +1.695 |
| $Ce^{4+} + e^- \rightleftharpoons Ce^{3+}$ (nitrate solution) | +1.61 |
| $BrO_3^- + 6H^+ + 5e^- \rightleftharpoons \frac{1}{2}Br_2 + 3H_2O$ | +1.52 |
| $MnO_4^- + 8H^+ + 5e^- \rightleftharpoons Mn^{2+} + 4H_2O$ | +1.51 |
| $Au^{3+} + 3e^- \rightleftharpoons Au$ | +1.50 |
| $ClO_3^- + 6H^+ + 5e^- \rightleftharpoons \frac{1}{2}Cl_2 + 3H_2O$ | +1.47 |
| $BrO_3^- + 6H^+ + 6e^- \rightleftharpoons Br^- + 3H_2O$ | +1.45 |
| $Ce^{4+} + e^- \rightleftharpoons Ce^{3+}$ (sulfate solution) | +1.44 |
| $Cl_2 + 2e^- \rightleftharpoons 2Cl^-$ | +1.359 |
| $Cr_2O_7^{2-} + 14H^+ + 6e^- \rightleftharpoons 2Cr^{3+} + 7H_2O$ | +1.33 |
| $Ce^{4+} + e^- \rightleftharpoons Ce^{3+}$ (chloride solution) | +1.28 |
| $MnO_2 + 4H^+ + 2e^- \rightleftharpoons Mn^{2+} + 2H_2O$ | +1.23 |
| $O_2 + 4H^+ + 4e^- \rightleftharpoons 2H_2O$ | +1.229 |
| $IO_3^- + 6H^+ + 5e^- \rightleftharpoons \frac{1}{2}I_2 + 3H_2O$ | +1.20 |
| $IO_3^- + 6H^+ + 6e^- \rightleftharpoons I^- + 3H_2O$ | +1.087 |
| $Br_2$ (liquid) $+ 2e^- \rightleftharpoons 2Br^-$ | +1.065 |
| $OP + e^- \rightleftharpoons OP'$ (orthophenanthroline) | +1.06 |
| $AuCl_4^- + 3e^- \rightleftharpoons Au + 4Cl^-$ | +1.00 |
| $NO_3^- + 4H^+ + 3e^- \rightleftharpoons NO + 2H_2O$ | +0.96 |
| $2Hg^{2+} + 2e^- \rightleftharpoons Hg_2^{2+}$ | +0.920 |
| $Cu^{2+} + I^- + e^- \rightleftharpoons CuI$ | +0.85 |
| DPAS $+ e^- \rightleftharpoons$ DPAS (red) (diphenylamine sulfonate) | +0.84 |
| $\frac{1}{2}O_2 + 2H^+(10^{-7}\,M) + 2e^- \rightleftharpoons H_2O$ | +0.815 |
| $Ag^+ + e^- \rightleftharpoons Ag$ | +0.799 |

| *Half-reaction* | *E* (V) |
|---|---|
| $Hg_2^{2+} + 2e^- \rightleftharpoons 2Hg$ | +0.789 |
| $Fe^{3+} + e^- \rightleftharpoons Fe^{2+}$ | +0.771 |
| $Sb^{5+} + 2e^- \rightleftharpoons Sb^{3+}$ | +0.75 |
| Q (saturated solution) + $2H^+ + 2e^- \rightleftharpoons H_2Q$ (saturated solution) (quinhydrone electrode) | +0.700 |
| $O_2 + 2H^+ + 2e^- \rightleftharpoons H_2O_2$ | +0.682 |
| $2HgCl_2 + 2e^- \rightleftharpoons Hg_2Cl_2 + 2Cl^-$ | +0.63 |
| $H_3AsO_4 + 2H^+ + 2e^- \rightleftharpoons H_3AsO_3 + H_2O$ | +0.559 |
| $I_2(I_3^-) + 2e^- \rightleftharpoons 2I^-(3I^-)$ | +0.535 |
| $Fe(CN)_6^{3-} + e^- \rightleftharpoons Fe(CN)_6^{4-}$ | +0.36 |
| $Cu^{2+} + 2e^- \rightleftharpoons Cu$ | +0.337 |
| $UO_2^{2+} + 4H^+ + 2e^- \rightleftharpoons U^{4+} + 2H_2O$ | +0.334 |
| $BiO^+ + 2H^+ + 3e^- \rightleftharpoons Bi + H_2O$ | +0.32 |
| $Hg_2Cl_2 + 2e^- \rightleftharpoons 2Hg + 2Cl^-$ (1 *M* KCl) (calomel cell) | +0.285 |
| $Hg_2Cl_2 + 2e^- \rightleftharpoons 2Hg + 2Cl^-$ (saturated KCl) (calomel cell) | +0.246 |
| $AgCl + e^- \rightleftharpoons Ag + Cl^-$ | +0.222 |
| $SbO^+ + 2H^+ + 3e^- \rightleftharpoons Sb + H_2O$ | +0.212 |
| $S_4O_6^{2-} + 2e^- \rightleftharpoons 2S_2O_3^{2-}$ | +0.17 |
| $SO_4^{2-} + 4H^+ + 2e^- \rightleftharpoons H_2SO_3 + H_2O$ | +0.17 |
| $Sn^{4+} + 2e^- \rightleftharpoons Sn^{2+}$ | +0.15 |
| $TiO^{2+} + 2H^+ + e^- \rightleftharpoons Ti^{3+} + H_2O$ | +0.1 |
| $AgBr + e^- \rightleftharpoons Ag + Br^-$ | +0.095 |
| $2H^+ + 2e^- \rightleftharpoons H_2$ | +0.000 |
| $Pb^{2+} + 2e^- \rightleftharpoons Pb$ | −0.126 |
| $Sn^{2+} + 2e^- \rightleftharpoons Sn$ | −0.136 |
| $AgI + e^- \rightleftharpoons Ag + I^-$ | −0.151 |
| $Ni^{2+} + 2e^- \rightleftharpoons Ni$ | −0.24 |
| $V^{3+} + e^- \rightleftharpoons V^{2+}$ | −0.26 |
| $Co^{2+} + 2e^- \rightleftharpoons Co$ | −0.28 |
| $Tl^+ + e^- \rightleftharpoons Tl$ | −0.34 |
| $PbSO_4 + 2e^- \rightleftharpoons Pb + SO_4^{2-}$ | −0.36 |
| $Cd^{2+} + 2e^- \rightleftharpoons Cd$ | −0.403 |
| $Cr^{3+} + e^- \rightleftharpoons Cr^{2+}$ | −0.41 |
| $2H^+(10^{-7}\ M) + 2e^- \rightleftharpoons H_2$ | −0.414 |
| $Fe^{2+} + 2e^- \rightleftharpoons Fe$ | −0.440 |
| $2CO_2 + 2H^+ + 2e^- \rightleftharpoons H_2C_2O_4$ | −0.49 |
| $S + 2e^- \rightleftharpoons S^{2-}$ | −0.51 |
| $AsO_4^{3-} + 3H_2O + 2e^- \rightleftharpoons H_2AsO_3^- + 4OH^-$ | −0.67 |
| $Cr^{3+} + 3e^- \rightleftharpoons Cr$ | −0.74 |
| $Zn^{2+} + 2e^- \rightleftharpoons Zn$ | −0.763 |
| $Mn^{2+} + 2e^- \rightleftharpoons Mn$ | −1.18 |
| $Al^{3+} + 3e^- \rightleftharpoons Al$ | −1.67 |
| $AlO_2^- + 2H_2O + 3e^- \rightleftharpoons Al + 4OH^-$ | −2.35 |
| $Mg^{2+} + 2e^- \rightleftharpoons Mg$ | −2.37 |
| $Na^+ + e^- \rightleftharpoons Na$ | −2.714 |
| $Ca^{2+} + 2e^- \rightleftharpoons Ca$ | −2.87 |
| $Sr^{2+} + 2e^- \rightleftharpoons Sr$ | −2.89 |
| $Ba^{2+} + 2e^- \rightleftharpoons Ba$ | −2.90 |
| $K^+ + e^- \rightleftharpoons K$ | −2.925 |

APPENDIX

# 8 Standard Deviation to Area Relationship for Gaussian Distribution

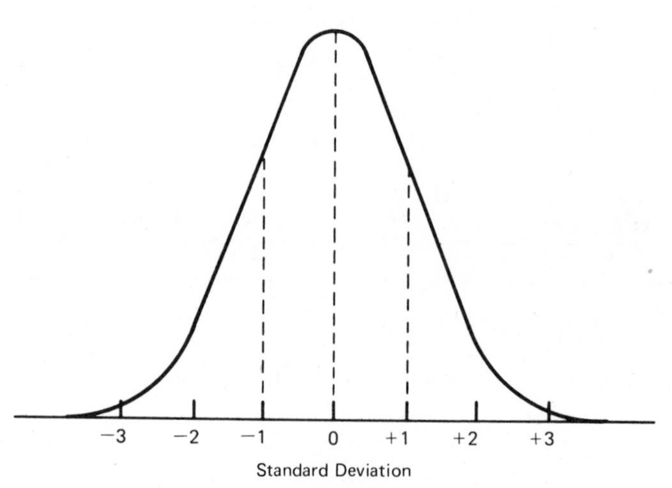


**Figure A8-1.** Typical Gaussian distribution showing relation of standard deviation to area under the curve.

| $t$ | % | $t$ | % |
|---|---|---|---|
| 0.0 | 0.00 | 2.0 | 95.44 |
| 0.1 | 7.96 | 2.1 | 96.42 |
| 0.2 | 15.86 | 2.2 | 97.22 |
| 0.3 | 23.58 | 2.3 | 97.86 |
| 0.4 | 31.08 | 2.4 | 98.36 |
| 0.5 | 39.30 | 2.5 | 98.76 |
| 0.6 | 45.14 | 2.6 | 99.06 |
| 0.7 | 51.60 | 2.7 | 99.30 |
| 0.8 | 57.62 | 2.8 | 99.48 |
| 0.9 | 63.18 | 2.9 | 99.62 |
| 1.0 | 68.26 | 3.0 | 99.730 |
| 1.1 | 72.86 | 3.1 | 99.806 |
| 1.2 | 76.98 | 3.2 | 99.862 |
| 1.3 | 80.64 | 3.3 | 99.902 |
| 1.4 | 83.94 | 3.4 | 99.932 |
| 1.5 | 86.64 | 3.5 | 99.952 |
| 1.6 | 89.04 | 3.6 | 99.968 |
| 1.7 | 91.08 | 3.7 | 99.978 |
| 1.8 | 92.82 | 3.8 | 99.984 |
| 1.9 | 94.26 | 4.0 | 99.992 |

APPENDIX

# 9 Use of Logarithms

The use of logarithms simplifies many calculations that must be performed in analytical chemistry, and, as a consequence, students who learn to use logarithms correctly will benefit greatly from the knowledge. A logarithm (often abbreviated log) of a number is the power (or exponent) of 10 that will give the number:

$$\text{number} = 10^{\text{logarithm of number}}$$

and

$$253 = 10^{\text{log of } 253}$$

In many cases it is advantageous to express numbers, especially very large or very small numbers, in the exponential form by showing them as numbers between 1 and 10 multiplied by 10 raised to the appropriate power, as

$$\begin{aligned} 25{,}300{,}000 &= 2.53 \times 10^{7} \\ 25{,}300 &= 2.53 \times 10^{4} \\ 253 &= 2.53 \times 10^{2} \\ 0.253 &= 2.53 \times 10^{-1} \\ 0.0000253 &= 2.53 \times 10^{-5} \end{aligned}$$

By expressing the number in this way, the magnitude of the number is shown by the exponent and thus is easy to determine. The exponent simply represents the number of times the number 1 must be multiplied or divided by 10, depending upon whether the exponent is positive or negative. Thus,

$$10^{2} = 1 \times 10 \times 10 = 100$$

and

$$10^{-3} = \frac{1}{10 \times 10 \times 10} = \frac{1}{1000} = 0.001$$

Since the log of a number is the power to which 10 must be raised to equal the number, the

$$\text{log of } 100 = \text{log of } 10^2 = 2$$

and

$$\text{log of } 0.001 = \text{log of } 10^{-3} = -3$$

Similarly, the log of 1 is 0, the log of 10 is 1, and any number between 1 and 10 will have a logarithm between 0 and 1. Since most numbers are not exact powers of 10, most logs are not whole numbers but are fractional numbers composed of an integer (known as the **characteristic**) and a decimal fraction (known as the **mantissa**). The mantissa is found in a logarithm table and the characteristic is obtained by inspection, since the characteristic serves only to locate the decimal point. The logarithms of all numbers that have the same integers in the same sequence all have the same mantissa but will have different characteristics depending upon the magnitude of the number. For example, using the integer sequence 3456, for which the mantissa is 0.5386,

$$\begin{aligned} \text{log of } 345{,}600 &= 5.5386 \\ \text{log of } 34.56 &= 1.5386 \\ \text{log of } 0.3456 &= \bar{1}.5386 \quad \text{or} \quad 9.5386 - 10 \quad \text{or} \quad -0.4614 \\ \text{log of } 0.0003456 &= \bar{4}.5386 \quad \text{or} \quad 6.5386 - 10 \quad \text{or} \quad -3.4614 \end{aligned}$$

For numbers less than 1, the negative sign is usually written over the integer to show that the characteristic is negative, whereas the mantissa is positive. The logs of small numbers of this type may also be written to show the addition of 10 followed by the subtraction of 10. In the last example above, the characteristic is calculated by the addition of $-4$ to $+10$, to obtain $+6$.

In some calculations it is desirable to show the logarithms of fractional numbers are completely negative numbers rather than as a negative integer followed by a positive decimal fraction. Expressed this way, the logarithm of 0.0003456 is $-3.4614$, which is obtained by subtracting the mantissa from 1.0000 and reducing the characteristic by 1. It is probably easier to express the logs in this way if the number is first written as a number between 1 and 10 multiplied by 10 with a whole-number negative exponent. Using the example above, first convert 0.0003456 to $3.456 \times 10^{-4}$ and then convert 3.456 to 10 with an exponent, or $10^{0.5386}$. Addition of exponents, equivalent to multiplication, gives $10^{-3.4614}$ and the log of $10^{-3.4614}$ is $-3.4614$.

The numerical operations often performed by logarithms to simplify calculations may be summarized as

| *Numerical Operation* | *Corresponding Manipulation with Logarithms* |
|---|---|
| Multiplication | Add logarithms |
| Division | Subtract logarithms |
| Raise to a power | Multiply logarithm by desired power |
| Extract a root | Divide logarithm by desired root |
| Addition | *Cannot be performed by logarithms* |
| Subtraction | *Cannot be performed by logarithms* |

## SIGNIFICANT FIGURES IN USING LOGARITHMS

In general, there should be the same number of significant figures in the mantissa of the logarithm as there are significant figures in the number being expressed. Thus, the log of 1.2300 is 0.08991, the log of 1.230 is 0.0899, the log of 1.23 is 0.090, and the log of 1.2 is 0.08. This rule has not been followed in the preceding examples because the primary purpose of the examples is to emphasize the manipulations involved in the use of logarithms.

## NEGATIVE LOGARITHMS

Some of the concepts used in analytical chemistry are defined in terms of negative logarithms. Examples are $\text{pH} = -\log(\text{H}^+)$, $\text{pK} = -\log K$, $A = -\log T$, and $\text{p}_c\text{M} = -\log[\text{M}^{n+}]$. In using negative logarithms, remember that the decimal part of the negative log is negative, whereas the decimal part (the mantissa) of a regular logarithm is positive and that the values found in log tables are thus positive. Consequently, it is not possible to treat the decimal part of a negative logarithm as a mantissa for location in a log table.

To express a number as a negative logarithm, it is usually advisable to write the number in the usual exponential form of a number between 1 and 10, multiplied by 10 with a whole-number exponent. The mantissa of the integer sequence is then looked up and the number expressed as 10, with the mantissa as the exponent. The exponents are then added algebraically and expressed as 10 raised to the proper power. The negative logarithm of this number is simply the exponent with the sign changed. Examples are shown below to illustrate the manipulations involved.

- **Example 1.** If $(\text{H}^+)$ is $4.5 \times 10^{-6}$, what is the pH?

$$4.5 \times 10^{-6} = 10^{0.65} \times 10^{-6} = 10^{-5.35}$$
$$\text{pH} = -\log(\text{H}^+) = -\log 10^{-5.35} = 5.35$$

Negative exponents of this type are usually expressed to either two or three decimal places, depending upon the number of significant figures in the number being converted. The rule to be followed is to use the same number of figures in the decimal part of the logarithm as is used in the number itself. In Example 1, the mantissa for 4.5 is 0.65321, which rounds off to 0.65 to give the proper number of significant figures.

- **Example 2.** If pH is $-0.36$, what is the hydrogen ion activity? Since

$$\text{pH} = -\log(\text{H}^+) = -\log 10^{-0.36} = \log 10^{0.36}$$
$$(\text{H}^+) = 10^{0.36} = 2.3 \times 10^0 = 2.3$$

In this case, the number corresponding to the mantissa 0.36 is 2.2902, which is rounded to 2.3 according to the significant-figure rule.

• **Example 3.** If the absorbance of a solution is 0.356, what is the percent transmittance of the solution?

$$T = 10^{-0.356} = 10^{0.644} \times 10^{-1} = 4.41 \times 10^{-1} = 0.441$$
$$\%T = 100T = 44.1\%$$

In this case the number corresponding to the mantissa 0.644 is 4.406, which is rounded off to 4.41 by the significant-figure rule.

• **Example 4.** What is the $p_c$Cu of a 0.1063 *M* solution of $CuCl_2$?

$$0.1063 = 1.063 \times 10^{-1} = 10^{0.0265} \times 10^{-1} = 10^{-0.9735}$$
$$p_c\text{Cu} = -\log 10^{-0.9735} = \log 10^{0.9735} = 0.9735$$

In this case, four places are carried in the mantissa, since there are four significant figures in the number.

In the preceding examples, enclosure of a symbol in parentheses refers to active concentration or activity, and enclosure in brackets represents molar concentration. The subscript *c* is used in the p notation to indicate that molar instead of active concentrations are being used.

APPENDIX

# 10 Four-Place Logarithm Table

## LOGARITHMS

| *t* | *0* | *1* | *2* | *3* | *4* | *5* | *6* | *7* | *8* | *9* |
|---|---|---|---|---|---|---|---|---|---|---|
| **1.0** | 0.0000 | 0.0043 | 0.0086 | 0.0128 | 0.0170 | 0.0212 | 0.0253 | 0.0294 | 0.0334 | 0.0374 |
| **1.1** | 0.0414 | 0.0453 | 0.0492 | 0.0531 | 0.0569 | 0.0607 | 0.0645 | 0.0682 | 0.0719 | 0.0755 |
| **1.2** | 0.0792 | 0.0828 | 0.0864 | 0.0899 | 0.0934 | 0.0969 | 0.1004 | 0.1038 | 0.1072 | 0.1106 |
| **1.3** | 0.1139 | 0.1173 | 0.1206 | 0.1239 | 0.1271 | 0.1303 | 0.1335 | 0.1367 | 0.1399 | 0.1430 |
| **1.4** | 0.1461 | 0.1492 | 0.1523 | 0.1553 | 0.1584 | 0.1614 | 0.1644 | 0.1673 | 0.1703 | 0.1732 |
| **1.5** | 0.1761 | 0.1790 | 0.1818 | 0.1847 | 0.1875 | 0.1903 | 0.1931 | 0.1959 | 0.1987 | 0.2014 |
| **1.7** | 0.2041 | 0.2068 | 0.2095 | 0.2122 | 0.2148 | 0.2175 | 0.2201 | 0.2227 | 0.2253 | 0.2279 |
| **1.7** | 0.2304 | 0.2330 | 0.2355 | 0.2380 | 0.2405 | 0.2430 | 0.2455 | 0.2480 | 0.2504 | 0.2529 |
| **1.8** | 0.2553 | 0.2577 | 0.2601 | 0.2625 | 0.2648 | 0.2672 | 0.2695 | 0.2718 | 0.2742 | 0.2765 |
| **1.9** | 0.2788 | 0.2810 | 0.2833 | 0.2856 | 0.2878 | 0.2900 | 0.2923 | 0.2945 | 0.2967 | 0.2989 |
| **2.0** | 0.3010 | 0.3032 | 0.3054 | 0.3075 | 0.3096 | 0.3118 | 0.3139 | 0.3160 | 0.3181 | 0.3201 |
| **2.1** | 0.3222 | 0.3243 | 0.3263 | 0.3284 | 0.3304 | 0.3324 | 0.3345 | 0.3365 | 0.3385 | 0.3404 |
| **2.2** | 0.3424 | 0.3444 | 0.3464 | 0.3483 | 0.3502 | 0.3522 | 0.3541 | 0.3560 | 0.3579 | 0.3598 |
| **2.3** | 0.3617 | 0.3636 | 0.3655 | 0.3674 | 0.3692 | 0.3711 | 0.3729 | 0.3747 | 0.3766 | 0.3784 |
| **2.4** | 0.3802 | 0.3820 | 0.3838 | 0.3856 | 0.3874 | 0.3892 | 0.3909 | 0.3927 | 0.3945 | 0.3962 |
| **2.5** | 0.3979 | 0.3997 | 0.4014 | 0.4031 | 0.4048 | 0.4065 | 0.4082 | 0.4099 | 0.4116 | 0.4133 |
| **2.6** | 0.4150 | 0.4166 | 0.4183 | 0.4200 | 0.4216 | 0.4232 | 0.4249 | 0.4265 | 0.4281 | 0.4298 |
| **2.7** | 0.4314 | 0.4330 | 0.4346 | 0.4362 | 0.4378 | 0.4393 | 0.4409 | 0.4425 | 0.4440 | 0.4456 |
| **2.8** | 0.4472 | 0.4487 | 0.4502 | 0.4518 | 0.4533 | 0.4548 | 0.4564 | 0.4579 | 0.4594 | 0.4609 |
| **2.9** | 0.4624 | 0.4639 | 0.4654 | 0.4669 | 0.4683 | 0.4698 | 0.4713 | 0.4728 | 0.4742 | 0.4757 |
| **3.0** | 0.4771 | 0.4786 | 0.4800 | 0.4814 | 0.4829 | 0.4843 | 0.4857 | 0.4871 | 0.4886 | 0.4900 |
| **3.1** | 0.4914 | 0.4928 | 0.4942 | 0.4955 | 0.4969 | 0.4983 | 0.4997 | 0.5011 | 0.5024 | 0.5038 |
| **3.2** | 0.5051 | 0.5065 | 0.5079 | 0.5092 | 0.5105 | 0.5119 | 0.5132 | 0.5145 | 0.5159 | 0.5172 |
| **3.3** | 0.5185 | 0.5198 | 0.5211 | 0.5224 | 0.5237 | 0.5250 | 0.5263 | 0.5276 | 0.5289 | 0.5302 |
| **3.4** | 0.5315 | 0.5328 | 0.5340 | 0.5353 | 0.5366 | 0.5378 | 0.5391 | 0.5403 | 0.5416 | 0.5428 |

| *t* | *0* | *1* | *2* | *3* | *4* | *5* | *6* | *7* | *8* | *9* |
|---|---|---|---|---|---|---|---|---|---|---|
| **3.5** | 0.5441 | 0.5453 | 0.5465 | 0.5478 | 0.5490 | 0.5502 | 0.5514 | 0.5527 | 0.5539 | 0.5551 |
| **3.6** | 0.5563 | 0.5575 | 0.5587 | 0.5599 | 0.5611 | 0.5623 | 0.5635 | 0.5647 | 0.5658 | 0.5670 |
| **3.7** | 0.5682 | 0.5694 | 0.5705 | 0.5717 | 0.5729 | 0.5740 | 0.5752 | 0.5763 | 0.5775 | 0.5786 |
| **3.8** | 0.5798 | 0.5809 | 0.5821 | 0.5832 | 0.5843 | 0.5855 | 0.5866 | 0.5877 | 0.5888 | 0.5899 |
| **3.9** | 0.5911 | 0.5922 | 0.5933 | 0.5944 | 0.5955 | 0.5966 | 0.5977 | 0.5988 | 0.5999 | 0.6010 |
| **4.0** | 0.6021 | 0.6031 | 0.6042 | 0.6053 | 0.6064 | 0.6075 | 0.6085 | 0.6096 | 0.6107 | 0.6117 |
| **4.1** | 0.6128 | 0.6138 | 0.6149 | 0.6160 | 0.6170 | 0.6180 | 0.6191 | 0.6201 | 0.6212 | 0.6222 |
| **4.2** | 0.6232 | 0.6243 | 0.6253 | 0.6263 | 0.6274 | 0.6284 | 0.6294 | 0.6304 | 0.6314 | 0.6325 |
| **4.3** | 0.6335 | 0.6345 | 0.6355 | 0.6365 | 0.6375 | 0.6385 | 0.6395 | 0.6405 | 0.6415 | 0.6425 |
| **4.4** | 0.6435 | 0.6444 | 0.6454 | 0.6464 | 0.6474 | 0.6484 | 0.6493 | 0.6503 | 0.6513 | 0.6522 |
| **4.5** | 0.6532 | 0.6542 | 0.6551 | 0.6561 | 0.6571 | 0.6580 | 0.6590 | 0.6599 | 0.6609 | 0.6618 |
| **4.6** | 0.6628 | 0.6637 | 0.6646 | 0.6656 | 0.6665 | 0.6675 | 0.6684 | 0.6693 | 0.6702 | 0.6712 |
| **4.7** | 0.6721 | 0.6730 | 0.6739 | 0.6749 | 0.6758 | 0.6767 | 0.6776 | 0.6785 | 0.6794 | 0.6803 |
| **4.8** | 0.6812 | 0.6821 | 0.6830 | 0.6839 | 0.6848 | 0.6857 | 0.6866 | 0.6875 | 0.6884 | 0.6893 |
| **4.9** | 0.6902 | 0.6911 | 0.6920 | 0.6928 | 0.6937 | 0.6946 | 0.6955 | 0.6964 | 0.6972 | 0.6981 |
| **5.0** | 0.6990 | 0.6998 | 0.7007 | 0.7016 | 0.7024 | 0.7053 | 0.7042 | 0.7050 | 0.7059 | 0.7067 |
| **5.1** | 0.7076 | 0.7084 | 0.7093 | 0.7101 | 0.7110 | 0.7118 | 0.7126 | 0.7135 | 0.7143 | 0.7152 |
| **5.2** | 0.7160 | 0.7168 | 0.7177 | 0.7185 | 0.7193 | 0.7202 | 0.7210 | 0.7218 | 0.7226 | 0.7235 |
| **5.3** | 0.7243 | 0.7251 | 0.7259 | 0.7267 | 0.7275 | 0.7284 | 0.7292 | 0.7300 | 0.7308 | 0.7316 |
| **5.4** | 0.7324 | 0.7332 | 0.7340 | 0.7348 | 0.7356 | 0.7364 | 0.7372 | 0.7380 | 0.7388 | 0.7396 |
| **5.5** | 0.7404 | 0.7412 | 0.7419 | 0.7427 | 0.7435 | 0.7443 | 0.7451 | 0.7459 | 0.7466 | 0.7474 |
| **5.6** | 0.7482 | 0.7490 | 0.7497 | 0.7505 | 0.7513 | 0.7520 | 0.7528 | 0.7536 | 0.7543 | 0.7551 |
| **5.7** | 0.7559 | 0.7566 | 0.7574 | 0.7582 | 0.7589 | 0.7597 | 0.7604 | 0.7612 | 0.7619 | 0.7627 |
| **5.8** | 0.7634 | 0.7642 | 0.7649 | 0.7657 | 0.7664 | 0.7672 | 0.7679 | 0.7686 | 0.7694 | 0.7701 |
| **5.9** | 0.7709 | 0.7716 | 0.7723 | 0.7731 | 0.7738 | 0.7745 | 0.7752 | 0.7760 | 0.7767 | 0.7774 |
| **6.0** | 0.7782 | 0.7789 | 0.7796 | 0.7803 | 0.7810 | 0.7818 | 0.7825 | 0.7832 | 0.7839 | 0.7846 |
| **6.1** | 0.7853 | 0.7860 | 0.7868 | 0.7875 | 0.7882 | 0.7889 | 0.7896 | 0.7903 | 0.7910 | 0.7917 |
| **6.2** | 0.7924 | 0.7931 | 0.7938 | 0.7945 | 0.7952 | 0.7959 | 0.7966 | 0.7973 | 0.7980 | 0.7987 |
| **6.3** | 0.7993 | 0.8000 | 0.8007 | 0.8014 | 0.8021 | 0.8028 | 0.8035 | 0.8041 | 0.8048 | 0.8055 |
| **6.4** | 0.8062 | 0.8069 | 0.8075 | 0.8082 | 0.8089 | 0.8096 | 0.8102 | 0.8109 | 0.8116 | 0.8122 |
| **6.5** | 0.8129 | 0.8136 | 0.8142 | 0.8149 | 0.8156 | 0.8162 | 0.8169 | 0.8176 | 0.8182 | 0.8189 |
| **6.6** | 0.8195 | 0.8202 | 0.8209 | 0.8215 | 0.8222 | 0.8228 | 0.8235 | 0.8241 | 0.8248 | 0.8254 |
| **6.7** | 0.8261 | 0.8267 | 0.8274 | 0.8280 | 0.8287 | 0.8293 | 0.8299 | 0.8306 | 0.8312 | 0.8319 |
| **6.8** | 0.8325 | 0.8331 | 0.8338 | 0.8344 | 0.8351 | 0.8357 | 0.8363 | 0.8370 | 0.8376 | 0.8382 |
| **6.9** | 0.8388 | 0.8395 | 0.8401 | 0.8407 | 0.8414 | 0.8420 | 0.8426 | 0.8432 | 0.8439 | 0.8445 |
| **7.0** | 0.8451 | 0.8457 | 0.8463 | 0.8470 | 0.8476 | 0.8482 | 0.8488 | 0.8494 | 0.8500 | 0.8506 |
| **7.1** | 0.8513 | 0.8519 | 0.8525 | 0.8531 | 0.8537 | 0.8543 | 0.8549 | 0.8555 | 0.8561 | 0.8567 |
| **7.2** | 0.8573 | 0.8579 | 0.8585 | 0.8591 | 0.8597 | 0.8603 | 0.8609 | 0.8615 | 0.8621 | 0.8627 |
| **7.3** | 0.8633 | 0.8639 | 0.8645 | 0.8651 | 0.8657 | 0.8663 | 0.8669 | 0.8675 | 0.8681 | 0.8686 |
| **7.4** | 0.8692 | 0.8698 | 0.8704 | 0.8710 | 0.8716 | 0.8722 | 0.8727 | 0.8733 | 0.8739 | 0.8745 |
| **7.5** | 0.8751 | 0.8756 | 0.8762 | 0.8768 | 0.8774 | 0.8779 | 0.8785 | 0.8791 | 0.8797 | 0.8802 |
| **7.6** | 0.8808 | 0.8814 | 0.8820 | 0.8825 | 0.8831 | 0.8837 | 0.8842 | 0.8848 | 0.8854 | 0.8859 |

| t | 0 | 1 | 2 | 3 | 4 | 5 | 6 | 7 | 8 | 9 |
|---|---|---|---|---|---|---|---|---|---|---|
| **7.7** | 0.8865 | 0.8871 | 0.8876 | 0.8882 | 0.8887 | 0.8893 | 0.8899 | 0.8904 | 0.8910 | 0.8915 |
| **7.8** | 0.8921 | 0.8927 | 0.8932 | 0.8938 | 0.8943 | 0.8949 | 0.8954 | 0.8960 | 0.8965 | 0.8971 |
| **7.9** | 0.8976 | 0.8982 | 0.8987 | 0.8993 | 0.8998 | 0.9004 | 0.9009 | 0.9015 | 0.9020 | 0.9025 |
| **8.0** | 0.9031 | 0.9036 | 0.9042 | 0.9047 | 0.9053 | 0.9058 | 0.9063 | 0.9069 | 0.9074 | 0.9079 |
| **8.1** | 0.9085 | 0.9090 | 0.9096 | 0.9101 | 0.9106 | 0.9112 | 0.9117 | 0.9122 | 0.9128 | 0.9133 |
| **8.2** | 0.9138 | 0.9143 | 0.9149 | 0.9154 | 0.9159 | 0.9165 | 0.9170 | 0.9175 | 0.9180 | 0.9186 |
| **8.3** | 0.9191 | 0.9196 | 0.9201 | 0.9206 | 0.9212 | 0.9217 | 0.9222 | 0.9227 | 0.9232 | 0.9238 |
| **8.4** | 0.9243 | 0.9248 | 0.9253 | 0.9258 | 0.9263 | 0.9269 | 0.9274 | 0.9279 | 0.9284 | 0.9289 |
| **8.5** | 0.9294 | 0.9299 | 0.9304 | 0.9309 | 0.9315 | 0.9320 | 0.9325 | 0.9330 | 0.9335 | 0.9340 |
| **8.6** | 0.9345 | 0.9350 | 0.9355 | 0.9360 | 0.9365 | 0.9370 | 0.9375 | 0.9380 | 0.9385 | 0.9390 |
| **8.7** | 0.9395 | 0.9400 | 0.9405 | 0.9410 | 0.9415 | 0.9420 | 0.9425 | 0.9430 | 0.9435 | 0.9440 |
| **8.8** | 0.9445 | 0.9450 | 0.9455 | 0.9460 | 0.9465 | 0.9469 | 0.9474 | 0.9479 | 0.9484 | 0.8489 |
| **8.9** | 0.9494 | 0.9499 | 0.9504 | 0.9509 | 0.9513 | 0.9518 | 0.9523 | 0.9528 | 0.9533 | 0.9538 |
| **9.0** | 0.9542 | 0.9547 | 0.9552 | 0.9557 | 0.9562 | 0.9566 | 0.9571 | 0.9576 | 0.9581 | 0.9586 |
| **9.1** | 0.9590 | 0.9595 | 0.9600 | 0.9605 | 0.9609 | 0.9614 | 0.9619 | 0.9624 | 0.9628 | 0.9633 |
| **9.2** | 0.9638 | 0.9643 | 0.9647 | 0.9652 | 0.9657 | 0.9661 | 0.9666 | 0.9671 | 0.9675 | 0.9680 |
| **9.3** | 0.9685 | 0.9689 | 0.9694 | 0.9699 | 0.9703 | 0.9708 | 0.9713 | 0.9717 | 0.9722 | 0.9727 |
| **9.4** | 0.9731 | 0.9736 | 0.9741 | 0.9745 | 0.9750 | 0.9754 | 0.9759 | 0.9763 | 0.9768 | 0.9773 |
| **9.5** | 0.9777 | 0.9782 | 0.9786 | 0.9791 | 0.9795 | 0.9800 | 0.9805 | 0.9809 | 0.9814 | 0.9818 |
| **9.6** | 0.9823 | 0.9827 | 0.9832 | 0.9836 | 0.9841 | 0.9845 | 0.9850 | 0.9854 | 0.9859 | 0.9863 |
| **9.7** | 0.9868 | 0.9872 | 0.9877 | 0.9881 | 0.9886 | 0.9890 | 0.9894 | 0.9899 | 0.9903 | 0.9908 |
| **9.8** | 0.9912 | 0.9917 | 0.9921 | 0.9926 | 0.9930 | 0.9934 | 0.9939 | 0.9943 | 0.9948 | 0.9952 |
| **9.9** | 0.9956 | 0.9961 | 0.9965 | 0.9969 | 0.9974 | 0.9978 | 0.9983 | 0.9987 | 0.9991 | 0.9996 |

# Index

## A

**D**

**E**

### K



### L



### M



### N



### O

P

## TABLE OF FORMULA WEIGHTS ($^{12}C$ = 12)

Rounded to two decimal places

|  |  |  |  |
|---|---|---|---|
| $AgCl$ | 143.32 | $HF$ | 20.01 |
| $Ag_2CrO_4$ | 331.74 | $HIO_4$ | 191.91 |
| $AgNO_3$ | 169.88 | $HNO_3$ | 63.02 |
| $AgSCN$ | 165.95 | $HSO_3NH_2$ (sulfamic) | 97.09 |
| $A1(OH)_3$ | 78.00 | $H_2O_2$ | 34.02 |
| $BaCl_2$ | 208.24 | $H_3PO_4$ | 97.99 |
| $BaSO_4$ | 233.40 | $H_2SO_4$ | 98.08 |
| $CaCO_3$ | 100.09 | $Hg_2Cl_2$ | 472.08 |
| $CaC_2O_4$ | 128.10 | $HgCl_2$ | 271.50 |
| $CaO$ | 56.08 | $KCN$ | 65.12 |
| $CO_2$ | 44.01 | $KCl$ | 74.56 |
| $CO(NH_2)_2$(urea) | 60.05 | $KClO_3$ | 122.55 |
| $Cu(NO_3)_2$ | 187.54 | $K_2CrO_4$ | 194.20 |
| $CuO$ | 79.54 | $K_2Cr_2O_7$ | 294.20 |
| $Cu_2O$ | 143.08 | $KHC_8H_4O_4$ | 204.23 |
| $CuSO_4$ | 159.60 | $KH_2PO_4$ | 136.09 |
| $Cu_2I_2$ | 380.88 | $K_2HPO_4$ | 174.19 |
| $Cu_2(SCN)_2$ | 243.24 | $KI$ | 166.00 |
| $Fe_2O_3$ | 159.69 | $KI_3$ | 419.81 |
| $Fe_3O_4$ | 231.54 | $KIO_3$ | 214.00 |
| $Fe\ (OH)_2$ | 89.87 | $KIO_4$ | 230.00 |
| $Fe(OH)_3$ | 106.87 | $KMnO_4$ | 158.04 |
| $Fe(NH_4)_2(SO_4)_2{\cdot}6H_2O$ | 392.14 | $KNO_2$ | 85.11 |
| $Fe_2(NH_4)_2(SO_4)_4{\cdot}24H_2O$ | 964.49 | $KOH$ | 56.11 |
| $HBO_2$ | 43.82 | $KSCN$ | 97.18 |
| $H_3BO_3$ | 61.83 | $K_2SO_4$ | 174.26 |
| $HC_2H_3O_2$ (acetic) | 60.05 | $MgCO_3$ | 84.32 |
| $H_2C_2O_4$ (oxalic) | 90.04 | $MgC_2O_4$ | 112.33 |
| $H_2C_2O_4{\cdot}2H_2O$ (oxalic) | 126.07 | $MgCl_2$ | 95.22 |
| $H_2C_4H_4O_6$ (tartaric) | 150.09 | $MgO$ | 40.31 |
| $H_2C_8H_4O_4$ (phthalic) | 166.14 | $Mg(OH)_2$ | 58.33 |
| $HC1$ | 36.46 | $NaC_2H_3O_2$ | 82.03 |
| $HClO_3$ | 84.46 | $NaCN$ | 49.01 |
| $HClO_4$ | 100.46 | $Na_2CO_3$ | 105.99 |